Conversion of Water and CO_2 to Fuels using Solar Energy

Conversion of Water and CO_2 to Fuels using Solar Energy

Science, Technology and Materials

Edited by

Oomman K. Varghese
University of Houston
Houston
USA

Flavio L. Souza
Universidade Federal do ABC
Santo Andre
Brazil

The American Ceramic Society

WILEY

Published by John Wiley & Sons, Inc., Hoboken, New Jersey.
Published simultaneously in Canada.

Library of Congress Cataloging-in-Publication Data applied for:

Hardback ISBN: 9781119600848

Cover Design and Image: Wiley

Set in 9.5/12.5pt STIXTwoText by Straive, Chennai, India

Contents

List of Contributors

Kirk H. Bevan
Division of Materials Engineering
Faculty of Engineering
McGill University
Montreal
QC
Canada

Kamran Dastafkan
School of Chemistry
The University of New South Wales
Faculty of Science
Sydney
NSW
Australia

Wanjae Dong
Department of Electrical Engineering and Computer Science
University of Michigan
Ann Arbor
MI
USA

Christoph Falter
Synhelion SA
Zurich
Switzerland

Ke Fan
State Key Laboratory of Fine Chemicals
School of Chemical Engineering
Dalian University of Technology
Dalian
China

Liniquer A. Fontana
Institute of Chemistry
University of Campinas (UNICAMP)
Campinas
SP
Brazil

Nicolas Gaillard
Hawaii Natural Energy Institute
University of Hawaii at Manoa
Honolulu
HI
USA

Bingchan Hu
State Key Laboratory of Multiphase Flow in Power Engineering (SKLMF)
School of Energy and Power Engineering
Xi'an Jiaotong University
Xi'an
Shaanxi
China

Asif Iqbal
Division of Materials Engineering
Faculty of Engineering
McGill University
Montreal
QC
Canada

Muhammad S. Khan
International Research Center for Renewable Energy (IRCRE)
State Key Laboratory of Multiphase Flow in Power Engineering (MPFE)
Xi'an Jiaotong University
Xi'an
China

Youjun Lu
State Key Laboratory of Multiphase Flow in Power Engineering (SKLMF)
School of Energy and Power Engineering
Xi'an Jiaotong University
Xi'an
Shaanxi
China

Jackson D. Megiatto Jr.
Institute of Chemistry
University of Campinas (UNICAMP)
Campinas
SP
Brazil

Zetian Mi
Department of Electrical Engineering and Computer Science
University of Michigan
Ann Arbor
MI
USA

Botong Miao
Division of Materials Engineering
Faculty of Engineering
McGill University
Montreal
QC
Canada

Catia Ornelas
Institute of Chemistry
University of Campinas
Campinas
SP
Brazil

Heng Pan
State Key Laboratory of Multiphase Flow in Power Engineering (SKLMF)
School of Energy and Power Engineering
Xi'an Jiaotong University
Xi'an
Shaanxi
China

Appu V. Raghu
Laser Molecular Photoscience Laboratory
Molecular Photoscience Research Center
Kobe University
Kobe
Japan

Vitor H. Rigolin
Institute of Chemistry
University of Campinas (UNICAMP)
Campinas
SP
Brazil

Ingrid Rodriguez-Gutierrez
Brazilian Nanotechnology National Laboratory (LNNANO)
Brazilian Center for Research in Energy and Materials (CNPEM)
Campinas
Brazil

Centro de Ciências Naturais e Humanas (CCNH)
Federal University of ABC (UFABC)
Santo André
Brazil

Shaohua Shen
International Research Center for Renewable Energy (IRCRE)
State Key Laboratory of Multiphase Q1Flow in Power Engineering (MPFE)
Xi'an Jiaotong University
Xi'an
China

Flavio L. Souza
Brazilian Nanotechnology National Laboratory (LNNANO)
Brazilian Center for Research in Energy and Materials (CNPEM)
Campinas
Brazil

Centro de Ciências Naturais e Humanas (CCNH)
Federal University of ABC (UFABC)
Santo André
Brazil

Mark T. Spitler
Department of Chemistry
University of North Carolina
Chapel Hill
NC
USA

Qian Sun
School of Chemistry
The University of New South Wales
Faculty of Science
Sydney
NSW
Australia

Takashi Tachikawa
Laser Molecular Photoscience Laboratory
Molecular Photoscience Research Center
Kobe University
Kobe
Japan
Department of Chemistry
Graduate School of Science
Kobe University
Kobe
Japan

Lianpeng Tong
School of Chemistry and Chemical Engineering
Guangzhou Key Laboratory for Clean Energy and Materials
Guangzhou University
Guangzhou
China

Oomman K. Varghese
Nanomaterials and Devices Laboratory
Department of Physics
University of Houston
Houston
TX
USA

Texas Center for Superconductivity
University of Houston
Houston
TX
USA

Lei Wang
Department of Chemical and Biomolecular Engineering
National University of Singapore
Singapore
Singapore

Chuan Zhao
School of Chemistry
The University of New South Wales
Faculty of Science
Sydney
NSW
Australia

Baowen Zhou
School of Mechanical Engineering
Research Center for Renewable Synthetic Fuel
Shanghai Jiao Tong University
Shanghai
China

Department of Electrical Engineering and Computer Science
University of Michigan
Ann Arbor
MI
USA

Peng Zhou
Department of Electrical Engineering and Computer Science
University of Michigan
Ann Arbor
MI
USA

Preface

The twenty-first century world is experiencing planetary and societal challenges that are unprecedented in human history. The United Nations identifies climate breakdown, air pollution, and biodiversity loss as a triple planetary crisis threatening the life on Earth. It is an irrefutable fact that fossil fuels, which form the backbone of the global energy system, are the primary cause of this crisis. The global average temperature is rising dramatically due to the changes in the planet's energy budget caused by the atmospheric accumulation of greenhouse gases originating from fossil fuel burning. The impact on the life on the planet is multifaceted. The changes happening in the environment are becoming irreversible. The window of opportunity to limit warming is rapidly closing. Our actions in the next couple of decades are crucial to securing a sustainable future.

Under these circumstances, the development of a viable alternative energy system has become imperative. It is no longer simply a matter of choice, but rather a collective responsibility to explore and implement innovative approaches that can power our societies without further compromising the delicate balance of the planet. In this scenario, solar fuel generation stands out as a promising avenue toward a green and sustainable future. Harnessing the power of the Sun, the most abundant and renewable energy source available, holds tremendous potential for mitigating the adverse impacts of traditional energy generation methods.

By converting water and carbon dioxide (CO_2) into fuels using energy from sunlight, the world can enjoy similar benefits as those provided by fossil fuels while minimizing their detrimental effects. The potential benefits of these synthetic fuels, called solar fuels, extend far beyond the realm of energy production and pollution control. The research, development, and implementation of solar fuel technologies stimulate innovation, drive technological advancements, and propel the growth of a new green economy. Moreover, incorporating solar fuel generation inspires a global shift towards a more harmonious relationship between humanity and the natural world, where environmental preservation and economic progress go hand in hand.

This book is a testament to the collective efforts of leading experts in the field who have devoted their time and energy over years and decades to unraveling the complexities of solar fuel generation. It gives the readers a ride through the historical, scientific, and environmental aspects of global warming, the establishment of an alternative energy system based on solar power and fuels, the primary approaches used for solar fuel generation, the scientific principles, technological and economic perspectives, and the key materials investigated for technology development and commercialization. By furnishing the fundamentals, discussing the latest advancements, and digging deep into the research on materials that enable solar fuel generation, this book is envisioned to be a resource for the knowledge accumulated through decades of research and a source of inspiration for further innovation in the quest for sustainable energy solutions.

The book is structured into an introductory chapter followed by two sections dealing with the techno-scientific and material aspects separately. The introductory chapter (Chapter 1) provides perspectives on global warming and climate change as consequences of excessive fossil fuel burning. It discusses briefly the current energy situation and sustainable options. The focus of this chapter is on introducing the two artificial fuel generation pathways, water splitting and CO_2 reduction, and the most promising approaches for accomplishing them. Section 1 (Chapters 2–7) entitled "Solar Fuel Generation Processes: Science and Technology," introduces the three promising approaches to fuel generation, namely, photoelectrochemical (PEC)/photocatalytic, concentrated solar thermochemical, and photovoltaic-assisted electrochemical. Certain unique perspectives on the natural and artificial photosynthetic systems are given in this section. In Section 2, entitled "Materials for Solar Fuel Generation," (Chapters 8–12), emphasis is given on discussing the key materials explored in the three fuel generation approaches mentioned above. Development of catalysts for photo-, thermo- and electro-conversion forms the heart of the discussion.

Chapter 2 discusses the basic science of artificial photosynthesis. It first introduces the catalyst complexes performing natural photosynthesis. This is followed by a brief discussion of bioinspired artificial systems for fuel generation. The remainder of the chapter is targeted at PEC and photocatalytic systems for water splitting and CO_2 reduction. The fundamental principles, device configurations, parameters of relevance for evaluating the performance of devices, and examples of materials used in these applications are given. Chapter 3 gives a comprehensive understanding of the scientific, technological, environmental, and economic aspects of the concentrated solar thermochemical fuel generation processes. It reviews the thermodynamics of the thermochemical cycles and the reactor designs for a two-step process. By taking a two-step thermochemical cycle as an example, the readers are taken deeper into a discussion of the energy and mass balance in the reactor and fuel plant scales, the techno-economic and life cycle aspects, water and land requirements for commercially implementing the technology, and the geographic locations around the world appropriate for producing fuels via solar thermochemical processes. Photovoltaics (PV) can be used for powering water electrolyzers and providing external bias for PEC cells. Both these technologies, from basic concepts to techno-economics, are discussed in detail in Chapter 4. Moreover, it covers the topic of monolithically stacked PV-PEC devices for spontaneous water splitting under solar irradiation.

Two different perspectives on artificial photosynthesis are given in Chapters 5 and 6. Chapter 5 takes the readers on a captivating journey through the history of molecular photosynthesis. It gives an overview of the evolution of natural photosynthetic systems, emphasizing the relevance of the photo-redox enzymes, and then connects the discussion to the development of molecular complexes and organic-inorganic hybrid systems for artificial photosynthesis. Chapter 6 gives a unique perspective on the kinetic designs for the solar reduction of CO_2. The catalytic reactions in natural photosynthesis and man-made systems are systematically analyzed. It points out the issues with the current approaches and identifies potential opportunities for widening the exploration of efficient and durable systems for fuel generation using CO_2 and water, following the pathway of natural photosynthesis. Chapter 7 presents the application of the band diagram approach to understanding the characteristics of PEC devices. Taking the case of photoanodes in PEC cells, both analytical and numerical modeling methods are discussed.

Section 2 begins with Chapter 8, which details the performance of redox pairs used in two-step thermochemical cycles for the conversion of CO_2 and water (H_2O) to fuels, while touching on the basis of thermolysis (single step) and multistep cycles. It also discusses the catalysts investigated for sulfur-iodine cycle, solar gasification, and reforming of methane and methanol. While PV-assisted water electrolysis has become a mature technology for commercial production of hydrogen, the electrocatalytic conversion of CO_2 to fuels and value-added chemicals is an emerging technology in which significant research efforts are being invested. Chapter 9 gives the fundamentals of CO_2 reduction reactions, electrochemical cell designs, electrolytes, and various catalysts, including compounds of metals, single-atom catalysts, and molecular catalysts. A feature of this chapter is the discussion of the in-situ characterization of electrocatalysts for CO_2 reduction.

Another very promising approach that has not yet reached the technology commercialization stage is PEC and photocatalytic water splitting and CO_2 conversion. Chapters 10–12 are dedicated to the materials researched for this approach. The semiconductors used for PEC fuel generation (called photocatalysts) are basically divided into oxides and non-oxides. Chapter 10 covers the ceramic materials with an emphasis on the oxide photocatalysts used for PEC water splitting. It discusses the properties and performance of carbides and nitrides including two-dimensional transition metal carbides and nitrides called MXenes. Among the non-oxides, gallium nitride (GaN)-based compounds form one of the most promising families of materials. Chapter 11 discusses GaN and its compounds used for artificial photosynthesis. Two dedicated sections focus on integrated devices for solar water splitting and CO_2 reduction separately. Nanostructured semiconductors are the most investigated class of materials in the area of PEC and photocatalytic fuel generation. Chapter 12 exposes the readers to the unique electronic, plasmonic, structural, morphological, and surface properties of nanomaterials and their heterostructures used for PEC and photocatalytic applications. It provides a comprehensive view of the characteristics of zero, one, and two-dimensional photocatalysts in performing solar water splitting and CO_2 reduction.

Through its compilation of various perspectives and insights, this book is expected to facilitate further innovations in solar fuel generation technologies and help speed up our journey toward a cleaner and more sustainable energy future. Recognizing our shared responsibility, embracing renewable energy sources, and prioritizing sustainability and responsible care for the environment are essential. It is our hope that this book will serve as an inspiring catalyst, sparking our collective efforts toward a world where solar fuel generation powers our societies while preserving the integrity of our planet.

Oomman K. Varghese (Editor)
Department of Physics & Texas Center for Superconductivity
University of Houston

Flavio L. Souza (Editor)
Centro de Ciências Naturais e Humanas (CCNH)
Federal University of ABC (UFABC)

1

Solar Fuel Generation: The Relevance and Approaches

Ingrid Rodriguez-Gutierrez[1,2], *Flavio L. Souza*[1,2], *and Oomman K. Varghese*[3,4]

[1] *Brazilian Nanotechnology National Laboratory (LNNANO), Brazilian Center for Research in Energy and Materials (CNPEM), Campinas, Brazil*
[2] *Centro de Ciências Naturais e Humanas (CCNH), Federal University of ABC (UFABC), Santo André, Brazil*
[3] *Nanomaterials and Devices Laboratory, Department of Physics, University of Houston, Houston, TX, USA*
[4] *Texas Center for Superconductivity, University of Houston, Houston, TX, USA*

1.1 Introduction

"Human activities, principally through emissions of greenhouse gases, have unequivocally caused global warming, with global surface temperature reaching 1.1 °C above [the average temperature in] 1850–1900 in 2011–2020", declared United Nations Intergovernmental Panel on Climate Change (IPCC) in the Synthesis Report of its 6th Assessment Report (IPCC, 2023). The "human activity" mentioned in the report as the primary cause of greenhouse gas accumulation in the atmosphere was not hidden to great scientific minds of the nineteenth century even though the consequences of the new energy system that emerged in conjunction with industrial revolution were not apparent. Professor Svante Arrhenius, in his book *Worlds in the Making: The Evolution of the Universe*, asserted: "We thus recognize that the percentage of carbonic acid in the air must be increasing at a constant rate as long as the consumption of coal, petroleum, etc., is maintained at its present figure and at a still more rapid rate if this consumption should continue to increase as it does now" (Arrhenius, 1908). He wrote this statement based on his calculation published in 1896. According to this calculation, a 50% increase in carbonic acid (historically a synonym for carbon dioxide, CO_2) concentration in the atmosphere would increase the global average temperature by about 4 °C (Arrhenius, 1896). Professor Arrhenius and many of his contemporary scientists knew that the blanket provided by the greenhouse gases, especially CO_2, was keeping the planet warm and hoped that it would not send the planet into another ice age. Apparently, they did not anticipate that the temperature increase due to excessive usage of fossil fuels would really go beyond safe limits and make catastrophic, potentially irreversible, changes in the climate and ecosystem. Until the second decade of the twenty-first century, the climate change due to anthropogenic global warming was a myth to many, including scientists, policy makers and corporate executives; however, now it is an undeniable truth.

It is clear that human beings have abused nature extensively. There is only a little time left to act prudently and reverse the changes to some extent, before the situation goes out of hand. Of course, going back to pre-industrial revolution era, burning wood for heat and using horse for work, is pretty naïve to think of. There has been an exponential growth in energy demands due to technological advancements and population growth in the post-industrial revolution period. According to the International Energy Agency (IEA) statistics (International Energy Agency, 2022), oil, coal, and natural gas currently account for roughly 80% of the global energy production. Hydro, biomass/biofuel, solar, and other renewables, and nuclear technologies bearing the remaining load are largely sustainable. The task of replacing fossil fuel-based energy system with a sustainable one is humongous and exceedingly complex. Fossil fuels are naturally available, do not need conversion to other forms for energy production, and can be stored or transported to almost any location. Moreover, plastic and a myriad of other products are derived from these fuels. No single energy vector has all these qualities. For example, electric power generated from renewable sources involves an energy conversion process and has limited options for transportation and storage. Hydrogen is another energy carrier. It can be transported, stored, and used as a raw material for chemicals; however, its production requires renewable energy and a renewable precursor such as water (H_2O). The same is true with other synthetic fuels such as ammonia and biomass products. Therefore, it is highly likely that the future energy system developed

Conversion of Water and CO_2 to Fuels using Solar Energy: Science, Technology and Materials, First Edition.
Edited by Oomman K. Varghese and Flavio L. Souza.

to replace the fossil fuel-oriented one will have contributions from electricity and different synthetic fuels generated using a wide range of sustainable pathways.

The sustainable pathways for electricity and fuel production should use, in principle, recycling approaches involving renewable energy sources while maintaining carbon neutrality. For example, hydrogen fuel generated via water splitting using renewable energy recycles water. Similarly, biofuels enable recycling of CO_2. A future energy system would quite likely use a range of renewable energy sources, especially solar, wind, hydro, geothermal, and tidal. Biomass has solar energy stored as potential energy in atomic and molecular bonds. Biomass and nuclear energy, along with a small fraction of fossil fuels, will continue contributing to the global energy supply for the foreseeable future. Among the renewable technologies, the solar power and fuel generation technologies are expected to have the highest share in the new energy system due to the abundance of solar energy.

In this rare time of energy system transition, it is important to accelerate the development of fuel generation technologies to make the transition fast and least disruptive, as renewable power generation alone cannot fill the gap created upon phasing out fossil fuels. There are significant challenges to overcome as the number of solar photons reaching Earth per unit area per second (called photon flux) is limited and the fuels should be produced within the limitations imposed by thermodynamics and kinetics. This chapter gives an introduction to the solar-driven technologies for synthetic fuel generation using H_2O and/or CO_2 as precursors. Prior to discussing these technologies, we would provide an overview of the effects of greenhouse gas emissions on global warming and associated climatc change.

1.2 The Nexus Between Fossil Fuels, Global Warming, and Climate Change

Even before the time of Svante Arrhenius, the scientific community was aware of the warming potentials of the constituents of the atmosphere. The seminal works of Foote and Tyndall published in 1856 and 1859, respectively, showing the infrared (IR) absorption ability of CO_2 and the resulting temperature rise are examples (Foote, 1856; Tyndall, 1861). Prior to these experiments, Joseph Fourier conducted studies on the role of atmosphere in keeping the surface of Earth warm. In his article published in 1824, Fourier stated: "[Earth's] temperature can be augmented by the interposition of the atmosphere, because heat in the state of light finds less resistance in penetrating the air, than in repassing into the air when converted into non-luminous heat"(Fleming, 1999; Fourier, 1824). Although the exact cause of the opacity of the atmosphere for the rays emitted by Earth was not clear to the scientific community two hundred years ago, they had insights about the role of atmosphere in preserving warmth in the planet.

Sunlight reaching Earth consists of wavelengths in the ultraviolet (UV) to IR region (about 200–4,500 nm). The incident light is partly reflected (this portion is called albedo) and partly absorbed. The atmosphere is generally transparent to the reflected wavelengths as its components do not absorb (but scatter) relatively high-energy radiations. On the other hand, Earth as a blackbody reemits the absorbed sunlight in the form of electromagnetic waves of longer wavelengths (mid-to-far IR region). In a steady state, the energy received by Earth and that released must be equal. Since nitrogen and oxygen are the primary constituents of the atmosphere (78.08% and 20.95% by volume, respectively, in dry air), the atmosphere must be transparent to IR rays. This is because N_2 and O_2 are homogenous diatomic molecules and, therefore, they are not IR-active. If N_2 and O_2 were the only constituents of the atmosphere, global average temperature would have stayed at about −18 °C (255 K). Nevertheless, the atmosphere contains water vapor (up to about 4% by volume), CO_2 and methane (CH_4). These molecules have IR-active modes of vibration and therefore, they absorb the IR waves reemitted by Earth causing the temperature to increase. This is termed the greenhouse effect because the glass enclosure of a greenhouse for growing plants traps outgoing long-wavelength IR rays and warms the environment (Florence et al., 1950).

The CO_2 level in the atmosphere has stayed between approximately 180 and 300 ppm since the last ice age about 800,000 years ago. The average CO_2 concentration during the pre-industrial revolution period was about 280 ppm. A blanket consisting primarily of CO_2 and water vapor formed via natural processes kept the atmosphere warmer by around 31 °C compared to −18 °C (taking the average temperature to ∼13 °C), with minor warming and cooling periods (approximately ±1 °C variation for over 5,000 years of human history) lasting a few hundred years. This pattern has been disrupted since the beginning of the twentieth century. In 1880, the average CO_2 concentration became 290 ppm and the global average temperature was about 13.6 °C. Since the middle of the twentieth century, both CO_2 concentration and global average temperature have increased consistently, as obvious from Figure 1.1,[1] which shows a graph between

1 The data for Figure 1.1 are taken from Our World in Data (https://ourworldindata.org/co2-and-greenhouse-gas-emissions).

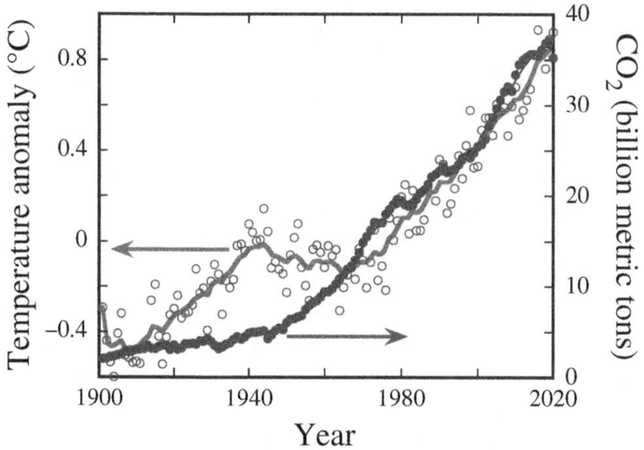

Figure 1.1 The trend in global average temperature change relative to the 1961–1990 average temperature and CO_2 emissions from fossil fuels and industry.

CO_2 released into the atmosphere and the global average temperature change with time. The average temperature in the period 1960–1990 is taken as reference for temperature change. Such a steep increase in CO_2 concentration has never been noticed since the last ice age. It has now been proven beyond doubt that increasing concentration of anthropogenic CO_2 in the atmosphere is primarily responsible for the observed rise in the global average temperature.

In addition to CO_2 and water vapor, CH_4, nitrous oxide (N_2O), and chlorofluorocarbons in the atmosphere also absorb IR radiation and contribute to greenhouse effect. Nevertheless, the contribution from CO_2 to global temperature rise is much higher than that from these compounds despite the fact that their global warming potential[2] is many folds higher than that of CO_2. This is because of the significantly higher amount of CO_2 in the atmosphere (~418 ppm of CO_2 vs. 1.9 ppm of CH_4 in 2022) and the associated higher radiative forcing.[3] Per the National Oceanic and Atmospheric Administration (NOAA) data (https://gml.noaa.gov/aggi/aggi.html), the radiative forcing caused by CO_2 was 2.17 W/m^2 while that due to methane was 0.65 W/m^2 in 2022. This means that the contribution of CO_2 in preventing the energy from escaping the Earth's atmosphere is over three times higher than that of methane. Fossil fuels are the sources of about 75% of the global greenhouse gases (CO_2 is ~90% in this composition) emitted to the atmosphere.

Among the three phases (solid, liquid, and gas) of fossil fuels, coal releases maximum CO_2 while burning. While some evidence of fossil fuel use is found in studies conducted on societies flourished before Christ, the modern era coal mining started in the sixteenth century, and drilling for gas and oil began in 1821 and 1859, respectively. Since the initiation of mechanization that occurred in conjunction with the invention of steam engine, coal has been used in place of wood due to its better heating value (two to three times). Coal consists of 70–90% carbon by mass and the rest hydrogen and impurities, including sulfur, oxygen, and oxides. The hydrogen-to-carbon ratio is lowest in coal. To compare, hydrogen-to-carbon molar ratio in anthracite coal is 0.5 : 1, whereas the ratio is 3 : 1 for methane, and this is the reason for higher CO_2 emissions from coal. The emission is lower for natural gas and oil products such as gasoline, diesel, and kerosene, but not substantially. While burning 1 ton of coal releases about 3.2 tons of CO_2, the corresponding numbers are 2.3 and 1.8 tons, respectively, for gasoline and natural gas. About 37 billion metric tons of CO_2 is emitted per year globally. The average annual CO_2 uptake by land is 25–30% and that by ocean is ~25%. The atmosphere retains the rest. The average CO_2 level in 2022 was 418 ppm, and it is increasing at a rate of around 3 ppm per year. Efforts are being carried out by several countries to control and reduce CO_2 production from fossil fuels. In 2015, the Paris agreement was adopted by 196 parties, outlining strategies to limit global warming to 1.5 °C by achieving net zero by 2050 (Bhore, 2016). Nevertheless, according to the UN Emissions Gap Report 2022, there has been very limited progress in reducing the large emissions gap for 2030 (United Nations Environment Programme, 2022). The gap between promised measures and those needed to keep the emissions per the Paris agreement is huge. The IPCC warns that unless sufficient measures are implemented per the time line of the Paris agreement, it will be difficult to keep the CO_2 concentration below 450 ppm and warming below 2 °C in this century (IPCC 2014, 2023). The

2 A measure of the ability of a greenhouse gas to trap heat and is defined (per IPCC report, 1990) as the ratio of the time-integrated radiative forcing from the instantaneous release of one kilogram of a trace substance (e.g., a greenhouse gas) to that of one kilogram of a reference gas (CO_2, generally).

3 The net change in the energy balance of the Earth system due to some imposed perturbations and is expressed in watts per square meter.

temperature rise of such magnitude would trigger catastrophic and irreversible changes in the climate and ecosystems, as well as extreme hardships, to vulnerable people (Lenton et al., 2023; McKay et al., 2022; Tollefson, 2021).

Apart from CO_2, release of CO produced due to incomplete combustion, unburnt hydrocarbons, oxides of nitrogen NO_x (primarily NO and NO_2), mercury (from coal), SO_2 produced via oxidation of sulfur, and particulate matter are other problems with fossil fuel burning. While CO_2 increases the temperature of the atmosphere, the small particulate matter dispersed in the atmosphere can scatter incoming light and cause cooling effect due to the reduction in energy reaching Earth. Although some particles absorb the infrared radiation from Earth and cause warming effect, global radiative forcing due to aerosol–radiation interactions and aerosol–cloud interactions is negative, indicating that their overall effect is cooling. Does this mean that atmospheric particulate concentrations should be increased to offset the global warming? An author of this chapter recalls the experience in traveling in coal-fueled trains with steam engines. The carbon particles from coal firing cover the body and nostrils of the passengers soon after the beginning of the journey. The relationship between climate change and air quality is very complex. Pollution makes the human population vulnerable to many chronic and terminal diseases including cancer (Turner et al., 2020). Benzene and polycyclic aromatic hydrocarbons are examples of carcinogens released from oil combustion. Although the use of catalytic converters, particulate filters, and unleaded fuel could help reduce the emission of hazardous pollutants, their concentrations in the atmosphere are increasing. The World Health Organization statistics associate about 7 million deaths every year with air pollution.

Nevertheless, some people still believe that global warming and its effect are not real. Some of the common arguments are the following: (a) CO_2 concentration has shown both increasing and decreasing characteristics several times in the history of the planet. In fact, it has been oscillating between about 280 and 180 ppm every 50,000–100,000 years for the past 800,000 years. Nevertheless, the steep increase in CO_2 concentration that the globe is currently experiencing is unprecedented. (b) A higher CO_2 level is desirable because plants would grow better and the food production would increase. While it is true that higher CO_2 favors the growth of plants, the overall effect of the global warming on the plants is complex. The extreme climatic conditions could adversely affect their growth. Moreover, the food crops would lose key nutrients and proteins at higher CO_2 levels. (c) The amount of anthropogenic greenhouse gas in the atmosphere is low compared to that from natural sources such as volcanos. The emission from natural sources are offset by the processes such as photosynthesis, absorption by ocean and creation of soil and peats. In the past, these processes prevented CO_2 concentrations from going above 300 ppm. (d) The CO_2 concentration in the atmosphere (0.042%) is too small to create an appreciable effect on the thermal budget of Earth. This is not correct. Every ppm of CO_2 is equivalent to dumping over 12 billion tons of CO_2 into the atmosphere (~37 billion tons @ ~3 ppm per year). The global warming potential of this amount of CO_2 is high. (e) Studies such as those conducted by the IPCC do not take into account all aspects of development, and, therefore, their reports do not portray the realistic picture of the situation. Even if the number of deaths due to climate change increases, the overall life expectancy will increase due to advancements in healthcare, sanitation, and other sectors (O'Neill, 2023). This is equivalent to stating that let the weak die and the strong survive. (6) The unusual wildfires and extreme weather conditions occurring in many parts of the world cannot be related to climate change. While all wildfire incidents and other weather extremes cannot be attributed to climate change, some disasters are indeed driven by global warming, as determined by various scientific agencies around the world, including NASA (https://climate.nasa.gov/), through daily monitoring of the atmosphere using multiple technologies.

It is a fact that there is a close agreement between the predictions of most climatic models and the observed changes. Regardless of the difference in opinion, it is important to unite and utilize the options available now, which may disappear for ever after a short time if appropriate actions are not promptly taken.

1.3 The Energy System Transformation

Fossil fuels account for about 80% of world's energy production (see Figure 1.2). Among the three forms of fossil fuels, natural gas has the highest heating value (~52 vs. 45 MJ/kg for crude oil and 24–30 MJ/kg for coal) and least CO_2 emission (about half vs. coal). A strategy followed by many nations is to increase the share of natural gas in the energy production and reduce that of coal and oil. While this is desirable, the global warming trend has reached a stage where a reversal may not happen by just halving the CO_2 emission. Moreover, there are concerns about using natural gas because natural gas typically contains 85–95% methane (the remaining consists of ethane, heavier hydrocarbons, and non-hydrocarbons such as CO_2), which is a greenhouse gas. Among the remaining major sources, nuclear, hydro, and biomass/biofuels are, in principle, carbon-neutral technologies. According to the IEA World Energy Outlook 2022 (https://www.iea.org/reports/world-energy-outlook-2022), while the contributions from all these sources would increase in the future, the increase in

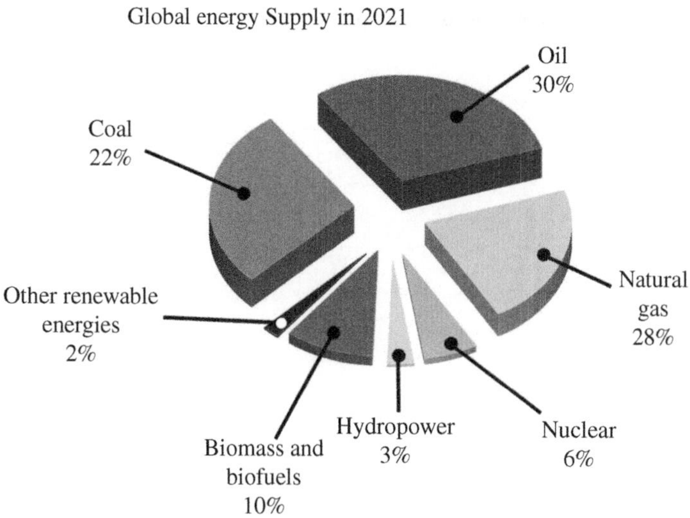

Figure 1.2 The world's total primary energy supply by fuels in 2021. The data are collected from the International Energy Agency.

the share of nuclear and biofuels in the energy production would be substantial by 2030; however, the growth is expected to slow down as 2050 approaches (International Energy Agency, 2022). Nuclear and biofuel technologies are mature, and these could help reduce the emissions rapidly. The concerns over radioactive waste produced during nuclear power production, food vs. fuel as well as low efficiency problems in biofuel-based processes, and challenges related to loss of forests and wild life habitats in establishing hydroelectric power projects are some of the factors that would lower the interest in the expansion of these technologies. Therefore, meeting the future energy demand efficiently and securely requires extensive utilization of renewable energy sources.

The renewable energy conversion technologies such as solar photovoltaics (PV), wind energy, geothermal energy, and ocean energy are the primary contributors to the renewable energy market (~3% of overall energy supply). A technology that is expected to replace the fossil fuel-based power production is solar PV. The contribution to global electricity production (~28,500 TWh) from solar PV reached 4.6% in 2022, and it is increasing consistently at a level envisioned for the period 2023–2030 under the net zero emission by 2050 scenario (International Energy Agency, 2022). The market share of solar PV is expected to grow at the rate of 30% in this period. Nevertheless, renewables still cannot replace fossil fuels completely unless radical improvements in establishing renewable source-based power plants are achieved. The implementation of sustainable energy technologies requires overcoming challenges such as the establishment of long-term contracts, ensuring priority grid access, and continually installing new renewable energy facilities. Other obstacles such as reduced electricity demand, supply chain disruptions, and construction delays in some regions, must be addressed for promoting the adoption of sustainable energy.

Despite the rapid growth of the renewable electric power market, the intermittency of the renewable energy supply related to diurnal cycles, seasons, and other weather-related factors makes it highly unreliable. The present energy storing technologies fall short, necessitating the development of strategies to overcome the limitations of sustainable energy technologies. One of the most promising approaches is the efficient conversion of the intermittent energy into chemical fuels. The high energy density of these fuels (e.g., hydrogen, methane, and methanol) makes them a viable option for solar energy storage.

1.4 Solar Fuels

The term solar fuel refers to the fuels generated using energy in sunlight, typically using water, carbon dioxide or both as sources. Solar fuels hold immense potential for transforming the global fuel production and driving the transition to a sustainable energy system. Figure 1.3 illustrates the two primary routes for solar fuel production: solar-driven water splitting for hydrogen production and synthesis of hydrocarbons, alcohols, or other fuels through solar reduction of carbon dioxide. Other potential fuels such as ammonia and hydrazine can also be produced with solar energy using nitrogen-abundant sources and solar hydrogen from water (Du et al., 2015; Patil et al., 2014; Wang et al., 2018); however, we limit the rest of our discussions to solar water splitting and carbon dioxide reduction routes.

Figure 1.3 Main routes of solar fuel production.

1.5 Solar Reduction of CO_2 for Fuel Production

The European Commission launched the European Strategic Energy Technology Plan (SET-Plan) in 2007 as a comprehensive scheme to combat global warming. This initiative was aimed at uniting national authorities, industry, civil society, and researchers in addressing the challenge of reducing carbon emissions (Commission of the European Communities, 2007). Within the SET-Plan strategy, carbon capture and storage (CCS) and carbon capture and utilization (CCU) are recognized as essential solutions for achieving this goal. CCS involves industrial methods for chemical capture of CO_2, conversion to pure CO_2, and its subsequent storage in geological formations. Nonetheless, CCS is limited by storage capacity constraints. CCU also involves capturing of CO_2, but the captured CO_2 is converted into products such as fuels, plastics, and so on. Utilization of captured carbon for the synthesis of fuels or value-added products cannot be viewed as a carbon dioxide removal (CDR) process to achieve a sudden reduction in the atmospheric CO_2 concentration (Smith et al., 2023). Nonetheless, it is a sustainable strategy that can facilitate CO_2 recycling and open opportunities toward achieving an everlasting greener economy if renewable energy is utilized for the processes. Solar fuel production from CO_2 is particularly appealing from economic and environmental perspectives, given the substantial demand for fuels compared to other value-added products (Herron et al., 2015).

In the past years, significant attention has been directed toward the reduction of CO_2 using H_2O. The technological utilization of CO_2 by reducing this molecule implies that the carbon atom changes its oxidation state from a +4 to +2 or lower. Such reactions require higher energy exchanges that can be provided by electricity, heat, or solar radiation (Zhu, 2019). For example, reducing CO_2 to CO (see Figure 1.4) involves a standard Gibbs energy of 257 kJ/mol and the process becomes sustainable if the energy comes from sunlight. Early studies by Halmann (1978) and Inoue et al. (1979) presented, respectively, photoelectrochemical (PEC) and photocatalytic (PC) methods for reducing CO_2 to organic compounds using H_2O (Halmann, 1978; Inoue et al., 1979). Subsequent research has focused on elucidating the fundamental mechanisms and finding the ideal semiconductor capable of harnessing sunlight to facilitate CO_2 conversion (Giménez & Bisquert, 2016; Huang et al., 2014; Low et al., 2015). Some typical products obtained are HCOOH, CO, H_2CO, CH_3OH, or hydrocarbons (e.g., CH_4), as illustrated in Figure 1.4. The underlying mechanism behind CO_2 photoreduction with H_2O is complex. Some researchers have suggested that H^+ photocatalyze this reaction via the proton-coupled multi-electron transfer process (He & Janáky, 2020; Li et al., 2021; Samanta & Srivastava, 2020; Wu et al., 2017). When water serves as an H^+ source, the active atomic hydrogen directly reduces CO_2 into solar products. On the contrary, when a strong reducing agent as H_2 is employed as an H source to promote the conversion, some peroxide intermediate states are the main ones responsible for performing the reduction (Goyal et al., 2020).

Although solar fuel production from CO_2 by sunlight-assisted processes has been widely explored, the reported conversion rates and efficiencies to date are rather low. Substantial improvements are necessary to develop scalable strategies that integrate existing nonrenewable technologies. In addition, since the electrochemical reactions from CO_2 to these compounds involve at least six electron transfer steps, a viable approach for CO_2 reduction involves the use of an H_2 source

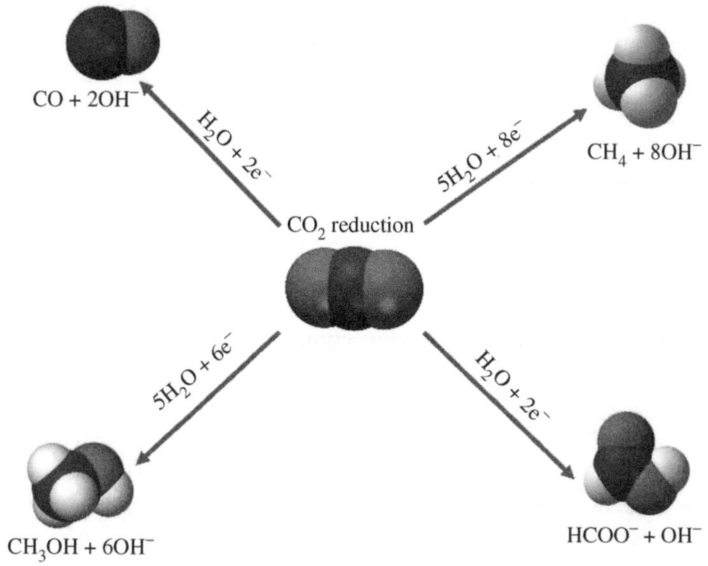

Figure 1.4 Possible CO_2 reduction reactions with their respective products, commonly considered as fuels or reactants for fuel production. The black, white, and dark gray spheres represent carbon, hydrogen, and oxygen atoms, respectively.

provided directly via solar water splitting. Contemplating these aspects, the solar-to-hydrogen conversion offers a faster route for a transition to the renewable fuels.

1.6 Solar Water Splitting for H_2 Generation

Hydrogen has gained significant attention as an ideal energy vector for creating a sustainable civilization. With its three times higher gravimetric energy density compared to fossil fuels and its clean combustion process that produces, in principle, only water as the product, hydrogen holds immense potential for various applications. It is an energy carrier similar to electricity; however, hydrogen is a chemical energy carrier consisting of molecules rather than only electrons. Therefore, it can be stored, transported, and used as a feedstock, just like fossil fuels. It can fuel industrial processes, provide electricity, and be used in transportation sector (Detz et al., 2018; Tuller, 2017). While H_2 can be employed in different areas, an energy system based on H_2 could help reduce greenhouse gas emissions, especially if it is generated from renewable sources, nuclear energy, or even fossil fuel-based systems with carbon sequestration.

Hydrogen produced using different methods is commonly identified by colors. The hydrogen produced from coal, natural gas, and lignite without carbon capture, utilization, and storage (CCUS) is labeled as Black, Gray, and Brown, respectively (International Energy Agency, 2019). Blue label is generally assigned to the hydrogen produced from fossil fuels, but CCUS applied. The hydrogen produced using renewable electricity is labeled as Green.

According to the IEA statistics, the global hydrogen production was 95 million metric tons, with ~62% of H_2 produced from natural gas without CCUS, ~21% using coal, and 16% as by-product from fossil fuel refining processes (International Energy Agency, 2023). Low-emission hydrogen, which is produced primarily from fossil fuels with CCUS, accounted for only 0.7% of the global production. Almost all hydrogen produced using natural gas is via steam methane reforming (SMR) without CCUS. From a commercial standpoint, steam reforming of methane remains the most economical route to produce H_2. However, the substantial CO_2 emissions associated with this process and the reliance on another greenhouse gas (natural gas) as a source impose limitations on using this technology. Renewable pathways for hydrogen production are being developed to generate long-term energy solutions and overcome the transition challenges toward a sustainable energy system. Countries such as the United States, Canada, Germany, and China have created their own initiatives to launch the H_2 economy, and the task aims at forming multidisciplinary collaborations to speed up the development and scale-up of H_2 technologies. The Hydrogen Shot program of the United States Department of Energy, launched in 2021, is such an initiative. It aims at reducing the cost of clean hydrogen by 80% to $1 per 1 kg in 1 decade ("111"). As of early 2021, the Hydrogen Council (Hydrogen Council, 2021) made known that over 30 countries released hydrogen roadmaps, the industry announced more than 200 hydrogen projects and ambitious investment plans, and governments worldwide committed more than US$ 70 billion in public funding to push the development of H_2 economy. The concept of a hydrogen-based economy is well-discussed in the literature (Bockris, 1999; Bonke et al., 2015; Kim et al., 2019; Tuller, 2017).

From all renewable routes, solar energy to hydrogen conversion with water as the H_2 source is considered as the most desirable option. Seeing that water covers about 70% of the Earth's surface, it can be contemplated as a convenient and abundant hydrogen source. Theoretically, 3.5×10^{13} L of water is needed to store the annual energy quote (4.7×10^{20} J) in the form of hydrogen, which corresponds to 0.01% of the annual rainfall, or 0.000,002% of the amount of water in the world. The overall water splitting reaction can be expressed as follows:

$$2H_2O + h\nu \rightleftharpoons 2H_2 + O_2 \quad \Delta G^0 = 237.2\,\text{kJ/mol} \tag{1.1}$$

where $h\nu$ represents the sunlight energy. This process corresponds to an oxidation–reduction reaction where two half-reactions occur separately. Under acidic media (pH = 0), the oxidation and reduction reactions can be written as follows:

$$2H^+ + 2e^- \rightleftharpoons H_2 \qquad \text{(reduction)} \tag{1.2}$$

$$2H_2O \rightleftharpoons O_2 + 4H^+ + 4e^- \qquad \text{(oxidation)} \tag{1.3}$$

From reactions (1.2) and (1.3), it can be noticed that the reduction half-reaction is a two-electron transfer reaction, while the oxidation reaction involves four electrons to obtain oxygen. Even though O_2 evolution is the limiting reaction in the water splitting process, water oxidation is significantly easier to accomplish compared to CO_2 reduction to fuels, since the latter needs at least six electrons to be transformed into fuels. However, alongside the challenges associated with successful hydrogen fuel production, the development of a safe, economical, and efficient hydrogen storage method represents another significant hurdle. Efforts are being directed toward the development of novel materials such as carbon nanotubes and metal–organic frameworks (MOFs) capable of storing large quantities of hydrogen for use in the transportation sector (Kaye et al., 2007; Li et al., 2009; Yang et al., 2012). The storage of hydrogen via chemical bonding by producing complex chemical and metallic hydrides is a recent development, and it presents an interesting opportunity in the field (Zacharia & Rather, 2015).

The development of efficient and scalable solar-driven processes for hydrogen and CO_2 reduction holds tremendous potential. Advanced catalysts, PEC systems, and PV technologies are being researched and optimized to enhance the performance and cost-effectiveness of these processes. In the upcoming section, an overview of various approaches and technologies employed for solar fuel production is discussed. This includes the utilization of solar energy for hydrogen generation through water splitting processes, as well as the conversion of CO_2 into fuels.

1.7 Solar to Fuel Conversion Pathways

Various pathways exist for converting solar energy into usable fuels from CO_2 and water, each employing unique approaches. Detz et al. (2018) have classified these pathways into four main categories, as illustrated in Figure 1.5: natural photosynthesis, concentration, transformation, and artificial photosynthesis. A parameter that is commonly used to compare different technologies is the energy efficiency. In general, it is defined as the ratio of the amount of energy stored in the fuel to the incident solar energy. The standard calculations involve use of work output from the fuel (the standard Gibbs energy change times the quantity of the product formed) in the numerator. For example, in the case of a water splitting experiment producing R moles of H_2 per second, the efficiency of the process is $(\Delta G^\circ R)/P_s$ where $\Delta G^\circ = 237.2\,\text{kJ/mol}$ at 25 °C and P_s is the solar power. Under AM 1.5 (air mass 1.5) conditions, which are the average atmospheric conditions in the United States, the solar power is taken as 1,000 W/m^2, which is generally termed as 1 sun.

1.7.1 Bioconversion

This route uses organisms such as algae, grasses, plants, and trees that harvest solar energy for producing biomass or biofuels (Varghese, 2013). Although these photosynthetic organisms can absorb light energy from most regions of the visible spectrum (400–700 nm), only the energy in the red region of the solar spectrum is effectively used for photosynthesis. The energy in excess of that in the red region is typically lost as heat. The near-infrared photons with energy lower than that of red photons are not absorbed. Another fraction is lost as reflected light. Furthermore, for each electron extracted

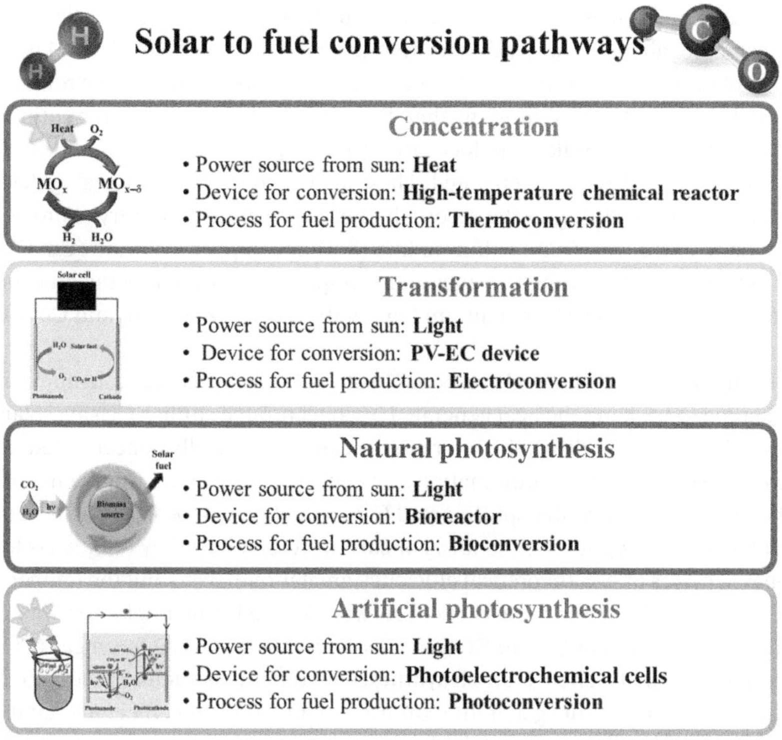

Figure 1.5 Solar to renewable fuel pathways according to Detz et al. (2018).

from water to reduce CO_2, two red photons are necessary to carry out the reaction. Considering these factors, the theoretical upper limit for the efficiency of solar energy to biomass conversion is about 4.5% for C3 plants (normal plants that produce 3-carbon compounds) and 6% for C4 plants (e.g., maize and sugarcane that form 4-carbon compounds) (Hankamer et al., 2007; Kruse et al., 2005; Skjånes et al., 2008). In reality, the efficiency of photosynthesis is <1%. Biogas, bioethanol, biokerosene, and biodiesel are examples of fuels derived from biomass.

A disadvantage of the solar energy to biomass/biofuel conversion strategy is the use of arable land and food crops that the growing population requires for survival. Besides the food scarcity and increase in prices this can cause, sustainability issues related to fertilizers, earth modifications, and water supplies are important concerns. Using lignocellulose instead of food crops such as maize and sugarcane for bioethanol production is a promising approach. Another approach is the conversion of waste biomass materials by direct photoconversion, which combines thermochemical and biological conversion routes. The production of methanol prepared by these routes could cost $0.48/kg (Zhu et al., 2011), while H_2 could be produced at $10/kg in a system with 2% of overall efficiency. These costs could be lowered by improving the efficiency (James et al., 2009).

The reduction in the emission of CO_2 also depends upon the biomass feedstock and the fuel produced. For example, ethanol derived from sugarcane could reduce CO_2 emissions by 90% (per life cycle analysis), whereas lignocellulose-based ethanol could lower the emissions by about 70% only. The emission reduction is still lower (60%) in the case of biodiesel (Varghese, 2013).

In recent years, there have been significant advancements in natural photosynthesis research. Efforts are underway to enhance light-harvesting capabilities, improve the efficiency of water oxidation and CO_2 reduction reactions, and optimize the overall performance of photosynthetic organisms. These developments aim to overcome the limitations of natural photosynthesis and increase its potential as a sustainable pathway for solar fuel production (Alexandropoulou et al., 2022; Cantera et al., 2022; Jin et al., 2023; Muñoz-Páez & Buitrón, 2022; Saravanan et al., 2022).

1.7.2 Thermoconversion

H_2O and CO_2 molecules can be dissociated using thermal energy. The process is called direct thermolysis. For the 100% dissociation of water and reduction of CO_2 to CO, temperatures of 4,300 and 3,350 K, respectively, are required. The Gibbs energy change in these reactions reduces as temperature increases and becomes zero at these temperatures. Partial dissociation can occur at lower temperatures. For example, about 9% water dissociation can take place at 2,500 K and 1 bar. Nevertheless, the reaction results in the formation of products as a mixture (H_2 and O_2 in the case of H_2O), and the reverse reaction can reduce the product yield (H_2 yield in the case of H_2O splitting) further. By performing the process in multiple steps, the useful work input (given by the Gibbs energy change) could be reduced to zero, the operating temperature could be lowered to manageable levels, and the products could be effectively separated. These are called thermochemical cycles.

Several thermochemical cycles were developed for water splitting in the second half of the twentieth century (Funk & Reinstrom, 1966; Kodama & Gokon, 2007). Zinc/zinc oxide (Zn/ZnO), sulfur–iodine (S–I) and copper–chlorine (Cu–Cl) are examples of cycles involving two, three, and four steps, respectively. The maximum temperature required in the process could be reduced by increasing the number of steps. For example, the maximum temperature in Zn/ZnO, S–I, and Cu–Cl cycles are 2,300, 1,123 and 803 K, respectively.

While the original motivation behind developing many thermochemical cycles was to utilize the heat wasted from nuclear power plants, in recent years the focus has been shifted to utilizing the heat generated by concentrated sunlight. Using solar concentrators in appropriate geometries (e.g., trough, tower, and dish), the solar energy can be practically concentrated to intensities in the range of 100–10,000 kW/m^2 from a 1 sun (1 kW/m^2) illumination (Leveque et al., 2017). The high temperatures required for even the two-step thermochemical cycles for water splitting and CO_2 reduction can be easily achieved from concentrated solar energy. Concentrated solar two-step cycles are the focus of current research as they offer ways to utilize concentrated sunlight efficiently (redox-based cycles offer a theoretical efficiency greater than 35%) and the temperatures are manageable with recently emerged redox materials. While the earlier efforts were focused almost solely on water splitting for H_2 generation, current research involves reduction of CO_2 to CO and co-splitting of CO_2-H_2O to syngas (CO and H_2 mix) also. Syngas could be used as a feedstock for the Fischer–Tropsch synthesis of liquid fuels such as methanol.

The challenges in utilizing the solar thermochemical cycles for fuel generation from an engineering and efficiency standpoint are enormous due to stability and radiative energy loss associated with high-temperature operation (Kim et al., 2012; Wang, 2020). Although a solar energy to chemical (fuel) energy conversion efficiency of over 20% can be practically achieved using the technology, the demonstrated efficiencies are in single digits. Solutions should also be found for technical problems such as those caused by the intermittency of solar energy, especially the frequent low light conditions due to the motion of clouds.

Nevertheless, there has been significant progress in recent years in developing advanced solar concentration technologies and utilizing them for scaling up the fuel generation using two-step thermochemical cycles. The conversion of CO_2 and H_2O to syngas and that of syngas to fuels could be accomplished in a single plant using energy from concentrated sunlight. For example, Steinfeld et al. demonstrated the operation of a pilot plant consisting of a solar concentrating tower, a solar reactor, and a gas-to-liquid conversion unit to synthesize kerosene (jet fuel) using concentrated solar energy (Zoller et al., 2022). The 50 kW reactor, utilizing a mean solar intensity of 2,500 suns, could perform a two-step thermochemical cycle involving cerium oxide (ceria; CeO_2) to produce syngas from CO_2 and H_2O at a solar to syngas efficiency of 4.1%. The syngas was then fed for Fischer–Tropsch synthesis of kerosene. The estimated cost of hydrogen produced from a two-step cycle varies between 3 and 17 \$/kg (Detz et al., 2018). Materials with improved thermal stability and optical properties such as ZnO/Zn (Bhosale et al., 2022a; Jiao et al., 2023), SnO_2/SnO (Bhosale et al., 2022b), CeO_2/Ce_2O_3 (Brugnoli et al., 2018; Bose et al., 2019), and perovskites (Gao et al., 2022, 2023; Pérez et al., 2022) are being explored, and advanced engineering solutions developed to optimize heat transfer and minimize energy losses. CeO_2 offers fast redox kinetics and long-term stability. It is currently considered the most promising material as it undergoes transition to a non-stoichiometric oxide state during the reduction process. The non-stoichiometric materials can perform better than the stoichiometric oxides in terms of yield and lower process temperature. The reactor designs for optimal operational parameters for highest solar-to-fuel efficiency are also being investigated (Abanades, 2022; Arribas et al., 2018; Boretti et al., 2022). These advancements have the potential to significantly enhance the efficiency and viability of solar fuel production by this approach.

1.7.3 Electroconversion

The splitting of water using an electric field was first reported in 1789 by Ardiaan Paets van Troostwijk and Jan Rudolph Deiman (Paets van Troostwijk & Deiman, 1789). Based on a finding made by Cavendish that hydrogen and oxygen in the ratio 2 : 1 could form water, they concluded that the gases that evolved were H_2 and O_2 (Cavendish, 1784). This electrochemical process of splitting water is called electrolysis and the device used for electrolysis is called an electrolyzer. By early 1900, many industrial electrolyzers were established. The technology is considered sustainable when electric power is generated using renewable sources and fuel is generated using water, CO_2, or both. Renewable H_2 can be produced by coupling water electrolysis with the power production technologies such as geothermal, hydro, wind, and solar PV. Water electrolysis is a mature technology that has been used for commercial hydrogen production for over a century. In 2022, only 0.2% of H_2 produced globally came from electrolysis (International Energy Agency, 2023). In an economic scope, the cost of electrolysis-based H_2 production oscillates between 2.91 and 10.92 $/kg according to the scale-up process (Schoots et al., 2008; U.S. Department of Energy Fuel Cell Technologies Office, 2013). Recent efforts have been focused on developing electrolyzers coupled with windfarms and PV installations.

In the electrolysis process, electrocatalysts perform redox processes under the influence of an electric field. An electrolyzer basically consists of a positive electrode (anode) and a negative electrode, both made of electrocatalysts, with an electrolyte between them. A potential difference of 1.229 V (corresponding to $\Delta G^\circ = 237.2$ kJ/mol or 1.229 eV/electron), applied between the electrodes, is theoretically required to split water. Oxygen evolution reaction (OER) takes place at the anode and hydrogen evolution reaction (HER) occurs at cathode. Cell stacks are used for high-volume production of H_2.

Water electrolysis efficiency is defined as the ratio of the high heating value of hydrogen (285.8 kJ/mol at 1 atm and 25 °C, which corresponds to 1.48 V, called thermoneutral voltage, V_{tn}) to the electrical energy expended to generate it. It can be calculated using the relation V_{tn}/V_{op}, where V_{op} is the operating voltage of the electrolyzer. Efficiencies as high as 98% have been reported for water electrolysis (Hodges et al., 2022). The overall efficiency of electrolytic hydrogen production is primarily limited by the efficiency of the system that powers it. For example, with a solar module efficiency of 20% (ratio of electrical power output to solar power input), an electrolyzer operating at an efficiency of 80% can produce hydrogen at a solar-to-hydrogen (STH) efficiency of 16% ($0.8 \times 0.2 \times 100$). PV-assisted electrolysis efficiency greater than 30% has been reported (Jia et al., 2016).

The electrolyzers are classified into three primary types: alkaline, polymer electrolyte membrane (PEM),[4] and solid oxide (SO) type based on the electrolyte used. The conventional electrolyzers use alkaline electrolytes (e.g. potassium hydroxide). PEM and SO electrolyzers use solid electrolytes that allow only the migration of ions between the electrodes. Examples of electrolytes used in PEM and SO electrolyzers are Nafion™ (perfluorosulfonic acid) and yttria-stabilized zirconia, respectively. While alkaline and PEM technologies are mature with an operating temperature typically less than 100 °C, SO electrolyzers operate at high-temperatures (in the range of 800 °C). The high-temperature operation permits SO electrolyzers to have lower electrical energy requirement and higher efficiency compared to the other two electrolyzer types. PEM electroyzers are currently used for commercial H_2 production. The SO electrolyzer technology is becoming ready for commercial implementation.

Simultaneous electrolysis of CO_2 and H_2O allows production of carbon-based gas and liquid-phase solar fuels. For example, a solid oxide electrolysis cell can produce syngas from a feed of CO_2 and H_2O (Gaudillere et al., 2014; Ni, 2012; Schäfer et al., 2019). Syngas could be used for making liquid fuels. The cost of methanol and gasoline so produced is estimated to be near 0.57 and 1.49 $/kg, respectively, if only electricity consumption is considered. Using alkaline and PEM electrolyzers, CO_2 dissolved in aqueous solutions could be directly converted into gas phase (e.g., CO and CH_4) and/or liquid phase (e.g., formic acid and methanol) fuels. The solar-to-fuel efficiency of a typical solar cell-powered CO-H_2O electrolyzer is currently lower compared to water electrolysis. Product selectivity is also a problem to be solved.

The advancements in PV technology and ongoing research to obtain highly efficient electrocatalysts (EC), mostly based on carbon-supported electrocatalysts such as platinum (Cui et al., 2023; Mensharapov et al., 2023; Muhyuddin et al., 2023; Najafli et al., 2023), nickel (Hao et al., 2023; Jia et al., 2023; Wei et al., 2023; Zhang et al., 2023a), and iridium (Huynh et al., 2023; Wang et al., 2023a; Xu et al., 2023; Zhang et al., 2023b), together with optimized PV-EC designs (Alhaj Omar, 2023; Gulzar et al., 2023; Holmes-Gentle et al., 2021; Mohammed et al., 2023), are shaping the state of the art in solar and renewable fuel technologies, offering promising systems for a sustainable and clean energy future. Many installations of electrolyzers powered by solar modules are emerging. By 2030, it is expected that 14 million tons of H_2 will be produced by electrolyzers with power from solar modules, other renewable sources and fossil fuels with CCUS and this is expected to account for about 10% of total hydrogen production.

4 These electrolyzers are also called solid polymer electrolytes (SPEs). PEM is also used as an acronym for proton exchange membrane. The term polymer electrolyte membrane is more generic as it covers both anion and cation exchange type polymer membranes and matches with that used for corresponding fuel cells.

1.7.4 Photoconversion

Solar PC and PEC fuel generation processes utilize sunlight as the energy source to split H_2O or co-split H_2O and CO_2 to produce fuels such as H_2, CO, hydrocarbons, alcohols, and so on. These processes are commonly known as artificial photosynthesis as they mimic mechanisms of natural photosynthesis. These processes involve a single unit that converts sunlight directly to fuels at normal atmospheric conditions, which is in contrast with a PV-assisted electrochemical fuel generation process that involves two separate systems, PV and electrolyzer units, or a thermochemical system that involves very high temperatures. They use organic, inorganic, or hybrid semiconductors (called photocatalysts) to absorb sunlight and generate charge carriers, electrons, and holes that perform redox reactions leading to fuel generation. PEC water splitting was demonstrated by Fujishima and Honda in 1969 using titania as the photocatalyst (Fujishima et al., 1969; Fujishima & Honda, 1972). A PEC process to reduce CO_2 was first reported by Halmann (1978).

In a PC (or photochemical) water splitting process the photocatalyst dissociates water molecules spontaneously upon solar illumination (Fujishima & Zhang, 2006; Lin et al., 2018; Tee et al., 2017). The OER and HER are not spatially separated and therefore, a hydrogen–oxygen mixture is generated in the same reaction chamber. The band gap of the photocatalyst must be at least 1.229 eV (Gibbs energy change per electron), in theory, to split water at 25 °C and 1 atm. The minimum band gap required is 1.33 eV for the reduction of CO_2 to CO. Considering the entropy change and overpotential losses, the band gap should be in the range of 2.0–2.2 eV for a photocatalyst to produce fuels most efficiently. Additionally, the conduction band minimum of the semiconductor must be at a potential above the reduction potential (0 V vs. normal hydrogen electrode, NHE, for H_2O at pH = 0) and the valence band maximum must be at a potential below the oxidation potential (1.23 V vs. NHE for H_2O at pH = 0) for a spontaneous fuel generation process upon illumination. For the gas phase conversion of CO_2 and H_2O to fuels, a photocatalyst bed or sheet is generally kept for illumination in a reaction chamber containing the reactants (Roy et al., 2010; Varghese et al., 2009). The photocatalyst particles are typically dispersed in water or in an aqueous solution containing dissolved CO_2 for conversion from liquid phase. Both gas phase and liquid phase fuels can be formed during such dissolved CO_2 conversion. The demonstrated efficiency of the process ($\Delta G°R/P$, as mentioned in the beginning of Section 1.7) is in single digits. Sustaining the fuel generation for a long time is a problem to be solved.

A solar PEC cell consists of an anode and a cathode separated by an electrolyte as in an electrochemical cell; however, in a PEC cell, semiconductors form the anode (called photoanode for OER) or cathode (called photocathode for HER) or both (Bosserez et al., 2015; Thongthep et al., 2021; Wang et al., 2014). If only one electrode is made of a photocatalyst, then the other electrode is usually an electrocatalyst such as platinum. The photovoltage generated between the electrodes is often insufficient to perform water splitting, so an external electrical bias is applied between the electrodes. Nevertheless, PEC cells with appropriate semiconductors forming both anode and cathode could generate sufficient photovoltage for water splitting without an external bias (Mor et al., 2008; Nozik, 1977). Unlike the case of PC reactors, in PEC cells the oxidation and reduction processes are spatially separated; therefore, product separation is spontaneous and the process is safe.

The practical sunlight to fuel efficiency of a PEC water splitting process involving a single semiconductor can reach about 18% (Varghese & Grimes, 2008). With the use of tandem configuration and more advanced strategies such as multiple carrier generation, hot carrier extraction, and so on, the efficiency can go significantly higher. Water splitting PEC cells with STH efficiency of about 14% have been reported (May et al., 2015). Nevertheless, the high-efficiency cells generally use materials that are expensive and prone to degradation over long-term operation.

There have been significant efforts to develop earth-abundant, low-cost, and nontoxic semiconductors as photocatalysts. The present research involves materials, especially many metal oxides (Bedin et al., 2023; Berardi et al., 2022; Gonzaga & Mercado, 2022; Ramakrishnan et al., 2023; Rodríguez-Gutiérrez et al., 2022; Wang et al., 2023b) and gallium nitride-based compounds (Cui et al., 2023; Luo et al., 2023; Xiao et al., 2022; Zheng et al., 2023), with different morphologies and dimensions or their combinations (Bai et al., 2022; Centurion et al., 2022; Zhao et al., 2022). This opens up new possibilities for efficient solar fuel generation. Furthermore, the state-of-the-art research in solar-to-renewable fuel technologies also demonstrates that the utilization of computational tools, operando technologies, and AI technologies plays a pivotal role in advancing the development of highly efficient materials for artificial photosynthesis (Bevan et al., 2023; Oral et al., 2022; Piccioni et al., 2023; Ramakrishnan et al., 2023; Sun et al., 2022; Wang et al., 2022; Yang et al., 2022). These advancements in PEC technologies are propelling the development of efficient and robust solar fuel production systems.

Economic analyses performed to assess the viability of the artificial photosynthesis systems showed that the estimated cost of hydrogen from artificial photosynthesis would be 9.2–11.4 \$/kg, considering the technological challenges that must be overcome for commercial-scale implementation (Pinaud et al., 2013; Shaner et al., 2016). The cost is expected to reduce with technology advancement. There are serious efforts to verify the large-scale implementation of the PEC technology. Domen et al. have recently demonstrated the operation of a large-area pilot plant (Nishiyama et al., 2021). The team developed a 100 m^2 reactor consisting of $SrTiO_3$:Al as the photocatalyst (Figure 1.6). This was the first demonstration of

Figure 1.6 Large-area deployment of PEC panels for solar hydrogen production. (Left) The 100 m^2 solar production system. (Right) The evolution of H_2 and O_2 upon exposing the panel to sunlight. This mixture is then introduced into an H_2 separation unit. Courtesy: Kazunari Domen, The University of Tokyo, Japan.

the feasibility of a PEC technology for large-scale fuel generation. Ongoing research also aims at optimizing the efficiency, stability, and cost-effectiveness of artificial photosynthesis systems, including reactor design (Arribas et al., 2018; Kan et al., 2022; Karaca & Dincer, 2023; Qureshy & Dincer, 2022), paving the way for their integration into installations for large-scale solar fuel production .

1.8 Conclusion

Renewable fuel technologies offer a transformative solution to tackle the pressing challenges of climate change and sustainability. The three most promising artificial fuel generation technologies utilizing sunlight with water, carbon dioxide, or both as precursors are thermochemical, PV-assisted electrolysis, and PEC. While PV-assisted electrolysis has started contributing to commercial level H_2 production, the other two technologies are in the development phase. While all technologies are advancing, there is an immediate need to scale-up the technologies to contribute to low-cost and large-volume solar fuel generation for achieving net zero emission by 2050. It is crucial for both government agencies and the private sector to actively participate by supporting initiatives and making investments that accelerate the global energy transition. This transition will not only help reduce greenhouse gas emissions but also promote energy independence and drive economic growth. It is also important to consider technological, environmental, and economic factors to ensure affordable, reliable, and sustainable energy access for all. By embracing the immense possibilities of solar to renewable fuels technologies, we pave the way for a brighter and greener future, ensuring a sustainable energy landscape for future generations to thrive upon.

References

Abanades, S. (2022). Redox cycles, active materials, and reactors applied to water and carbon dioxide splitting for solar thermochemical fuel production: A review. *Energies 15* (19): 7061.

Alexandropoulou, M., Antonopoulou, G., and Lyberatos, G. (2022). Modeling of continuous dark fermentative hydrogen production in an anaerobic up-flow column bioreactor. *Chemosphere 293*: 133527.

Alhaj Omar, F. (2023). A new approach for improving the efficiency of the indirectly coupled photovoltaic-electrolyzer system. *International Journal of Hydrogen Energy 48* (24): 8768–8782.

Arrhenius, S. (1896). On the influence of carbonic acid in the air upon the temperature of the ground. *Philosophical Magazine and Journal of Science 41*: 237–276.

Arrhenius, S. (1908). *Worlds in the making: the evolution of the universe*. New York: Harper & Brothers Publishers.

Arribas, L., González-Aguilar, J., and Romero, M. (2018). Solar-driven thermochemical water-splitting by cerium oxide: determination of operational conditions in a directly irradiated fixed bed reactor. *Energies 11* (9): 2451.

Bai, J., Shen, R., Chen, W. et al. (2022). Enhanced photocatalytic H_2 evolution based on a $Ti_3C_2/Zn_{0.7}Cd_{0.3}S/Fe_2O_3$ Ohmic/S-scheme hybrid heterojunction with cascade 2D coupling interfaces. *Chemical Engineering Journal 429* (September 2021): 132587.

Bedin, K.C., Rodríguez-Gutiérrez, I., Peregrino, L.R.P. et al. (2023). On electron loss lowering at hematite photoelectrode interfaces. *Journal of the American Ceramic Society 106* (1): 79–92.

Berardi, S., Cristino, V., Bignozzi, C.A. et al. (2022). Hematite-based photoelectrochemical interfaces for solar fuel production. *Inorganica Chimica Acta 535*: 120862.

Bevan, K.H., Miao, B., and Iqbal, A. (2023). SLJCompact: A semiconductor-liquid junction solver for rapid band diagram insights into photoelectrochemical devices. *Computer Physics Communications 286*: 108638.

Bhore, S. (2016). Paris agreement on climate change: A booster to enable sustainable global development and beyond. *International Journal of Environmental Research and Public Health 13* (11): 1134.

Bhosale, R.R., Gupta, R.B., and Shende, R.V. (2022a). Thermodynamic evaluation of solar assisted ZnO/Zn thermochemical CO_2 splitting cycle. *Environmental Research 212* (Pt B): 113266.

Bhosale, R.R., Shende, R.V., and Gupta, R.B. (2022b). SnO_2/SnO based redox thermochemical CO_2 splitting cycle: Effect of inert gas flowrate, reduction temperature, and gas separation on the solar-to-fuel energy conversion efficiency. *International Journal of Energy Research 46* (7): 9267–9280.

Bockris, J.O. (1999). Hydrogen economy in the future. *International Journal of Hydrogen Energy 24* (1): 1–15.

Bonke, S.A., Wiechen, M., MacFarlane, D.R., and Spiccia, L. (2015). Renewable fuels from concentrated solar power: Towards practical artificial photosynthesis. *Energy & Environmental Science 8* (9): 2791–2796.

Boretti, A., Castelletto, S., and De Angelis, F. (2022). A solar concentrator/receiver/storage/reactor system for thermochemical splitting cycles based on perovskites. *International Journal of Hydrogen Energy 47* (7): 4970–4975.

Bose, A., Farooqui, A., Ferrero, D. et al. (2019). Thermodynamic assessment of non-catalytic Ceria for syngas production by methane reduction and $CO_2 + H_2O$ oxidation. *Materials for Renewable and Sustainable Energy 8* (1): 5.

Bosserez, T., Rongé, J., van Humbeeck, J. et al. (2015). Design of compact photoelectrochemical cells for water splitting. *Oil & Gas Science and Technology – Revue d'IFP Energies nouvelles 70* (5): 877–889.

Brugnoli, L., Ferrari, A.M., Civalleri, B. et al. (2018). Assessment of density functional approximations for highly correlated oxides: The case of CeO_2 and Ce_2O_3. *Journal of Chemical Theory and Computation 14* (9): 4914–4927.

Cantera, S., Tamarit, D., Strong, P.J. et al. (2022). Prospective CO_2 and CO bioconversion into ectoines using novel microbial platforms. *Reviews in Environmental Science and Bio/Technology 21* (3): 571–581.

Cavendish, H. (1784). Experiments on air. *Philosophical Transactions of Royal Society of London 74*: 119–153.

Centurion, H.A., Rabelo, L.G., Rodriguez-Gutierrez, I. et al. (2022). Constructing particulate p–n heterojunction Mo:$SrTiO_3$/NiO@$Ni(OH)_2$ for enhanced H_2 evolution under simulated solar light. *ACS Applied Energy Materials 5* (10): 12727–12738.

Commission of the European Communities (2007). A European strategic energy technology plan (SET-PLAN). *Commission of the European Communities.*

Cui, P., Zhao, L., Long, Y. et al. (2023). Carbon-based electrocatalysts for acidic oxygen reduction reaction. *Angewandte Chemie International Edition 62* (14): 2300911.

Detz, R.J., Reek, J.N.H., and van der Zwaan, B.C.C. (2018). The future of solar fuels: When could they become competitive? *Energy & Environmental Science 11* (7): 1653–1669.

Du, Z., Denkenberger, D., and Pearce, J.M. (2015). Solar photovoltaic powered on-site ammonia production for nitrogen fertilization. *Solar Energy 122*: 562–568.

Fleming, J.R. (1999). Joseph Fourier, the 'greenhouse effect', and the quest for a universal theory of terrestrial temperatures. *Endeavour 23*: 72–75.

Florence, J.M., Allshouse, C.C., Glaze, F.W., and Hahner, C.H. (1950). Absorption of near-infrared energy by certain glasses. *Journal of the Research of the National Bureau of Standards 45*: 121–128.

Foote, E. (1856). On the heat in the Sun's rays. *The American Journal of Science and Arts 22*: 377–383.

Fourier, J. (1824). Remarques générales sur les températures du globe terrestre et des espaces planétaires. *Annales de Chimie et de Physique 27*: 136–167. English translation: Burgess, E. (1837). General remarks on the temperature of the terrestrial globe and the planetary spaces. *American Journal of Sciences, 32*, 1–20.

Fujishima, A. and Honda, K. (1972). Electrochemical photolysis of water at a semiconductor electrode. *Nature 238*: 37–38.

Fujishima, A. and Zhang, X. (2006). Titanium dioxide photocatalysis: present situation and future approaches. *Comptes Rendus Chimie 9* (5–6): 750–760.

Fujishima, A., Honda, K., and Kikuchi, S. (1969). Photosensitized electrolytic oxidation on TiO_2 semiconductor electrode. *Journal of the Chemical Society of Japan (Kogyo Kagaku Zasshi) 72*: 108–109.

Funk, J.E. and Reinstrom, R.M. (1966). Energy requirements in the production of hydrogen from water. *I&EC Process Design and Development 5*: 336–342.

Gao, K., Liu, X., Jiang, Z. et al. (2022). Direct solar thermochemical CO_2 splitting based on Ca- and Al-doped $SmMnO_3$ perovskites: Ultrahigh CO yield within small temperature swing. *Renewable Energy 194*: 482–494.

Gao, K., Liu, X., Wang, Q. et al. (2023). Remarkable solar thermochemical CO_2 splitting performances based on Ce- and Al-doped $SrMnO_3$ perovskites. *Sustainable Energy & Fuels 7* (4): 1027–1040.

Gaudillere, C., Navarrete, L., and Serra, J.M. (2014). Syngas production at intermediate temperature through H_2O and CO_2 electrolysis with a Cu-based solid oxide electrolyzer cell. *International Journal of Hydrogen Energy 39* (7): 3047–3054.

Giménez, S. and Bisquert, J. (2016). *Photoelectrochemical solar fuel production: From basic principles to advanced devices.* Cham: Springer International Publishing.

Gonzaga, I.L.E. and Mercado, C.C. (2022). Copper ternary oxides as photocathodes for solar-driven CO_2 reduction. *Reviews On Advanced Materials Science 61* (1): 430–457.

Goyal, A., Marcandalli, G., Mints, V.A., and Koper, M.T.M. (2020). Competition between CO_2 reduction and hydrogen evolution on a gold electrode under well-defined mass transport conditions. *Journal of the American Chemical Society 142* (9): 4154–4161.

Gulzar, M.M., Iqbal, A., Sibtain, D., and Khalid, M. (2023). An innovative converterless solar PV control strategy for a grid connected hybrid PV/wind/fuel-cell system coupled with battery energy storage. *IEEE Access 11* (March): 23245–23259.

Halmann, M. (1978). Photoelectrochemical reduction of aqueous carbon-dioxide on p-type gallium-phosphide in liquid junction solar-cells. *Nature 275*: 115–116.

Hankamer, B., Lehr, F., Rupprecht, J. et al. (2007). Photosynthetic biomass and H_2 production by green algae: from bioengineering to bioreactor scale-up. *Physiologia Plantarum 131* (1): 10–21.

Hao, J., Wu, K., Lyu, C. et al. (2023). Recent advances in interface engineering of Fe/Co/Ni-based heterostructure electrocatalysts for water splitting. *Materials Horizons 10*: 2312–2342.

He, J. and Janáky, C. (2020). Recent advances in solar-driven carbon dioxide conversion: expectations versus reality. *ACS Energy Letters 5* (6): 1996–2014.

Herron, J.A., Kim, J., Upadhye, A.A. et al. (2015). A general framework for the assessment of solar fuel technologies. *Energy and Environmental Science 8* (1): 126–157.

Hodges, A., Hoang, A.L., Tsekouras, G. et al. (2022). A high-performance capillary-fed electrolysis cell promises more cost-competitive renewable hydrogen. *Nature Communications 13*: 1304.

Holmes-Gentle, I., Tembhurne, S., Suter, C., and Haussener, S. (2021). Dynamic system modeling of thermally-integrated concentrated PV-electrolysis. *International Journal of Hydrogen Energy 46* (18): 10666–10681.

Huang, Z.-F., Pan, L., Zou, J.-J. et al. (2014). Nanostructured bismuth vanadate-based materials for solar-energy-driven water oxidation: a review on recent progress. *Nanoscale 6* (23): 14044–14063.

Huynh, T.B.N., Song, J., Bae, H.E. et al. (2023). Ir–Ru electrocatalysts embedded in N-doped carbon matrix for proton exchange membrane water electrolysis. *Advanced Functional Materials 33*: 2301999.

Hydrogen Council (2021). Hydrogen insights (February), 58.

Inoue, T., Fujishima, A., Konishi, S., and Honda, K. (1979). Photoelectrocatalytic reduction of carbon dioxide in aqueous suspensions of semiconductor powders. *Nature 277* (5698): 637–638.

International Energy Agency (2019). The future of hydrogen: Seizing today's opportunities. Retrieved from https://www.iea.org/reports/global-hydrogen-review-2023

International Energy Agency (2022). World Energy Outlook 2022, IEA. License: Creative Commons Attribution CC BY-NC-SA 4.0. https://www.iea.org/reports/world-energy-outlook-2022

International Energy Agency (2023). Global hydrogen review 2023. Retrieved from https://www.iea.org/reports/global-hydrogen-review-2023

IPCC (2014). *Climate change 2014: Synthesis report. Contribution of Working Groups I, II and III to the Fifth Assessment Report of the Intergovernmental Panel on Climate Change* [Core Writing Team, (ed. R.K. Pachauri and L.A. Meyer). Geneva, Switzerland: IPCC, 151 pp.

IPCC (2023). Summary for policymakers. In: *Climate change 2023: Synthesis report. Contribution of Working Groups I, II and III to the Sixth Assessment Report of the Intergovernmental Panel on Climate Change* [Core Writing Team, (ed. H. Lee and J. Romero), 1–34. Geneva, Switzerland: IPCC https://doi.org/10.59327/IPCC/AR6-9789291691647.001.

James, B.D., Baum, G.N., Perez, J., and Baum, K.N. (2009). Technoeconomic boundary analysis of biological pathways to hydrogen production. *DOE Contract Number SR-560-46674*, **22201** (September), 1–193.

Jia, J., Seitz, L.C., Benck, J.D. et al. (2016). Solar water splitting by photovoltaic-electrolysis with a solar-to-hydrogen efficiency over 30%. *Nature Communications 7*: 13237.

Jia, L., Li, J., Yu, X. et al. (2023). In situ synthesis of nanorod arrays of nickel–molybdenum nitrides as stable electrocatalysts for hydrogen evolution reactions. *ACS Applied Nano Materials 6* (2): 1050–1058.

Jiao, F., Lu, B., Chen, C. et al. (2023). In-depth energy and irreversibility analysis in the solar driven two-step thermochemical water splitting cycle for hydrogen production. *Journal of Thermal Science 32* (1): 1–16.

Jin, T., Wang, Y., Yao, S. et al. (2023). Bioconversion of carbon dioxide to succinate by Citrobacter. *Chemical Engineering Journal 452* (P4): 139668.

Kan, M., Wang, Q., Hao, S. et al. (2022). System engineering enhances photoelectrochemical CO_2 reduction. *The Journal of Physical Chemistry C 126* (4): 1689–1700.

Karaca, A.E. and Dincer, I. (2023). Development and evaluation of a solar based integrated hydrogen energy system for mobile applications. *Energy Conversion and Management 280* (November 2022): 116808.

Kaye, S.S., Dailly, A., Yaghi, O.M., and Long, J.R. (2007). Impact of preparation and handling on the hydrogen storage properties of $Zn_4O(1,4\text{-benzenedicarboxylate})_3$ (MOF-5). *Journal of the American Chemical Society 129* (46): 14176–14177.

Kim, J., Johnson, T.A., Miller, J.E. et al. (2012). Fuel production from CO_2 using solar-thermal energy: System level analysis. *Energy & Environmental Science 5* (9): 8417.

Kim, J.H., Hansora, D., Sharma, P. et al. (2019). Toward practical solar hydrogen production – An artificial photosynthetic leaf-to-farm challenge. *Chemical Society Reviews 48* (7): 1908–1971.

Kodama, T. and Gokon, N. (2007). Thermochemical cycles for high temperature solar hydrogen production. *Chemical Reviews 107*: 4048–4077.

Kruse, O., Rupprecht, J., Bader, K.-P. et al. (2005). Improved photobiological H_2 production in engineered green algal cells. *Journal of Biological Chemistry 280* (40): 34170–34177.

Lenton, T., Xu, C., Abrams, J.F. et al. (2023). Quantifying the human cost of global warming. *Nature Sustainability 6*: 1237–1247.

Leveque, G., Bader, R., Lipinski, W., and Haussener, S. (2017). High-flux optical systems for solar thermochemistry. *Solar Energy 156*: 133–148.

Li, J., Cheng, S., Zhao, Q. et al. (2009). Synthesis and hydrogen-storage behavior of metal–organic framework MOF-5. *International Journal of Hydrogen Energy 34* (3): 1377–1382.

Li, Y., Hui, D., Sun, Y. et al. (2021). Boosting thermo-photocatalytic CO_2 conversion activity by using photosynthesis-inspired electron-proton-transfer mediators. *Nature Communications 12* (1): 123.

Lin, Z., Li, W., and Yang, G. (2018). Hydrogen-interstitial $CuWO_4$ nanomesh: A single-component full spectrum-active photocatalyst for hydrogen evolution. *Applied Catalysis B: Environmental 227*: 35–43.

Low, J., Yu, J., and Ho, W. (2015). Graphene-based photocatalysts for CO_2 reduction to solar fuel. *Journal of Physical Chemistry Letters 6* (21): 4244–4251.

Luo, Y., Wang, D., Kang, Y. et al. (2023). Reprogrammable binary and ternary optoelectronic logic gates composed of nanostructured GaN photoelectrodes with bipolar photoresponse characteristics. *Advanced Optical Materials 11* (13): 2300129.

May, M.M., Lewerenz, H.-J., Lackner, D. et al. (2015). Efficient direct solar-to-hydrogen conversion by in situ interface transformation of a tandem structure. *Nature Communications 6*: 8286.

McKay, D.I.A., Staal, A., Abrams, J.F. et al. (2022). Exceeding 1.5 °C global warming could trigger multiple climate tipping points. *Science 377*: 7950.

Mensharapov, R.M., Spasov, D.D., Ivanova, N.A. et al. (2023). Screening of carbon-supported platinum electrocatalysts using Frumkin adsorption isotherms. *Inorganics 11* (3): 103.

Mohammed, A., Ghaithan, A.M., Al-Hanbali, A., and Attia, A.M. (2023). A multi-objective optimization model based on mixed integer linear programming for sizing a hybrid PV-hydrogen storage system. *International Journal of Hydrogen Energy 48* (26): 9748–9761.

Mor, G.K., Varghese, O.K., Wilke, R.H.T. et al. (2008). p-Type Cu-Ti-O nanotube arrays and their use in self-biased heterojunction photoelectrochemical diodes for hydrogen generation. *Nano Letters 8*: 1906–1911.

Muhyuddin, M., Friedman, A., Poli, F. et al. (2023). Lignin-derived bimetallic platinum group metal-free oxygen reduction reaction electrocatalysts for acid and alkaline fuel cells. *Journal of Power Sources 556* (November 2022): 232416.

Muñoz-Páez, K.M. and Buitrón, G. (2022). Bioconversion of H_2 and CO_2 from dark fermentation to methane: Effect of operating conditions on methane concentration. *Chemosphere 308* (August): 136305.

Najafli, E., Ratso, S., Ivanov, Y.P. et al. (2023). Sustainable CO_2-derived nanoscale carbon support to a platinum catalyst for oxygen reduction reaction. *ACS Applied Nano Materials 6* (7): 5772–5780.

Ni, M. (2012). 2D thermal modeling of a solid oxide electrolyzer cell (SOEC) for syngas production by H_2O/CO_2 co-electrolysis. *International Journal of Hydrogen Energy 37* (8): 6389–6399.

Nishiyama, H., Yamada, T., Nakabayashi, M. et al. (2021). photocatalytic solar hydrogen production from water on a 100 m^2 scale. *Nature 598*: 304–307.

Nozik, A.J. (1977). Photochemical diodes. *Applied Physics Letters 30*: 567–569.

O'Neill, B.C. (2023). Envisioning a future with climate change. *Nature Climate Change 13*: 874–876.

Oral, B., Can, E., and Yildirim, R. (2022). Analysis of photoelectrochemical water splitting using machine learning. *International Journal of Hydrogen Energy 47* (45): 19633–19654.

Paets van Troostwijk, A. and Deiman, J.R. (1789). Sur une mani'ere de d'ecomposer l'Eau en Air inflammable et en Air vital. *Observations et Mémoires sur la physique, sur l'histoire naturelle et sur les arts et Métiers. Journal de Physique, de Chimie et d'Histoire Naturelle 35*: 369. English translation: Pearson, G. (1797). Experiments and observations, made with the view of ascertaining the nature of the gas produced by passing electric discharges through water. *Philosophical Transactions of Royal Society of London*, 87, 142–158.

Patil, A., Laumans, L., and Vrijenhoef, H. (2014). Solar to ammonia – via proton's NFuel units. *Procedia Engineering 83*: 322–327.

Pérez, A., Orfila, M., Linares, M. et al. (2022). Hydrogen production by thermochemical water splitting with $La_{0.8}Al_{0.2}MeO_{3-\delta}$ (Me = Fe, Co, Ni and Cu) perovskites prepared under controlled pH. *Catalysis Today 390–391* (July 2021): 22–33.

Piccioni, A., Vecchi, P., Vecchi, L. et al. (2023). Distribution of relaxation times based on lasso regression: a tool for high-resolution analysis of IMPS data in photoelectrochemical systems. *The Journal of Physical Chemistry C 127* (17): 7957–7964.

Pinaud, B.A., Benck, J.D., Seitz, L.C. et al. (2013). Technical and economic feasibility of centralized facilities for solar hydrogen production via photocatalysis and photoelectrochemistry. *Energy & Environmental Science 6* (7): 1983.

Qureshy, A.M.M.I. and Dincer, I. (2022). A new photoelectrochemical reactor with special photocathode design for hydrogen production. *Advanced Engineering Materials 24* (8): 2101509.

Ramakrishnan, V., Tsyganok, A., Davydova, E. et al. (2023). Competitive photo-oxidation of water and hole scavengers on hematite photoanodes: Photoelectrochemical and operando Raman spectroelectrochemistry study. *ACS Catalysis 13* (1): 540–549.

Rodríguez-Gutiérrez, I., Bedin, K.C., Mouriño, B. et al. (2022). Advances in engineered metal oxide thin films by low-cost, solution-based techniques for green hydrogen production. *Nanomaterials 12* (12): 1957.

Roy, C.R., Varghese, O.K., Paulose, M., and Grimes, C.A. (2010). Toward solar fuels: Photocatalytic conversion of carbon dioxide to hydrocarbons. *ACS Nano 4*: 1259–1278.

Samanta, S. and Srivastava, R. (2020). Catalytic conversion of CO_2 to chemicals and fuels: the collective thermocatalytic/photocatalytic/electrocatalytic approach with graphitic carbon nitride. *Materials Advances 1* (6): 1506–1545.

Saravanan, A., Kumar, P.S., Mat Aron, N.S. et al. (2022). A review on bioconversion processes for hydrogen production from agro-industrial residues. *International Journal of Hydrogen Energy 47* (88): 37302–37320.

Schäfer, D., Fang, Q., Blum, L., and Stolten, D. (2019). Syngas production performance and degradation analysis of a solid oxide electrolyzer stack. *Journal of Power Sources 433*: 126666.

Schoots, K., Ferioli, F., Kramer, G., and Vanderzewaan, B. (2008). Learning curves for hydrogen production technology: An assessment of observed cost reductions. *International Journal of Hydrogen Energy 33* (11): 2630–2645.

Shaner, M.R., Atwater, H.A., Lewis, N.S., and McFarland, E.W. (2016). A comparative technoeconomic analysis of renewable hydrogen production using solar energy. *Energy and Environmental Science 9* (7): 2354–2371.

Skjånes, K., Knutsen, G., Källqvist, T., and Lindblad, P. (2008). H_2 production from marine and freshwater species of green algae during sulfur deprivation and considerations for bioreactor design. *International Journal of Hydrogen Energy 33* (2): 511–521.

Smith, S.M., Geden, O., Nemet, G. et al. (2023). *The State of Carbon Dioxide Removal*, 1e. Retrieved from https://www.stateofcdr.org.

Sun, X., Tiwari, D., Li, M., and Fermin, D.J. (2022). Decoupling the impact of bulk and surface point defects on the photoelectrochemical properties of $LaFeO_3$ thin films. *Chemical Science 13* (37): 11252–11259.

Tee, S.Y., Win, K.Y., Teo, W.S. et al. (2017). Recent progress in energy-driven water splitting. *Advanced Science 4* (5): 1600337.

Thongthep, P., Moonmangmee, S., and Ponchio, C. (2021). Solar/photoelectrocatalytic cell development for H_2 production and simultaneous organic dye degradation. *Materials Science in Semiconductor Processing 124* (June 2020): 105597.

Tollefson, J. (2021). Top climate scientists are skeptical that nations can rein in global warming. *Nature 599*: 22–24.

Tuller, H.L. (2017). Solar to fuels conversion technologies: A perspective. *Materials for Renewable and Sustainable Energy 6* (1): 3.

Turner, M.C., Andersen, Z.J., Baccarelli, A. et al. (2020). Outdoor air pollution and cancer: an overview of the current evidence and public health recommendations. *CA: A Cancer Journal for Clinicians 70*: 460–479.

Tyndall, J. (1861). The Bakerian Lecture: On the absorption and radiation of heat by gases and vapors and on the physical connexion of radiation, absorption and conduction. *Philosophical Transactions of the Royal Society of London 151*: 1–36.

United Nations Environment Programme (2022). Emissions Gap Report 2022: The Closing Window – Climate crisis calls for rapid transformation of societies. Nairobi. Retrieved from https://www.unep.org/emissions-gap-report-2022.

U.S. Department of Energy Fuel Cell Technologies Office (2013) PEM electrolysis H2A production case study documentation. *PEM Electrolysis H2A Production Case Study Documentation* (December).

Varghese, O.K. (2013). Sunlight for fuel generation by way of carbon dioxide recycling. *Nanomaterials and Energy 5*: 244–255.

Varghese, O.K. and Grimes, C.A. (2008). Appropriate strategies for determining the photoconversion efficiency of water photoelectrolysis cells: A review with examples using titania nanotube array photoanodes. *Solar Energy Materials and Solar Cells 92*: 374–384.

Varghese, O.K., Paulose, M., LaTempa, T.J., and Grimes, C.A. (2009). High-rate solar photocatalytic conversion of CO_2 and water vapor to hydrocarbon fuels. *Nano Letters 9*: 731–737.

Wang, H. (2020). Solar thermochemical fuel generation. In: *Wind solar hybrid renewable energy system* (ed. K.E. Okedu, A. Tahour, and A.G. Aissaou), 13. IntechOpen.

Wang, C., Chen, Z., Jin, H. et al. (2014). Enhancing visible-light photoelectrochemical water splitting through transition-metal doped TiO_2 nanorod arrays. *Journal of Materials Chemistry A 2* (42): 17820–17827.

Wang, H., Zhou, Z., Long, R., and Prezhdo, O.V. (2023b). Passivation of hematite by a semiconducting overlayer reduces charge recombination: An insight from nonadiabatic molecular dynamics. *The Journal of Physical Chemistry Letters 14* (4): 879–887.

Wang, L., Xia, M., Wang, H. et al. (2018). Greening ammonia toward the solar ammonia refinery. *Joule 2* (6): 1055–1074.

Wang, Y., Guo, X., Wang, X. et al. (2023a). Construction of steady-active self-supported porous Ir-based electrocatalysts for the oxygen evolution reaction. *Chemical Communications 59* (13): 1813–1816.

Wang, Z., Gu, Y., Zheng, L. et al. (2022). Machine learning guided dopant selection for metal oxide-based photoelectrochemical water splitting: The case study of Fe_2O_3 and CuO. *Advanced Materials 34* (10): 2106776.

Wei, Z., Guo, M., and Zhang, Q. (2023). Scalable electrodeposition of NiFe-based electrocatalysts with self-evolving multi-vacancies for high-performance industrial water electrolysis. *Applied Catalysis B: Environmental 322* (68): 122101.

Wu, J., Huang, Y., Ye, W., and Li, Y. (2017). CO_2 reduction: From the electrochemical to photochemical approach. *Advanced Science 4* (11): 1700194.

Xiao, Y., Vanka, S., Pham, T.A. et al. (2022). Crystallographic effects of GaN nanostructures in photoelectrochemical reaction. *Nano Letters 22* (6): 2236–2243.

Xu, H., Han, Y., Wu, Q. et al. (2023). Iridium-based electrocatalysts for the acidic oxygen evolution reaction: engineering strategies to enhance the activity and stability. *Materials Chemistry Frontiers 7* (7): 1248–1267.

Yang, J., Qi, H., Li, A. et al. (2022). Potential-driven restructuring of cu single atoms to nanoparticles for boosting the electrochemical reduction of nitrate to ammonia. *Journal of the American Chemical Society 144* (27): 12062–12071.

Yang, S.J., Kim, T., Im, J.H. et al. (2012). MOF-derived hierarchically porous carbon with exceptional porosity and hydrogen storage capacity. *Chemistry of Materials 24* (3): 464–470.

Zacharia, R. and Rather, S.U. (2015). Review of solid state hydrogen storage methods adopting different kinds of novel materials. *Journal of Nanomaterials 2015*: 1–18.

Zhang, K., Wang, F., Li, X. et al. (2023a). Connected design of Ni foam-supported porous Ni and FeOOH/porous Ni electrocatalysts for overall water splitting in alkaline media. *Journal of Alloys and Compounds 942*: 169014.

Zhang, S., Sun, M., Yin, L. et al. (2023b). Tailoring the electronic structure of Ir alloy electrocatalysts through lanthanide (La, Ce, Pr, and Nd) for acidic oxygen evolution enhancement. *Advanced Energy and Sustainability Research 4* (9): 2300023.

Zhao, D., Guan, X., and Shen, S. (2022). Design of polymeric carbon nitride-based heterojunctions for photocatalytic water splitting: A review. *20* (6): 3505–3523.

Zheng, H., Zhao, C., Hu, S. et al. (2023). Nanoporous GaN photoanode decorated with plasmonic Au nanoparticles for enhanced solar water splitting. *Physica Status Solidi A 220* (11): 2300037.

Zhu, Q. (2019). Developments on CO_2-utilization technologies. *Clean Energy 3* (2): 85–100.

Zhu, Y., Tjokro Rahardjo, S., Valkenburg, C., Snowden-Swan, L., Jones, S., and Machinal, M. (2011). Techno-economic analysis for the thermochemical conversion of biomass to liquid fuels (U.S. DOE). *DOE (USA)* (June), 152.

Zoller, S., Koepf, E., Nizamian, D. et al. (2022). A solar tower fuel plant for the thermochemical production of kerosene from H_2O and CO_2. *Joule 6*: 1606–1616.

Section 1

Solar Fuel Generation Processes: Science and Technology

2

Introduction to Photocatalytic/Photoelectrochemical Fuel Generation: Science and Technology Perspective

Ke Fan[1], Lei Wang[2], and Lianpeng Tong[3]

[1] State Key Laboratory of Fine Chemicals, School of Chemical Engineering, Dalian University of Technology, Dalian, China
[2] Department of Chemical and Biomolecular Engineering, National University of Singapore, Singapore, Singapore
[3] School of Chemistry and Chemical Engineering, Guangzhou Key Laboratory for Clean Energy and Materials, Guangzhou University, Guangzhou, China

2.1 Introduction

To mitigate the pressing climate issues and meet our increasing needs for energy and chemical materials, sustainable alternatives to fossil resources must be developed. Recently, renewable electricity produced using solar cells, wind turbines, and other carbon-emission-free processes has become relatively cost-competitive (De Luna et al., 2019). One common nature of renewable electricity is intermittency, which requires the development of economically viable, large-scale, and seasonal (i.e., typically over a period of several months, in order to balance the seasonal variations in energy supply and demand) energy storage technologies. While there have been significant advancements in battery technologies, due to their insufficient energy density they are currently not compatible with many energy-intensive industries such as cement, steel, and transportation (e.g., aviation and long-distance land carriage). On the other hand, production of fossil fuel-derived carbon-based chemical feedstocks significantly contributes to global CO_2 emissions (Agency, 2018). Taken together, it is crucial to establish a technology that can directly convert intermittent renewable energy into stable and high-energy-density fuels and chemical feedstocks with low and even negative carbon footprints.

Artificial photosynthesis has been recognized as one of the most promising technologies to address the challenges mentioned above, as it can store solar energy directly into a broad range of chemicals and fuels with high energy densities. As shown in Scheme 2.1, the basis of using artificial photosynthesis is to utilize abundant molecules, e.g., H_2O, CO_2, and N_2, to produce valuable fuels and chemicals through several key photocatalytic/photoelectrochemical (PEC) reactions. Examples are hydrogen generation via water splitting, production of carbon-based fuels and chemicals through CO_2 reduction, and synthesis of ammonia via N_2 reduction. Developing cost-effective and robust photocatalysts that can efficiently convert solar energy into chemical energy is essential for the delivery of this technology. A breakthrough in the area of artificial photosynthesis is expected to facilitate a faster transition to a sustainable energy system based on sunlight.

The pioneering work of Fujishima and Honda (Fujishima & Honda, 1972) on water photolysis based on titanium dioxide (a wide bandgap n-type semiconductor) laid the foundation of the research on photocatalytic/PEC fuel production. The major technical obstacles preventing this technology from reaching the commercialization stage are poor long-term stability of the semiconductor/catalytic interfaces, high cost, low activity and stability of the catalysts, and the low overall energy efficiency resulting from high overpotential losses associated with oxygen evolution. In addition, photocatalytic/PEC CO_2 and N_2 reductions also face challenges from poor selectivity of products due to the intricate reaction networks and the kinetically facile competing hydrogen evolution reaction (Andersen et al., 2019; Nitopi et al., 2019). To help readers understand the origins of these challenges, we start this chapter with the introduction to the current knowledge about natural photosynthesis (e.g., the oxygen evolution complex), hydrogenase, and CO_2 fixation enzymes, and how such knowledge inspires the artificial systems for hydrogen production and CO_2 reduction. We then discuss the basic physical and chemical principles of photocatalytic and PEC systems for both water splitting and CO_2 reduction processes. Finally, we analyze the key scientific needs and future opportunities in this research area.

Conversion of Water and CO_2 to Fuels using Solar Energy: Science, Technology and Materials, First Edition.
Edited by Oomman K. Varghese and Flavio L. Souza.

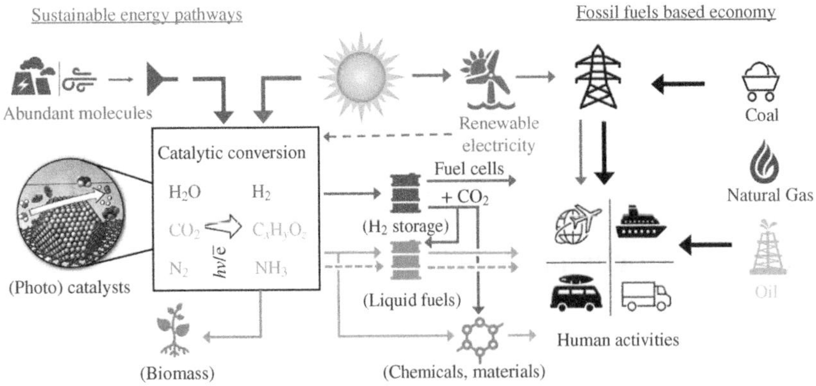

Scheme 2.1 Proposed sustainable energy pathways. In addition to the direct use of renewable electricity, these clean energies can be stored in the form of high energy density fuels, such as hydrogen and carbon-based fuels. These clean energies can also be utilized to produce important chemical feedstocks, such as NH_3, and reduce their carbon footprint. To enable all these sustainable alternatives, the development of the corresponding cost-effective (photo)catalytic systems is the key.

2.2 The Natural Photosynthetic Water Splitting and CO_2 Reduction

Natural photosynthesis involves a series of complex physical and chemical processes, which cannot be reproduced by modern technologies, at least not to the extent of elegance and efficiency of the processes at the catalytic center. Understanding the mechanisms of these complex biologic processes will undoubtedly extend our knowledge of dealing with these complex photoreactions involving multi-electron/proton transfer and multiple bond-breaking/forming steps.

2.2.1 Oxygen-Evolving Complex (OEC)

Oxygen-evolving photosynthetic-bacteria have existed on our planet since around three billion years ago (Blankenship, 2010). Evolution has enabled enzyme photosystem II (PSII) to directly split water molecules into dioxygen and protons. PSII possesses a complex protein structure (around 650 kDa) and is located in the thylakoid membranes of the oxygen-evolving photosynthetic organisms, including algae, cyanobacteria, and other higher plants. So far, the most precise PSII crystal structure has been reported by Jian-Ren Shen and Nobuo Kamiya at a resolution of 1.9 Å based on samples from the thermophilic cyanobacterium *Thermosynechococcus vulcanus* (Umena et al., 2011). This atomic-resolution picture allows people to obtain unambiguous positions of critical components in the PSII enzyme. The active center for oxidizing water in PSII is a Mn_4CaO_5 cluster, usually called oxygen-evolving complex (OEC). The resolved crystal structure of PSII reveals that the OEC adopts a distorted-chair structure with an organized $[Mn_3CaO_4]$ cubic core and a fourth Mn atom attached to the core via two oxo-bridges (Figure 2.1) (Suga et al., 2015). The OEC cluster also includes four water ligands (W1–W4), two of which directly coordinate with the Ca atom and the other two chelate with the dangling Mn (Mn_4A). Meanwhile, there are seven amino acid residues in the inner coordinating sphere of the OEC cluster. These amino acid residues stabilize the OEC structure in the enzyme matrix of PSII via metal-carboxylate and metal-imidazole coordinating bonds. A few more amino acid residues reside in the outer coordinating sphere. These residues do not directly ligate to the OEC but play key roles in the substrate (proton and oxygen) transportation and charge compensation during the catalytic oxygen-evolving reaction.

Oxygen evolution occurring at PSII goes through a periodic cycle of five states denoted by S_0 to S_4. This famous S-state cycle was initially proposed by Bessel Kok and therefore is also called Kok cycle (Kok et al., 1970). It has been proposed that two Mn ions of the OEC are trivalent at the dark-stable S_1 state, and the other two Mn ions are tetravalent (Suga et al., 2015; Yano & Yachandra, 2014; Zaharieva et al., 2016). Multiple deprotonation and oxidation events occur in a stepwise sequence of the cycle. The OEC as well as its surrounding protein residues dynamically change their structures at different S-states as the biological oxygen-evolving reaction proceeds. Enormous efforts in experimental and theoretical investigations have elucidated important information about the biological water oxidation process, yet some mechanism details in the reaction pathway remain unclear (Cox et al., 2013; Najafpour et al., 2020; Pantazis, 2018). Despite some dispute about certain steps in the catalytic cycle, it is now largely accepted that Mn(III) ions of the OEC are oxidized to Mn(IV) through the S_0–S_3 state,

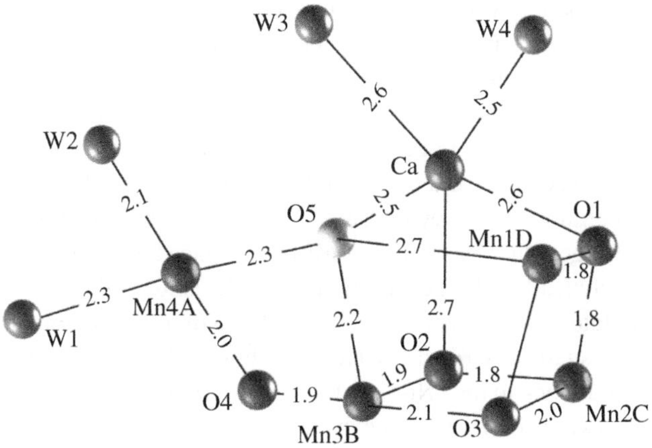

Figure 2.1 Structure of the OEC (bond distances in Å) from *Thermosynechococcus vulcanus* at the S1 state. W denotes water molecule.

and the critical O—O bond forms during the S_4–S_0 state transition. There are two main hypotheses about the mechanism of O—O bond formation: (i) nucleophilic attack of the metal (Ca or Mn)-bound water molecule on the adjacent terminal or bridging oxygen ligand and (ii) radical coupling between the bridging O atom and an oxyl radical (Najafpour et al., 2020; Shen, 2015; Yano & Yachandra, 2014). The upper limit turnover frequency of OEC was estimated to be 100–500 s^{-1} in vivo (Vinyard et al., 2013). Current understanding of the OEC structure may serve as a blueprint for the development of artificial photosynthetic model compounds.

2.2.2 Hydrogenase

In nature, proton reduction in certain microorganisms belonging mainly to the bacteria and archaea domains of life, and some eukarya, is catalyzed by hydrogenases (Vignais & Billoud, 2007). These enzymes can be classified into three distinct categories: [FeFe]-, [NiFe]-, and [Fe]-hydrogenases, according to the metal compositions of their active sites (Figure 2.2) (Lubitz et al., 2014). The [FeFe]- and [NiFe]-hydrogenases catalyze protons to hydrogen conversion or its reverse reaction with a negligible kinetic barrier. In contrast, [Fe]-hydrogenase only catalyzes the heterolytic cleavage of H_2 into H^+/H^- in the presence of specific H^- acceptors (Corr & Murphy, 2011).

The catalytic site of [FeFe]-hydrogenase is a di-iron metal center bridged by an azadithiolate ligand and ligated by auxiliary inorganic groups (CN^- and CO, Figure 2.2a) (Peters et al., 1998). One of the iron atoms at the so-called "proximal" position connects to a [4Fe-4S] cubane cluster and the cysteine thiolate donor. The [FeFe]-hydrogenases from the most known organisms are proposed to possess similar electronic configurations (Vignais and Billoud, 2007). Current investigation suggests that the [FeFe]-hydrogenases catalyze H_2 evolution/oxidation via a highly reduced Fe^I–Fe^I intermediate (Adamska et al., 2012). The amine moiety of the azadithiolate bridge ligand has been believed to be involved in the catalytic process by transferring protons to the open coordination site of iron. Purified [FeFe]-hydrogenase from *Clostridium pasteurianum* has achieved a catalytic hydrogen evolution frequency of 3400 $\mu mol \cdot min^{-1} \cdot mg^{-1}$ (Lubitz et al., 2014).

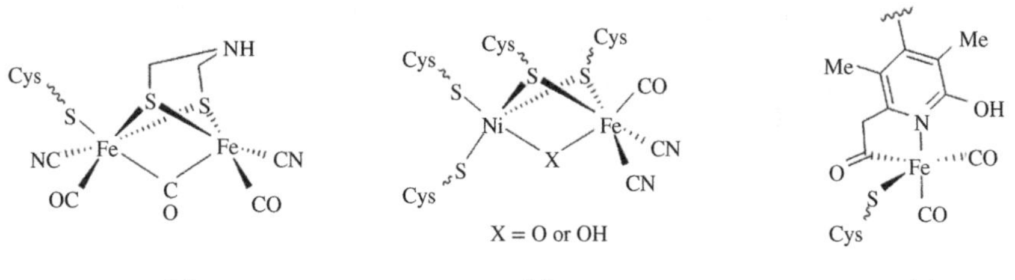

Figure 2.2 Chemical structures of (a) [FeFe]-hydrogenase active site from *Clostridium pasteurianum;* (b) [NiFe]-hydrogenase active site from *D. v.* Miyazaki F. (Ogata et al., 2005); (c) [Fe]-hydrogenase active site from *Methanocaldococcus jannaschii* (Hiromoto et al., 2009; Shima et al., 2008).

The [NiFe]-domains have almost identical structures among all the types of O_2-sensitive or O_2-tolerant [NiFe]-hydrogenases. The crystal structure of [NiFe]-hydrogenase from *Desulfovibrio* strains has been resolved at atomic resolution, revealing a bimetallic Ni–Fe cluster as the active site of the hydrogenase (Ogata et al., 2005). Two cysteine residue thiolates bridge the Ni and Fe atoms, and another two cysteine residues are bound to the nickel in a terminal fashion (Figure 2.2b). The low spin Fe(II) adopts an octahedral coordination geometry with one CO and two CN^- auxiliary ligands. A third bridge ligand connecting the Ni and the Fe changes the identity between oxyl and hydroxyl depending on the redox state of the enzyme. The basic catalytic cycle of [NiFe]-hydrogenases contains three key intermediate states (Lubitz & Tumas, 2007; Yagi et al., 2014). The initial "Ni-SI_a" does not have the X bridge, but it has open coordination space between Ni(II) and Fe(II). By reacting with the Ni site, H_2 is polarized and heterolytically split, generating the "Ni-R" state of the [NiFe] center with a H^- bridge. The resulting H^+ is attached to the thiolate of one cysteine residue. Release of a proton and an electron to the environment affords the "Ni-C" state intermediate usually described as a Ni^{III}–Fe^{II} bimetallic center with a hydride bridge ligand. Loss of another electron and a proton proceeds back to the "Ni-SI_a" state and completes the catalytic cycle. Catalytic H_2 evolution reaction undergoes the reverse process of the H_2 splitting cycle. Insertion of oxygen (from O_2) between the Ni and Fe atoms may deactivate the catalytic activity of certain hydrogenases. The turnover frequencies of [NiFe]-hydrogenases have been reported to be more than 20,000 s^{-1} (Frey, 2002; Madden et al., 2012).

[Fe]-hydrogenases are found only in methanogenic archaea (Thauer et al., 2010). They catalyze the reaction between H_2 and methylene-H_4MPT to generate the methenyltetrahydromethanopterin (methenyl-H_4MPT^+) (Thauer et al., 1996). Resolved crystal structure of the [Fe]-hydrogenase from *Methanocaldococcus jannaschii* displays a single Fe active site in an octahedral coordination geometry (Figure 2.2c) (Hiromoto et al., 2009; Shima et al., 2008). The iron coordinates with the nitrogen atom, and the formylmethyl group of pyridinol is linked with a guanylylpyridinol molecule. Two CO and one cysteine residue thiolate occupy another three coordination positions, leaving one open site in the coordination sphere. The catalytic H_2 conversion cycle of [Fe]-hydrogenase differs from that of [NiFe]- or [FeFe]-hydrogenases. The initial reaction step of [Fe]-hydrogenases requires binding of the second substrate methenyl-H_4MPT^+ to the open site of the enzyme. The H_2 molecule is then transferred to the active site and heterolytically split at the Fe core. The resulting proton and hydride are accepted by the thiol of cysteine and the adjacent methenyl-H_4MPT^+.

2.2.3 Enzymes that Reduce CO_2

There are two classes of enzymes in nature that catalyze the direct reduction of CO_2: one class is formate dehydrogenases (FDHs) that reversibly convert CO_2 to formate (HCO^{2-}), and the other one is CO dehydrogenases (CODHs) that reversibly convert CO_2 to CO. The metal-based FDHs contain molybdenum or tungsten centers as their active sites. The W(VI) or Mo(VI) center possesses a coordination geometry of distorted trigonal prism with sulfur or selenium donors of protein residues (Dobbek, 2011). These Mo- and W-formate dehydrogenases are capable of catalyzing the reduction of CO_2 to formate at rates as high as 280 s^{-1} (Reda et al., 2008).

The CO dehydrogenases include two types of enzymes. One type of enzyme arising mainly from obligate anaerobes possesses a [Fe_4S_4Ni] active center (Figure 2.3) and is sensitive to O_2 (Jeoung & Dobbek, 2007). These enzymes catalyze the CO_2/CO reduction at a rate of 45 s^{-1} with a trivial overpotential (Svetlitchnyi et al., 2001). The other type is air-stable and is found in aerobes such as *Oligotropha carboxidovorans*. These enzymes contain a [MoSCu] active center and are able to catalyze the oxidation of CO to CO_2, not the reverse reduction of CO_2/CO (Zhang et al., 2010).

The catalytic CO_2 reduction mechanism by the [Fe_4S_4Ni] active site of CODHs has been extensively investigated using spectroscopic methods and X-ray diffraction techniques (Appel et al., 2013). Currently proposed catalytic cycle is initiated

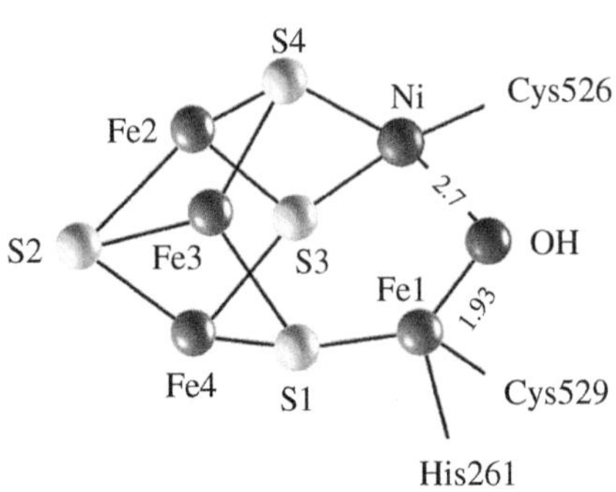

Figure 2.3 Chemical structure of the [Fe_4S_4Ni] active cluster of CODHs in a reduced state (Jeoung and Dobbek, 2007).

by a one-electron reduction of the Ni ion to a monovalent state. The carbon atom of CO_2 then binds to the Ni(I) site, and another one-electron transfer occurs simultaneously. One oxygen atom of the binding CO_2 coordinates with the Fe site and weakens the corresponding C—O bond. Cleavage of the C—O(Fe) bond and loss of water afford a Ni^{II}–CO species that readily releases the CO production. Protons and water molecules involved in the catalytic pathway are transported through a hydrogen-binding network within the enzyme.

2.3 Artificial Systems for Solar-Driven Chemical Fuels Production

Along with the advancement of understanding the natural photosynthesis system, various artificial systems have been established to mimic the active center structures of natural enzymes, and/or their working principles for synthesizing chemicals and fuels using solar energy. The related research has been named as "artificial photosynthesis" and a more recent term "solar fuels" has been adopted by the global research community with respect to the increasing interest in expanding diversity of energy-storing chemical fuels. We primarily divide the field of "solar fuels" research into the following sub-areas: (i) bioinspired synthetic systems, (ii) bioinorganic hybrid systems, (iii) PEC systems, and (iv) particulate photocatalytic systems. In the following sections, we will introduce the working principles, current research scopes, and major challenges of these research areas.

2.3.1 Bioinspired Synthetic Systems

2.3.1.1 Synthetic Molecular Catalysts

Elucidating the catalytic water splitting and CO_2 reduction processes in nature has inspired people to synthesize model compounds to mimic the structures and functions of key components, especially the active sites, of corresponding enzymes. So far, there have been several synthetic Mn complexes that closely mimic the core structure of OEC (Sun, 2015; Tsui et al., 2013). As examples of synthetic compounds shown in Figure 2.4, the classic propanedithiolate Fe–Fe complex (Winter et al., n.d.), the heterodinuclear Ni–Fe complex (Ahmed et al., 2018; Brazzolotto et al., 2016; Brazzolotto et al., 2018), and the mononuclear Fe complex with a pincer ligand are biomimic motifs of the [FeFe]-, [NiFe]-, and [Fe]-hydrogenases active sites, respectively. Despite the highly structural similarities between the model complexes and enzyme active sites, the synthetic compounds usually show little or much inferior catalytic performance to the enzymes in nature. One reason is that a slight discrepancy in the structure may lead to completely different chemical properties. Another reason is that the molecular complexes, of which the catalytic activity is assessed in a homogeneous medium, do not possess a protein environment that stabilizes the active sites of enzymes and effectively transports the substrates.

Considering the difference in the reaction microenvironments between synthetic model compounds and enzymes, chemists have designed and prepared transition metal complexes with an emphasis on their catalytic ability rather than structural resemblance to the enzymatic sites. While certain features of these synthetic molecular complexes might be inspired by the natural enzymes, a variety of their organic ligands have a far broader scope than that in the coordination sphere of the active sites of enzymes (Kärkäs & Åkermark, 2016; Tong et al., 2020). Metal elements without any biological origins are also employed in order to acquire great efficiency and stability of the synthetic catalysts (Appel et al., 2013; Francke et al., 2018; Tong & Thummel, 2016). Moreover, organic ligands of these synthetic complexes can be elaborately tailored and integrated in a PEC cell.

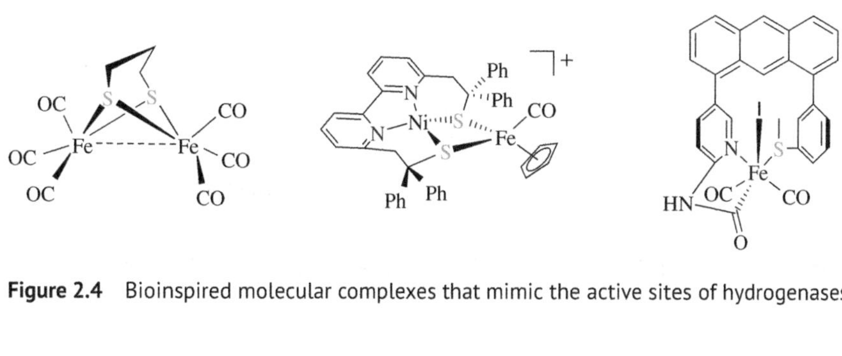

Figure 2.4 Bioinspired molecular complexes that mimic the active sites of hydrogenases.

2.3.1.2 Application of Synthetic Model Compounds in PEC Cells

The synthetic catalysts anchored to mesoporous semiconductors can form PEC electrodes that carry out light-driven water splitting or CO_2 reduction reactions (Ashford et al., 2015; Sekizawa et al., 2018). Integration of other molecule components together with the catalysts in the PEC electrodes can improve the light absorption and electron transfer of the system, thus increasing the efficiency of solar-to-chemical energy conversion. Figure 2.5, for example, describes a sophisticated architecture of a PEC anode that consists of a molecular assembly linked to nanoITO (indium tin oxide) particles (Wang et al., 2019). The methyl viologen electron acceptor (MV^{2+}), Ru^{II} tris-bipyridine sensitizer, Fe^{II} terpyridyl electron donor, and the [Ru^{II}(2,2′-bipyridine-6,6′-carboxylic acid)-((3-(pyridine-4-yloxy)propyl)phosphonic acid)$_2$] water oxidation catalyst correspond to the key functional components of PSII accordingly. Under solar light illumination, the excited Ru sensitizer first injects electrons into ITO via MV^{2+}. The oxidized Ru sensitizer then accepts electrons from the iron donor and Ru catalysts. The PEC water oxidation by the PEC anode was conducted in a pH 4.65 acetate buffer (0.1 M) with a Pt cathode. A maximum photocurrent of about 250 μA/cm^2 of this PEC anode was achieved under an applied 0.5 V bias voltage. The photocurrent, however, decreased to 75 μA/cm^2 after 10 min of illumination. The overall Faradaic efficiency for O_2 production was 67% over a 60 min illumination period. While this assembly notably mimics the key functions in PSII, it currently does not possess enough stability to meet the requirements of practical application.

2.3.2 Bioinorganic Hybrid Systems

The fact that enzymes can retain their function in vitro allows people to construct semi-artificial PEC cells (Zhang and Reisner, 2020). Enzymes immobilized onto electrodes may play the role of a light harvester that generates charges under irradiation, or the role of a catalyst that drives fuel-forming reactions, or both (in the case of PSII). Earlier research efforts were focused on either the anode or cathode of the semi-artificial PEC cell. Recent advancements have facilitated the development of semi-artificial photosynthetic systems performing complete light-driven H_2O splitting with the use of biocatalysts alone (Figure 2.6) (Mersch et al., 2015). This PEC cell consisted of a PSII-decorated anode wired to a hydrogenase-decorated cathode, separated by a glass frit. Both enzymes were supported on hierarchically structured indium tin oxide electrodes (IO-mesoITO) with large porosities that match the sizes of PSII and hydrogenases. Under irradiation of red light (λ = 660 nm, 10 mW/cm^2) upon the anode and a bias voltage of 0.9 V applied to the system, H_2O splitting took place with a light-to-H_2 efficiency of up to 5.4% and a nearly 100% Faradaic yield. This bias voltage application

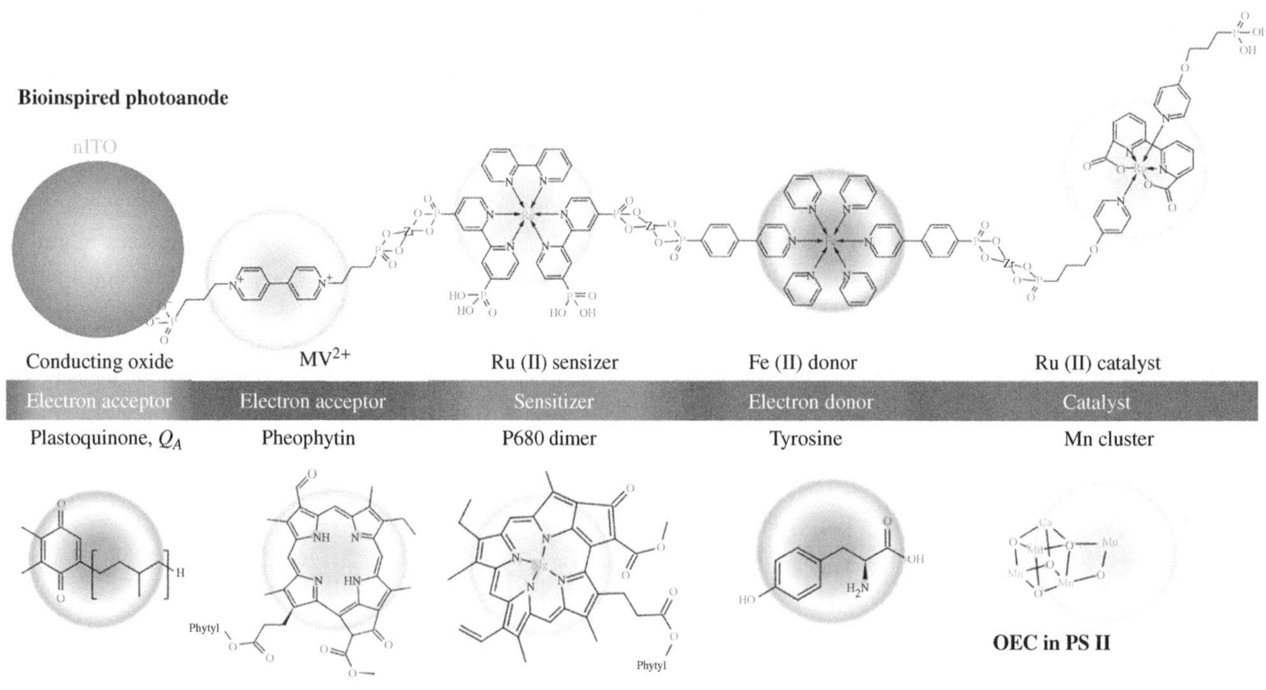

Figure 2.5 Comparison between the molecular assembly, nanoITO|-MV^{2+}-S-Fe^{II}-Ru^{II}, and the key components of PS II (Wang et al., 2019).

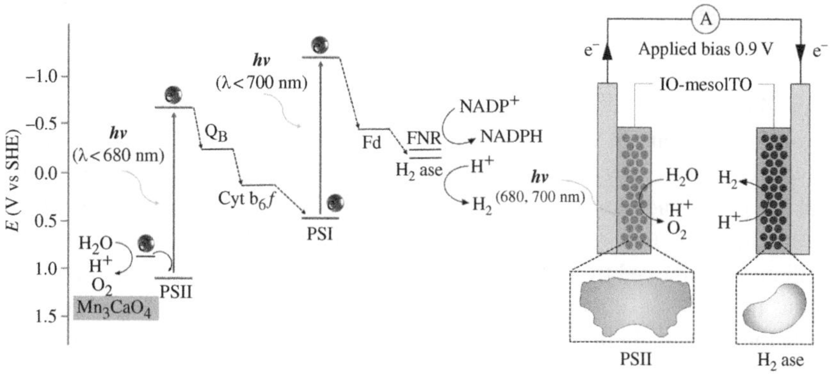

Figure 2.6 Comparison of energy diagrams between the biological photosynthetic system (left) and a semi-artificial photoelectrochemical setup for water splitting (right) (Mersch et al., 2015). NADP = nicotinamide adenine dinucleotide phosphate; FNR = ferredoxin-nicotinamide adenine dinucleotide phosphate reductase. Cyt b_6f = Cytochrome b_6/f complex.

corresponds to the photoexcitation process in PSI that compensates the energy gap between Q_b (secondary plastoquinone electron acceptor) and Fd (ferredoxin) in the naturally occurring photosynthesis (Figure 2.4). This energy gap can also be overcome by integrating semiconductor light harvester in the semi-artificial PEC cell (Nam et al., 2018; Sokol et al., 2018).

Besides isolated enzymes, there is a growing interest in directly coupling whole live cells with the electrode of a semi-artificial PEC. The direct employment of microorganisms avoids the costly purification process of enzymes and takes advantage of their potential self-healing capability. Meanwhile, certain microorganisms have evolved to have both light-harvesting function and efficient catalytic site that drives water splitting, CO_2 fixation, or even N_2 reduction reactions (Ding et al., 2019). The wildtype and engineered bacterium *Ralstonia eutropha*, for instance, has been immobilized on the anode of semi-artificial PECs, which could convert CO_2 to polyhydroxybutyric acid, C^{3-} alcohol, or C^{4+} alcohol under light irradiation (Cestellos-Blanco et al., 2020; Smith et al., 2020). A recent work showed that a close-packed silicon nanowire/*Sporomusa ovata* anode in a PEC cell could generate a photocurrent of 0.65 ± 0.11 mA/cm^2 originating from CO_2 to acetate reduction under AM 1.5 G illumination without any applied bias voltage (Su et al., 2020). This system maintained a solar-to-CO_2 fixation efficiency of ~3.6% for more than one week, indicating the robust bonding between the microorganism and the inorganic material. Current research about semi-artificial PEC has centered around the enhancement of current density that converts CO_2, H_2O, and N_2 into value-added products. The photocurrent response and overall solar-to-chemical conversion efficiency are influenced by a number of factors, including the activity of microorganisms, biotic/abiotic interface, the quality of the electronic wiring between the electrodes, the electrode architecture, the number of bio-assemblies on the surface, and the light intensity. Despite significant developments in this field of bioinorganic hybrid PEC, the technology has not yet reached the commercialization stage.

2.3.3 Photoelectrochemical Water Splitting and CO_2 Reduction

PEC conversion of solar light into fuels and chemicals has been investigated as a sustainable pathway over a century. Honda and Fujishima discovered water splitting based on electrochemical cell with TiO_2-photoanode illuminated by UV light (Fujishima & Honda, 1972). The ultimate goal of Honda's PEC cell is to store the solar light energy into high-energy-density chemical, in this case, hydrogen. Research following this principle has extended the reactants and target products, e.g., carbon-based fuels and chemicals, by reducing CO_2 molecules. Consequently, the term "solar fuels" has been widely adopted by the research community.

2.3.3.1 Some Basic Concepts of Semiconductors

Semiconductor is normally referred to as a kind of material whose electrical conductivity is between conductors and insulators. According to the molecular bonding theory, the energy levels of semiconductors consist of the conduction band (CB) and valence band (VB), which are formed by a combination of anti-bonding orbitals and bonding orbitals, respectively. The gap between the CB minimum and VB maximum is called a bandgap. In insulators, the electrons in the VB are separated from the CB by a large bandgap; in conductors, such as metals, the VB overlaps the CB; and in semiconductors, the bandgap between the valence and CBs is so small that it can be filled by thermal or other excitations. The bandgap of semiconductors can range from less than 1 eV (0.67 eV, germanium) to more than 5 eV (5.47 eV, diamond), which is between that of conductors and insulators, making semiconductors possess unique PEC properties. Conventionally, in semiconductors, the CB is considered mostly empty (completely empty at absolute zero), while VB is mostly fully filled with electrons. When a semiconductor is excited using external energy, such as light and thermal, the electrons in the VB could be excited and transferred to the CB, leaving holes at the VB. Such externally energy-generated electron–hole pairs are important for the photocatalytic/photoelectrocatalytic application of semiconductors. Photocatalysis is a type of catalysis that results in the modification of the rate of a photoreaction—a chemical reaction that involves the absorption of light by one or more reacting species—by adding substances (catalysts) that participate in the chemical reaction without being consumed. Photoelectrocatalysis is a powerful method derived from the combination of heterogeneous photocatalysis and electrochemical techniques. For photocatalysis to occur, the energy of photons in the incident light must be equal to or larger than the bandgap energy (E_g), to enable the electron–hole pair formation. That means, one necessary condition for photocatalysis is the following:

$$E_g \leq h\nu = \frac{hc}{\lambda}$$

where $h = 6.63 \times 10^{-34}$ *J·s, the Planck constant;* $c = 3 \times 10^8$ *m/s, the speed of light;* λ*, the wavelength of light;* ν*, the frequency of light*. When λ is expressed in nanometers, this condition can be written in the form,

$$\text{Band gap}(E_g, \text{eV}) \leq \frac{1240}{\lambda(\text{nm})}$$

Generally, a pure semiconductor has very poor electrical conductivity. To increase the conductivity, doping the semiconductors with foreign atoms is a widely used strategy. Such impurity atoms can act as electron donors or acceptors. When electron donors, which contain more valence electrons than that of the host semiconductor, are used as dopants, the semiconductor becomes electron-rich. Such a material is called a p-type semiconductor. For instance, doping phosphorus (number of valence electrons: 5) to the semiconductor of Si (number of valence electrons: 4) (Figure 2.7a) makes Si a p-type semiconductor. On the other hand, if the dopant atom has fewer valence electrons than the host atom, the dopant atom will act as an electron acceptor, which makes the semiconductor hole-rich. This is the case of boron (number of valence electrons: 3) -doped Si (Figure 2.7b), which is called an n-type semiconductor. Thus, p- and n-type semiconductors have different functions in the photoelectrocatalytic field.

When a semiconductor contacts with a liquid electrolyte containing a redox couple, if there is a difference between the Fermi levels (E_F) of the semiconductor and the electrochemical potential $-qE°$ ($E°$, Nernst potential) of the redox couple in the electrolyte (Figure 2.8a), electrons will flow through the interface of the semiconductor/electrolyte until an equilibrium is reached. Such electron transfer re-distributes the charges at the surface of the semiconductor, causing the band to bend and forming a space charge region. If the Fermi level of the semiconductor equals the Nernst potential of the redox couple, neither electron transfer nor band bending will occur at the interface. If the E_F of the semiconductor is more positive

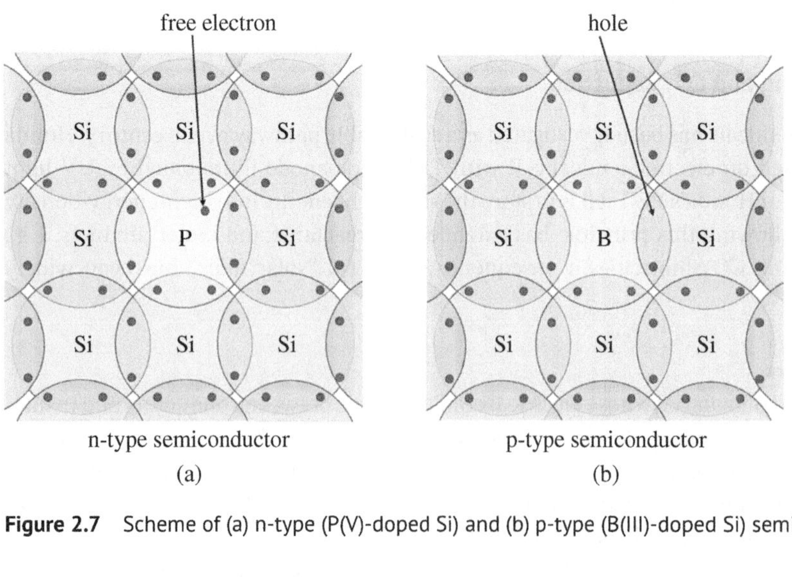

Figure 2.7 Scheme of (a) n-type (P(V)-doped Si) and (b) p-type (B(III)-doped Si) semiconductors.

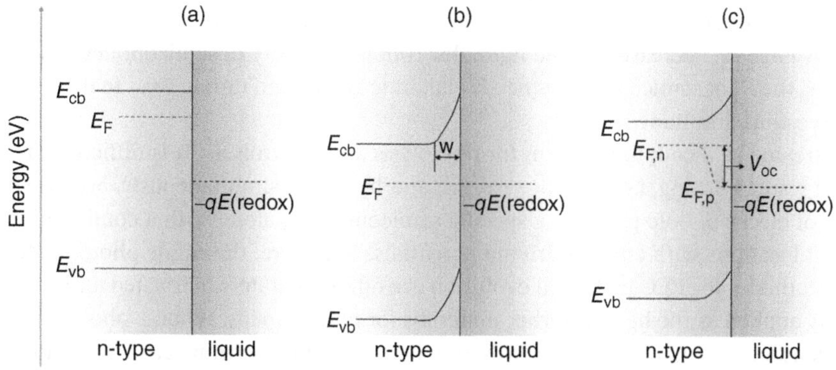

Figure 2.8 Scheme of three cases of band energetics of a semiconductor/liquid contact (taking n-type semiconductors as the example): (a) before equilibrium, (b) after equilibrium in dark, and (c) in quasi-equilibrium under illumination.

(vs. normal hydrogen electrode (NHE)) than the $E°$ of the redox couple, electrons will transfer from the solution to the semiconductor until the surface E_F and the $E°$ establish an equilibrium. A downward band bending of the semiconductor is generated in this case. Conversely, if the E_F is more negative than $E°$, the majority of charge carriers, i.e., electrons, will transfer from the semiconductor to the solution, producing excess positive charge in the semiconductor, and forming an upward band bending and depletion layer (W, the width of the depletion layer, Figure 2.8b). Under illumination, the photoexcited electrons disequilibrate the system. The E_F of n-type semiconductors, for instance, moves to more negative quasi-Fermi level ($E_{F,n}$) and form the hole quasi-Fermi level ($E_{F,p}$) (Figure 2.8c). The difference between $E_{F,n}$ and $-qE$(redox) determines the photogenerated voltage (V_{oc}).

2.3.3.2 Photoelectrochemical (PEC) Water Splitting

PEC water splitting is considered an efficient means to produce the energy carrier–hydrogen (H_2). The theoretical maximum efficiency for a tandem PEC water splitting system could reach over 22.5% (Haussener et al., 2013). An overall water splitting reaction includes two half reactions: oxygen evolution reaction (OER) and hydrogen evolution reaction (HER). Under the standard conditions, the standard free energy change of $\Delta G = 237.2$ kJ/mol (at 273.15 K and 1 atm pressure) is required to convert one molecule of H_2O to H_2 and $1/2O_2$, which corresponds to standard redox potential $\Delta E° = 1.23$ V. Therefore, to initiate an overall water splitting reaction merely under illumination of incident light, light absorbers (normally semiconductors) should thermodynamically produce a minimum photovoltage of 1.23 V between anode and cathode. In such a case, the bandgap energy E_g of semiconductors should be large enough to straddle the reduction and oxidation potentials of water (0 and 1.23 V vs. reversible hydrogen electrode, RHE, respectively), as shown in Figure 2.9. A few semiconductors such as TiO_2, ZnO, and $SrTiO_3$ are able to meet this requirement on the bandgap, (Figure 2.9). Taking into account the kinetics loss

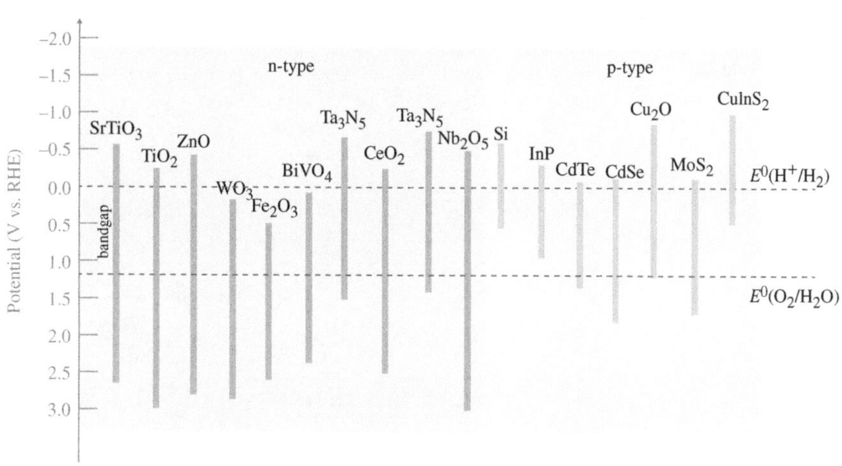

Figure 2.9 Conduction and valence bands of some n-type and p-type semiconductors.

and overpotentials during reactions, however, a larger energy bandgap of at least 2.0 V is necessary to create a photovoltage sufficient for practical water splitting. A large E_g decreases visible light absorption capability of semiconductors and lowers their light-harvesting efficiency. Thus, the E_g of semiconductors needs elaborate choice and tuning so as to meet the requirements of photovoltage and light harvesting simultaneously.

A PEC device for water splitting can be treated as a coupling system for photolysis and electrolysis. It is difficult for a single semiconductor to perform an overall water splitting, i.e., produce oxygen and hydrogen simultaneously, because it requires effective separation and transport of electron–hole pairs for a specific semiconductor, along with a combination of an appropriate bandgap energy and band locations with enough driving potentials. Therefore, the single photoanodes for PEC oxygen evolution and single photocathodes for PEC hydrogen evolution are often separately fabricated and studied. n-Type semiconductors are traditionally applied as the light absorber materials for photoanodes. When a photoanode contacts an electrolyte, an upward band bending occurs (Figure 2.8), which is beneficial for the transfer of the photogenerated holes and electrons to the surface and bulk of the photoelectrode, respectively. The holes at the surface react with water molecules to produce oxygen. On the contrary, a photocathode normally consists of p-type semiconductors, and the downward band bending formed upon contact with the electrolyte facilitates the transfer of the photogenerated holes to the bulk and electrons to the surface of the photocathode. The electrons at the surface reduce the protons to generate hydrogen. A two-photoelectrode system constructed by coupling the photoanode and photocathode can be used for bias voltage-free water splitting. In this configuration, the majority of charge carriers driven by the light-induced electric field flow to the counter photoelectrode, and the photogenerated minority charge carriers transfer to the surface of the photoelectrode and participate in the catalytic reaction. This is reminiscent of the Z-scheme process of photosynthesis in nature. The current density-potential measurements for each photocathode and photoanode, which are each independently evaluated, can be used to estimate the overall water splitting efficiency for photoelectrolysis cells (Figure 2.9). The maximum operational current density for the entire cell can be calculated from the intersection of the two curves. Unfortunately, such a two-photoelectrode configuration has so far exhibited very low performance. A photocurrent density of only a few mA/cm^2 has been achieved for the overall water splitting. More detailed device configurations for PEC water splitting will be discussed in the following sections (Figure 2.10).

There are three main processes in a semiconductor photoelectrode that are considered the efficiency determining factors in a PEC water splitting device. Process **1:** the incident light causes electrons in the VB of a semiconductor photoelectrode, to get photoexcited to the CB, generating photogenerated electron–hole pairs. Process **2**: the photogenerated holes (in n-type semiconductor photoanodes) or electrons (in p-type semiconductor photocathodes) transfer from the bulk to the surface of the semiconductor. Process **3**: the injection of the surface electrons to the electrolyte for proton reduction (hydrogen evolution) or surface holes for water oxidation (oxygen evolution). These three processes are summarized as light absorption, charge transfer from the bulk to the surface, and charge injection from the surface to water molecules/protons, respectively, as illustrated in Figure 2.11 (taking the case of an n-type semiconductor photoanode for PEC water oxidation as an example). Correspondingly, the light harvesting efficiency (η_{abs}), the charge transfer efficiency (η_{trans}), and charge injection efficiency (η_{inj}) determine the incident photon to current efficiency (IPCE) of the photoelectrodes. The IPCE of water

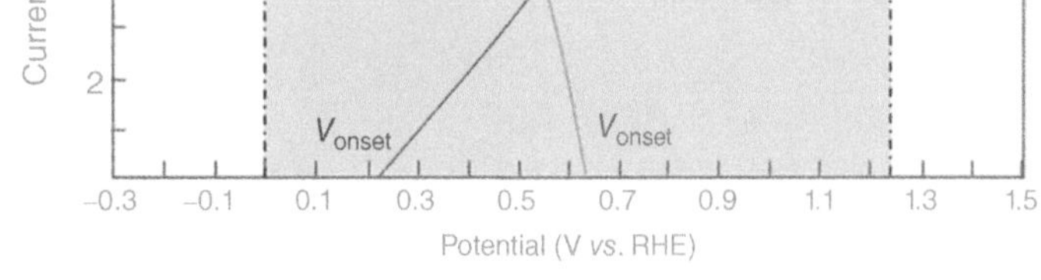

Figure 2.10 Performance curves of photoanodes and photocathodes for water splitting.

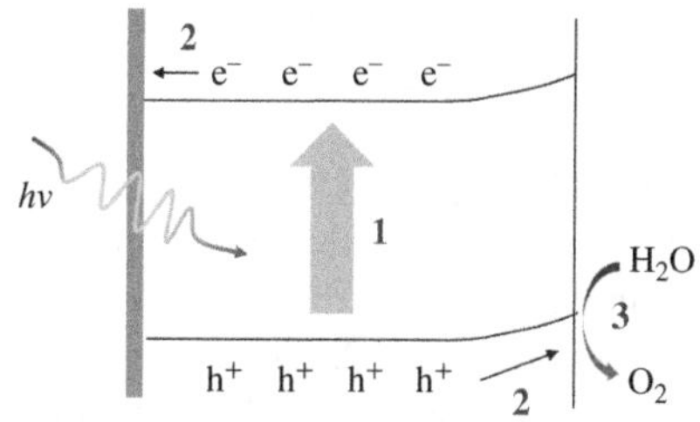

Figure 2.11 Scheme of the three main processes occurring in a semiconductor photoelectrode in a PEC water splitting (taking n-type semiconductors as the example) device under illumination. Process **1:** the incident light causes electrons in the VB of a semiconductor photoelectrode, to get photoexcited to the CB, generating photogenerated electron–hole pairs; Process **2**: the photogenerated holes (in n-type semiconductor photoanodes) or electrons (in p-type semiconductor photocathodes) transfer from the bulk to the surface of the semiconductor; Process **3**: the injection of the surface electrons to the electrolyte for proton reduction (hydrogen evolution) or surface holes for water oxidation (oxygen evolution).

splitting can be expressed by the equation: IPCE $(\lambda) = \eta_{abs}(\lambda) \times \eta_{trans} \times \eta_{inj}$, where λ signifies the wavelength of the incident light. Enhancing the value of η_{abs}, η_{trans}, or η_{inj} will increase the overall PEC efficiency.

There have been several specific strategies developed to improve η_{abs}, tuning the morphologies of the photoelectrodes in a nanoscale (e.g., implementation of 3D ordered nanophotonic structures) (Singh & Sarkar, 2019), applying the materials with surface plasmon resonance effect (such as Au (Gao et al., 2012; Thimsen et al., 2011) and Pd (Mattos et al., 2021)) to the semiconductors, building dual-absorber systems (such as Fe_2O_3/WO_3 photoanode for PEC water oxidation (Sivula et al., 2009), and doping foreign atoms to the semiconductors to modulate the electronic structures for better light harvesting (like doping F to TiO_2 to make the latter visible-light responsible (Dozzi et al., 2013) are a few among such approaches. Applying these methods to the semiconductors could broaden the light absorption range, especially for the utilization of the visible light energy range in the solar spectra.

With respect to η_{trans} of Process **2**, the recombination time (t_1) and the carrier drift time (t_2, the time needed for the charge carriers to travel through the materials) are highly relevant concerning the transfer of the charge carriers from the bulk to the surface of the semiconductor. A longer t_1 and/or shorter t_2 is beneficial for a greater η_{trans}, and strategies on optimization focusing on these two time-related parameters have emerged. To increase t_1, an ideal approach is passivation of the recombination center and/or reduction of the electron (for photoanodes)/hole (for photocathodes) concentration in the space charge region. This can be realized by surface passivation by adding passivation layer to semiconductors (such as coating TiO_2 layer to Cu_2O) (Trang et al., 2019) and building type II junctions for improved charge separation (Moniz et al., 2015) (the type II junction is introduced in Figure 2.12). To reduce t_2, shortening the distance covered by the charge carriers traveling through the materials and enhancing the charge carrier's mobility are two major effective approaches.

The carrier injection efficiency, η_{inj}, related to Process 3, can be affected by the surface state of the semiconductors. It can be improved by decreasing the density of the surface recombination sites and improving the kinetics at the photoelectrode/electrolyte interface. Common strategies are coating thin passivation overlayers (such as TiO_2 (Ahmed et al., 2015), Al_2O_3 (Le Formal et al., 2011), and Ga_2O_3 (Steier et al., 2014)) onto the semiconductors and modifying the surface of the semiconductors by appropriate cocatalysts, such as Pt (Dubale et al., 2016), CoPi (Haydous et al., 2018), and IrO_2 (Collomb et al., 2020).

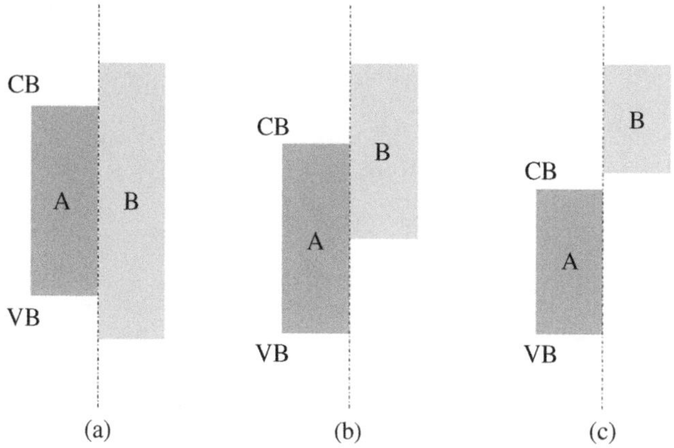

Figure 2.12 Three types of junctions: (a) Type I: the CB edge of semiconductor A is lower than that of semiconductor B, while the VB edge of A is higher than that of B; (b) Type II: the CB edge of A is lower than that of B, but higher than the VB edge of B, while the VB edge of A is lower than that of B; (c) Type III: the CB edge and VB edge of A are lower than VB edge of B. Different types of junctions lead to varied charge carrier migrations.

On a lab scale, the performance of a PEC device for water oxidation or hydrogen evolution can be quantitatively described by the following parameters:

Current Density (J) J is the measured current density (expressed in mA/cm^2) in the external circuit of the PEC devices. For a single photoelectrode PEC cell (e.g., a cell with a cathode and a photoanode), J is typically generated using a combination of an external electrical bias and incident light. For a two-photoelectrode PEC cell consisting of a photoanode and a photocathode (Figure 2.13) J can be generated with incident light alone (without the applied bias) if sufficient photovoltage is generated between the electrodes to meet the thermodynamic requirements and overpotential losses.

Onset Voltage (V_{onset}) V_{onset} is the highest (for cathodic reactions) or lowest (for anodic reactions) voltage value at which a reaction product is produced at a given electrode and defined conditions. For photoanodes, a more negative V_{onset} is desired, while for photocathodes, a more positive V_{onset} is preferred (vs. NHE).

Applied Bias Photon-to-Current Efficiency (ABPE) ABPE = $J\times(1.23\,V$ (standard reversible voltage) $- V_{app})/P_{in}$, where J is the measured external current density, V_{app} is the applied potential referred to the counter electrode, and P_{in} is the power density of the incident light (in most research, $P_{in} = 100\,mW/cm^2$ with AM 1.5G solar spectrum).

Solar-to-Hydrogen Conversion Efficiency (η_{STH}) η_{STH} defines the overall efficiency of the total H_2 energy produced by the input light power. Note η_{STH} can be only applied for the two-photoelectrode configuration. In other words, the measurement of meaningful η_{STH} only applies to systems where sunlight and water are the only inputs. The amount of H_2 produced can be measured using gas chromatography and similar tools. The STH efficiency can be determined using the relation

$$\eta_{STH} = (\Delta G \times r_{H_2})/(P_{in} \times S),$$

where standard Gibbs energy $\Delta G = 237.2\,kJ/mol$, r_{H_2} is the rate of hydrogen produced (mol per second), and S is the illuminated area (cm^{-2}). The most intense band of solar spectrum locates at the visible region (from 400 to 700 nm, Figure 2.13), which accounts for about 28% of the total photon flux that reaches the Earth's surface. As illustrated in Figure 2.13, the ideal STH is theoretically capped at 2% assuming only ultraviolet ($\lambda < 400$ nm) band absorbed by the semiconductors/photocatalysts (neglecting entropic and any energy losses). Although fundamentally unachievable, the theoretical maximum STH, considering absorption of 100% incident photons and their conversion to hydrogen (with all losses ignored), extends to 25% and 47% if the photoelectrode absorbs visible light up to 700 and 1,000 nm, respectively (Wang & Domen, 2020). Thus, to obtain high STHs in practical applications, it is crucial to develop photoactive materials that can harvest as much visible light as possible.

Faradaic Efficiency (FE) Faradaic efficiency describes the efficiency with which charge (electrons) is transferred in a system facilitating an electrochemical reaction. In an ideal circumstance, FE can be close to 1, but in some cases, because of problems such as self-oxidation or self-reduction of the semiconductors, or the existence of a capacitance, the FE could be significantly lower than 1.

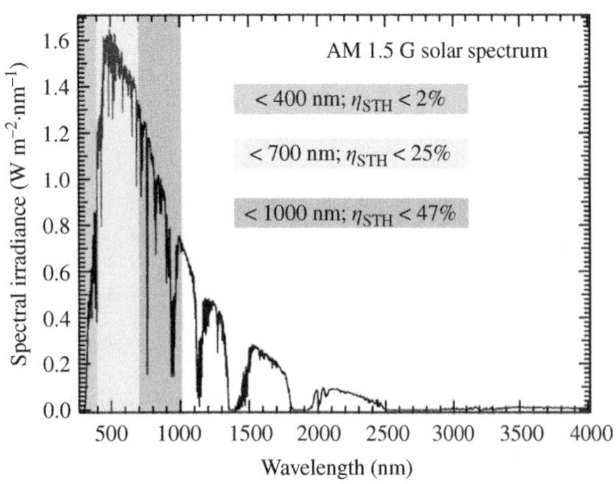

Figure 2.13 Illustration of the dependence of STHs on incident photon wavelengths. Estimation based on AM 1.5 G solar spectrum.

IPCE IPCE is the number of electrons driven to the external circuit per incident photon at a given wavelength.

Absorbed Photon-to-Current Conversion Efficiency (APCE) APCE is defined by the following equation: APCE (λ) = IPCE(λ)/(1–$10^{-A(\lambda)}$), where A represents the optical absorbance. Unlike IPCE, APCE takes into account the optical loss due to reflection and transmission and considers only the net absorbed photons contributing to the photocurrent. Therefore, APCE > IPCE. In some studies, it is also termed as internal quantum efficiency.

2.3.3.3 Configurations of PEC Cell for Water Splitting

Several established configurations of the PEC devices for water splitting, such as the aforementioned single-photoelectrode and two-photoelectrode (photoanode–photocathode coupling) systems, are summarized in Figure 2.14. The photovoltaic (PV) cell-coupled PEC configurations are also shown.

Single-photoelectrode Configuration The single-photoelectrode system consists of one photoactive electrode as the working electrode and a counter electrode normally based on metals (e.g., Pt is widely used) or carbon materials (e.g., graphite rods) to complete the circuit. For example, $BiVO_4$ is a very promising photoactive semiconductor for a photoanode that catalyzes oxygen evolution. With the help of a bias voltage from an external source applied between the anode and cathode, the hydrogen evolution can happen on the Pt counter electrode, as shown in Figure 2.14a. In recent years, heterojunction photoelectrodes were investigated extensively. These photoelectrodes are formed by adding one or more secondary semiconductors to the primary one. As shown in Figure 2.14b, such a configuration has many merits, including broadened light absorption range and/or enhanced driving force for charge carrier separations. The energy band structures of the two semiconductors should be carefully chosen/modulated so that the constructed heterojunction can meet the energy band requirements of type II junctions. Otherwise, the interface between the two semiconductors may promote unwanted recombination.

Two-photoelectrode configuration: The two-photoelectrode system is formed by coupling a photoanode and a photocathode. In this configuration, if the photoanode and photocathode materials are chosen appropriately, the oxygen and hydrogen evolution can take place simultaneously at respective electrodes under incident light illumination without bias.

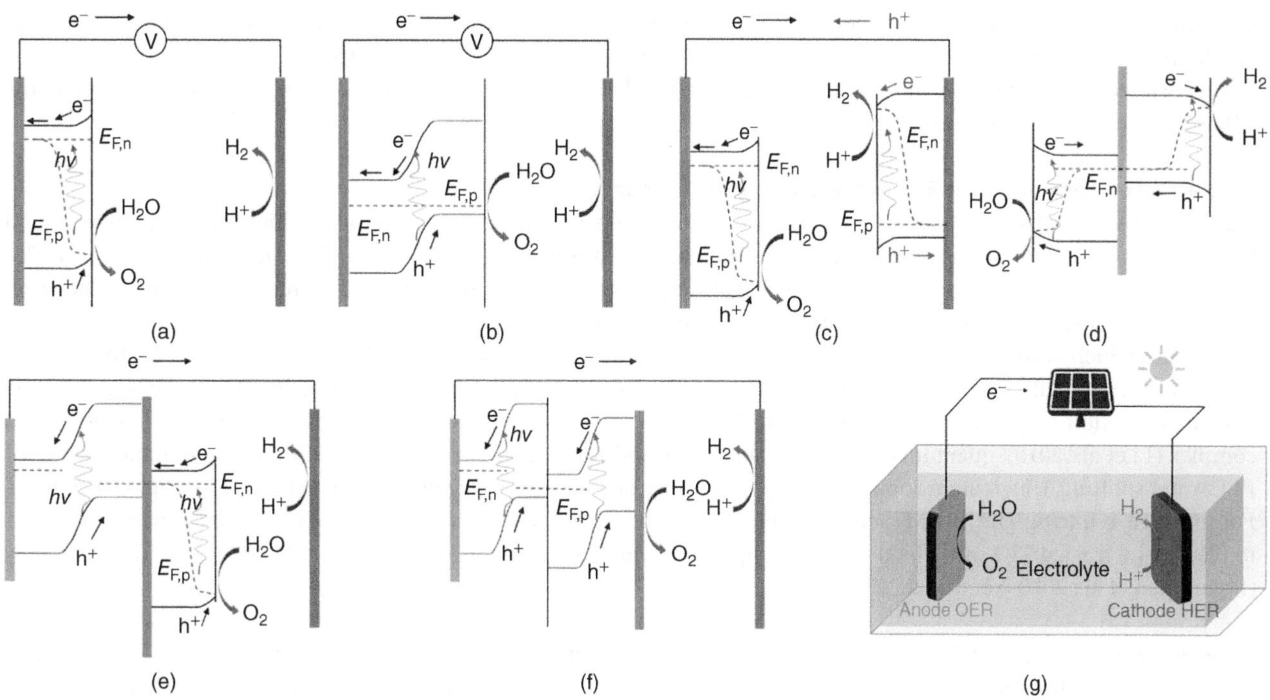

Figure 2.14 Scheme of the configurations of PEC devices for water splitting: (a) single-photoelectrode configuration with one light absorber and (b) heterojunction (taking a photoanode as the example); (c) wired two-photoelectrode configuration; (d) wireless two-photoelectrode configuration; (e) PV-PEC cells; (f) and (g) PV-electrolyzer cells. The incident light typically enters from the back of photoanodes in (a–f).

Such a configuration can be divided into two classes: a photoanode and a photocathode connected by a conducting wire (Figure 2.14c) and an integrated structure consisting of a transparent conductive substrate directly forming an ohmic contact between the photoanode and photocathode (Figure 2.14d). This wireless configuration is more compact compared with the former one with wires, but the transparent conductive substrate could lead to more recombination. In both cases, maximizing light harvesting efficiency requires careful band gap alignment between the photoanode and photocathode, as well as desirable transmittance properties of both photoelectrodes to enable high transmittance of incident light. Many p-type semiconductors for photocathodes have high absorption in the visible region of the solar spectrum, but their VB-edge potentials are not sufficiently positive to oxidize water. Therefore, the energy band structures of the semiconductors for both photoanode and photocathode must be appropriate. Typically, the photogenerated minority carrier holes in the n-type semiconductor photoanode migrate to the surface to participate in the water oxidation. The majority carrier, electrons, are oppositely directed and recombine with holes from the photocathode. Similarly, the minority carriers, which are electrons in the p-type semiconductor photocathode, drift to the surface and reduce protons or water for HER, and the majority carriers, holes, flow to the back contact and recombine with electrons from the photoanode (Smith et al., 2015). To give the surface collected holes in photoanodes enough oxidation power, the VB maximum needs to be more positive than the redox potential of H_2O/O_2. The CB minimum for the photocathode must be more negative than the corresponding reduction potential of H_2/H_2O. Meanwhile, the CB minimum of the photoanode should be more negative (higher) than the VB maximum of the photocathode. However, due to the tandem circuit, the operating photocurrent of the two-photoelectrode configuration is decided by the photoelectrode that generates the smaller current under illumination. The electrode with inferior current limits the performance of the devices.

PV included PEC devices: PV devices (such as Si and perovskite solar cells) are also used to supply the required voltage bias for PEC water splitting. In such devices, the PV device can be single or multi-junction. Typically, those with very high power conversion efficiency (beyond 20%) are used. The incident light passes through the front photoelectrode and then reaches the back PV to generate the photovoltage required to drive the water splitting (PV-PEC cells, Figure 2.14e). In the case of PEC cells with PV-backed photoanodes, hydrogen evolves on the metal cathode. Wide-bandgap semiconductors are preferred for photoelectrodes to ensure sufficient optical transmission. Notably, the ohmic contact between the PV and photoelectrode may cause more charge carrier recombination. Another example of PV assisted device for bias-free water splitting is shown in Figure 2.14f. This configuration is termed as "buried" photoelectrolysis due to the absence of photoelectrodes. The PV cells are directly integrated not on either photoelectrode but in the circuit. It can be regarded as an identical configuration to the PV-electrolyzer that directly wires the PV to the electrolyzer, as shown in Figure 2.14g. In this case, no semiconductor/liquid junction exists, so the driving force to split water is not necessarily correlated to the CB or VB of the semiconductors.

2.3.3.4 A Few Semiconductors Extensively Studied for Water Splitting

TiO$_2$ The merits of TiO_2 (in anatase phase) semiconductor as a photoelectrode material are high stability, low cost, and non-toxicity. Its CB minimum and VB maximum straddle the redox potentials of H^+/H_2 (0 V vs. RHE) and O_2/H_2O (1.23 V vs. RHE), which enables the ability of overall water splitting, as shown in Figure 2.3. However, the relatively large bandgap of TiO_2, ~3.2 eV, enables it to possess only narrow band absorption in UV region of the solar spectrum. The utilized ultraviolet light band, which accounts for less than ~5% of photons in the solar spectrum, results in a very poor solar-to-chemical energy conversion efficiency (Fujishima & Honda, 1972). Therefore, sensitization is normally mandatory to extend the optical absorption range of TiO_2 to visible light. To this end, some organic and inorganic sensitizers, such as Ru-based complex (Li et al., 2010), quantum dots (Han et al., 2022), and Au (Celebi et al., 2021) have been used to modify TiO_2 for PEC water splitting. Heteroatom doping and creating appropriate vacancies also enable TiO_2 to have visible light absorption. For instance, reducing TiO_2 with H_2 creates sufficient vacancy in the material, and turns TiO_2 from the initial white color to black. The treated TiO_2 absorbs the incident light in a wide energy range and achieves relatively high STH efficiency (Rajaraman et al., 2020; Yoon et al., 2022).

Hematite α-Fe_2O_3 (hematite), known as rust in daily life, is a very cheap and environmentally friendly semiconductor for PEC water splitting in alkaline conditions. The CB minimum of hematite is more positive than the redox potential of H^+/H_2 (0 V vs. RHE), and its VB maximum is more positive than the redox potential of O_2/H_2O (1.23 V vs. RHE), as shown in Figure 2.9. Such energy band structure makes it feasible for water oxidation rather than proton reduction. The bandgap of hematite is approximately 2.0–2.2 eV, which corresponds to theoretical absorption of 40% photons in the solar spectrum. Thus, the theoretical photocurrent density of hematite for water oxidation can be as high as 12.6 mA/cm^2, but so far, the highest photocurrent reported in the literature is only ~4.32 mA/cm^2 (at 1.23 V vs. RHE without sacrificial agents) (Kim et al., 2013). The short hole-diffusion length (2–20 nm) and low carrier mobility are the main reasons for the low photocurrent density in labs (Barroso et al., 2013). Methods to improve the PEC performance of hematite include loading cocatalysts, nanostructure modulation, and heteroatom doping (Zhou & Fan, 2021).

Bismuth Vanadate Monoclinic $BiVO_4$, which usually exhibits yellow color, is also a low-cost and non-toxic semiconductor for PEC water splitting. Its bandgap is ~2.4 eV, and the VB maximum and CB minimum are more positive than those of the redox potentials of H^+/H_2 (0 V vs. RHE) and O_2/H_2O (1.23 V vs. RHE), respectively (Figure 2.9), showing great potential as water oxidation catalysts. Its theoretical photocurrent density is 7.5 mA/cm^2 at 1.23 V vs. RHE for OER, and in experiments, 6.73 mA/cm^2 at 1.23 V vs. RHE has been achieved (Zhang, Huang, et al., 2022), and the separation efficiency of the photogenerated hole–electron pairs can be more than 85% in the $BiVO_4$ photoanodes. Therefore, the key to increasing the STH efficiency of $BiVO_4$ is to reduce the recombination loss. The reported strategies include effectively tuning the morphologies, compositions, and cell structures (Gaikwad et al., 2022).

Tungsten Oxide WO_3 is also a low-cost and environmentally friendly semiconductor for PEC water oxidation. Its bandgap is 2.6–2.7 eV, which means it can absorb 12% of the solar spectrum, in the wavelength region up to 450 nm. Its VB maximum is positive enough to drive the holes for oxygen evolution. Almost 150 nm hole diffusion length of WO_3, which is longer than that of hematite, made it attract the attention of researchers; however, it is only suitable in acidic and neutral environments (moderate pH range) due to its instability in alkaline electrolytes. As a semiconductor for photoanodes, the maximum theoretical STH efficiency of WO_3 is 4.8%, but only a far lower value has been reported (Kalanur et al., 2018), because of the serious recombination of the photogenerated hole–electron pairs. Morphology control has been proven as an efficient means to boost the PEC performance of WO_3, but there is still considerable room for improvement (Kalanur et al., 2018).

Cu_2O p-Type semiconductor cuprous oxide with a bandgap ~2.0 eV has a negative enough CB minimum to reduce the protons for hydrogen evolution. Its low cost and low toxicity attract intensive attention in PEC devices. For water splitting, the calculated theoretical photocurrent density of ~15 mA/cm^2 (at the bias of 0 V vs. RHE) and 18% STH efficiency make Cu_2O very promising (Paracchino et al., 2011). However, only moderate photocurrent densities of Cu_2O have been reported so far and the long-term stability is still a great challenge for its practical application due to photocorrosion (Paracchino et al., 2011). Constructing a p–n junction (for instance Cu_2O–ZnO) (Luo et al., 2016) or adding cocatalysts (like MoS_2) (Zhang et al., 2021) is supposed to be effective means of improving the photocurrent density. Coating a protection layer, such as TiO_2 layer, (Pan et al., 2018) can also promote the long-term stability of Cu_2O photocathodes for hydrogen evolution. But to date, the highest STH efficiency of Cu_2O photocathodes in labs is not more than 3% (Pan et al., 2018). Enormous efforts are still needed for causing further improvements.

Sb_2Se_3 Sb_2Se_3 is a newly emerged p-type semiconductor for PEC proton reduction. The small bandgap of Sb_2Se_3, ~1.1–1.2 eV, guarantees sufficient absorption of the solar spectra. The cost of Sb is as low as that of Cu, and its high mobility (~10 cm^2/Vs) and excellent optical property ($\alpha > 10^5$ cm^{-1}) make it a very promising material in PEC devices. Meanwhile, it shows better stability in terms of photocorrosion than Cu_2O. Regarding Sb_2Se_3 photocathodes, the theoretical photocurrent density at 0 V vs. RHE can be as high as ~40, and ~30 mA/cm^2 has been achieved to date (Yang et al., 2020). To improve the PEC performance for proton reduction, unique nanostructures, particularly one-dimensional nanoarrays such as Sb_2Se_3 nanoarray, have been developed (Liu et al., 2020). p–n Junction construction (e.g., Sb_2Se_3–In_2S_3 photocathodes) (Yang et al., 2020) and co-catalyst addition (e.g., MoS_x) (de Araújo et al., 2022) are also effective ways to boost the charge transfer in Sb_2Se_3 photocathodes.

2.3.3.5 Photoelectrochemical (PEC) CO_2 reduction

PEC CO_2 reduction research originated in the year 1978, when Halmann successfully converted CO_2 to formic acid and methanol using p-type GaP photocathodes (Halmann, 1978). Since then, CO_2 reduction has shown great potential to alleviate carbon emissions and obtain high-value-added products. Multiple hydrocarbons and oxygenates, like monocarbon (such as CO, HCOOH, CH_3OH, and CH_4) and C_{2+} products (such as C_2H_4 and C_2H_5OH), have been obtained via CO_2 reduction reactions. The thermodynamic potentials of CO_2 reduction reactions are different depending on the specific products, as shown in Table 2.1.

Table 2.1 Some thermodynamic redox potentials of CO_2 reduction.

CO_2 reduction (half reaction)	V vs. RHE
$CO_2 + H_2O + 2e^- \rightarrow CO + 2OH^-$	−0.10
$CO_2 + H_2O + 2e^- \rightarrow HCOO^- + OH^-$	−0.03
$CO_2 + 5H_2O + 6e^- \rightarrow CH_3OH + 6OH^-$	0.03
$CO_2 + 6H_2O + 8e^- \rightarrow CH_4 + 8OH^-$	0.17
$2CO_2 + 8H_2O + 12e^- \rightarrow C_2H_4 + 12OH^-$	0.08
(HER) $2H_2O + 2e^- \rightarrow H_2 + 2OH^-$	0.00

Obviously, to thermodynamically conduct the desired CO_2 reduction to the specific product, the CB minimum of the p-type semiconductor should be more negative than the redox potential of corresponding half reactions. For example, $2e^-$ transfer is needed in CO_2 reduction reaction to produce one molecule of HCOOH and the CB edge of the semiconductor (normally p-type) must be negative than −0.03 V (vs. RHE), while $8e^-$ transfer in CO_2 reduction to one molecular CH_4 requires a CB edge negative than 0.17 V (vs. RHE). Thus, it is easy to note that when the applied potential (photovoltage) is sufficiently negative, multiple CO_2 reduction reactions for varied products could happen simultaneously. If the production of CO is the desired reaction, the conversion of CO_2 to CH_4 and HCOOH is very likely to occur as the companion reactions due to their more positive redox potentials than that of CO_2/CO. Therefore, high selectivity is one of the main challenges for CO_2 reduction. Since the reaction is always carried out in aqueous systems, the hydrogen evolution by proton reduction (as described in the above PEC water splitting sections) normally becomes the most severe competing reaction of CO_2 reduction.

With the assistance of theoretical calculations (density function theory, DFT), several pathways of CO_2 reduction to different products were proposed, and a fraction of them could be experimentally verified. As shown in Figure 2.15, these pathways can be classified into two categories according to the products of the reactions: monocarbon and multicarbon (C_{2+}). For monocarbon products, the adsorption of the intermediates on the photocathodes should be modest. Too strong or weak adsorption could lower the efficiency. For C_{2+}, such as ethylene and ethanol, the difficulty of the reactions is the formation of C—C bonds, which need more steps and intermediates than the formation of monocarbon. In these processes, the intermediates could be desorbed from the surface, normally resulting in more by-products and low selectivity of the target product.

With respect to the configuration of PEC devices for CO_2 reduction, technically, the configuration of the device for PEC water splitting in an aqueous environment can be used directly for PEC CO_2 reduction, as shown in Figure 2.13, except that the photocathodes should be replaced to reduce CO_2 rather than to reduce protons. It should be noted that water oxidation reaction (OER) is usually the counter-reaction of the CO_2 reduction reaction in a PEC device, therefore, OER also plays a critical role in CO_2 reduction reactions, due to its sluggish kinetics for the overall reaction. For a single-photoelectrode (photocathode) system, the counter electrodes for OER are normally Pt, carbon materials, and IrO_2. For a two-photoelectrode configuration driven by bias-free light illumination (Figure 2.16a), the CO_2 conversion efficiency is determined by the overlaid performance curves of the photocathode (for CO_2 reduction) and photoanode (for OER) (Figure 2.16b), similar to the bias-free photo-driven overall water splitting (Figure 2.13).

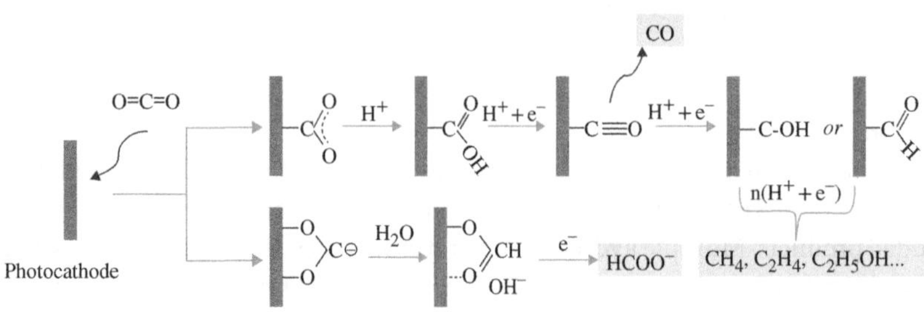

Figure 2.15 Simplified scheme of the pathways of CO_2 reduction to different products.

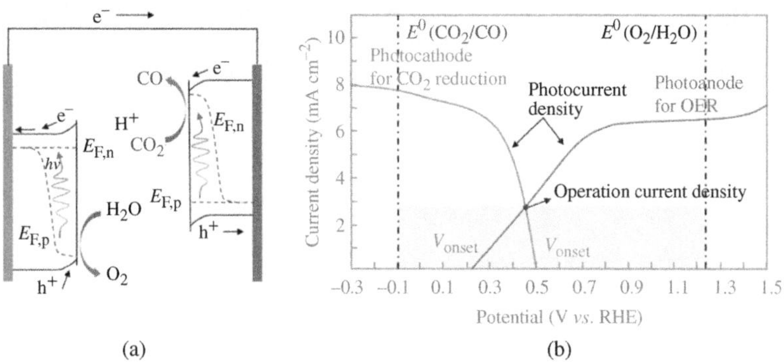

Figure 2.16 (a) Two-photoelectrode configuration for CO_2 reduction, and (b) the overlaid performance curves of photocathodes and photoanodes for overall CO_2 reduction. The conversion of CO_2 to CO is taken as an example here.

The photocurrent density in PEC devices for CO_2 reduction reaction deserves additional discussion. The total photocurrent density expresses the overall reduction activity of the photocathodes. It is the sum of the photocurrent density for all reduction reactions, including CO_2 reduction reaction of various products and proton reduction. Therefore, only partial photocurrent density is responsible for certain reduction products, and the Faradaic efficiency for the specific product can be obtained by dividing the amount of the product by the theoretical amount produced by the sum photocurrent.

Semiconductors useful for CO_2 reduction are normally p-type. Thus, when the photocathodes contact the electrolyte, the downwards band bending takes place, as described in the previous PEC hydrogen evolution cases (Figure 2.14c). Cu_2O and C_3N_4 are examples of materials extensively investigated as novel p-type semiconductors for PEC CO_2 reduction reactions. Cu_2O was introduced in the former Section 2.3.3.4 of HER photocathodes. It has been discovered that by carefully tuning the morphology (Zeng et al., 2019), junction design (Wang et al., 2022), cocatalysts selection (Chang et al., 2019; Zhang, Pan, et al., 2022), and pH environment (de Brito et al., 2015), HER could be suppressed and CO_2 reduction reaction could be made the primary catalytic reaction under illumination and bias. The products of PEC CO_2 reduction include CH_3OH, CH_2O, CH_3CH_2OH, and CH_3CHO. But the photocorrosion of Cu_2O leading to its poor stability is still a significant issue in PEC CO_2 reduction reaction. Graphite-like C_3N_4 has also gained considerable attention due to ease of preparation, low cost, non-toxicity, and existence in two-dimensional nanostructure geometry. The CB minimum of C_3N_4 is sufficiently negative to reduce CO_2 (and water reduction as well, of course). Its bandgap is around ~2.7 eV, which enables it to absorb visible light under 460 nm. By doping heteroatoms (Riyadh Atta et al., 2023) and loading cocatalysts on C_3N_4 (Xu et al., 2018), the products of PEC CO_2 reduction reaction could be HCOOH, CH_4, and ethanol. But the photocurrent for CO_2 reduction reaction is still very low up to now.

2.3.3.6 Particulate Photocatalytic Systems for Water Splitting/CO_2 Reduction

Particulate photocatalysts offer another straightforward route for producing solar fuels without an external electrical bias (Wang & Domen, 2019). The working principle and evaluation criteria of photocatalysis based on particulate photocatalysts are similar to those discussed in the previous sections for the PEC-based electrode reactions. In short, there are three sequential steps involved in the particulate photocatalytic systems (Figure 2.17): first, the photocatalysts absorb incident photons with energy greater than their bandgap, generating the electron–hole pairs; second, the separated charges migrate to the surface/cocatalysts; third, occurrence of the surface chemical reactions based on potential energies of the charge carriers. Similar to the case of PEC systems, multistep and coherently connected processes occur during photocatalytic water splitting and CO_2 reduction, and they all contribute to the overall energy efficiency of the system. Thus, it is crucial to take into consideration a number of different intrinsic physical and chemical properties to design efficient particulate photocatalytic systems. The important parameters relevant to photocatalysts include the visible light absorption coefficients (determine the maximum STH), the lifetimes and diffusion lengths of the carriers, concentrations of defects that may affect the distribution of the charge carriers, and the corresponding recombination rate, the identity, and density of the active sites on the particle surfaces. Many specific design principles developed based on these intrinsic properties of the photocatalysts exist in the literature.

Particulate photocatalytic systems have the advantage that they can be easily scaled up while retaining their intrinsic activity. The poor overall solar-to-fuel efficiency, however, has put constraints on the practical implementation of this technology. The low solar-to-fuel efficiency is due to the low utilization efficiency of the solar light spectrum.

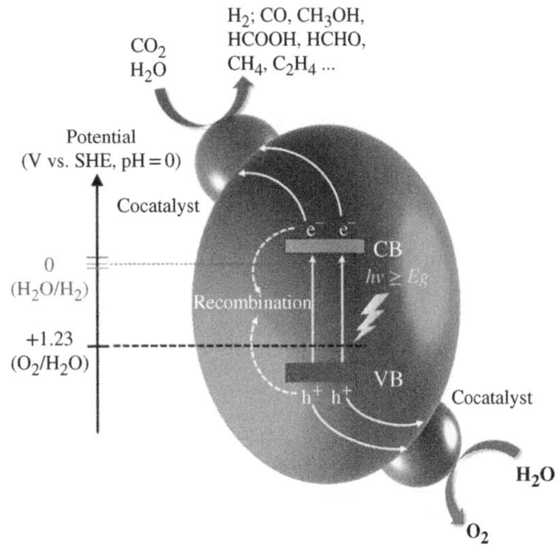

Figure 2.17 Schematic of a single particulate photocatalyst system and its working principle for water splitting and carbon dioxide reduction. The dotted lines denote the pathways for undesired electron transfer reactions.

Designing photoactive semiconductor materials with suitable bad gaps can effectively fill this gap. Therefore, extensive research efforts were spent on exploring metal oxide-based semiconductors with tunable band structures and gaps aiming to improve the utilization efficiency of the solar light spectrum. In general, the precursor of these metal oxides usually consists of d^0 (Ti^{4+}, V^{5+}, W^{6+}, Ce^{4+}, etc.) and/or d^{10} (Zn^{2+}, In^{3+}, Sn^{4+}, Sb^{5+}, etc.) cations (Wang & Domen, 2019). The unsatisfactory charge separation/transfer efficiency, particularly at the reaction interface, is another major constraint for achieving high solar-to-fuel efficiency. The bandgaps, structures, and edge positions, together with the symmetry of the coordination environments of the metal cations are crucial factors that can affect the interfacial charge separation/transfer. A handful of strategies were established to improve the charge separation/transfer toward the desired reactions, including VB potential engineering and control (Goga et al., 1999, Goto et al., 2016 Nandy et al., 2018 Ogisu et al., 2008 Sato et al., 2003), solid solution formation to tune band structure and electronic properties (Sahoo & Maggard, 2013; Takata et al., 1997); doping with foreign atoms to optimize the physicochemical properties of the photocatalysts (An et al., 2018; Arai et al., 2006; Arai et al., 2007; Pasumarthi et al., 2019; Yuan et al., 2007); quantum confinement engineering (e.g., quantum dots-based materials, QDs) (Holmes et al., 2012; Zhao et al., 2013); dye sensitization to extend the usable wavelength range (Chen et al., 2010; Grätzel, 2001; Youngblood et al., 2009); and introducing local surface plasmon resonance to mitigate the surface charge transfer reactions (Hou & Cronin, 2013; Liu et al., 2017).

Inspired by the two-step photoexcitation process in natural photosynthesis (photosystem II–photosystem I, as shown in Figure 2.6), a so-called Z-scheme approach was proposed to achieve higher solar-to-fuel conversion efficiency (Wang et al., 2018). Taking water splitting as an example, a Z-scheme system consists of a hydrogen evolution photocatalyst (HEP), an oxygen evolution photocatalyst (OEP), and a redox mediator between to shuttle the electron, as shown in Figure 2.18. In such a Z-scheme system, the photogenerated H_2 and O_2 gases can be spatially separated when a suitable reactor is carefully designed. Neither HEP nor OEP requires to have bandgaps that straddle both the water oxidation and reduction potentials at the same time. This would broaden the range of materials useful as photocatalysts for PEC fuel production, which is exceptionally beneficial in terms of materials development. This also gives Z-scheme systems a higher freedom in choosing suitable band positions for improving product selectivity. Moreover, it would increase the theoretical solar energy conversion efficiency in comparison to a single-particle photocatalytic system, since a wider range of the solar spectrum can be harnessed for PEC reactions. Notably, both HEP and OEP generate electron–hole pairs when exposed to light, requiring two incident photons to split the water molecule into H_2 and O_2. A redox shuttle reaction with a suitable chemical potential is also necessary to combine the charges effectively. Despite the advantages of developing Z-scheme systems, the system design and optimization are much more complex, because more surface reactions cause greater chances of undesired reactions.

Overall, both solar energy conversion efficiency and the system durability of particulate photocatalytic systems must undergo dramatic progress before practical implementation of this technology. To achieve high efficiency of overall solar energy conversion, one must maximize the visible light absorption and charge separation abilities of the photocatalysts. It is also vital to design facile and stable redox shuttles in terms of the Z-scheme systems and to develop effective cocatalysts that extract the charge carriers at multiple interfaces.

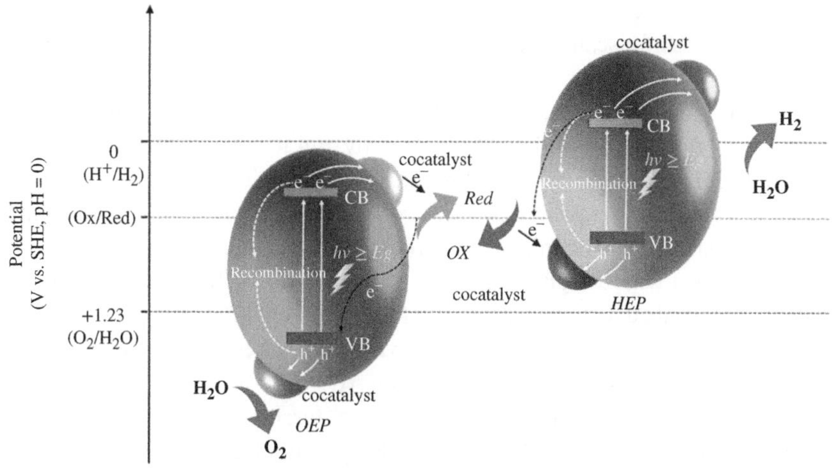

Figure 2.18 Schematic of a two-step photoexcitation Z-scheme system for water splitting. The dotted lines denote the pathways for undesired electron transfer reactions.

2.4 Challenges and Outlook

Solar fuel production based on the aforementioned routes requires synergistic functioning of multiple yet interconnected components within the system: visible light absorbers, surface cocatalysts, redox shuttles, ion conductors, and counter electrode/reactions. Each component might dictate activity, selectivity, stability, and overall energy conversion efficiency of the system. To maximize the overall solar-to-fuel conversion efficiency, the performance of every component has to be optimized. Enhanced light absorbers, for example, may increase quantum efficiency and capture broader solar spectrum; novel catalysts may show high activity and selectivity for the desired electrode reactions; suppressing side reactions and undesired electron transfer pathways may improve stability of both light absorber and cocatalysis. To achieve these goals, it is necessary to comprehensively understand the thermodynamics and kinetics of various processes within different components and at different interfaces. Advanced *operando* spectroscopy (i.e., vibrational spectroscopy that can probe the reaction intermediates, X-ray based spectroscopy that can probe the material structural changes during the PEC reaction) with high time resolutions, combined with the capability of real-time product analysis can help understand and reveal the bottleneck that controls the activity, selectivity, and stability of the system. In particular, it is critical to bridge the different time and length gaps between the single or multi-photon excitation processes and the desired surface chemical reactions. Model systems designed using first-principle computations are beneficial to rationalize the mechanistic pathways, desired and unwanted, within the multicomponent systems.

Selectivity toward desired products or pathways remains a grand challenge for solar fuel production through photocatalytic and PEC processes. While the intrinsic properties of the photo absorbers and cocatalysts are critical, microenvironment around the active center also plays a significant role in determining the efficiencies of both photoinduced charge separations and surface catalytic reactions, particularly the photo(electro) CO_2 reduction. The microenvironment factors include secondary coordination sphere (that can facilitate the charge separation and the reaction kinetics), solvent/electrolyte identity, pH, substrate manifold, and electrical field. They control the transportation of electrons, protons, reactants, and products. Consequently, they influence the overall system activity, selectivity, durability, and overall energy efficiency. Optimization of the microenvironments around different components within the system provides further opportunities to facilitate efficient solar fuel technologies.

The success of solar fuel production not only relies on the optimization of the light-absorbing layers and catalysts but also depends on reactor engineering targeting industrial-level performance. Identifying practical challenges associated with the integration of different components is crucial in reactor design, because unexpected problems may be encountered as multiple chemical and physical processes are involved. These processes occurring at multiple interfaces include the light absorption, electrons and holes separation, and transportation to the surface to promote the interfacial reactions. Moreover, these interfacial reactions involve multielectron and proton transfers, multiple bond breaking, and formation steps at the reaction interface. Additionally, the transport of reactants to the reaction interface and collection and separation of products have to be considered as well. Nevertheless, these processes happening at different time and length scales within the reactor need to be investigated. They must be well coordinated so as to achieve the maximum efficiency. The recent advances in vapor-fed devices provide further opportunities for scaling up solar fuel production in both PEC and particulate photocatalytic systems.

Additionally, the development of efficient semiconductor materials, active and selective catalysts, and the suitable operating conditions have been largely done on empirical approaches. This strategy is costly and tedious although a number of materials were developed in this way. Emerging approaches such as high-throughput materials screening and quantum computing-assisted materials discovery have shown promising capabilities. However, a large portion of the reported data has not been well utilized, partially due to the lack of standard benchmarks and protocols for conducting experiments and reporting data. Artificial intelligence (AI) tools equipped with designed algorithms for big data analysis are able to decipher hidden trends and descriptors underlying the established data, inspiring solar fuels research. Efforts must be made by researchers working in both experimental and computational fields to establish benchmarks for collecting and reporting data, in order to create high-accuracy and compatible databases that can be utilized by AI.

In summary, to mitigate the pressing climate change issues, sustainable alternatives to fossil-based resources must be established to meet our needs for energy and chemical stocks. As one of the most promising technologies, solar fuel research aims to develop cost-effective photocatalytic and PEC processes that can enable efficient conversion of solar energy to fuels and chemicals directly. In this chapter, we introduced the basic physical and chemical principles of commonly studied photocatalytic and PEC systems and presented the corresponding scientific challenges. The core challenges lie in optimization and organization of multi-component and multi-photo(electro)chemical processes within the systems at different time and length scales. The employment of the state-of-art *operando* technologies, first principle computation, and AI technologies would provide new opportunities for the solar fuel research. While the advancements in scientific research are crucial for the practical implementation of these sustainable alternatives, the awareness and support from the public, major stakeholders, and policymakers can greatly promote the transition toward renewable and sustainable future.

References

Adamska, A., Silakov, A., Lambertz, C., Rüdiger, O., Happe, T., Reijerse, E., & Lubitz, W. (2012). Identification and characterization of the "super-reduced" state of the H-cluster in [FeFe] hydrogenase: A new building block for the catalytic cycle? *Angewandte Chemie International Edition*, *51*, 11458–11462.

Agency, I. E. (2018). *Oil, 2018.*

Ahmed, M. E., Chattopadhyay, S., Wang, L., Brazzolotto, D., Pramanik, D., Aldakov, D., Fize, J., Morozan, A., Gennari, M., Duboc, C., Dey, A., & Artero, V. (2018). Hydrogen evolution from aqueous solutions mediated by a heterogenized [NiFe]-hydrogenase model: Low pH enables catalysis through an enzyme-relevant mechanism. *Angewandte Chemie International Edition*, *57*, 16001–16004.

Ahmed, M. G., Kretschmer, I. E., Kandiel, T. A., Ahmed, A. Y., Rashwan, F. A., & Bahneman, D. W. (2015). A facile surface passivation of hematite photoanodes with TiO_2 overlayers for efficient solar water splitting. *ACS Applied Materials & Interfaces*, *7*, 24053–24062.

An, L., Kitta, M., Iwase, A., Kudo, A., Ichikuni, N., & Onishi, H. (2018). Photoexcited electrons driven by doping concentration gradient: Flux-prepared $NaTaO_3$ photocatalysts doped with strontium cations. *ACS Catalysis*, *8*, 9334–9341.

Andersen, S. Z., Čolić, V., Yang, S., Schwalbe, J. A., Nielander, A. C., McEnaney, J. M., Enemark-Rasmussen, K., Baker, J. G., Singh, A. R., Rohr, B. A., Statt, M. J., Blair, S. J., Mezzavilla, S., Kibsgaard, J., Vesborg, P. C. K., Cargnello, M., Bent, S. F., Jaramillo, T. F., Stephens, I. E. L., … Chorkendorff, I. (2019). A rigorous electrochemical ammonia synthesis protocol with quantitative isotope measurements. *Nature*, *570*, 504–508.

Appel, A. M., Bercaw, J. E., Bocarsly, A. B., Dobbek, H., DuBois, D. L., Dupuis, M., Ferry, J. G., Fujita, E., Hille, R., Kenis, P. J., & Kerfeld, C. A. (2013). Frontiers, opportunities, and challenges in biochemical and chemical catalysis of CO_2 fixation. *Chemical Reviews*, *113*, 6621–6658.

Arai, N., Saito, N., Nishiyama, H., Domen, K., Kobayashi, H., Sato, K., & Inoue, Y. (2007). Effects of divalent metal ion (Mg^{2+}, Zn^{2+} and Be^{2+}) doping on photocatalytic activity of ruthenium oxide-loaded gallium nitride for water splitting. *Catalysis Today*, *129*, 407–413.

Arai, N., Saito, N., Nishiyama, H., Inoue, Y., Domen, K., & Sato, K. (2006). Overall water splitting by RuO_2-dispersed divalent-ion-doped GaN photocatalysts with d10 electronic configuration. *Chemistry Letters*, *35*, 796–797.

de Araújo, M. A., Costa, M. B., & Mascaro, L. H. (2022). Improved photoelectrochemical hydrogen gas generation on Sb_2S_3 films modified with an earth-abundant MoS_x co-catalyst. *ACS Applied Energy Materials*, *5*, 1010–1022. https://pubs.acs.org/doi/10.1021/acsaem.1c03374

Ashford, D. L., Gish, M. K., Vannucci, A. K., Brennaman, M. K., Templeton, J. L., Papanikolas, J. M., & Meyer, T. J. (2015). Molecular chromophore–catalyst assemblies for solar fuel applications. *Chemical Reviews*, *115*, 13006–13049.

Barroso, M., Pendlebury, S. R., Cowan, A. J., & Durrant, J. R. (2013). Charge carrier trapping, recombination and transfer in hematite (α-Fe_2O_3) water splitting photoanodes. *Chemical Science*, *4*, 2724–2734.

Blankenship, R. E. (2010). Early evolution of photosynthesis. *Plant Physiology*, *154*, 434–438.

Brazzolotto, D., Gennari, M., Queyriaux, N., Simmons, T. R., Pécaut, J., Demeshko, S., Meyer, F., Orio, M., Artero, V., & Duboc, C. (2016). Nickel-centred proton reduction catalysis in a model of [NiFe] hydrogenase. *Nature Chemistry*, *8*, 1054–1060.

Brazzolotto, D., Wang, L., Tang, H., Gennari, M., Queyriaux, N., Philouze, C., Demeshko, S., Meyer, F., Orio, M., Artero, V., & Hall, M. B. (2018). Tuning reactivity of bioinspired [NiFe]-hydrogenase models by ligand design and modeling the CO inhibition process. *ACS Catalysis*, *8*, 10658–10667.

de Brito, J. F., Araujo, A. R., Rajeshwar, K., & Zanoni, M. V. B. (2015). Photoelectrochemical reduction of CO_2 on Cu/Cu_2O films: Product distribution and pH effects. *Chemical Engineering Journal*, *264*, 302–309.

Celebi, N., Aydin, M. Y., Soysal, F., Ciftci, Y. O., & Salimi, K. (2021). Ligand-free fabrication of Au/TiO_2 nanostructures for plasmonic hot-electron-driven photocatalysis: Photoelectrochemical water splitting and organic-dye degredation. *Journal of Alloys and Compounds*, *860*, 157908.

Cestellos-Blanco, S., Zhang, H., Kim, J. M., Shen, Y.-X., & Yang, P. (2020). Photosynthetic semiconductor biohybrids for solar-driven biocatalysis. *Nature Catalysis*, *3*, 245–255.

Chang, X., Wang, T., Yang, P., Zhang, G., & Gong, J. (2019). The development of cocatalysts for photoelectrochemical CO_2 reduction. *Advanced Materials*, *31*, 1804710.

Chen, X., Shen, S., Guo, L., & Mao, S. S. (2010). Semiconductor-based photocatalytic hydrogen generation. *Chemical Reviews*, *110*, 6503–6570.

Collomb, M.-N., Morales, D. V., Astudillo, C. N., Dautreppe, B., & Fortage, J. (2020). Hybrid photoanodes for water oxidation combining a molecular photosensitizer with a metal oxide oxygen-evolving catalyst. *Sustainable Energy & Fuels*, *4*, 31–49.

Corr, M. J., & Murphy, J. A. (2011). Evolution in the understanding of [Fe]-hydrogenase. *Chemical Society Reviews*, *40*, 2279–2292.

Cox, N., Pantazis, D. A., Neese, F., & Lubitz, W. (2013). Biological water oxidation. *Accounts of Chemical Research*, *46*, 1588–1596.

De Luna, P., Hahn, C., Higgins, D., Jaffer, S. A., Jaramillo, T. F., & Sargent, E. H. (2019). What would it take for renewably powered electrosynthesis to displace petrochemical processes? *Science*, *364*, eaav3506.

Ding, Y., Bertram, J. R., Eckert, C., Bommareddy, R. R., Patel, R., Conradie, A., Bryan, S., & Nagpa, P. (2019). Nanorg microbial factories: light-driven renewable biochemical synthesis using quantum dot-bacteria nanobiohybrids. *Journal of the American Chemical Society*, *141*, 10272–10282.

Dobbek, H. (2011). Structural aspects of mononuclear Mo/W-enzymes. *Coordination Chemistry Reviews*, *255*, 1104–1116.

Dozzi, M. V., D'Andrea, C., Ohtani, B., Valentini, G., & Selli, E. (2013). Fluorine-doped TiO_2 materials: photocatalytic activity vs time-resolved photoluminescence. *The Journal of Physical Chemistry C*, *117*, 25586–25595.

Dubale, A. A., Tamirat, A. G., Chen, H.-M., Berhe, T. A., Pan, C.-J., Su, W.-N., & Hwang, B.-J. (2016). A highly stable CuS and CuS–Pt modified Cu_2 O/CuO heterostructure as an efficient photocathode for the hydrogen evolution reaction. *Journal of Materials Chemistry A*, *4*, 2205–2216.

Francke, R., Schille, B., & Roemelt, M. (2018). Homogeneously catalyzed electroreduction of carbon dioxide—methods, mechanisms, and catalysts. *Chemical Reviews*, *118*, 4631–4701.

Frey, M. (2002). Hydrogenases: Hydrogen-activating enzymes. *ChemBioChem*, *3*, 153–160.

Fujishima, A., & Honda, K. (1972). Electrochemical photolysis of water at a semiconductor electrode. *Nature*, *238*, 37–38.

Gaikwad, M. A., Suryawanshi, U. P., Ghorpade, U. V., Jang, J. S., Suryawanshi, M. P., & Kim, J. H. (2022). Emerging surface, bulk, and interface engineering strategies on $BiVO_4$ for photoelectrochemical water splitting. *Small*, *18*, 2105084.

Gao, H., Liu, C., Jeong, H. E., & Yang, P. (2012). Plasmon-enhanced photocatalytic activity of iron oxide on gold nanopillars. *ACS Nano*, *6*, 234–240.

Goga, M., Seshadri, R., Ksenofontov, V., Gütlich, P., & Tremel, W. (1999). $Ln_2Ti_2S_2O_5$ (Ln= Nd, Pr, Sm): a novel series of defective ruddlesden–popper phases. *Chemical Communications, 979–980.*

Goto, Y., Seo, J., Kumamoto, K., Hisatomi, T., Mizuguchi, Y., Kamihara, Y., & Katayama, M. (2016). Crystal structure, electronic structure, and photocatalytic activity of oxysulfides: $La_2Ta_2ZrS_2O_8$, $La_2Ta_2TiS_2O_8$, and $La_2Nb_2TiS_2O_8$. *Inorganic Chemistry*, *55*, 3674–3679.

Grätzel, M. (2001). Photoelectrochemical cells. *Nature, 414*, 338–344.

Halmann, M. (1978). Photoelectrochemical reduction of aqueous carbon dioxide on p-type gallium phosphide in liquid junction solar cells. *Nature, 275*, 115–116.

Han, Y., Wu, J., Li, Y., Gu, X., He, T., Zhao, Y., Huang, H., Liu, Y., & Kang, Z. (2022). Carbon dots enhance the interface electron transfer and photoelectrochemical kinetics in TiO_2 photoanode. *Applied Catalysis B: Environmental, 304*, 120983.

Haussener, S., Hu, S., Xiang, C., Weber, A. Z., & Lewis, N. S. (2013). Simulations of the irradiation and temperature dependence of the efficiency of tandem photoelectrochemical water-splitting systems. *Energy & Environmental Science, 6*, 3605–3618. https://doi.org/10.1039/C3EE41302K

Haydous, F., Si, W., Guzenko, V. A., Waag, F., Pomjakushina, E., El Kazzi, M., Sévery, L., Wokaun, A., Pergolesi, D., & Lippert, T. (2018). Improved photoelectrochemical water splitting of $CaNbO_2N$ photoanodes by CoPi photodeposition and surface passivation. *The Journal of Physical Chemistry C, 123*, 1059–1068.

Hiromoto, T., Ataka, K., Pilak, O., Vogt, S., Stagni, M. S., Meyer-Klaucke, W., Warkentin, E., Thauer, R. K., Shima, S., & Ermler, U. (2009). The crystal structure of C176A mutated [Fe]-hydrogenase suggests an acyl-iron ligation in the active site iron complex. *FEBS Letters, 583*, 585–590.

Holmes, M. A., Townsend, T. K., & Osterloh, F. E. (2012). Quantum confinement controlled photocatalytic water splitting by suspended CdSe nanocrystals. *Chemical Communications, 48*, 371–373.

Hou, W., & Cronin, S. B. (2013). A review of surface plasmon resonance-enhanced photocatalysis. *Advanced Functional Materials, 23*, 1612–1619.

Jeoung, J.-H., & Dobbek, H. (2007). Carbon dioxide activation at the Ni, Fe-cluster of anaerobic carbon monoxide dehydrogenase. *Science, 318*, 1461–1464.

Kalanur, S. S., Duy, L. T., & Seo, H. (2018). Recent progress in photoelectrochemical water splitting activity of WO_3 photoanodes. *Topics in Catalysis, 61*, 1043–1076.

Kärkäs, M. D., & Åkermark, B. (2016). Water oxidation using earth-abundant transition metal catalysts: Opportunities and challenges. *Dalton Transactions, 45*, 14421–14461.

Kim, J. Y., Magesh, G., Youn, D. H., Jang, J.-W., Kubota, J., Domen, K., & Lee, J. S. (2013). Single-crystalline, wormlike hematite photoanodes for efficient solar water splitting. *Scientific Reports, 3*, 2681.

Kok, B., Forbush, B., & McGloin, M. (1970). Cooperation of charges in photosynthetic O_2 evolution–I. A linear four step mechanism. *Photochemistry and Photobiology, 11*, 457–475.

Le Formal, F., Tetreault, N., Cornuz, M., Moehl, T., Grätzel, M., & Sivula, K. (2011). Passivating surface states on water splitting hematite photoanodes with alumina overlayers. *Chemical Science, 2*, 737–743.

Li, L., Duan, L., Xu, Y., Gorlov, M., Hagfeldt, A., & Sun, L. (2010). A photoelectrochemical device for visible light driven water splitting by a molecular ruthenium catalyst assembled on dye-sensitized nanostructured TiO_2. *Chemical Communications, 46*, 7307–7309.

Liu, C., Liu, T., Li, Y., Zhao, Z., Zhou, D., Li, W., Zhao, Y., Yang, H., Sun, L., Li, F., & Li, Z. (2020). A dendritic Sb_2Se_3/In_2S_3 heterojunction nanorod array photocathode decorated with a MoS_x catalyst for efficient solar hydrogen evolution. *Journal of Materials Chemistry A, 8*, 23385–23394.

Liu, L., Zhang, X., Yang, L., Ren, L., Wang, D., & Ye, J. (2017). Metal nanoparticles induced photocatalysis. *National Science Review, 4*, 761–780.

Lubitz, W., Ogata, H., Rudiger, O., & Reijerse, E. (2014). Hydrogenases. *Chemical Reviews, 114*, 4081–4148.

Lubitz, W., & Tumas, W. (2007). Hydrogen: An overview. *Chemical Reviews, 107*, 3900–3903.

Luo, J., Steier, L., Son, M.-K., Schreier, M., Mayer, M. T., & Grätze, M. (2016). Cu_2O nanowire photocathodes for efficient and durable solar water splitting. *Nano Letters, 16*, 1848–1857. https://doi.org/10.1021/acs.nanolett.5b04929

Madden, C., Vaughn, M. D., Díez-Pérez, I., Brown, K. A., King, P. W., Gust, D., Moore, A. L., & Moore, T. A. (2012). Catalytic turnover of [FeFe]-hydrogenase based on single-molecule imaging. *Journal of the American Chemical Society, 134*, 1577–1582.

Mattos, G. J., Salamanca-Neto, C. A. R., Barbosa, E. C. M., Camargo, P. H. C., Dekker, R. F. H., Barbosa-Dekker, A. M., & Sartori, E. R. (2021). A photoelectrochemical enzyme biosensor based on functionalized hematite microcubes for rutin determination by square-wave voltammetry. *Microchimica Acta, 188*, 1–11.

Mersch, D., Lee, C.-Y., Zhang, J. Z., Brinkert, K., Fontecilla-Camps, J. C., Rutherford, A. W., & Reisner, E. (2015). Wiring of photosystem II to hydrogenase for photoelectrochemical water splitting. *Journal of the American Chemical Society, 137*, 8541–8549.

Moniz, S. J., Shevlin, S. A., Martin, D. J., Guo, Z.-X., & Tang, J. (2015). Visible-light driven heterojunction photocatalysts for water splitting–a critical review. *Energy & Environmental Science, 8*, 731–759.

Najafpour, M. M., Zaharieva, I., Zand, Z., Hosseini, S. M., Kouzmanova, M., Hołyńska, M., Tranca, I., Larkum, A. W., Shen, J.-R., & Allakhverdiev, S. I. (2020). Water-oxidizing complex in Photosystem II: Its structure and relation to manganese-oxide based catalysts. *Coordination Chemistry Reviews, 409*, 213183.

Nam, D. H., Zhang, J. Z., Andrei, V., Kornienko, N., Heidary, N., Wagner, A., Nakanishi, K., Sokol, K. P., Slater, B., Zebger, I., & Hofmann, S. (2018). Solar water splitting with a hydrogenase integrated in photoelectrochemical tandem cells. *Angewandte Chemie International Edition, 57*, 10595–10599.

Nandy, S., Hisatomi, T., Sun, S., Katayama, M., Minegishi, T., & Domen, K. (2018). Effects of Se incorporation in $La_5Ti_2CuS_5O_7$ by annealing on physical properties and photocatalytic H_2 evolution activity. *ACS Applied Materials & Interfaces, 11*, 5595–5601.

Nitopi, S., Bertheussen, E., Scott, S. B., Liu, X., Engstfeld, A. K., Horch, S., Seger, B., Stephens, I. E., Chan, K., Hahn, C., & Nørskov, J. K. (2019). Progress and perspectives of electrochemical CO_2 reduction on copper in aqueous electrolyte. *Chemical Reviews, 119*, 7610–7672.

Ogata, H., Hirota, S., Nakahara, A., Komori, H., Shibata, N., Kato, T., Kano, K., & Higuchi, Y. (2005). Activation process of [NiFe] hydrogenase elucidated by high-resolution X-ray analyses: Conversion of the ready to the unready state. *Structure, 13*, 1635–1642.

Ogisu, K., Ishikawa, A., Shimodaira, Y., Takata, T., Kobayashi, H., & Domen, K. (2008). Electronic band structures and photochemical properties of La–Ga-based oxysulfides. *The Journal of Physical Chemistry C, 112*, 11978–11984.

Pan, L., Kim, J. H., Mayer, M. T., Son, M.-K., Ummadisingu, A., Lee, J. S., Hagfeldt, A., Luo, J., & Grätzel, M. (2018). Boosting the performance of Cu_2O photocathodes for unassisted solar water splitting devices. *Nature Catalysis, 1*, 412–420.

Pantazis, D. A. (2018). Missing pieces in the puzzle of biological water oxidation. *ACS Catalysis, 8*, 9477–9507.

Paracchino, A., Laporte, V., Sivula, K., Grätzel, M., & Thimsen, E. (2011). Highly active oxide photocathode for photoelectrochemical water reduction. *Nature Materials, 10*, 456–461.

Pasumarthi, V., Liu, T., Dupuis, M., & Li, C. (2019). Charge carrier transport dynamics in W/Mo-doped $BiVO_4$: First principles-based mesoscale characterization. *Journal of Materials Chemistry A, 7*, 3054–3065.

Peters, J. W., Lanzilotta, W. N., Lemon, B. J., & Seefeldt, L. C. (1998). X-ray crystal structure of the Fe-only hydrogenase (CpI) from Clostridium pasteurianum to 1.8 angstrom resolution. *Science, 282*, 1853–1858.

Rajaraman, T., Parikh, S. P., & Gandhi, V. G. (2020). Black TiO_2: A review of its properties and conflicting trends. *Chemical Engineering Journal, 389*, 123918.

Reda, T., Plugge, C. M., Abram, N. J., & Hirst, J. (2008). Reversible interconversion of carbon dioxide and formate by an electroactive enzyme. *Proceedings of the National Academy of Sciences, 105*, 10654–10658. https://doi.org/10.1073/pnas.0801290105

Riyadh Atta, M., Shaharun, M. S., Khan, M. R., Abdullah, B., Al-Mahmodi, A. F., Ridzuan, N. D. M., Munusamy, T. D., & Wei, L. J. (2023). Enhancing the photo-electrocatalytic properties of g-C_3N_4 by boron doping and ZIF-8 hybridization. *Inorganic Chemistry Communications, 148*, 110235. https://doi.org/10.1016/j.inoche.2022.110235

Sahoo, P. P., & Maggard, P. A. (2013). Crystal chemistry, band engineering, and photocatalytic activity of the $LiNb_3O_8$–$CuNb_3O_8$ solid solution. *Inorganic Chemistry, 52*, 4443–4450.

Sato, J., Kobayashi, H., & Inoue, Y. (2003). Photocatalytic activity for water decomposition of indates with octahedrally coordinated d10 configuration. II. Roles of geometric and electronic structures. *The Journal of Physical Chemistry B, 107*, 7970–7975.

Sekizawa, K., Sato, S., Arai, T., & Morikawa, T. (2018). Solar-driven photocatalytic CO_2 reduction in water utilizing a ruthenium complex catalyst on p-type Fe_2O_3 with a multiheterojunction. *ACS Catalysis, 8*, 1405–1416.

Shen, J.-R. (2015). The structure of photosystem II and the mechanism of water oxidation in photosynthesis. *Annual Review of Plant Biology, 66*, 23–48.

Shima, S., Pilak, O., Vogt, S., Schick, M., Stagni, M. S., Meyer-Klaucke, W., Warkentin, E., Thauer, R. K., & Ermler, U. (2008). The crystal structure of [Fe]-hydrogenase reveals the geometry of the active site. *Science, 321*, 572–575.

Singh, A. K., & Sarkar, D. (2019). Enhanced light absorption and charge carrier management in core-shell Fe_2O_3@ nickel nanocone photoanodes for photoelectrochemical water splitting. *ChemCatChem, 11*, 6355–6363.

Sivula, K., Formal, F. L., & Gratzel, M. (2009). WO_3–Fe_2O_3 photoanodes for water splitting: A host scaffold, guest absorber approach. *Chemistry of Materials, 21*, 2862–2867.

Smith, P. T., Nichols, E. M., Cao, Z., & Chang, C. J. (2020). Hybrid catalysts for artificial photosynthesis: merging approaches from molecular, materials, and biological catalysis. *Accounts of Chemical Research, 53*, 575–587.

Smith, W. A., Sharp, I. D., Strandwitz, N. C., & Bisquert, J. (2015). Interfacial band-edge energetics for solar fuels production. *Energy & Environmental Science, 8*, 2851–2862. https://doi.org/10.1039/C5EE01822F

Sokol, K. P., Robinson, W. E., Warnan, J., Kornienko, N., Nowaczyk, M. M., Ruff, A., Zhang, J. Z., & Reisner, E. (2018). Bias-free photoelectrochemical water splitting with photosystem II on a dye-sensitized photoanode wired to hydrogenase. *Nature Energy*, *3*, 944–951.

Steier, L., Herraiz-Cardona, I., Gimenez, S., Fabregat-Santiago, F., Bisquert, J., Tilley, S. D., & Grätzel, M. (2014). Understanding the role of underlayers and overlayers in thin film hematite photoanodes. *Advanced Functional Materials*, *24*, 7681–7688.

Su, Y., Cestellos-Blanco, S., Kim, J. M., Shen, Y.-x., Kong, Q., Lu, D., Liu, C., Zhang, H., Cao, Y., & Yang, P. (2020). Close-packed nanowire-bacteria hybrids for efficient solar-driven CO_2 fixation. *Joule*, *4*, 800–811.

Suga, M. Fusamichi Akita, Kunio Hirata, Go Ueno, Hironori Murakami, Yoshiki Nakajima, Tetsuya Shimizu, Keitaro Yamashita, Masaki Yamamoto, Hideo Ago, Jian-Ren Shen Native structure of photosystem II at 1.95 Å resolution viewed by femtosecond X-ray pulses. *Nature 517*, 99-103 (2015).

Sun, L. (2015). A closer mimic of the oxygen evolution complex of photosystem II. *Science*, *348*, 635–636.

Svetlitchnyi, V., Peschel, C., Acker, G., & Meyer, O. (2001). Two membrane-associated NiFeS-carbon monoxide dehydrogenases from the anaerobic carbon-monoxide-utilizing eubacterium Carboxydothermus hydrogenoformans. *Journal of Bacteriology*, *183*, 5134–5144.

Takata, T., Furumi, Y., Shinohara, K., Tanaka, A., Hara, M., Kondo, J. N., & Domen, K. (1997). Photocatalytic decomposition of water on spontaneously hydrated layered perovskites. *Chemistry of Materials*, *9*, 1063–1064.

Thauer, R. K., Kaster, A. K., Goenrich, M., Schick, M., Hiromoto, T., & Shima, S. (2010). Hydrogenases from methanogenic archaea, nickel, a novel cofactor, and H_2 storage. *Annual Review of Biochemistry*, *79*, 507–536.

Thauer, R. K., Klein, A. R., & Hartmann, G. C. (1996). Reactions with molecular hydrogen in microorganisms: eEvidence for a purely organic hydrogenation catalyst. *Chemical Reviews*, *96*, 3031–3042.

Thimsen, E., Le Formal, F., Gratzel, M., & Warren, S. C. (2011). Influence of plasmonic Au nanoparticles on the photoactivity of Fe_2O_3 electrodes for water splitting. *Nano Letters*, *11*, 35–43.

Tong, L., Duan, L., Zhou, A., & Thummel, R. P. (2020). First-row transition metal polypyridine complexes that catalyze proton to hydrogen reduction. *Coordination Chemistry Reviews*, *402*, 213079.

Tong, L., & Thummel, R. P. (2016). Mononuclear ruthenium polypyridine complexes that catalyze water oxidation. *Chemical Science*, *7*, 6591–6603.

Trang, T., Tu, L. T. N., Man, T. V., Mathesh, M., Nam, N. D., & Thu, V. T. H. (2019). A high-efficiency photoelectrochemistry of Cu_2O/TiO_2 nanotubes based composite for hydrogen evolution under sunlight. *Composites Part B: Engineering*, *174*, 106969.

Tsui, E. Y., Kanady, J. S., & Agapie, T. (2013). Synthetic cluster models of biological and heterogeneous manganese catalysts for O_2 evolution. *Inorganic Chemistry*, *52*, 13833–13848.

Umena, Y., Kawakami, K., Shen, J.-R., & Kamiya, N. (2011). Crystal structure of oxygen-evolving photosystem II at a resolution of 1.9 Å. *Nature*, *473*, 55–60.

Vignais, P. M., & Billoud, B. (2007). Occurrence, classification, and biological function of hydrogenases: An overview. *Chemical Reviews*, *107*, 4206–4272.

Vinyard, D. J., Ananyev, G. M., & Charles Dismukes, G. (2013). Photosystem II: The reaction center of oxygenic photosynthesis. *Annual Review of Biochemistry*, *82*, 577–606.

Wang, D., Sampaio, R. N., Troian-Gautier, L., Marquard, S. L., Farnum, B. H., Sherman, B. D., Sheridan, M. V., Dares, C. J., Meyer, G. J., & Meyer, T. J. (2019). Molecular photoelectrode for water oxidation inspired by photosystem II. *Journal of the American Chemical Society*, *141*, 7926–7933.

Wang, Q., & Domen, K. (2019). Particulate photocatalysts for light-driven water splitting: Mechanisms, challenges, and design strategies. *Chemical Reviews*, *120*, 919–985.

Wang, Q., & Domen, K. (2020). Particulate photocatalysts for light-driven water splitting: Mechanisms, challenges, and design strategies. *Chemical Reviews*, *120*, 919–985. https://doi.org/10.1021/acs.chemrev.9b00201

Wang, Q., Zhang, Y., Liu, Y., Wang, K., Qiu, W., Chen, L., Li, J., & Li, W. (2022). Core-shell In/Cu_2O nanowires Schottky junction for enhanced photoelectrochemical CO_2 reduction under visible light. *Industrial & Engineering Chemistry Research*, *61*, 16470–16478. https://doi.org/10.1021/acs.iecr.2c02579

Wang, Y., Suzuki, H., Xie, J., Tomita, O., Martin, D. J., Higashi, M., Kong, D., Abe, R., & Tang, J. (2018). Mimicking natural photosynthesis: Solar to renewable H_2 fuel synthesis by Z-scheme water splitting systems. *Chemical Reviews*, *118*, 5201–5241.

Winter, A., Zsolnai, L. & Huttner, G. (n.d. Zweikernige und dreikernige Carbonyleisenkomplexe mit 1.2-und 1.3-Dithiolatobrückenliganden.

Xu, Y., Wang, S., Yang, J., Han, B., Nie, R., Wang, J., Dong, Y., Yu, X., Wang, J., & Jing, H. (2018). Highly efficient photoelectrocatalytic reduction of CO_2 on the Ti_3C_2/gC_3N_4 heterojunction with rich Ti^{3+} and pyri-N species. *Journal of Materials Chemistry A*, *6*, 15213–15220.

Yagi, T., Ogo, S., & Higuchi, Y. (2014). Catalytic cycle of cytochrome-c3 hydrogenase, a [NiFe]-enzyme, deduced from the structures of the enzyme and the enzyme mimic. *International Journal of Hydrogen Energy*, *39*, 18543–18550.

Yang, W., Kim, J. H., Hutter, O. S., Phillips, L. J., Tan, J., Park, J., Lee, H., Major, J. D., Lee, J. S., & Moon, J. (2020). Benchmark performance of low-cost Sb_2Se_3 photocathodes for unassisted solar overall water splitting. *Nature Communications*, *11*, 861.

Yano, J., & Yachandra, V. (2014). Mn_4Ca cluster in photosynthesis: Where and how water is oxidized to dioxygen. *Chemical Reviews*, *114*, 4175–4205.

Yoon, D.-H., Biswas, M. R. U. D., & Sakthisabarimoorthi, A. (2022). Impact of crystalline core/amorphous shell structured black TiO_2 nanoparticles on photoelectrochemical water splitting. *Optical Materials*, *133*, 113030.

Youngblood, W. J., Lee, S.-H. A., Maeda, K., & Mallouk, T. E. (2009). Visible light water splitting using dye-sensitized oxide semiconductors. *Accounts of Chemical Research*, *42*, 1966–1973.

Yuan, Y., Lv, J., Jiang, X., Li, Z., Yu, T., Zou, Z., & Ye, J. (2007). Large impact of strontium substitution on photocatalytic water splitting activity of $BaSnO_3$. *Applied Physics Letters*, *91*, 094107.

Zaharieva, I., Dau, H., & Haumann, M. (2016). Sequential and coupled proton and electron transfer events in the S2→S3 transition of photosynthetic water oxidation revealed by time-resolved X-ray absorption spectroscopy. *Biochemistry*, *55*, 6996–7004.

Zeng, Z., Yan, Y., Chen, J., Zan, P., Tian, Q., & Chen, P. (2019). Boosting the photocatalytic ability of Cu_2O nanowires for CO_2 conversion by MXene quantum dots. *Advanced Functional Materials*, *29*, 1806500.

Zhang, B., Hemann, C. F., & Hille, R. (2010). Kinetic and spectroscopic studies of the molybdenum-copper CO dehydrogenase from *Oligotropha carboxidovorans*. *Journal of Biological Chemistry*, *285*, 12571–12578.

Zhang, J. Z., & Reisner, E. (2020). Advancing photosystem II photoelectrochemistry for semi-artificial photosynthesis. *Nature Reviews Chemistry*, *4*, 6–21.

Zhang, M., Wang, J., Wang, Y., Zhang, J., Han, X., Chen, Y., Wang, Y., Karim, Z., Hu, W., & Deng, Y. (2021). Promoting the charge separation and photoelectrocatalytic water reduction kinetics of Cu_2O nanowires via decorating dual-cocatalysts. *Journal of Materials Science and Technology*, *62*, 119–127.

Zhang, Y., Pan, D., Tao, Y., Shang, H., Zhang, D., Li, G., & Li, H. (2022). Photoelectrocatalytic reduction of CO_2 to syngas via SnO_x-enhanced Cu_2O nanowires photocathodes. *Advanced Functional Materials*, *32*, 2109600.

Zhang, Z., Huang, X., Zhang, B., & Bi, Y. (2022). High-performance and stable $BiVO_4$ photoanodes for solar water splitting via phosphorus–oxygen bonded FeNi catalysts. *Energy & Environmental Science*, *15*, 2867–2873. https://doi.org/10.1039/D2EE00936F

Zhao, J., Holmes, M. A., & Osterloh, F. E. (2013). Quantum confinement controls photocatalysis: A free energy analysis for photocatalytic proton reduction at CdSe nanocrystals. *ACS Nano*, *7*, 4316–4325.

Zhou, D., & Fan, K. (2021). Recent strategies to enhance the efficiency of hematite photoanodes in photoelectrochemical water splitting. *Chinese Journal of Catalysis*, *42*, 904–919.

3

Solar Thermochemical Fuels

Christoph Falter

Laboratory of Aviation and the Environment, Massachusetts Institute of Technology, Cambridge, MA, USA
Synhelion SA, Zurich, Switzerland

Nomenclature

A	Area, annuity factor
C	Solar flux concentration ratio, cost
CHP	Combined heat and power
CSP	Concentrated solar power
DNI	Direct normal irradiation
ERP	Equity risk premium
FT	Fischer–Tropsch
GHG	Greenhouse gas
GIS	Geographical information system
GWP	Global warming potential
HEFA	Hydrotreated esters and fatty acids
HHV	Higher heating value
I	Normal beam insolation, investment costs
i	Interest rate
j	Year
LCOE	Levelized cost of energy
LHV	Lower heating value
MED	Mediterranean region
n	Lifetime
N	Fuel volume
NPV	Net present value
PtL	Power to liquid
PV	Photovoltaic
Q	Solar energy, annual fuel volume
T	Temperature
TLCC	Total life cycle cost
V	Volume
WACC	Weighted average cost of capital
WTT	Well-to-tank
WTW	Well-to-wake
ΔG	Gibbs energy change
ΔH	Enthalpy change
σ	Stefan–Boltzmann constant (5.6705×10^{-8} $Wm^{-2} \cdot K^{-4}$)
$\eta_{absorption}$	Solar energy absorption efficiency
η_{Carnot}	Efficiency of a Carnot heat engine operating between T_H and T_L

Conversion of Water and CO_2 to Fuels using Solar Energy: Science, Technology and Materials, First Edition.
Edited by Oomman K. Varghese and Flavio L. Souza.

$\eta_{\text{land coverage}}$	Ground coverage factor
$\eta_{\text{solar - to - chemical energy}}$	Energy conversion efficiency from incident solar energy to chemical energy of the products
$\eta_{\text{solar - to - jet fuel}}$	Energy conversion efficiency from incident solar energy to jet fuel

3.1 Thermodynamics

A comprehensive overview of the basic principles of solar thermochemical processes and thermodynamics is given in (Fletcher, 2000; Romero & Steinfeld, 2012; Steinfeld & Palumbo, 2003). In this subsection, the basic principles are shown. The fundamental concept of solar thermochemical processes is the use of high-temperature heat to drive endothermal chemical reactions. For the provision of high-temperature heat, solar energy is concentrated by either line-focusing systems in parabolic troughs or point-focusing in solar towers or parabolic dishes (Figure 3.1).

Troughs use 2D parabolic mirrors and one-axis tracking to focus the sunlight onto a linear tubular receiver, while tower systems reflect the sunlight from a field of heliostats to the top of a tower, where a receiver converts the radiation into heat. Dish systems consist of a 3D paraboloidal mirror and a central receiver above the mirror surface. Tower and dish systems use two-axis tracking and are capable of reaching higher temperatures because of a higher concentration ratio, which is a measure of their ability to concentrate solar radiation. It is measured as the mean flux over an aperture in the focal plane and normalized to the direct normal irradiation (DNI). Using a DNI of 1 kW/m^2, the unit of "suns" is often used. Using secondary, non-imaging optics, the concentration ratio can be enhanced, where the upper theoretical limits of 46,200 and 215 apply for 3D and 2D reflectors, respectively (Romero & Steinfeld, 2012). Due to losses, e.g., by sunshape, tracking errors, and surface imperfections, the achievable concentration ratios are lower and typically on the order of 100 for the trough, 1,000 for the tower, and 10,000 for the dish system (Steinfeld, 2014). Toward higher concentration ratios, the receiver temperatures increase, as well, because of lower heat losses from smaller aperture areas. Linear focusing systems achieve temperatures of up to 500 °C, which is well suited for the use of a Rankine cycle to produce electricity. Tower and dish systems, on the other hand, reach temperatures of up to about 2,000 °C, which enables advanced thermodynamic and thermochemical cycles. These concentration devices are already used in demonstrations and also commercially for the generation of electricity from heating of a working fluid and driving Rankine, Brayton, or Sterling engines.

For effective absorption of concentrated solar radiation in thermochemical processes, a cavity reactor design is chosen. Solar radiation enters a cavity through a small aperture and is reflected multiple times off the walls, whereas the effective absorptivity of the cavity greatly exceeds the surface absorption of the walls. A black body absorber can be approached by increasing the ratio of characteristic length of the cavity to its aperture. The absorption efficiency, $\eta_{\text{absorption}}$, is defined as the solar energy being absorbed by the receiver divided by the radiative energy entering through the aperture, Q_{solar}. For a perfectly insulated cavity receiver without losses through convection and conduction, which operates at the temperature T, the absorption efficiency is (Fletcher & Moen, 1977)

$$\eta_{\text{absorption}} = 1 - \frac{\sigma T^4}{IC}, \tag{3.1}$$

where C is the mean concentration ratio of solar radiation, I is the DNI of 1 kW/m^2, and σ is the Stefan–Boltzmann constant.

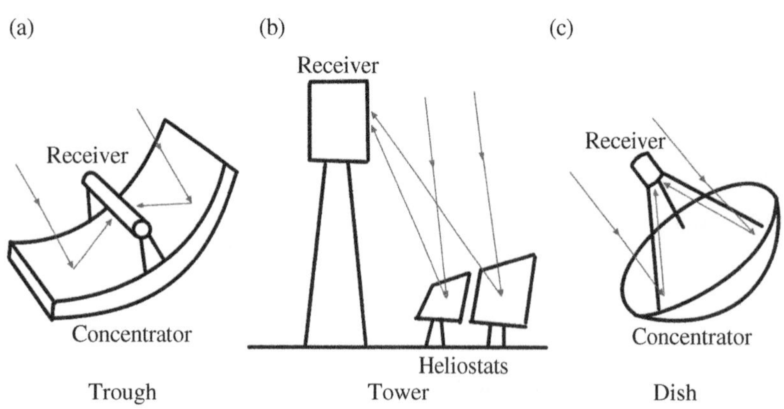

Figure 3.1 Overview of main optical concentration configurations of sunlight: (a) parabolic trough, (b) tower, (c) parabolic dish.

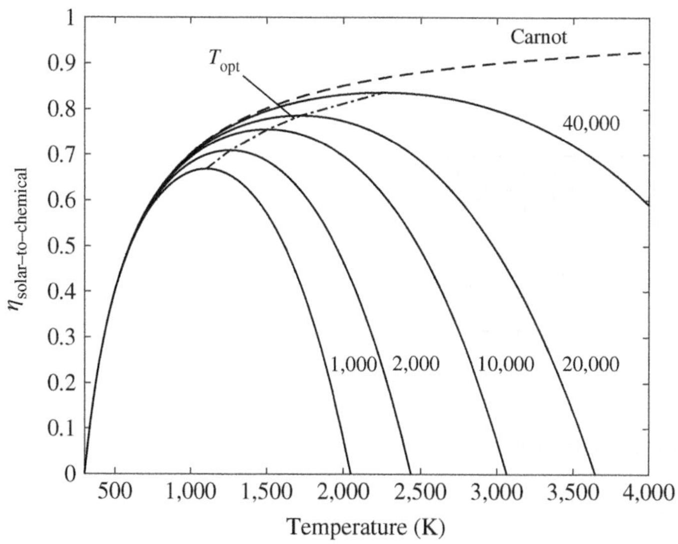

Figure 3.2 Variation of optimum energy conversion efficiency of ideal black body absorber as a function of operating temperature T_H for the conversion of concentrated solar radiation into chemical energy. Indicated is also the Carnot efficiency limit and the optimum temperature at the maximum efficiency.

As a measure of the conversion of solar energy into chemical energy, the reactor energy conversion efficiency is defined as

$$\eta_{\text{solar-to-chemical energy}} = \frac{-\Delta G}{Q_{\text{solar}}}, \tag{3.2}$$

where ΔG denotes the maximum work that can be extracted from the chemical products. Alternatively, the efficiency can also be measured with ΔH (Romero & Steinfeld, 2012).

For an ideal process, the maximum conversion efficiency is derived by multiplying the absorption efficiency by the Carnot efficiency for a heat engine.

$$\eta_{\text{solar-to-chemical energy,ideal}} = \left[1 - \left(\frac{\sigma T^4}{IC}\right)\right] \times \left[1 - \frac{T_L}{T_H}\right] \tag{3.3}$$

T_L and T_H are the lower and higher process temperatures of the equivalent heat engine. The ideal reactor efficiency is shown as a function of the reactor temperature in Figure 3.2 and is a trade-off between higher efficiencies of the heat engine and lower absorption efficiencies at higher operating temperatures. The optimum efficiency is found between the ambient temperature, where the heat engine efficiency is zero, and the stagnation temperature, the maximum attainable temperature of the receiver, where the absorber efficiency is zero. The optimum temperature at which the energy conversion efficiency is maximized is found by setting $\partial\eta_{\text{solar-to-chemical energy, ideal}}/\partial T = 0$.

3.2 Solar Thermochemical Processes and Reactor Concepts

Six main routes to solar thermochemical fuels can be distinguished (Figure 3.3). All processes use concentrated solar radiation to convert either water and CO_2 or carbonaceous feedstocks, such as natural gas, biomass, coal, or oil into solar fuels, where among the latter are hydrogen, syngas, or hydrocarbons.

3.2.1 Thermolysis of H_2O

The simplest way to thermally split water is by direct dissociation. The process only requires a single step and is conceptually simple; however, very high temperatures are required to achieve a reasonable degree of dissociation. This is demonstrated by observing the enthalpy, entropy, and Gibbs energy of the reaction $H_2O \rightleftarrows H_2 + \frac{1}{2}O_2$ (Figure 3.4). While the enthalpy is rather constant over the plotted range, the term $T\Delta S^0$ increases with temperature. The reaction is thermodynamically favorable for negative values of the Gibbs energy ($\Delta G^0 = \Delta H^0 - T\Delta S^0$). This is achieved at temperatures exceeding about 4,300 K.

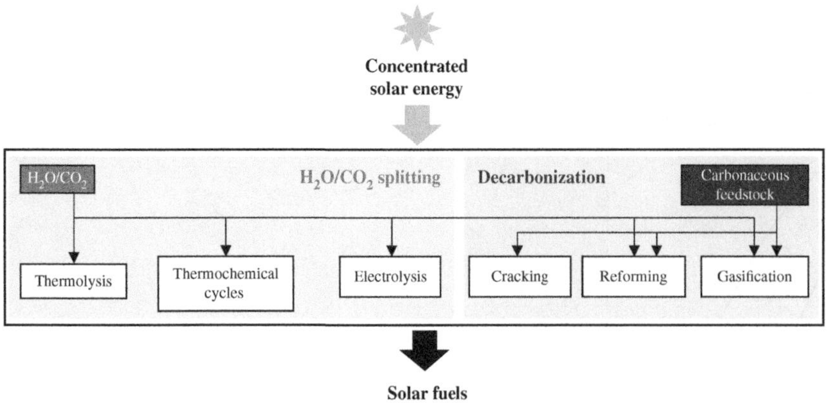

Figure 3.3 Overview of pathways from concentrated solar energy to solar fuels.

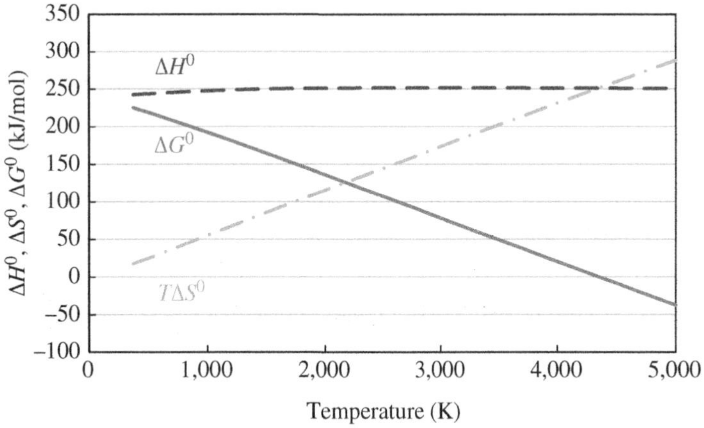

Figure 3.4 Standard Gibbs energy, standard reaction enthalpy, and the product $T\Delta S^0$ as a function of temperature for the direct thermolysis of water.

Assuming the required operating temperature can be reached, the choice of a suitable insulation and reactor material proves to be a considerable engineering problem since the temperature of 4,300 K exceeds the melting point of most common materials.

Furthermore, hydrogen and oxygen are produced in a single process step, which makes a gas separation at high temperatures necessary in order to avoid an explosive mixture (Romero & Steinfeld, 2012). The high temperatures will also lead to significant reradiation from the cavity and thus reduce the absorption efficiency. Different means of gas separation, such as effusion or electrolytic separation, or the use of a semipermeable membrane have been tested at temperatures up to 2,500 K (Steinfeld, 2014) but have often failed due to the severe thermal shocks that can occur when operating with concentrated solar radiation. Another way is rapid quenching, i.e., introducing a large volume of cold inert gas to quickly reduce the temperature of the mixture of hydrogen and oxygen to prevent their recombination. Quenching, however, leads to an explosive gas mixture and significantly reduces the efficiency of the process. A viable way to reduce the operating temperatures and prevent high-temperature gas separation is the introduction of thermochemical cycles with multiple steps.

3.2.2 H_2/CO From H_2O/CO_2 Using Thermochemical Cycles

The introduction of multiple steps allows the reduction of the process temperature and the production of H_2 and O_2 in separate steps, which circumvents the challenge of gas separation at high temperatures. Some of the first cycles to be studied were the sulfur-iodine process with three steps and the UT3-cycle with four steps. Both processes operate at relatively moderate temperatures of <1,200 K that can be reached using waste heat from nuclear power plants or other sources. These cycles have the disadvantage of low efficiencies due to heat transfer and product separation in each step

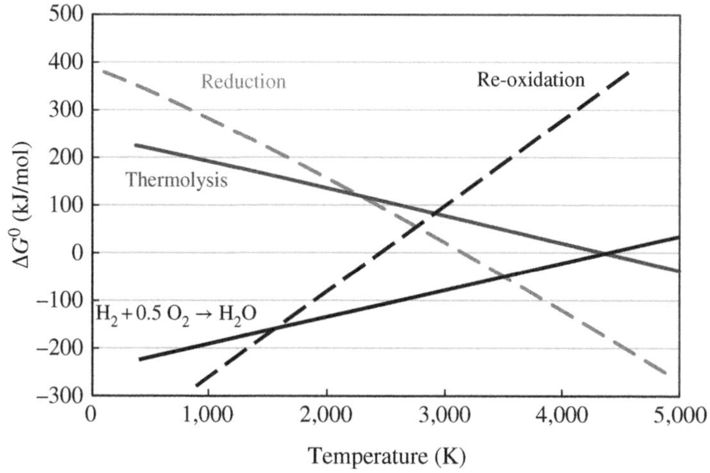

Figure 3.5 Gibbs energy as a function of temperature for thermolysis and the two steps of reduction and re-oxidation of a generic metal oxide. Thermolysis is thermodynamically favorable only at very high temperatures. The introduction of a multi-step thermochemical cycle (here shown is a generic two-step cycle consisting of reduction and re-oxidation) can reduce the temperature requirement and circumvent the production of an explosive gas mixture.

(Steinfeld, 2014). The UT3-cycle (developed at the University of Tokyo) is a four-step process that operates with the hydrolysis of $CaBr_2$ and $FeBr_2$ at temperatures of <1,033 K (Sakurai et al., 1996; Steinfeld, 2014). While the temperatures are available from certain nuclear reactions, the process has the challenges of slow reaction rates (hydrolysis of hydrogen bromide is the rate-limiting step), high-temperature gas separation, and handling of corrosive environments (Lee, 2008). An economical process is thought to be possible only if the efficiency loss of a conventional condensation method can be avoided (Lee, 2008; Teo et al., 2005). The sulfur–iodine process, on the other hand, also has the issue of corrosive reactants that may be dangerous to apply close to a nuclear reactor. Furthermore, the value of iodine that has to be handled per kg of produced hydrogen is on the order of 20,000 $, which means that even small losses on the order of 10–5 could result in significant cost increases (Vitart et al., 2008). Finally, there is also the question of sufficient abundance of iodine (Vitart et al., 2008).

Due to the difficulties of practical implementation, multi-step processes are used. Multiple steps allow a reduction of the maximum operating temperature and separate the production of hydrogen and oxygen. For an assumed generic two-step redox cycle of a metal oxide, Figure 3.5 shows the Gibbs energy of thermolysis, the recombination reaction of oxygen and hydrogen to water, and the reduction and re-oxidation of the metal oxide. The reduction reaction reaches $\Delta G = 0$ at lower temperatures than the thermolysis reaction and thus relaxes the required upper process temperature. The production of hydrogen occurs during the re-oxidation of the material with water, which is favored at temperatures of roughly <1,500 K (for $\Delta G_{\text{re-oxidation}} < \Delta G_{H_2+0.5O_2\rightarrow H_2O}$). Since hydrogen and oxygen are produced in different steps, gas separation of these two gases is not necessary. The required reduction temperatures are about 1,500 K and more. Progress in the development of solar concentrating devices enabled reaching these temperatures and led to the demonstration of several two-step cycles. Among them are the cycles based on ZnO/Zn, Fe_3O_4/FeO, and $CeO_2/CeO_{2-\delta}$.

The general working principle of the two-step process is the endothermal reduction of a metal oxide to a lower valence state producing oxygen, and the subsequent re-oxidization of the material with water, which brings the material back into its initial oxidation state and produces hydrogen. Overall, water is then split into hydrogen and oxygen in two separate process steps, which avoids the challenges of separating the two gases at a high temperature and opens up the possibility of decoupling the process steps geographically and temporally. This was especially proposed as a way to avoid hydrogen transportation in the zinc cycle by reducing ZnO at a sunny location and oxidizing it at the demand site (Werder & Steinfeld, 2000). By using H_2O and CO_2 to oxidize the reactive material, syngas can be produced, a mixture of H_2 and CO, which is a precursor for many fuels and chemicals and thus has a wide range of possible uses. Specifically, alternative fuels such as jet fuel, diesel, gasoline, or methanol can be produced from syngas via Fischer–Tropsch (FT) synthesis. A schematic of the two-step process with a metal oxide undergoing a nonstoichiometric change in oxidation state is shown in Figure 3.6.

An early implementation of the redox pair was suggested to be Fe_3O_4/FeO, where the reduction step requires temperatures above 2,300 K. Having studied the process both thermodynamically (Steinfeld et al., 1999) and experimentally

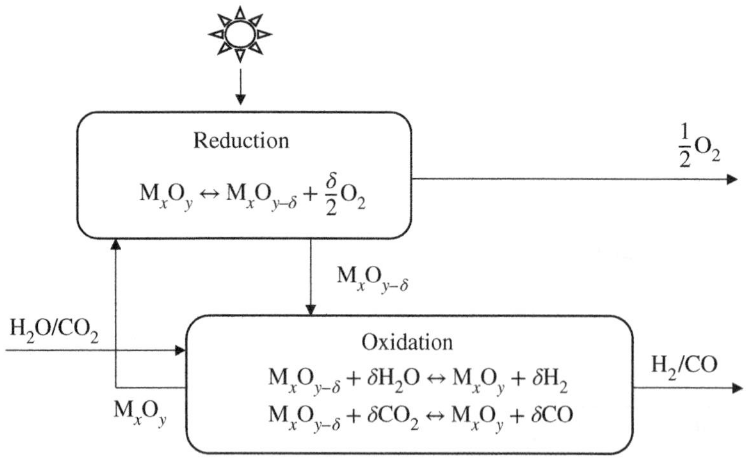

Figure 3.6 Schematic of two-step solar thermochemical cycle for splitting H_2O and CO_2 into H_2 and CO (syngas). In the first step, a metal oxide is reduced to a lower-valence state producing oxygen, which requires the input of high-temperature solar energy. In the second step, the lower-valence metal oxide is re-oxidized with H_2O or CO_2 or both to produce H_2 and or CO, bringing the metal oxide back into its original state. A nonstoichiometric change in oxidation state is shown here, as is encountered, in e.g., the CeO_2 cycle.

(Sibieude et al., 1982), quenching the gaseous products was found to be necessary, which however introduces a significant energy penalty. This is also required for the effective separation of gaseous Zn and O_2 evolving from the thermal dissociation of ZnO at high temperatures. Theoretical and experimental work has been performed on the ZnO/Zn oxide pair with research focusing on the gaseous separation processes through dilution and rapid quenching below the solidification temperature of Zn (Lédé et al., 2001; Loutzenhiser et al., 2010; Palumbo et al., 1998; Steinfeld, 1998; Steinfeld, 2002), or electrolytic methods (Fletcher, 1999). The reoxidation of Zn with H_2O was analyzed in aerosol flow reactors that allowed the rapid and effective chemical conversion to ZnO and H_2 (Abu Hamed et al., 2009; Melchior et al., 2009; Wegner et al., 2006).

In the past decade, the two-step CeO_2 cycle has been the focus of theoretical and experimental work (Brendelberger et al., 2015; Chueh et al., 2010; Chueh & Haile, 2010; Ermanoski et al., 2013; Falter et al., 2014; Falter et al., 2015; Falter et al., 2016; Falter et al., 2017; Falter & Pitz-Paal, 2017; Falter & Pitz-Paal, 2018; Falter, Valente, et al., 2020; Furler et al., 2014; Furler et al., 2017; Furler, Scheffe, Gorbar, et al., 2012; Furler, Scheffe, & Steinfeld, 2012; Geissbühler, 2019; Lapp, 2013; Le Gal et al., 2011; Marxer et al., 2015; Marxer et al., 2017; Miller et al., 2014; Scheffe & Steinfeld, 2012; Siegel et al., 2013; Suter et al., 2014). As a fundamental difference from volatile reactions, the reactive material remains in its solid state throughout both process steps, retaining its fluorite-type structure. The material is cycled between different nonstoichiometric states of oxygen in its lattice with typical values of $\delta < 0.05$ (Chueh et al., 2010; Furler, Scheffe, Gorbar, et al., 2012; Marxer et al., 2017). All reactions, therefore, occur between solid ceria and the gas phase. High-temperature gas separation processes, including energy-intensive quenching, can therefore be avoided, which crucially simplifies the reactor design and leads to comparatively benign experimental procedures that can mainly focus on efficient heat transfer and reaction rates at the surface of the reactive material. Ceria as a reactive material has high oxygen ion diffusion enabling rapid chemical reaction rates (Chueh & Haile, 2010), it is stable in its fluorite-type structure in a wide range of nonstoichiometries, and has a high melting point. Despite its classification as a rare earth metal, cerium is about as abundant in the Earth's crust as copper (Lumen Learning, 2022), with China being the main supplier of rare earth oxides at the moment. Having a naturally high occurrence in certain regions, ceria is in some cases the main product of the mining activities and has a market price currently below 5 \$/kg (Statista, 2019). The supply is presently mainly dependent on Chinese mines; however, there are other resources in the *i.a.* United States of America and Australia that may be leveraged to increase and secure the supply in the future.

Experiments with ceria redox cycles have demonstrated significantly enhanced cycle efficiencies in the past decade with the first reported values below 1% (Chueh et al., 2010). Chueh et al. operated a cavity-type solar reactor at the Paul-Scherrer Institute with ceria porous bricks supplied by the California Institute of Technology (Chueh et al., 2010) (Figure 3.7). In the experiments, a solar-to-syngas efficiency of 0.8% was measured, which was the highest value of this cycle up to that point in time. As an alternative to the brick structures, porous ceria felt was also tested and achieved lower efficiencies due to the high thermal resistance across the optically thick material, which led to a larger temperature drop across the material

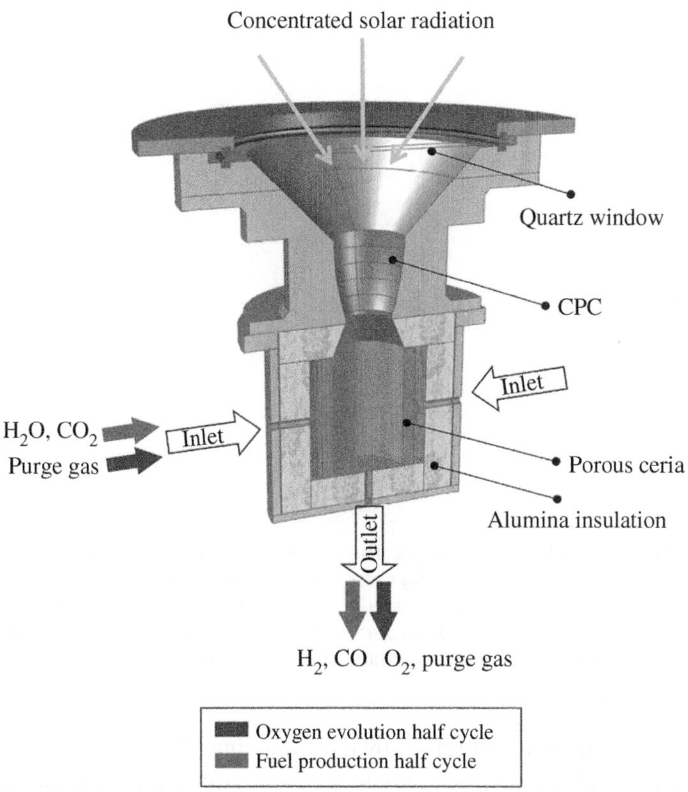

Figure 3.7 Schematic representation of the solar reactor configuration used for two-step thermochemical syngas production. The cavity-type reactor contains the reactive material ceria, which is in the form of a porous monolithic cylinder. Concentrated solar radiation enters the cavity through the opening at the top, is further concentrated by a compounded parabolic concentrator, and heats up the reactive material in the cavity from the inside. Reacting gases are introduced radially through openings in the cavity, while product gases exit through an opening at the bottom of the cavity. The reduction is indicated by the black arrows, and the oxidation by the gray arrows. Adapted from (Chueh et al., 2010).

and prevented equally high heating rates. On the other hand, the bricks had a higher large-scale porosity, which allowed a more effective heat transfer by radiation through the structure at a larger mass loading of the reactor.

The realization of the limitation of higher efficiencies with respect to heat transfer to and across the reactive material led to efforts to enhance the radiative part, the dominant heat transfer mechanism at these high temperatures. Marxer et al. describe the introduction and experimental testing of a dual-scale porosity structure in ceria, where the large-scale porosity is intended to enhance heat transfer rates and the small-scale porosity to allow for rapid kinetics (Furler et al., 2014). In experimental runs in a solar simulator at ETH, solar-to-syngas an energy conversion efficiency of 1.72% was reported (Furler et al., 2014). In a subsequent experimental campaign, the largest up-to-date number of completed cycles in a solar thermochemical reactor was reported to be close to 300 (Marxer et al., 2015). The produced syngas was compressed and for the first time converted to liquid hydrocarbons in the SOLAR-JET project (Bauhaus Luftfahrt e.V, 2011-2015; Marxer et al., 2015). Up to this time, the required reduction of the oxygen partial pressure in the solar reactor was achieved by applying inert gas flushing to the reactor prior to and during the reduction reaction. An improvement on this method was achieved by the introduction of vacuum pumping, which in combination with the dual-scale porosity ceria structures, allowed to reach solar-to-syngas energy conversion efficiencies of 5.25% (Marxer et al., 2017). A schematic of the reactor design is shown in Figure 3.8.

The nonstoichiometric cycling of the material has the advantage of simpler reactor design but also has the disadvantage of a relatively small fuel productivity per cycle, which is on the order of a few percent of the oxygen stored in the material. Furthermore, all of the material has to be thermally cycled, which introduces the challenge of heat recuperation to reach high cycle efficiencies. To solve this issue, it was recently suggested in (Geissbühler, 2019) to make use of the thermocline energy storage concept in a dual-tank storage approach. The system consists of two vertically packed beds of inert storage material and two reaction zones of reactive material that are located on top of the storage material. Inside the packed beds,

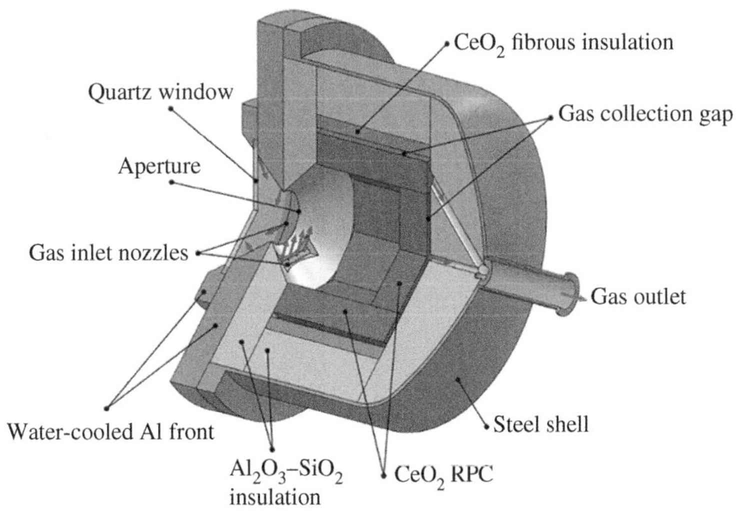

Figure 3.8 Schematic of a 4-kW solar thermochemical reactor for the production of syngas. The cavity-type receiver-reactor receives concentrated solar radiation through the aperture to operate a temperature–vacuum swing cycle with CeO_2. A window prevents convective heat losses and mixing with the atmosphere. The cavity contains the reactive material, a CeO_2 reticulated porous ceramic (RPC), and insulation to limit heat losses. For the reduction, the reactor is subjected to a vacuum pressure at high temperatures, while during oxidation, CO_2 and H_2O are injected into the cavity at lower temperatures. Figure reproduced from (Marxer, 2016) by permission of Prof. Steinfeld, ETH Zurich.

axial temperature profiles are established (thermoclines) that can be shifted up or down by pumping a heat transfer fluid through the packed beds. In this way, the reactive material is cyclically exposed to reduction and oxidation temperatures, and sensible heat is stored and recuperated in the storage areas. Concentrated solar energy is used to supply the reaction enthalpy and balance heat losses to the environment. The indirect introduction of solar radiation, therefore, avoids local hotspots, a challenge in earlier reactor designs, and enables more freedom in the design of the system. Geissbühler estimates that an energy conversion efficiency of about 35% can be reached with the dual-storage concept (Geissbühler, 2019), which would be a significant enhancement of the technology.

3.3 Energy and Mass Balance

The ability of the solar reactor to convert concentrated solar energy into chemical energy stored in the product gases strongly influences the economic and environmental performance of the process. Both the size of the reactor and that of the solar concentrator are scaled by the efficiency of the reactor, which is therefore a fundamental variable in the process design. Reactor efficiencies (concentrated solar energy to higher heating value of gaseous products) of greater than 5% have been achieved both at laboratory-scale (Marxer et al., 2017) and in-field demonstration (Bauhaus Luftfahrt e.V, 2011-2015). It is believed that efficiencies of about 20% are required to make solar thermochemical liquid fuel production economically viable (Falter et al., 2016; Falter, Scharfenberg, & Habersetzer, 2020; Falter, Valente, et al., 2020; Stechel & Miller, 2013), where the threshold of economic feasibility depends on the equipment cost, process design, and political boundary conditions, i.e., potential mandates, subsidies, or carbon prices that reduce the price gap between fossil and alternative fuels.

In the following, the energy balance of a solar thermochemical reactor and of the associated plant producing jet fuel is discussed together with possible plant integration efforts to increase the efficiency of fuel production.

3.3.1 Thermochemical Reactor

A solar thermochemical reactor receives concentrated radiation and converts it in a redox cycle into chemical energy stored in the lower heating value (LHV) of hydrogen and carbon monoxide. In principle, many different reactor technologies exist that differ in their design and each has different loss mechanisms. For example, a ZnO/Zn cycle may have a fundamentally different energy balance than a CeO_2 cycle because of the nature of the volatile or nonvolatile reactions during reduction of the reactive material and the requirement of product separation. For nonvolatile processes, heat losses have a crucial importance due to the comparably small fuel productivity per cycle. Here, a nonvolatile ceria cycle is discussed, which has received attention in research groups recently.

In (Marxer et al., 2017), a 4-kW reactor is presented and experimentally tested in a high-flux solar simulator at ETH Zurich in a temperature–vacuum swing cycle. The reactor achieves an energy conversion efficiency of 5.25% from concentrated solar energy to chemical energy stored in the product gases. In (Marxer, 2016), Marxer discusses the energy balance of the reactor and identifies the main loss mechanisms. Of the total heat input to the reactor, 36.2% is lost to heating inert parts and conduction to water cooling, 26.4% is used for heating of the reactive material, 8.8% is used to reduce ceria (this value is higher than the final efficiency because of the exothermal oxidation reaction), 8.2% is lost by reradiation, 6.6% is reflected from the window shielding the cavity, 5.0% and 4.3% are reradiated and lost by convection from the shell, respectively, and smaller parts are lost due to vacuum pumping, reflection from the cavity, inert gas production and heating. Vacuum pumping requires only a small amount of energy due to the moderate pressure level of 10^{-2} bar applied, where the energy is attained by converting heat to electricity or mechanical energy. It should be noted that the system is a small-scale reactor designed for experimental work in the laboratory and is not fully optimized. It is to be expected that a larger optimized reactor will perform better, especially in terms of heat losses.

An advanced design including sensible heat recuperation and storage is presented in (Geissbühler, 2019). Geissbühler indicates the potential to reach 35% energy conversion efficiency in a larger-scale system. The dual storage concept combines a solar receiver with two reactors and two heat storages that operate with a thermocline, i.e., a distribution of different temperature levels in the storage. The receiver is heated continually beyond the reduction temperature, which provides the heat input to the reaction zones and heat storages. On top of the heat storages, a reaction zone is established, where the reactive material is cyclically reduced and oxidized, producing syngas. The temperature swing is achieved by pumping a heat transfer fluid through the storage, and due to the low temperatures of the storage at the lower end, a normal compressor operating at room temperature can be used to move the heat transfer fluid. By cyclically reversing the direction of the fluid, the thermocline is moved up and down in the storage medium and the reaction zone is periodically exposed to the reduction temperature (~1,800 K) and the oxidation temperature (~1,000 K). According to the prevalent cycle, either a vacuum is created using a vacuum pump during reduction and the evolving oxygen is removed from the reaction zone, or oxidants (CO_2 and/or H_2O) are introduced for the production of syngas. The implementation of a heat storage is crucial for achieving high energy conversion efficiencies, especially in nonvolatile cycles such as the CeO_2 cycle.

From the solar heat input to the cycle, 42.7% is extracted at 1,373 K to provide energy for vacuum pumping and for pumping the heat transfer fluid through the system. Reradiation losses are relatively large at 18.7%, and thermal energy losses have a relative weight of 3.1%. Compared to the cavity design discussed above, the reactive medium is not heated directly and most of the system can be efficiently insulated from the environment. This allows for very low thermal heat losses. Reradiation losses could be higher than for the batch reactor if a constant high temperature was held in a common cavity-type receiver, while the batch receiver-reactor is cyclically heated and thus reradiates also at lower temperatures, which incurs smaller losses. However, a modified receiver design for indirect heating of the reactant, making use of an absorbing gas as the heat transfer fluid, could enable very high absorption efficiencies (Ambrosetti & Good, 2019), alleviating this issue. Continuous on-sun operation is a precondition for cost-efficient fuel production because it makes the maximum use of the concentration infrastructure.

3.3.2 Energy and Mass Balance of Solar Thermochemical Fuel Plant

The production of liquid hydrocarbon fuels requires the supply of water and carbon dioxide, the concentration of solar energy for the provision of high-temperature heat, gas storage, FT synthesis of hydrocarbons from syngas, and refining of raw FT products into the final liquid fuels. The thermochemical reactor is the centerpiece of this concept, and being at a relatively low TRL today, it has the highest potential for enhancements of efficiency. In the following, the energy and mass balance of a solar thermochemical fuel plant for the production of jet fuel are discussed. The analysis of the plant including energy and mass balances has been published in (Falter & Pitz-Paal, 2018) and is shown in Figure 3.9.

In total, 1.1 GJ of incident solar energy (unconcentrated) is required for the production of one functional unit (f.u.), i.e., one liter of jet fuel and 0.87 L of naphtha. Of this total primary energy, 78% is converted into heat and 22% into electricity. The required water of 13.5 L/f.u. is produced from seawater desalination at a distance of 500 km from the plant and is mainly required for the cleaning of the mirrors of the solar concentrators (42%), for the generation of electricity in a concentrated solar power (CSP) plant on-site (36%), and for the thermochemical production of hydrogen (22%). At a specific energy requirement of 3 kWh/m^3 and adding the energy demand of pumping the water to the plant site,

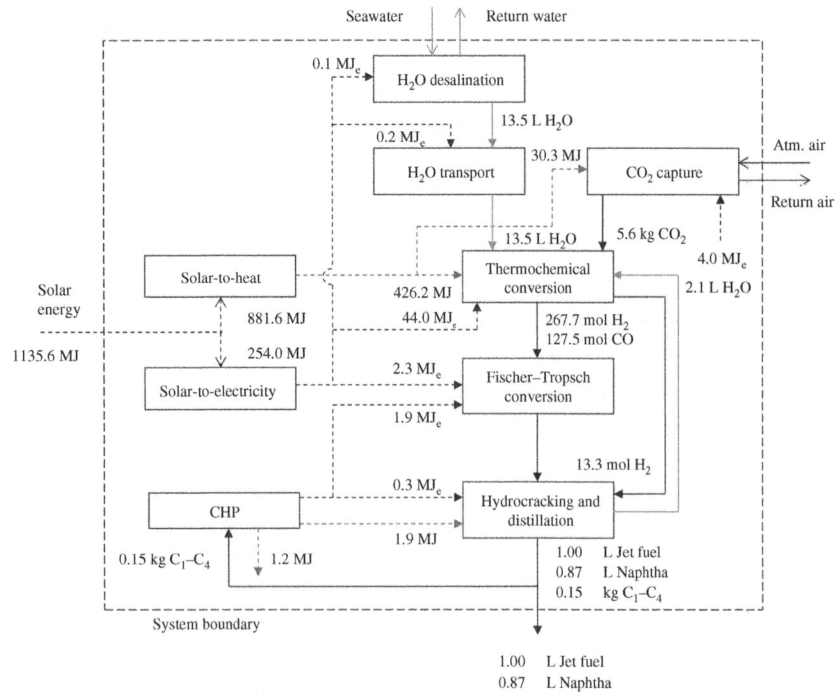

Figure 3.9 Energy and mass balance of a solar thermochemical fuel plant for the production of jet fuel with naphtha as a by-product. The thermochemical reactor has an energy conversion efficiency of 21.6% (concentrated solar energy to chemical energy (HHV) stored in syngas) and operates with vacuum pumping for the reduction of the oxygen partial pressure. All values are normalized to 1 L of jet fuel. Mass flows are represented by solid lines and energy flows by dashed lines. The system has not been optimized for internal heat recuperation. Reproduced from (Falter & Pitz-Paal, 2018)/with permission Elsevier.

water provision does not account for a significant part of the total energy requirement (<1%). For the thermochemical conversion, 5.6 kg of CO_2 is used, which is provided via direct air capture (DAC) through chemical adsorption to an ammine-functionalized sorbent as pursued by, for example, the Swiss company Climeworks (Climeworks, 2020). This process requires 2,000 kWh/t of low-temperature heat at about 100 °C and 200 kWh/t of electricity. The thermochemical reaction is driven by high-temperature heat provided by the concentration of solar energy using a field of heliostats and a tower, and electricity for auxiliary processes such as vacuum pumping. A location with a solar irradiation of 2,500 $kWh{\cdot}m^{-2}{\cdot}y^{-1}$ is chosen with 51.7% conversion efficiency from incident sunlight to concentrated radiation (Sargent and Lundy LLC Consulting Group, 2003) and 20% for the conversion to electricity. The thermochemical reactor has an efficiency of 21.6% for combined H_2O and CO_2 splitting at a reduction temperature of 1,900 K and operates with a solid–solid heat exchanger of 70% effectiveness. Sunlight is concentrated by a factor of 5,000, and vacuum pumping is assumed to operate at a 50% higher efficiency than what was described in an analysis in the recent scientific literature (Brendelberger et al., 2017). The evolving gases from the thermochemical reactor are pressurized to 30 bar for further conversion in the FT synthesis unit. The synthesis reactions themselves are exothermal and thus do not require the input of energy, but the heat could be used to cover energy demands of other process steps in the plant. Refining of the raw FT products requires the input of hydrogen for hydrocracking of the long-chained hydrocarbons and of 1.9 MJ/kg_{C5+} thermal and 0.2 MJ/kg_{C5+} electrical energy (Beiermann, 2007). The light fraction of the FT products (C_1–C_4 hydrocarbons) is assumed to be converted to heat and electricity in a combined heat and power (CHP) plant.

The pathway efficiency is then calculated as the LHV of the liquid products, jet fuel and naphtha, divided by the sum of the primary energy input in the form of incident sunlight. The LHV of FT-derived jet fuel is 33.4 MJ/L and that of FT-derived naphtha is 31.1 MJ/L (Stratton et al., 2010). At an overall energy input of 1,135.6 MJ/f.u., the pathway efficiency is defined as in Equation (3.4) below.

$$\eta_{\text{system}} = \frac{V_{\text{jet fuel}} \cdot \text{LHV}_{\text{jet fuel}} + V_{\text{naphtha}} \cdot \text{LHV}_{\text{naphtha}}}{Q_{\text{primary solar}}} \tag{3.4}$$

For the derived energy inputs and outputs, the efficiency of the unoptimized system is 5.3%, which means that about one in twenty units of the incident energy is converted into liquid fuels. At the assumed thermochemical conversion efficiency of above 20%, the system efficiency is thus about one quarter of this number. A seemingly low number, this efficiency should be compared to other alternative fuel pathways using the same boundary conditions. Biofuels usually have a very low system efficiency <1% (from sunlight to LHV of biomass) being limited by photosynthesis (Blankenship et al., 2011; Hall & Rao, 1999). Power-to-liquid (PtL) fuels can achieve conversion efficiencies (from electricity to LHV of fuels) of 38%–46% (Schmidt et al., 2016), which corresponds to 7.6–9.2% using the same conversion efficiency of solar primary energy to electricity of 20%.

As can be seen from the discussion of energy inputs, a solar thermochemical fuel plant is a system of subcomponents that each have specific energy requirements. Some of them require the input of heat and electricity, while others are exothermal processes that emit waste heat. The efficient operation of such a plant then depends on the integration of heat demand and heat requirements between the components. The plant efficiency can be enhanced if an energy demand can be satisfied by waste heat from another plant component instead of using more primary energy. In the following, the possibilities for increasing the plant efficiency by integrating heat sources and sinks are discussed.

3.3.3 Possibilities of Enhancing Plant Efficiency

In Section 3.3.2, a solar thermochemical plant for the production of jet fuel and naphtha was presented. The plant design was, however, not optimized for the recuperation of waste heat between the system components, which could play a significant role in enhancing the plant's economics and reducing the life-cycle emissions of the products. By using waste heat from another plant subunit, the amount of concentrated solar energy coming from the heliostat field can be reduced (at constant plant output), which decreases its required size. This directly lowers the investment and operating costs of the solar concentrator and the associated emissions because the size is reduced and the material input can be decreased.

When analyzing the system for waste heat recovery, both the energy flow rates and the temperature levels of heat source and sink have to be taken into account. A heat demand can only be covered by a source at least equal in energy and temperature (not accounting for energy losses during heat transfer). An example of possible heat recovery is the supply of the heat demand of DAC of CO_2 by the waste heat from the FT synthesis: the FT reaction is exothermal at 170 kJ/mol, which leads to an available waste heat of 21.7 MJ at >200 °C. This can cover part of the heat demand of DAC of CO_2, which is 30.3 MJ at 100 °C.

The analysis of energy recovery therefore requires a comprehensive overview of energy demands and waste energy together with the respective temperature levels for all components of the plant. This is shown in Figure 3.10, where energy inputs have positive values and energy losses have negative values. The largest thermal energy requirements are the reduction of ceria at 1,900 K, the heating of ceria from the exit temperature of the heat exchanger to the reduction temperature, the heating of the oxidants to the oxidation temperature of 1,000 K, CO_2 capture at 373 K, and gas separation of CO and CO_2 at 473 K. The most important energy losses are from reradiation at 1,900 K, incomplete heat recuperation of the reactive material between the oxidation and reduction temperatures, the exothermal oxidation of ceria at 1,000 K, spillage of concentrated radiation around the reactor, conduction and convection losses, and the exothermal FT synthesis reaction.

Electrical energy is needed for vacuum pumping, CO_2 capture from air, the separation of product gases, and the compression of syngas to the operating pressure of the FT synthesis reaction.

The total losses from the system sum to 4.9 J/J of liquid fuel output. If all of these losses can be recuperated without losses, the energy demand for CO_2 capture, H_2O desalination and transport, heating of oxidants H_2O and CO_2, compression of syngas, and FT raw product refining could be covered. This would consequently increase the energy conversion efficiency of the process from 5.3% to 8.6%, a relative increase of over 60%. Of course, this is an upper limit for the process enhancements through waste heat recovery because no losses in the recuperation and transfer of waste heat have been assumed. As these losses occur in different steps of the process and at different temperature levels, it is not clear that all the energy can be recovered with high effectiveness. Nevertheless, this analysis demonstrates that a large potential lies within the recovery of energy lost at different places in the system. Even though the specific recovery systems and their efficiencies and costs are not analyzed in further detail, it is clear that the reuse of energy losses can contribute to cost reductions, an enhanced environmental performance, and can relax the required increase in thermochemical efficiency. Especially limiting the required thermochemical reactor efficiency can be very important for further development of the technology considering that published efficiencies are still significantly below what is assumed to be required to make the process economical (Falter et al., 2016; Falter, Scharfenberg, & Habersetzer, 2020; Falter, Valente, et al., 2020; Marxer et al., 2017; Stechel & Miller, 2013).

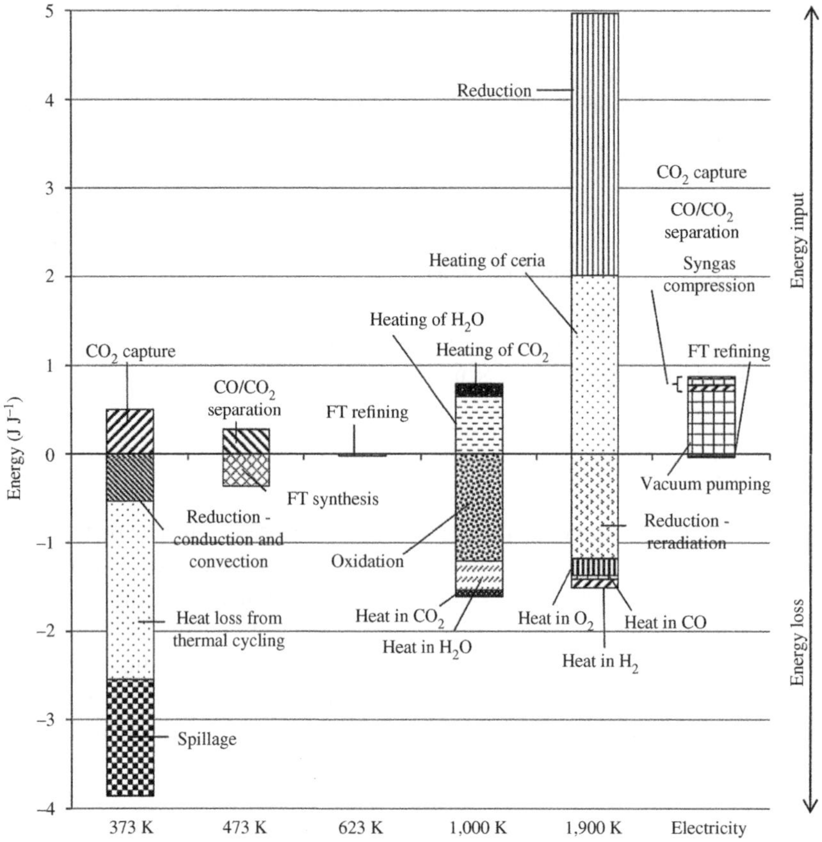

Figure 3.10 Energy demands (positive values) and energy losses (negative values) from solar thermochemical fuel production. Solid heat recuperation effectiveness is 70% in the thermochemical reactor. The corresponding temperature levels are shown for thermal energy (373–1,900 K). All values are normalized with respect to the lower heating values of the products (1 L jet fuel and 0.87 L naphtha). The value for "FT refining" is composed of the energy demands for refining minus the energy from the combustion of the gaseous products and has a negative value at a temperature of 623 K and for electricity, whereas "FT synthesis" indicates the exothermal chemical reaction. The heat lost from the thermal cycling of ceria and the spillage of solar concentration is assumed to occur at a temperature of 373 K. Reproduced with permission from (Falter & Pitz-Paal, 2018)/Elsevier.

3.4 Techno-Economic Analysis

The goal of the introduction of solar thermochemical fuels into the transportation sector is to reduce its climate impact by reducing the net emission of greenhouse gases. The environmental impact of this shift to renewable fuels is then determined by the specific fuel emissions and the availability of the fuel, i.e., by the share of conventional fuels that are replaced by alternative ones. The specific emissions of solar thermochemical fuels are discussed in Section 3.5 and the availability in Section 3.7. Another integral part of the feasibility of solar thermochemical fuels is their economic performance: as we are limited by finite financial resources, the most cost-efficient route to limiting global warming should be chosen. A reliable estimate of the production costs of different alternative fuel pathways and their sensitivity to important variables is therefore required for an informed decision on the required investments into different technologies.

In this subsection, a techno-economic analysis of the solar thermochemical fuel pathway is presented: the system to be analyzed is discussed, and the economic model is introduced before the results of the analysis are shown. The following analysis has been published in the scientific literature (Falter, Scharfenberg, & Habersetzer, 2020; Falter, Valente, et al., 2020) and is reproduced here in adapted form.[1]

1 Reproduced from (Falter, Valente, et al., 2020) with permission from The Royal Society of Chemistry.

3.4.1 System Description

The system as described in the following is identical to the one used in the life-cycle analysis. The analysis is performed for solar thermochemical jet fuel production from a baseline plant design with a capacity of 1,054 barrels of jet fuel per day and 888 bpd of naphtha as a by-product. The heat and electricity demands are covered by concentrated solar energy from separate tower plants on-site. An arid and sunny location in Morocco is chosen for the plant with a level of annual DNI of 2,500 kWh/m^2 and year and a distance to the sea of 250 km. Water and carbon dioxide are captured onsite from the air, via chemical adsorption (Climeworks LLC, 2014; Wurzbacher et al., 2012). Water is captured concurrently with CO_2 from the air, where it is assumed here that no further source is necessary. Supply via seawater desalination has been shown to have a negligible energy demand (Falter & Pitz-Paal, 2017). The plant is assumed to be a solar stand-alone plant, i.e., all heat and electricity demands are covered by conversion of solar energy on-site.

3.4.2 Economic Model

Different models exist for the assessment of the economic performance of a product or process (Short et al., 1995). A rather simple instrument is the payback period, which expresses the time it takes for the revenues to match the investment costs and does not give consideration to the changing value of money over time. Another model is based on the net present value (NPV), which is a discounted cash flow technique that subtracts the present value of all expenses from the present value of the revenues. If the NPV is positive, the project will create benefit, while when negative a loss is created and the project should not be pursued. Another related financial instrument is the internal rate of return, which indicates the interest created by the project at which the NPV is equal to zero.

For the derivation of production costs of solar thermochemical fuel, the levelized cost of energy (LCOE) method is chosen here (Short et al., 1995), which sums all costs over the plant lifetime taking into account the time value of money. From the total life-cycle cost (TLCC), an equivalent cost per year is derived with the annuity method, and finally, the costs per liter of fuel are found by dividing by the annual production of fuel. The TLCC is found from the following Equation (3.5) (Short et al., 1995).

$$\mathrm{TLCC} = \frac{I - (T \times \mathrm{PV}_{\mathrm{DEP}}) + (1 - T)\mathrm{PV}_{\mathrm{O\&M}}}{1 - T} \tag{3.5}$$

I are the investment costs, T the tax rate, $\mathrm{PV}_{\mathrm{DEP}}$ the present value of recurring depreciation, and $\mathrm{PV}_{\mathrm{O\&M}}$ the present value of the operational costs of the plant. The present value of a series of costs or revenues occurring over a series of years is found from Equation (3.6) (Short et al., 1995).

$$\mathrm{PV} = \sum_{j=1}^{n} \frac{C_j}{(1 + i)^j} \tag{3.6}$$

For constant annual payments, the equation can be simplified and the present value expressed as $\mathrm{PV} = A \times C_j$ with the annuity factor $A = i(1 - (1 + i)^{-n})^{-1}$, where i is the interest rate and n is the lifetime of the plant (Short et al., 1995).

The production costs of the fuel are then found by dividing the TLCC by the annual production of fuel multiplied by the annuity factor (Equation (3.7)) (Falter, Valente, et al., 2020; Short et al., 1995).

$$\text{Production costs} = \frac{\mathrm{TLCC}}{Q \times A} \tag{3.7}$$

The method uses the costs accumulated over the lifetime of the plant, discounted to their value at the start of the project, and divides them by the annual amount of fuel produced. In the following, it is assumed that the plant is supported by the state and that no taxes have to be paid. This is a possible scenario for the introduction of alternative fuels to the market. In the absence of an effective carbon tax and at higher production costs of alternative fuels than conventional fuels, incentives by the government, such as purchasing agreements or quotas for the former, are required for a start of fuel production. One way of such a support could be a tax rate of zero for the production of alternative fuels.

The determination of production costs with the LCOE method therefore requires the definition of investment costs and operational costs of the plant in addition to its lifetime and a plausible interest rate for the discounting of the payments occurring in different years. For the latter, it is assumed that the plant is financed through 60% debt and 40% equity (SunShot Initiative: U.S. Department of Energy, SunShot Vision Study. U.S. Department of Energy, 2012). However, the definition of a meaningful interest rate in the absence of experience from actual commercial projects is a challenge. To solve this problem, economy-wide proxy indicators (in this case, for Morocco) are used to determine suitable debt and equity interest rate estimates. Debt interest rates are derived from official statistics of the IMF (International Monetary Fund, 2018) expressing bank prime lending rates. The debt interest rate is then taken to be the bank lending rate. The equity interest rate is defined as the sum of government bond yields and equity risk premiums (ERPs). The bond yields are taken from publicly accessible information (Trading Economics, 2018; World Government Bonds, 2018), and the risk premium is determined after Damodoran (Damodoran, 2018). More detail on the methodology can be found in the literature (Falter, Scharfenberg, & Habersetzer, 2020; Falter, Valente, et al., 2020). This method yields interest rates for debt of 5.7% and for equity of 11.8% for Morocco. With the inflation taken to be 1.2% as the five-year average of the years 2013–2017, the nominal weighted average cost of capital (WACC) is 8.1%.

Depending on the location of the plant, labor costs can be a significant part of the operational cost. For high-income countries such as the United States of America or Australia, which also have an excellent solar resource, labor costs play an important role in the overall costs (see Section 3.7 and (Falter, Scharfenberg, & Habersetzer, 2020; Falter, Valente, et al., 2020) for more detail). Labor costs are estimated by assuming a fixed number of employees in different occupations working at the plant, where the total number of employees lies within the range of estimates of CSP plants on an equal power basis (Falter, Scharfenberg, & Habersetzer, 2020): 5 managers, 30 engineers, 15 clerks, 100 technicians, and 246 workers (Falter, Scharfenberg, & Habersetzer, 2020; Falter, Valente, et al., 2020). Country-specific labor rates are collected from the International Labor Organization (ILO) (ILO, 2018), where occupation classes from the database are matched to the classes defined for the plant. The labor for the construction of the plant is assumed to be covered by the investment costs.

3.4.3 Production Costs

For the defined baseline case of a solar thermochemical fuel production plant located in Morocco, the investment costs and operational costs are described in the following. It should be considered that the cost estimate is performed for a future implementation of the plant when the thermochemical reactor efficiency has improved and the cost of the solar concentrator and the CO_2 capture units have been reduced. A list of the investment and operational costs is shown in Table 3.1. All values are converted from 2020 Euro to 2021 US Dollar based on an exchange rate of 1.14 (Internal Revenue Service, 2022) and an inflation rate of 5.39% (CPI, 2021).

The DAC units are assumed to cost 420 $ per ton per year at operational costs of 48 $ per ton plus electricity and heat consumption, resulting in total costs slightly above 120 $ per ton, which is the long-term cost target of the technology.

Table 3.1 Overview of Investment Costs and O&M Costs for Baseline Plant.

Subsystem	Investment costs (M$)	O&M costs (M$/y)
Heliostats	1,038	20.8
Thermochemical reactors	151	1.00
Syngas storage	132	3.97
CO_2 capture units	124	14.2
Solar tower	107	–
Vacuum pumps	70.3	2.11
FT	53.6	3.41
CO_2 storage	37.4	1.13
Product pipeline	26.9	1.34
Ship transport	–	1.02
CO/CO_2 separation	15.5	0.47
Buildings	13.0	–
Water storage	9.28	0.28
Electricity	–	13.8
Labor	–	2.56
Total	**1,778**	**65.9**

Adapted from (Falter, Valente, et al., 2020).

Concurrently with CO_2, water is captured from the atmosphere at a relative molar ratio of approximately two to six (Wurzbacher et al., 2012). It is assumed here that the water captured from the atmosphere is sufficient for the operation of the plant, whereas in case of additional water demand seawater desalination and transport via pipeline to the plant site have been shown to be feasible at a small cost increase (Falter et al., 2016). In 2013, costs of heliostats were in the range 150–200 $/m^2$ (Coventry & Pye, 2013), while the goal of the SunShot project in the United States of America is a cost of 50–75 $/m^2$ (U.S. DOE, 2017; SunShot Initiative: U.S. Department of Energy, SunShot Vision Study. U.S. Department of Energy, 2012) in the future. Here, it is assumed that the solar concentrator has specific costs of 120 $/m^2$. At a required area of 8.6 million m^2 of heliostats for the provision of heat for the process, the investment costs are 1,038 million \$. The tower has a cost of 20 \$ per kW_{th} (Mancini et al., 2011) and the solar reactors are estimated to have a cost of 14.4 \$ per kW_{th}. The cost for the latter is estimated with material demands of recent experiments and assuming a scaling factor of 0.6 toward a large-scale production (see the supporting information of (Falter, Valente, et al., 2020) for more information). In the reactor costs, the reactive material ceria is not yet included, which is conservatively assumed to have a unit cost of 6 \$/kg (Statista, 2019). As market prices are lower than this value, further upgrading steps to the final ceria structure could at least partially be covered by the assumed value. Unlike price peaks in the past, which were due to export constraints from the largest producer China, the market prices are assumed to decline in the coming years (Statista, 2019). At the required amount of 16.9 kt for the plant, the cost for ceria is 101 M\$, which corresponds to two-thirds of the total cost of the reactor system. For the reduction of the oxygen partial pressure in the reactors, jet vacuum pumps are used. Their cost estimates are taken from (Brendelberger et al., 2017) and are a function of the established reduction pressure, where costs increase strongly toward low pressures. In the baseline case, at a reduction pressure of 1,100 Pa, the specific costs of the pumps are 50,640 \$ per mol per second. At an operation of eight hours per day, the required flow rate of oxygen is 930 mol per second. For the derivation of installed costs, additional cost factors are taken into account that increase the delivered costs by 0.2 for installation and 0.23 for additional costs, the investment costs are 70.3 M\$ for the vacuum pumps. The investment costs of the separation of CO and CO_2 at the exit of the thermochemical reactor are modeled based on (Kim et al., 2011) using a scaling factor of 0.7. At the given mass flow rates, the specific separation costs are 54 \$ per ton of CO_2 at a capture efficiency of 90%. CO_2 is captured from the air also at times when the thermochemical reactors are not operational and thus need to be stored. The optimal storage size and pressure need to be defined in a dedicated plant layout. At a storage size equivalent to one day of plant operation and a pressure of 10 bar, the investment costs for vessels and compressors are 37.4 M\$. Likewise, syngas is stored after the thermochemical reactors before entering the FT synthesis plant to decouple both subsystems. As H_2 and CO can be produced in separate reactors, storage in separate vessels is assumed. At a specific storage cost of 70 \$ per standard cubic meter, the cost of the storage vessels are 70.1 M\$ for the hydrogen tanks, 37.2 M\$ for the CO tanks, and 25.2 M\$ for the compressors, where the cost for the latter is taken from (Apt et al., 2008).

Operation and maintenance of the heliostat field are assumed to cost 2.4 $/(m^2 y)$, which is based on an estimate for central receiver plants (Fichtner, 2010). The operational cost of CSP electricity produced on-site is derived from the specific demand of 2.8 kWh_{el}/L jet fuel and a cost of 0.05 $/kWh_{el}$. The maintenance of all systems is taken into account with 3% of the investment costs annually, except for the thermochemical reactors, where a value of 2% is chosen plus the labor costs associated with the refurbishment of ceria. An overview of the investment costs and O&M costs of the baseline plant is shown in Table 3.1 and their relative weight in Figure 3.11.

The main influence on the O&M costs stems from the heliostat field (also the major contribution to the investment costs), the provision of water and CO_2, and the generation of electricity.

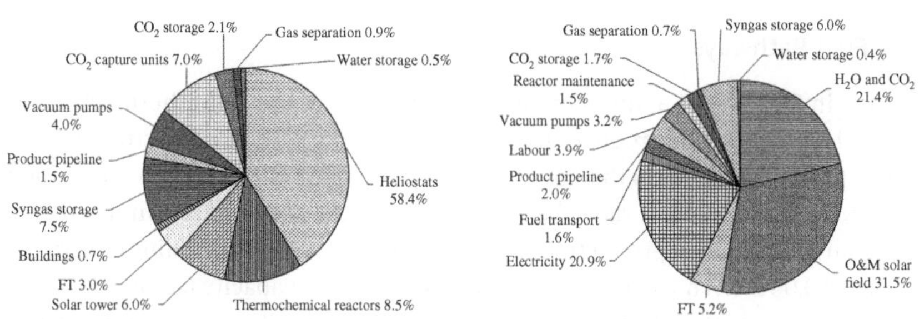

Figure 3.11 Distribution of investment costs (left) and operational cost (right) of the baseline plant. Figure taken from (Falter, Valente, et al., 2020) – reproduced with permission from The Royal Society of Chemistry.

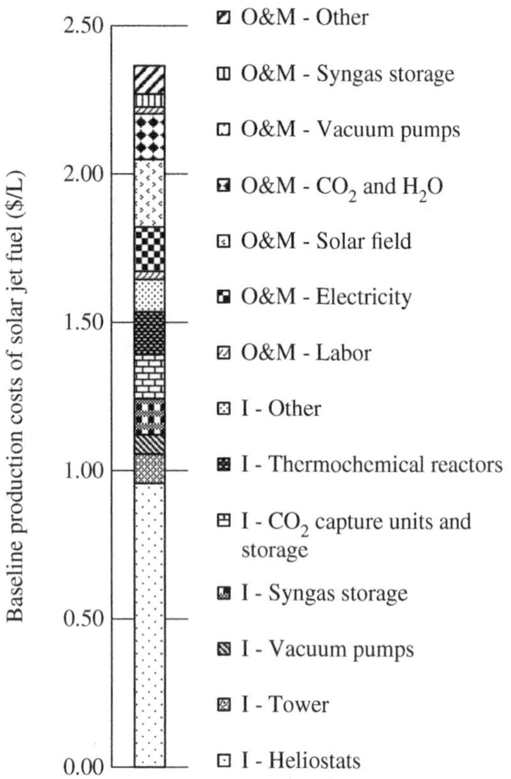

Figure 3.12 Overview of contributions to production costs of 1 L of solar thermochemical jet fuel from a baseline plant located in Morocco. Water and CO_2 are captured from the air at specific costs of 120 \$/t CO_2, the thermochemical energy conversion efficiency is 19%, and electricity is produced at a cost of 0.05 \$ per kWh_{el} on-site. Reproduced from (Falter, Valente, et al., 2020)/with permission of The Royal Society of Chemistry.

The production costs of 1 L of jet fuel from the solar thermochemical plant are then 2.36 \$ (nominal) or 2.10 \$ (real). An overview of the contributions to the final cost is shown in Figure 3.12. 70% of the total cost can be attributed to the investment cost and 30% to the operation of the plant. This fundamental cost structure is characteristic of many renewable energy technologies with a large investment at the beginning of the project and relatively low O&M costs over the lifetime of the plant.

The cost estimate for the baseline case was derived from a set of assumptions that are in parts ambitions, i.e., for the achievement of a thermochemical reactor efficiency of 19% (which is well below the thermodynamic limit) or the reduction of costs for CO_2 capture of 120 \$ per ton (the long-term target of commercial suppliers). In the following, to investigate the cost potential, more aggressive assumptions are made such as a thermochemical reactor efficiency of 30%, heliostat costs of 60 \$/m^2 (U.S. DOE, 2017), and CO_2 supply costs of 48 \$ per ton. For the baseline case in Morocco, the production costs under these new assumptions drop to 1.45 \$/L (nominal) or 1.30 \$/L (real). If the plant were built in Chile, the costs would be 1.39 \$/L (nominal) or 0.98 \$/L (real), and if built in Arizona, 1.5 \$/L (nominal) or 1.3 \$/L (real). Significant cost reductions are thus possible if the process can be improved over the assumptions of the baseline case. However, it should be noted that the achievement of these conditions requires the mass production of heliostats at very low cost and a considerable increase in process efficiency over the current state of the art. Furthermore, a low-cost renewable source of CO_2 has to be available, which may present an issue with respect to availability.

3.4.4 Comparison with Other Alternative Fuel Pathways

Based on the estimated production costs shown in Section 3.4.3, the economic performance of the solar thermochemical fuel pathway is compared to other alternative fuel pathways. The production costs of the PtL fuel pathway are estimated to be between 0.96 and 5.4 \$/L in a range of different studies that differ in their assumptions of the year of production, investment and O&M costs, costs of electricity, capacity factors, and interest rates, besides others. These studies are summarized in (Brynolf et al., 2018), where also a reference scenario for PtL fuel production in 2030 is shown: assuming alkaline electrolysis at investment costs of 840 \$/kW, electricity costs of 0.06 \$/kWh_{el}, carbon capture costs of 36 \$/t, a capacity factor of 80%, and an interest rate of 5%, production costs of 2.0 \$/L are estimated. The goal of the US DoE Earthshot Initiative is a renewable hydrogen production cost of 1 \$/kg, which corresponds to 0.3 \$/L of jet fuel equivalent. This value appears to be significantly lower than what is found here and discussed in the literature.

Different short-term biofuel pathways are discussed in (de Jong et al., 2015), and cost estimates are 1.2 $/L for HEFA conversion of used cooking oil, 1.6–2.4 $/L for FT conversion, 0.8–1.2 $/L for hydrothermal liquefaction (HtL), 2.0–3.2 $/L for alcohol-to-jet conversion (AtJ), and 4.4–5.8 $/L for direct-sugar-to-hydrocarbons conversion (DSHC), each for the same set of feedstock of forestry residues and wheat straw. In (Bann et al., 2017), a Monte Carlo analysis of production costs is performed, and the following most likely values of middle distillates are estimated: 0.8–1.2 $/L for HEFA conversion of yellow grease and soybean oil, 1.3–2.5 $/L for advanced fermentation (AE) of sugarcane and herbaceous biomass, 2.75 $/L for HTL conversion of woody biomass, and 1.2 $/L for FT conversion of municipal solid waste.

When comparing cost estimates of different fuel pathways, the time of the cost estimate should be taken into account, i.e., a point in the future for solar thermochemical fuels and PtL fuels, and today or in the short-term future for biofuels. The technology readiness level of solar thermochemical fuels is today at four to five with demonstration in the field (Bauhaus Luftfahrt e.V, 2016–2019), where technological development is expected from the thermochemical reactor and cost reductions from CO_2 DAC and heliostat production for eventual commercialization. The PtL is further advanced and mainly requires cost reductions of main system components such as the electrolyzer, the CO_2 DAC, and electricity generation. Furthermore, an increase in capacity factor of the electrical energy supply for PV or wind generation will be required for a reduction in production cost, e.g., through cost-efficient storage. Biofuels are technologically ready and available in small quantities (especially in the HEFA pathway). There have been commercial flights with biofuels for demonstration and testing purposes, e.g., by Lufthansa in the BurnFAIR project (Gröngröft et al., 2014).

Alternative fuels, if needed in a short timeframe, are therefore likely to be biofuels from waste streams and residues that achieve low costs and emissions. In a longer timeframe, a shift toward PtL and solar thermochemical fuels could allow for a higher scalability at lower cost.

3.5 Life-Cycle Analysis

The goal of a life-cycle analysis is the quantification of environmental impacts during all phases in the life of a product or process. To formalize the methodology and to improve the results of LCAs, the International Organization for Standardization has issued two norms, ISO 14040 and 14044. In these norms, the general methodology for the execution of a LCA is described and recommendations are given. The fundamental steps are first the definition of goal and scope of the analysis, which comprises the system boundaries, the functional unit, cut-off criteria, impact categories, and the allocation method. Second, in the inventory analysis, an energy and mass balance is defined with flows to and from the environment into the system. Third, the impacts in the chosen impact categories by the product or process are quantified in the impact assessment phase. Fourth, the obtained results are checked for data quality and consistency and interpreted with respect to the goal of the analysis. Furthermore, a sensitivity study can be performed, limitations discussed, and conclusions drawn. In general, these four phases of a LCA are not performed strictly sequentially but one may need to go back and forth between the single phases to adjust the system boundary, cut-off criteria, e.g., judge the quality of the results with respect to the input data, or reevaluate the chosen set of impact categories. Especially the recommendations on the outcomes of the study should be chosen carefully since a LCA usually contains a large number of assumptions, methodological choices, omissions (e.g., in case of the cut-off criterion), and details that are difficult to take into account without carefully studying the analysis. This is supported by the fact that the result of a LCA is usually a set of numbers that are quite easily understood, but their context and origins reveal themselves only by getting to know the particulars of the study.

In the following section, the recommended phases of a LCA in ISO 14040 and 14044 are applied to the case of solar thermochemical fuel production, and the specific result of the greenhouse warming potential (GWP) is discussed. In the following subsection, other impact criteria of water demand and land requirement of this fuel pathway are analyzed in more detail. The energy and mass balance is shown in Section 3.3. The material in this section is based on a prior journal publication (Falter, Valente, et al., 2020).[2]

3.5.1 Goal and Scope

The system analyzed is identical to the one used for the techno-economic analysis in Section 3.4. The LCA is performed for solar thermochemical jet fuel production from a baseline plant design with a capacity of 1,054 barrels of jet fuel per

2 Reproduced from (Falter, Valente, et al., 2020) with permission from The Royal Society of Chemistry.

day and 888 bpd of naphtha as a by-product. The heat and electricity demands are covered by concentrated solar energy from separate tower plants on-site. An arid and sunny location in Morocco is chosen for the plant with a level of annual DNI of 2,500 kWh/m^2 and a distance to the sea of 250 km. Water and carbon dioxide are captured onsite from the air, via chemical adsorption (Climeworks LLC, 2014; Wurzbacher et al., 2012). Water is captured concurrently with CO_2 from the air, where it is assumed here that no further source is necessary. Supply via seawater desalination has been shown to have a negligible energy demand (Falter & Pitz-Paal, 2017). The plant is assumed to be a solar stand-alone plant, i.e., all heat and electricity demands are covered by conversion of solar energy on-site. The functional unit is chosen to be 1 L of jet fuel and 0.84 L of naphtha and well-to-tank (WTT) and well-to-wake (WTW) boundaries are chosen for the determination of environmental impacts in the category of GWP. The analysis is limited to globally averaged effects and does therefore not take into account GWP from non-CO_2 effects. To allocate the emissions between the two products, an energy basis (LHV) is used. This leads to 56% of the emissions from the plant being allocated to jet fuel and 44% to naphtha. For the calculation of the life-cycle impacts of solar thermochemical fuel production, the software GaBi Professional (Thinkstep AG, 2012) in combination with the ecoinvent database V3.5 (Wernet et al., 2016) is used.

3.5.2 Inventory Analysis

The inventory analysis comprises a detailed description of mass and energy flows to and from the baseline case of the solar thermochemical fuel facility. The reader is referred to Section 3.3 to find this information and a discussion on the energy and mass balance. This is the basis for the quantification of environmental impacts as performed in the following section on impact assessment.

3.5.3 Impact Assessment

In the baseline case, for the production and transportation of the reference flow, 4.8 kg of CO_2 are captured from the air and stored in the products, while 1.7 kg of CO_2-eq. are emitted from the process, resulting in net 3.1 kg of CO_2-eq. being removed from the atmosphere. The WTT emissions are thus negative, which means that CO_2 is removed from the air and stored in the fuels. Of course, when they are combusted, CO_2 is re-emitted (4.1 kg of CO_2-eq. per functional unit), making the overall emissions positive: 1.0 kg of CO_2-eq. per functional unit. According to the allocation approach and the specific use-phase emissions of each product, the WTW emissions of jet fuel are then 0.61 kg CO_2-eq. L^{-1} and for naphtha 0.54 kg CO_2-eq. L^{-1}. In Figure 3.13, an overview of the emissions for 1 L of jet fuel is shown.

In the following paragraph, the emissions for the production of 1 L of jet fuel are analyzed in more detail. The capture of CO_2 for the process leads to negative emissions of 2.71 kg CO_2-eq. L^{-1}, whereas the production of the capture units, their operation including replacement of materials, and their decommissioning, lead to positive emissions of 0.06 kg CO_2-eq. L^{-1}.

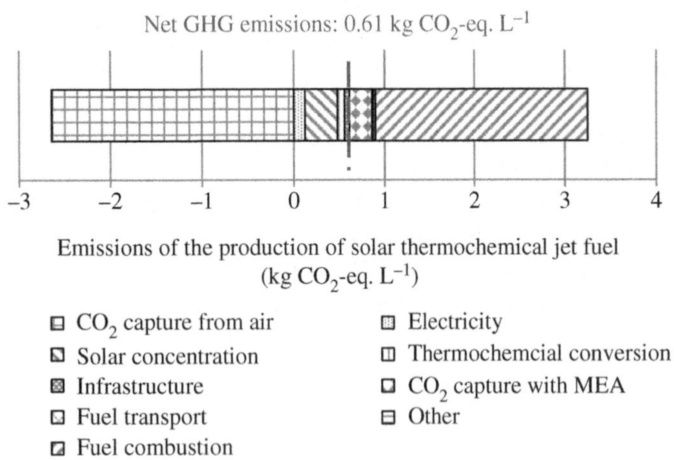

Figure 3.13 Overview of the emissions from the production and combustion of solar thermochemical jet fuel. Reproduced from (Falter, Valente, et al., 2020)/with permission of The Royal Society of Chemistry.

The production of CSP electricity on-site is connected to emissions of 0.13 kg CO_2-eq. L^{-1}, which is mainly due to the construction of the infrastructure including heliostats and receivers. Likewise, the concentration infrastructure for the provision of solar heat creates emissions of 0.37 kg CO_2-eq. L^{-1}, whereas the thermochemical reactors have a contribution of 0.08 kg CO_2-eq. L^{-1}, which is due to the materials used such as steel, aluminum oxide, or ceria. During the capture of CO_2 with MEA from the combined streams from the thermochemical reactors and from the recycle stream from the reformer, the capture efficiency is assumed to be 90%, resulting in an emission of 0.25 kg CO_2-eq. L^{-1}. All other emissions from the infrastructure of, e.g., compressors and the storage of gas and water combined have a sum of 0.06 kg CO_2-eq. L^{-1}. The transport of the fuel over 250 km with a pipeline and 1,000 km with a ship leads to emissions of 0.01 kg CO_2-eq. L^{-1}. The combustion of jet fuel eventually emits 2.35 kg CO_2-eq. L^{-1} and that of naphtha emits 2.16 kg CO_2-eq. L^{-1} (Stratton et al., 2010). For the purpose of comparing the results to GWP of conventional fuels, the data for the latter are taken from the literature and the ecoinvent database (Gómez et al., 2006; Stratton et al., 2010; Wernet et al., 2016). The largest emissions are associated with the capture of CO_2 from the atmosphere (negative emissions), with the concentration of solar energy, with losses during CO_2 capture with MEA, and with the combustion of fuel itself. The WTW emissions of solar jet fuel and naphtha achieve a reduction of about 80% with respect to their conventional counterparts (Stratton et al., 2010).

3.5.4 Interpretation

In this section, scenario analyses are presented for the supply of CO_2 and electricity from fossil point sources (compared to the renewable supply of the baseline case), and the results are discussed in the context of GWP of other fuel pathways. Furthermore, conclusions are drawn from the analysis and limitations are discussed.

3.5.4.1 Scenario Analysis–CO_2 From Natural Gas Combustion

There is an ongoing debate about the adequacy of different sources of CO_2 for carbon capture and use. In this work, a straightforward mass balance approach of carbon atoms was chosen at the system boundary. In this sense, CO_2 taken from the atmosphere creates "negative" emissions that partly compensate the emissions incurred by the other process steps, while the use of fossil carbon leads (after proper allocation between the products) to an increase in the CO_2 concentration in the atmosphere.

In the baseline case shown above, it was assumed that CO_2 is captured from the atmosphere via chemical adsorption. This process is currently in a demonstration and early commercial phase (Climeworks LLC, 2014), and the assumed cost of 120 $ per ton is a target for future commercial implementation. If solar thermochemical fuel were to be produced with CO_2 from a point source, there is in principle a broad range of options such as fossil point sources (e.g., coal and gas power plants), industrial processes (e.g., from steel or cement production), or biological processes (e.g., from biomass conversion). It was shown previously (Falter et al., 2016; van der Giesen et al., 2014) that the associated life-cycle emissions of the CO_2 used in the process are crucial for the GWP of the fuel produced. This is due to the fact that the total emissions mainly comprised those of CO_2 supply, of the construction and operation of the plant, and of the combustion of the fuel. The emissions of combustion are similar to those of a conventional fuel and account for about 70% of the emissions without CO_2 supply. To produce an environmentally friendly fuel with significantly lower CO_2 emissions, a negative source term has to be introduced via capture of CO_2 from a renewable origin, e.g., from the atmosphere. If fossil CO_2 from a natural gas combined cycle power plant is used for the process, the emissions rise from 0.61 to 2.9 kg CO_2-eq. L^{-1} and are thus almost as high as the emissions from conventional jet fuel. The use of biomass-based CO_2 leads to low emissions (as in the baseline case), but this source may not be scalable or available at the locations of high solar irradiation and in dry climates.

3.5.4.2 Scenario Analysis–Grid Electricity

Instead of generating electricity in a dedicated solar thermal plant on-site as in the baseline case, it could also be taken from the local grid. At a consumption of about 2.8 kWh/L of jet fuel, the use of Moroccan grid electricity at a specific carbon intensity of 0.73 kg CO_2-eq. per kWh_{el} would increase the WTW GHG emissions from 0.61 kg to 2.6 kg CO_2-eq. L^{-1} jet fuel. The improvement of GHG emissions with respect to conventional jet fuel would consequently be reduced from 80% to about 15%. It should be noted that the solar thermochemical fuel pathway as described in this work requires only a small amount of electricity when compared to the PtL pathway that has a consumption of >15 kWh_{el}/L jet fuel (König et al., 2015). The findings of higher GHG emissions when using grid electricity therefore hold to an even larger degree for the PtL pathway (and other alternative fuel pathways requiring significant amounts of electricity). Moving to a different country with a more favorable carbon intensity than Morocco could reduce the emissions of the fuel. However, for a reduction of 70% of the emissions, as foreseen to be the lower limit to be considered as an alternative fuel by the EU (European Union, 2018) from 2021 onwards, the grid should not have a carbon intensity higher than 0.15 kg CO_2 eq. kWh^{-1}, which may not be reached in many sunny locations today but could be in the future, when electricity grids become more and more decarbonized. In any case, the use of renewable electricity maximizes the emission savings potential and is a precondition for a renewable fuel.

3.5.4.3 Comparison with Published GWP Values of Other Fuel Pathways

The GWP of solar thermochemical fuels is compared to that of the PtL pathway, which is similar to the former but uses hydrogen from water electrolysis, and to different biofuels. Among the biofuels are BtL fuels from switchgrass (gasification of biomass and FT conversion of syngas) and HEFA fuels from soybean oil, palm oil, rapeseed oil, and algae oil (hydrotreatment of plant oil for deoxygenation and hydrocracking). LCA results for the PtL fuel are taken from (Schmidt et al., 2016) and for the biofuels from (Stratton et al., 2010). The comparison is shown in Table 3.2.

A reduction of GHG emissions from alternative jet fuel production can be achieved with biofuels, solar thermochemical fuels, and PtL fuels. For biofuels, a wide range of estimates exists depending on the inclusion of land-use change emissions into the estimate, which may lead to significantly higher emissions than conventional jet fuel. The best biofuel pathway in the literature sources was found to be a biomass-to-liquid pathway based on the conversion of forest residues. Solar thermochemical and PtL pathways can reduce emissions by about 80% and 70–90% compared to conventional jet fuel, respectively. As discussed above, the use of renewable sources of CO_2 and electricity is imperative if emissions are to be significantly lowered. The conversion of coal and natural gas to jet fuel leads to higher emissions than from the conventional sources because of higher process energy inputs. Another important aspect is the availability of the primary energy source, which in case of biomass-based processes could present a challenge for a large-scale supply of fuels (Suresh et al., 2018). As biomass is usually grown in regions with a moderate climate, natural rainfall, and thus a high biodiversity, a conflict with uses for food and fodder production could arise. Furthermore, the scalability of waste streams to large demands is not easily possible. On the other hand, the use of residues and wastes for fuel production is technically feasible, commercially available, and could reduce emissions by over 90% compared to conventional jet fuel. Especially for the near term, the use of these resources is seen to be a desirable solution for the transition to alternative fuels.

Table 3.2 Overview of Well-to-Wake GHG Emissions of Different Fuel Pathways.

Fuel pathway	GWP without land-use change (g CO_2-eq./MJ)	GWP with direct land-use change (g CO_2-eq./MJ)	Data source
Conventional jet fuel	87.5		(Stratton et al., 2010)
Coal-to-liquid	97.2–194.8[a)]		(Stratton et al., 2010)
Natural gas-to-liquid	101.0		(Stratton et al., 2010)
BtL (switchgrass)	17.7	−2.0[b)]	(Stratton et al., 2010)
BtL (forest residue)	7.7		(Suresh et al., 2018)
BtL (municipal solid waste)	27.6		(Suresh et al., 2018)
HEFA (soybean oil)	37.0	97.8–564	(Stratton et al., 2010)
HEFA (rapeseed oil)	54.9	97.9	(Stratton et al., 2010)
HEFA (palm oil)	30.1	39.8–698	(Stratton et al., 2010)
HEFA (jatropha oil)	39.4		(Stratton et al., 2010)
HEFA (algae oil)	50.7		(Stratton et al., 2010)
Solar thermochemical fuel	18.3		(Falter, Valente, et al., 2020)
PtL (Wind/PV in Germany)	11–28		(Schmidt et al., 2016)

a) CtL: the smaller value denotes a process with carbon capture and the larger one a process without.
b) BtL: the negative value occurs due to increased carbon stored in soil with growth of switchgrass.

3.6 Land and Water Demand

On a fundamental level, the production of synthetic fuels can be expressed as the conversion of water and CO_2 to synthetic hydrocarbons with the input of energy. If renewable energy is used as the primary energy for the process, the usually dilute primary energy is converted into heat or electricity by, e.g., wind turbines, photovoltaic modules, or heliostats and receivers. If biomass is converted, photosynthesis produces hydrocarbons, which are subsequently converted to the desired chain lengths for liquid fuels. In these cases, the amount of energy converted is directly proportional to the areas used, while the different conversion technologies have varying efficiencies, i.e., the demand for area for the production of one joule of heat or electricity will be different for each of the conversion processes used. However, the availability of land is primarily an issue for biomass-based fuel processes as the best locations are found in areas of high biodiversity, where a competition with food and fodder production could arise. The most favorable locations for the conversion of solar energy are found in sunny and arid regions that also have low levels of relative humidity in the air. In these locations, the land being used for other purposes is usually much less a challenge than in regions of high biomass growth. On the other hand, availability of clean drinking water is one of the major challenges for many economically challenged regions of the Earth and presents a serious risk to human health. Until 2025, it is expected that 1.8 billion people will live in regions of absolute water scarcity (United Nations, 2013; United Nations, 2013). The availability of land and water can therefore present an obstacle with respect to the feasibility and scalability of a fuel production pathway. In this subsection, the demand for land and water for the solar thermochemical fuel pathway is quantified and compared to that of other fuel pathways. Parts of the contents of this section have been published in a journal article (Falter & Pitz-Paal, 2017).[3]

3.6.1 Water Footprint

For the analysis, a plant producing 1,000 barrels per day (bpd) of jet fuel and 865 bpd of naphtha as a by-product is considered. CO_2 is captured on-site from the atmosphere via chemical adsorption, and water is supplied via desalination of seawater and subsequent transport via pipeline to the plant site located at a distance of 500 km and at an altitude difference of 500 m. It is assumed that an existing pipeline can be used. The plant is located in a region with a DNI of 2,500 $kWh \cdot m^{-2} \cdot y^{-1}$ and heat and electricity are assumed to be supplied by converting solar energy in a CSP plant. The energy conversion efficiency of the thermochemical reactor is 20%, which is a value well within the thermodynamic limits of the technology (Scheffe & Steinfeld, 2012). The baseline plant design requires an area of 31 km^2, which is mainly composed of the solar concentrator (26.1 km^2) and the CSP plant (5.3 km^2), where for the derivation of the areas a concentration efficiency of 51.6% (Pitz-Paal et al., 2011) and a conversion efficiency of 20% for the generation of solar electricity are assumed.

The on-site flows of energy and water are shown in Figure 3.14. Water is required for the thermochemical process and the generation of heat and electricity (i.a. for mirror cleaning and the steam turbine). In the FT and refining steps, water is generated and assumed to be recycled in the process plant. Heat and electricity are generated by conversion of sunlight.

The functional unit is chosen to be 1 L of jet fuel and 0.87 L of naphtha, which is a by-product of the process. A WTW boundary is chosen that includes provision of resources, concentration of solar energy, thermochemistry, and the FT conversion. For the plant components, construction, manufacturing, and disassembly are taken into account. The construction of the desalination plant is estimated to have a relative contribution below 1% of the total water footprint and is therefore not taken into account (Raluy et al., 2006). For the same reason, the impact on the water footprint of the CO_2 capture plants, which is estimated from publicly available data (Climeworks, 2015), is neglected. For the FT unit, the impact associated with the plant construction is estimated based on material requirements of a large-scale GtL plant in Qatar, and it was equally found that the contribution is well below 1% of the overall water footprint. The water demand for the transport of the materials to the plant location and the deconstruction is not assumed to exceed that of the manufacturing phase and is neglected except for the main infrastructure component of the solar concentration field.

In general, a distinction has to be made between water used at the plant site (e.g., for the chemical synthesis or mirror cleaning) and off-site during the production of the materials such as steel, concrete, or ceria. In the following analysis, both direct and indirect water demand is quantified for the baseline case plant design outlined above. Depending on the material intensity and the types of materials used, the indirect water demand can surpass the direct water demand, while the latter is usually the main focus of attention. In the following, the origins of water consumption are identified throughout the

3 For rights and permissions of further use of this material, the reader is referred to the journal Environmental Science and Technology of the American Chemical Society https://pubs.acs.org/doi/10.1021/acs.est.7b02633.

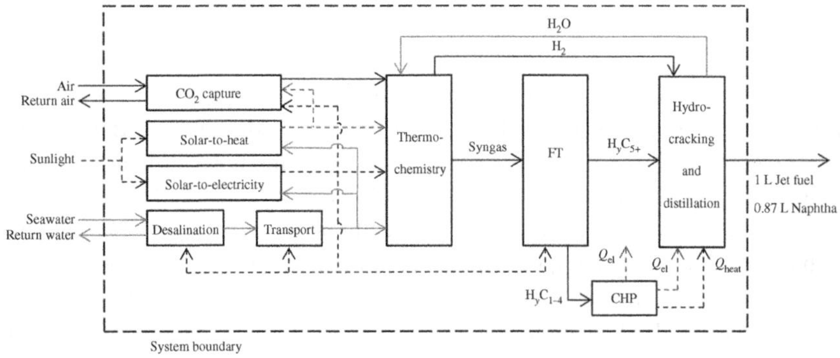

Figure 3.14 Flowsheet of solar thermochemical fuel production as discussed for the demand for land and water. CO_2 is captured by direct air capture (DAC), and H_2O is provided by seawater desalination and transported over a distance of 500 km and 500 m altitude to the plant. Electricity is provided by concentrated solar power on-site. Gaseous hydrocarbons are combusted in a combined heat and power (CHP) for the provision of heat and electricity to the process, while long-chain hydrocarbons are hydrocracked and distilled into the final products jet fuel and naphtha. Solid lines depict material flows and dashed lines energy flows. Reproduced from (Falter & Pitz-Paal, 2017)/with permission of American Chemical Society.

production process and the respective values are indicated that serve as a basis for the calculation of the water footprint. The results are normalized to the amount of product (jet fuel or naphtha) with an allocation between the products based on their energy content (LHV).

The contributions to the direct water footprint are discussed in the following. Sand and dirt accumulate with time on the heliostats, which makes a cleaning procedure necessary to maintain high reflectivity. In the literature, different values for the water demand of the cleaning have been published. Here, 58.0 $L{\cdot}m^{-2}{\cdot}y^{-1}$ from a recent life-cycle analysis of a tower CSP plant (Whitaker et al., 2013) are used. For the solar thermochemical fuel plant, this corresponds to 6.5 L per functional unit or 3.62 L/L of jet fuel. This water demand could be further reduced through the filtration of the cleaning water, which is, however, not further pursued at this point. Assuming stoichiometric conversion of water into hydrogen and oxygen, 5.1 L of water has to be supplied to the thermochemical reaction. 3.1 L of water are produced in the FT unit that can be recycled, reducing the required water input for thermochemical conversion to 3.0 L, which corresponds to 1.7 L/L jet fuel. For the production of solar thermochemical fuels, electricity is required for the purification of inert gases (80% of total electricity requirement), for CO_2 capture (10%), for FT synthesis (6%), for the separation of CO and CO_2 coming out of the thermochemical reactors (3%), and for the desalination and transport of water (2%). The production of inert gases requires 16 kJ/mol based on cryogenic rectification (Haering, 2008), whereas the thermochemical reactors could also be operated under a vacuum (Marxer et al., 2017). The DAC of CO_2 requires 200 kWh of electricity and 1,500 kWh of low-temperature heat (Climeworks LLC, 2014). For a more detailed description of the energy demands, the reader is referred to (Falter & Pitz-Paal, 2017).

The contributions to the indirect water demand are listed and explained in the following. One major contribution to the water demand is the materials used for the construction of the solar concentrator, where the heliostats consist mainly of a foundation holding a steel frame with a mirror structure made from a coated glass, and gear drives that perform two-axis tracking to maintain the reflected radiation on the target. The specific water demand for manufacturing, construction, dismantling, and disposal of the heliostats is estimated to be 2,848 L/m^2 from a study on CSP plants (Whitaker et al., 2013) and that of the tower to be 381 L/m^2 of heliostat field area. At a reflective area of 6.5 million m^2, the water demand for field and tower is estimated to be $2.11{\times}10^{10}$ L or 8.1 L/L jet fuel at an assumed lifetime of 25 years. The thermochemical reactors are mainly made from steel, alumina, glass, and the reactive material ceria, where mostly the latter is responsible for the water footprint of the reactor at a specific water demand of 11,830 L/kg (Koltun & Tharumarajah, 2014) for supply from the currently largest mine in China. The corresponding water demand of steel, on the other hand, is 3 orders of magnitude lower at 15 L/kg (Institute for Applied Ecology, 2008), rendering the impact of the reactor materials other than ceria negligible. The construction of the CSP plant is indicated to require 0.43 L/kWh (Whitaker et al., 2013), which corresponds to 2.7 L/L jet fuel. The construction of the chemical plant, including FT conversion and refining, is estimated on the basis of the Pearl GtL plant in Qatar and is a negligible contribution to the total water footprint.

In total, the direct water footprint is 13.4 L of water per functional unit or 7.4 L/L of jet fuel, and the indirect water footprint is 76.5 L per functional unit or 42.4L/L of jet fuel. The contributions to both the direct and indirect water footprint

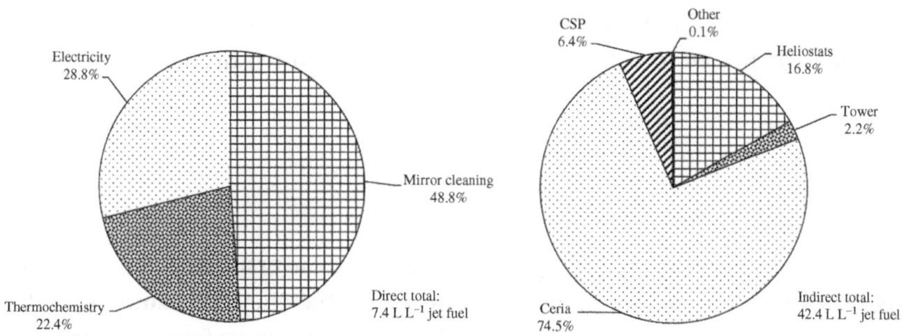

Figure 3.15 Direct (left) and indirect (right) contributions to total water footprint of solar thermochemical fuel production. Figure reproduced upon permission from (Falter & Pitz-Paal, 2017), https://pubs.acs.org/doi/10.1021/acs.est.7b02633. Further permissions to reuse this figure should be directed to the ACS.

are shown in Figure 3.15. The overall water footprint is dominated by indirect contributions, which are responsible for more than five times the amount of water used on-site. Ceria mining has a relative weight of 75% of the indirect water use, followed by the water demand of the heliostat field with 17% and the CSP facility with 6%. The other items only have minor contributions. The single largest contributor to the water demand is therefore the provision of the rare earth oxide ceria, which is currently almost exclusively provided by China. If ceria is replaced by a reactive material with a low specific water demand (such as Fe_3O_4 or ZnO), the water footprint of solar fuel production can be decreased from 49.8 to about 18 H_{L_2O} L^{-1} of jet fuel. About half of the direct water footprint is due to cleaning of the heliostat surfaces, and about a quarter each for the production of CSP electricity (which also requires water for mirror cleaning) and for thermochemistry. As can be expected from an efficient chemical process, the water demand for the actual synthesis of the fuels is on the order of the volume of the produced fuels and small compared to the overall water footprint.

The overall water footprint is the sum of direct and indirect water demands and is 49.8 L/L jet fuel. In the following, the water footprint is compared to that of other pathways. For this comparison, published data for biofuels and conventional jet fuels are used, while the PtL pathway is modeled based on an energy conversion efficiency of 44.6% (Haije & Geerlings, 2011) and referenced water demands of electricity production. The reader is referred to (Falter & Pitz-Paal, 2017) for more information on the data sources. The solar thermochemical pathway has a similar water footprint compared to the fossil-based options and one that is lower by orders of magnitude than the biomass-based options. Even the consideration of the higher value of the solar thermochemical fuel pathway does not change this result. However, the water demand for biofuels occurs mainly in less arid regions and may therefore be less critical. Nevertheless, the large water footprint of biofuels may have a negative impact on their feasibility and large-scale scalability. Compared to the competing PtL pathway based on water electrolysis, reverse water gas shift, and FT synthesis, the direct water footprint is expected to be about equal. The reason for this is an efficient chemical synthesis of hydrocarbons that does not include water losses from biomass-based plant growth with the associated evapotranspiration. Additionally, water is required for cleaning the mirrors or PV modules and for electricity taken from different sources such as coal, CSP, PV, or wind.

3.6.2 Land Demand

Land requirement is defined here as the total area of land used for the production of a defined amount of jet fuel. This metric can be used to compare different fuel production pathways, like PtL fuels, biofuels, or other alternatives. A smaller land requirement is advantageous because the environmental and social impact will be reduced. However, the plant location is also very important because a plant construction in a desert region is likely to have a much smaller impact on the environment and the regional population than a construction in areas of large population density and rich flora and fauna. In the metric of land requirement this is not reflected. In addition, the quality of the land coverage is decisive: while biological plants are perceived to be a more natural environment, industrial facilities may be seen to be more critical.

If the system boundaries are chosen such that primary solar energy, water and CO_2 are converted into liquid fuels, land requirement is directly proportional to system efficiency. A lower efficiency will require a larger mirror area to supply the same fuel output. The reference area for the solar thermochemical efficiency is the reflective area of the mirrors. To derive the total covered land area of the facility, a land coverage factor has to be defined. In the case of concentrated solar power plants, this factor is approximately 25% (Trieb et al., 2009), i.e., the total covered land area is four times the reflective area of the mirrors. The area requirement for the solar tower and the fuel conversion plant is neglected because they cover a small area compared to the heliostat field and may be placed in between the mirrors. The land requirement of the solar thermochemical plant is then defined as

$$A_{\text{total ground}} = \frac{N_{\text{fuel,daily}} \cdot 365 \frac{\text{days}}{\text{year}} \cdot \text{LHV}_{\text{jet fuel}}}{\text{DNI}_{\text{annual}} \cdot \eta_{\text{land coverage}} \cdot \eta_{\text{solar–to–jet fuel}}} \tag{3.8}$$

$\text{DNI}_{\text{annual}}$ is the annual DNI per unit area at the plant location, $\eta_{\text{solar–to–jet fuel}}$ is the energy conversion efficiency of solar primary energy to jet fuel [$\eta_{\text{solar–to–jet fuel}}$ is equal to 55.4% of the overall energy conversion efficiency from sunlight to 1 L of jet fuel and 0.87 L of naphtha based on an energy allocation (LHV)], $\eta_{\text{land coverage}}$ is the land coverage factor (assumed to be 25%), $N_{\text{fuel,daily}}$ is the daily jet fuel production from the plant in liters, and $\text{LHV}_{\text{jet fuel}}$ is the LHV of jet fuel (33.4 MJ/L) (Stratton et al., 2010).

Overall, 1.22 GJ of solar primary energy is captured and converted into the intermediates heat and electricity, producing 1 L of jet fuel and 0.87 L of naphtha. The overall energy conversion efficiency based on the LHV of jet fuel and naphtha is thus (1 L × 33.4 MJ/L + 0.87 L × 31.1 MJ/L)/1.22 GJ = 5.0%. This value includes the provision of heat and electricity used in the process. While in other publications, higher numbers are mentioned for the overall efficiency, this more conservative estimate is based on a thermochemical efficiency of 20%, which is well below the thermodynamic limit. At experimental values that are at about 5% today (Marxer et al., 2017), the achievement of 20% seems to be an ambitious but realistic target for the midterm future and was therefore selected here. Using Equation (3.8) for the facility with a total jet fuel production of 1,000 bpd (and 865 bpd naphtha) located in a favorable region with a solar irradiation of 2,500 $\text{kWh·m}^{-2}\text{·y}^{-1}$ and an efficiency of 5.0% gives an area of 31.40 km^2, i.e., a square with a length of about 5.6 km. The corresponding annual production per hectare is 18,480 L of jet fuel and 15,985 L of naphtha or 1.85 L of jet fuel and 1.60 L of naphtha per square meter per year. As for the water demand, the land requirement is compared to other fuel pathways using published values for conventional fuels and biofuels. For PtL fuels, area demands of renewable electricity production are used together with the above-indicated energy conversion efficiency of 44.6%. More information on the data sources is found in (Falter & Pitz-Paal, 2017). An overview of fuel productivities per area (the inverse of the area demand) is shown in Figure 3.16.

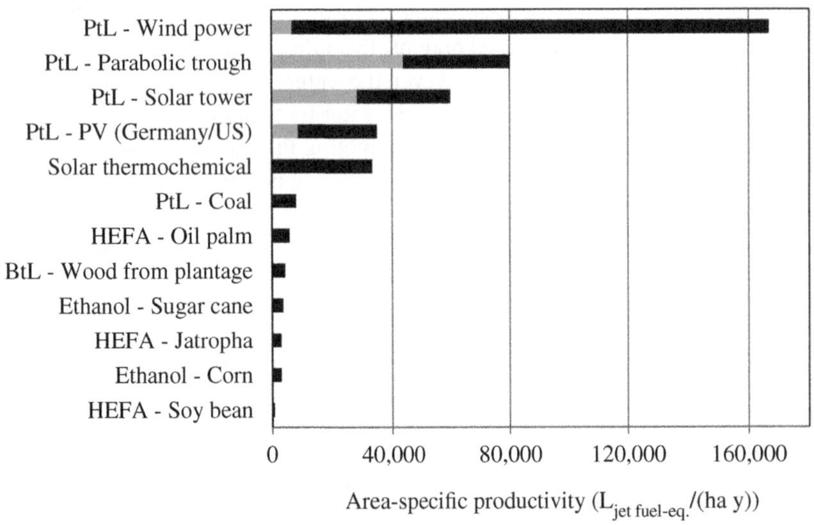

Figure 3.16 Area-specific productivities of the solar thermochemical fuel pathway in comparison with biomass-based pathways (BtL, HEFA, and ethanol; sources (Gröngröft et al., 2014) and (Riegel, 2014)) and power-to-liquid pathways (PtL, based on different sources of electricity: (Eberle, 2010; European Academies Science Advisory Council, 2011; Haije & Geerlings, 2011; McDonald et al., 2009) and own computations). The results are given in liters of jet fuel equivalent per hectare per year, where the conversion is done on an energy basis (LHV). The gray and black bars denote the actually covered land area (gray) and the total affected land area for wind power (black), plants in Spain (gray), and under a higher solar resource of 2,500 $\text{kWh·m}^{-2}\text{·y}^{-1}$ (black) for solar trough and tower, and plant locations in Germany (gray) and the United States (black) for solar PV. The assumptions are ideal conditions for plant growth of the biomass-based pathways and favorable developments of the thermochemical conversion efficiency and the energy penalty of carbon dioxide capture for the solar thermochemical pathway. Figure reproduced upon permission from (Falter & Pitz-Paal, 2017), https://pubs.acs.org/doi/10.1021/acs.est.7b02633. Further permissions to reuse this figure should be directed to the ACS.

Solar thermochemical fuels in the baseline case have an area-specific productivity of 33,362 L of jet fuel-equivalent ha^{-1} $year^{-1}$ and therefore achieve a higher value than some of the PtL pathways, i.e., those based on electricity from coal combustion, photovoltaics in Germany, and a solar tower in Spain. The reason for the higher value of solar thermochemistry over the PtL pathway with a solar tower in Spain is the higher solar irradiation of the baseline case plant over the location in Spain. The PtL pathway based on coal combustion has a lower productivity than the solar thermochemical fuel pathway due to a relatively high area demand for coal mining in Germany (European Academies Science Advisory Council, 2011). The production of biofuels relies on photosynthesis, which in theory achieves efficiencies of up to 5% (Hall & Rao, 1999), but in practice has much lower values <1% (Blankenship et al., 2011). As the efficiency directly translates into the required cultivation area, in general higher values are noted for biofuels. The preferred areas of biomass cultivation are located in mild and humid climate zones, while the best locations for solar fuels are found in arid climate zones with high solar irradiation. Biofuels and solar fuels therefore do not compete for the same land and could be implemented complementarily. The solar thermochemical fuel pathway therefore achieves lower area-specific productivities (but on the same order of magnitude) as electrochemical pathways, where both the final values depend on the specific assumptions made, e.g., the primary energy source or plant location.

3.7 Geographical Potential

The following material is based on a prior publication in the scientific literature (Falter, Scharfenberg, & Habersetzer, 2020).[4] As can be shown by a simple energy balance of solar irradiation on Earth and our global energy demand, solar energy is in principle easily able to cover the global energy demand. Therefore, its conversion into liquid fuels could also easily cover the fuel demand of global aviation. As the production of solar thermochemical fuels requires only sunlight, water, and carbon dioxide, it could give a range of countries the possibility to produce their own environmentally friendly fuels without having to rely on imports from oil-producing countries. However, as sunlight is unevenly distributed over the surface of the Earth, there are regions that are more suitable for solar fuel production than others. It is therefore interesting to analyze the dependency of the fuel production costs on geographical location and to quantify the production potential in different regions of the Earth.

The focus of this subsection is on the geographical variability of solar thermochemical fuel production characteristics. The theoretical volumes of fuels that can be produced along with the estimated production costs for the most interesting regions of the Earth are presented. To this end, a geographic information system (GIS)-based approach is used in combination with technical and economic models of the fuel production process. Prior to the following analysis, studies have been performed on the geographical potential of solar technologies for electricity production (Ennaceri et al., 2018; Huld et al., 2017; Trieb et al., 2005), and on the potential of biogenic sustainable energy (Deng et al., 2015; Long et al., 2013; Perea-Moreno et al., 2019; Searle & Malins, 2015), but none on the particular case of solar thermochemical fuels. After the exclusion of unsuitable areas due to sustainability criteria, the linkage between production potential and production cost is shown on a regional and national level with cost–supply curves. With the information shown in this subsection, the regional and global potential for solar fuel production is quantified, and the life-cycle costs for single regions and countries can be estimated, establishing a useful tool for decision-makers in the area of alternative fuels.

In the following, the methodology for the determination of suitable areas and production costs of solar thermochemical fuels is described. In general, the analysis was limited to regions with a high level of DNI, i.e., the United States of America, the Andes region in South America (including Chile, Bolivia, Argentina, and Peru), the Mediterranean region (MED; including Southern Europe, the Middle East, and North Africa), Southern Africa (including South Africa, Botswana, Angola, and Namibia), and Australia. First, a list of exclusion criteria is defined for the determination of the available areas. Then the remaining net areas are combined with the available DNI and a conversion factor to determine the production potential. In a second step, an economic model is applied to express the production costs of solar thermochemical fuels as a function of the location.

3.7.1 Determination of Suitable Areas for Solar Thermochemical Fuel Production

The following areas are excluded from consideration (similarly to the calculations performed in the MED CSP study (Trieb et al., 2005) for the determination of suitable areas for CSP plants in the Mediterranean): areas with existing ground

4 The article is published under the CC-BY 4.0 license https://creativecommons.org/licenses/by/4.0/legalcode.

structures, water bodies, shifting sands, slopes ≥ 5%, protected areas, as well as areas covered by forest, closed shrubland, woody savannas, wetland, cropland, urban settlements, or snow/ice. Consequently, allowed land types are open shrubland, savannas, and barren or sparsely vegetated land that do not fall under the restrictions listed above. For the net area calculations, the software QGIS (QGIS Development Team, 2017) is used as a GIS-based tool. For an overview of the data sources used for the analysis, the reader is referred to (Falter, Scharfenberg, & Habersetzer, 2020).

A map with the net available areas is created by subtracting all excluded areas described above. On the remaining land, the DNI is mapped with data from the Global Solar Atlas (World Bank, 2019) with a resolution of 1 km^2. In the Global Solar Atlas model, local DNI values are derived from a combination of a clear sky model that calculates the local solar energy flux based on the position of the sun, altitude, concentration of aerosols, water vapor content, ozone, and satellite images that detect the cloud cover. The model uses a primary grid resolution of 3–7 km, which is downscaled to a resolution of 1 km.

In the following, the results for the available areas are shown exemplarily for the United States, while the other regions mentioned above can be found in (Falter, Scharfenberg, & Habersetzer, 2020). In the United States of America (Figure 3.17), most of the land is being used for agriculture and livestock farming or is protected from other use in national parks. In mountainous regions such as the Appalachian and Rocky Mountains, the terrain may be too steep for the construction of a fuel production plant. Most of the suitable areas are found in the southwest, where the solar irradiation is highest. While most of the country is unavailable for solar thermochemical fuel production, the remaining suitable areas are nevertheless considerable in size and could enable large fuel production.

Having identified suitable areas for fuel production, the production potential can be derived from the solar irradiation and the conversion efficiency of the plant (DNI to LHV of products). This conversion factor comprises the energy conversion efficiency of the plant and the land-use factor, whereas the latter is defined as the mirror area divided by total covered ground area, and is assumed to be 25% (Kim et al., 2013). The energy conversion efficiency of concentrated solar energy to LHV of jet fuel is found to be 2.46% (Falter, Scharfenberg, & Habersetzer, 2020), where naphtha is a by-product. The area-based conversion factor is then found to be 25% × 2.46% = 0.61%. The jet fuel production on one square meter of land with a DNI of 2,500 $kWh{\cdot}m^{-2}{\cdot}y^{-1}$ is then 2,500 $kWh{\cdot}m^{-2}{\cdot}y^{-1}$ × 0.61% × 3.6 MJ/kWh/33.4 MJ/L = 1.66 $L{\cdot}m^{-2}{\cdot}y^{-1}$. Additionally, 1.40 L naphtha are produced. For the purposes of this study, it is assumed that the conversion factor is constant and thus independent of solar irradiation.

The suitable areas and associated production volumes are shown in Table 3.3.

The largest suitable areas for solar fuel production are found in the MED region with close to 6 million km^2, where the largest areas ($\geq 10^6$ km^2) are found in North Africa and the Middle East, i.e., in Algeria, Western Sahara, and Saudi Arabia. As a single country, Australia has by far the largest potential with about 4.5 million km^2 of suitable area corresponding to more than half of the country's area. In the MED region, the share of suitable areas in the respective country areas is also

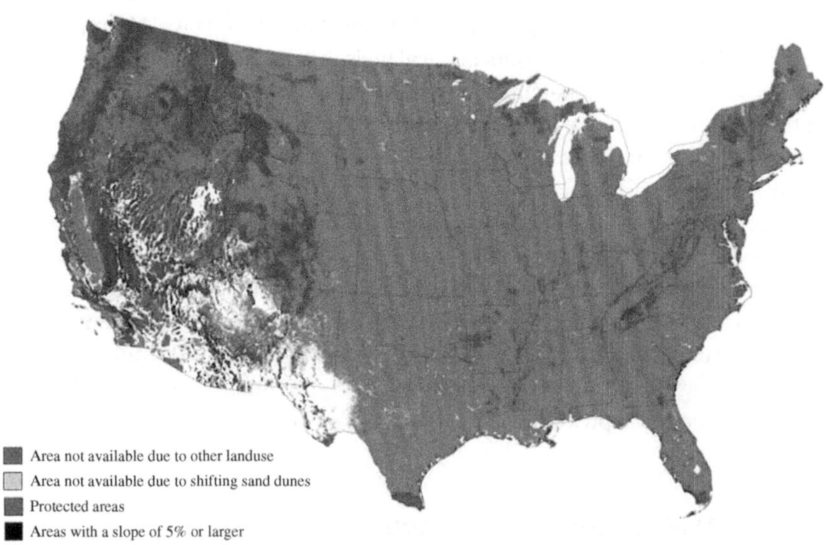

Figure 3.17 Suitable areas for solar thermochemical fuel production in the United States of America (shown in white). Excluded areas comprise other land use, protected areas, ≥ 5% slope, and shifting sands. The majority of the areas is excluded based on other land use (such as agricultural land) and areas with a slope of 5% or more. Reproduced from (Falter, Scharfenberg, & Habersetzer, 2020)/MDPI/Public domain CC BY 4.0.

Table 3.3 Suitable Areas and Production Potentials in Different Regions.

	Suitable area for fuel production ($\times 10^6$ km^2)	Suitable area/calculated country area (%)	Production volume on suitable areas (10^3 Mt/y)	Production volume/World demand (–)
MED region	5.71	52	6.13	20.4
Australia	4.49	59	5.81	19.4
South Africa, Botswana, Namibia, Angola	1.87	48	2.38	7.9
Chile, Argentina, Peru, Bolivia	1.24	23	1.47	4.9
United States of America	0.599	8	0.777	2.6

a) The deviation from the country area as reported by the FAO is largely due to the data being restricted to latitudes between 60° N and 45 °S.
b) Current global jet fuel demand is about 300 Mt/y (IATA, 2018).

high with >50% on average, whereas this value is mainly achieved in North Africa and the Middle East, where large regions are available in the deserts. The value found for the countries in southern Africa is also close to 50%. In the United States of America, however, below 10% of the country is suitable for solar fuel production, which is due to large areas being used for agriculture and farming. The suitable and most interesting sites are found in California, Nevada, Arizona, New Mexico, and Texas, which is also where solar irradiation is highest. In the analyzed countries in South America, only about 20% of the total areas are found to be suitable, which is mainly due to high slopes in the Andes region, agricultural land use, pastureland, and protected areas.

The largest volumes can be produced in the MED region with 6.1×10^3 Mt/y, which is more than 20 times the current value of global jet fuel consumption (about 300 Mt/y (IATA, 2018)). The other regions achieve smaller production volumes corresponding to respectively 19.4 times the world demand in Australia, 4.9 in southern America, 7.9 in southern Africa, and 2.6 in the United States of America, which is still more than twice the global jet fuel consumption. With these large volumes that could be produced, a concentration on the most attractive areas for fuel production is possible, which will reduce costs and land use.

3.7.2 Determination of Life-Cycle Production Costs

To determine the production costs of 1 L of solar thermochemical jet fuel, an economic model is developed based on the LCOE (Short et al., 1995) (see Section 3.4 for more detail). Due to the long lifetime of the plant of 25 years, the time value of money has to be taken into account. This is achieved through the annualization of the investment and operational costs, which requires the definition of an interest rate. This interest rate is equivalent to the WACC and comprises an equity rate and a debt rate. Consequently, the calculation of an annual value of the total costs of the plant is the sum of the annualized investment costs and the yearly operation and maintenance costs. The total annual costs are then divided by the annual production volume of fuel. Importantly, in the current analysis, the WACC is differentiated based on geographical location on a country level. This is not commonly done in comparable studies, where often a standard value is used across all countries. However, the impact on LCOE is potentially very large. For comparison, WACC values for solar projects in the literature are 7% (nominal) in the US SunShot Vision Study (Mehos et al., 2016; SunShot Initiative: U.S. Department of Energy, SunShot Vision Study. U.S. Department of Energy, 2012), 7.5% for OECD countries and China, and 10% (real) for other countries in a study by the IRENA (IRENA, 2018). Labordena et al. used default values for the real equity rate of return recommended by the UNFCCC to calculate country-specific WACC (Labordena et al., 2017) for CSP projects in sub-Saharan Africa. In the following, the methodology for the derivation of local WACC values is described.

The equity interest rate is assumed to be the sum of the government bond yields and an ERP. Essentially, the ERP is an indication of "the compensation investors require to make them indifferent at the margin between holding the risky market portfolio and a risk-free bond" (Duarte and Rosa, 2015). More specifically, one can assume that government bonds represent the benchmark for risk-free bonds. Thus, to estimate equity interest rates, one can sum up country-specific government bond yields with country-specific ERPs. The debt interest rate is taken to be the nominal bank lending rate. A detailed description of the databases used for the determination of the interest rates for equity and debt can be found in (Falter, Scharfenberg, & Habersetzer, 2020), whereas for some countries and indicators data were missing. As multiple imputation

Table 3.4 Interest Rates, Equity Risk Premiums, Inflation Rates, and WACC Across a List of Representative Countries.

	Debt interest rate (%)	Government bond yields (%)	Equity risk premium (%)	Inflation (5-year average) (%)	Nominal weighted average cost of capital (%)
United States of America	4.0	3.1	5.1	1.3	5.7
Spain	2.2	1.6	7.3	0.5	4.9
Chile	4.6	5.0	5.8	3.3	7.1
Israel	3.4	2.2	5.9	0.2	5.3
Egypt	18.8	17.5	12.6	12.3	23.3
South Africa	10.4	9.5	7.6	5.6	13.1

methods prove to be a useful method to adequately deal with missing data (especially in comparison to simpler methods, like dropping observations with missing values), multiple imputation using chained equations is employed here (White et al., 2011).

The resulting interest rates and the WACC for different countries are shown in Table 3.4. As is evident from the assumed rates, costs of capital vary significantly across the countries. This can be attributed to the fact that some regions provide a more stable environment for investments than others, which is expressed in the interest rate estimates and inflation. High costs of capital represent the risk involved for the investor and thus increase the total cost of the investment. If interest rates and inflation are high, the resulting production costs of jet fuel will be high as well. This may render a very sunny location unfeasible for solar fuel production because of the resulting high costs of capital. An example is Egypt, which has the sunniest locations in the MED region but has very high estimates of interest rates, which increase the production costs over those of Israel and even Spain, which have a smaller solar resource.

It is important to note that these estimated WACCs are only a snapshot for the year 2017 and rely on aggregated proxy indicators. Thus, the cost of capital for a specific project in the future may be quite different from the estimates presented here. The aim here is not to give a precise prediction of future investment environments but to sensitize for the importance of country-specific variations in the cost of capital. The sensitivity of the results with respect to the inflation rate and WACC can be found in the appendix of (Falter, Scharfenberg, & Habersetzer, 2020).

3.7.3 Production Cost

The description of the model used for the estimation of investment and operational costs is found in Section 3.4.2. All values are converted from 2019 Euro to 2021 US Dollar based on an exchange rate of 1.14 (Internal Revenue Service, 2022) and an inflation rate of 5.39% (CPI, 2021). For a plant with a capacity of 1,050 barrels per day (bpd), located in Morocco with a DNI of 2,500 $kWh{\cdot}m^{-2}{\cdot}y^{-1}$ and at a distance of 250 km to the sea, the total investment costs are 1.80 billion \$. Of this sum, 59% are for the heliostat field, 8% for the thermochemical reactors, 7% for the CO_2 capture units and syngas storage, respectively, and other smaller contributions. The operational costs sum up to 66.2 million \$ per year, whereas the largest cost contributors are the heliostat field (32%), the capture of H_2O and CO_2 (21%), the generation of electricity (21%), and syngas storage (6%). A total workforce of 396 is assumed, distributed into workers, technicians, clerks, and managers. While this number is constant across countries, the salaries are location-dependent and thus change the operational costs of the plant. As a thermochemical fuel plant has yet to be built and also the experience with CSP plants of the same size is limited, the estimation of the workforce has an inherent uncertainty. It was chosen to lie in the range of estimates for CSP plants on an equivalent energy basis (Falter, Scharfenberg, & Habersetzer, 2020; Falter, Valente, et al., 2020).

Further variables, which are location-dependent are the level of solar irradiation (DNI) and the distance to the sea, where the former scales the size of the solar concentrator and the latter the length of the water pipeline to supply fresh water from desalination at the sea. In Figure 3.18, the sensitivity of the production costs with respect to a change in DNI and distance to the sea is shown. Additionally, a change in solar reactor efficiency (concentrated solar energy to chemical energy stored in syngas) is shown to demonstrate a change in reactor technology.

A change of location from the south of Spain with a DNI of 2,000 $kWh{\cdot}m^{-2}{\cdot}y^{-1}$ to the best locations in the Andes with 3,500 $kWh{\cdot}m^{-2}{\cdot}y^{-1}$ will decrease the production costs of solar thermochemical jet fuel by 25%. If the reactor efficiency is

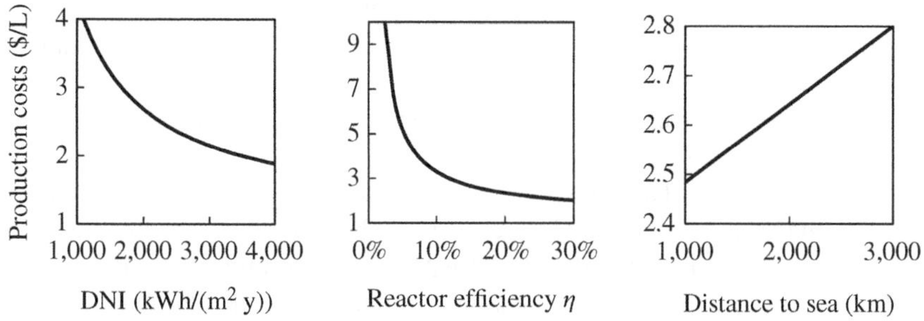

Figure 3.18 Sensitivity of production costs of solar thermochemical jet fuel from baseline case plant with respect to changes in direct solar irradiation (DNI, left), thermochemical reactor efficiency (center), and distance to sea (right). The distance to the sea determines the length of the water supply pipeline from a desalination facility at the sea. Reproduced from (Falter, Scharfenberg, & Habersetzer, 2020)/MDPI/Public domain CC BY 4.0.

increased from 10%, a value that should be reached soon, to 25%, which could be a target value for commercial production in the future, the production costs would drop by 35%. Finally, as the production facility moves away from the sea, the length of the required water pipeline will increase the costs, e.g., by 15% for a distance of 2,000 km. As however, these distances are unlikely to be covered by a pipeline, the strongest influence of the location-dependent variables on production costs is expected to be the DNI.

The economic model described in Section 3.4.2 is used to derive production costs of solar thermochemical jet fuel production from a baseline case plant at different locations, using the country-specific variables discussed in Section 3.7.2. Again, the results are shown here exemplarily for the United States of America and more maps and data can be found in (Falter, Scharfenberg, & Habersetzer, 2020).

In (Falter, Scharfenberg, & Habersetzer, 2020), a map for the local production costs of jet fuel in the United States of America is shown, where the costs range roughly between 1.8 $/L at the best locations in the southwest and exceed 3.0 $/L at the most unfavorable. The best locations are found in desert regions with a high level of direct solar radiation and at a location close to the sea, while those with high production costs are in less sunny regions that are also further away from the sea.

The combination of location-specific production potentials with the respective production costs allows for their expression in cost–supply curves. In Figure 3.19, the cost–supply curve of solar thermochemical jet fuel production in the United States of America is shown for three different thermochemical reactor efficiencies. Also indicated are the national and global jet fuel demand with vertical lines. The cost–supply curves for the selected regions are shown for reactor efficiencies of 0.15, 0.19, and 0.25. In general, a change in efficiency leads to a shift of the cost–supply curves along the *y*-axis, where a higher efficiency reduces the costs due to the smaller heliostat field required. Among the interesting process parameters, reactor efficiency has been chosen to be varied here due to its large influence on costs and its large potential improvement over the current state of the art. In the following, the results are discussed only for the case of a reactor efficiency of 0.19 in the United States of America.

Figure 3.19 Cost–supply curve of solar thermochemical jet fuel production in the United States of America. The national demand and the global demand of jet fuel are indicated with vertical lines. Three different efficiencies of the thermochemical reactor are shown. Reproduced from (Falter, Scharfenberg, & Habersetzer, 2020)/MDPI/Public domain CC BY 4.0.

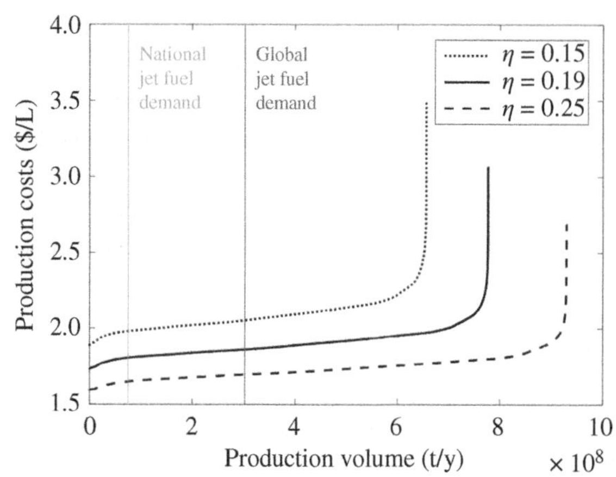

The United States of America has highly irradiated areas in the southwest and favorable costs of capital (at a nominal WACC of 5.7%), enabling low-cost production of solar fuels. At the best locations, production costs of under 2.1 \$/L can be achieved, whereas the national demand for jet fuel could be covered at costs between 2.1 and 2.18 \$/L, and at average costs of 2.15 \$/L. If the global jet fuel demand were to be covered, the costs would be between 2.1 and 2.26 \$/L at average cost of 2.21 \$/L. The progression of the supply curves is quite flat, indicating a large availability of fuels at relatively constant cost.

On a global level, there is an interesting interplay between solar irradiation and costs of capital: while the highest DNI occurs in the Andes mountains in South America, the local costs of capital are expected to be above those for more mature markets in the United States of America or the European Union. Both DNI and WACC have an important influence on the production costs, the former through scaling the investment costs and the latter through higher costs for the provision of capital. Furthermore, the salaries of the workforce may present a significant contribution to the operational cost of the plant. The ideal location would therefore have high solar irradiation and low levels of both WACC and salaries, while in reality it is rather unlikely that these variables reach their respective optimums at the same location. In Chile, the DNI values reach their global maximum at up to 3,500 kWh $m^{-2} \cdot y^{-1}$ but the WACC is estimated to be 7.1%, which is somewhat higher than in the best locations such as France (3.5%), Spain (4.9%), the United States of America (5.7%) or Israel (5.3%). In Egypt, for example, DNI values are excellent, as well, but the WACC reaches values of 23%, leading to high production costs. The monthly salary of a worker, normalized to the one in the United States of America, varies from 6% in Egypt, about 40% in Israel and Spain, to 110% in Australia. Salaries are found to be comparably low in less-developed economies, which may also have a high DNI but are also likely to have a higher WACC at the same time. On a global level, the lowest production costs are found in Israel and Spain at close to 1.9 \$/L. Chile and the United States of America also have very low minimal production costs (<2.0 \$/L and <2.2 \$/L, respectively), while Australia, the country with the largest production potential, has slightly higher costs of capital (WACC: 6.2%) and the highest salaries of the analyzed countries, which leads to minimal production costs of <2.3 \$/L.

3.8 Conclusions

Solar thermochemical fuels are an attractive option to decarbonize the transportation sector, especially for aviation, shipping, and heavy-duty road transport. These subsectors cannot easily be electrified due to the comparably low energy density of batteries and will thus also depend on liquid fuels in the future. At currently >5% thermochemical energy conversion efficiency, the reactor technology has to be significantly improved to make this pathway viable. However, there exist promising developments in implementing heat recuperation from the redox cycle of the reactive material, which could improve the performance greatly. Furthermore, research is being conducted to reduce the required reduction temperature and increase the oxygen partial pressure to reach a certain level of oxygen nonstoichiometry of the reactive material. Especially the material class of perovskites was the focus of attention for the next-generation family of materials. At the time of writing this Chapter, ceria was however still the material achieving the highest demonstrated efficiencies.

By using renewable CO_2, e.g., as derived from DAC, the solar thermochemical fuel pathway is able to save 80% of the emissions compared to conventional jet fuel. The remaining emissions are embedded in the infrastructure of the plant and will decrease even further if the manufacturing of steel, cement, and glass is decarbonized. The water footprint of solar thermochemical jet fuel production is 7.4 L/L on-site in the production plant. Additionally, if ceria is used as the reactive material, 42.4 L/L is consumed off-site mainly for the mining and purification of ceria and the manufacturing of the heliostats for the concentration of sunlight. This is on the same order of magnitude as the water footprint of PtL fuels and significantly lower than that of many biofuels, which sometimes consume several thousand liters per liter of fuel.

The land demand is found to be on the order of 33,300 L of jet fuel-eq./(ha y) in a location with an average direct solar irradiation of 2,500 kWh $m^{-2} \cdot y^{-1}$. At this specific productivity, and excluding areas based on sustainability criteria (e.g., prohibited use of forests, water bodies, protected areas, and agricultural land), many countries in the sunny regions of the Earth have the potential to cover their national jet fuel demand. Some of them (such as Australia, United States of America, and South Africa) could even produce the world demand for the available locations.

In terms of production cost, a mature plant with an energy conversion efficiency of about 19% could achieve 2.4 \$/L. This requires a significant enhancement of the reactor technology over the current state of the art. Promising concepts have been suggested that incorporate heat recuperation from the thermochemical cycle and may achieve this target in the future. For even more optimistic assumptions on the energy conversion efficiency and capital cost of the plant components, production costs of 1.4 $/L could be reached. While still more than double the current cost of conventional jet fuel, this value would place the pathway among the most economical alternative pathways.

References

Abu Hamed, T., Venstrom, L., Alshare, A., Brülhart, M., & Davidson, J. H. (2009). Study of a quench device for the synthesis and hydrolysis of Zn nanoparticles: Modeling and experiments. *Journal of Solar Energy Engineering*, *131*. https://doi.org/10.1115/1.3142825

Ambrosetti, G., & Good, P. (2019). A novel approach to high temperature solar receivers with an absorbing gas as heat transfer fluid and reduced radiative losses. *Solar Energy*, *183*, 521–531. https://doi.org/10.1016/j.solener.2019.03.004

Apt, J., Newcomer, A., Lave, L.B., Douglas, S., & Dunn, L.M.. (2008). *An engineering-economic analysis of syngas storage* (DOE/NETL-2008/1331). US Dep Energy's Natl Energy Technol Lab: 143.

Bann, S. J., Malina, R., Staples, M. D., Suresh, P., Pearlson, M., Tyner, W. E., Hileman, J. I., & Barrett, S. (2017). The costs of production of alternative jet fuel: A harmonized stochastic assessment. *Bioresource Technology*, *227*, 179–187. https://doi.org/10.1016/j.biortech.2016.12.032

Bauhaus Luftfahrt e.V. (2011-2015). Solar chemical reactor demonstration and optimization for long-term availability of renewable JET fuel. FP 7-project SOLAR-JET: 2011–2015. Grant agreement number 285098. www.solar-jet.aero

Bauhaus Luftfahrt e.V. (2016–2019). SUNlight-to-LIQUID: Integrated solar-thermochemical synthesis of liquid hydrocarbon fuels. Horizon 2020-project SUN-to-LIQUID: 2016–2019. Grant agreement number 654408. www.sun-to-liquid.eu

Beiermann, D. (2007). Development of an upgrading model and application to XtL processes: Integration of the upgrading unit into the BtL process chain. Opportunities and Challenges at the Interface between Petrochemistry and Refinery: DGMK/SCI-Conference, (pp. 199–206). Hamburg, Germany: DGMK-Tagungsbericht.

Blankenship, R. E., Tiede, D. M., Barber, J., Brudvig, G. W., Fleming, G., Ghirardi, M., Gunner, M. R., Junge, W., Kramer, D. M., Melis, A., Moore, T. A., Christopher, C. M., Daniel, G. N., Arthur, J. N., Donald, R. O., William, W. P., Roger, C. P., & And Richard, T. S. (2011). Comparing photosynthetic and photovoltaic efficiencies and recognizing the potential for improvement. *Science*, *332*, 805–809. https://doi.org/10.1126/science.1200165

Brendelberger, S., Roeb, M., Lange, M., & Sattler, C. (2015). Counter flow sweep gas demand for the ceria redox cycle. *Sol Energy*, *122*, 1011–1022. https://doi.org/10.1016/j.solener.2015.10.036

Brendelberger, S., von Storch, H., Bulfin, B., & Sattler, C. (2017). Vacuum pumping options for application in solar thermochemical redox cycles—Assessment of mechanical-, jet- and thermochemical pumping systems. *Solar Energy*, *141*, 91–102. https://doi.org/10.1016/j.solener.2016.11.023

Brynolf, S., Taljegard, M., Grahn, M., & Hansson, J. (2018). Electrofuels for the transport sector: A review of production costs. *Renewable and Sustainable Energy Reviews*, *81*, 1887–1905. https://doi.org/10.1016/j.rser.2017.05.288

Chueh, W. C., Falter, C., Abbott, M., Scipio, D., Furler, P., Haile, S. M., & Steinfeld, A. (2010). High-flux solar-driven thermochemical dissociation of CO_2 and H_2O using nonstoichiometric ceria. *Science (80-)*, *330*, 1797–1801. https://doi.org/10.1126/science.1197834

Chueh, W. C., Falter, C., Abbott, M., Scipio, D., Furler, P., Haile, S. M., & Steinfeld, A. (2010). High-flux solar-driven thermochemical dissociation of CO_2 and H_2O using nonstoichiometric ceria. *Science (80-)*, *330*, 1797–1801. https://doi.org/10.1126/science.1197834

Chueh, W. C., & Haile, S. M. (2010). A thermochemical study of ceria: Exploiting an old material for new modes of energy conversion and CO_2 mitigation. *Philosophical Transactions of the Royal Society A Mathematical, Physical and Engineering Sciences*, *368*, 3269–3294. https://doi.org/10.1098/rsta.2010.0114

Chueh, W. C., & Haile, S. M. (2010). A thermochemical study of ceria: exploiting an old material for new modes of energy conversion and CO_2 mitigation. *Philosophical Transactions of the Royal Society A Mathematical, Physical and Engineering Sciences*, *368*, 3269–3294. https://doi.org/10.1098/rsta.2010.0114

Climeworks. (2015). Capturing CO_2 from air. Manufacturing of Green Fuels from Renew Energy Work DTU Risø, Roskilde, 14 April 2015.

Climeworks. (2020) Fight climate change with direct air capture www.climeworks.ch (accessed January 7, 2020).

Climeworks LLC. (2014). CO_2 air capture demonstration plant. http://www.climeworks.com/co2-capture-plants.html (accessed June 1, 2014).

Coventry, J., & Pye, J. (2013). Heliostat cost reduction—Where to now? *Energy Procedia*, *49*, 60–70. https://doi.org/10.1016/j.egypro.2014.03.007

CPI. (2021). Inflation calculator https://www.in2013dollars.com/us/inflation/2020?amount=1.

Damodoran, A. (2018). Estimated country risk premiums. Stern School of Business, New York University. http://pages.stern.nyu.edu/~adamodar/ (accessed July 3, 2019).

Deng, Y. Y., Koper, M., Haigh, M., & Dornburg, V. (2015). Country-level assessment of long-term global bioenergy potential. *Biomass and Bioenergy*, *74*, 253–267. https://doi.org/10.1016/j.biombioe.2014.12.003

Duarte, F., & Rosa, C. (2015). The equity risk premium: A review of models. Economic Policy Review, pp. 39–57. New York.

Eberle, M.K.. (2010). Sonnenkraft aus der Steckdose—Concentrating Solar Power (CSP)—eine Renaissance. ETH Zürich, Treffpunkt Science City 9. Mai 2010.

Ennaceri, H., Ghennioui, A., Benyoussef, A., Ennaoui, A., & Khaldoun, A. (2018). Direct normal irradiation-based approach for determining potential regions for concentrated solar power installations in Morocco. *International Journal of Ambient Energy*, *39*, 78–86. https://doi.org/10.1080/01430750.2016.1237888

Ermanoski, I., Siegel, N. P., & Stechel, E. B. (2013). A new reactor concept for efficient solar-thermochemical fuel production. *Journal of Solar Energy Engineering*, *135*, 031002. https://doi.org/10.1115/1.4023356

European Academies Science Advisory Council. (2011). *Concentrating solar power: Its potential contribution to a sustainable energy future.* Halle, Germany: EASAC Secretariat.

European Union. (2018). Directive (Eu) 2018/2001 of the European Parliament and of the Council of 11 December 2018 on the promotion of the use of energy from renewable sources. Official Journal of the European Union 2018.

Falter, C., Batteiger, V., & Sizmann, A. (2014). The SOLAR-JET project. In *ILA Berlin Air Show*, Berlin.

Falter, C., Batteiger, V., & Sizmann, A. (2016). Climate impact and economic feasibility of solar thermochemical jet fuel production. *Environmental Science & Technology*, *50*, 470–477. https://doi.org/10.1021/acs.est.5b03515

Falter, C., Batteiger, V., & Sizmann, A. (2016). Climate impact and economic feasibility of solar thermochemical jet fuel production. *Environmental Science & Technology*, *50*, 470–477. https://doi.org/10.1021/acs.est.5b03515

Falter, C., & Pitz-Paal, R. (2017). Water footprint and land requirement of solar thermochemical jet-fuel production. *Environmental Science & Technology*, *51*. https://doi.org/10.1021/acs.est.7b02633

Falter, C., & Pitz-Paal, R. (2018). Energy analysis of solar thermochemical fuel production pathway with a focus on waste heat recuperation and vacuum generation. *Solar Energy*, *176*, 230–240. https://doi.org/10.1016/j.solener.2018.10.042

Falter, C., Scharfenberg, N., & Habersetzer, A. (2020). Geographical potential of solar thermochemical jet fuel production. *Energies*, *13*, 1–32. https://doi.org/10.3390/en13040802

Falter, C., Sizmann, A., & Pitz-Paal, R. (2017). Perspectives of advanced thermal management in solar thermochemical syngas production using a counter-flow solid-solid heat exchanger. *AIP Conference Proceedings*, *1850*. https://doi.org/10.1063/1.4984462

Falter, C., Valente, A., Habersetzer, A., Iribarren, D., & Dufour, J. (2020). An integrated techno-economic, environmental and social assessment of the solar thermochemical fuel pathway. *Sustain Energy & Fuels*, *4*, 3992–4002. https://doi.org/10.1039/d0se00179a

Falter, C. P., & Pitz-Paal, R. (2017). A generic solar-thermochemical reactor model with internal heat diffusion for counter-flow solid heat exchange. *Solar Energy*, *144*, 569–579. https://doi.org/10.1016/j.solener.2017.01.063

Falter, C. P., & Pitz-Paal, R. (2018). Modeling counter-flow particle heat exchangers for two-step solar thermochemical syngas production. *Applied Thermal Engineering*, *132*, 613–623. https://doi.org/10.1016/j.applthermaleng.2017.12.087

Falter, C. P., Sizmann, A., & Pitz-Paal, R. (2015). Modular reactor model for the solar thermochemical production of syngas incorporating counter-flow solid heat exchange. *Solar Energy*, *122*, 1296–1308. https://doi.org/10.1016/j.solener.2015.10.042

Fichtner. (2010) Assessment of technology options for development of concentrating solar power in South Africa for the World Bank. Presentation, Johannesburg, 9–10 December 2010.

Fletcher, E. A. (1999). Solarthermal and Solar Quasi-Electrolytic Processing and Separations: Zinc from Zinc Oxide as an Example. *Ind Eng Chem Res*, *38*, 2275–2282. https://doi.org/10.1021/ie990053t

Fletcher, E. A. (2000). Solarthermal processing: A review. *Journal of Solar Energy Engineering*, *123*, 63–74. https://doi.org/10.1115/1.1349552

Fletcher, E. A., & Moen, R. L. (1977). Hydrogen- and oxygen from water. *Science*, *197*, 1050–1056. https://doi.org/10.1126/science.197.4308.1050

Furler, P., Marxer, D., Scheffe, J., Reinalda, D., Geerlings, H., Falter, C., Batteiger, V., Sizmann, A., & Steinfeld, A. (2017). Solar kerosene from H_2O and CO_2. *AIP Conference Proceedings*, *1850*. https://doi.org/10.1063/1.4984463

Furler, P., Scheffe, J., Gorbar, M., Moes, L., Vogt, U., & Steinfeld, A. (2012). Solar thermochemical CO_2 splitting utilizing a reticulated porous ceria redox system. *Energy & Fuels*, *26*, 7051–7059. https://doi.org/10.1021/ef3013757

Furler, P., Scheffe, J., Marxer, D., Gorbar, M., Bonk, A., Vogt, U., & Steinfeld, A. (2014). Thermochemical CO_2 splitting via redox cycling of ceria reticulated foam structures with dual-scale porosities. *Phys Chem Chem Phys*, *16*, 10503–10511. https://doi.org/10.1039/c4cp01172d

Furler, P., Scheffe, J. R., & Steinfeld, A. (2012). Syngas production by simultaneous splitting of H_2O and CO_2 via ceria redox reactions in a high-temperature solar reactor. *Energy & Environmental Science*, *5*, 6098. https://doi.org/10.1039/c1ee02620h

Geissbühler, L. (2019). *Thermocline thermal energy storage: advances and applications to CSP, compressed air energy storage, and solar fuels* (Doctoral thesis), ETH Zürich.

van der Giesen, C., Kleijn, R., & Kramer, G. J. (2014). Energy and climate impacts of producing synthetic hydrocarbon fuels from CO_2. *Environmental Science & Technology*, *48*, 7111–7121. https://doi.org/10.1021/es500191g

Gómez, D. R., Watterson, J. D., Americanohia, B. B., Ha, C., Marland, G., Matsika, E., Namayanga, N., Osman-Elasha, B., Kalenga Saka, J. D., & Treanton, K. (2006). Chapter 2: Stationary Combustion. In *2006 IPCC Guidelines for National Greenhouse Gas Inventories* (pp. 2.1–2.47). https://doi.org/10.1016/S0166-526X(06)47021-5

Gröngröft A, Meisel K, Hauschild S, Grasemann E, Peetz D, Meier K, Roth, A., Riegel, F., Endres, C. *Abschlussbericht Projekt BurnFAIR*. 2014.

Haering, H.-W.. (Ed.) (2008). The air gases nitrogen, oxygen, and argon. In *Industrial Gases Processing* (pp. 9–109). Wiley-VCH. https://doi.org/10.1002/9783527621248.ch2

Haije, W., & Geerlings, H. (2011). Efficient production of solar fuel using existing large scale production technologies. *Environmental Science & Technology*, *45*, 8609–8610. https://doi.org/10.1021/es203160k

Hall, D. O., & Rao, K. K. (1999). *Photosynthesis* (6th ed.). Cambridge University Press.

Huld, T., Moner-Girona, M., & Kriston, A. (2017). Geospatial analysis of photovoltaic mini-grid system performance. *Energies*, *10*, 218. https://doi.org/10.3390/en10020218

IATA. (2018). Economic performance of the airline industry 2018. Cambridge: Cambridge University Press. https://www.iata.org/contentassets/f88f0ceb28b64b7e9b46de44b917b98f/iata-economic-performance-of-the-industry-end-year-2018-report.pdf.

ILO. (2018). Mean nominal monthly earnings of employees by sex and occupation—Harmonized series. ILOSTAT. https://www.ilo.org/shinyapps/bulkexplorer51/?lang=en&segment=indicator&id=EAR_4MTH_SEX_ECO_CUR_NB_A (accessed October 5, 2019).

Institute for Applied Ecology, Gemis software, http://iinas.org/gemis-de, Darmstadt, Germany, (2008).

Internal Revenue Service. (2022). Yearly average currency exchange rates. https://www.irs.gov/individuals/international-taxpayers/yearly-average-currency-exchange-rates.

International Monetary Fund. (2018). Lending rates, Percent per annum. International Financial Statistics. https://data.imf.org/regular.aspx?key=61545867 (accessed July 3, 2019).

IRENA. (2018). *Renewable power generation costs in 2017*. IRENA.

de Jong, S., Hoefnagels, R., Faaij, A., Slade, R., Mawhood, R., & Junginger, M. (2015). The feasibility of short-term production strategies for renewable jet fuels—a comprehensive techno-economic comparison. *Biofuels, Bioproducts & Biorefining*, *9*, 778–800. https://doi.org/10.1002/bbb.1613

Kim, J., Henao, C. A., Johnson, T. A., Dedrick, D. E., Miller, J. E., Stechel, E. B., & Maravelias, C. T. (2011). Methanol production from CO_2 using solar-thermal energy: process development and techno-economic analysis. *Energy & Environmental Science*, *4*, 3122. https://doi.org/10.1039/c1ee01311d

Kim, J., Miller, J. E., Maravelias, C. T., & Stechel, E. B. (2013). Comparative analysis of environmental impact of S2P (Sunshine to Petrol) system for transportation fuel production. *Applied Energy*, *111*, 1089–1098. https://doi.org/10.1016/j.apenergy.2013.06.035

Koltun, P., & Tharumarajah, A. (2014). Life cycle impact of rare earth elements. *ISRN Metallurgy*, *2014*, 1–10. https://doi.org/10.1155/2014/907536

König, D. H., Freiberg, M., Dietrich, R.-U., & Wörner, A. (2015). Techno-economic study of the storage of fluctuating renewable energy in liquid hydrocarbons. *Fuel*, *159*, 289–297. https://doi.org/10.1016/j.fuel.2015.06.085

Labordena, M., Patt, A., Bazilian, M., Howells, M., & Lilliestam, J. (2017). Impact of political and economical barriers for concentrating solar power in Sub-Saharan Africa. *Energy Policy*, *102*, 52–72. https://doi.org/10.1016/j.enpol.2016.12.008

Lapp, J. (2013). *Thermal modeling and design of a solar non-stoichiometric redox reactor with heat recovery* (Doctoral thesis). University of Minnesota.

Le Gal, A., Abanades, S., & Flamant, G. (2011). CO_2 and H_2O splitting for Thermochemical production of solar fuels using nonstoichiometric ceria and ceria/zirconia solid solutions. *Energy & Fuels, 25*, 4836–4845. https://doi.org/10.1021/ef200972r

Lédé, J., Elorza-Ricart, E., & Ferrer, M. (2001). Solar thermal splitting of zinc oxide: A review of some of the rate controlling factors. *Journal of Solar Energy Engineering, 123*, 91–97. https://doi.org/10.1115/1.1351815

Lee, M.S. (2008). *Theoretical studies of UT-3 thermochemical hydrogen production cycle and developmment of calcium oxide reactant for UT-3 cycle and carbon dioxide capture* (Doctoral thesis). Available from ProQuest Dissertations and Theses database. (UMI No.3381457).

Long, H., Li, X., Wang, H., & Jia, J. (2013). Biomass resources and their bioenergy potential estimation: A review. *Renewable and Sustainable Energy Reviews, 26*, 344–352. https://doi.org/10.1016/j.rser.2013.05.035

Loutzenhiser, P. G., Meier, A., & Steinfeld, A. (2010). Review of the two-step H_2O/CO_2-splitting solar thermochemical cycle based on Zn/ZnO redox reactions. *Materials (Basel), 3*, 4922–4938. https://doi.org/10.3390/ma3114922

Lumen Learning. (2022). Abundance of Elements in Earth's Crust. *Geology*. https://courses.lumenlearning.com/geology/chapter/reading-abundance-of-elements-in-earths-crust/

Mancini, T.R., Gary, J.A., Kolb, G.J., Ho, C.K. (2011). *Power tower technology roadmap and cost reduction plan* (Report No. SAND2011-2419). Albuquerque, NM, and Livermore, CA (United States): Sandia National Laboratories. https://doi.org/10.2172/1011644.

Marxer, D. (2016). *4 kW solar reactor technology for splitting of H_2O and CO_2 with a temperature–pressure swing redox cycle* (Doctoral thesis). ETH.

Marxer, D., Furler, P., Scheffe, J., Geerlings, H., Falter, C., Batteiger, V., Sizmann, A., & Steinfeld, A. (2015). Demonstration of the entire production chain to renewable kerosene via solar thermochemical splitting of H_2O and CO_2. *Energy & Fuels, 29*, 3241–3250. https://doi.org/10.1021/acs.energyfuels.5b00351

Marxer, D., Furler, P., Takacs, M., & Steinfeld, A. (2017). Solar thermochemical splitting of CO_2 into separate streams of CO and O_2 with high selectivity, stability, conversion, and efficiency. *Energy & Environmental Science, 10*, 1142–1149. 10.1039/C6EE03776C

Marxer, D., Furler, P., Takacs, M., & Steinfeld, A. (2017). Solar thermochemical splitting of CO_2 into separate streams of CO and O_2 with high selectivity, stability, conversion, and efficiency. *Energy & Environmental Science, 10*, 1142–1149. 10.1039/C6EE03776C

Marxer, D., Furler, P., Takacs, M., & Steinfeld, A. (2017). Solar thermochemical splitting of CO_2 into separate streams of CO and O_2 with high selectivity, stability, conversion, and efficiency. *Energy & Environmental Science, 10*, 1142–1149. 10.1039/C6EE03776C

McDonald, R. I., Fargione, J., Kiesecker, J., Miller, W. M., & Powell, J. (2009). Energy sprawl or energy efficiency: climate policy impacts on natural habitat for the United States of America. *PLoS One, 4*, e6802. https://doi.org/10.1371/journal.pone.0006802

Mehos, M., Turchi, C., Jorgenson, J., Denholm, P., Ho, C., Armijo, K. (2016). On the path of SunShot: Advancing Concentrating Solar Power Technology, Performance, and Dispatchability (NREL/TP-5500-65688). Golden, CO: National Renewable Energy Laboratory. 1–66. doi.org/10.2172/1344199.

Melchior, T., Perkins, C., Lichty, P., Weimer, A. W., & Steinfeld, A. (2009). Solar-driven biochar gasification in a particle-flow reactor. *Chemical Engineering and Processing: Process Intensification, 48*, 1279–1287. https://doi.org/10.1016/j.cep.2009.05.006

Miller, J. E., McDaniel, A. H., & Allendorf, M. D. (2014). Considerations in the design of materials for solar-driven fuel production using metal-oxide thermochemical cycles. *Adv Energy Mater, 4*, 1300469. https://doi.org/10.1002/aenm.201300469

Palumbo, R., Léde, J., Boutin, O., Elorza Ricart, E., Steinfeld, A., Möller, S., Weidenkaff, A., Fletcher, E. A., & Bielicki, J. (1998). The production of Zn from ZnO in a high-temperature solar decomposition quench process—I. The scientific framework for the process. *Chemical Engineering Science, 53*, 2503–2517. https://doi.org/10.1016/S0009-2509(98)00063-3

Perea-Moreno, M.-A., Samerón-Manzano, E., & Perea-Moreno, A.-J. (2019). Biomass as renewable energy: worldwide research trends. *Sustainability, 11*, 863. https://doi.org/10.3390/su11030863

Pitz-Paal, R., Botero, N. B., & Steinfeld, A. (2011). Heliostat field layout optimization for high-temperature solar thermochemical processing. *Solar Energy, 85*, 334–343. https://doi.org/10.1016/j.solener.2010.11.018

QGIS Development Team. (2017). QGIS Geographic Information System. Open Source Geospatial Foundation Project. http://qgis.osgeo.org (accessed February 7, 2019).

Raluy, G., Serra, L., & UChe, J. (2006). Life cycle assessment of MSF, MED and RO desalination technologies. *Energy, 31*, 2361–2372. https://doi.org/10.1016/j.energy.2006.02.005

Riegel, F. (2014). Das nachhaltige Potenzial von Flüssigkraftstoffen aus Biomasse: Eine globale Abschätzung auf der Basis von hochaufgelösten Geodaten. Dissertation, submitted, Ludwig-Maximilians-Universität München.

Romero, M., & Steinfeld, A. (2012). Concentrating solar thermal power and thermochemical fuels. *Energy & Environmental Science, 5*, 9234. https://doi.org/10.1039/c2ee21275g

Sakurai, M., Bilgen, E., Tsutsumi, A., & Yoshida, K. (1996). Solar UT-3 thermochemical cycle for hydrogen production. *Solar Energy, 57*, 51–58. https://doi.org/10.1016/0038-092X(96)00034-5

Sargent & Lundy LLC Consulting Group. (2003). Assessment of Parabolic Trough and Power Tower Solar Technology Cost and Performance Forecasts; SL-5641. Chicago, IL, USA.

Scheffe, J. R., & Steinfeld, A. (2012). Thermodynamic analysis of cerium-based oxides for solar thermochemical fuel production. *Energy and Fuels, 26*, 1928–1936. https://doi.org/10.1021/ef201875v

Schmidt, P., Weindorf, W., Roth, A., Batteiger, V., & Riegel, F. (2016). *Power-to-liquids potentials and perspectives for the future supply of renewable aviation fuel.* German Environment Agency.

Searle, S., & Malins, C. (2015). A reassessment of global bioenergy potential in 2050. *GCB Bioenergy, 7*, 328–336. https://doi.org/10.1111/gcbb.12141

Short, W., Packey, D.J., & Holt, T. (1995). *A manual for the economic evaluation of energy efficiency and renewable energy technologies* (No. NREL/TP-462-5173). National Renewable Energy Lab (NREL), Golden, CO (United States).

Sibieude, F., Ducarroir, M., Tofighi, A., & Ambriz, J. (1982). High temperature experiments with a solar furnace: The decomposition of Fe_3O_4, Mn_3O_4, CdO. *International Journal of Hydrogen Energy, 7*, 79–88. https://doi.org/10.1016/0360-3199(82)90209-9

Siegel, N. P., Miller, J. E., Ermanoski, I., Diver, R. B., & Stechel, E. B. (2013). Factors affecting the efficiency of solar driven metal oxide thermochemical cycles. *Industrial & Engineering Chemistry Research, 52*, 3276–3286. https://doi.org/10.1021/ie400193q

Statista. (2019). Cerium oxide price worldwide from 2009 to 2020 with a forecast from 2021 to 2030. https://www.statista.com/statistics/450146/global-reo-cerium-oxide-price-forecast/

Stechel, E. B., & Miller, J. E. (2013). Re-energizing CO_2 to fuels with the sun: Issues of efficiency, scale, and economics. *Journal of CO_2 Utilization, 1*, 28–36. https://doi.org/10.1016/j.jcou.2013.03.008

Steinfeld, A. (1998). Solar-processed metals as clean energy carriers and water-splitters. *International Journal of Hydrogen Energy, 23*, 767–774. https://doi.org/10.1016/S0360-3199(97)00135-3

Steinfeld, A. (2002). Solar hydrogen production via a two-step water-splitting thermochemical cycle based on Zn/ZnO redox reactions. *International Journal of Hydrogen Energy, 27*, 611–619. https://doi.org/10.1016/S0360-3199(01)00177-X

Steinfeld, A. (2014). Solar thermochemical production of hydrogen. In *Handbook of Hydrogen Energy* (pp. 421–443). CRC Press.

Steinfeld, A., & Palumbo, R. (2003). Solar Thermochemical Process Technology. In *Encyclopedia of Physical Science and Technology*, Elsevier Reference Collection in Earth Systems and Environmental Sciences (pp. 237–256). Elsevier. https://doi.org/10.1016/B0-12-227410-5/00698-0

Steinfeld, A., Sanders, S., & Palumbo, R. (1999). Design aspects of solar thermochemical engineering—a case study: Two-step water-splitting cycle using the Fe_3O_4/FeO redox system. *Solar Energy, 65*, 43–53. https://doi.org/10.1016/S0038-092X(98)00092-9

Stratton, R.W., Wong, H.M., & Hileman, J.I. (2010). *Life cycle gas emissions from alternative jet fuels* (Report No. Partner-Coe-2010-001); PARTNER Project 28 report.

SunShot Initiative: U.S. Department of Energy, SunShot Vision Study. U.S. Department of Energy. (2012). *Energy efficiency and renewable energy* (NREL Report No. BK5200-47927; DOE/GO-102012-3037), U.S. Department of Energy.

Suresh, P., Malina, R., Staples, M. D., Lizin, S., Olcay, H., Blazy, D., Pearlson, M. N., & Barrett, S. R. H. (2018). Life cycle greenhouse gas emissions and costs of production of diesel and jet fuel from municipal solid waste. *Environmental Science & Technology, 52*, 12055–12065. https://doi.org/10.1021/acs.est.7b04277

Suter, S., Steinfeld, A., & Haussener, S. (2014). Pore-level engineering of macroporous media for increased performance of solar-driven thermochemical fuel processing. *International Journal of Heat and Mass Transfer, 78*, 688–698. 10.1016/j.ijheatmasstransfer.2014.07.020

Teo, E., Brandon, N., Vos, E., & Kramer, G. (2005). A critical pathway energy efficiency analysis of the thermochemical UT-3 cycle. *International Journal of Hydrogen Energy, 30*, 559–564. https://doi.org/10.1016/j.ijhydene.2004.08.003

Thinkstep AG. (2012). GaBi Software-System and Database for Life Cycle Engineering.

Trading Economics. (2018). Government Bond yields. Trading Economics. https://tradingeconomics.com/bonds (accessed August 2, 2019).

Trieb, F., Schillings, C., Kronshage, S., Klann, U., Viebahn, P., May, N., Wilde, R., & Paul, C. (2005). Concentrating solar power for the Mediterranean region. *MED-CSP Final Rep.* https://doi.org/10.1155/1995/27197

Trieb, F., Schillings, C., O'Sullivan, M., Pregger, T., & Hoyer-Klick, C. (2009). Global potential of concentrating solar power. In the 15th *SolarPACES Conference* (15–18 September 2009), Berlin, Germany.

U.S. DOE. (2017). The SunShot 2030 Goals. https://www.energy.gov/sites/prod/files/2020/09/f79/SunShot%202030%20White %20Paper.pdf.

United Nations. (2013). Water Security & the Global Water Agenda. Hamilton, Ontario, Canada: The United Nations University.

United Nations. (2013). UN water scarcity fact sheet.

Vitart, X., Carles, P., & Anzieu, P. (2008). A general survey of the potential and the main issues associated with the sulfur–iodine thermochemical cycle for hydrogen production using nuclear heat. *Progress in Nuclear Energy*, *50*, 402–410. https://doi.org/10.1016/j.pnucene.2007.11.023

Wegner, K., Ly, H., Weiss, R., Pratsinis, S., & Steinfeld, A. (2006). In situ formation and hydrolysis of Zn nanoparticles for H_2 production by the 2-step ZnO/Zn water-splitting thermochemical cycle. *International Journal of Hydrogen Energy*, *31*, 55–61. https://doi.org/10.1016/j.ijhydene.2005.03.006

Werder, M., & Steinfeld, A. (2000). Life cycle assessment of the conventional and solar thermal production of zinc and synthesis gas. *Energy*, *25*, 395–409. https://doi.org/10.1016/S0360-5442(99)00083-3

Wernet, G., Bauer, C., Steubing, B., Reinhard, J., Moreno-Ruiz, E., & Weidema, B. (2016). The ecoinvent database version 3 (part I): overview and methodology. *The International Journal of Life Cycle Assessment*, *2*, 1218–1230.

Whitaker, M. B., Heath, G. A., Burkhardt, J. J., & Turchi, C. S. (2013). Life cycle assessment of a power tower concentrating solar plant and the impacts of key design alternatives. *Environmental Science & Technology*, *47*, 5896–5903. https://doi.org/10.1021/es400821x

White, I. R., Royston, P., & Wood, A. M. (2011). Multiple imputation using chained equations: Issues and guidance for practice. *Statistics in Medicine*, *30*, 377–399. https://doi.org/10.1002/sim.4067

World Bank. (2019). This report was prepared by Solargis, under contract to the World Bank. *Global Solar Atlas 2016* (Project ID: P161130). https://globalsolaratlas.info/ (accessed January 15, 2019).

World Government Bonds. (2018). 10 Year Bond Yields. http://www.worldgovernmentbonds.com/ (accessed October 8, 2019).

Wurzbacher, J. A., Gebald, C., Piatkowski, N., & Steinfeld, A. (2012). Concurrent separation of CO_2 and H_2O from air by a temperature-vacuum swing adsorption/desorption cycle. *Environmental Science & Technology*, *46*, 9191–9198. https://doi.org/10.1021/es301953k

4

Principles, Operations, and Techno-Economics of Photovoltaic-Electrolysis and Photoelectrochemical Water Splitting Processes

Nicolas Gaillard

Hawaii Natural Energy Institute, University of Hawaii at Manoa, Honolulu, HI, USA

4.1 Introduction

With new fuel cell vehicle models introduced every year to the market, hydrogen has emerged as a feasible energy-carrier alternative to fossil fuels in the transportation sector to help decarbonize our society. However, a *green* mode of transportation is as *green* as the means of producing its fuel, and with 96% of hydrogen currently produced from nonrenewable steam–methane reforming (SMR), the road to truly *green* hydrogen-powered vehicles will be long. Nevertheless, hydrogen from SMR is a necessary short-term enabler to support the deployment of a hydrogen-based transportation infrastructure, and with time, more *green* hydrogen will be added into the mix.[1] It is worth noting that the need for hydrogen in transportation represents only a small fraction of the world's current demand, as it is also a crucial chemical commodity used in many industrial sectors. For instance, it serves as a fundamental building block in the production of ammonia, 90% of which eventually goes into fertilizer. Hydrogen also plays a vital role in the refinement of crude oil into gasoline and diesel, a process consuming alone nearly 75% of world's hydrogen supply. As we see, the hydrogen economy is real, and it is in total more than 70,000,000,000 kg which are produced worldwide annually primarily from SMR, accounting for approx. 830,000,000 tons of CO_2 emission per year. Therefore, replacing SMR with *green* hydrogen could have a significant impact on greenhouse gas emissions, even if the application in which hydrogen is used is not itself renewable.

With over 180,000 TW of power irradiating Earth continuously, the Sun is undeniably the go-to renewable energy source for future production of *green* hydrogen. Several light-driven thermodynamic pathways exist, some requiring more conversion steps than others. Before focusing our attention on *artificial* means of producing hydrogen with solar power, it is worth reviewing what nature has come up with. In the well-known photosynthesis process (Figure 4.1a), plants are capable of storing sunlight energy as carbon–carbon bonds via a complex chain of chemical ($NADP^+$ to NADPH and ADP to ATP) and electronic exchanges between several specialized light-harvesting proteins (PSII and PSI). As a result, water and CO_2 feedstocks are converted into O_2 and carbohydrates (typically glucose, $C_6H_{12}O_6$). Interestingly, some algae and cyanobacteria actually produce hydrogen for a brief period as a means to get rid of excess energy right before the photosynthetic carbon fixation process. Finding methods to delay the latter and extend the production of hydrogen in these microorganisms is of great interest to the scientific community (Ghirardi & Mohanty, 2010). So far the solar-to-hydrogen (STH) efficiency achieved from photobiological processes is rather modest (<1%), yet recent research efforts have demonstrated progress toward improving microorganisms' light harvesting efficiency and their activity with respect to hydrogen production.

We now turn our attention to solar-thermochemical cycles (STCs, Figure 4.1b), our first example of *artificial* paths for *green* hydrogen production. In this approach, sunlight is concentrated onto a multi-chamber reactor filled with specific chemical compounds (solid or liquid). Using metal oxide (MO_x) particles as an example, a two-step direct STC process can be described as follows. In the high-temperature chamber (~2,000 K), heat reduces the metal oxide particles to a lower oxidation state (MO_{x-d}), releasing oxygen gas. The reduced particles are then conveyed to a low-temperature chamber (~1,000 K) and exposed to steam. Follows a re-oxidation of the particles by water, liberating hydrogen. Water splitting via STC has been demonstrated with over 300 chemical compounds so far, including high-temperature Ce-, Zn-, Fe- and La-based oxides, as well as low-temperature $CuCl_2$ and hybrid sulfur (HyS) (Gorensek et al., 2017). Some of the key challenges in the

1 Some states including California have already passed laws to include at least 30% of *green* hydrogen in their refueling stations.

Conversion of Water and CO_2 to Fuels using Solar Energy: Science, Technology and Materials, First Edition.
Edited by Oomman K. Varghese and Flavio L. Souza.

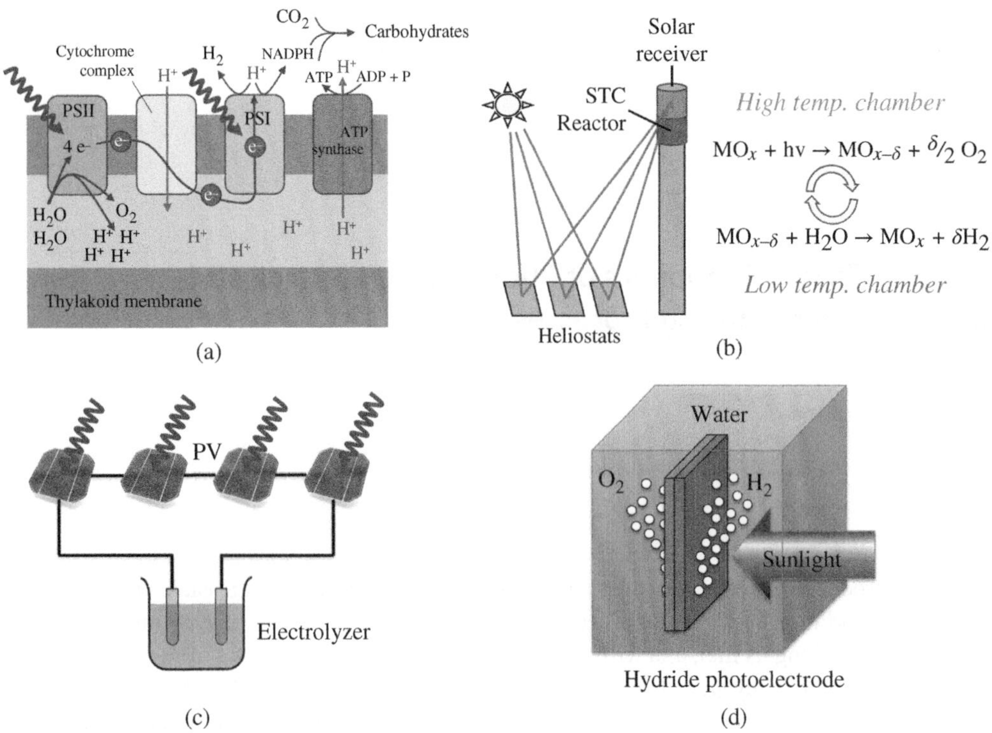

Figure 4.1 Examples of natural and artificial paths to *green* hydrogen production using (a) photobiological, (b) solar thermal cycle, (c) PV-electrolysis, and (d) photoelectrochemical processes.

deployment of STC *green* H_2 production include discovering materials which can undergo multiple redox cycles with minimal degradation, designing robust reactors capable of handling extreme temperatures and heat cycling, as well as reducing the cost of solar concentrators.

The aforementioned STC process relies on a *solar–thermal* conversion route, with some "hybrid" conversion cycles requiring up to four thermal and electrochemical steps (e.g., $CuCl_2$). In contrast, *solar-potential* routes such as photovoltaic-electrolysis (PV-E, Figure 4.1c) or photoelectrochemical (PEC, Figure 4.1d) water splitting offer more direct paths for the storage of photon energy into hydrogen–hydrogen bonds. Many semiconductor-based PEC technologies exist, but in this chapter, we will focus exclusively on bulk and thin film-based water splitting devices. In PV-E, a solar-to-electric step is followed by an electric-to-chemical process, both taking place in two discrete and distant apparatuses. The same two processes take place in PEC systems; however, the length scale (few microns) is considerably reduced compared to PV-E, and as such, PEC water splitting is often referred to as a single-step conversion.

As we will see in this chapter, both PV-E and PEC water splitting processes require essentially the same fundamental components, which are, semiconducting solar absorbers for light capture and charge carrier generation, diode-like junctions for charge separation, and thin film or nanoparticulate catalysts for hydrogen evolution reaction (HER) and oxygen evolution reaction (OER), respectively. For the untrained eye, the two technologies could appear to be rather identical, but are they? What are the similarities and differences between PV-E and PEC water splitting? Is one fundamentally more efficient than the other? Are they equivalent in terms of operation and *green* hydrogen production cost? Will one supplant the other in the long-term, or will both technologies be complementary?

In this chapter, we review the fundamentals, working principles, and techno-economics of PV-E and PEC processes, providing the reader with the necessary information to better understand and assess the potential and pitfalls of each technology. We first discuss the basics of the STH conversion process, define essential metrics, and provide key material and device considerations researchers should have in mind when investigating new semiconductor-based water splitting concepts. Then, we present the fundamentals of photovoltaics and water electrolysis and the engineering solutions commonly used to integrate the two into efficient PV-E systems. The concept of PEC water splitting is then introduced and additional information regarding the solid-electrolyte interface is provided. Examples of standalone PEC water splitting devices developed by the University of Hawaii are introduced. Finally, techno-economic analyses (TEA) of PV-E and PEC concepts are presented and discussed.

4.2 The Solar-to-Hydrogen Conversion Process

4.2.1 Fundamental Concepts

Regardless of the absorbing material class or the device architecture used, solar-assisted water splitting with semiconductor-based materials requires four fundamental steps: charge carrier generation, separation, transport, and transfer at the solid/electrolyte interface (Figure 4.2). The overall external quantum efficiency (EQE) of the water splitting process is given by (Jacobsson et al., 2014):

$$\text{EQE}(\lambda) = \phi_{\text{gen.}}(\lambda) \times \phi_{\text{sep.}}(\lambda) \times \phi_{\text{trans.}} \times \phi_{\text{cat.}} \tag{4.1}$$

where $\phi_{\text{gen.}}$ is the quantum efficiency of charge generation as a function of photon wavelength λ, $\phi_{\text{sep.}}$ is the yield of charge separation (also wavelength-dependent), $\phi_{\text{trans.}}$ is the yield of charge transport, and $\phi_{\text{cat.}}$ is the yield of catalytic charge transfer to the specific redox species. The first three terms describe solid-state processes taking place in the semiconducting material that are required to generate and shuttle carriers to the solid/liquid interface, while the last one represents the actual electrochemical process taking place at the catalytic sites.

Equation 4.1 provides valuable information with respect to material selection and integration, which must be taken into consideration when designing water splitting devices. Clearly, reaching high EQE requires semiconductors with excellent photoconversion efficiency coupled with efficient HER and OER catalysts. Also, to minimize electric losses and maximize the yield of charge transport, one may favor *all-in-one* PEC devices over PV-E systems, since the length scale in the former (few microns) is considerably shorter than in the latter (few meters for centralized plants to several kilometers for distributed solutions).

Figure 4.2b represents the energetic diagram of the solar-powered water splitting process with a single absorber. Briefly, an electron is promoted from the valence band maximum (VBM) to the conduction band minimum (CBM) upon absorption of a photon with energy ($h\nu$) greater than the semi-conductor bandgap (E_G), leaving a hole in the valence band. Under the action of a solid–solid (for PV) or solid–liquid (for PEC) diode, both electron and hole are separated and directed toward catalytic reaction sites where the HER and the OER take place. Although the reversible potential ($V_{\text{REV.}}$) for water splitting is 1.23 V, in practice additional energy is required to overcome losses occurring during carriers separation ($\eta_{\text{sep.}}$) and transport ($\eta_{\text{trans.}}$) in the solar absorber as well as the overpotentials to drive the HER (η_{HER}) and the OER (η_{OER}) at the catalytic sites. The relationship between the single absorber bandgap and the aforementioned losses can be expressed as:

$$E_G > 1.23 + \eta_{\text{sep.}} + \eta_{\text{trans.}} + \eta_{\text{HER}} + \eta_{\text{OER}} \tag{4.2}$$

Equation 4.2 outlines the thermodynamic requirements for water splitting to occur, yet it does not provide any information about the rate at which hydrogen is being produced. This metric, arguably the most important for ranking

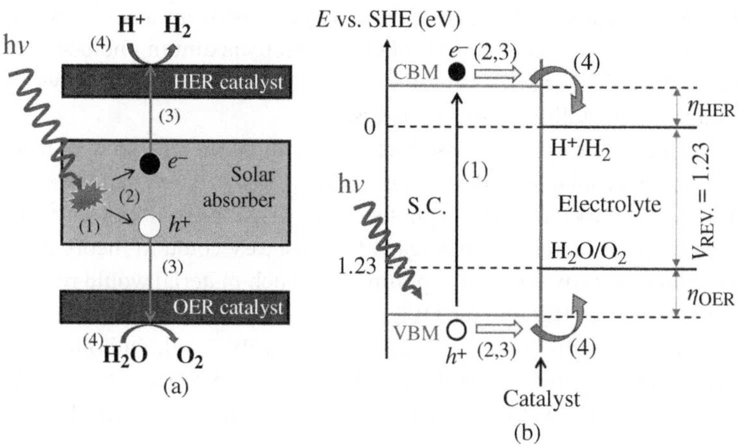

Figure 4.2 (a) Basic components and (b) energetic diagram of semiconductor-based solar-driven water splitting process with a single junction absorber. The four essential steps are photo-carrier (1) generation, (2) separation, (3) transport, and (4) transfer into the redox species at the catalytic sites.

solar-powered water splitting technologies, is captured in the STH efficiency. It is described as the ratio of the useable chemical energy in the generated hydrogen gas to the total solar energy provided to the system (Chen et al., 2010), as given by the ratio of the power output to the power input provided to the system:

$$\frac{P_{out}}{P_{in}} = \frac{(\text{hydrogen production rate}) \times (\text{hydrogen energy density})}{\text{solar flux integrated over the illumination area}} \tag{4.3}$$

For bulk or thin-film PV-E and PEC water splitting devices, the hydrogen production rate is tied directly to the steady-state short-circuit photocurrent (I_{SC}) flowing between the (photo-) anode and (photo-) cathode, generating oxygen and hydrogen, respectively. At the (photo-) cathode surface, 2 protons will be reduced by 2 electrons to form one molecule of hydrogen. Thus, the hydrogen production rate is:

$$R(H_2) = \frac{I_{SC} \times \eta_F}{2e} = \frac{(J_{SC} \times A) \times \eta_F}{2e} \tag{4.4}$$

where $R(H_2)$ is the hydrogen production rate (s^{-1}), I_{SC} is short-circuit photocurrent (A), e is the charge of an electron (C), A is the area of the illuminated photoelectrode (m^2), J_{SC} is short-circuit photocurrent density (A/m^2), and η_F the Faradaic efficiency. Using the expression of hydrogen production rate from Equation 4.4 and the standard Gibbs energy change ($\Delta G°$) as the useful energy density of hydrogen, the STH efficiency for a PV-E or PEC water splitting system can be defined as:

$$\text{STH}(\%) = \frac{\Delta G \times R(H_2)}{P_{solar} \times A} = \frac{\Delta G° \times \left(\frac{J_{SC} \times A \times \eta_F}{2e}\right)}{P_{solar} \times A} \tag{4.5}$$

where P_{solar} is the solar flux of energy in W/m^2. Under reference illumination as defined by air mass global condition ($AM1.5_G$), where the Sun is approximately 41° above the horizon (total irradiance: $100\,mW/cm^2$), Equation 4.5 can be simplified to:

$$\text{STH}(\%) = \frac{\left| J_{SC}\left(\frac{mA}{cm^2}\right)\right| \times 1.23(V) \times \eta_F}{100\left(\frac{mW}{cm^2}\right)} \tag{4.6}$$

4.2.2 Material and Device Considerations

Equation 4.6 emphasizes that the STH efficiency of PV-E and PEC water splitting devices depends primarily on two main variables: the Faradaic efficiency, which is the fraction of photogenerated electrical charges participating in the HER, and the short circuit photocurrent density produced by the device, which describes the ability of a semiconducting material to effectively absorb photons, use their energy to generate electron–hole pairs, and transport these charge carriers across the layer. Assuming a Faradaic efficiency of unity and the optical limit (i.e., 100% quantum efficiency for photons with energy greater than the semiconductor bandgap), it is then possible to establish the STH efficiency *upper limit* of any absorber based on its bandgap using Equation 4.6, as presented in Figure 4.3. Also included in this figure are the maximum photocurrent densities reported for various material classes integrated in PV (J_{SC}) or PEC (saturation photocurrent density: $J_{SAT.}$) devices, including metal oxides, chalcopyrites, perovskites, and monocrystalline semiconductors.

The minimum bandgap required to drive the water splitting process and associated STH can be then estimated using Equations 4.2 and 4.6. Considering typical free energy losses in solid-state devices ($\eta_{sep.} + \eta_{trans.} \sim 0.4$–$0.6$ eV) and overpotentials for the HER (Pt, ~0.05 eV) and the OER (RuO_2, ~0.35 eV) at relevant photocurrent densities (>10 mA/cm^2) for high STH efficiency (>10%) (Seitz et al., 2014), a single semiconductor with bandgap of at least 2 eV could in theory drive the water splitting process. It is worth noting that achieving spontaneous water splitting with such material would require ideal band edge position with both CBM and VBM straddling the HER and the OER potentials, respectively (more on this latter in this chapter). In practice, such situation has not been achieved with a single junction PV-E yet, and only PEC materials with bandgaps significantly larger than 2.0 eV, such as 3.4 eV GaN (Kibria et al., 2015), 3.7 eV $SrTiO_3$ (Goto et al., 2018), or 3.2 eV TiO_2 (Fujishima & Honda, 1972), have been demonstrated to spontaneously split water but with limited STH efficiencies.

Connecting two or more cells in series is a simple and practical approach to overcome the fundamental conversion limits of single junction devices. In its most simple form, such device comprises cells of identical bandgap connected in a *coplanar* fashion: the device open circuit voltage (V_{OC}) is equal to the sum of each individual cell's V_{OC}, whereas its J_{SC} is limited by the photocurrent produced by each cell and a function of the *relative area* it occupies in the device. In other

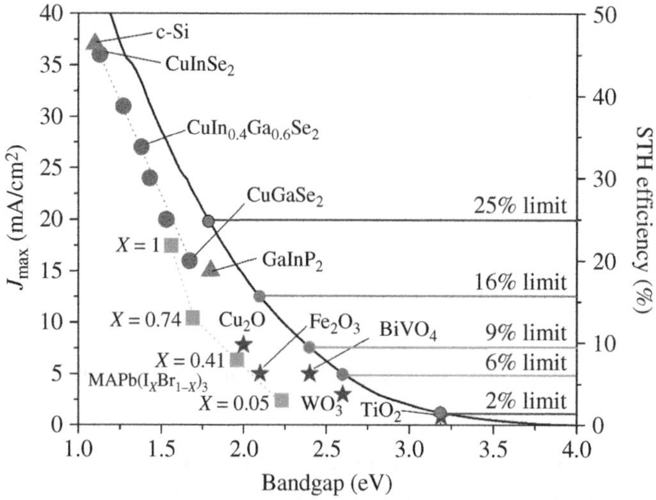

Figure 4.3 Maximum photocurrent density and STH efficiency as a function of the semiconductor bandgap. The maximum photocurrent density was calculated using the optical limit, while the STH efficiency limits were calculated using Equation 4.6, assuming 100% Faradaic efficiency. Also included are experimental photocurrent densities measured on various material classes integrated in PV or PEC devices, including metal oxides (stars), perovskites (rectangles), chalcopyrites (circles), and mono-crystalline semiconductors (triangles).

words, the more the cells that are connected in series in the coplanar device, the lower the J_{SC}. In first approximation, the open circuit voltage, $J_{SC,}$ and maximum achievable STH efficiency of coplanar PV-E made of four individual c-Si cells (V_{OC} = 0.55 V, J_{SC} = 42 mA/cm^2) connected to an external electrolyzer are 2.2 V, 10.5 mA/cm^2, and 12.9%, respectively. Coplanar PV-E devices can be constructed simply using individual cells connected in series (Figure 4.1c); however, in practice, laboratory-scale demonstrations often comprise cells integrated onto the same substrate (Luo et al., 2014) (Figure 4.4a). To further reduce transmission losses, both the anode and cathode can be integrated monolithically with the solar absorbers onto the same substrate (Jacobsson et al., 2013), as presented in Figure 4.4b.

It is worth reiterating that coplanar designs do not make the best use of device real-estate, as their maximum achievable photocurrent density is intrinsically reduced by the number of cells used, and the more cells are needed to drive the water splitting process, the lower the photocurrent density. In contrast, the multi-junction (MJ) stacked architecture (Nozik, 1976; Yoneyama et al., 1975) (Figure 4.4c,d) constitutes a more practical and efficient approach to integrate multiple absorbers into a single system. This design relies on the concept of "spectrum splitting" where each junction absorbs a specific fraction of the solar spectrum. This concept is depicted in Figure 4.5 with a triple-junction device made of 1.8, 1.4, and 1.0 eV semiconductors stacked on top of each other. Photons with energy equal to or greater than 1.8 eV (λ < 690 nm range) are

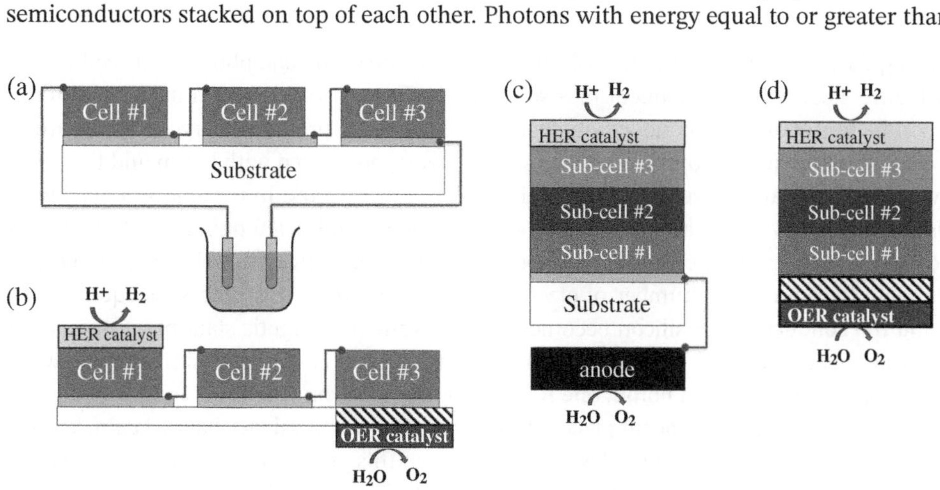

Figure 4.4 Examples of multi-cell semiconductor-based water splitting devices: (a) coplanar PV-E with separated components, (b) integrated coplanar PV-E, (c) "wired" multi-junction tandem connected in series with a dark anode, and (d) "wireless" fully integrated multi-junction tandem.

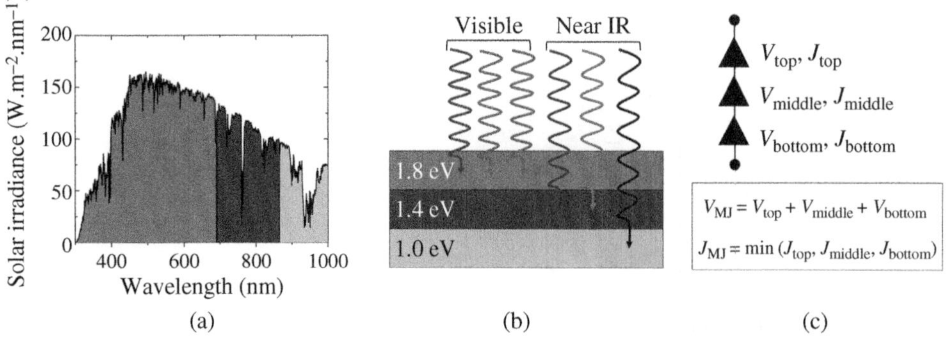

Figure 4.5 (a) AM1.5$_G$ spectrum, (b) schematic of a multi-junction device made of 1.8 eV top, 1.4 eV middle, and 1.0 eV bottom solar absorbers, and (c) equivalent circuit of a multi-junction device.

absorbed by the top junction, generating its own photocurrent density J_{top} and a photovoltage V_{top}. The photons with energy lower than 1.8 eV are transmitted to the second (middle) junction located underneath. Photons with energy lower than 1.8 eV but greater than 1.4 eV (690–885 nm range) are absorbed, and the middle junction generates a photocurrent density J_{middle} and a photovoltage V_{middle}. Finally, photons with energy in the 1.0–1.4 eV range (885–1240 nm) are absorbed by the bottom junction, generating J_{bottom} and V_{bottom}. With the three junctions electrically in series, the overall photovoltage of the triple-junction device is equal to $V_{top} + V_{middle} + V_{bottom}$, while the net photocurrent density corresponds to the smaller of the three photocurrents generated by each of the sub-cells ($J = \min[J_{top}, J_{middle}, J_{bottom}]$). When designing MJ devices, it is thus critical not only to select absorbers with complementary bandgaps but also to fine-tune each film's thickness to ensure that all sub-cells produce equivalent photocurrent densities to prevent "bottleneck" effects.

In PV-E, MJ structures are simply connected to the external electrolyzer (Nakamura et al., 2015; Rocheleau et al., 1998), a situation identical to the one depicted in Figure 4.4a. For direct PEC water splitting, MJ cells may consist of a bare photoelectrode mechanically stacked (Gaillard, Chang, et al., 2010) or monolithically integrated (Hu et al., 2014) on top of one or more *PV drivers*. To further improve the photocatalytic water splitting process, HER or OER catalysts can be directly deposited on top of the photoelectrode (Khaselev & Turner, 1998), as presented in Figure 4.4c. Ultimately, both HER and OER catalysts can be integrated at both the photoelectrode and the substrate surfaces (Figure 4.4d), a structure often referred to as *hybrid photoelectrode* (Miller et al., 2005) or *artificial leaf* (Reece et al., 2011).

4.3 PV-Electrolysis Water Splitting

4.3.1 The Photovoltaic Process

At the heart of PV-E device powering the electrochemical water splitting process is the solid-state photovoltaic cell whose basic schematic is presented in Figure 4.6a. A key component of a solar cell is the *p–n junction*, a built-in asymmetry obtained by contacting two semiconductors of different conductivity types and designed to promote charge carriers extraction. In the following section, we focus on a p–n junction made of two silicon layers, one doped with boron and the other doped with phosphorus. For now, the Si layers are not yet in contact and are placed in the dark. In the Si layer doped with pentavalent phosphorus, one P is surrounded by four Si atoms in the lattice: one electron in P is left unbonded. Calculation shows that the average thermal energy of atomic vibration at room temperature is enough to free the fifth electron in the Si conduction band. Assuming complete ionization, the number of electrons originating from this process is equal to the number of phosphorus *donors*, and the conductivity of silicon becomes *n-type*. From an energetic standpoint, the *Fermi level* (E_F), which is defined as the energy at which the probability of finding an electron is ½ at 0 K, shifts "up" toward the conduction band. In the case of doping with trivalent boron, one B has a missing electron when surrounded by four Si atoms. A nearby electron bond to Si can tunnel into this vacant place, creating a *hole* in the Si valence band. Again, assuming complete ionization at room temperature, the number of holes originating from this process is equal to the number of boron *acceptors*, and the silicon conductivity becomes *p-type*. This time, E_F shifts "down" toward the valence band.

Still in the dark, the p-side is now placed in contact with the n-side. Due to the difference in the concentrations of electrons and holes on either side of the junction, charge carriers will migrate across it: holes will *diffuse* from the p-side to the n-side, while electrons will *diffuse* from the n-side to the p-side. Once equilibrium is reached, the Fermi level of the n-side aligns with that of the p-side. As this process takes place, unbalanced ionized donors (P^-) and acceptors (B^+) are left behind: a depletion region, or space charge region (SCR), is created at the interface and a built-in electric field arises. Any electron or hole "close enough" to this electric field will *drift* back to its place of origin (n-type region for electrons, p-type region for holes). In dark and in absence of any external driving force (i.e., in short circuit condition), the *diffusion* ($J_{\text{diff.}}$) and *drift* (J_{drift}) currents flow in opposite directions but with the same magnitude, and as such the net total current is zero.

We now illuminate the junction from the n-side using photons with energy greater than the silicon bandgap. Under such conditions, electrons are promoted from the VBM to the CBM anywhere in the junction, as illustrated in Figure 4.6; however, only carriers generated in the SCR will be swept away by the built-in electric field, generating a photocurrent (J_L). In the neutral regions outside of SCR, electrons and holes quickly recombine. In practice, SCR must extend the entire solar cell's thickness in order to maximize photocarrier extraction.

Using the equivalent circuit presented in Figure 4.7a, the current output of the solar cell can be expressed as:

$$J = J_L - J_D - J_{SH} \tag{4.7}$$

where J_L, J_D, and J_{SH} are the photocurrent density, the diode current density, and the shunt current density, respectively. Using the expressions of the diode current as defined by the Shockley equation and the current passing through the shunt resistor as defined by Ohm's law, the equation characterizing the total current of the solar cell can be expressed as (Sze & Ng, 2006):

$$J = -J_L + J_s\left[\exp\left(\frac{q(V - JR_S)}{nk_BT}\right) - 1\right] + \frac{V - JR_S}{R_{SH}} \tag{4.8}$$

with J_S the reverse saturation current density, n the ideality factor, q the elementary charge, k_B the Boltzmann constant, and T the absolute temperature.

The current density–voltage (J–V) characteristic of the cell under dark conditions can also be expressed using Equation 4.8, with $J_L = 0$. The J–V characteristics of such solar cells measured in dark and under illumination are presented in Figure 4.7b, along with the associated energy band diagrams under reverse, short circuit, and forward bias

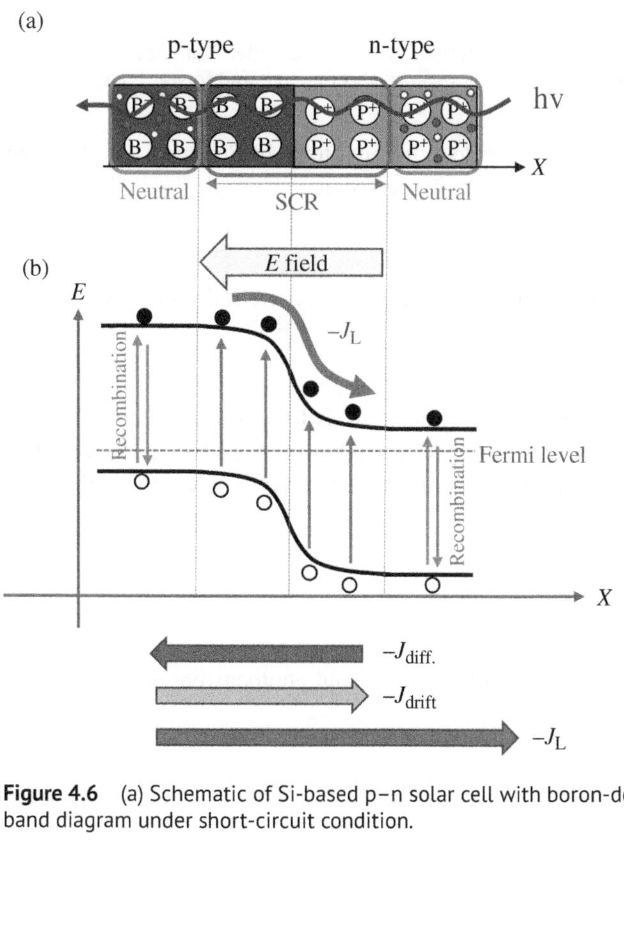

Figure 4.6 (a) Schematic of Si-based p–n solar cell with boron-doped p-side and phosphorous-doped n-side, and (b) its energetic band diagram under short-circuit condition.

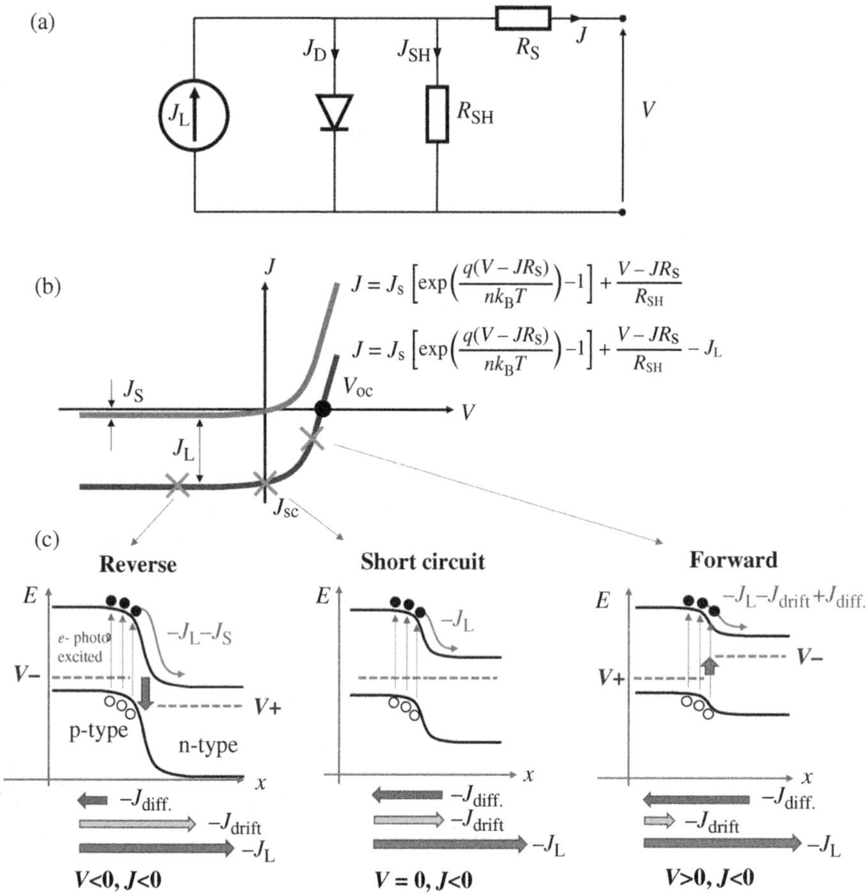

Figure 4.7 (a) Equivalent circuit diagram and (b) schematic current–voltage characteristics in dark and under illumination of a p–n solar cell. (c) Energy band diagrams of the p–n junction under illumination and under reverse, short circuit, and forward bias conditions.

conditions in Figure 4.7c. Assuming a shunt resistance high enough to neglect the final term in Equation 4.8, the open circuit voltage of the solar cell, as defined by the voltage at zero current, can be expressed as:

$$V_{OC} \approx \frac{nk_BT}{q}\left(\frac{J_L}{J_S} + 1\right) \tag{4.9}$$

Similarly, for high-quality solar cells (low R_S and J_S and high R_{SH}), the short-circuit photocurrent density at 0 V can be expressed as:

$$J_{SC} \approx J_L \tag{4.10}$$

The efficiency of p–n solar cell, defined as the ratio of power output to power input, is formalized in Equation 4.11. The output power vs. voltage characteristic of the cell is presented in Figure 4.8. A maximum is reached at the so-called maximum power point (MPP), and power decreases rapidly to reach zero at J_{SC} and V_{OC}. In practice, a variable solid-state load is required to maintain the cell's operation at (or near) MPP.

$$\eta_{PV} = \frac{P_{electrical}\,(\text{out})}{P_{light}\,(\text{in})} = \frac{\left|\text{Voltage(V)} \times J\left(\frac{\text{mA}}{\text{cm}^2}\right)\right|}{P_{light}\left(\frac{\text{mW}}{\text{cm}^2}\right)} \tag{4.11}$$

The cell's maximum efficiency can then be expressed as function of the voltage (V_{MP}) and photocurrent density (J_{MP}) at maximum power by:

$$\eta_{PV,\,Max} = \frac{\left|V_{MP}(\text{V}) \times J_{MP}\left(\frac{\text{mA}}{\text{cm}^2}\right)\right|}{P_{light}\left(\frac{\text{mW}}{\text{cm}^2}\right)} \tag{4.12}$$

Figure 4.8 Schematic photocurrent density and power vs. voltage curves of a p–n solar cell.

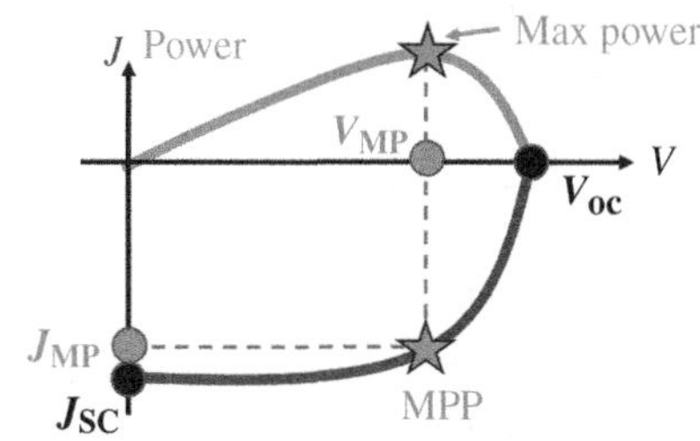

4.3.2 Fundamentals of Water Electrolysis

We now turn our attention to water splitting device in PV-E system: the electrolyzer. The basic schematics of an acidic and an alkaline cell are presented in Figure 4.9. Both systems comprise a cathode and an anode, where hydrogen and oxygen are respectively produced. Between both electrodes reside an ionic conductor and electric insulator: the electrolyte. For acidic electrolyzers (Figure 4.9a), also known as polymer electrolyte membrane (PEM) cells, protons H^+ are transferred between the cathode and the anode through a proton-conductive membrane (typically sulfonated tetrafluoroethylene polymers, such as Dupont's Nafion), closing the *electrochemical circuit* between the two half reactions. PEM electrolyzers typically operate at 2 A/cm^2 and 1.8 V, leading to efficiencies as high as 72% (Phillips & Dunnill, 2016); however, such performance comes with high costs for both the membrane and catalysts used. In addition to being good proton conductors, the membranes must also possess excellent mechanical properties to effectively separate the hydrogen and oxygen gas formed on either side, support the associated pressure buildup and be durable over tens of thousands of hours of continuous operation. Also, because of the harsh environment in which PEM electrolyzers operate (low pH and high oxygen concentration at the anode), additional constraints apply to the catalysts used, and costly platinum group metal (PGM)-based catalysts (cathode: platinum, anode: iridium oxide) are currently necessary to drive the water splitting process with minimum losses. Although PGM catalysts in small-scale (kW range) PEM electrolyzers account for only 5% of the total cell cost, this share is projected to significantly increase in large-scale (MW) systems (Bernt et al., 2020). As such, a great deal of research is currently focused on finding non-PGM alternatives (e.g., transition metal-based catalysts) or high-structured designs (e.g., nanostructured materials with increased number of catalytic sites) to reduce PGM loading in PEM electrolyzers, especially on the anode side.

In contrast, the operating conditions of alkaline electrolyzer (AEL, Figure 4.9b) allow for the use of cheaper catalysts, such as Ni, Fe, or Co-based alloys (Xiang et al., 2016). Two main differences between conventional AEL and PEM systems include the ionic conductor (OH^-) and use of an aqueous H_2O/KOH electrolyte, while product separation is typically achieved by a thin, porous foil (i.e., a diaphragm), such as AGFA's Zirfon (zirconia/polysulfone composites). A drawback of this design is the low current density generated by the electrolyzer (<0.25 mA/cm^2), leading to efficiency of approx. 65%. Zero-gap AEL design has been proposed to reduce cell resistance from both the electrolyte and gas bubbles (Phillips & Dunnill, 2016), leading to current density as high as 0.5 Acm^2 (Kim et al., 2015). The slower ionic transport in liquid electrolyte makes AEL

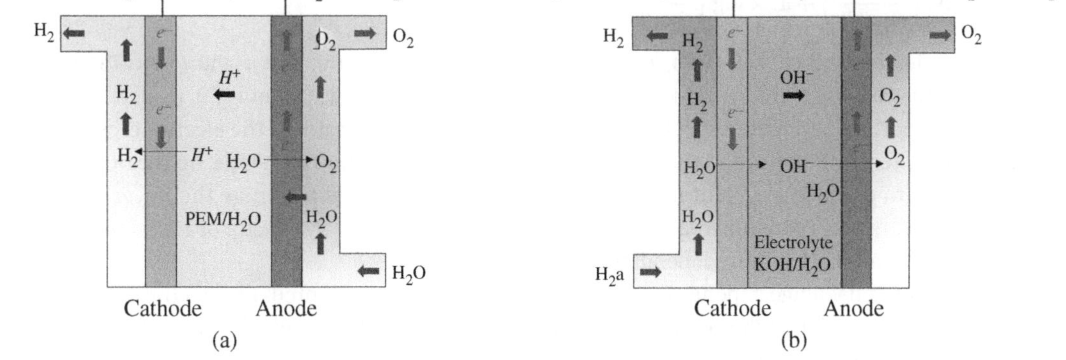

Figure 4.9 Schematic diagrams of (a) acidic (PEM) and (b) alkaline (AEL) water splitting cells. Reproduced from (Gallandat et al., 2017)/Scientific Research Publishing Inc./Public Domain CC BY 4.0.

Table 4.1 Components, Typical Operating Conditions, and Performances of PEM and AEL Systems.

-	PEM	AEL
Membrane, electrolyte, ionic species	Proton exchange membrane, Nafion®, H^+.	Diaphragm, KOH, Electrolyte, OH^-.
Cathode catalyst	Carbon-supported platinum (Pt/C)	Ni-based
Anode catalyst	Ir/IrO_2	Ni-based
Operating temperature/(°C)	50–80	60–80
Cell voltage/(V)	1.8–2.2	1.8–2.4
Operating current density/(A/cm^2)	>1.5	<0.25 (standard design) 0.5 (zero-gap)
Efficiency (HHV)/(%)	67–82	62–82
Cell pressure/(10^5 Pa)	<30	<30
Typical size of commercially available systems	kW	MW
Response to power input variation (e.g., renewables)	Fast	Slow
Current research thrusts	- Non-PGM catalysts or reduced PGM loading, - Increasing material lifetime, - Proof of scale-up concepts for stacks.	- Faster dynamics (e.g., anion exchange membranes), - Increasing current density, - Increasing pressure tightness.

slower to respond compared to PEM systems to the power variations typically observed with renewable energy sources. Recent advances in OH^- anion exchange membranes could solve this issue. A summary of the components, operating conditions, and performances of conventional PEM and AEL systems is included in Table 4.1, along with the main research trusts currently undertaken for each technology.

As discussed previously in this chapter, no system is capable of operating at the reversible voltage of 1.23 V due to inevitable losses occurring at the many steps of the water splitting process. In an electrolyzer, those losses are expressed as overpotentials at the electrodes, the membrane, and in the electrolyte. Thus, the cell voltage V_{cell} required to operate at a given current density can be expressed as:

$$V_{cell} = V_{REV.} + V_{act.} + V_{Ohm} \tag{4.13}$$

with $V_{REV.}$, $V_{act.}$, and V_{Ohm} the reversible water splitting voltage (1.23 V at 298 K and 1 atm.), the sum of the activation overpotentials at the cathode and anode, and the ohmic overpotential associated with the resistances of the membrane, the potential drop across the Helmholtz double layer, and mass transfer in the electrolyte.

Using the Tafel region of the Butler–Volmer relation for anodic and cathodic overpotentials, Equation 4.13 can be expressed as function of the current density:

$$V_{cell} = V_{REV.}(T) + \alpha_a \times \log\left(\frac{J}{j_{0,a}(T)}\right) + \alpha_c \times \log\left(\frac{J}{j_{0,c}(T)}\right) + JR_{EC}(T) \tag{4.14}$$

with R_{EC} total resistance of the electrochemical cell (EC), T the operating cell temperature, and α and j_0 the Tafel slope and exchange current density of the catalyst of interest at the anode (α_a and $j_{0,a}$) and cathode (α_c and $j_{0,c}$), respectively. The exchange current density is an important parameter which depicts the ease of charge transfer at the electrode/electrolyte interface, and the larger j_0 is, the faster the kinetics and the higher the electron transfer. The Tafel slope, on the other hand, indicates the electrode's efficiency to produce a current in response to a change in applied potential: the lower the slope (expressed in mV/decade), the lower the overpotential required to increase the reaction rate by a factor of 10. It is important to note that the $V_{REV.}$, j_0, and R_{EC} are all temperature-dependent.

As described in Equation 4.13, the cell voltage depends on the water splitting reverse potential. From a thermodynamic standpoint, V_{REV} can be expressed as function of the standard Gibbs energy change ($\Delta G°$), Faraday's constant (F), and the number of electrons transferred during the reaction (n):

$$V_{REV.} = -\frac{\Delta G°}{n\mathrm{F}} \tag{4.15}$$

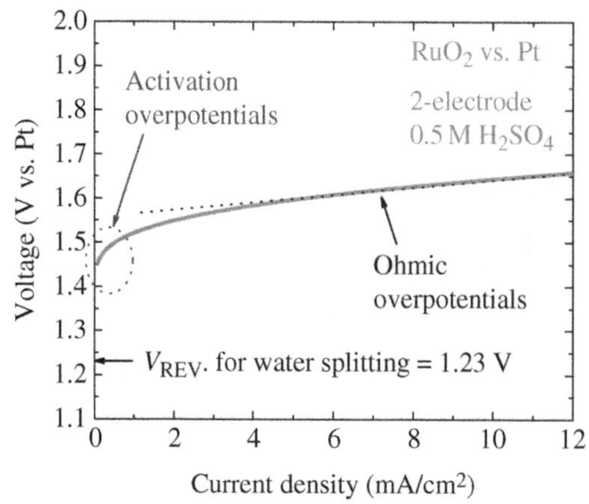

Figure 4.10 Polarization curve (V–J) measured between a thin film RuO_2 (anode) deposited by reactive sputtering onto an FTO substrate and a platinum foil (cathode) in a 2-electrode configuration in 0.5 M H_2SO_4 aqueous electrolyte.

For water electrolysis, the minimum amount of electrical energy required to produce H_2, that is, the change in Gibbs energy (ΔG), can be formally calculated using the enthalpy of formation (ΔH = 286.03 kJ/mol) and ideal gas entropy ($\Delta S = 0.163$ kJ·mol^{-1}·K^{-1}) of gaseous water and the cell temperature (T):

$$\Delta G = \Delta H - T\Delta S \tag{4.16}$$

Equation 4.16 implies that ΔG, and thus V_{REV} are temperature-dependent, and the higher the temperature, the lower the amount of electrical energy required to split water. Likewise, the kinetic overpotentials at the cathode and the anode decrease exponentially with increased temperature. Finally, since ion mobility increases with temperature, the resistance of the electrolyte is reduced as temperature increases. All these aforementioned facts justify why PEM and AEL systems are typically operated at an elevated temperature (Table 4.1). Also, depending on the end use of H_2, the pressure at which the electrolyzer operates can be significantly greater than atmospheric, reducing bubble size and the associated ohmic losses. However, the efficiency of high-pressure cells is not necessarily greater than that of ambient systems (Zeng & Zhang, 2010).

A typical voltage vs. current density characteristic of an electrolyzer, or *polarization curve*, is presented in Figure 4.10. The electrochemical system consists of a RuO_2 thin film anode reactively sputtered onto SnO_2:F (FTO, 25 × 25 mm) substrate and a platinum foil cathode (5 cm^2) immersed in a 0.5 M H_2SO_4 aqueous electrolyte and approx. 2 cm apart. Each term of Equation 4.13 is clearly observed in the polarization curve, with the activation overpotentials corresponding to the non-linear regime at low current densities (<1 mA/cm^2 in Figure 4.10) and the ohmic overpotentials at higher J values (linear portion of the polarization curve).

It should be noted that the polarization curve presented in Figure 4.10 was measured in a 2-electrode configuration (Chen et al., 2013), and as such, the activation and ohmic overpotentials include losses across the entire system. In order to differentiate which component, for example, the cathode or the anode, is the most limiting element, it is convenient to measure polarization curves in a 3-electrode configuration using a reference electrode. Also, new catalyst candidates are commonly benchmarked in 3-electrode configuration, with the resistance drop between the working and reference electrodes subtracted (i.e., *iR corrected*), and their performance assessed by the overpotential (measured vs. the redox potential of interest, e.g., 1.23 V vs. NHE for OER catalysts) required to achieve a current density relevant to a particular application (e.g., 10 mA/cm^2 for PEC water splitting) (McCrory et al., 2015).

4.3.3 PV-E Operating Principles

The solar converter is now connected to the electrochemical system to form the PV-E water splitting system. Unlike PV, where a *variable* solid-state load is used to ensure maximum power delivery, the electrochemical load in PV-E is *fixed*. Since the two components are series-connected, both the current density and voltage must be identical. Such condition can be numerically established using the expression for the current–voltage characteristics of a solar cell (*J–V curve*, Equation 4.8) and an electrolyzer (*polarization curve*, Equation 4.14). Graphically, this condition can be determined using the so-called *load line analysis*, as presented in Figure 4.11, with V_{OP} and J_{OP} defined as the voltage and current density at the operating point where the two curves intersect. The STH efficiency for such PV-E system can be formally described

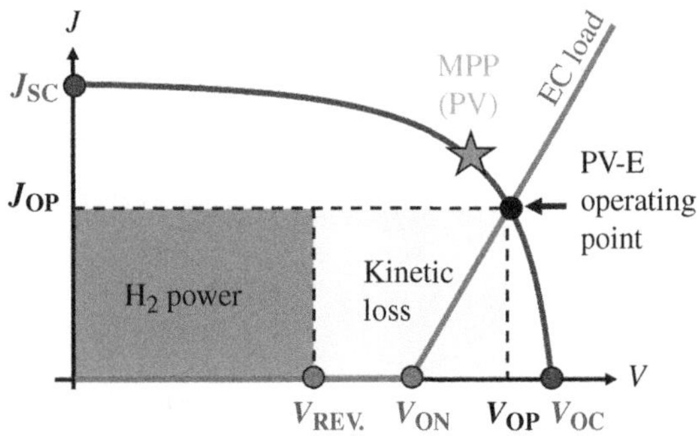

Figure 4.11 Schematic load line analysis of a PV-E water splitting system. The operating voltage and current density are defined by the intersection of the solar cell J–V and the electrolyzer polarization curves. The hydrogen power per unit of area is represented by $J_{OP} \times V_{REV.}$, assuming a Faradaic efficiency of 1.

as:

$$\mathrm{STH}_{\mathrm{PV}-E} = \eta_{\mathrm{PV}} \times \eta_C \times \eta_{\mathrm{EC}} \times \eta_{\mathrm{F}} \tag{4.17}$$

where η_{PV}, η_C, η_{EC}, and η_{F} are the PV efficiency, the coupling efficiency between PV and the electrolyzer, the energy loss owing to the overpotential of the EC, and the Faradaic efficiency, respectively. The $\mathrm{STH}_{\mathrm{PV}\text{-}E}$ can be further expressed as function of V_{MP}, J_{MP}, V_{Op}, and J_{Op}:

$$\mathrm{STH}_{\mathrm{PV}-E} = \frac{V_{\mathrm{MP}} \times J_{\mathrm{MP}}}{P_{\mathrm{light}}} \times \frac{V_{\mathrm{Op}} \times J_{\mathrm{Op}}}{V_{\mathrm{MP}} \times J_{\mathrm{MP}}} \times \frac{1.23\,\mathrm{V}}{V_{\mathrm{Op}}} \times \eta_{\mathrm{F}} \tag{4.18}$$

Some important considerations can be derived from Figure 4.11 and Equation 4.18. Clearly, V_{OC} must be greater than the onset potential V_{ON} ($V_{\mathrm{REV.}}$ + overpotentials > 1.5 V) for the water splitting process to take place, confirming once again that, to this date, no single junction solar cell can drive this electrochemical process; the highest V_{OC} are currently achieved with wide bandgap lead-halide perovskites (Liu et al., 2019): 1.26 V with 1.6 eV $CH_3NH_3PbI_3$ (Kirchartz, 2019), 1.30 V with 1.8 eV $MAPb(I_{0.6}Br_{0.4})_3$ (Rajagopal et al., 2018). Also, since the electrochemical load is fixed, the operating point of a PV-E does not necessarily correspond to the PV's maximum power point (MPP), implying power losses on the PV side of the system. Even if the system was designed to operate at MPP under specific conditions (e.g., 100 mW/cm^2, 25 °C), any change in irradiance or temperature occurring under real-life operations would inevitably *move* the operating point away from MPP (Reuß et al., 2019). Those power losses are encapsulated in Equation 4.17 by the coupling efficiency factor η_C between the PV and electrolyzer systems.

From a practical standpoint, a real PV-E system may comprise a PV module made of multiple series-connected single junction cells and generating a voltage well over V_{ON} (e.g., module made of 36 Si cells, $V_{\mathrm{OC,module}} = 36 \times V_{\mathrm{OC,Si}} = 21.6\,\mathrm{V} \gg 1.5\,\mathrm{V}$). In order to operate closer to the module's MPP and increase the STH efficiency, multiple ECs can be combined as a stack to *shift* V_{ON} closer to the module's MPP (Figure 4.12a). The voltage of the electrochemical stack is then equal to the voltage of the cell times the number of cells in the stack (series-connected case). However, since the number of cells is an integer, matching exactly a predefined operating point is nearly impossible. Alternatively, a PV-E system can be assisted by a DC/DC converter connected electrically between the module(s) and the EC(s) (Dahbi et al., 2016). The converter's task is to adjust the output voltage and current density of the module to ensure maximum utilization of the electrical power by the EC connected to it, as presented in Figure 4.12b.

4.3.4 Evolution of PV-E Systems and Current State-of-the-Art

4.3.4.1 PV-E Systems with Planar Photovoltaics

Early research on PV-E was focused on system engineering/optimization and used photovoltaic modules and electrolyzers available commercially. The first report of a PV-E water splitting system was published in 1977 by Costogue and Yasui from the California Institute of Technology's Jet Propulsion Laboratory (Costogue & Yasui, 1977). The PV system consisted of c-Si-based modules generating 143 W at 35 V near Earth at air mass zero illumination (originally developed for NASA's Mariner 4 spacecraft) and rewired to reduce their output voltage down to ~8 V. A commercially available "Elhygen-R" electrolyzer manufactured by Milton Roy Co. and equipped with palladium cathode tubes was used for the water splitting operation (V_{ON}~2 V). Although the PV and EC components were not matched (as depicted in Figure 4.12a), a volumetric

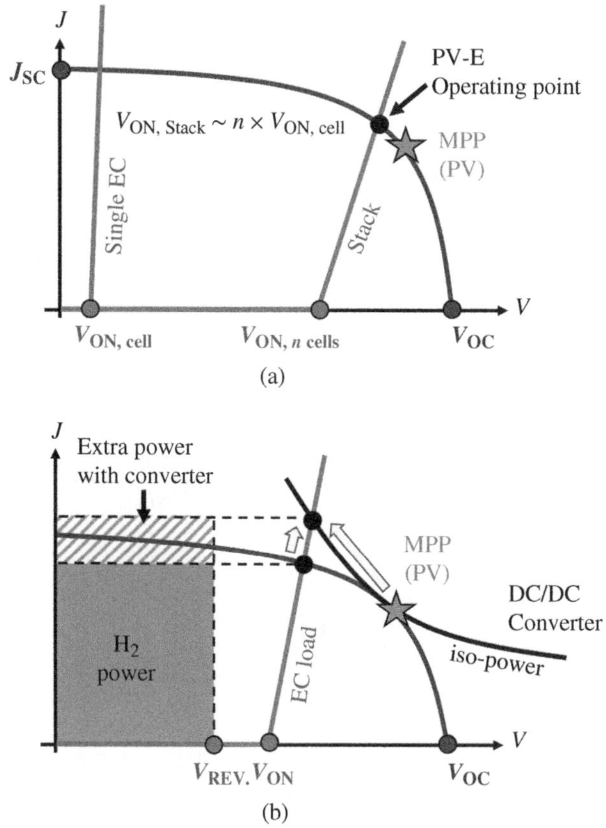

Figure 4.12 Schematic load line analyses of (a) a PV system and either a single EC or a stack and (b) PV-E system assisted by a DC/DC converter. The hydrogen power per unit of area is represented by $J_{OP} \times V_{REV.}$, assuming a Faradaic efficiency of 1.

production rate of 390 mL_{H2}/A hr was demonstrated. Further optimization was proposed, including increasing the number of series-connected EC cells (Figure 4.13a) or using a MPP tracking (MPPT) system to maximize solar power usage.

Siegle and Schott later reported on the economic impact such MPPT would have on PV-E system's *specific investment costs*, that is, the ratio of the initial capital investment to the net power produced by the system (Siegel & Schott, 1988). This analysis was based on hydrogen production data gathered from 10 and 350 kW PV-E plants installed in Stuttgart (Germany) and Riyadh (Saudi Arabia), respectively. Considering the regional solar variations at each location, the authors concluded that high efficiencies could be achieved without DC/DC converters, as long as the electrolyzer characteristic (adjusted by the number of cells) matched closely the annual MPP frequency partition of the solar generator (Figure 4.13b). This conclusion was later supported by several reports, including one by Paul and Andrews in 2008 using laboratory-scale systems (Paul & Andrews, 2008). Nonetheless, it became progressively apparent that future large-scale PV-E farms will require precise control systems to adjust for periodic (changes in ambient temperature, insolation, or partial shading) and long-term (degradation) variations (Fischer, 1986). As such, research groups have continued to focus on improving DC/DC conversion to optimize PV-E operation and safety, developing new algorithms capable of tracking fast dynamic variations in both the PV and EC response curves (García-Valverde et al., 2008).

Some recent models indicate that DC/DC converters can lead to significant boost in hydrogen production with PV-E, as reported by Chang et al. (Chang et al., 2017). The authors calculated that the amount of "extra H_2 power" gained with DC/DC conversion outweighs the power loss from additional kinetic overpotentials (arising from the system operating at higher voltage and current densities), especially when high-efficiency PV (η_{PV} = 30%) and catalysts ($\eta_{10\ mA/cm^2}$ = 300 mV) are used (Figure 4.13c). The use of such high-efficiency systems was reported by Licht et al. in 2001, where a STH efficiency of over 18% was reported with a AlGaAs/Si MJ tandem cell directly coupled to Pt black (HER) and RuO_2 (OER) electrodes immersed in 1 M $HClO_4$ (Licht et al., 2001).

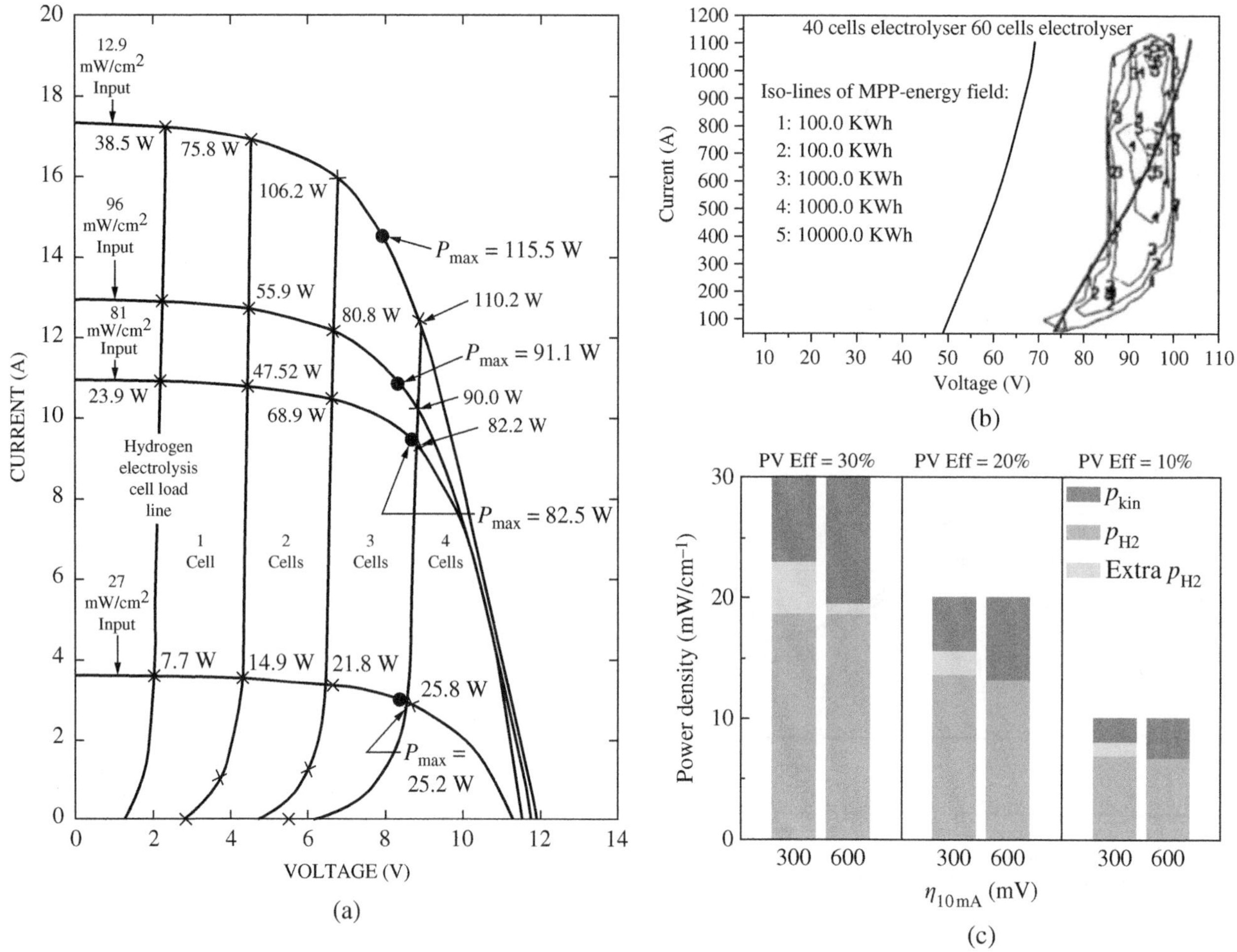

Figure 4.13 (a) Load line analysis of solar photovoltaic (measured under different illumination conditions) and 1, 2, 3, or 4 series-connected hydrogen electrolysis cells. Reproduced from ref. (Costogue & Yasui, 1977) with permission from Elsevier, copyright 1977. (b) Electrolyzer characteristics and MPP-energy field. The annual maximum power point frequency partition of the solar generator (100 kW, Riyadh climate 1978), weighted by appropriate power values, forms an MPP-energy field within the current–voltage diagram. Reproduced from ref. (Siegel & Schott, 1988) with permission from Elsevier, copyright 1988. (c) Hydrogen power (p_{H2}) and kinetic power loss (p_{kin}) of the PV-Conv-EC system at each maximum PV efficiency ($p_{PV,max}$) and the overpotential of reaching the 10 mA/cm^2 ($\eta_{10\,mA}$) condition. Extra p_{H2} gain by the application of the converter is also displayed. Reproduced from ref. (Chang et al., 2017) with permission from the American Chemical Society, copyright 2017.

4.3.4.2 PV-E Systems with Concentrated Photovoltaics

Increasing the STH efficiency of PV-E systems with solar concentration was initially proposed in 1982 by Estève et al. (1982). Their 1 kW 2-axis tracking heliostat system was equipped with Fresnel lenses providing 40× optical concentration to silicon solar cells (passive cooling, $\eta_{PV} = 13\%$ at 45 °C). Water splitting was carried out with an AEL (Ni electrodes, asbestos-based membrane, 30% KOH electrolyte, atm. pressure, 70–80 °C operation temperature, $\eta_{EC} = 70\%$), and an overall STH efficiency of 7% was reported. The authors also provided guidelines for the development of future large-scale hydrogen production with concentrated PV-E systems, and included projected cost based on contemporary (year 1986: Si PV, classical electrolysis, $\$_{Nm^3H_2} = 1.4$ or $\$_{kgH_2} = 15.5$) and future (year 1996: GaAs PV, advanced electrolysis, $\$_{Nm^3H_2} = 0.5$ or $\$_{kgH_2} = 5.6$) technology advancements.

It should be noted that heliostats and solar tracking add significant cost to the overall solar conversion systems (PV or PV-E), and as such concentration is economically sound only when high-cost/high-efficiency photovoltaics are used, as it allows to reduce material utilization. This concept is embodied in the work of Peharz et al., published in 2007 on PV-E with III-V MJ devices. A total of six parallel-connected 1.7 eV GaInP/1.2 eV GaInAs double junction devices were placed under 500× Fresnel lenses and directly connected to a PEM electrolyzer (Peharz et al., 2007) (Figure 4.14a). A STH efficiency value of 15.1% was calculated based on the averaged current generated by the system over the course of 2.3 hr.

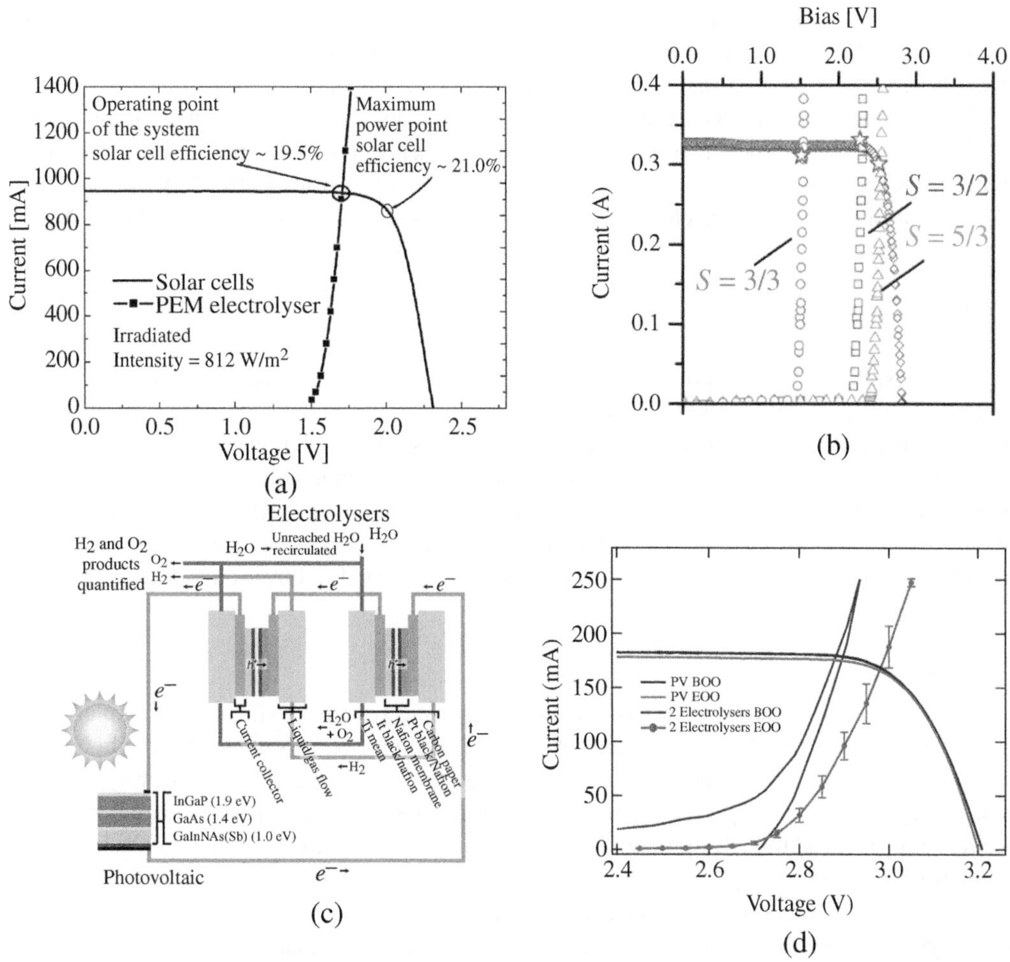

Figure 4.14 (a) Load line analysis of a PEM electrolyzer and six parallel connected solar cells measured outdoor under the Fresnel lenses and at a direct irradiation intensity of 812 W/m^2. Reproduced from Peharz et al. (2007) with permission from Elsevier, copyright 2007. (b) Load line analysis of InGaP/GaAs/Ge CPV modules and PEM EC cells. The lateral voltage axis is normalized so that the curve for a single CPV module is shown equivalently. The number $S = x/y$ represents the series connection configuration of PV (x) and ECs (y). The stars represent the operation points (V_{OP} and I_{OP}) measured in the course of hydrogen production. Reproduced from Nakamura et al., (2015)/Springer Nature/CC BY 4.0 (c) Schematic and (d) load line analysis of a InGaP/GaAs/GaInNAs (Sb) triple junction PV connected in series with two ECs. Reproduced from ref. (Jia et al., 2016).

Eight years later, Nakamura et al. exceeded the 20% STH efficiency barrier using CPV GaInP/GaAs/Ge triple junction modules (device area: 2.5 mm^2, 23× light focusing elements) and PEM ECs (Nakamura et al., 2015). No DC/DC conversion was used, and the operating conditions were adjusted by connecting multiple PV and ECs in series. A STH efficiency of 24.4% was reported using a total of three PV cells connected to five ECs, with a nearly perfect match between the EC polarization curve and the PV MPP (Figure 4.14b).

Finally, in 2016, Jia et al. reported a STH efficiency of 30% using a high-efficiency triple junction cell (InGaP/GaAs/GaInAs(Sb), area = 0.316 cm^2, η_{PV} = 39%) connected in series with two PEM (Pt black, Ir black, T_{EC} = 80 °C) ECs (Figure 4.14c) under 42 suns concentrated illumination (Jia et al., 2016). This value was averaged over the course of 48-hr testing. The authors reported a slight reduction in efficiency at the end of the experiment, due in the most part to a reduction in activity of the ECs from loss of catalyst material and/or membrane degradation (Figure 4.14d).

4.4 Photoelectrochemical Water Splitting

4.4.1 Energetics of the Semiconductor/Liquid Junction

So far, we have covered exclusively the situation where water splitting took place at the surface of metallic electrodes in electrolyzers: the photogenerated carriers by the PV elements were shuttled to the electrode/electrolyte interface and injected into the electrolyte to drive the HER and the OER. Any energetic considerations were discussed solely for the solid/solid interface at the junction between the n-type and p-type semiconductors in the PV device (Figure 4.6), while the electrode/liquid interface was treated as a metal/metal junction with no energetic mismatch. In reality, the dynamics at the solid/electrolyte interface are much more complex, especially with semiconducting electrodes, and merit further discussion. We have introduced earlier the concept of Fermi level for electrons in a semiconductor. Another term often used to describe this quantity for solids is the *chemical potential of electrons,* which derives from the electronic density and density of states in the solid. In the case of electrolytes, the electron energy in a particular redox couple is given by the Nernst equation as:

$$E_{\text{Redox}} = E^0_{\text{Redox}} + \frac{\text{RT}}{\text{nF}} \ln \left[\frac{C_{\text{ox}}}{C_{\text{red}}}\right] \tag{4.19}$$

where $E^0{}_{\text{Redox}}$, C_{ox}, and C_{red} are the standard potential of the redox species involved and the concentration of the oxidized and reduced species, respectively. From an energetic point of view, E_{Redox} is nothing else than the *Fermi level of the electrolyte* $E_{\text{F,redox}}$. Conventionally, E_{Redox}, is measured with respect to an arbitrarily scale based on a reference electrode, commonly the normal hydrogen electrode (NHE); however, reporting this value against a level common with solid-state physics, that is, *the absolute vacuum level*, greatly simplifies energetic assessments at the semiconductor/electrolyte interface in photoelectrochemistry (Bär et al., 2010). Using estimates of the position of NHE with respect to the absolute vacuum level ($E_{\text{abs}} = E_{\text{NHE}} + 4.5\,\text{eV}$) (Trasatti, 1982; Trasatti, 1990), the E_{Redox} (measured vs. NHE) can be converted to $E_{\text{F,redox}}$ (reported against vacuum):

$$E_{\text{F,Redox}} = -4.5\text{eV} - eU_{\text{Redox}} \tag{4.20}$$

Fundamental models developed by Gerischer allow to treat the semiconductor/electrolyte interfaces using similar methods employed for solid-state junctions (Gerischer, 1979; Gerischer, 1990). As an example, we consider an n-type photoanode connected in series with a cathode and immersed in an aqueous electrolyte (Figure 4.15a). As described for the n-type/p-type junction in PV in Section 4.3.1, the Fermi levels of the photoanode ($E_{\text{F},n}$, located close to the conduction band), the electrolyte ($E_{\text{F,redox}}$, located between the OER and the HER potentials), and the metallic cathode ($E_{\text{F},M}$) align to reach thermal equilibrium. With $E_{\text{F},n}$ originally higher than $E_{\text{F,redox}}$, free electrons in the photoanode will migrate toward the back contact, leaving behind ionized donors: a SCR, or depletion region, is formed in the sub-surface of the semiconductor.

Upon illumination, this rectifying junction drives photogenerated electrons to the photoanode back contact and eventually to the cathode, while holes are accumulated at the semiconductor/electrolyte interface. Under these conditions, the thermal equilibrium is now disturbed, and the former description of energy levels in the semiconductor is no longer valid. The new *quasi-equilibrium* condition gives rise to a split in the Fermi level between electrons ($E_{\text{F},e-}$) and holes ($E_{\text{F},h+}$) *quasi-Fermi levels.* Since the electron concentration in the n-type photoanode (majority carriers) was much greater than that of the holes (minority carriers) to begin with, the newly photogenerated electrons barely affect the global electron concentration, and as such $E_{\text{F},e-}$ is nearly identical to $E_{\text{F},n}$. The situation for $E_{\text{F},h+}$ is quite different, though, since the excess hole population appreciably alters the initial minority carrier distribution before illumination, and as a result, a significant $E_{\text{F},h+}$ (down) shift takes place.

The *quasi-Fermi level* split is a critical parameter for all solar absorbers as it defines the photovoltage, or *usable energy*, produced by the cell. In the example presented in Figure 4.15a, $E_{\text{F},h+}$ at the semiconductor/electrolyte interface is lower than the OER potential (+ overpotentials), and as such photogenerated holes are in favorable energetic position to drive the water oxidation process. However, with no external bias supplied to the system, $E_{\text{F},M}$ on the cathode remains aligned with $E_{\text{F},n}$ on the photoanode back contact (and $E_{\text{F},e-}$ for that matter), and electrons do not have sufficient energy to drive the HER. At this point, no water splitting takes place, as both the OER and the HER must happen simultaneously to complete the *electrochemical circuit.*

This situation changes with the use of an external bias ($U_{\text{B}} = E_{\text{B}}/e$), as presented in Figure 4.15a, moving $E_{\text{F},M}$ above the HER potential (+ overpotentials). A more detailed description of the energy diagram evolution at the semiconductor/electrolyte interface upon immersion, illumination, and biasing can be found in Nozik & Memming (1996).

It is worth reemphasizing the important role of the quasi-Fermi level split and its implications in PEC water splitting. It is often mentioned that a single semiconductor with conduction and valence band edges straddling the HER and the OER potentials (+ overpotentials) can spontaneously split water. Although necessary, this condition is not sufficient, and it is in reality the electrons and holes quasi-Fermi level positions with respect to the water splitting potentials that dictate whether such process can take place. Considering that typical photovoltages in proven PV material classes correspond to

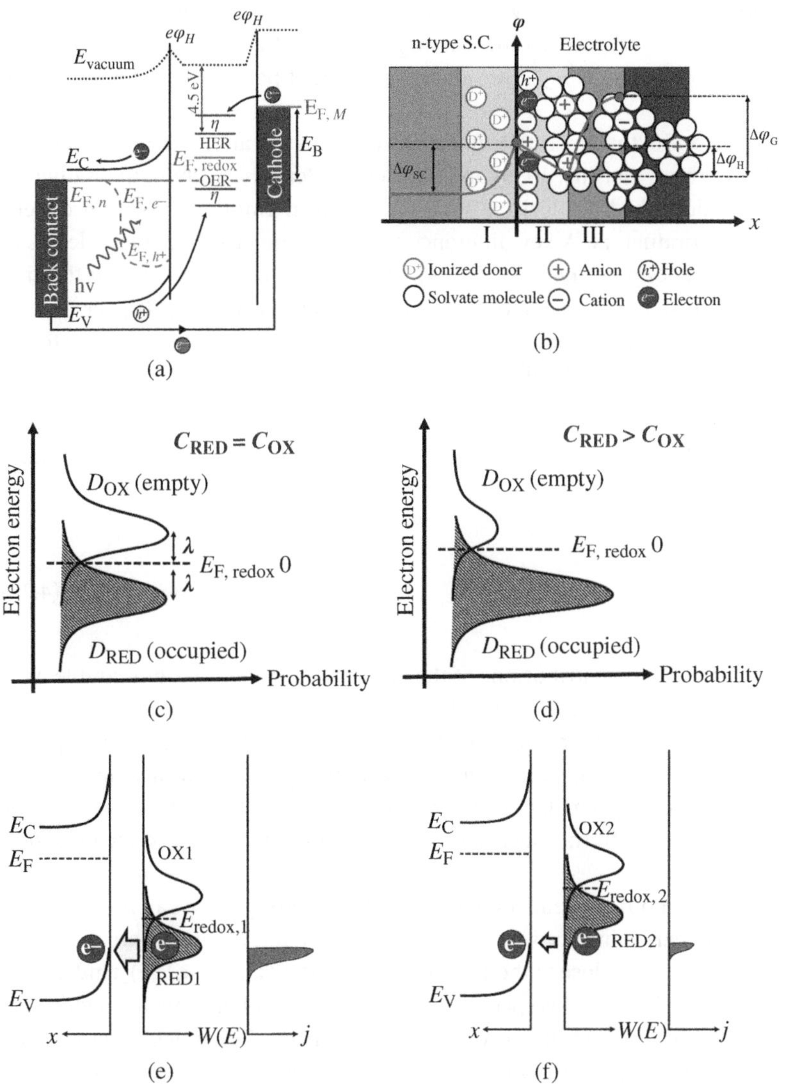

Figure 4.15 (a) Schematic energy band diagram of an n-type semiconductor/aqueous electrolyte/cathode system under illumination. (b) Schematic diagram of the n-type semiconductor/electrolyte interface under equilibrium conditions and associated potential profile across. $\Delta\varphi_{SC}$, $\Delta\varphi_H$, and $\Delta\varphi_G$ are the potential drops in the space charge layer (region I), the Helmholtz layer (region II), and the Gouy layer (region III), respectively. Schematic probability distributions of electron energy in a redox system as described by the Marcus and Gerischer models for the case where (c) $C_{RED} = C_{OX}$ and (d) $C_{RED} > C_{OX}$. Energy level model between an n-type semiconductor and two redox couples with (e) high and (f) low electron transfer rates.

60–80% of the absorbers bandgap at best (e.g., $V_{OC} = 0.706$ V for 1.1 eV Si, $V_{OC} = 1.12$ V for 1.4 eV GaAs, $V_{OC} = 0.713$ V for 1.1 eV $CuInGaSe_2$) (Green et al., 2013), it is clear why very large bandgaps (>3 eV) far greater than 1.23 eV are required to spontaneously split water with single junction PEC devices.

4.4.2 Charge Transfer Dynamics at the Semiconductor/Liquid Junction

The charge carriers transfer occurring in the semiconductor depletion region also gives rise to ionic reorganization in the electrolyte near the solid/liquid interface. A schematic of this process is presented in Figure 4.15a. The charges at the surface of the photoanode consist mostly of trapped electrons and holes in surface states, as well as adsorbed H^+ and OH^- ions created continuously via protonation and deprotonation reactions. Other non-absorbed H^+ and OH^- ions in solution are surrounded by a cloud of water molecules, preventing further accumulation of ions at the electrode surface; the so-called *Helmholtz layer* forms between the absorbed ions and the closest ions in solution (Region II in Figure 4.15b). A potential drop $\Delta\varphi_H$ of 100–500 mV within only few Ångström takes place in the *Helmholtz layer*, creating a capacitance in the 10–20 μF/cm^2 range. Outside of the *Helmholtz layer* is the *Gouy layer*, which is induced by mobile solvated ions in the vicinity of the outer Helmholtz plane (Region III in Figure 4.15b). For highly concentrated electrolytes, such as those used for water splitting (strong acids or strong base), the *Gouy layer* and its associated potential drop $\Delta\varphi_G$ can be neglected.

Now that we have covered the energetics of the semiconductor/electrolyte interface, we focus our attention on the electron transfer process itself. Following pioneer work by Marcus (Markus, 1956) and Gerischer (Gerischer, 1990), the reduction and oxidation energy levels in an electrolyte can be described as electron-filled and electron-empty bands, respectively, similarly to how energetic levels are depicted in semiconductors. A key difference resides in the fact that energy levels in the electrolyte are not fixed but fluctuating, since the energy of the ions themselves varies continuously due to thermal fluctuations in the electrolyte. After the reduction or oxidation process, the ion-modified electrical charge triggers a reorganization of solvate molecules surrounding it. Both concepts of *fluctuating energy* and *reorganization* are encapsulated in the probability distribution of the electron energy in a redox system defined as:

$$W_{\mathrm{Red}}(E) = (4k_{\mathrm{B}}T\lambda)^{-\frac{1}{2}} \exp\left(\frac{-\left(E - E^0_{\mathrm{F,redox}} - \lambda\right)^2}{4k_{\mathrm{B}}T\lambda}\right) \tag{4.21}$$

$$W_{\mathrm{Ox}}(E) = (4k_{\mathrm{B}}T\lambda)^{-\frac{1}{2}} \exp\left(\frac{-\left(E - E^0_{\mathrm{F,redox}} + \lambda\right)^2}{4k_{\mathrm{B}}T\lambda}\right) \tag{4.22}$$

where λ represents the reorganization energy. The probability distributions of the electron energy for a redox couple with identical C_{red} and C_{ox} concentrations are presented in Figure 4.15c. The density of electronic states $D(E)$ is simply defined by the product of the species concentration and the probability distribution $W(E)$. The electrolyte Fermi level corresponds graphically to the intersection of the two distributions, and the energy difference between the maximum of each distribution and the electrolyte Fermi level is the reorganization energy (λ). The latter can be as large as 1 eV (Cell, 2012). The case for a redox couple with $C_{\mathrm{red}} > C_{\mathrm{ox}}$ is presented in Figure 4.15d, with the electrolyte Fermi level located closer to the oxidation level.

Knowing the energetic positions of electrical charges in both the semiconductor and the electrolyte, we can now describe the charge transfer at the interface. The energetics between an n-type semiconductor and two redox couples are presented in Figure 4.15e,f. We assume first that Redox 1 has a Fermi level close to the semiconductor VBM (Figure 4.15e). Under such conditions, a large overlap exists between the energy levels in the redox species and the valence band, leading to a significant electron transfer via tunneling from the electrolyte to the semiconductor (or hole transfer from the semiconductor to the electrolyte). The current originating from this charge transfer process can be formally described as (Deangelis et al., 2018):

$$J_{\mathrm{VB}\rightarrow E_{\mathrm{Red}}} \propto \int_{-\infty}^{E_V} f(E, E_{\mathrm{F}}) \cdot g_{\mathrm{VB}}(E) \cdot D_{\mathrm{Red}}(E) \cdot dE \tag{4.23}$$

where $J_{\mathrm{VB}\rightarrow E_{\mathrm{RED}}}$ is the anodic current due to hole transferring from the valence band into the electrolyte, $g_{\mathrm{VB}}(E)$ is the density of states in the semiconductor valence band, $f(E)$ is the Fermi--Dirac function, and $D_{Red}(E)$ is the distribution of reduction states in the electrolyte. A qualitative representation of such anodic current between the photoanode and Redox 1 is presented on the right side of Figure 4.15e. Then, Redox 1 is replaced with Redox 2 in which the Fermi level sits closer to the semiconductor conduction band (further away from the valence band). In this condition, presented in Figure 4.15f, the overlap between the energy levels in the Redox couple and the valence band is reduced when compared to Redox 1: the charge transfer is lower and the overall anodic current is reduced.

4.4.3 Current–Potential Behavior of a Photoelectrode

We now formally establish the expression of the current–potential characteristic of majority carriers for an n-type semiconductor immersed into a redox electrolyte, starting with dark conditions. The general expression of net current can be expressed as (Rajeshwar, 2007):

$$i_c = -e_0 A k_{\mathrm{et}} C_{\mathrm{ox}}(n_s - n_{\mathrm{so}}) \tag{4.24}$$

where k_{et}, A, n_s, and n_{so} are the rate constants for electron transfer, the electrode area, and the surface concentration of electrons under dynamic and equilibrium conditions, respectively. Without external perturbations (i.e., light or bias), $n_s = n_{\mathrm{so}}$

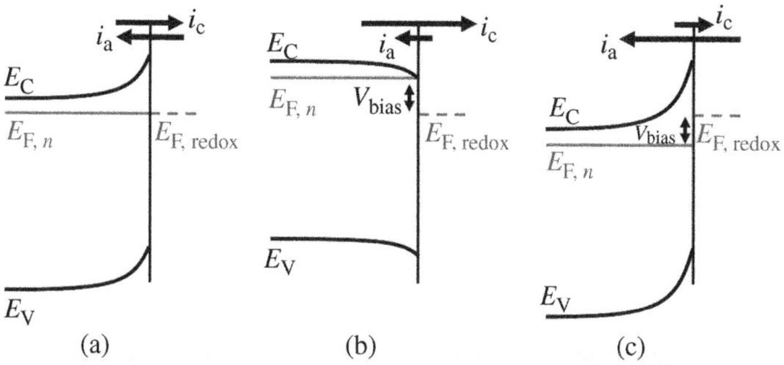

Figure 4.16 Schematic band diagram of an n-type semiconductor/electrolyte interface in dark under (a) equilibrium, (b) forward bias, and (c) reverse bias. The size of the arrows indicates the magnitude of the anodic (i_a) and cathodic (i_c) currents.

and no current exists in the system. In reality, the semiconductor/electrolyte interface is in constant dynamic equilibrium with charges continuously tunneling back and forth with same magnitude but in opposite directions (Figure 4.16a), a situation similar to the one described for the p–n junction under similar conditions (see Section 4.3.1). The surface concentration of electrons under equilibrium can be further expressed as function of the bulk majority carrier concentration n and the built-in voltage V_{SC}:

$$n_{so} = n \exp\left(-\frac{eV_{SC}}{k_B T}\right) \tag{4.25}$$

Still under dark conditions, a bias is applied to the system: the relative position of the Fermi levels will be altered and a SCR will be created in the semiconductor sub-surface. With the conduction and valence bands bending *down*, electrons will accumulate at the interface: n_s is now greater than n_{so} and a cathodic current is flowing (Figure 4.16b), reducing oxidized species (Ox → Red). Conversely, if the bias is such that energetic bands are bending *up*, n_s is lower than n_{so} and an anodic current flows at the interface (Figure 4.16c), oxidizing reduced species (Red → Ox). Under such dynamic conditions, Equation 4.25 can be re-written as:

$$n_s = n \exp\left(-\frac{e(V_{SC} + V_{\text{bias}})}{k_B T}\right) \tag{4.26}$$

Using the expression of n_s and n_{so} established in Equations 4.23 and 4.24, the cathodic current for the majority carriers defined in Equation 4.24 can finally be expressed as:

$$i_c = -i_0 \exp\left(-\frac{eV_{\text{bias}}}{k_B T} - 1\right) \tag{4.27}$$

with i_0 the exchange current defined as:

$$i_0 = -e_0 A k_{et} C_{ox} n_{so} \tag{4.28}$$

Following the same concepts applied to solid-state solar cells (Equations 4.7 and 4.8), the total current density flowing at the semiconductor/electrolyte junction under illumination can be expressed as:

$$J = -J_0 \exp\left(-\frac{eV_{\text{bias}}}{k_B T} - 1\right) + J_L \tag{4.29}$$

A schematic representation of the current density–potential characteristics under dark and illumination is presented in Figure 4.17. It is worth reiterating that the cathodic current measured in dark on the n-type semiconductor under forward bias condition stems from the injection of majority carriers (electrons), while the anodic current observed under illumination and reverse bias condition originates from photogenerated minority carriers (holes). The typical dark and light J–V characteristics measured in a 3-electrode configuration in 0.5 M H_2SO_4 on an n-type 2.6 eV bandgap WO_3 photoanode reactively sputtered onto a FTO substrate (Gaillard, Cole, et al., 2010) are presented in Figure 4.18. Also included in this figure is the J–V characteristic of a platinum cathode measured under identical conditions. The contribution of the majority carriers to the current density is clearly observed under cathodic potential conditions. As the potential becomes more positive, less electrons are injected into the redox system to a point where cathodic and anodic currents balance each other at the *flat-band potential* ($V_{FB} = 0.82\ V_{RHE}$). As the potential becomes more anodic, the current density increases as

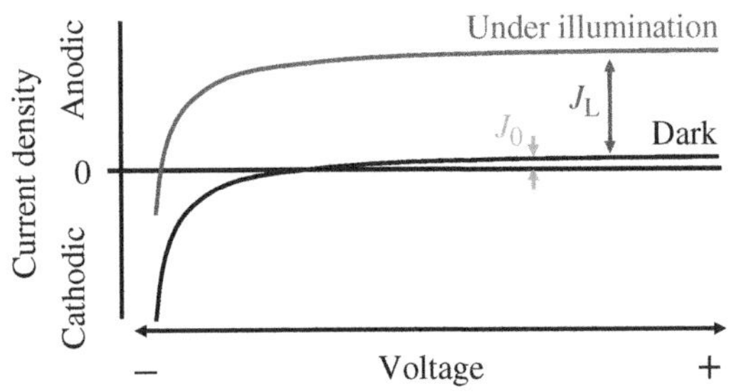

Figure 4.17 Ideal current density–voltage characteristics of a photoanode in dark and under illumination. Adapted from (Rajeshwar, 2007).

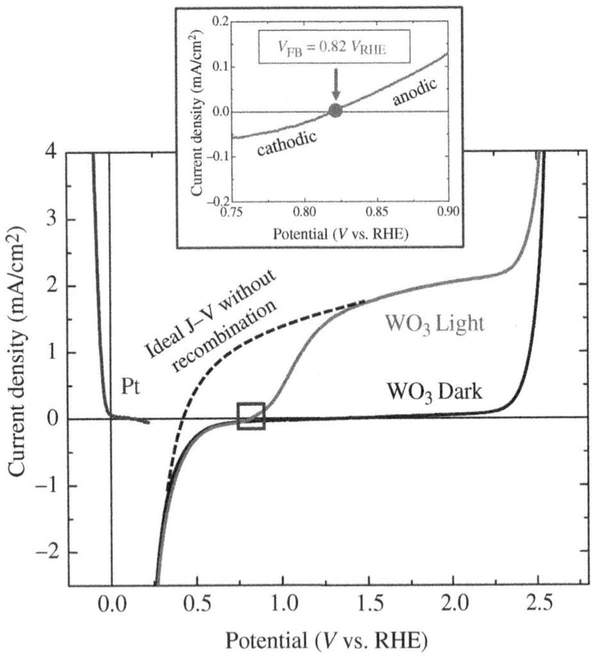

Figure 4.18 Photocurrent–potential characteristics measured with a 3-electrode setup in 0.5 M H_2SO_4 in dark and light (AM1.5_G) conditions on a reactively sputtered 2.6 eV bandgap WO_3 photoanode and platinum cathode. The ideal illuminated *J–V* characteristic the photoanode should have in absence of surface recombination is represented by a broken line. (Inset) Light *J–V* characteristic zoomed in near the flat-band potential.

a result of the increased charge separation in the SCR, until a plateau (or saturation) is reached. Eventually, a dark current arises under higher reverse bias condition (>2.25 V_{RHE}). Several phenomena could cause this sharp anodic current increase, including (in order of probability for WO_3) a lowering of the Schottky barrier height, shunting of majority carriers at surface defect states, or reverse bias breakdown.

A notable difference between the *J–V* characteristic of the WO_3 photoanode and the idealized one materialized by the broken curve in Figure 4.18 is the reduction in electrical charge carrier extraction under low anodic potentials ($V_{FB} < V < 1.2\, V_{RHE}$). In this particular experiment, no OER catalyst was deposited on top of the WO_3 photoanode, and hole transfer was most likely too slow to compete with surface recombination (Chu et al., 2017). Redox couples with high rate constant for oxidation compared to O_2/H_2O, also known as *hole scavengers* (e.g., O_2/H_2O_2), can be added to the aqueous electrolyte to assess if a slow carrier injection is indeed responsible for the nonideal behavior of metal oxide photoanodes (Abdi & Van De Krol, 2012; Dotan et al., 2011; Gaillard, Chang, et al., 2013). Finally, it is worth noting that the flat-band potential of 2.6 eV bandgap WO_3 photoanode ($V_{FB} = 0.82\, V_{RHE}$) is too positive with respect to the HER potential for spontaneous water splitting to occur, a situation clearly illustrated in Figure 4.18 by the absence of any intersection point between the photoanode and the platinum cathode *J–V* characteristics.

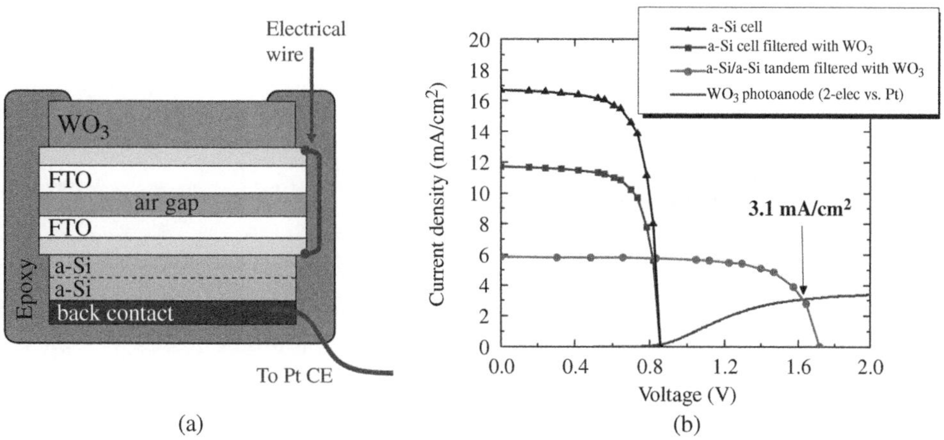

Figure 4.19 (a) Schematic representation of a mechanically stacked standalone PEC device comprising a WO_3 photoanode and an a-Si/a-Si tandem PV cell, both deposited on 2 distinct FTO-coated glass substrates. (b) *J–V* characteristics measured on the components in (a) under various conditions. All data was measured under AM1.5_G illumination. PEC data was gathered in 0.33 M H_3PO_4 using a 2-electrode setup and a Pt cathode.

4.4.4 Spontaneous Water Splitting with Multi-Junction PEC Devices

As discussed previously, combining a photoabsorber with one or more PV *drivers* in a MJ device is a proven and effective approach for efficient standalone water splitting. Multiple MJ PEC architectures have been reported over the years, including dual photoanode/photocathode systems (Jang et al., 2015) as well as photoanode/PV(s) (Abdi et al., 2013) and photocathode/PV(s) (Zhu et al., 2009) devices, the most emblematic example of the latter being the Turner cell comprising a Pt-coated p-type $GaInP_2$ PEC cell monolithically integrated over a GaAs PV and capable of splitting water with an STH efficiency of 12.4% under 11× concentrated light (Khaselev & Turner, 1998). In this section, we will review some of the MJ PEC concepts recently proposed by the University of Hawaii's Hawaii Natural Energy Institute (HNEI) using 1.6 eV amorphous silicon (a-Si) PV paired with either WO_3 photoanodes or $CuGaSe_2$ photocathodes. The reader is invited to examine Ager et al., (2015) and Kim et al. (2019) for comprehensive reviews of other standalone water splitting devices.

Tungsten oxide-based photoanodes have been a central focus of HNEI's PEC research for nearly a decade. Miller, Marsen, et al. first reported on the solid-state properties (primarily electrical conductivity) of reactively sputtered films in 2005 (Miller et al., 2005), followed by their PEC performance in 2007 (Marsen et al., 2007). The team demonstrated that large grain size, which was influenced by the substrate temperature during deposition and the condition of the tungsten sputter target, was a key indicator for high performance. Cole et al. later investigated nitrogen "doping" as a means to reduce WO_3 bandgap in 2008 (Cole et al., 2008). Despite an apparent bandgap reduction, no increase in photocurrent density was reported. Finally, Gaillard et al. published a bi-layer approach to improve photogenerated hole extraction using Mo incorporation in the photoanode top sub-layer (Gaillard, Cole, et al., 2010). Subsequent photoelectron spectroscopy analysis performed by the Heske group at the University of Nevada, Las Vegas concluded that Mo incorporation shifted the Fermi level in WO_3, leading to the formation of a built-in potential at the WO_3/WO_3:Mo interface (Bär et al., 2010). Finally, Chang et al. (Chang et al., 2011) and Gaillard et al. (Gaillard, Chang, et al., 2013) reported on WO_3 alloying with Cu to form 2.2 eV bandgap $CuWO_4$. Although thermodynamically stable, this quaternary material could not generate sufficient photocurrent densities (<0.5 mA/cm^2) to compete with WO_3, and PEC research at HNEI was re-directed onto more promising material classes (i.e., chalcopyrites, more on this later). To this date, the WO_3/WO_3:Mo bi-layer architecture has led to the highest photocurrent density (3.6 mA/cm^2) reported by the group for the tungsten oxide material class.

We present now the results achieved by HNEI on WO_3-based standalone PEC water splitting. All PEC data presented herein were measured in 0.33 M H_3PO_4 using a 2-electrode configuration with a Pt cathode. First, a mechanically stacked device comprising a 3 μm thick WO_3 reactively sputtered onto a FTO glass substrate, and an a-Si/a-Si tandem superstrate PV cell (Zhu et al., 2009) deposited by plasma enhanced chemical vapor deposition (PECVD) was evaluated (Figure 4.19a). The two samples were pressed back-to-back against each other and connected with an electrical wire. One additional wire was soldered onto the tandem cell back contact and epoxy was finally applied to protect all parts of the device but the WO_3 surface. The *J–V* characteristics of each sub-components of the device are presented in Figure 4.19b. A single a-Si PV could generate a J_{SC} up to 16.4 mA/cm^2 and a V_{OC} of 0.83 V. Shading this cell with a WO_3/SnO_2:F substrate reduced J_{SC}

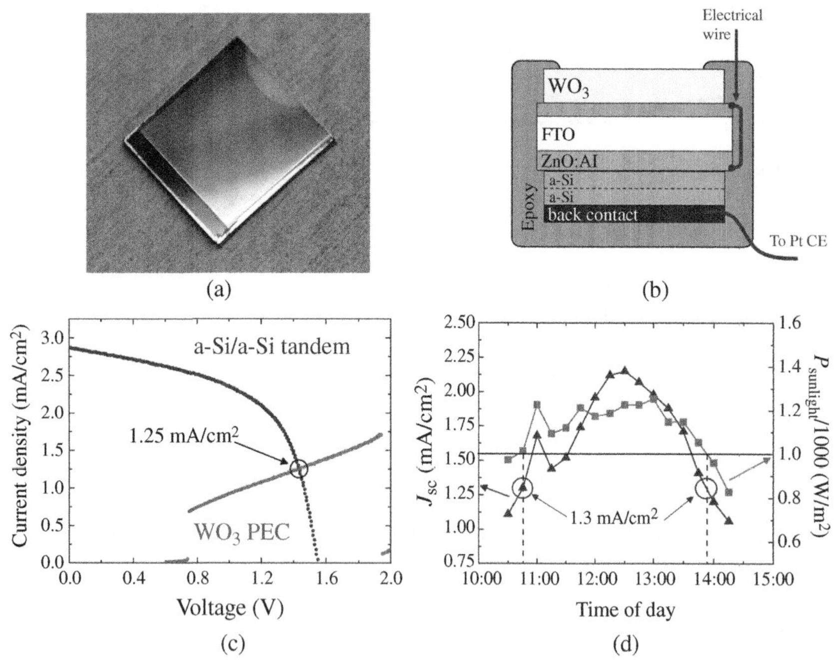

Figure 4.20 (a) Picture and (b) schematic representation of a monolithically stacked standalone PEC device comprising a WO_3 photoanode and an a-Si/a-Si tandem PV cell, both integrated on either side of a SnO_2:F substrate. (c) J–V characteristics were measured under AM1.5$_G$ illumination of a WO_3 photoanode/Pt cathode system and an a-Si/a-Si tandem optically filtered by the WO_3 cell. (d) Short-circuit photocurrent density measured on the monolithic PEC device presented in (a) under outdoor illumination at different times of the day. All PEC data were measured in a 2-electrode configuration in 0.33 M H_3PO_4 using a Pt counter electrode.

down to 11.8 mA/cm^2, while V_{OC} remained largely unaffected. Nonetheless, the total photovoltage produced by a single a-Si was not enough to lead to any measurable water splitting, although an intersection point exists between a-Si and WO_3 (vs. Pt, 2-electrode) J–V curves. It is only with addition of a second a-Si cell that an operating point could be defined, that is, 3.1 mA/cm^2 at 1.6 V. Chronoamperometry analysis was then carried out with the specimen, and a stabilized short-circuit photocurrent density of 2.4 mA/cm^2 was achieved after 100 min, leading to a maximum achievable STH efficiency of approx. 3% (no gas collection was performed to confirm this efficiency) (Gaillard, Chang, et al., 2010).

Similar materials were integrated into a monolithic fashion. After reactive sputtering of a 2 μm thick WO_3 layer onto the SnO_2:F substrate, the specimen was subjected to the a-Si/a-Si tandem integration. A transparent, conductive Al-doped ZnO layer was first sputtered onto the back side of the SnO_2:F glass substrate, followed by the two p-i-n a-Si sub-cells (Figure 4.20a,b). Similar to the mechanical stacked device, a metal wire was used to connect the PEC SnO_2:F back contact to the PV ZnO: Al top contact and a wire was soldered to the PV driver prior to epoxy encapsulation. The load line analysis between the PEC and PV components measured individually in a 2-electrode configuration (vs. Pt for PEC) under simulated AM1.5$_G$ illumination is presented in Figure 4.20c. A lower operating photocurrent density (1.25 mA/cm^2) was achieved compared to that obtained with the mechanical stack device (3.1 mA/cm^2), presumably due to the photoanode being thinner (2 μm for monolithic device vs. 3 μm for the mechanical stack). The PEC performance of the monolithic device measured outdoors is presented in Figure 4.20d. The photocurrent flowing between the PEC device and a Pt cathode was measured at different times of the day, and the solar illumination power was tracked using a calibrated Si diode provided by the National Renewable Energy Laboratory (NREL). Standard AM1.5$_G$ conditions (100 mW/cm^2) were reached around 10:45 a.m. and 1:55 p.m., with the device generating a short-circuit photocurrent density of 1.3 mA/cm^2, setting the STH efficiency upper limit for this device at 1.6% (no gas collection was performed to confirm the device efficiency).

Photoelectrochemical research at HNEI was then re-centered on copper chalcopyrites, a material class usually identified by its most popular PV-grade alloy Cu(In,Ga)Se_2. A key asset of this band gap tunable, direct absorber, thin film semiconductor material is its outstanding photon-to-electron conversion efficiency, as demonstrated with the short-circuit photocurrent densities (J_{SC}, 35 mA/cm^2) achieved with low band gap Cu(In, Ga)Se_2-based PV cells (Septina et al., 2021). HNEI demonstrated that chalcopyrite-based PEC systems can generate photocurrent densities comparable to those achieved by CuInGaSe_2-based solar cells and evolve H_2 with Faradaic efficiencies greater than 85% (Gaillard, Prasher, et al., 2013). However, chalcopyrite photocathodes notoriously suffer from nonideal interfacial band-edge mismatch with respect to the HER potential, requiring both surface treatments (Gunawan et al., 2014; Moriya et al., 2013) and additional PV cells to drive the water splitting process.

In 2013, HNEI reported on $CuGaSe_2$-based PEC device capable of standalone water splitting (Gaillard, Prasher, et al., 2013). The coplanar device comprised a 1.6 eV $CuGaSe_2$ photocathode co-evaporated onto a SnO_2:F substrate and series connected to three a-Si PV drivers. In this experiment, only the photocathode and the RuO_2 anode were immersed in the 0.5 M H_2SO_4 electrolyte (Figure 4.21a). The duration of testing (typically less than 60 min) was well within the stability range of the chalcopyrite photocathodes (150 hr (Palm et al., 2021) to 2,500 hr (Muzzillo et al., 2018) depending on copper concentration). The J–V of each sub-components was measured under simulated AM1.5$_G$ illumination to identify the optimum area ratio between the PV drivers and the photocathode to be used in the coplanar device. To do so, raw I–V curves were normalized to the various relative areas that would be occupied by the a-Si/a-Si/a-Si and $CuGaSe_2$ cells in the coplanar device (Figure 4.21b). A short-circuit photocurrent density of 3.76 mA/cm^2 was established for a PV/PEC relative area ratio of 70/30. Such device was then constructed for zero-bias outdoor testing. Each sub-cell area was adjusted with opaque epoxy until the 70/30 optimum area ratio was achieved. A similar test procedure as the one presented earlier for the monolithically integrated WO_3/a-Si/a-Si cell was used. A short-circuit photocurrent density of 3.53 mA/cm^2 was achieved, leading to an STH efficiency of 3.7% (calculated using the 85% Faradic efficiency previously measured on bare $CuGaSe_2$). Since, the Ångström Laboratory at Uppsala University, Sweden, has reported a 10% STH efficiency using three series-connected $CuInGaSe_2$ solar cells with the same 1.1 eV narrow band gap (Jacobsson et al., 2013). The use of coplanar schemes in both HNEI and the Ångström Laboratory's experiments was dictated by the band gaps of Cu(In, Ga)Se_2 compounds that were too narrow for integration into efficient MJ cells. The development of novel wide band gap chalcopyrites has been the central focus of HNEI since, primarily sulfur-based alloys.

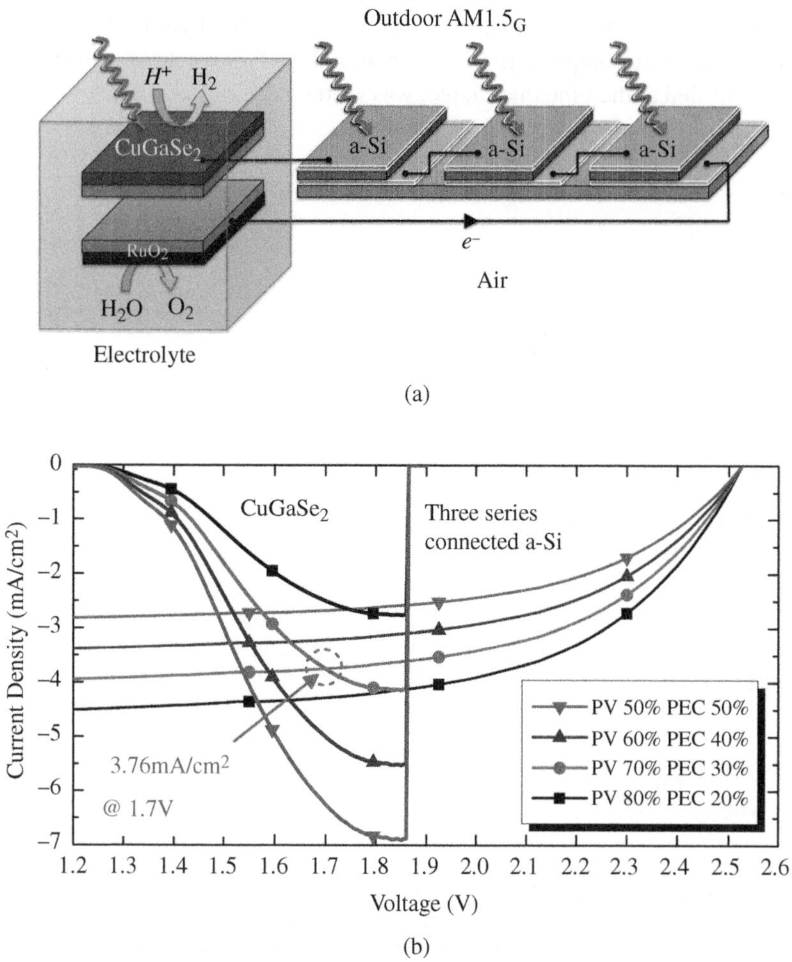

Figure 4.21 (a) Schematic diagram and (b) load-line analysis of a $CuGaSe_2$/a-Si/a-Si/a-Si/RuO_2 coplanar standalone PEC device.

Widening the optical bandgap of $Cu(In,Ga)Se_2$ via incorporation of sulfur has been reported for over three decades; however, the vast majority of these studies focus on surface modification of polycrystalline $Cu(In,Ga)Se_2$ only to *open up* the absorber bandgap near the interface with the buffer and increase the solar cell V_{OC} (Nakada et al., 2003). Quite interestingly, subjecting a polycrystalline chalcopyrite (conventionally Cu-poor for improved PV performance) to a sulfur source (either elemental sulfur vapor or H_2S) has never led to any complete bulk conversion of $Cu(In,Ga)Se_2$ into $Cu(In,Ga)S_2$ (Nakada et al., 1997; Zou et al., 2013). Some reports do exist on 1.5 eV $CuInS_2$ absorbers for single junction PV, but their synthesis requires Cu-rich precursors (Merdes et al., 2011; Merdes et al., 2013; Siemer et al., 2001), which limits their sub-bandgap optical transmission (more on this point later). Also, most reports on $CuInS_2$ use Mo as electrical back contact, rendering the device unusable for MJ-stacked applications. Integrating $Cu(In,Ga)S_2$ on a transparent conductive SnO_2:F substrate is in fact a much more challenging task, as this material tends to react readily with sulfur to form resistive SnS_x. As such, direct co-evaporation of $Cu(In,Ga)S_2$ on SnO_2:F may not be applicable, as SnS_x would inevitably form at the absorber/back contact interface during the initial stage of deposition.

To address this barrier, Deangelis et al. from HNEI first reported on a two-stage process to synthesize bandgap-tunable $CuGa(S,Se)_2$ on SnO_2:F with no loss in substrate conductivity (Figure 4.22a) (Deangelis et al., 2018). The authors first co-evaporated Cu-rich $CuGaSe_2$ precursors using the high-temperature co-evaporation techniques typically used for chalcopyrite synthesis (3-stage process, 550 °C max temperature). The samples were then subjected to two annealing steps: a *dosing phase* for 1 hr at 300 °C in presence of elemental sulfur, followed by an *alloying stage* for 10 min at 500 °C under inert atmosphere. Sequencing the sulfurization process into these two stages permitted the synthesis of 1.72 eV photocathodes capable of producing photocurrent densities over 12 mA/cm^2 (Figure 4.22b). It was speculated that the excess copper used during the precursor growth segregated to form a Cu_2Se top layer, which further reacted with sulfur during the *dosing phase* to form Cu_2S. The latter could then diffuse in $CuGaSe_2$ during the *alloying stage* and drive sulfur with it. The bandgap of $CuGa(S,Se)_2$ was eventually determined by the amount of sulfur used. Unfortunately, the need for excess copper was found to be detrimental to the absorber sub-bandgap transmission ($\%T_{E<EG}$), as Cu-rich films ($\%T_{E<EG} = 30\%$) would block nearly twice as many photons with energy lower than their bandgap compared to Cu-poor materials ($\%T_{E<EG} = 80\%$). A definitive explanation for this phenomenon had not been identified at the time this chapter was written.

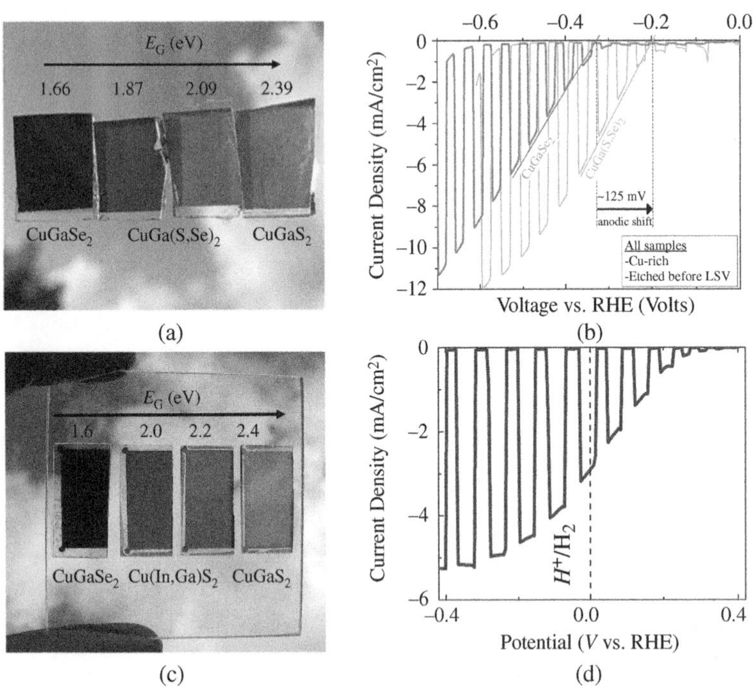

Figure 4.22 (a) Picture of bandgap-tunable $CuGa(S,Se)_2$ integrated onto SnO_2:F substrates and (b) *J*–*V* characteristics measured on 1.6 eV $CuGaSe_2$ and 1.72 eV $CuGa(S,Se)_2$ photocathodes. Reproduced from Deangelis et al. (2018) with permission from the American Chemical Society, copyright 2018. (c) Picture of bandgap-tunable $Cu(In,Ga)S_2$ absorbers integrated onto SnO_2:F substrates and (d) *J*–*V* characteristics measured on a 2.0 eV $CuInGaS_2$ photocathode. Adapted from Gaillard et al. (2019).

At this point, it was apparent that Cu-rich layers were required to sulfurize polycrystalline Cu(In,Ga)Se_2, yet the need for this very Cu-rich layer would render Cu(In,Ga)(S,Se)$_2$ absorbers unusable for tandem applications. The HNEI team later theorized that reducing the crystallographic ordering in Cu-poor Cu(In,Ga)Se_2 should allow for sulfur diffusion, and a low-temperature co-evaporation process was developed to synthesize such *defective* precursors (Gaillard et al., 2019). A two-stage annealing process similar to the one reported by Deangelis et al. was used, leading to absorbers with tunable bandgap ranging from 2.0 to 2.4 eV depending on the initial In and Ga contents in the precursor films (Figure 4.22c). Subsequent Raman analysis confirmed the lack of crystallographic ordering in the low-temperature co-evaporated Cu(In,Ga)Se_2 precursors, while X-ray diffraction spectra confirmed the substitution of selenium with sulfur after the *dosing* and *alloying* steps. Finally, a maximum photocurrent density of 5.25 mA/cm^2 was reported on a 2.0 eV $CuInGaS_2$ photoanode (Figure 4.22d). Unfortunately, the onset potential achieved with this material (500 mV_{RHE}) was found to be too low for spontaneous water splitting, even if paired with a high-efficiency PV driver. Other research efforts were focused on the development of MgZnO buffer layers with tunable energetics (Khan et al., 2021) to increase the photovoltage of wide bandgap chalcopyrites. We invite the reader to review recent publications from HNEI and collaborators on the modeling (Sharan et al., 2020; Varley et al., 2018) and development (Muzzillo et al., 2018) of novel chalcopyrite materials and interfaces (Hellstern et al., 2019; Palm et al., 2021) for standalone PEC water splitting devices.

4.5 Techno-Economics of PV-E and PEC Water Splitting

4.5.1 Similarities and Differences Between PV-E and PEC Water Splitting Technologies

We have discussed so far the fundamentals and basic operations of PV-E and PEC water splitting systems separately, and several examples of reported technologies were discussed. At this point, major questions still remain unanswered: which of PV-E or PEC water splitting processes is the most technically sound? Are they equivalent in terms of hydrogen production cost? Will one supplant the other in the long term, or will both technologies be complementary? Answering these questions requires detailed TEA of the two technologies that will be presented in this section. Before this, we briefly compare them side-by-side to highlight the key similarities and differences and see if one is clearly more advantageous than the other.

From a device integration standpoint, PV-E comprises two independent components, whereas PEC is an *all-in-one* solution. This fact alone imposes some important constraints regarding production plant sizing and the balance of systems. While the electrolyzer component can produce H_2 nearly 24 hr a day (97% capacity factor), the PV component operates at maximum power for only 5 hr per day (20% capacity factor). Thus, the electrolyzers, which make up approximately 40% of the PV-E system cost (James et al., 2009), are greatly underutilized, and are essentially on idle 80% of the time. In PEC systems, the catalytic components can be integrated directly over (in the form of nanoparticles) and under (dark electrode) the solar absorber (e.g., *artificial leaf*, Figure 4.4d), rationalizing their utilization. Another benefit of PEC over PV-E relies on the energy carrier being produced at the device surface. As it turns out, transferring electrical current between the PV modules and the electrolyzers in PV-E has a significant cost, about $1 MM per GW-mile (Shaner et al., 2016). In PEC, H_2 is the only energy carrier transferred between the PEC reactors and the plant gate through pipelines, which is much cheaper ($0.1 MM per GW-mile).

Another key difference is found in the power plant layout options offered by both technologies. Two terminologies are commonly used to define the arrangement of solar-powered hydrogen production plants: *centralized*, where both solar capture and hydrogen production take place at the same site, and *distributed*, where both processes can take place at distant locations (Ardo et al., 2018). By design, PECs are inherently centralized systems, while PV-E could be either centralized (with both components few meters from each other) or distributed (and linked via the electric grid). A distributed approach gives PV-E the edge over PEC, since water splitting can be used as "plan B" to store excess PV-generated power produced when the demand for electricity from the grid is low. Materials durability is another area where PV-E has a clear advantage over PEC: both PV and electrolyzers have been studied for decades and materials lifetime exceed years, compared to only few weeks (at best) for PEC systems. It is worth noting that once a PEC device is exhausted, the entire system must be replaced (at least with the integration schemes reported thus far). Also, the use of separated components and conversion devices (e.g., DC/DC converters) in PV-E allows for maximizing power delivery to the electrochemical system. As such, the STH efficiencies reported with PV-E (Jia et al., 2016) are roughly twice (30%) those reported with PEC (Young et al., 2017) (16.2%). Another important metric to consider is the *technology readiness level* (TRL) of these hydrogen production methods (Grube et al., 2020). In other words: are any of these techniques ready for commercialization yet? Although PEC

Table 4.2 Comparison of PV-E vs. PEC Technologies with Some of Their Advantages and Drawbacks.

-	PV-E	PEC
Key advantages	• Commercially available components, • Components developed independently.	• All-in-one wireless technology, • One-step conversion process, • Cooling by water feedstock.
TRL (higher is better)	6–7 out of 8	3–4 out of 8
Max reported STH	30% (Jia et al., 2016)	16.2% (Young et al., 2017)
Typical durability	Years	Weeks
Current density during electrolysis	1–2 A/cm^2	10–20 mA/cm^2
Technological options	Distributed or centralized	Centralized
Cost of transport of energy carrier (Shaner et al., 2016)	$1 MM per GW-mile	$0.1 MM per GW-mile

water splitting with devices as large as 100 m^2 has been reported (Nishiyama et al., 2021), more research is needed to develop systems that can meet the durability targets required for cheap renewable hydrogen production. In contrast, all the necessary components for PV-E have been well studied and improved for decades, and large-scale production plants could be implemented today if proven to be economically sustainable.

Also, we have discussed in Section 4.3.2 the interplay between current density and overpotentials. Since PV-E is operating at high current densities (1–2 A/cm^2), highly efficient (and often expensive) catalysts are required to reduce kinetic losses. In contrast, PEC systems operate at current densities 100× lower (10–20 mA/cm^2), reducing the needs (and loading) of precious catalysts. Being directly immersed in the electrolyte, PEC systems can be also cooled directly by the water feedstock. As we see from these few examples summarized in Table 4.2, neither of the two technologies has a clear edge over the other, although PV-E appears more mature for short/mid-term renewable hydrogen production. Ultimately, only one metric will crown the winning technology: *the cost per kilogram of* H_2 (kg_{H2}).

4.5.2 Independent Assessments of PEC Technologies

We present in this section the results of a TEA performed by the University of Hawaii on chalcopyrite-based PEC water splitting devices using the U.S. Department of Energy's H2A model.[2] In *Scenario #1*, a coplanar system comprising three narrow bandgaps $CuInGaSe_2$ capable of splitting water with already demonstrated STH efficiency of 10% (Jacobsson et al., 2013) was considered. Materials cost was set at $100/m^2, a value reported for thin film-based PV technologies. *Scenario #1* represents essentially the current state-of-the-art in chalcopyrite-based PEC water splitting. In *Scenario #2*, a MJ stacked device with hypothetical STH efficiency of 20% is used, a value theoretically achievable with chalcopyrite-based MJ devices (Seitz et al., 2014). Materials cost was further reduced to $60/m^2, exploiting recent advances in liquid-based chalcopyrite manufacturing (Septina et al., 2021). For the base case in both scenarios, a 25× concentration factor was chosen to reduce constraints on the durability target, and a 2-yr PEC material lifetime was used (equivalent to 4,400 hr of operation assuming a 25% capacity factor).

All calculations were performed using so-called "Type 4" hydrogen production reactors equipped with optical concentrators and solar tracking systems (James et al., 2009) in which the coplanar or MJ devices are integrated (Figure 4.23a). Each individual concentrator array is 6 m long and 3 m wide. The 50 ton per day (TPD) centralized hydrogen production plant is composed of fifty 1 TPD sub-plants. The number of reactors and total light capture area of each sub-plant depend on the PEC device efficiency, the amount of H_2 produced per day per sub-plant (here 1 TPD), the concentration factor, and the yearly average solar energy (6.55 kWh·m^{-2}·d^{-1}). As an example, 50,300 m^2 is required to produce 1 TPD using reactors equipped with 25× concentrations and 10% efficient PEC devices (Figure 4.23b). Reactors are aligned in rows with 0.83 m of clearance between them to allow for movements during tracking. Two consecutive rows are 14.3 m apart to prevent shading. The electrolyte reservoir and the H_2 and O_2 gases are pressurized by the inlet water pump at 300 psi (Figure 4.23c). This obviates the need for a separate compressor, minimizes water vapor loss by the reactor, reduces H_2 gas bubble size,

2 Case studies and templates can be downloaded at https://www.nrel.gov/hydrogen/h2a-production-models.html

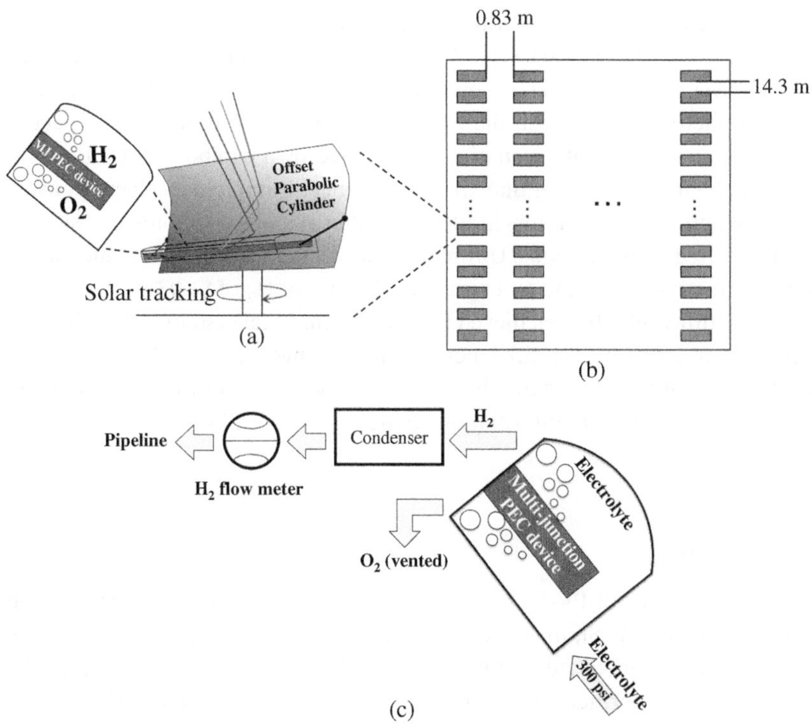

Figure 4.23 Schematic diagrams of PEC-based hydrogen production plant. (a) The PEC reactors comprise the water splitting device integrated into parabolic mirrors and is equipped with solar tracking systems. (b) The reactors are arranged into arrays, forming one sub-plant. (c) Simplified process flow of the hydrogen production system.

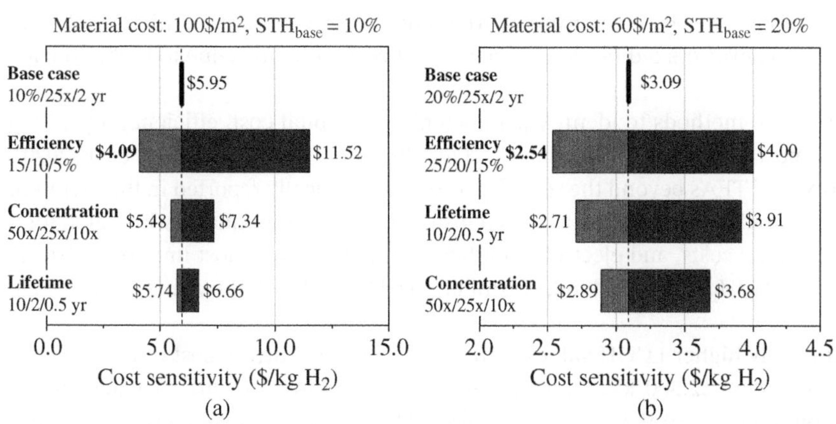

Figure 4.24 Effect of efficiency, panel cost, optical concentration, and component lifetime on the cost of H_2 for PEC reactors equipped either (a) coplanar or (b) stacked chalcopyrite-based PEC water splitting devices. Each calculation represents the variation of a single parameter from the base case.

and minimizes light scattering. Only hydrogen gas is collected and oxygen gas is vented. Hydrogen gas is then fed into the condenser, with a final purity of 99.999%.

The results of these TEAs are shown in Figure 4.24. The so-called Tornado plots present the effect of the STH efficiency, the material lifetime, and the optical concentration factor on the cost of 1 kg of hydrogen in both scenarios. Each calculation reported in these figures represents the variation of a single parameter from the base case, which is STH = 10%, 2-yr durability, \$100/m^2 for scenario #1, and STH = 20%, 2-yr durability, \$60/m^2 for scenario #2. Under these assumptions, a cost of \$5.95/kgH$_2$ and \$3.09/kgH$_2$ are obtained for the coplanar and MJ architectures, respectively. Regardless of the technology, the STH efficiency is the variable with the most leverage on PEC hydrogen production costs.

In Scenario #1 (low-efficiency coplanar device), increasing the STH efficiency from 10% to 15%, a value close to the theoretical maximum for any coplanar device, leads to a cost of \$4.09/kg_{H2}. In Scenario #2 (high-efficiency MJ device), increasing the STH efficiency from 20% to 25% reduces the cost per kg_{H2} down to \$2.54 (/unit). The relative impact and leverage of the material lifetime and concentration factor vary between the two architectures. The rather low impact of the PEC material lifetime on the hydrogen production costs in both scenarios is explained by a capital cost for the PEC electrode (approx. \$200,000 for a 1 TPD sub-plant) that is ten times lower than that of the concentrator and receiver (typically \$2,000,000 for a 1 TPD sub-plant). As such, the use of optical concentrators limits the burden of material photo-corrosion on the cost of hydrogen when compared to non-concentrated planar designs. Ultimately, the calculations show that increasing both STH efficiency to 25% and material durability to 10 yr could lead to a cost per kilogram of H_2 of \$1.95. Although the STH and durability values used in this TEA exceed significantly those achieved by any PEC material system to date, it is the strong economic lever that comes from having a significantly higher efficiency that makes investigating MJ PEC devices a particularly relevant route to achieve low-cost water splitting. We invite the reader to review the work from James et al. (James et al., 2009) and Pinaud et al. (Pinaud et al., 2013) for more information regarding TEA of PEC technologies with the H2A model.

4.5.3 Independent Assessments of PV-E Technology

Rodriguez et al. reported in 2014 on a detailed cost analysis of PV-E systems (Rodriguez et al., 2014), focusing on the impact of the cost of solar absorbers (thin film a-Si, crystalline Si, unconcentrated and concentrated III-V) and catalysts (e.g., PGM Pt/IrO_2 or low-cost NiMo/Co_3O_4). The authors concluded that hydrogen production costs were dominated by the PV components and not the catalytic materials, since the former occupy an area 100× larger than the latter. Cost as low as \$2.90/kg_{H2}, excluding compression, installation cost, operation, and maintenance, were reported in this study.

Grube et al. analyzed PV-E systems equipped or not with DC/DC conversion and located in regions of low (Oldenburg, Germany) and high (Dagget, United States of America) solar irradiance distribution (Grube et al., 2020). Hydrogen production costs were found to be nearly twice as high in low irradiance locations (approx. €9/kg_{H2}) compared to regions with high irradiance (approx. €5/kg_{H2}). The levelized cost of hydrogen (LCOH), that is the ratio between the total annual cost and annual hydrogen production, was found to be 58–84% greater in low irradiation areas compared to high irradiation areas. The authors also determined that power converters did not lead to significant cost reductions due to higher related investments.

Yates et al. reported on the use of Monte Carlo methods to identify parameters (e.g., capital cost, efficiency, operating costs, stack lifetime, discount rates) with the most significant impact on LCOH for off-grid PV-E systems (Yates et al., 2020). This randomized approach was selected to expand TEAs beyond the specific case studies typically reported in the literature and based on predetermined parameters (e.g., fixed capacity factor, averaged insulation data, or unique yearly degradation rate). The team concluded that system size, capital costs, and electrolyzer efficiency are the three most important drivers for LCOH. Averaged LCOH ranging from \$3.38 kg/H_2 (Port Hedland, Australia) to \$4.72 kg/H_2 (Fukushima, Japan) were calculated with this model.

In contrast, Hinkley et al. reported significantly higher LCOH values (\$18.70 kg/H_2) using current cost of PV and electrolyzer components, assuming a total system cost of \$2,300/kW for a large-scale, non-stacking, PV with a capacity factor of 20.5% (Hinkley et al., 2016). Cost reductions down to \$9.10 kg/H_2 were projected by 2030 from the reduction in balance of system cost.

4.5.4 Comparative Assessments of PV-E and PEC Technologies

The spread in dollar figures reported in independent TEA of PV-E on the one hand, and PEC on the other, highlights the difficulty to forecast hydrogen production costs, even when mature designs are considered. Part of the issue is in the selection of predetermined parameters, such as solar irradiance, capacity factors, device efficiency, or lifetime, which are not consistent across the various studies, making the comparison of TEAs difficult. Therefore, it is best if both PEC and PV-E technologies are compared on a level playing field in a single study.

In their TEA published in 2016, Shanner et al. (Shaner et al., 2016) analyzed four technologies: (a) planar thin film PEC with concentration and tracking ("Type 4", STH = 20%, 10× optical concentration, III-V tandems, Pt/IrO_x catalysts, capacity factor = 18.6%, 7-yr lifetime), (b) planar PEC without solar concentration and tracking ("Type 3", STH = 9.76%, c-Si coplanar,

Pt/IrO_x catalysts, 7-yr lifetime), as well as (c) "standard" and (d) grid-supplemented[3] PV-E (GSPV-E) systems (STH = 9.76%, η_{PV} = 16%, η_{PEM} = 61%, capacity factor = 20.4%, 25-yr lifetime). The LCOH and capital expenditure (calculated per m^2 of solar collection) for the "standard" PV-E were found to be \$12.1 kg_{H2} and \$371/m^2, respectively. For comparison, the LCOH of direct grid-electrolysis was \$5.5 kg_{H2}. The plant efficiency was found to have the greatest leverage on hydrogen cost, and a LCOH value of \$9.4 kg_{H2} was calculated by replacing the PV component (η_{PV} = 16%) with state-of-the-art more expensive cells (η_{PV} = 46%). Also, the authors found that reducing LCOH down to \$3 kg_{H2} while keeping the PV capital cost of \$0.5/W_P would require an electrolyzer capital cost of \$0.1/W_P, a value one order of magnitude lower than that currently achieved with PEM electrolyzers. The LCOH and capital expenditure calculated for GSPV-E were \$6.1 kg_{H2} and \$441/m^2, respectively. The PV efficiency and cost of grid electricity were the two most sensitive parameters for this solution. Again, using high-efficiency PV would reduce LCOH, down to \$5.6 kg_{H2}. The LCOH and capital expenditure of Type 3 PEC scheme (9.76% STH efficiency, no concentration) were \$11.4 kg_{H2} and \$293/m^2, respectively. The two most impactful parameters were plant efficiency and the capital expenses of the active component (i.e., cost of semiconducting and catalytic materials). The LCOH and capital expenditure of Type 4 PEC system (20% STH efficiency, 10× optical concentration) were \$9.2 kg_{H2} and \$428/m^2, respectively. Similar to the "Type 3" system, plant efficiency and active component capital expenses were found to have the greatest impact on hydrogen production costs. Achieving an STH efficiency of 25% would further reduce the LCOH to \$7.4 kg_{H2}. Ultimately, it was the lower cost of transporting hydrogen (\$0.1 MM/GW-mile) compared to electricity (\$1 MM/GW-mile), which gave the advantage of PEC design ("Type 3": \$11.4 kg_{H2}, "Type 4": \$9.2 kg_{H2}) over "standard" PV-E (\$12.1 kg_{H2}).

Grimm et al. reported on another TEA comparison between PV-E and PEC (none-concentrated) technologies (Grimm et al., 2020). The PV-E system concept comprised 18% efficient c-Si PV modules (\$0.3/W, 20-yr lifetime) connected to PEM electrolyzers (\$0.4/W, 7-yr lifetime) with efficiency of 61%. Since DC/DC converter can yield to higher operating costs, hydrogen production was maximized by altering the PV-to-electrolyzer ratio (as shown in Figure 4.12a), leading to a capacity factor increase from 28% to 31%. The overall plant area was 4× that of the PV surface. The PEC cell concept (\$153/m^2, 7-yr lifetime) with STH of 10% comprised one anode fully immersed in solution and a photoelectrode that only had its back contact (cathode) in solution, keeping the photoactive region (a-Si triple junction) in contact with ambient air. A Nosepta® branded membrane was used to prevent oxygen and hydrogen crossover. The LCOH calculated for the PV-E scheme was \$6.22 kg_{H2}, including \$2.30 kg_{H2} for the PV modules and \$3.92 kg_{H2} for the PEM electrolyzer. For the PEC system, a LCOH of \$8.43 kg_{H2} was calculated, 50% of which was attributed to the cost of the PEC module. Out of the \$4.61 kg_{H2} attributed to the PEC module, the membrane (\$1.50 kg_{H2}) and photoactive materials (\$1.35 kg_{H2}) were the biggest contributors.

A summary of all TEA presented in this section is provided in Table 4.3 for reference. One should keep in mind that since PV-E and PEC have different TRLs, TEAs should not be used necessarily to rank the technologies but rather be used as guidelines to help researchers identify the parameters with the greatest leverage on hydrogen cost.

4.6 Conclusion and Outlook

With a production cost below \$2/kg_{H2}, SMR remains to this date the most economical path for hydrogen production. Although not a green technology by any stretch of the imagination, SMR is still a necessary enabler that will help propel the deployment of a hydrogen-based transportation infrastructure. In this context, fuel cell car owners are expected to pay \$4/kgH_2 at the pump in a near future (\$2/kg_{H2} for storage and transport added to \$2/kg_{H2} for production), a price competitive with oil considering 1 kg_{H2} contains the same energy as 1 gallon of gasoline.

Electrolysis powered by renewables has been identified as *green* alternatives to SMR, with PV-E being the most promising with STH efficiencies of over 30%. Recent TEA estimate hydrogen production costs with PV-E in the ~\$4–12 kg_{H2} range, compared to ~\$5–7 kg_{H2} for grid-electrolysis. PEC water splitting is another key contender for *green* hydrogen, with production costs as low as ~\$1.95 kg_{H2}. Unfortunately, no PEC device has simultaneously reached the efficiency, cost target, and durability required to meet this value. Above all, materials limited lifetime under operations is arguably the most critical issue that needs to be solved by the community to convince stakeholders and investors of the value of PEC water splitting.

---

3 A hybrid of grid-electrolysis and PV-E. This design offers the possibility to boost water electrolysis with grid electricity whenever the PV components are not operating at their maximum capacity.

Table 4.3 Hydrogen Production Costs Reported for PV-E and PEC Technologies in Several Independent and Combined Techno-Economic Analyses.

PV-E		PEC		
Key Assumptions	**cost/kg$_{H2}$**	**Key Assumptions**	**cost/kg$_{H2}$**	**References**
PV: a-Si/a-Si/μc-Si PV (16%). E: various. Costs related to compression, installation, and O&M are excluded.	\$2.9	–	–	Rodriguez et al. (2014)
PV: c-Si (\$0.8/W_P). E: PEM (\$0.6/W_P).	€5–€9	–	–	Grube et al. (2020)
Monte Carlo-assisted TEA PV: various, \$0.818/W_{AC} nominal. E: various, \$0.784/W_{AC} nominal.	\$3.38–\$4.72	–	–	Yates et al. (2020)
–	–	PEC: coplanar (STH_{max} = 15%) or MJ (STH_{max} = 25%) CIGSe-based devices. Lifetime: 0.5–10 yr	\$2.54–\$11.52	This chapter
PV efficiency: 16%. E efficiency: 61%. Lifetime: 25 yr	\$12.1	PEC: 9.76–20%. Lifetime: 7 yr	\$9.2–\$11.4	Shaner et al. (2016)
PV: 18% efficient c-Si, \$0.3/W, 20-yr lifetime. E: 61% efficient, \$0.4/W, 7-yr lifetime.	\$6.22	PEC: 10% STH efficiency, c-Si, 7-yr lifetime.	\$8.43	Grimm et al. (2020)

In this chapter, we have presented PV-E and PEC as two technologies competing for large-scale hydrogen production, and clearly, PV-E currently has the edge both in terms of maturity and cost. Instead, it appears more sensible to look for new opportunities where the *value proposition* of PEC is intrinsically higher than PV-E. In military operations, for example, portability and ease of utilization are far more critical than raw performance, and one could envision all-in-one PEC devices being the lightweight solution military troops need to produce hydrogen for fuel cell-equipped autonomous vehicles. The value of high-purity oxygen has thus far not been assessed in any TEAs of PEC devices. Being able to produce medical-grade oxygen on demand and on the spot with simple PEC systems could be extremely valuable during disaster relief operations. Finally, PEC devices could be part of storage energy solutions to buffer long-term solar fluctuations in small islands and other remote off-grid locations. Other niches not yet envisioned probably exist for PEC water splitting.

Acknowledgments

The author would like to acknowledge and express his gratitude to the U.S. Department of Energy's Hydrogen and Fuel Cell Technologies Office for the financial support. The author also thanks the current and former members of the Thin Films Laboratory at the University of Hawaii's Hawaii Natural Energy Institute, including Drs. Richard Rocheleau, Eric Miller, Dixit Prasher, Marina Chong, Yuancheng Chang, Jess Kaneshiro, and Wilman Septina, as well as Alexander Deangelis, Joshua Crunk, and Kai Outlaw-Spruell for their encouragement and support. Finally, the authors would like to recognize Drs. Clemens Heske, Thomas Jaramillo, Kai Zhu, Christopher Muzzillo, Heli Wang, Andriy Zakutayev, Tadashi Ogitsu, Joel Varley, Jason Cooper, and coworkers for their outstanding contributions to the discovery of unique chalcopyrite absorbers and interfaces.

References

Abdi, F. F., Han, L., Smets, A. H. M., Zeman, M., Dam, B., & van de Krol, R. (2013). Efficient solar water splitting by enhanced charge separation in a bismuth vanadate-silicon tandem photoelectrode. *Nature Communications*, *4*(1), 2195. Available from: http://www.nature.com/articles/ncomms3195

Abdi, F. F., & Van De Krol, R. (2012). Nature and light dependence of bulk recombination in Co-Pi-catalyzed $BiVO_4$ photoanodes. *Journal of Physical Chemistry C*, *116*(17), 9398–9404. Available from: http://pubs.acs.org/doi/abs/10.1021/jp3007552

Ager, J. W., Shaner, M. R., Walczak, K. A., Sharp, I. D., & Ardo, S. (2015). Experimental demonstrations of spontaneous, solar-driven photoelectrochemical water splitting. *Energy & Environmental Science*, *8*(10), 2811–2824. Available from: 10.1039/C5EE00457H

Ardo, S., Fernandez Rivas, D., Modestino, M. A., Schulze Greiving, V., Abdi, F. F., Alarcon Llado, E., Artero, V., Ayers, K., Battaglia, C., Becker, J. P., & Bederak, D. (2018). Pathways to electrochemical solar-hydrogen technologies. *Energy & Environmental Science*, *11*(10), 2768–2783.

Bär, M., Weinhardt, L., Marsen, B., Cole, B., Gaillard, N., Miller, E., & Heske, C. (2010). Mo incorporation in WO_3 thin film photoanodes: Tailoring the electronic structure for photoelectrochemical hydrogen production. *Applied Physics Letters*, *96*(3), 032107. Available from: http://link.aip.org/link/APPLAB/v96/i3/p032107/s1&Agg=doi

Bernt, M., Hartig-Weiß, A., Tovini, M. F., El-Sayed, H. A., Schramm, C., Schröter, J., Gebauer, C., & Gasteiger, H. A. (2020). Current challenges in catalyst development for PEM water electrolyzers. *Chemie-Ingenieur-Technik*, *92*(1–2), 31–39.

Cell, T. P. (2012). Chapter 2: Principle of photoelectrochemical cells. In R. van de Krol & M. Grätzel (Eds.), *Photoelectrochemical hydrogen production*, Electronic Materials: Science & Technology (Vol. *102*). Springer. Available from: http://link.springer.com/10.1007/978-1-4614-1380-6

Chang, W. J., Lee, K. H., Ha, H., Jin, K., Kim, G., Hwang, S. T., Lee, H. M., Ahn, S. W., Yoon, W., Seo, H., & Hong, J. S. (2017). Design principle and loss engineering for photovoltaic-electrolysis cell system. *ACS Omega*, *2*(3), 1009–1018.

Chang, Y., Braun, A., Deangelis, A., Kaneshiro, J., & Gaillard, N. (2011). Effect of thermal treatment on the crystallographic, surface energetics, and photoelectrochemical properties of reactively cosputtered copper tungstate for water splitting. *The Journal of Physical Chemistry*, *115*(51), 25490–25495. Available from: http://pubs.acs.org/doi/abs/10.1021/jp207341v

Chen, Z., Dinh, H. N., & Miller, E. (2013). *Photoelectrochemical water splitting* (pp. 49–62). Springer. Available from: http://link.springer.com/10.1007/978-1-4614-8298-7

Chen, Z., Jaramillo, T. F., Deutsch, T. G., Kleiman-Shwarsctein, A., Forman, A. J., Gaillard, N., Garland, R., Takanabe, K., Heske, C., Sunkara, M., & McFarland, E. W. (2010). Accelerating materials development for photoelectrochemical hydrogen production: Standards for methods, definitions, and reporting protocols. *Journal of Materials Research*, *25*(1), 3–16.

Chu, S., Li, W., Yan, Y., Hamann, T., Shih, I., Wang, D., & Mi, Z. (2017). Roadmap on solar water splitting: Current status and future prospects. *Nano Futures*, *1*(2), 022001.

Cole, B., Marsen, B., Miller, E., Yan, Y., To, B., Jones, K., & Al-Jassim, M. (2008). Evaluation of nitrogen doping of tungsten oxide for photoelectrochemical water splitting. *Journal of Physical Chemistry C*, *112*(13), 5213–5220. Available from: http://pubs.acs.org/doi/abs/10.1021/jp077624c

Costogue, E. N., & Yasui, R. K. (1977). Performance data for a terrestrial solar photovoltaic/water electrolysis experiment. *Solar Energy*, *19*(2), 205–210.

Dahbi, S., Aboutni, R., Aziz, A., Benazzi, N., Elhafyani, M., & Kassmi, K. (2016). Optimised hydrogen production by a photovoltaic-electrolysis system DC/DC converter and water flow controller. *International Journal of Hydrogen Energy*, *41*(45), 20858–20866. Available from:. https://doi.org/10.1016/j.ijhydene.2016.05.111

Deangelis, A. D., Horsley, K., & Gaillard, N. (2018). Wide band gap $CuGa(S,Se)_2$ thin films on transparent conductive fluorinated tin oxide substrates as photocathode candidates for tandem water splitting devices. *Journal of Physical Chemistry C*, *122*(26), 14304–14312.

Dotan, H., Sivula, K., Grätzel, M., Rothschild, A., & Warren, S. C. (2011). Probing the photoelectrochemical properties of hematite (α-Fe_2O_3) electrodes using hydrogen peroxide as a hole scavenger. *Energy & Environmental Science*, *4*(3), 958.

Esteve, D., Ganibal, C., Steinmetz, D., & Vialaron, A. (1982). Performance of a photovoltaic electrolysis system. *International Journal of Hydrogen Energy*, *7*(9), 711–716.

Fischer, M. (1986). Review of hydrogen production with photovoltaic electrolysis systems. *International Journal of Hydrogen Energy*, *11*(8), 495–501.

Fujishima, A., & Honda, K. (1972). Electrochemical photolysis of water at a semiconductor electrode. *Nature*, *238*(5358), 37–38. Available from:. https://doi.org/10.1038/238037a0

Gaillard, N., Chang, Y., Deangelis, A., Higgins, S., & Braun, A. (2013). A nanocomposite photoelectrode made of 2.2 eV band gap copper tungstate ($CuWO_4$) and multi-wall carbon nanotubes for solar-assisted water splitting. *International Journal of Hydrogen Energy, 38*(8), 3166–3176.

Gaillard, N., Chang, Y., Kaneshiro, J., Deangelis, A., Miller, E.L. (2010). Status of research on tungsten oxide-based photoelectrochemical devices at the University of Hawai'i. In H. Idriss & H. Wang (Eds.), *Proceedings of SPIE—The International Society for Optical Engineering* [*Internet*]. *77700V*, *Solar Hydrogen and Nanotechnology V*, vol. 7770, pp. 100-113. San Diego, CA, US: SPIE. Available from: http://proceedings.spiedigitallibrary.org/proceeding.aspx?doi=10.1117/12.860970

Gaillard, N., Cole, B., Kaneshiro, J., Miller, E. L., Marsen, B., Weinhardt, L., Bär, M., Heske, C., Ahn, K. S., Yan, Y., & Al-Jassim, M. M. (2010). Improved current collection in WO_3:Mo/WO_3 bilayer photoelectrodes. *Journal of Materials Research, 25*, 45–51.

Gaillard, N., Prasher, D., Chong, M., Deangelis, A., Horsley, K., Ishii, H. A., Bradley, J. P., Varley, J., & Ogitsu, T. (2019). Wide-bandgap Cu(In,Ga)S_2 photocathodes integrated on transparent conductive F:SnO_2 substrates for chalcopyrite-based water splitting tandem devices. *ACS Applied Energy Materials, 2*(8), 5515–5524.

Gaillard, N., Prasher, D., Kaneshiro, J., Mallory, S., Chong, M. (2013). Development of chalcogenide thin film materials for photoelectrochemical hydrogen production. *Materials Research Society Symposium Proceedings: Vol. 1558 Symposium Z—Nanotechnology and Sustainability*: mrss13-1558-z02-07.

Gallandat, N., Romanowicz, K., & Züttel, A. (2017). An analytical model for the electrolyser performance derived from materials parameters. *Journal of Power and Energy Engineering, 05*(10), 34–49.

García-Valverde, R., Miguel, C., Martínez-Béjar, R., & Urbina, A. (2008). Optimized photovoltaic generator water electrolyser coupling through a controlled DC-DC converter. *International Journal of Hydrogen Energy, 33*(20), 5352–5362.

Gerischer, H. (1979). Solar photoelectrolysis with semiconductor electrodes. In B. O. Seraphin (Ed.), *Solar energy conversion: Solid-state physics aspects* (pp. 115–172. Available from:). Springer. https://doi.org/10.1007/3-540-09224-2_4

Gerischer, H. (1990). The impact of semiconductors on the concepts of electrochemistry. *Electrochimica Acta, 35*(11), 1677–1699. Available from: https://www.sciencedirect.com/science/article/pii/001346869087067C

Ghirardi, M. L., & Mohanty, P. (2010). Oxygenic hydrogen photoproduction—current status of the technology. *Current Science, 98*(4), 499–507.

Gorensek, M. B., Corgnale, C., & Summers, W. A. (2017). Development of the hybrid sulfur cycle for use with concentrated solar heat. I. Conceptual design. *International Journal of Hydrogen Energy, 42*(33), 20939–20954. Available from:. https://doi.org/10.1016/j.ijhydene.2017.06.241

Goto, Y., Hisatomi, T., Wang, Q., Higashi, T., Ishikiriyama, K., Maeda, T., Sakata, Y., Okunaka, S., Tokudome, H., Katayama, M., & Akiyama, S. (2018). A particulate photocatalyst water-splitting panel for large-scale solar hydrogen generation. *Joule, 2*(3), 509–520. Available from:. https://doi.org/10.1016/j.joule.2017.12.009

Green, M. A., Emery, K., Hishikawa, Y., Warta, W., & Dunlop, E. D. (2013). Solar cell efficiency tables (version 41). *Progress in Photovoltaics: Research and Applications, 21*(1), 1–11.

Grimm, A., de Jong, W. A., & Kramer, G. J. (2020). Renewable hydrogen production: A techno-economic comparison of photoelectrochemical cells and photovoltaic-electrolysis. *International Journal of Hydrogen Energy, 45*(43), 22545–22555. Available from:. https://doi.org/10.1016/j.ijhydene.2020.06.092

Grube, T., Reul, J., Reuß, M., Calnan, S., Monnerie, N., Schlatmann, R., Sattler, C., Robinius, M., & Stolten, D. (2020). A techno-economic perspective on solar-to-hydrogen concepts through 2025. *Sustainable Energy & Fuels, 4*(11), 5818–5834.

Gunawan, G., Septina, W., Ikeda, S., Harada, T., Minegishi, T., Domen, K., & Matsumura, M. (2014). Platinum and indium sulfide-modified $CuInS_2$ as efficient photocathodes for photoelectrochemical water splitting. *Chemical Communications, 50*(64), 8941–8943. Available from: http://xlink.rsc.org/?DOI=C4CC03634D

Hellstern, T. R., Palm, D. W., Carter, J., Deangelis, A. D., Horsley, K., Weinhardt, L., Yang, W., Blum, M., Gaillard, N., Heske, C., & Jaramillo, T. F. (2019). Molybdenum disulfide catalytic coatings via atomic layer deposition for solar hydrogen production from copper gallium diselenide photocathodes. *ACS Applied Energy Materials, 2*(2), 1060–1066.

Hinkley, J., Hayward, J., Mcnaughton, R., Gillespie, R., Matsumoto, A., Watt, M. and Lovegrove, K. (2016). *Cost assessment of hydrogen production from PV and electrolysis.* Report to ARENA as part of Solar Fuels Roadmap, Project A-3018. Autralia, CSIRO. Available from: https://arena.gov.au/assets/2016/05/Assessment-of-the-cost-of-hydrogen-from-PV.pdf

Hu, J., Zhu, F., Kunrath, A., Gaillard, N. (2014). Development of a-SiC thin film photoelectrodes and hybrid PV/a-SiC devices for photoelectrochemical water splitting. In J. Turner (Ed.), *Materials Research Society Symposium Proceedings. Vol. 1669. Symposium D—Materials for Photoelectrochemical and Photocatalytic Solar-Energy Harvesting and Storage.* San Francisco, CA, US: Cambridge University Press, 13–18.

Jacobsson, T. J., Fjällström, V., Edoff, M., & Edvinsson, T. (2014). Sustainable solar hydrogen production: From photoelectrochemical cells to PV-electrolyzers and back again. *Energy & Environmental Science*, *7*(7), 2056–2070.

Jacobsson, T. J., Fjällström, V., Sahlberg, M., Edoff, M., & Edvinsson, T. (2013). A monolithic device for solar water splitting based on series interconnected thin film absorbers reaching over 10% solar-to-hydrogen efficiency. *Energy & Environmental Science*, *6*(12), 3676. Available from: https://pubs.rsc.org/en/content/articlelanding/2013/ee/c3ee42519c

James, B.D., Baum, G.N., Perez, J., Baum, K.N. (2009). *Technoeconomic analysis of photoelectrochemical (PEC) hydrogen production*. Vol. 22201, DOE Contract Number: GS-10F-009J. Arlington, VA, US: Directed Technologies Inc.Available from: http://www.directedtechnologies.com/publications/fuel_options/PEC_Technoeconomic_Analysis-DTI.pdf

Jang, J. W., Du, C., Ye, Y., Lin, Y., Yao, X., Thorne, J., Liu, E., McMahon, G., Zhu, J., Javey, A., & Guo, J. (2015). Enabling unassisted solar water splitting by iron oxide and silicon. *Nature Communications*, *6*(May), 1–5.

Jia J, Seitz LC, Benck JD, Huo Y, Chen Y, Ng JWD, Bilir, T., Harris, J.S., and Jaramillo, T.F. Solar water splitting by photovoltaic-electrolysis with a solar-to-hydrogen efficiency over 30%. *Nature Communications* 2016;*7*(May):1–6. Available from: 10.1038/ncomms13237

Khan, I. S., Muzzillo, C. P., Perkins, C. L., Norman, A. G., Young, J. L., Gaillard, N., & Zakutayev, A. (2021). $Mg_xZn_{1-x}O$ contact to $CuGa_3Se_5$ absorber for photovoltaic and photoelectrochemical devices. *Journal of Physics: Energy*, *3*(2), 024001. Available from: https://iopscience.iop.org/article/10.1088/2515-7655/abd3b3

Khaselev, O., & Turner, J. A. (1998). A monolithic photovoltaic-photoelectrochemical device for hydrogen production via water splitting. *Science (80-)*, *280*(5362), 425–427. Available from: http://www.sciencemag.org/cgi/doi/10.1126/science.280.5362.425

Kibria, M. G., Chowdhury, F. A., Zhao, S., AlOtaibi, B., Trudeau, M. L., Guo, H., & Mi, Z. (2015). Visible light-driven efficient overall water splitting using p-type metal-nitride nanowire arrays. *Nature Communications*, *6*, 1–8.

Kim, J. H., Hansora, D., Sharma, P., Jang, J. W., & Lee, J. S. (2019). Toward practical solar hydrogen production-an artificial photosynthetic leaf-to-farm challenge. *Chemical Society Reviews*, *48*(7), 1908–1971.

Kim, J.-H., Lee, J.-N., Yoo, C.-Y., Lee, K.-B., & Lee, W.-M. (2015). Low-cost and energy-efficient asymmetric nickel electrode for alkaline water electrolysis. *International Journal of Hydrogen Energy*, *40*(34), 10720–10725. Available from: https://www.sciencedirect.com/science/article/pii/S036031991501753X

Kirchartz, T. (2019). High open-circuit voltages in lead-halide perovskite solar cells—experiment, theory and open questions. *Philosophical Transactions of the Royal Society A*, *377*(2152), 20180286.

Licht, S., Wang, B., Mukerji, S., Soga, T., Umeno, M., & Tributsch, H. (2001). Over 18% solar energy conversion to generation of hydrogen fuel; theory and experiment for efficient solar water splitting. *International Journal of Hydrogen Energy*, *26*(7), 653–659.

Liu, Z., Krückemeier, L., Krogmeier, B., Klingebiel, B., Márquez, J. A., Levcenko, S., Öz, S., Mathur, S., Rau, U., Unold, T., & Kirchartz, T. (2019). Open-circuit voltages exceeding 1.26 V in planar methylammonium lead iodide perovskite solar cells. *ACS Energy Letters*, *4*(1), 110–117.

Luo, J., Im, J. H., Mayer, M. T., Schreier, M., Nazeeruddin, M. K., Park, N. G., Tilley, S. D., Fan, H. J., & Grätzel, M. (2014). Water photolysis at 12.3% efficiency via perovskite photovoltaics and Earth-abundant catalysts. *Science (80-)*, *345*(6204), 1593–1596.

Markus, R. A. (1956). On the theory of oxidation-reduction reactions involving electron transfer. I. *The Journal of Chemical Physics*, *24*(5), 966–978.

Marsen, B., Miller, E. L., Paluselli, D., & Rocheleau, R. E. (2007). Progress in sputtered tungsten trioxide for photoelectrode applications. *International Journal of Hydrogen Energy*, *32*(15), 3110–3115. Available from: http://linkinghub.elsevier.com/retrieve/pii/S0360319907000547

McCrory, C. C. L., Jung, S., Ferrer, I. M., Chatman, S. M., Peters, J. C., & Jaramillo, T. F. (2015). Benchmarking hydrogen evolving reaction and oxygen evolving reaction electrocatalysts for solar water splitting devices. *Journal of the American Chemical Society*, *137*(13), 4347–4357.

Merdes, S., Abou-Ras, D., Mainz, R., Klenk, R., Lux-Steiner, M. C., Meeder, A., Schock, H. W., & Klaer, J. (2013). $CdS/Cu(In,Ga)S_2$ based solar cells with efficiencies reaching 12.9% prepared by a rapid thermal process. *Progress in Photovoltaics: Research and Applications*, *21*(1), 88–93. Available from: http://doi.wiley.com/10.1002/pip.2165

Merdes, S., Mainz, R., Klaer, J., Meeder, A., Rodriguez-Alvarez, H., Schock, H. W., Lux-Steiner, M. C., & Klenk, R. (2011). 12.6% efficient $CdS/Cu(In,Ga)S_2$-based solar cell with an open circuit voltage of 879 mV prepared by a rapid thermal process. *Solar Energy Materials & Solar Cells*, *95*(3), 864–869. Available from: http://linkinghub.elsevier.com/retrieve/pii/S092702481000646X

Miller, E. L., Paluselli, D., Marsen, B., & Rocheleau, R. E. (2005). Development of reactively sputtered metal oxide films for hydrogen-producing hybrid multijunction photoelectrodes. *Solar Energy Materials & Solar Cells, 88*(2), 131–144. Available from: http://linkinghub.elsevier.com/retrieve/pii/S0927024804004702

Moriya, M., Minegishi, T., Kumagai, H., Katayama, M., Kubota, J., & Domen, K. (2013). Stable hydrogen evolution from CdS-modified $CuGaSe_2$ photoelectrode under visible-light irradiation. *Journal of the American Chemical Society, 135*(10), 3733–3735. Available from: http://www.ncbi.nlm.nih.gov/pubmed/23437875

Muzzillo, C. P., Klein, W. E., Li, Z., Deangelis, A. D., Horsley, K., Zhu, K., & Gaillard, N. (2018). Low-cost, efficient, and durable H_2 production by photoelectrochemical water splitting with $CuGa_3Se_5$ photocathodes. *ACS Applied Materials & Interfaces, 10*(23), 19573–19579.

Nakada, T., Matsumoto, K., Okumura, M. (2003). Improved efficiency of Cu(In,Ga)Se_2 thin film solar cells by surface sulfurization using wet process. In: *Conference Record of the Twenty-Ninth IEEE Photovoltaic Specialists Conference, 2002* (p. 527–530). New Orleans, LA, USA: IEEE. Available from: http://ieeexplore.ieee.org/document/1190618/

Nakada, T., Ohbo, H., Watanabe, T., Nakazawa, H., Matsui, M., & Kunioka, A. (1997). Improved Cu(In,Ga)(S,Se)2 thin film solar cells by surface sulfurization. *Solar Energy Materials & Solar Cells, 49*(1–4), 285–290. Available from: http://linkinghub.elsevier.com/retrieve/pii/S0927024897000548

Nakamura, A., Ota, Y., Koike, K., Hidaka, Y., Nishioka, K., Sugiyama, M., & Fujii, K. (2015). A 24.4% solar to hydrogen energy conversion efficiency by combining concentrator photovoltaic modules and electrochemical cells. *Applied Physics Express, 8*(10), 107101. Available from:. https://doi.org/10.7567/APEX.8.107101

Nishiyama, H., Yamada, T., Nakabayashi, M., Maehara, Y., Yamaguchi, M., Kuromiya, Y., Nagatsuma, Y., Tokudome, H., Akiyama, S., Watanabe, T., & Narushima, R. (2021). Photocatalytic solar hydrogen production from water on a 100-m^2 scale. *Nature, 598*(7880), 304–307. Available from:. https://doi.org/10.1038/s41586-021-03907-3

Nozik, A. J. (1976). p-n photoelectrolysis cells. *Applied Physics Letters, 29*(3), 150–153. Available from:. https://doi.org/10.1063/1.89004

Nozik, A. J., & Memming, R. (1996). Physical chemistry of semiconductor—liquid interfaces. *The Journal of Physical Chemistry, 3654*(95), 13061–13078.

Palm, D. W., Muzzillo, C. P., Ben-Naim, M., Khan, I., Gaillard, N., & Jaramillo, T. F. (2021). Tungsten oxide-coated copper gallium selenide sustains long-term solar hydrogen evolution. *Sustainable Energy & Fuels, 5*(2), 384–390.

Paul, B., & Andrews, J. (2008). Optimal coupling of PV arrays to PEM electrolysers in solar—hydrogen systems for remote area power supply. *International Journal of Hydrogen Energy, 33*(2), 490–498.

Peharz, G., Dimroth, F., & Wittstadt, U. (2007). Solar hydrogen production by water splitting with a conversion efficiency of 18%. *International Journal of Hydrogen Energy, 32*(15), 3248–3252.

Phillips, R., & Dunnill, C. W. (2016). Zero gap alkaline electrolysis cell design for renewable energy storage as hydrogen gas. *RSC Advances, 6*(102), 100643–100651.

Pinaud, B. A., Benck, J. D., Seitz, L. C., Forman, A. J., Chen, Z., Deutsch, T. G., James, B. D., Baum, K. N., Baum, G. N., Ardo, S., & Wang, H. (2013). Technical and economic feasibility of centralized facilities for solar hydrogen production via photocatalysis and photoelectrochemistry. *Energy & Environmental Science, 6*(7), 1983–2002. Available from: 10.1039/C3EE40831K

Rajagopal, A., Stoddard, R. J., Jo, S. B., Hillhouse, H. W., & Jen, A. K. Y. (2018). Overcoming the photovoltage plateau in large bandgap perovskite photovoltaics. *Nano Letters, 18*(6), 3985–3993.

Rajeshwar, K. (2007; Available from:). Fundamentals of semiconductor electrochemistry and photoelectrochemistry. *Encyclopedia of Electrochemistry*. https://doi.org/10.1002/9783527610426.bard060001

Reece, S. Y., Hamel, J. A., Sung, K., Jarvi, T. D., Esswein, A. J., Pijpers, J. J. H., & Nocera, D. G. (2011). Wireless solar water splitting using silicon-based semiconductors and earth-abundant catalysts. *Science (80-), 334*(6056), 645–648. Available from: http://science.sciencemag.org/content/334/6056/645.abstract

Reuß, M., Reul, J., Grube, T., Langemann, M., Calnan, S., Robinius, M., Schlatmann, R., Rau, U., & Stolten, D. (2019). Solar hydrogen production: A bottom-up analysis of different photovoltaic-electrolysis pathways. *Sustainable Energy & Fuels, 3*(3), 801–813.

Rocheleau, R. E., Miller, E. L., & Misra, A. (1998). High-efficiency photoelectrochemical hydrogen production using multijunction amorphous silicon photoelectrodes. *Energy & Fuels, 12*(1), 3–10. Available from:http://pubs.acs.org/doi/abs/10.1021/ef9701347

Rodriguez, C. A., Modestino, M. A., Psaltis, D., & Moser, C. (2014). Design and cost considerations for practical solar-hydrogen generators. *Energy & Environmental Science, 7*(12), 3828–3835. Available from: 10.1039/C4EE01453G

Seitz, L. C., Chen, Z., Forman, A. J., Pinaud, B. A., Benck, J. D., & Jaramillo, T. F. (2014). Modeling practical performance limits of photoelectrochemical water splitting based on the current state of materials research. *ChemSusChem*, *7*(5), 1372–1385.

Septina, W., Muzzillo, C. P., Perkins, C. L., Giovanelli, A. C., West, T., Ohtaki, K. K., Ishii, H. A., Bradley, J. P., Zhu, K., & Gaillard, N. (2021). In situ Al_2O_3 incorporation enhances the efficiency of $CuIn(S,Se)_2$ solar cells prepared from molecular-ink solutions. *Journal of Materials Chemistry A*, *9*(16), 10419–10426. Available from: 10.1039/D1TA00768H

Shaner, M. R., Atwater, H. A., Lewis, N. S., & McFarland, E. W. (2016). A comparative technoeconomic analysis of renewable hydrogen production using solar energy. *Energy & Environmental Science*, *9*(7), 2354–2371. Available from: 10.1039/C5EE02573G

Sharan, A., Sabino, F. P., Janotti, A., Gaillard, N., Ogitsu, T., & Varley, J. B. (2020). Assessing the roles of Cu- and Ag-deficient layers in chalcopyrite-based solar cells through first principles calculations. *Journal of Applied Physics*, *127*(6), 065303. Available from: http://aip.scitation.org/doi/10.1063/1.5140736

Siegel, A., & Schott, T. (1988). Optimization of photovoltaic hydrogen production. *International Journal of Hydrogen Energy*, *13*(11), 659–675. Available from: https://linkinghub.elsevier.com/retrieve/pii/0360319988900766

Siemer, K., Klaer, J., Luck, I., Bruns, J., Klenk, R., & Bräunig, D. (2001). Efficient $CuInS_2$ solar cells from a rapid thermal process (RTP). *Solar Energy Materials & Solar Cells*, *67*(1–4), 159–166. Available from: http://www.sciencedirect.com/science/article/pii/S0927024800002762

Sze, S. M., & Ng, K. K. (2006). *Physics of semiconductor devices* (pp. 568). Wiley Interscience.

Trasatti, S. (1982). The concept and physical meaning of absolute electrode potential. *Journal of Electroanalytical Chemistry*, *139*(1), 1–13.

Trasatti, S. (1990). The absolute electrode potential—the end of the story. *Electrochimica Acta*, *35*(1), 269–271.

Varley, J. B., Lordi, V., Ogitsu, T., Deangelis, A., Horsley, K., & Gaillard, N. (2018). Assessing the role of hydrogen in Fermi-level pinning in chalcopyrite and kesterite solar absorbers from first-principles calculations. *Journal of Applied Physics*, *123*(16), 161408. Available from: http://aip.scitation.org/doi/10.1063/1.5006272

Xiang, C., Papadantonakis, K. M., & Lewis, N. S. (2016). Principles and implementations of electrolysis systems for water splitting. *Materials Horizons*, *3*(3), 169–173.

Yates, J., Daiyan, R., Patterson, R., Egan, R., Amal, R., Ho-Baille, A., & Chang, N. L. (2020). Techno-economic analysis of hydrogen electrolysis from off-grid stand-alone photovoltaics incorporating uncertainty analysis. *Cell Reports Physical Science*, *1*(10), 100209. Available from:. https://doi.org/10.1016/j.xcrp.2020.100209

Yoneyama, H., Sakamoto, H., & Tamura, H. (1975). A photo-electrochemical cell with production of hydrogen and oxygen by a cell reaction. *Electrochimica Acta*, *20*(5), 341–345. Available from:https://www.sciencedirect.com/science/article/pii/001346867590016X

Young, J. L., Steiner, M. A., Döscher, H., France, R. M., Turner, J. A., & Deutsch, T. G. (2017). Direct solar-to-hydrogen conversion via inverted metamorphic multi-junction semiconductor architectures. *Nature Energy*, *2*(4), 17028. Available from:. https://doi.org/10.1038/nenergy.2017.28

Zeng, K., & Zhang, D. (2010). Recent progress in alkaline water electrolysis for hydrogen production and applications. *Progress in Energy and Combustion Science*, *36*(3), 307–326. Available from:. https://doi.org/10.1016/j.pecs.2009.11.002

Zhu, F., Hu, J., Matulionis, I., Deutsch, T., Gaillard, N., Kunrath, A., Miller, E., & Madan, A. (2009). Amorphous silicon carbide photoelectrode for hydrogen production directly from water using sunlight. *Philosophical Magazine*, *89*(28–30), 2723–2739. Available from: http://www.tandfonline.com/doi/abs/10.1080/14786430902740729

Zou, P., Wan, L., Pan, S., Meng, M., Guo, Z., & Xu, J. (2013). Effect of sulfurization in hydrogen sulfide on the properties of $Cu(In,Ga)Se_2$ thin-film absorbers. *Journal of Materials Science: Materials in Electronics*, *24*(11), 4530–4534. Available from: http://link.springer.com/10.1007/s10854-013-1437-2

5

A Brief History of Molecular Photosynthesis: The Quest for the Bridge Between Light and Chemistry

Liniquer A. Fontana, Vitor H. Rigolin, Catia Ornelas, and Jackson D. Megiatto Jr.

Institute of Chemistry, University of Campinas (UNICAMP), 13083-970, Campinas, SP, Brazil

Following the light of the Sun, we left the Old World behind.

Christopher Columbus.

5.1 Introduction

The Sun is the brightest object in Earth's sky. Naturally, humans have been attracted by our nearest star from the dawn of our species. For most of human history, the Sun has been an object of mystery, reverence, and, sometimes, fear. Our curiosity has driven several attempts to understand the functions and properties of this "ball of fire" in the sky. Obviously, the first attempts to understand the Sun were exclusively based on our senses and gave birth to a series of myths and traditions in virtually all cultures and early civilizations. With its unblinking all-seeing eye, together with the universal connection of light with enlightenment, the Sun was viewed by our ancestors as the supreme deity, who was the stern guarantor of justice, source of wisdom, and bestower of life on Earth. These qualities led many cultures to worship the Sun, which later laid the foundations for many religions and became source of political power as kings ruled by the power of the Sun. Besides myths and religions, early civilizations also learned that the regular motion of the Sun through the sky could be used to predict seasonal changes, which profoundly impacted the development of agriculture and, ultimately, the survival of our species. Without knowing as much as we do nowadays, early humans harvested about 100 times more solar energy by transitioning from hunting and gathering activities to domesticating plants and animals (Phené, 1875; von Frischi, 1940).

As the presence of the Sun gives us life, the absence of the Sun in broad daylight has terrified humans through the ages. Rudimentary methods were devised to predict when eclipses would happen. Those who learned to predict eclipses became enormously powerful, and some of them turned out to be the early scientists. With the renascence and the further development of the scientific method, humans started to layer out the subtle mechanisms by which the Sun powers life on Earth (Close, 2019). The understanding of such mechanisms has been a cumulative process over our history, with many scientists and philosophers giving important contributions. In the following sections, we give a historical account of the most important findings that have enriched our understanding of how solar energy is converted into chemical energy.

5.2 Historical Context and Early Findings

Over a period of about 200 years after the Renaissance epoque, key discoveries revealed the existence of natural photosynthesis. Worth mentioning are the experiments of the Belgian chemist Jan Baptiste van Helmont in 1600s, who scientifically demonstrated what ancient farmers were at some level aware of: water is central to plant growth (Hershey, 1991). Later in the 1700s, Joseph Priestley's experiments informed that there was also something in the air that influenced the biological functions of plants. In his famous experiments of the candle flame in the jar, Priestley discovered that plants were able to restore what he called the "injury" air inside the jar. Although Priestley did not know at the time that his experiments

Conversion of Water and CO_2 to Fuels using Solar Energy: Science, Technology and Materials, First Edition.
Edited by Oomman K. Varghese and Flavio L. Souza.

proved the existence of oxygen gas and the "injury" air was carbon dioxide produced by the combustion of the candle, the link between plant physiology and gases was established (Priestley, 2013).

Soon after the publication of Priestley's findings, the Dutch scientist Jan Ingenhousz in 1779 proved that the green parts of plants produced oxygen gas only under sunlight, which was a landmark discovery in our understanding of natural photosynthesis (Ingenhousz, 1779). In 1796, the Swiss botanist Jean Senebier demonstrated that plants absorb carbon dioxide and release oxygen with the assistance of sunlight, upon which the Swiss chemist Nicolas-Théodore de Saussure concluded after further experiments that the increased mass of growing plants was the result of absorption of carbon dioxide and water (Farmer, 2010). The advent of the thermodynamic laws during the industrial revolution allowed the German physicist Julius Robert Mayer to establish that plants use sunlight to transform carbon dioxide and water into organic compounds, therefore defining photosynthesis as the process that transforms sunlight energy into chemical energy (Devlin & Barker, 1971). Further support for Mayer's conclusions came from the investigations by the German botanist Julius von Sachs in the 1860s, which provided unequivocal evidence for the relationship between starch production, sunlight exposure, and the greenness of the plants (Kutschera, 2015). Last but not least, the American biochemist Cornelis Van Niel proposed the general chemical equation for natural photosynthesis in the 1930s, which upon further elaboration by others led to the photosynthesis' chemical equation that we are all familiar with (Eq. (5.1)) (Niel, 1932).

$$6CO_2 + 6H_2O + \text{sunlight} \rightarrow C_6H_{12}O_6 + 6O_2 \tag{5.1}$$

The broad outlines of natural photosynthesis understood in conjunction with the rapid and exciting scientific developments of the 1800s and early 1900s provided scientists with encouragement to explore the idea of artificial photosynthesis, i.e., to convert sunlight energy into other forms of more useful energy by man-made systems. The first steps into artificial photosynthesis were made with solar concentrators, which were designed to provide heat for boiling water. The steam produced was then used to drive steam engines. Among the first designs, the solar steam engine invented by French mathematician Augustin Mouchot dragged a lot of attention. He demonstrated, at the 1878 World's Fair in Paris, that his machine could harness and use sunlight to produce ice. Such innovation must have indeed astonished his audience, as Mouchot stated in his presentation that heat provided by the Sun could be used to cool things. However, the typically low efficiency and the impossibility to run solar machines when the Sun is not up in the sky rendered solar energy expensive when compared to coal, the primary energy source of the time. A little later, in the 1910s, American inventor Frank Schuman devised solar thermal power plants to demonstrate that sunlight could be used at industrial scales to power mechanical engines able to pump water to agricultural fields. His works proved the commercial viability of solar power against coal in tropical regions. However, the break of World War I and the discovery of even cheaper and easier-to-process and transport liquid fossil fuel, namely petroleum, halted the possibility of the sun-powered society imagined by Schuman (Braga et al., 2014).

In the academic world, the concept of artificial photosynthesis to make fuels able to replace natural fossil fuels also became evident. Petroleum and coal are nothing more than stored solar energy into chemical bonds. The problem is that the geothermal processes take several millions of years to yield coal and petroleum. So, why not try to speed up the process? Among the pioneers hooked by this idea was Italian chemist Giacomo Ciamician (1912). In his exposition at New York City in 1912, he imagined forests of glass tubes deployed in arid lands, in which photochemical processes like those performed by plants would take place, but with even more abundant fruits than nature's ones, as nature is not in hurry and mankind is. He went further and stated, "if such photochemical processes are possible, our civilization will thrive as long as the Sun shines." However, Ciamician was a man ahead of his time as our understanding of natural photosynthesis and the development of Photochemistry and related sciences were not even close to the level necessary to design a practical artificial photosynthetic device at that time. Furthermore, the success achieved by the petrochemical industry in providing society with cheap fuels and raw materials at large scales in the first half of the twentieth century led to a gradual loss of interest in artificial photosynthesis. Accordingly, research efforts were mostly directed toward refining our understanding of natural photosynthesis rather than applications of photochemical concepts in functional artificial photosynthetic systems.

However, another scientific history that seemed not connected to that of photosynthesis was unfolding parallelly, which would ultimately reveal the dangers of burning fossil fuels at large scales. In 1824, the famous French mathematician Jean-Baptiste Joseph Fourier described that visible light finds less resistance in penetrating the air than the reflected invisible heat "light" when re-entering the same air. He also coined for the first time the term greenhouse effect. A little later, the Irish physicist John Tyndall showed in 1861 that water vapor and other gases present in our atmosphere acted as a planetary blanket; therefore, such gases were the main cause of the greenhouse effect. The Swedish chemist Svante Arrhenius (the same as the acid–base theory) concluded from his works that the burning of fossil fuels indeed enhanced the greenhouse effect. This landmark work suggested for the first time that human activities would lead to a man-made greenhouse effect, which consequently would raise global temperature. In 1900, the Swedish physicist Knut Ångströn, while investigating the heat radiation coming from the Sun, established that carbon dioxide strongly absorbs infrared radiation, thus confirming Arrhenius' works. In 1938, the British engineer Guy Callendar finally connected all the pieces together. Using reliable records from weather stations around the globe, Callendar unequivocally showed that the observed increase in Earth's temperature was proportional to the rise in the carbon dioxide level in the atmosphere. In the following decades, scientists refined the measurements and models to realize that global warming was indeed mostly caused by a man-made greenhouse effect, which was the direct result of burning fossil fuels (Bolin, 2007; Braga et al., 2014).

In the early 1970s, the global warming issue gained political attention with the first United Nations Climate Conference in 1972 in Sweden, which resulted in the United Nations Environment Program. In the same year, Japanese chemists Akira Fujishima and Keinich Honda (1972) reported the first photoelectrochemical cell (PEC) able to split water molecules into hydrogen and oxygen gases upon irradiation of a TiO_2 semiconducting electrode. Although their apparatus required UV light, which is almost absent in the sunlight that reaches the Earth's surface, for the excitation of the semiconducting electrode, and the quantum yield of their PEC was extremely low, this experiment demonstrated for the first time that photochemical splitting of water to produce fuels (in this case, hydrogen gas) could be achieved. This landmark contribution associated with the political debate about the greenhouse effect brought back some of the interest in artificial photosynthesis.

5.3 The Beginning of the Modern Understanding of Photosynthesis

The 1970s were also the time in which the techniques for isolation of proteins and enzymes reached maturity with the extensive research works done in the biological sciences. Physicists and engineers also invented and perfected powerful spectroscopic techniques at the same time. X-ray crystallographic methods for structural determinations of complex biomolecules associated with the development of ultra-fast spectroscopic tools enabled pioneering investigations of the structure/function relationships in the biological world, including the photosynthetic proteins. Our understanding of natural photosynthesis increased exponentially because of such remarkable scientific advances.

To understand the photosynthetic process and fully appreciate its complexity and beauty, a brief general description of plant's anatomy (Figure 5.1) along with a general description of the plant's photosynthetic process are necessary. Plant cells in the leaf contain several membrane-bounded organelles called chloroplasts. The chloroplasts also contain membrane-bounded stacked subunits called thylakoids, which are surrounded by the chloroplast's fluid called stroma. Zooming in on the thylakoids, we find that the active photo-redox enzymes are directly embedded and span the thylakoid membrane, which in turn separates the fluid inside the thylakoid, called lumen, from the stroma. The active photo-redox enzymes drive the light-dependent reactions of photosynthesis, and the primary ones are called photosystems (PS) I and II,

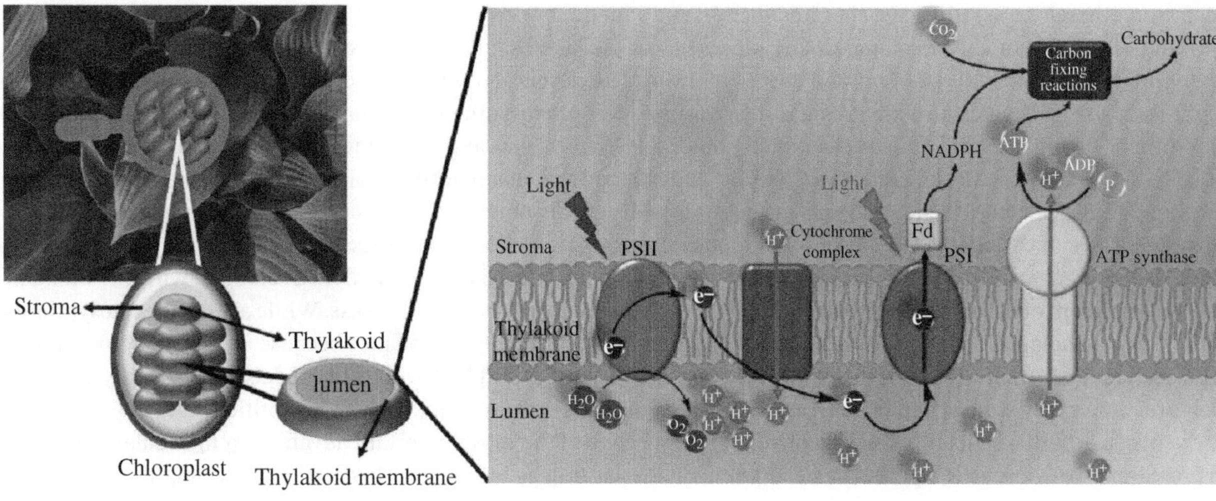

Figure 5.1 Schematic representation of the main components of a leaf and a simplified version of the photosynthetic conversion of water and carbon dioxide into oxygen and carbohydrates. PSII = photosystem II, PSI = photosystem I, Fd = ferredoxins, ADP = adenosine diphosphate, ATP = adenosine triphosphate, P = inorganic phosphate. Source: Jackson Megiatto.

which work cooperatively powered by sunlight. The interactions between PSI and PSII are mediated by the molecular pump referred to as cytochrome complex, while the subunit called adenosine triphosphate (ATP) synthase is a molecular machine that carries out the synthesis of the biochemical energy "currency" ATP molecules from adenosine diphosphate (ADP) and inorganic phosphate (P) substrates (Blankenship, 2002).

The functions of such elaborated set of transmembrane photo-redox enzymes and molecular machines are remarkable. Photoinduced electron and proton transfer reactions are the heart of the photosynthetic process. The process begins with absorption of visible light by PSII[1], which carries out a sequence of electron transfers that ultimately oxidize water molecules in the lumen into protons, excited electrons, and oxygen gas. The latter is released into the atmosphere as a byproduct. The cytochrome complex uses the energy stored in the excited electrons afforded by the PSII photoinduced processes to transport protons (H^+) from the stroma to the lumen. As that proton transport is against a proton concentration gradient (higher concentration in the lumen than in the stroma), the cytochrome complex functions as a molecular proton pump. The relatively low energy-containing electron in the cytochrome complex is then transferred by soluble electron carriers in the lumen (called plastocyanins) to the PSI, which absorbs visible light to drive its own chain of electron transfer reactions. The "reenergized" electron in PSI is transferred to iron–sulfur-containing proteins called as ferredoxins (Fd), which in conjunction with some H^+ present in the stroma promote the reduction of the cofactor nicotinamide adenine dinucleotide phosphate ($NADP^+$) into NADPH molecules. Meanwhile, the ATP synthase molecular machine uses the proton gradient (proton motive force) created between the stroma and lumen by the cytochrome complex and the water oxidation reaction as energy sources to drive the thermodynamically uphill phosphorylation reaction of ADP into ATP molecules. The NADPH and ATP molecules finally enter the Calvin Cycle, in which they are combined with CO_2 to produce carbohydrates in light-independent reactions. Therefore, natural photosynthesis carries out both light-dependent and light-independent reactions in the "day shift," but only the light-independent reactions in the "night shift" (Blankenship, 2002; Braga et al., 2014).

Although the last two paragraphs give a fair overview of the structural/function relationships of the main enzymatic components of the natural photosynthetic apparatus, the reader should have noticed that such description is not at the molecular level. To understand the photosynthetic process at molecular dimensions, we must zoom in even closer, analyze the organic chromophores and metal complexes that make up the photo-redox enzymes, and investigate in detail the mechanisms of light absorption, electronic excitation, energy and electron transfers, charge transport, and catalysis that drive photosynthesis. The molecular view also offers the possibility to study the thermodynamics and kinetics of such photophysical, photochemical, and chemical processes and how they operate in a highly synchronized manner to ensure the smoothness of the photosynthetic process.

As we shall see, nature has devised ingenious solutions to fine-tune the thermodynamic and kinetic properties of the photosynthetic apparatus. For example, noncovalent interactions (hydrogen bonds, π–π stacking, solvophobic forces, and ionic interactions) between the active photo-redox subunits and the protein scaffold drive the structural organization of the photosynthetic apparatus at the molecular level. Such protein–cofactor interactions also modulate the redox power of the cofactors, therefore adjusting the thermodynamic and kinetic properties of the set of photoinduced reactions. Such fine-tuning of the thermodynamic and kinetic properties of the photosynthetic processes ensures that each step yields the right products in the right amounts and at the right time. However, photosynthetic organisms are exposed to variable sunlight intensities in the environment, and a redox overload of photosynthetic apparatus eventually occurs. Therefore, photosynthetic organisms have evolved elaborated photoprotective systems and repair mechanisms to limit damages when exposed to bright sunlight. Such features can only be understood from the molecular point of view (Blankenship, 2002).

The study of the photosynthetic processes at the molecular level is much more difficult to carry out using the natural photosystems due to the complexity of the photo-redox enzymes as well as difficulties in extracting and isolating them without structural changes/damages. Here is where artificial photosynthesis research offers a scientific win-win situation for the advancement of our molecular understanding of the solar energy conversion process. While artificial photosynthesis uses the natural analog as a blueprint, the former helps elucidate the details of the latter because artificial photosynthetic models abstract important aspects of the natural processes into simpler and better-defined models. Congruently, artificial models are more easily studied and offer systems, in which parameters may be more easily quantified and varied.

As the purpose of this chapter is to give a historical perspective of the most important advances in molecular photosynthesis research, we chose to adopt the artificial photosynthesis strategy in our text. Instead of describing the structural/

1 Unfortunately, Photosystems I and II were names given to the enzymes according to the order of their discovery in the biological sciences and not according to their functions in the natural photosynthetic apparatus.

functional relationships of each subunit of the natural system, we will describe the main structural properties and mechanisms of the natural system that have inspired the most important achievements in artificial photosynthesis research, which we review in the next section. With this approach, it will become easier to understand the fruitful scientific marriage between research in natural and artificial photosynthesis.

5.4 Molecular Photosynthesis: Human Ingenuity Enters the Game

5.4.1 Biomimetic Reaction Centers

The primary electron transfer reactions in photosynthesis occur in the photosynthetic sub-compartment called reaction center, which is the subunit where the electronic excitation energy stored in the chromophores is used to produce the charge-separated state (CSS). By the mid-1970s, photochemists had investigated a large family of electron donors and acceptors free in solution and established the general framework for our understanding of photoinduced electron transfer reactions. Furthermore, a quantitative theory for electron transfer reactions, published (Marcus, 1956; Marcus & Sutin, 1985) by the Canadian-born chemist Rudolph Marcus in 1956 and further elaborated by others, provided experimentalists with the right tool for designing artificial photosynthetic models from the thermodynamic and kinetic points of view (Figure 5.2). Marcus theory basically predicts that an increase in thermodynamic driving force ($-\Delta G°$) for electron transfer will result in an increase in rate up to a maximum value. This is called the normal region of the Marcus rate vs. free energy change curve. At the maximum, the driving force equals the sum of the internal and solvent reorganization energies (λ) for electron transfer. When the driving force exceeds the total reorganization energy, further increases result in a decrease in rate constant. This region of the Marcus curve is called inverted region. In practice, the driving forces for the desirable forward-electron transfer (ΔG_{ET}) as well as that for the wasteful back-electron transfer analog (charge recombination [CR] process, ΔG_{CR}) are calculated from simple absorption/emission and cyclic voltammetry experiments. For example, the energy level of the first singlet excited state ($^1D^*$–A, Figure 5.2) is calculated from the average of the frequencies of the longest-wavelength absorption maxima and shortest-wavelength emission maxima of the electron donor, while the energy content of the CSS ($D^{\cdot+}$–$A^{\cdot-}$, Figure 5.2) is estimated from the first oxidation of the electron donor (D) and first reduction potentials of the acceptor (A).

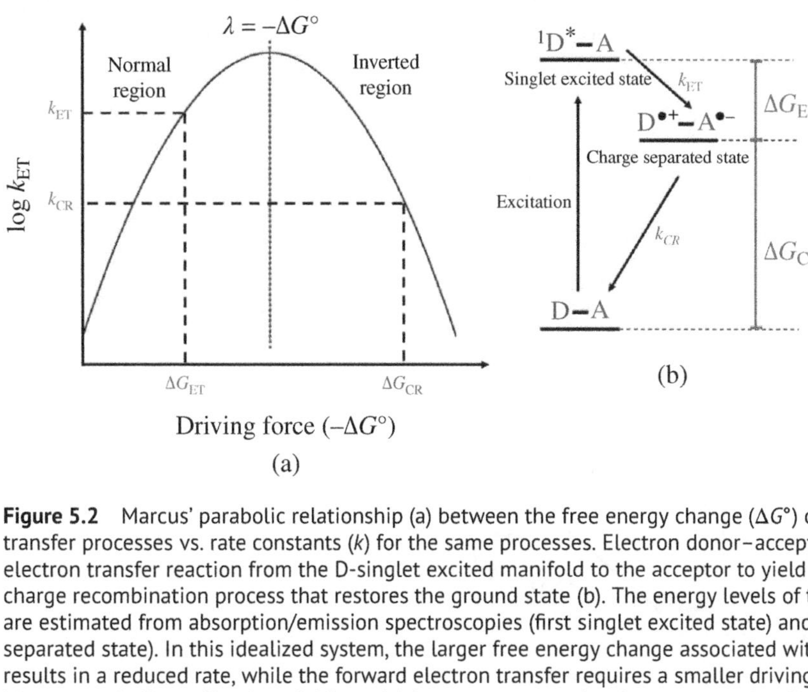

Figure 5.2 Marcus' parabolic relationship (a) between the free energy change ($\Delta G°$) of the forward- (ET) and back- (CR) electron transfer processes vs. rate constants (k) for the same processes. Electron donor–acceptor photosynthetic system (D–A) describing an electron transfer reaction from the D-singlet excited manifold to the acceptor to yield a charge separated state, which decays by a charge recombination process that restores the ground state (b). The energy levels of the intermediate states afforded upon excitation are estimated from absorption/emission spectroscopies (first singlet excited state) and cyclic voltammetry experiments (charge separated state). In this idealized system, the larger free energy change associated with the charge recombination reaction (ΔG_{CR}) results in a reduced rate, while the forward electron transfer requires a smaller driving force (ΔG_{ET}) to approach the apex of the Marcus parabolic profile, thus yielding a higher rate constant for charge separation. Because of such kinetics, the CSS lives for a longer time, therefore allowing the use of its chemical potential to drive further useful reactions. Source: Jackson Megiatto.

Electron transfer reactions require overlap of molecular orbitals. Therefore, the efficiency of the process strictly depends on the separation and orientation of electron donors and acceptors. When the chromophores are free in solution, such important structural features of the photosynthetic models are poorly defined. Furthermore, electron transfers require collisions between donors and acceptors to take place if the chromophores are free in solution. That also restricts the time scale of electron transfers to the molecular diffusion rate limit. In natural photosynthesis, electron transfer between donors and acceptors takes place from the chromophore's singlet excited states. As the singlet excited states of organic chromophores have short lifetimes, ET between the chromophores free in solution must occur rapidly to afford CSSs in high yields. However, the lifetimes of the singlet excited state of organic chromophores are usually shorter than the molecular diffusion rate limit. Congruently, chromophores free in solution afford CSSs in low yields. Thus, such approach was not appropriate to mimic the ET processes occurring in the reaction centers of natural photosynthesis (Blankenship, 2002; Gust, 2016).

It then became obvious that by covalently linking the electron donors and acceptors, such limitations could be overcome to yield more robust artificial reaction centers. The chemical linkage defines or at least limits the relative separation of the moieties, and the time constraints of diffusion are no longer a problem. In addition, electronic orbitals of the linkage itself play an important role in the overlap of the orbitals of the donor and acceptor and thus in determining the rate of electron transfer. By the end of 1970s, chemists began to synthesize covalently linked donor/acceptor "dyads" for the study of photoinduced electron transfers. The first electron donor–acceptor artificial photosynthetic dyads (Figure 5.3) were reported by Kong and Loach (Kong & Loach, 1978) in 1978 and by Tabushi (Tabushi et al., 1979) in 1979.

In both dyads, porphyrin moieties (Figure 5.3) act as electron donors, while quinones (Figure 5.3) are the acceptor counterparts. Such choice of chromophores was inspired by the natural photosynthetic apparatus, in which the electron donors are chlorophyll molecules and the acceptors are plastoquinones. However, chlorophyll molecules are not stable without the protein environment and readily decompose in solution upon irradiation. Therefore, chemists have used porphyrin molecules instead. Porphyrins are useful synthetic alternatives to chlorophylls as they are much easier to synthesize, are relatively stable in solution, absorb sunlight with reasonable efficiencies, and their redox and optical properties can be tuned by introduction of chemical functionalities at the peripheral positions of their macrocyclic structures. Accordingly, porphyrins are the most common light absorbers and electron donors in artificial photosynthesis research.

Due to the rich variety of covalent bonds that could be used to link porphyrins (P) to quinones (Q), as well as the many possible porphyrin and quinone derivatives that could be synthetically prepared, hundreds of P–Q dyads were reported in the 1980s. Steady-state and time-resolved spectroscopic experiments revealed the photophysical processes of the dyads published in the literature. The P–Q dyads reported (Kuciauskas et al., 1997) by GMM will be described to exemplify the

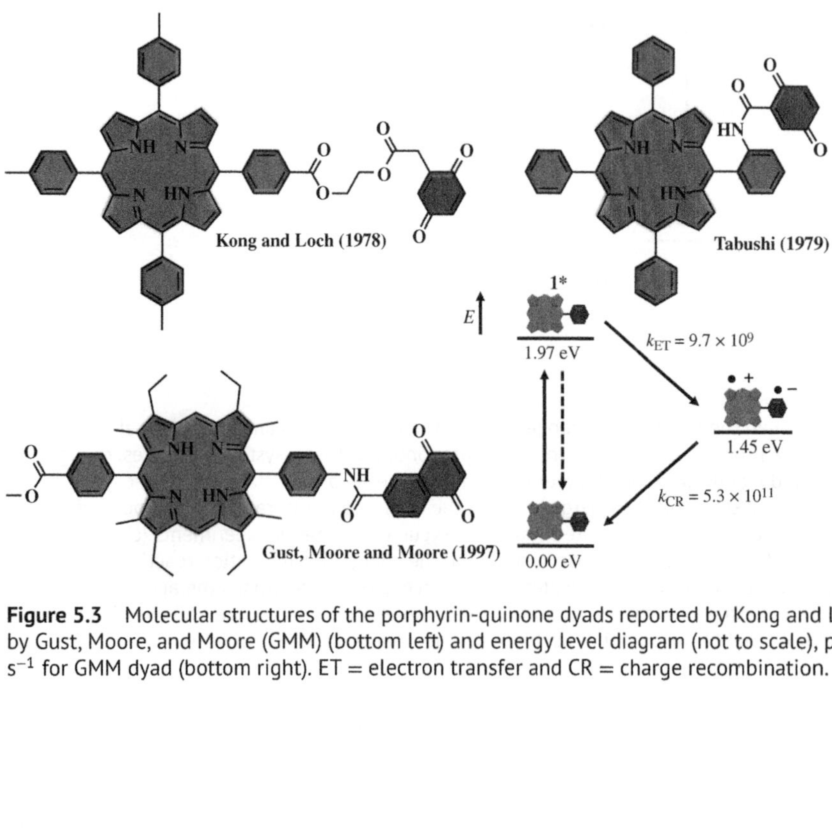

Figure 5.3 Molecular structures of the porphyrin-quinone dyads reported by Kong and Loch, and Tabushi (top). Similar dyad reported by Gust, Moore, and Moore (GMM) (bottom left) and energy level diagram (not to scale), proposed decay pathway and rate constants in s^{-1} for GMM dyad (bottom right). ET = electron transfer and CR = charge recombination. Source: Jackson Megiatto.

light response (Figure 5.3, bottom right). Excitation of the porphyrin subunit in the dyad produces the respective singlet excited state $^1P^*$–Q, which lies 1.97 eV above the ground state. The main pathway for the decay of the porphyrin excited state is by electron transfer (ET) to the quinone to generate the $P^{\cdot+}$–$Q^{\cdot-}$ CSS. From electrochemical measurements on the P–Q dyad, it was established that $P^{\cdot+}$–$Q^{\cdot-}$ CSS lies about 1.45 eV above the ground state. Therefore, the thermodynamic driving force for this electron transfer is $\Delta G^\circ = -0.54$ eV. The rate constant for this electron transfer, afforded from the short lifetime of the $^1P^*$–Q excited state (102 ps), was $k = 9.7 \times 10^9\ s^{-1}$ with a virtual quantitative quantum yield of $\phi = 0.99$. Despite such quantum efficiency and fast ET, P–Q dyads usually yield $P^{\cdot+}$–$Q^{\cdot-}$ CSS with truly short lifetimes. In GMM's dyad, the $P^{\cdot+}$–$Q^{\cdot-}$ CSS has a lifetime of only 1.9 ps ($k = 5.3 \times 10^{11}\ s^{-1}$). That means that the wasteful CR process, which occurs through back electron transfer, is also kinetically competent in P–Q dyads. Such short lifetime renders it almost impossible to use the chemical potential stored in $P^{\cdot+}$–$Q^{\cdot-}$ CSS to drive further reactions, hence revealing the limitations of P–Q dyads as artificial photosynthetic reaction centers.

From the large set of data afforded from the investigation of P–Q dyads, it was observed that temporal stabilization of the CSSs was critical. Therefore, artificial models should be designed to avoid recombination of the separated charges and the consequent loss of the stored sunlight energy as heat. To address this challenge, researchers looked once again at natural photosynthesis for inspiration. Nature's solution to this problem is to isolate the charges at large distances by moving the photogenerated charges produced in the primary ET reaction in opposite directions to secondary donor and acceptor moieties. These charge shift (CS) reactions (also ET processes) among the chromophores are directional, short-range, and thereby fast, and ultimately yield CSSs with large distance between the charges. For example, in the reaction center of PSI in cyanobacteria, excitation of the chlorophyll dimers labeled as P_A and P_B in Figure 5.4 (the dimer is also called as P700 complex), triggers an electron transfer reaction to the nearby primary chlorophyll electron acceptors A_{0A} and A_{0B}, a process that is assisted by the accessory chlorophyll molecules A_{-1A} and A_{-1B} to generate the intermediate $P700^{\cdot+}$–$A_{0A,B}{}^{\cdot-}$ CSS. The electron in the reduced primary acceptor $A_{0A,B}{}^{\cdot-}$ is then transferred to the phylloquinone secondary acceptors $A_{1A,B}$ to afford the CSS $P700^{\cdot+}$–$A_{1A,B}{}^{\cdot-}$, which practically spans the thylakoid membrane. This larger spatial separation leads to weak electronic coupling between the photogenerated charges, thereby increasing the driving force and slowing down the CR processes (Blankenship, 2002).

Inspired by such mechanisms, scientists began to covalently connect primary and secondary electron donors and acceptors to afford the so-called triads, tetrads, and higher D–A models able to mimic the sequential electron transfers upon irradiation. A myriad of such systems was reported in the literature by the end of 1980s and early 1990s. The main photophysical decay of the pentad reported (Gust et al., 1990, 1993) by GMM (Figure 5.5) in 1990 will be described to highlight the concept of kinetic competition between charge separation/recombination processes in multicomponent systems to afford large-distance and long-living CSSs. The molecular structure of the pentad consists of the primary donor free-base porphyrin (P) connected to a pair of quinones (Q–Q), which are the primary and secondary acceptors, on one of its side and on the other to the zinc(II)porphyrinate (ZnP), which functions as secondary donor. The ZnP subunit is in turn connected to a carotene molecule (C), which is the final electron donor.

Excitation of P (step 1) produces the corresponding singlet excited state C–ZnP–$^1P^*$–Q–Q, which decays through ET (step 2) to yield the first C–ZnP–$P^{\cdot+}$–$Q^{\cdot-}$–Q CSS. Two charge shift reactions operating in parallel (steps 3 and 4) occur to produce the C–ZnP–$P^{\cdot+}$–Q–$Q^{\cdot-}$ and C–$ZnP^{\cdot+}$–P–$Q^{\cdot-}$–Q CSSs, respectively. The charge shift reaction in step 5 yields the C–$ZnP^{\cdot+}$–P–Q–$Q^{\cdot-}$ CSS, while the charge shift reaction from the C moiety to the oxidized $ZnP^{\cdot+}$ (step 6) produces the $C^{\cdot+}$–ZnP–P–$Q^{\cdot-}$–Q CSS. Two final charge shift reactions in these two intermediates (steps 8 and 9) afford

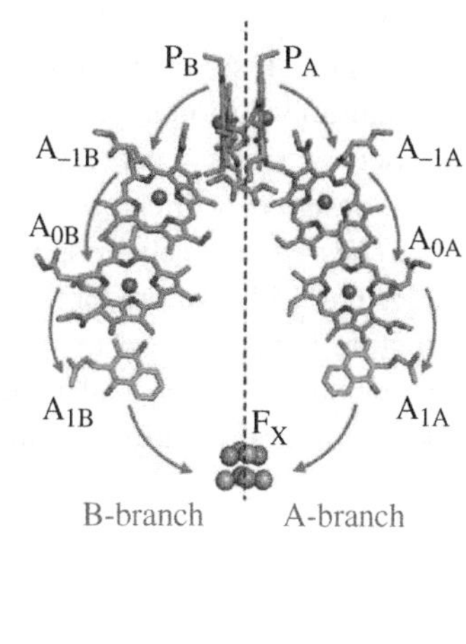

Figure 5.4 Spatial organization of the main active photo-redox subunits in the reaction center of the photosystem I extracted from the thermophilic cyanobacterium *Synechococcus elongatus* according to the crystal structure at 2.5 Å resolution (Jordan et al., 2001). Arrows indicate the electron transfer chain in the photosynthetic reaction center, while the dotted line indicates the pseudo-C_2 axe symmetry of the chromophoric array. Source: Adapted from the Protein Data Bank Archive (PDB ID 1JB0). Public domain.

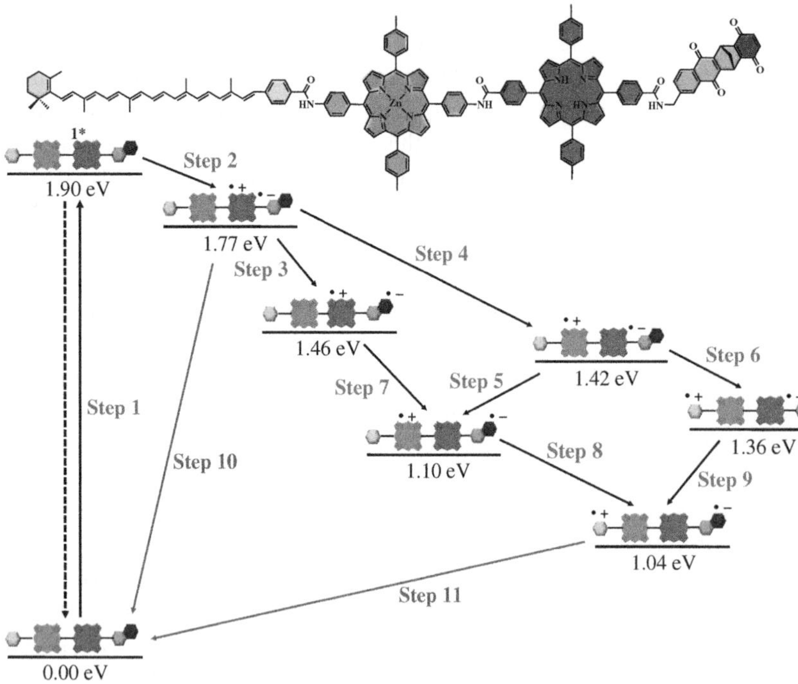

Figure 5.5 Molecular structure, energy level diagram (not to scale), and proposed decay pathways for the pentad reported by GMM group, following excitation of the porphyrin (P) moiety. Source: Adapted from Gust et al. (1993). Reproduced with permission of the American Chemical Society.

the final $C^{\cdot+}$–ZnP–P–Q–$Q^{\cdot-}$ CSS. Competing with all the charge shift reactions occurring in the intermediate CSSs are the respective CR processes. The most rapid CR process is that of step 10 as the photogenerated charges in the C–ZnP–$P^{\cdot+}$–$Q^{\cdot-}$–Q CSS are strongly coupled at this stage of the charge separation process. However, the quantum yield of the final $C^{\cdot+}$–ZnP–P–Q–$Q^{\cdot-}$CSS measured from the first C–ZnP–$P^{\cdot+}$–$Q^{\cdot-}$–Q analog is $\phi = 0.97$. Such high efficiency unequivocally demonstrates that all charge shift reactions (steps 2–9) occur faster than the respective CR processes, in particular that of step 10. The remarkable photophysical properties of the pentad are not only attributed to the proper tuning of the thermodynamic driving force for each charge shift reaction but also to the fact that the critical CR in step 10 competes against two charge shift reactions operating in parallel (steps 3 and 4). The lifetime of the final CSS $C^{\cdot+}$–ZnP–P–Q–$Q^{\cdot-}$ was 55 μs. The elegant work by GMM undoubtably demonstrated that molecular engineering based on the theoretical concepts devised by Marcus to favor charge separation events (forward ET) over competing recombination (back ET) processes in electron transfer chains is a powerful approach to designing artificial photosynthetic reaction centers that mimic the natural ones in terms of quantum yield and lifetime of CSSs.

5.4.2 Artificial Reaction Centers with Nonnatural Electron Donors and Acceptors

The biomimetic approach has driven most part of the artificial photosynthesis research. However, the natural photosynthetic apparatus evolved under evolutionary pressures to allow organisms to use a new power source, namely solar energy. Therefore, evolution might not have explored electron donors and acceptors other than chlorophyll and quinone molecules to assemble reaction centers. Here, human ingenuity can enter the game and explore more robust synthetic chromophores in the design of artificial reaction centers. One class of exceptional artificial donors and acceptors are ferrocenes (Fcs) and fullerenes, respectively. Fc is a sandwich organometallic complex that has valence electrons in an essentially non-bonding molecular orbital of predominant *d*-character. Therefore, Fc has a low oxidation potential. Furthermore, the Fc sandwich structure is maintained upon oxidation. Consequently, the eventual photogenerated positive charge is well-distributed throughout the complex, while the iron atom is protected from solvent molecules by the cyclopentadienyl ligands. Such structural features provide the ferrocenium species with extra stability (Astruc, 2007). Fullerenes with their large spherical rigid structure and poor solvation properties yield small values of reorganization energies (λ) for ET reactions. Therefore, a small thermodynamic driving force is necessary to reach the apex of the Marcus parabolic curve, while the large driving force associated with the CR is buried in the inverted region (Guldi, 2002; Hirsch & Brettreich, 2005). Thus, artificial photosynthetic models bearing fullerenes as acceptors instead of quinones afford CSSs with much longer lifetimes.

In the 1990s, a multitude of dyads, triads, and related systems bearing fullerenes as electron acceptors were reported, which demonstrated the superior properties of the bulky balls as electron acceptors when compared to quinones. The first triad of this large family of fullerene artificial photosynthetic models was reported (Liddell et al., 1994) by GMM in 1997 (Figure 5.6, top). However, one of the best artificial photosynthetic models with fullerenes in terms of quantum yield and lifetime of the final CSS was the pentad reported (Imahori et al., 2004) by Imahori et al. (Figure 5.6, bottom) in 2004, which is composed of a Fc secondary electron donor linked to a zinc(II)porphyrinate-trimer subunit (ZnP_3) that serves as both light harvesting moieties and primary donor. The trimer is attached to the fullerene derivative C_{60}, which is the acceptor. Excitation of the ZnP chromophores yields the Fc–(^{1}ZnP*)$_3$–C_{60} excited state, which decays to the $Fc^{\cdot+}$–ZnP–$C_{60}{}^{\cdot-}$ CSS after a sequence of ETs. The lifetime of this CSS in DMF at 163 K was 0.53 s and was formed with a quantum yield of 0.83. This pentad is an extreme example of lifetimes and efficiency with artificial reaction centers containing synthetic electron donors and acceptors.

In the 1990s and the 2000s, research on artificial reaction centers flourished with many interesting designs and photo-redox active units. Transition metal complexes had become popular chromophores in artificial photosynthetic models, an area that was pioneered by Meyer with his investigations of polypyridyl Ru-complexes in the 1980s. Following Meyer's works, several groups, specially Balzani's, Armaroli's, Scandolla's, and Campagna's groups in Italy (Balzani et al., 1996) and the group of Sauvage in France (Collin et al., 1991), extensively investigated metal complexes as chromophores in artificial reaction centers. Complexes with second and third-row transition metals exhibit low-lying excited states with metal-to-ligand charge transfer (MLCT) characters, which can act as both electron donors and acceptors once populated through photoexcitation. Furthermore, transition metal complexes can be molecularly engineered to accept or donate more than one electron upon successive photoexcitation steps (Astruc, 2007). As we shall discuss later, accumulation of multiple charges by chromophores upon successive absorption of photons is key to the multielectron chemistry necessary to produce fuel from solar energy. An elegant example of such strategy was reported (Molnar et al., 1994) by Brewer et al. in 1994. They investigated a series of triads related to the one shown in Figure 5.7. Photoexcitation of either ruthenium complex is followed by ET for the iridium complex. If the oxidized ruthenium complex is reduced by a sacrificial donor in solution (dimethylaniline), absorption of a second photon by a ruthenium complex leads to a second ET to the iridium complex, thereby producing a doubly reduced Ir-species from two successive photon absorptions, albeit in low quantum yield and lifetime of a few nanoseconds.

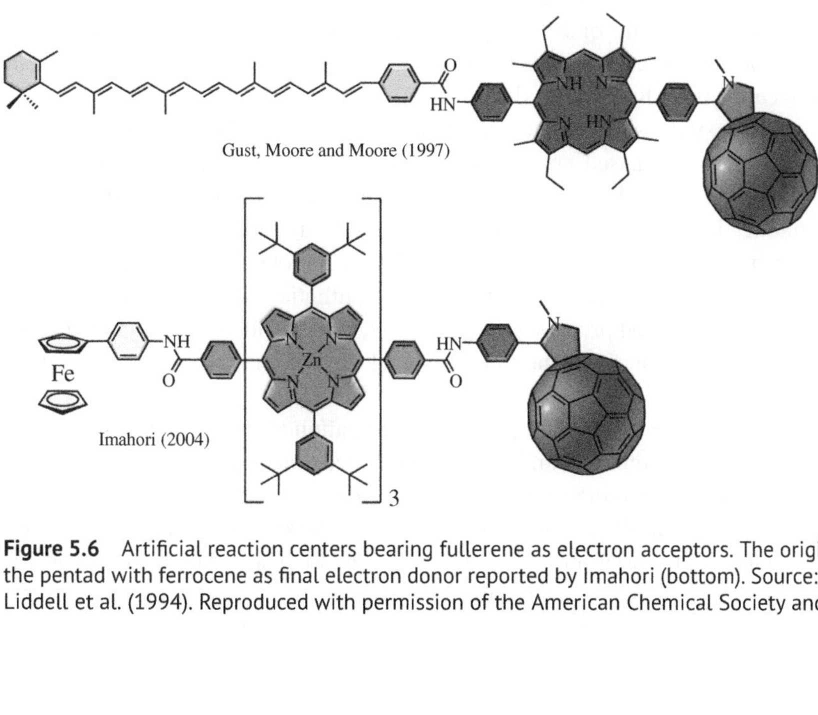

Figure 5.6 Artificial reaction centers bearing fullerene as electron acceptors. The original triad reported by GMM's group (top) and the pentad with ferrocene as final electron donor reported by Imahori (bottom). Source: Adapted from Hirsch and Brettreich (2005) and Liddell et al. (1994). Reproduced with permission of the American Chemical Society and WILEY-VCH Verlag GmbH & Co.

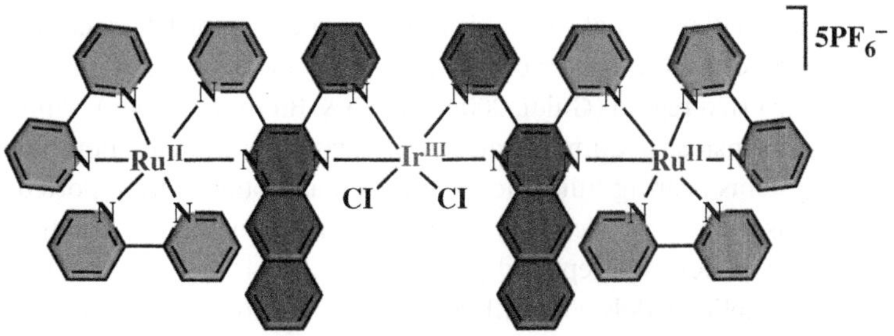

Figure 5.7 Transition metal complex triad able to accumulate two electrons from two consecutive photoinduced electron transfer reactions. Source: Adapted from Molnar et al. (1994). Reproduced with permission of the American Chemical Society.

5.4.3 Supramolecular Assembly of Artificial Reaction Centers

Unfortunately, similar complexes based on first-row transition metals suffer from the presence of non-emissive low-lying metal-centered *d–d* transitions that quickly quench the excited state emission. However, such low-lying *d–d* transitions are not present in complexes of d^{10} transition metal ions, such as those based on Cu(I) and Zn(II) ones (Astruc, 2007). Therefore, complexes based upon d^{10} ions exhibit luminescence and are attractive as chromophores in artificial photosynthesis research. Another important property of transition metal complexes is their ability to gather and position ligands in well-defined geometries, which renders them excellent template motifs to assemble supramolecular systems (Alcantara et al., 2018, 2020; Fontana et al., 2020; Lehn, 2011). So far in our discussion, we have described artificial reaction centers, whose molecular structures are "linear" and the photoactive subunits are organized by covalent bonds. However, in natural photosynthesis, the protein backbone directs and spatially positions the primary and secondary electron donors and acceptors in the reaction center to control the yield and kinetics of the sequential electron transfer. Organization of the active cofactors is mainly driven by noncovalent interactions between the protein scaffold and the chromophores. Therefore, molecular topology, which is hard to achieve using covalently linked D–A systems, plays an important role in the light energy conversion process (Blankenship, 2002).

The concepts of Supramolecular Chemistry (Lehn, 2011), such as host–guest chemistry and molecular recognition through secondary interactions, can be used to assemble large D–A photosynthetic arrays with well-defined structures, in which higher-order features such as molecular topology naturally emerge (Balzani & Scandola, 1991). Among the supramolecular systems available, mechanically linked molecules are excellent platforms to organize active photo-redox groups. Furthermore, the effectiveness of transition metal ion template synthesis is mostly appreciated in formation of mechanical bonds (Lehn, 2011). A mechanical bond is an entanglement in space between two or more component parts, such that they cannot be separated without distorting or breaking chemical bonds between atoms. Rotaxanes (a dumbbell-shaped molecular component in which a molecular ring is trapped) and catenanes (two or more interlocked rings) are prime examples of mechanically linked systems (Burns & Stoddart, 2017).

The 1,10-phenanthroline-copper(I) metal template synthesis technique developed by Sauvage in the 1980s is a powerful strategy to assemble rotaxanes and catenanes. This idea is based on tetrahedral coordination of a Cu(I) ion with a pair of bidentate ligands, which creates the entwined $[Cu(phen)_2]^+$ complex. The physical overlapping of the ligands in the entwined complex orients chemical functionalities in an orthogonal fashion to favor introduction of stoppering groups (to afford rotaxanes) or to effect macrocyclizations (to afford catenanes) (Sauvage, 1990). As rotaxanes and catenanes are topologically distinct from the molecular point of view, they are appealing as photosynthetic models to investigate the effects of molecular topology on the dynamics of photoinduced processes. Furthermore, the $[Cu(phen)_2]^+$ complex used to assemble the mechanically linked systems is a d^{10} complex, thereby functioning as an electronic relay upon photoexcitation (Balzani & Scandola, 1991).

Sauvage et al. early recognized such features and pioneered the photophysical investigation of rotaxanes and catenanes assembled through formation of a central $[Cu(phen)_2]^+$ complexes and bearing appended zinc(II)porphyrinate moieties (ZnP) as the electron donors and gold(III)porphyrinate groups (AuP) as the electron acceptor. Unfortunately, overlapping of the spectroscopic signals, as well as ultrafast CR processes, prohibited the determination of the kinetic parameters for many of the expected intermediates formed upon excitation of these systems (Flamigni et al., 2006). Inspired by Sauvage's work, Megiatto and Schuster prepared analogous $[Cu(phen)_2]^+$-based rotaxanes and catenanes bearing ZnP electron donors, but with C_{60} as the ultimate electron acceptor in place of AuP (Kirner et al., 2014, 2015; Megiatto Jr. et al., 2009a, 2009b;

Megiatto Jr. et al., 2020; Megiatto Jr. & Schuster, 2008; Megiatto Jr. & Schuster, 2009; Megiatto Jr. & Schuster, 2010; Megiatto Jr. & Schuster, 2011; Megiatto Jr., Li, et al., 2010; Megiatto Jr., Schuster, et al., 2010; Megiatto Jr., Schuster, et al., 2012). Although a few rotaxane-based artificial photosynthetic models decorated with C_{60} had been reported by several groups at that time, the catenane analogs were extremely rare, probably due to the intrinsic synthetic hurdles imposed by catenane synthesis when compared to the more accessible rotaxanes (Armaroli et al., 1998; Megiatto Jr. and Schuster, 2011b, 2012).

Megiatto and Schuster developed (Megiatto Jr. et al., 2009a; Megiatto Jr. et al., 2009b; Megiatto Jr. and Schuster, 2008; Megiatto Jr. and Schuster, 2009; Megiatto Jr. and Schuster, 2010; Megiatto Jr. and Schuster, 2011a; Megiatto Jr., Schuster, et al., 2010; Megiatto Jr., Schuster, et al., 2012) a powerful synthetic strategy at the end of the 2000s, which combines the virtues of Sauvage's $[Cu(phen)_2]^+$ template technique (Sauvage, 1990) with those of "click" chemistry protocols (Meldal & Tornøe, 2008) to enable the syntheses of rotaxanes and catenanes whose size and shape of the ring and thread components as well as the chemical linkages that connect the photo-redox active subunits are identical (Figure 5.8). Therefore, the effects of the distinct molecular topology provided by the two interlocked systems on the ET could be investigated for the first time. A detailed spectroscopic and electrochemical investigation revealed that both interlocked systems presented an identical main sequence of photoinduced processes and functioned as a triad. Briefly, exclusive excitation of the ZnP produces the respective singlet excited state $^1ZnP^*–[Cu(phen)_2]^+–C_{60}$ (step 1) that decays through energy transfer (step 2) to the $[Cu(phen)_2]^+$ complex to yield the respective MLCT state. The MLCT manifold is oxidatively quenched by the C_{60} via ET (step 3) to afford the intermediate $ZnP–[Cu(phen)_2]^{2+}–C_{60}^{\bullet-}$ CSS. A charge shift reaction (step 4), which involves ET from the ZnP moiety to the oxidized $[Cu(phen)_2]^{2+}$ complex, yields the final CSS $ZnP^{\bullet+}–[Cu(phen)_2]^+–C_{60}^{\bullet-}$. The common decay pathway of rotaxanes and catenanes reported in Figure 5.8 was also detected by electron paramagnetic resonance spectroscopy (EPR) (Jakob et al., 2011a; Jakob et al., 2011b).

However, the most dramatic effect of the distinct rotaxane and catenane molecular topologies was the rates for the CR reactions of the final $ZnP^{\bullet+}–[Cu(phen)_2]^+–C_{60}^{\bullet-}$ states (step 6). The lifetime of the final CSS in the rotaxane

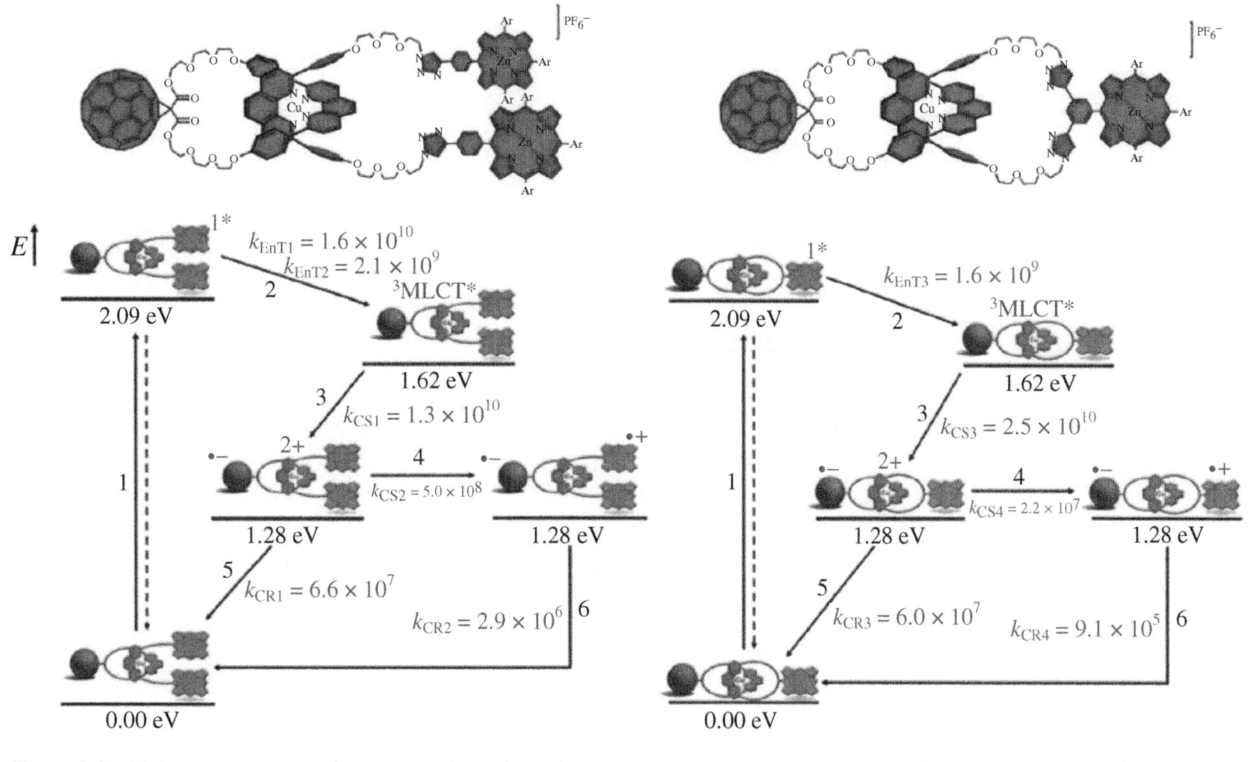

Figure 5.8 Molecular structure of rotaxanes (top left) and catenanes (top right) reported by Megiatto and Schuster, and their energy level diagram (not to scale), proposed decay pathways, and rate constants in s^{-1} following exclusive excitation of the ZnP subunit in benzonitrile (bottom). Ar at the meso positions of the porphyrin moieties is abbreviation for 3,5-di-*tert*-butylphenyl groups. EnT = energy transfer, CS = charge shift, and CR = charge recombination. Source: Jackson Megiatto.

was 0.24 μs (Megiatto Jr. et al., 2020; Megiatto Jr., Schuster, et al., 2012), while it was four times longer (1.1 μs) in the catenane analog (Megiatto Jr. et al., 2020; Megiatto Jr., Schuster, et al., 2010) under the same conditions. From a detailed structural investigation, the authors concluded that the catenane adopted an extended molecular conformation with the chromophores far apart, while the rotaxane was conformationally flexible and folded into a more compact conformation, thereby bringing the chromophores closer together and favoring CR. Accordingly, the rigid catenane structure keeps the chromophores further apart, decreasing the electronic coupling between the photogenerated charges and retarding their recombination when compared to the structurally flexible rotaxane analog (Megiatto Jr. et al., 2009b, 2020). Those findings unequivocally demonstrated the importance of molecular topology in photoinduced processes in artificial reaction center models. In further works (Kirner et al., 2015; Megiatto Jr. et al., 2020), Megiatto and Schuster reported a family of multichromophoric rotaxanes also assembled through formation of a central $[Cu(phen)_2]^+$ entwined complex and with C_{60} as acceptor, but bearing several combinations of zinc(II)phtalocyanate, Fc and zinc(II)porphyrinate electron donors as the rotaxanes' stoppers. Rotaxane Fc–ZnP–$[Cu(phen)_2]^+$–C_{60} (Figure 5.9) exemplifies the light response of the family of rotaxanes. Accordingly, excitation of the ZnP moiety leads to a sequence of EnT and ET processes that yield the expected final $Fc^{\cdot+}$–ZnP–$[Cu(phen)_2]^+$–$C_{60}^{\cdot-}$ CSS with good quantum yield and lifetime of 61 μs; the longest lifetime ever reported for CSSs in mechanically linked artificial reaction centers (Megiatto Jr. and Schuster, 2012). These works clearly demonstrated the versatility of rotaxanes and catenanes as molecular scaffolds for the organization of multichromophoric artificial photosynthetic reaction centers.

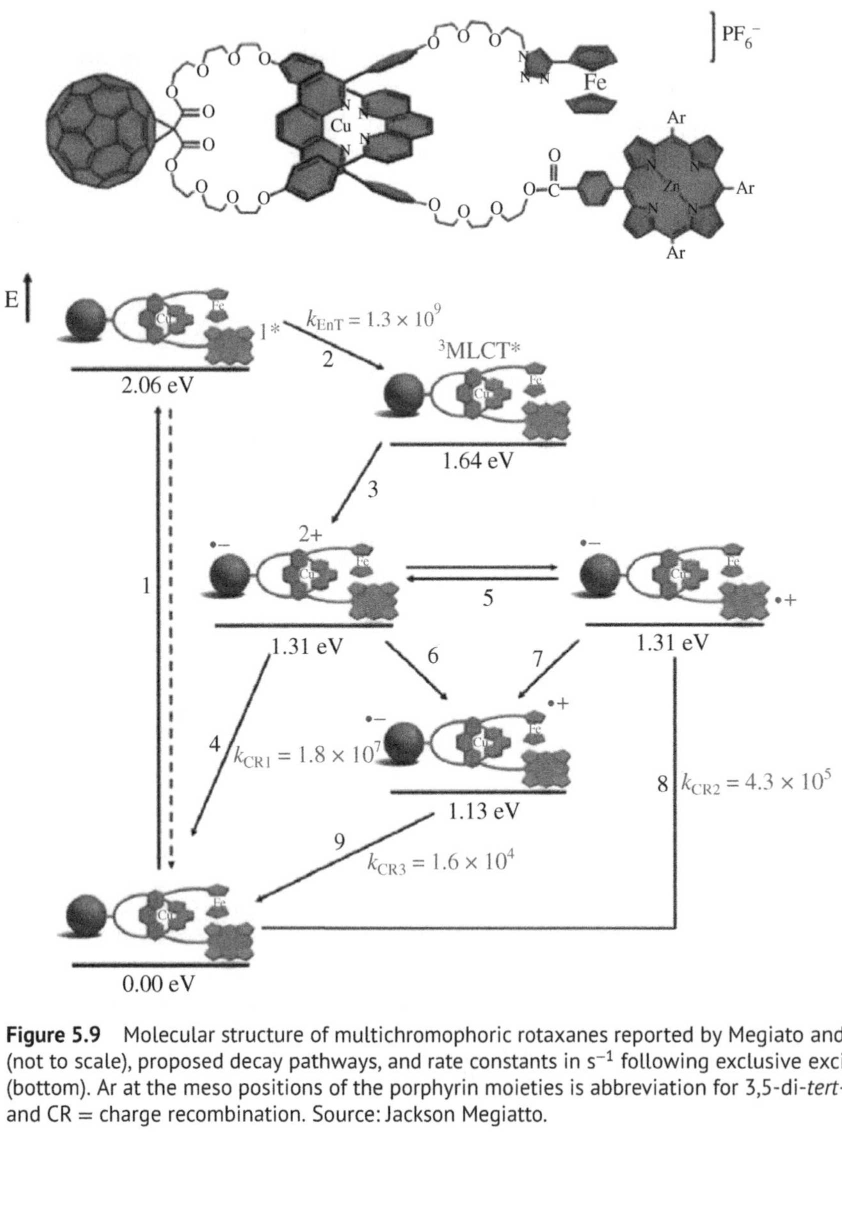

Figure 5.9 Molecular structure of multichromophoric rotaxanes reported by Megiato and Schuster (top) and its energy level diagram (not to scale), proposed decay pathways, and rate constants in s^{-1} following exclusive excitation of the ZnP subunit in tetrahydrofuran (bottom). Ar at the meso positions of the porphyrin moieties is abbreviation for 3,5-di-*tert*-butylphenyl groups. EnT = energy transfer and CR = charge recombination. Source: Jackson Megiatto.

5.4.4 Artificial Antenna

As described above, the first step of the photophysical decay in the catenanes and rotaxanes reported by Megiatto and Schuster occurs through an EnT process from the first singlet excited state $^{1}ZnP^{*}$–$[Cu(phen)_2]^{+}$–C_{60} to the $[Cu(phen)_2]^{+}$ MLCT excited state. Similar energy transfer processes also occur in natural photosynthesis, in which most of sunlight is absorbed by protein complexes called antenna that contain hundreds of chromophores, rather than by the reaction center itself. To approach full spectral coverage and maximize sunlight absorption, antennas in photosynthetic organisms are composed of a variety of chromophores, such as carotenoids, phycoerythrins, and phycocyanins. Antenna functions depend on singlet-to-singlet energy transfers between the chromophores. As the lifetimes of singlet excited states in organic chromophores are usually short, the chromophores are packed close together to optimize the transfer of the electronic excitation energy in the antenna (Blankenship, 2002). Furthermore, that electronic excitation energy must be funneled to the reaction center to start the ET reactions that afford CSSs. The protein's globular shape warrants the efficient packing, and consequent close contact, of the chromophores, while organizing them in a configuration that creates the necessary thermodynamic gradient within the antenna to funnel the excitation energy to the reaction center.

Dendrons and dendrimers belong to a class of synthetic macromolecules that can be used to mimic the globular shape of proteins. Dendrimers have spatially controlled globular structures in solution, which are prepared by iterative stepwise processes. Therefore, dendrimers are monodispersing macromolecules with well-defined structures, which contrasts with linear polymers that are usually polydisperse and can adopt several conformations in solution. Accordingly, dendrimers are useful systems to investigate the resulting properties emerging from structural higher-order molecular organizations such as those afforded by the protein environment. Many photo-responsive dendritic systems have been reported in the literature, in which the dendron or dendrimer architecture arranges multiple energy donor chromophores in close contact with each other at the dendrimer's periphery while surrounding the energy acceptor group located at the center of the globular structure. Extensive spectroscopic investigation of photoactive dendrimers has revealed that excitation of the peripheral chromophores results in exclusive emission from the core subunit. Accordingly, dendrons/dendrimers can be viewed as molecular light concentrators or molecular lenses (Astruc et al., 2010).

One elegant example of the use of dendrons in the design of artificial antennas was reported (Devadoss et al., 1996) in 1996 by J. S. Moore et al. (Figure 5.10). In their design, a perylene moiety is located at the focal point of the conjugated phenylacetylene dendron scaffold, which has 3,5-di-*tert*-butylbenzene peripheral groups. They reported several dendron generations, whose structures are increased by introduction of tri-connected phenylacetylene junctions from one generation to the next. Time-resolved fluorescence experiments revealed that all generations have similar light responses. Excitation of the dendron backbone at 310 nm leads to exclusive emission of the perylene moiety at λ_{max} = 484 and 518 nm, thus confirming the expected singlet-to-singlet EnT processes from the periphery to the focal point. However, the authors found an interesting trend, in which the light absorption efficiency increases with increasing the dendron generation, while the quantum yield of the EnT from the dendron scaffold to the perylene focal point decreases. Accordingly, the first-generation performed best in terms of EnT efficiency, with quantum yield of 98% and a time constant $k = 1.9 \times 10^{11}$ s^{-1} (lifetime of the singlet excited state, τ = 270 ps). On the other hand, the last-generation yielded 54% EnT and a biexponential fluorescence decay with an average $k = 4.3 \times 10^{8}$ s^{-1} (τ = 2.35 ns), suggesting that the chromophores were not uniformly

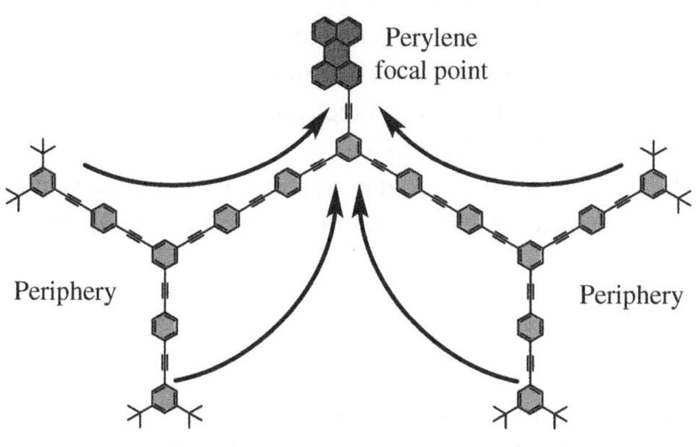

Figure 5.10 Artificial antenna device with chromophores spatially organized by a dendron architecture that illustrates the molecular lens effect. EnT processes (arrows) focus the electronic excitation energy collected by the periphery subunits at the perylene central point. Source: Adapted from Devadoss et al. (1996). Reproduced with permission of the American Chemical Society.

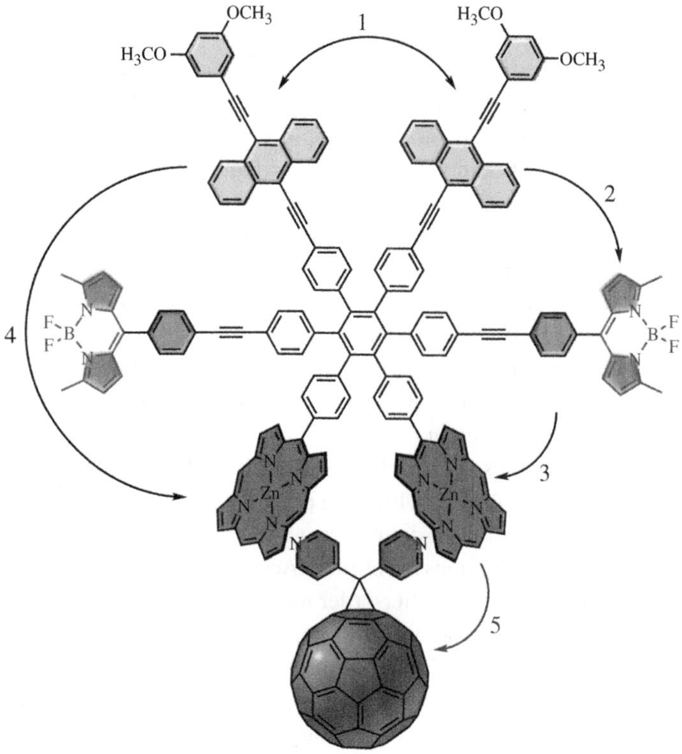

Figure 5.11 Artificial multiantenna-reaction center reported by GMM group with the expected pathway decay, which was observed among others in the spectroscopic investigation of the heptad. Steps 1–4 denote singlet-to-singlet energy transfers, and step 5 is electron transfer reaction. Source: Adapted from Terazono et al. (2009). Reproduced with permission of the American Chemical Society.

distributed around the periphery of the large dendron. However, the biggest reported dendron collected about 100 times more photons than the smaller one at 290 nm, according to UV-Vis absorption experiments. Thus, it was demonstrated that there is a trade-off between the efficiency of photon absorption by the antenna and the quantum yield of the EnT process within the system. This work unequivocally shows that dendron molecular engineering is useful to create thermodynamic gradients for capture and concentration of sunlight in artificial antenna devices.

The dendron antenna reported by Moore et al. cannot produce CSSs due to the lack of a reaction center in the system. An elegant biomimetic model containing an antenna connected to a reaction center (Figure 5.11) was reported (Terazono et al., 2009) by GMM in 2009. The system is a synthetic molecular heptad that features two bis(phenylethynyl)anthracene (absorption in the 400–500 nm region) and two borondipyrromethene (absorption in the 450–550 and 330–430 nm regions) antennas linked to a hexaphenylbenzene core that also bears two zinc porphyrins (ZnP). A fullerene electron acceptor self-assembles to both porphyrins via formation of coordinative bonds to yield the reaction center. Excitation of the antenna chromophores results in rapid energy transfer from the antenna chromophores to the porphyrin moieties via a singlet-to-singlet mechanism with near quantitative yield. Transient absorption measurements informed that the excitation energy is transferred from the antennas to the ZnP moieties through both direct and multistep pathways. The resulting ^{1}ZnP* state donates an electron to the C_{60} acceptor to produce the CSS $ZnP^{\cdot+}$–$C_{60}{}^{\cdot-}$ with a lifetime of 230 ps. Although the lifetime of the CSS was relatively short, the heptad demonstrated that by incorporating antennas, it is possible for a molecular system to efficiently harvest light throughout the visible wavelengths and to funnel the excitation energy to the reaction center to produce CSSs.

5.4.5 Photo-Regulation

Using a similar antenna-reaction center system (Figure 5.12), GMM also demonstrated (Straight et al., 2008) the important phenomenon of photo-regulation in artificial photosynthesis in 2008. Under conditions of bright sunlight, the natural

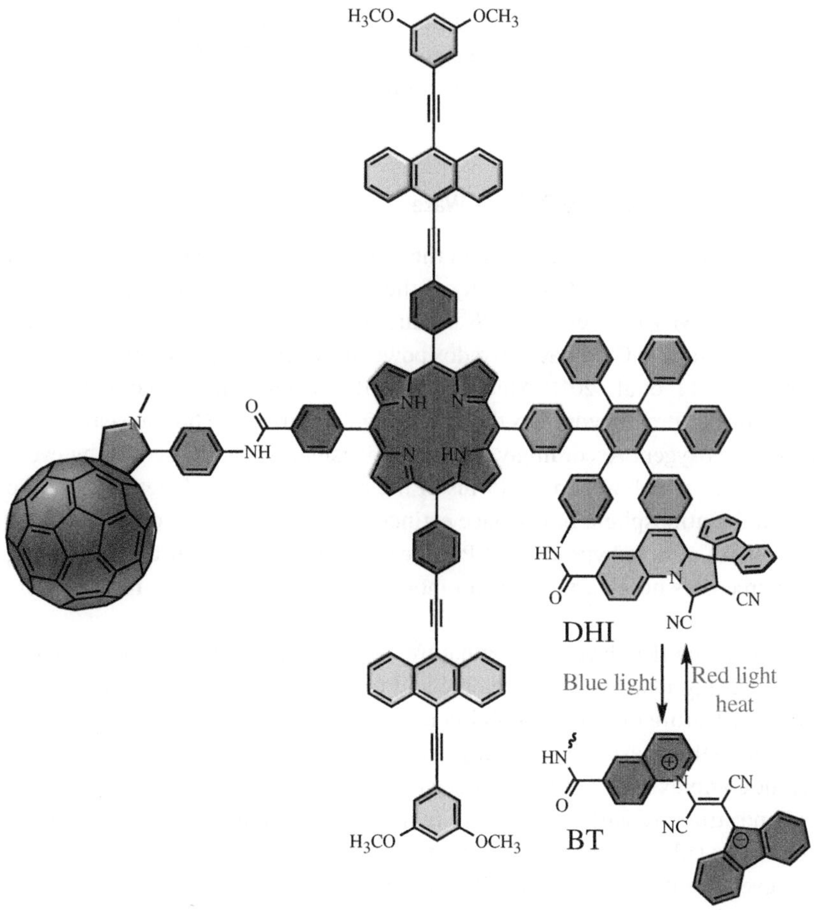

Figure 5.12 Self-regulated artificial photosynthetic pentad, which mimics the nonphotochemical processes in cyanobacteria. DHI = dihydroindolazine form and BT = betaine form. Source: Courtesy of Jackson Megiatto.

antenna absorbs far more solar energy than the reaction center can manage to produce the suitable reactants for the following chemical reactions. This causes a redox overload of the reaction center that starts producing harmful high-energy intermediates via secondary photochemical processes that can ultimately kill the organism. Therefore, photosynthetic organisms are equipped with photo-protection molecular devices that regulate the photosynthetic process as a function of sunlight intensity. For example, cyanobacteria respond to potentially damaging light levels by down-regulating the fraction of excitation energy that drives ET using a carotenoid protein. Under low-intensity light, this protein is orange in color and has no effect on the photosynthetic processes. Under high-intensity light, this protein undergoes a photoisomerization process that yields a red carotenoid analog. The red form quenches the electronic excitation collected by the antenna before it reaches the reaction center, therefore down-regulating photosynthesis. The red form can be thermally converted back to the inactive orange analog. The net result is that the reversible orange/red carotenoid protein photoisomerization allows the natural photosynthetic machinery to harmlessly dissipate excess excitation energy harnessed by the antenna as heat (Blankenship, 2002).

The self-regulated artificial photosynthetic system designed by GMM is a pentad composed of two anthracene antennas, a porphyrin-fullerene reaction center, and a photochromic dye. Under low-intensity light levels, excitation of the antenna leads to energy transfer to the porphyrin unit with quantitative quantum yield. The porphyrin singlet excited state decays through ET to the fullerene to produce the respective CSS with 0.82 quantum yield and lifetime of 2.4 ns. Thus, at low light levels, the pentad functions as an artificial antenna-reaction center system, and the photochromic dye, which is in the dihydroindolazine form (DHI), has no effect on the pentad light response. However, when light intensity increases, the photochrome undergoes photoisomerization to generate the colorful betaine form (BT), which rapidly quenches the porphyrin singlet excited state by energy transfer (time constant of 33 ps), thus reducing the quantum yield of charge separation to as low as 37%. The resulting BT excited state returns to the ground state by intersystem crossing, therefore dissipating the absorbed light energy as heat (Straight et al., 2008). This beautiful work clearly mimics the functions of the cyanobacteria carotenoid protein complex (Blankenship, 2002).

5.4.6 Artificial Reaction Centers Thermodynamically Poised to Oxidize Water

As can be contemplated from the artificial photosynthetic systems described in the previous section, fullerenes are indeed exceptional electron acceptors and have allowed chemists to investigate many of the natural photosynthetic processes in artificial systems (Guldi, 2002; Hirsch & Brettreich, 2005). However, the CSSs produced from fullerene-containing models are not suitable to mimic the functions of PSII as the resulting CSSs have no redox power to activate water splitting catalysts (Guldi, 2002; Hirsch & Brettreich, 2005; Megiatto Jr., Li, et al., 2010; Moore et al., 2012; Schuster & Megiatto Jr, 2009). PSII is the photo-redox membrane-bound heterodimeric protein complex that nature evolved about 3.5 billion years ago to catalyze the water oxidation reaction into molecular oxygen. Accordingly, PSII is responsible for the most striking event in Earth's atmosphere, namely the great oxidation, in which anaerobic forms of life existing on the planet at that time were forced to adapt to increasing levels of O_2 in the atmosphere or else face extinction. This dramatical event has been nicknamed as biology's big bang as the availability of a mere byproduct of PSII became the primary source of oxidizing power for the high-energy-requiring metabolic processes necessary to sustain more complex life forms (Gumsleya et al., 2017).

PSII is one of the most wonderful molecular machineries developed by nature, and it is the only enzyme known to promote water oxidation reactions. Such water splitting reaction is endergonic and produces O_2, four protons, and four reducing equivalents (electrons). Therefore, the key feature of PSII is its ability to combine single-photon excitation with the multielectron process of water oxidation. To drive this thermodynamically challenging reaction, PSII is equipped with the oxygen-evolving complex (OEC), an inorganic complex with a core composed of three manganese and one calcium ion held together by oxo bridges in a distorted cubane structure with a fourth external manganese ion linked to the cubane. The whole structure is embedded in a protein matrix (Blankenship, 2002). Despite huge research efforts, the OEC water oxidation mechanism is still under debate. However, extensive studies on PSII using XAS, XANES, and EXAFS X-ray spectroscopic techniques along with EPR investigations have allowed the description of Mn oxidation states and established that the OEC is photooxidized four times to reach the manganese high-oxidation states necessary for O—O bond formation and release of O_2. Therefore, the OEC functions as a molecular charge capacitor, able to accumulate the four oxidizing equivalents required for water splitting (Blankenship, 2002; Ferreira et al., 2004; Limburg et al., 1999; Umena et al., 2011).

In PSII, the oxidizing equivalents are produced through photoexcitation of the chlorophyll dimer pairs labeled Chl_{D1} and Chl_{D2} in Figure 5.13 (the whole structure is called P680 complex), which decay by electron transfer to the protein-bound quinone cofactor Q_A via a pheophytin primary electron acceptor. Q_A further transfers the electron to the loosely bound quinone electron acceptor Q_B, which functions as a mobile electron carrier in the thylakoid membrane and ultimately

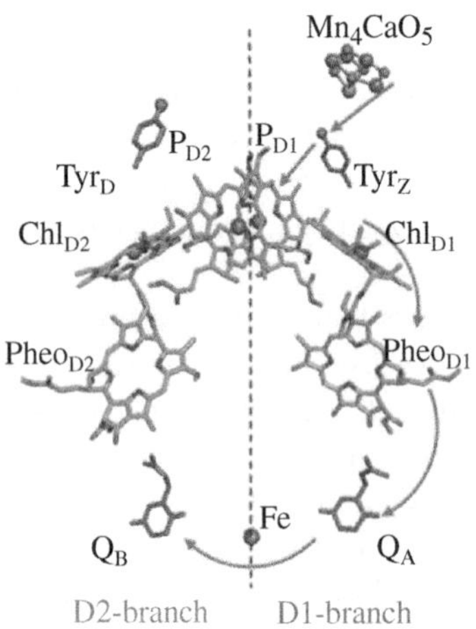

Figure 5.13 Spatial organization of the main active photo-redox subunits in the reaction center of the Photosystem II extracted from the *Thermosynechococcus vulcanus* thermophilic cyanobacterium *Synechococcus elongatus* cyanobacteria according to the crystal structure at 1.9 Å resolution (Umena et al., 2011) (PDB code 3WU2). Arrows indicate the electron transfer chain in the photosynthetic reaction center, while the dotted line indicates the pseudo-C_2 axe symmetry of the chromophoric array. Source: Adapted from the Protein Data Bank Archive (PDB ID 3WU2). Public domain.

delivers the acquired electron to the cytochrome complex. The photogenerated $P680^{\cdot+}$ is a highly oxidizing species, which removes an electron from the OEC labeled as $CaMn_4O_5$ in Figure 5.13. This electron transfer reaction from the OEC to the $P680^{\cdot+}$ species is one of the secrets of PSII and is mediated by a redox-active tyrosine–hystidine pair protein residue (TyrZ-D1His190) that features a short hydrogen bond between the phenolic and the imidazole moieties. In PSII chain electron transfer, the $P680^{\cdot+}$ species is reduced by the TyrZ residue, which loses its phenolic proton most likely to the nearby D1His190 via a proton-coupled electron transfer (PCET) mechanism. The resulting neutral TyrZ· radical then removes one electron from the OEC increasing its oxidation state. After four photoinduced charge separation in the PSII chain electron transfer, the OEC oxidizes two water molecules (Barry & Babcock, 1987; Blankenship, 2002; Debus et al., 1988; Groot et al., 2005; Hammarstrom & Styring, 2011; Rappaport et al., 2002; Tommos & Babcock, 1998).

The high redox power of the chromophores in PSII as well as the critical short hydrogen bond in the TyrZ-D1His190 pair that mediates the electron transfer reaction coupled to the proton translocation are warranted by the protein environment. Designing synthetic models able to mimic intricate protein radical interactions like those in PSII that ultimately allow the generation of highly energetic but long-lived radicals through PCET reactions is a challenging step in the development of artificial photosynthetic devices. In 2012, Megiatto, Gust, Moore, and Moore reported (Megiatto Jr., Antoniuk-Pablant, et al., 2012) a bioinspired molecular triad (Figure 5.14) engineered to functionally mimic the initial charge-separation events of PSII. The three units were designed and covalently assembled to simulate the specific interactions between P680, pheophytin, and the TyrZ-D1His190 pair of PSII.

In their design (Antoniuk-Pablant et al., 2016; Megiatto Jr., Antoniuk-Pablant, et al., 2012; Megiatto Jr., Patterson, et al., 2012; Ravensbergen et al., 2015), the tetracyanoporphyrin (TCNP) subunit acts as a powerful electron acceptor due to the

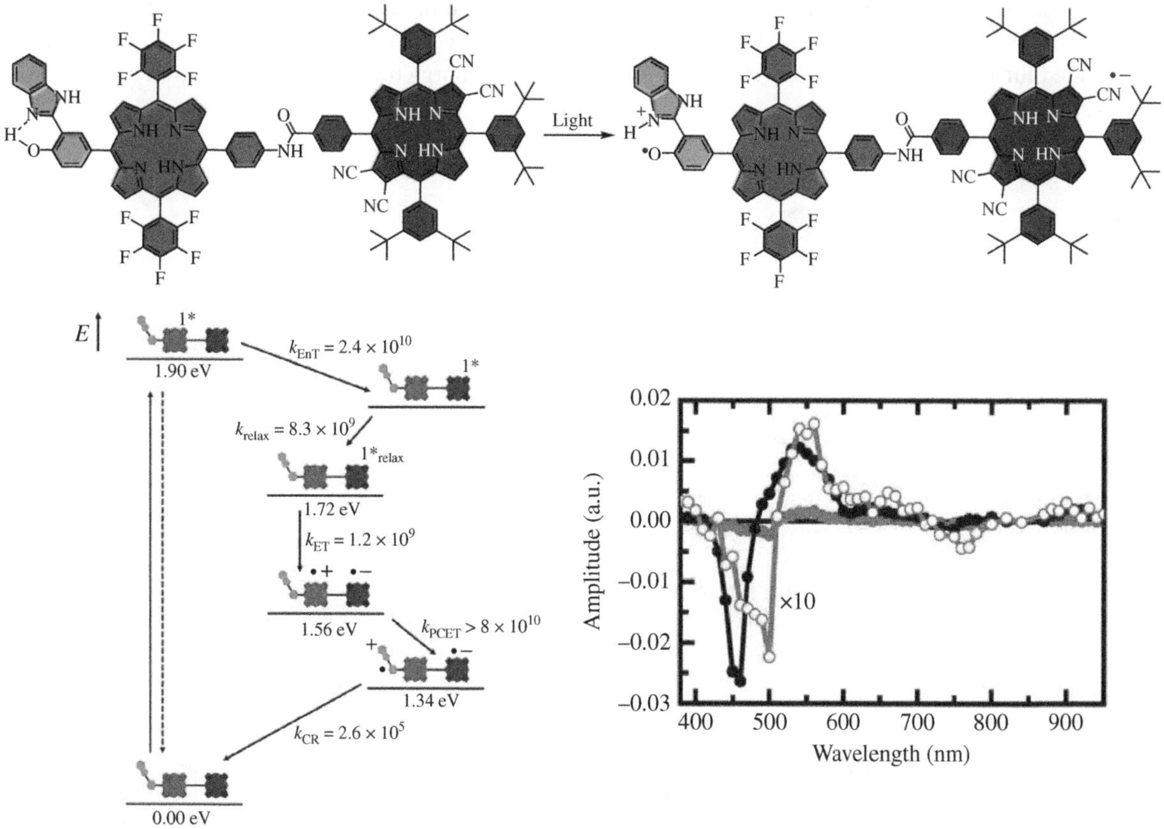

Figure 5.14 Molecular structure of PSII biomimetic triad and observed charge-separated state afforded from sequential ET and PCET reactions upon irradiation of the triad (top). Energy level diagram (not to scale), proposed decay pathways, and rate constants in s^{-1} following exclusive excitation of the PF_{10} subunit in benzonitrile (bottom left). Transient absorption measurements presented in the form of DAS confirmed formation of the expected CSS BiH^+-PhO·–PF_{10}–$TCNP^{\cdot-}$. Source: Adapted from Barry and Babcock (1987). Source: Jackson Megiatto.

electron withdrawing effects of the four cyano groups at the β-positions of the tetrapyrrolic core. The two pentafluorophenyl groups provide the primary electron-donor porphyrin (PF_{10}) with the right redox balance to transfer one electron to the TCNP acceptor after irradiation of the complex, resulting in the $PF10^{\cdot+}$ cation, which is thermodynamically competent to oxidize the benzimidazole-phenol (Bi-PhOH) secondary electron-donor component. The Bi-PhOH unit features an intramolecular hydrogen bond between the phenolic proton and the lone pair of the nitrogen atom in the benzimidazole moiety. During oxidation of Bi-PhOH by the photogenerated $PF10^{\cdot+}$, the phenol synthon transfers its proton to the benzimidazole group by a PCET mechanism to produce a neutral phenoxyl radical ($E = 1.06$ V vs. SCE) suitably poised for water oxidation ($E° = +0.58$ V vs. SCE, pH 7), thus mimicking the role of the TyrZ-D1His190 in PSII.

A complete spectroscopic investigation (Megiatto Jr., Antoniuk-Pablant, et al., 2012) revealed that excitation of PF_{10} in benzonitrile is followed by singlet energy transfer to TCNP ($\tau = 41$ ps), whose excited state structurally relaxes and then decays by ET ($\tau = 830$ ps) to yield the primary CSS Bi-PhOH–$PF_{10}^{\cdot+}$–$TCNP^{\cdot-}$. A second electron transfer reaction follows ($\tau < 12$ ps), giving a final state postulated as BiH^+-PhO·–PF_{10}–$TCNP^{\cdot-}$, in which the phenolic proton now resides on the benzimidazole group. This final state decays with a time constant of 3.8 μs. The relatively long-lifetime of the highly energetic final state is consistent with the involvement of PCET in the second step of the charge separation and suggests that coupling of the triad to water oxidation catalysts might yield useful systems for solar-fuel production. In a related study (Megiatto Jr. et al., 2014), the same authors investigated an organic/inorganic hybrid system, in which the Bi-PhOH–PF_{10} organic moiety was attached to TiO_2 semiconductor nanoparticles that function as electron acceptors (Figure 5.15). Using EPR spectroscopy, it was demonstrated that the hybrid system functions as a molecular triad to produce the target BiH^+-PhO·–PF_{10}–$TiO_2^{\cdot-}$ CSS. Most importantly, the EPR investigation unambiguously demonstrated that formation of the phenoxyl radical species in the PSII biomimetic at 13 K results in an energetic intermediate radical state, which at 100 K undergoes a change that can be interpreted in terms of solvation and conformational reorganization to yield a relatively more stable phenoxyl radical. This relaxation was also observed in EPR investigations of PSII. Therefore, that finding provided further support to the key role of PCET in mediating charge transfer between natural and artificial reaction centers and water oxidizing catalysts.

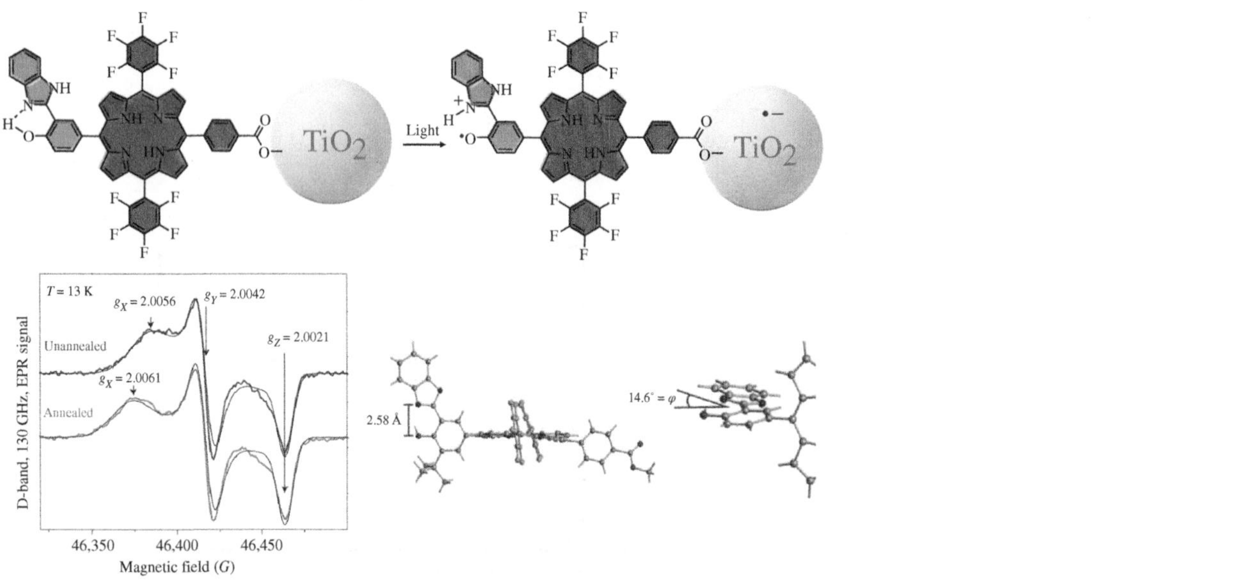

Figure 5.15 Molecular structure of the inorganic-organic hybrid system that mimics the radical interactions between protein backbone and chromophores in PSII (top). Photoinduced D-band (130 GHz) EPR spectra of an acetonitrile suspension of the system recorded at 13 K (bottom left). The unannealed curve means spectrum measured immediately after sample illumination in the cavity of the spectrometer, while the annealed curve means the spectrum afforded after the illuminated sample has been annealed at 100 K in the dark for 10 min and cooled back to 13 K. Annealing the sample shifts the components of the phenoxyl radical's g-tensors, especially the g_x component (from 2.0056 to 2.0061), which is spectroscopic evidence for PCET reactions. Crystal structure of the organic part of the system, which reveals formation of the critical short H-bond between the phenolic proton and the imidazole moiety that facilitates the proton transfer in the PCET process (bottom right). Source: Jackson Megiatto.

5.5 Harvesting the Energy of Charge-Separated States for Solar Fuel Production

5.5.1 Solar-Sensitized Photoelectrochemical Cells

Artificial photosynthesis research has made impressive progress in the last 50 years. The field has matured to the point where construction of complete systems, in which the redox equivalents photogenerated by reaction centers can be used to activate catalysts, can be contemplated. A complete photosynthetic system must include a chemical species that can be oxidized to serve as a source of electrons, another species that can be reduced to a useful fuel by these electrons, a reaction center that can use sunlight to carry out photoinduced electron transfer to generate oxidation and reduction equivalents, and catalysts to promote the oxidation and reduction processes. Assemble of a biomimetic complete functional photosynthetic device that integrates all components able to oxidize water molecules and reduce CO_2 to sugars is still impossible at our current technological level. Therefore, this research field is in its infancy, but significant progress has been made with encouraging results. Furthermore, a thermodynamic analysis of the water oxidation and CO_2 reduction reactions (Eqs. (5.2) and (5.3), respectively) informs that the former is the reaction that drives natural photosynthesis, while the latter is just nature's way to store the hydrogen produced from the water splitting process. In fact, carbohydrate production in natural photosynthesis is required to produce cellulose for plant growth and NADPH is essentially hydrogen in solid form. Accordingly, the chief goal of artificial photosynthesis is to achieve water splitting by a direct method to produce oxygen and hydrogen gases as fuels. The two gases can then be recombined into water and electricity as output in a conventional fuel cell to close the cycle (Gust et al., 2008).

$$H_2O \rightarrow H_2 + \frac{1}{2}O_2 \quad E^0 = 1.23\,V \tag{5.2}$$

$$CO_2 + H_2O \rightarrow \frac{1}{6}C_6H_{12}O_6 + O_2 \quad E^0 = 1.24\,V \tag{5.3}$$

In this context, Megiatto, Gust, Moore, and Moore in collaboration with Mallouk's group reported in 2012 (Zhao et al., 2012) the PEC schematically shown in Figure 5.16. The cell is composed of two electrodes immersed in water buffer solution. The photoanode is a transparent glass substrate coated with semiconducting TiO_2 nanoparticles (NPs) sensitized with a ruthenium tris-bipyridyl-type metal complex. The water oxidation catalyst, a colloidal iridium oxide nanoparticle, is decorated with the benzimidazole-phenol PCET electron transfer mediator that mimics the TyrZ-D1His190 pair in PSII and the 2-carboxyethylphosphonic acid anchoring molecule. The iridium oxide water catalyst is co-adsorbed onto the porous sensitized TiO_2 semiconducting electrode, which is connected to the metallic platinum cathode by an insulated cooper wire.

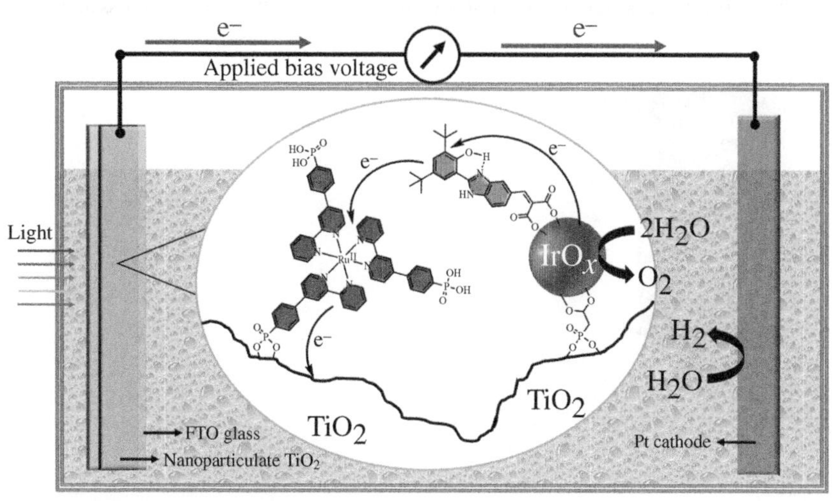

Figure 5.16 Photoelectrochemical cell and electron transfer reactions for solar splitting of water to hydrogen and oxygen gases. Nanoparticulate TiO_2 is deposited over a fluorine-dopped tin oxide film (FTO) on a transparent glass electrode. The deposited TiO_2 film is sensitized with a polypyridyl-Ru(II)complex followed by co-adsorption of the nanoparticulate IrO_x catalyst decorated with the biomimetic redox mediator and the 2-carboxyethylphosphonic acid anchoring molecule. Source: Jackson Megiatto.

Operationally, excitation of the ruthenium dye leads to injection of an electron into the TiO_2 conduction band and oxidation of the ruthenium dye, which in turn oxidizes the iridium oxide catalyst. After four photoinduced successive electron transfers, the catalyst accumulates the four oxidizing equivalents to split water molecules into oxygen and protons. The latter are reduced to hydrogen gas using the electrons that migrate from the TiO_2 electrode to the platinum cathode through the wire. Although the technique of co-adsorption to assemble the photoanode inevitably led to uncontrolled distances between the dye and the biomimetic mediator, the quantum yield for water splitting was more than double (2.3%) in the PEC with the biomimetic electron relay when compared to that of the PEC without it (1.0%). Despite the low efficiency, this work demonstrated the possibility of solar fuel production in biomimetic artificial photosynthesis.

5.5.2 Artificial Leaf

A breakthrough in water splitting photosynthetic devices was achieved by Nocera and collaborators in the first-half of the 2010s (Nocera, 2012). In PECs, the catalytic centers are usually noble transition metal complexes and/or nanoparticles. In a series of impressive works, Nocera et al. reported oxygen- and hydrogen-evolving catalysts made of earth-abundant elements, thus paving the way for the construction of inexpensive solar-to-fuel systems. Nocera's water oxidation catalyst is a cobalt-phosphate (Co-OEC) cluster with ions arranged in a cubane-like structure akin to the Mn_3CaO_4-Mn OEC in PSII. From an extensive investigation, the authors demonstrated that the Co-OEC operates via Co^{II}/Co^{IV} oxidation states, with the Co(IV) ion as the active catalytic species that oxidizes water molecules via a PCET mechanism (Figure 5.17). Furthermore, the inherent lability of Co(II) species, which usually leads to dissolution of its oxides in neutral water, is overcome as the catalyst has the property of self-repair through a reversible corrosion process. Therefore, the Co-OEC can operate in water under benign conditions. Consequently, nonprecious metals may be used in place of Pt in the cathode to produce hydrogen. Nocera et al. developed the NiMoZn ternary alloy, which is electrodeposited from ${Ni_{(aq)}}^{2+}$ solution, sodium molybdate, and anhydrous zinc chloride in the presence of pyrophosphate, bicarbonate, and hydrazine, followed by a base leaching solution. Molybdenum leaches from the alloy to furnish a high surface area material. The alloy was shown to achieve current densities as high as 1000 mA/cm^2 at an overpotential of 35 mV.

From these developments, Nocera et al. went a step further and constructed an artificial leaf (Figure 5.17) (Nocera, 2012). In their design, the photosynthetic membrane is replaced by a triple silicon junction that absorbs sunlight to generate CSSs. The OEC in PSII and the ferredoxin of the natural photosynthetic apparatus are replaced by Co-OEC and NiMoZn oxygen- and hydrogen-evolving catalysts, respectively, to perform water splitting. More specifically, the triple silicon wafer is a 3jn-a-Si solar cell, in which the *p*-side was protected with an ITO layer on which the Co-OEC is electrodeposited. The NiMoZn HEC alloy was electrodeposited onto the stainless steel support of the 3jn-a-Si cell. When this wireless artificial

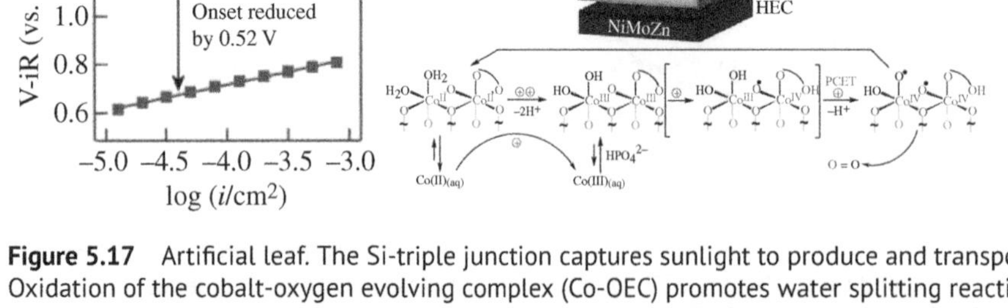

Figure 5.17 Artificial leaf. The Si-triple junction captures sunlight to produce and transport charges to the catalysts. Top left: Oxidation of the cobalt-oxygen evolving complex (Co-OEC) promotes water splitting reactions to yield oxygen gas and protons. The latter are reduced by the hydrogen-evolving complex (HEC). Bottom right: The artificial leaf (oxidation site) for onset of water oxidation in the dark (triangles) and at 100 mW/cm^2 illumination (squares). Under illumination, the onset potential needed to be applied to the cell for water oxidation is decreased by 0.52 V. Top right: Schematic view of the components of the multilayer artificial leaf. Bottom right: Proposed mechanism for the water splitting reaction promoted by the Co-OEC. Source: Adapted from Nocera (2012). Reproduced with permission of the American Chemical Society.

leaf is immersed in an electrolyte (1 M potassium borate, pH 9.2) and illuminated with 1 sun (AM 1.5, 100 mA cm^{-2}), O_2 bubbles evolve from the anode at the front face and bubbles of H_2 evolve from the cathode at the back of the cell at an efficiency of 2.5%. Although this apparent low efficiency seems to doom this pioneered artificial leaf, it should be noted that most efficient terrestrial plants barely convert 1% of sunlight into biofuel. Therefore, this scientific discovery has set a course for the large-scale deployment of solar energy, and further developments of the artificial leaf are expected to increase the cell efficiency to 10%.

5.6 Conclusions

Since the 1970s, scientists have tried to produce fuel from water and sunlight more efficiently than natural photosynthesis. As this brief overview of the artificial photosynthesis research field demonstrated, science has made slow but considerable progress on the critical steps of solar-to-fuel conversion process. The basic outlines of such a process have been demonstrated. Although there is not a commercially viable artificial photosynthetic technology yet, and whether the molecule-based approach will be practically fruitful, rapid progress has been made toward the scientific holy grail of assembling an artificial leaf in recent years. One of the biggest challenges in the field is to design artificial photosynthetic systems that are robust with longer lifetimes under operational conditions. Photodegradation of the molecular components is still a technological issue that needs to be overcome for practical applications. Many materials are currently under investigation to address the photodegradation problem with promising results. If solar fuel research continues at this pace, an artificial photosynthetic device able to produce 100% clean, drop-in replacements for fossil fuels from sunlight and water may be just around the corner. If such technology becomes available at large scales, the transition from a black and nervous civilization based on oil and coal to a quieter civilization based on solar energy will surely follow, as envisioned by Ciamician in 1912.

References

Alcantara, A. F. P., Fontana, L. A., Almeida, M. P., Rigolin, V. H., Ribeiro, M. A., Barros, W. P., & Megiatto, J. D., Jr. (2020). Control over the redox cooperative mechanism of radical carbene transfer reactions for the efficient active-metal-template synthesis of [2]rotaxanes. *Chemistry – A European Journal*, *26*, 7808–7822.

Alcantara, A. F. P., Fontana, L. A., Rigolin, V. H., Andrade, Y. F., Ribeiro, M. A., Barros, W. P., Ornelas, C., & Megiatto, J. D., Jr. (2018). Olefin cyclopropanation by radical carbene transfer reactions promoted by cobalt(II)/porphyrinates: Active-metal-template synthesis of [2]rotaxanes. *Angewandte Chemie International Edition*, *57*, 8979–8983.

Antoniuk-Pablant, A., Terazono, Y., B. J. B., Sherman, B. D., Megiatto, J. D., Brudvig, G. W., Moore, A. L., Moore, T. A., & Gust, D. (2016). A new method for the synthesis of β-cyano substituted porphyrins and their use as sensitizers in photoelectrochemical devices. *Journal of Materials Chemistry A*, *4*, 2976–2985.

Armaroli, N., Diederich, F., Dietrich-Buchecker, C. O., Flamigni, L., Marconi, G., Nierengarten, J. F., & Sauvage, J. P. (1998). A copper(I)-complexed rotaxane with two fullerene stoppers: Synthesis, electrochemistry, and photoinduced processes. *Chemistry – A European Journal*, *4*, 406–416.

Astruc, D. (2007). *Organometallic chemistry and catalysis*. Springer.

Astruc, D., Boisselier, E., & Ornelas, C. (2010). Dendrimers designed for functions: From physical, photophysical, and supramolecular properties to applications in sensing, catalysis, molecular electronics, photonics, and nanomedicine. *Chemical Reviews*, *10*, 1857–1959.

Balzani, V., Juris, A., Venturi, M., Campagna, S., & Serroni, S. (1996). Luminescent and redox-active polynuclear transition metal complexes. *Chemical Reviews*, *96*, 759–833.

Balzani, V., & Scandola, F. (1991). *Supramolecular photochemistry*. Horwood.

Barry, B. A., & Babcock, G. T. (1987). Tyrosine radicals are involved in the photosynthetic oxygen-evolving system. *Proceedings of the National Academy of Sciences of the United States of America*, *84*, 7099–7103.

Blankenship, R. E. (2002). *Molecular mechanisms of photosynthesis*. Blackwell Science.

Bolin, B. (2007). *A history of the science and politics of climate change: The role of the intergovernmental panel on climate change*. U.S. Department of Energy, Office of Scientific and Technical Information, USA.

Braga, B., Pope, B., Druyan, A., Tyson, N. D., & Silvestri, A. (2014). *Cosmos: A spacetime odyssey*. Twentieth Century Fox Home Entertainment Ltd.

Burns, C. J., & Stoddart, J. F. (2017). *The nature of the mechanical bond: From molecules to machines*. John Wiley & Sons, Inc.

Ciamician, G. (1912). The photochemistry of the future. *Science, 36*, 385–394.

Close, F. (2019). *Eclipses: What everyone needs to know*. Oxford University Press.

Collin, J. P., Guillerez, S., Sauvage, J. P., Barigelletti, F., De Cola, L., Flamigni, L., & Balzani, V. (1991). Photoinduced processes in dyads and triads containing a ruthenium(II) bis(terpyridine) photosensitizer covalently linked to electron-donor and acceptor groups. *Inorganic Chemistry, 30*, 4230–4238.

Debus, R. J., Barry, B. A., Babcock, G. T., & Mcintosh, L. (1988). Site-directed mutagenesis identifies a tyrosine radical involved in the photosynthetic oxygen-evolving system. *Proceedings of the National Academy of Sciences of the United States of America, 85*, 427–430.

Devadoss, C., Bharathi, P., & Moore, J. S. (1996). Energy transfer in dendritic macromolecules: Molecular size effects and the role of an energy gradient. *Journal of the American Chemical Society, 118*, 9635–9644.

Devlin, R. M., & Barker, A. V. (1971). *Photosynthesis*. Van Nostrand Reinhold Company.

Farmer, E. E. (2010). *Jean Senebier's thoughts on experimentation and their relevance for today's researcher. Science Archives, 63*, 185–192.

Ferreira, K. N., Iverson, T. M., Maghlaoui, K., Barber, J., & Iwata, S. (2004). Architecture of the photosynthetic oxygen-evolving center. *Science, 303*, 1831–1838.

Flamigni, L., Heitz, V., & Sauvage, J.-P. (2006). Porphyrin rotaxanes and catenanes: Copper(I)-templated synthesis and photoinduced processes. In E. Alessio (Ed.), *Non-covalent multi-porphyrin assemblies*, Structure and bonding (Vol. 121, pp. 217–261). Springer.

Fontana, L. A., Almeida, M. P., Alcântara, A. F., Rigolin, V. H., Ribeiro, M. A., Barros, W. P., & Megiatto, J. D., Jr. (2020). Ru(II)Porphyrinate-based molecular nanoreactor for carbene insertion reactions and quantitative formation of rotaxanes by active-metal-template syntheses. *Nature Communications, 11*, 6370.

von Frischi, K. (1940). *Life and you*. Scientific Book Club; Book Club (BCE/BOMC)

Fujishima, A., & Honda, K. (1972). Electrochemical photolysis of water at a semiconductor electrode. *Nature, 238*, 37–38.

Groot, M. L., Pawlowicz, N. P., van Wilderen, L. J., Breton, J., van Stokkum, I. H., & van Grondelle, R. (2005). Initial electron donor and acceptor in isolated photosystem II reaction centers identified with femtosecond mid-IR spectroscopy. *Proceedings of the National Academy of Sciences of the United States of America, 102*, 13087–13092.

Guldi, D. M. (2002). Fullerene–porphyrin architectures; photosynthetic antenna and reaction center models. *Chemical Society Reviews, 31*, 22–36.

Gumsleya, A. P., Chamberlain, K. R., Bleeker, W., Söderlund, U., De Kock, M. O., Larsson, E. R., & Bekker, A. (2017). Timing and tempo of the Great Oxidation Event. *Proceedings of the National Academy of Sciences of the United States of America, 114*, 1811–1816.

Gust, D. (2016). An illustrative history of artificial photosynthesis. In B. Robert (Ed.), *Advances in botanical research* (pp. 1–40). Elsevier.

Gust, D., Kramer, D., Moore, A., Moore, T. A., & Vermaas, W. (2008). Engineered and artificial photosynthesis: Human ingenuity enters the game. *MRS Bulletin, 33*, 383–387.

Gust, D., Moore, T. A., Moore, A. L., Lee, S. J., Bittersmann, E., Luttrull, D. K., Rehms, A. A., DeGraziano, J. M., Ma, X. C., Gao, F., Belford, R. E., & Trier, T. T. (1990). Efficient multistep photoinduced electron transfer in a molecular pentad. *Science, 248*, 199–201.

Gust, D., Moore, T. A., Moore, A. L., Macpherson, A. N., Lopez, A., DeGraziano, J. M., Gouni, I., Bittersmann, E., & Seely, G. R. (1993). Photoinduced electron and energy transfer in molecular pentads. *Journal of the American Chemical Society, 115*, 11141–11152.

Hammarstrom, L., & Styring, S. (2011). Proton-coupled electron transfer of tyrosines in photosystem II and model systems for artificial photosynthesis: The role of a redox-active link between catalyst and photosensitizer. *Energy & Environmental Science, 4*, 2379–2388.

Hershey, D. R. (1991). *Digging deeper into Helmont's famous willow tree experiment. The American Biology Teacher, 53*, 458–460.

Hirsch, A., & Brettreich, M. (2005). *Fullerenes*. Wiley-VCH.

Imahori, H., Sekiguchi, Y., Kashiwagi, Y., Sato, T., Araki, Y., Ito, O., Yamada, H., & Fukuzumi, S. (2004). Long-lived charge-separated state generated in a ferrocene-meso, meso-linked porphyrin trimer-fullerene pentad with a high quantum yield. *Chemistry – A European Journal, 10*, 3184–3196.

Ingenhousz, J. (1779). *Experiments upon vegetables, discovering their great power of purifying the common air in the sun-shine, and of injuring it in the shade and at night. To which is joined, A new method of examining the accurate degree of salubrity of the atmosphere*. Printed for Elmsly, P., Payne, H.

Jakob, M., Berg, A., Levanon, H., Schuster, D. I., & Megiatto, J. D., Jr. (2011a). Photoexcited state properties of H_2-porphyrin/C_{60}-based rotaxanes as studied by time-resolved EPR spectroscopy. *The Journal of Physical Chemistry A, 115*, 5044–5052.

Jakob, M., Berg, A., Levanon, H., Schuster, D. I., & Megiatto, J. D., Jr. (2011b). Photoinduced processes in some mechanically interlocked supramolecules as studied by time-resolved electron paramagnetic resonance. *The Journal of Physical Chemistry C, 115*, 24555–24563.

Jordan, P., Fromme, P., Witt, H. T., Klukas, O., Saenger, W., & Krauß, N. (2001). Three-dimensional structure of cyanobacterial photosystem I at 2.5 Å resolution. *Nature, 411*, 909–917.

Kirner, S. V., Guldi, D. M., Megiatto, J. D., Jr., & Schuster, D. I. (2014). Synthesis and photophysical properties of new catenated electron donor–acceptor materials with magnesium and free base porphyrins as donors and C_{60} as the acceptor. *Nanoscale, 7*, 1145–1160.

Kirner, S. V., Henkel, C., Guldi, D. M., Megiatto, J. D., Jr., & Schuster, D. I. (2015). Multistep energy and electron transfer processes in novel rotaxane donor–acceptor hybrids generating microsecond-lived charge separated states. *Chemical Science, 6*, 1–12.

Kong, J. L., & Loach, P. A. (1978). In P. L. Dutton & H. Scarpa (Eds.), *Frontiers of biological energetics: From electrons to tissues* (pp. 73–82). Academic Press.

Kuciauskas, D., Liddell, P. A., Hung, S. C., Lin, S., Stone, S., Seely, G. R., Moore, A. L., Moore, T. A., & Gust, T. A. (1997). Structural effects on photoinduced electron transfer in carotenoid-porphyrinquinone triads. *The Journal of Physical Chemistry B, 101*, 429–440.

Kutschera, U. (2015). Basic versus applied research: Julius Sachs (1832–1897) and the experimental physiology of plants. *Plant Signaling & Behavior, 10*, e1062958.

Lehn, J.-M. (2011). *Supramolecular chemistry: Concepts and perspectives*. John Wiley & Sons.

Liddell, P. A., Kuciauskas, D., Sumida, J. P., Nash, B., Nguyen, D., Moore, A. L., Moore, T. A., & Gust, D. (1994). Photoinduced charge separation and charge recombination to a triplet state in a carotene–porphyrin–fullerene triad. *Journal of the American Chemical Society, 119*, 1400–1405.

Limburg, J., Vrettos, J. S., Liable-Sands, L. M., Rheingold, A. L., Crabtree, R. H., & Brudvig, G. W. (1999). A functional model for O–O bond formation by the O-2-evolving complex in photosystem II. *Science, 283*, 1524–1527.

Marcus, R. A. (1956). The theory of oxidation-reduction reactions involving electron transfer. *The Journal of Chemical Physics, 24*, 966–978.

Marcus, R. A., & Sutin, N. (1985). Electron transfers in chemistry and biology. *Biochimica et Biophysica Acta, 811*, 265–322.

Megiatto, J. D., Jr., Antoniuk-Pablant, A., Sherman, B. D., Kodis, G., Gervaldo, M., Moore, T. A., Moore, A. L., & Gust, D. (2012). Mimicking the electron transfer chain in photosystem II with a molecular triad thermodynamically capable of water oxidation. *Proceedings of the National Academy of Sciences of the United States of America, 109*, 15578–15583.

Megiatto, J. D., Jr., Guldi, D. M., & Schuster, D. I. (2020). Design, synthesis and photoinduced processes in molecular interlocked photosynthetic [60]fullerene systems. *Chemical Society Reviews, 49*, 8–20.

Megiatto, J. D., Jr., Li, K., Schuster, D. I., Palkar, A., Herranz, M. Á., Echegoyen, L., Abwandner, S., de Miguel, G., & Guldi, D. M. (2010). Convergent synthesis and photoinduced processes in multi-chromophoric rotaxanes. *The Journal of Physical Chemistry B, 114*, 14408–14419.

Megiatto, J. D., Jr., Méndez-Hernández, D. D., Tejeda-Ferrari, M. E., Teillout, A. L., Llansola-Portolés, M. J., Kodis, G., Poluektov, O. G., Rajh, T., Mujica, V., Groy, T. L., & Gust, D. (2014). A bioinspired redox relay that mimics radical interactions of the Tyr–His pairs of photosystem II. *Nature Chemistry, 6*, 423–428.

Megiatto, J. D., Jr., Patterson, D., Sherman, B. D., Moore, T. A., Gust, D., & Moore, A. L. (2012). Intramolecular hydrogen bonding as a synthetic tool to induce chemical selectivity in acid catalyzed porphyrin synthesis. *Chemical Communications, 48*, 4558–4560.

Megiatto, J. D., Jr., & Schuster, D. I. (2008). General method for synthesis of functionalized macrocycles and catenanes utilizing "click" chemistry. *Journal of the American Chemical Society, 130*, 12872–12873.

Megiatto, J. D., Jr., & Schuster, D. I. (2009). "Click" methodology for synthesis of functionalized [3]catenanes: Toward higher interlocked structures. *Chemistry – A European Journal, 15*, 5444–5448.

Megiatto, J. D., Jr., & Schuster, D. I. (2010). Introduction of useful peripheral functional groups on [2]catenanes by combining Cu(i) template synthesis with "click" chemistry. *New Journal of Chemistry, 34*, 276–286.

Megiatto, J. D., Jr., & Schuster, D. I. (2011a). Alternative demetalation method for Cu(I)-phenanthroline-based catenanes and rotaxanes. *Organic Letters*, *13*, 1808–1811.

Megiatto, J. D., Jr., & Schuster, D. I. (2011b). Synthesis of electron donor-[60]fullerene multi-ring interlocked systems. In F. D'Souza & K. M. Kadish (Eds.), *Handbook of carbon nanomaterials* (pp. 207–244). World Scientific Press.

Megiatto, J. D., Jr., & Schuster, D. I. (2012). Interlocked artificial photosynthetic model systems composed of electron-donor and [60]fullerene units. In F. L. de la Puente & J.-F. Nierengarten (Eds.), *Fullerenes: Principles and applications* (pp. 354–385). Royal Society of Chemistry.

Megiatto, J. D., Jr., Schuster, D. I., Abwandner, S., de Miguel, G., & Guldi, D. M. (2010). [2]Catenanes decorated with porphyrin and [60]fullerene groups: Design, convergent synthesis, and photoinduced processes. *Journal of the American Chemical Society*, *132*, 3847–3861.

Megiatto, J. D., Jr., Schuster, D. I., de Miguel, G., Wolfrum, S., & Guldi, D. M. (2012). Topological and conformational effects on electron transfer dynamics in porphyrin-[60]fullerene interlocked systems. *Journal of Materials Chemistry*, *24*, 2472–2485.

Megiatto, J. D., Jr., Spencer, R., & Schuster, D. I. (2009a). Efficient one-pot synthesis of rotaxanes bearing electron donors and [60] fullerene. *Organic Letters*, *11*, 4152–4155.

Megiatto, J. D., Jr., Spencer, R., & Schuster, D. I. (2009b). Optimizing reaction conditions for synthesis of electron donor-[60] fullerene interlocked multiring systems. *Journal of Materials Chemistry*, *21*, 1544–1550.

Meldal, M., & Tornøe, C. W. (2008). Cu-catalyzed azide–alkyne cycloaddition. *Chemical Reviews*, *108*, 2952–3015.

Molnar, S. M., Nallas, G., Bridgewater, J. S., & Brewer, K. J. (1994). Photoinitiated electron collection in a mixed-metal trimetallic complex of the form $\{[(bpy)_2Ru(dpb)]_2IrCl_2\}(PF_6)_5$ (bpy = 2,2′-bipyridine and dpb = 2,3-bis(2-pyridyl)benzoquinoxaline). *Journal of the American Chemical Society*, *116*, 5206–5210.

Moore, G. F., Megiatto, J. D., Hambourger, M., Gervaldo, M., Kodis, G., Moore, T. A., Gust, D., & Moore, A. L. (2012). Optical and electrochemical properties of hydrogen-bonded phenol-pyrrolidino[60]fullerenes. *Photochemical & Photobiological Sciences*, *11*, 1018–1025.

Niel, C. B. (1932). On the morphology and physiology of the purple and green sulphur bacteria. *Archives of Microbiology*, *3*, 1–112.

Nocera, D. G. (2012). The artificial leaf. *Accounts of Chemical Research*, *45*, 767–776.

Phené, J. S. (1875). *On prehistoric traditions and customs in connection with sun and serpent*. R. Hardwicke.

Priestley, J. (2013). *Experiments and observations on different kinds of air*. Cambridge University Press.

Rappaport, F., Guergova-Kuras, M., Nixon, P. J., Diner, B. A., & Lavergne, J. (2002). Kinetics and pathways of charge recombination in photosystem II. *Biochemistry*, *41*, 8518–8527.

Ravensbergen, J., Antoniuk-Pablant, A., Sherman, B. D., Kodis, G., Megiatto, J. D., Jr., Méndez-Hernández, D. D., Frese, R. N., Van Grondelle, R., Moore, T. A., Moore, A. L., & Gust, D. (2015). Spectroscopic analysis of a biomimetic model of TyrZ function in PSII. *The Journal of Physical Chemistry B*, *119*, 12156–12163.

Sauvage, J.-P. (1990). Interlacing molecular threads on transition metals: Catenands, catenates, and knots. *Accounts of Chemical Research*, *23*, 319–327.

Schuster, D. I., & Megiatto, J. D., Jr. (2009). Carbon nanotubes: Nanotubes reveal all in solution. *Nature Chemistry*, *1*, 182–183.

Straight, S. D., Kodis, G., Terazono, Y., Hambourger, M., Moore, T. A., Moore, A. L., & Gust, D. (2008). Self-regulation of photoinduced electron transfer by a molecular nonlinear transducer. *Nature Nanotechnology*, *3*, 280–283.

Tabushi, I., Koga, N., & Yanagita, M. (1979). Efficient intramolecular quenching and electron transfer in tetraphenylporphyrin attached with benzoquinone or hydroquinone as a photosystem model. *Tetrahedron Letters*, *20*, 257–260.

Terazono, Y., Kodis, G., Liddell, P. A., Garg, V., Moore, T. A., Moore, A. L., & Gust, D. (2009). Multiantenna artificial photosynthetic reaction center complex. *The Journal of Physical Chemistry B*, *113*, 7147–7155.

Tommos, C., & Babcock, G. T. (1998). Oxygen production in nature: A light-driven metalloradical enzyme process. *Accounts of Chemical Research*, *31*, 18–25.

Umena, K., Kawakami, J., Shen, R., & Kamiya, N. (2011). Crystal structure of oxygen-evolving photosystem II at a resolution of 1.9 angstrom. *Nature*, *473*, 55–60.

Zhao, Y., Swierk, J. R., Megiatto, J. D., Jr., Sherman, B., Youngblood, W. J., Qin, D., Lentz, D. M., Moore, A. L., Moore, T. A., Gust, D., & Mallouk, T. E. (2012). Improving the efficiency of water splitting in dye-sensitized solar cells by using a biomimetic electron transfer mediator. *Proceedings of the National Academy of Sciences of the United States of America*, *109*, 15612–15616.

6

The Competitive Kinetics of Solar-Driven CO_2 Reduction

Mark T. Spitler

Department of Chemistry, University of North Carolina, Chapel Hill, NC, USA

6.1 Introduction

Research has accelerated over the past decade into the technical problems of the solar reduction of CO_2 to fuels or value-added chemicals (He & Janáky, 2020). Wide-ranging explorations have been made of the homogeneous or heterogeneous catalysis involved in CO_2 reduction and how it can be integrated within light-driven systems consisting of semiconductors, metals, particles, and nanostructures. Much of the recent work in this area is an extension of previous efforts for solar hydrogen fuel production where reduced CO_2 has supplanted molecular hydrogen. However, the challenges in the CO_2 area are much greater and the response of scientific community to this problem has been much larger in scope and effort. The thousands of publications on this topic and many reviews cover not only the theory, materials, physics, and chemistry of the process but also the commercial aspects and societal impact of such technology (Goeppert et al., 2012; Spitler et al., 2020; Stolarczyk et al., 2018).

This work examines the science and technology involved in photo-initiated CO_2 reduction from a design perspective of representative prototypes from the literature and how their form is strongly intertwined with the complexity of the chemical kinetics of CO_2 reduction to fuels or value-added chemicals. The analysis is made within the context of a market-based definition of the prototype, its size, and how this is tied to fundamental science of its active components. This is similar to an earlier examination of solar hydrogen production (Spitler et al., 2020), but differs in that the complicated interplay between the prototype design and its component catalytic functions dominates as a topic.

A framework for the marketplace for CO_2 reduction is set up here that encompasses the full range of chemistry being explored in research efforts in this area around the world. It is evident that the market for solar-powered CO_2 reduction to make multi-carbon products differs from that for hydrogen as a fuel. Whereas solar H_2 production provides a fuel with a widespread use in the economy, the solar reduction of CO_2 can also be seen to serve a number of specific uses over a broader range. In a future world of zero carbon emissions, or en route to such a world, there may be a social carbon tax imposed upon the emission of CO_2, and this will create a market need for on-site solar-driven systems to reduce this emission. For an electric power plant burning coal, natural gas, or liquid hydrocarbons and emitting an effluent of ~10% CO_2 (Goeppert et al., 2012), solar CO_2 reduction systems can serve as a form of recycling where CO_2 emissions can be captured and returned to a high energy and useful form of carbon. This counteracts any such tax through creation of revenue. It is also evident that there is general interest in reducing CO_2 produced by industrial processes at specific locations as a form of environmental remediation. A good example is concrete formation, which currently accounts for about 5% of anthropogenic sources of CO_2 emissions (Nitopi et al., 2019). At another limit is simple fuel harvesting to create fuels or chemical feedstocks, much like photosynthetic organisms on Earth do. This could be a distributed product that would use the low-level and dispersed CO_2 in the atmosphere at its current 0.04% level. It is here that solar-powered CO_2 reduction will have to be cost-competitive in the energy marketplace with other technologies for these fuels and high-value chemicals. However, as has been shown for the relation between photovoltaic (PV) cost and market size over the past 50 years (Spitler et al., 2020), there will be an initial market demand for even more expensive, higher-priced CO_2 reduction products.

The maturity of technology differs in these markets. For a recycling CO_2 reduction system at the fixed location of a power plant, a desire to avoid a carbon tax and simultaneously create value from the CO_2 in the gaseous effluent of the power

Conversion of Water and CO_2 to Fuels using Solar Energy: Science, Technology and Materials, First Edition.
Edited by Oomman K. Varghese and Flavio L. Souza.

plant can motivate the installation of a centralized electrolyzer system for CO_2 reduction that is powered by a solar PV farm. PV and electrolyzer technologies are well known, although the particular form of a CO_2 reduction electrolyzer would have to be determined where both liquid-phase and vapor-phase catalytic designs are possible (Modestino et al., 2015; Xiang et al., 2013). In any case, economies of scale come into play here, involving a significant capital investment. Whereas the concept of using expensive electronic-grade semiconductors in PV systems to produce commodity fuels or chemicals would not be cost-effective in an energy-based market, in this context of imposed social costs, it may be justified. The other limit of the market would be the fuel harvesting of CO_2 from the atmosphere. It may be possible that the cost of a solar PV farm/electrolyzer could be reduced to make its application usable in this market. However, the miniaturization of all of its capital-intensive, active components into an integrated design would be a more certain pathway to lower costs. Within this system design would be light absorption, charge transport, catalysis, and fuel harvesting. This has been the goal of most of the experimental work on CO_2 reduction and would be most suitable for a product application that is distributed, much like present solar PV rooftop installations. The remediation market would be intermediate between these two limits and would contain elements of both. It may be required to be portable from site to site.

As has been illustrated in prior work, there are a number of technological steps that connect the marketplace definition of a system for solar fuel production to the internal chemistry and chemical engineering of a device prototype (Spitler et al., 2020). These include the need for an independent confirmation of efficiency and yield from a device, the material challenges in making large-area photoactive solids, and the details of ensuring a durable fuel production system. However important these may be, the focus here will be on the relationship of the prototype design of the photosystem to the kinetics of the CO_2 reduction reaction and how to categorize these relationships using models drawn from the literature.

This analysis of light-driven CO_2 reduction will be done in three parts. First, we look at examples from natural photosynthesis since it contains a functioning system for the light-driven fixation of CO_2. Here in this work, a different perspective is taken on the models and kinetics of photosynthesis and the lessons to be drawn from them, highlighting the important role of the coupling between the different elements involved in CO_2 fixation in the chloroplast. Second, within the context of these photosynthetic kinetic structures, a review will be of man-made light-driven water oxidation efforts and the designs of the kinetic systems with which they operate. Finally, a similar examination will be made of CO_2 reduction research, with the conclusion that present research on solar-driven CO_2 reduction is yet narrow in its scope and that many opportunities for research remain to be explored.

6.2 Photosynthetic Systems

In this section, the operation of the photosynthetic system for fixation of CO_2 from the atmosphere will be considered from a device perspective. The design characteristics will be identified and analyzed with respect to their utility in man-made systems for CO_2 reduction.

Photosynthetic systems for the light-driven fixation of CO_2 have existed in one form or another for several billion years (Blankenship, 2010). They are found in bacteria and eukaryotes in the myriad evolutionary variants of such systems that have developed during that period. Research into these various photosynthetic systems has provided a wealth of knowledge on their design, the structure of their components, and the dynamics of their mode of operation. Given that an essential function is to fix CO_2 from the atmosphere into multi-carbon compounds, these natural systems and the scientific studies of them can serve as useful references in the design and construction of man-made systems for CO_2 reduction.

There are a number of design elements in photosynthesis that will be of use to those assembling a man-made system and a number that are to be avoided in certain circumstances. Given the many evolutionary variations on photosynthetic systems, the list given here of such features is not complete but represents the main elements of concern for eventual man-made systems. The intent is to take advantage of the billions of years of trial and error by nature in developing CO_2 fixation process in order to identify tools and concepts that can be of utility in artificial systems for CO_2 reduction.

6.2.1 General

The location of the photosynthetic system for CO_2 reduction in eukaryotes is within the chloroplast organelle (Blankenship, 2014). Two distinct elements are involved, and they are found in different locations within the stroma,

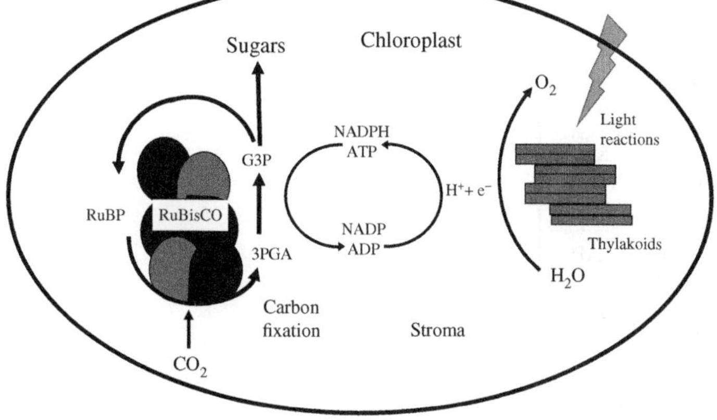

Figure 6.1 Within the chloroplast are found the light harvesting and reaction centers of the thylakoid membrane, which oxidize water and produce NADPH and ATP for use in CO_2 fixation by RuBisCO at another location in the stroma. M. T. Spitler.

or interior volume, of the chloroplast. This is depicted schematically in Figure 6.1 and consists of the thylakoid membranes, where the light reactions occur to oxidize water, and the subcellular structures and chemistry for CO_2 fixation, where ribulose-1,5-biphosphate carboxylase/oxygenase (RuBisCO) operates. The light-driven reactions release H^+, nicotinamide adenine dinucleotide phosphate hydrogen (NADPH), and adenosine triphosphate (ATP) to the stroma, which are utilized by RuBisCO in the Calvin–Benson Cycle to fix CO_2 into a carbohydrate.

The structure and function of the photosystems within the thylakoid membrane will be addressed first, with a schematic depiction of its form for higher plants being presented in Figure 6.2 (Turmo et al., 2017). Prominent within this picture are the two well-known reaction centers Photosystem I (PSI) and Photosytem II (PSII) imbedded within the membrane. Although the various components in this membrane are not always present in an exact one-to-one ratio (Fan et al., 2007), these two photosystems and their associated cofactors act in coordination to extract an electron from water and then elevate its reducing power through absorption of light energy from sunlight to enable the reduction of ferredoxin on the membrane periphery. This then leads to reduction of ferredoxin-NADP$^+$ reductase in the stroma to make NADPH.

Both reaction centers initiate an electron transfer reaction upon excitation of their central chromophores, labeled P680 for PSII and P700 for PSI, and both have acceptor electron transfer chains that serve to separate electrons from oxidized reaction centers P680$^+$ and P700$^+$. A distinction of note between these two reaction centers is that PSII sources its electrons through a four-electron catalyzed oxidation of water and gives up these electrons through two, two-electron quinone reductions at the end of its electron transfer chain, whereas the electrons in PSI both enter and exit through one electron reactions. It has been observed that only natural photosynthetic organisms with this combination of two photosystems are able to use water as a source of electrons (Blankenship, 2010).

PSII is an evolutionary form of the photosynthetic system from over 2.5 billion years ago that uses water as a sacrificial molecule to be oxidized (Blankenship, 2010). This is done through an oxygen-evolving catalyst (OEC) integrated into PSII. For PSII-containing organisms, this donor was very convenient because its oxidation produced not only electrons, but also protons, which the organism is built to manage, and gaseous oxygen, which it can throw away to the atmosphere, producing as a by-product the conditions for life as we know it today. One downside of oxygen evolution for the plant is that its carbon fixation process is very sensitive to O_2 and protection had to be created to minimize the O_2 interference with RuBisCO (Hayer-Hartl & Hartl, 2020).

An effective operation of the photosynthetic CO_2 fixation apparatus requires an even flow of electrons and protons through its various elements from water oxidation in the OEC through PSII, PSI, and then to the site of RuBisCO (Derks et al., 2015). This relatively constant flow of charge requires a coordinated matching of the various concentration and potential gradients driving the reactions between photosystems and solution species. From the perspective of the intent to construct man-made systems for CO_2 reduction, it will be seen how the most significant characteristics of the photosynthetic system for CO_2 fixation are found in the different types of coupling involved in this charge transfer matching between the specific elements of the photosynthetic process. There are three major chemical coupling points and coupling types connecting these elements and one physical one in the light-harvesting center (LHC), and they bear close inspection.

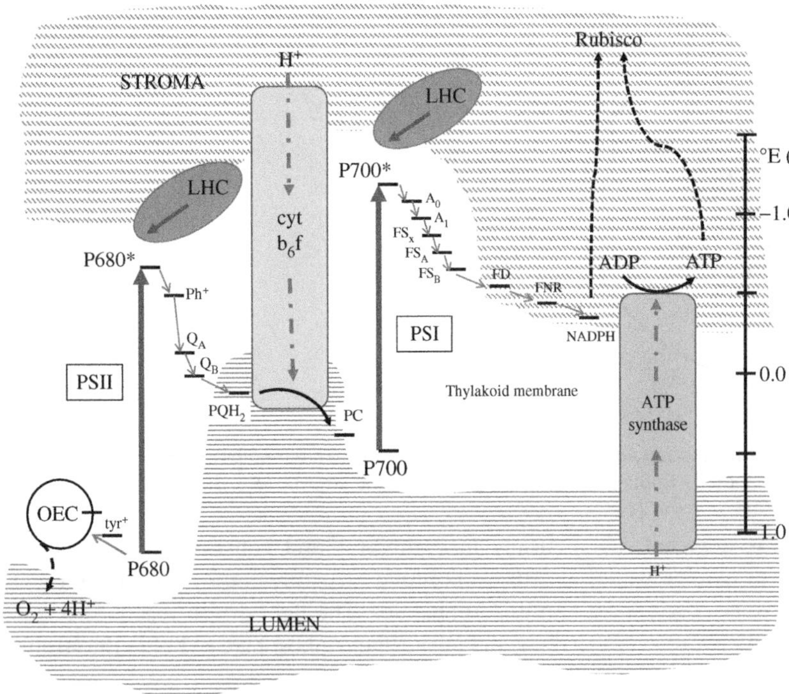

Figure 6.2 An energy level diagram is presented for the components of the reaction centers embedded within the thylakoid membrane between its interior lumen and its exterior stroma. M. T. Spitler.

They are circled in Figure 6.3 and will be addressed in turn in the following discussion. They include the direct coupling found between the OEC and PSII, the indirect coupling between PSI and RuBisCO and between PSI and PSII, and the direct coupling of the LHCs with PSI and PSII. A description of the function of RuBisCO is also provided to provide more detail on the indirect coupling with the output of PSI.

6.2.2 PSII Coupling to the OEC

The first type of chemical coupling is at one extreme and is the direct photocoupling of the OEC with the excited state reactivity of the chlorophyll molecule P680 in PSII. The structure of this Mn-based OEC complex has been a subject of research for many decades, and high-resolution crystallographic determinations of its physical structure are now available (Kern et al., 2018). Four oxidizing equivalents are required to effect water oxidation to O_2, which can only result from four successive excitations and oxidations of the P680 chromophore and charge transport down the electron transfer chain of charge to reduce plastiquinones, which are then released into the lumen as plastoquinol QH_2 molecules.

In the energy level diagram of Figure 6.2 the excitation of the chlorophyll molecule P680 results within 3 ps in electron transfer to a pheophytin (Ph) species with a subsequent transfer in less than 1 ps to a quinone in the chain (Renger & Renger, 2008). A recombination of this electron with the hole state in the $P680^+$ species is possible at each step, but it becomes slower with each subsequent step, having been measured at 200 ps from the initial $P680^+/Ph^-$ charge separation and 100 μs from the reduced quinone Q_{A-}. After the passage of two electrons down the chain, the quinone Q_B is doubly reduced and extracts two protons from the thylakoid lumen to make QH_2 which detaches into solution in the lumen in a 95% efficient charge separation of the electron from the hole state (de Wijn & van Gorkom, 2001). The release of this reduced QH_2 into solution effectively precludes a recombination reaction for the two electrons involved and allows the OEC to proceed or complete the water oxidation reaction.

The $P680^+$ hole state initiates its own sequence of reactions where the positive charge is transferred through oxidized tyrosine tyr^+ to the OEC, as is depicted in Figure 6.2. The sequence of catalytic steps for water oxidation, known as the Kok cycle, is well defined and is labeled schematically as S_0 through S_4 in Figure 6.4 (Joliet et al., 1969; Kok et al., 1970). Four-hole transfers are required to effect the evolution of O_2. Each successive injection of the four holes from $P680^+$ via tyr^+

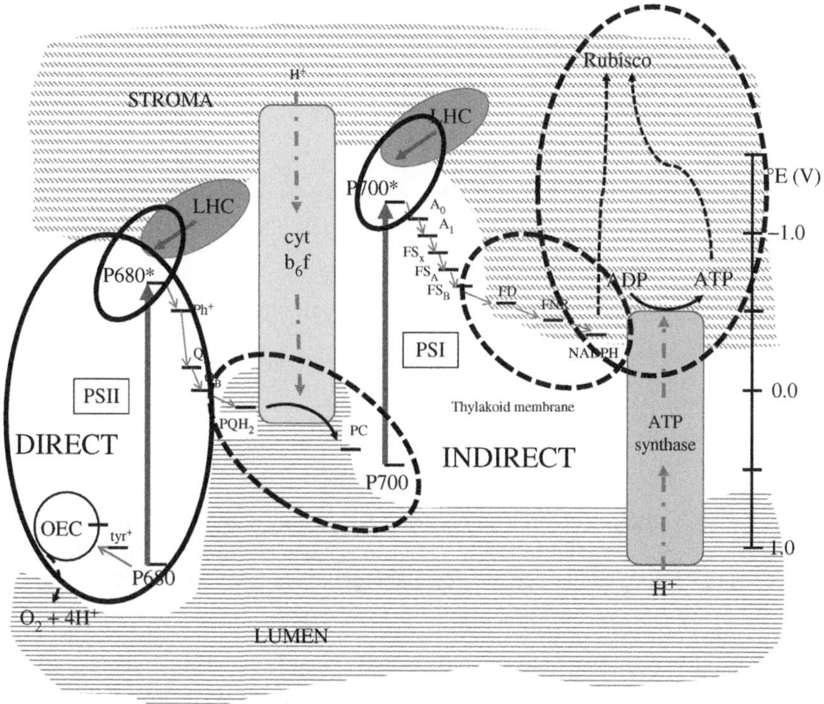

Figure 6.3 The flow of charge through the reaction centers from water oxidation to CO_2 reduction by RuBisCO runs through coupling points between its elements, designated here as direct (solid circles) in their coupling, or indirect (dashed circles). M. T. Spitler.

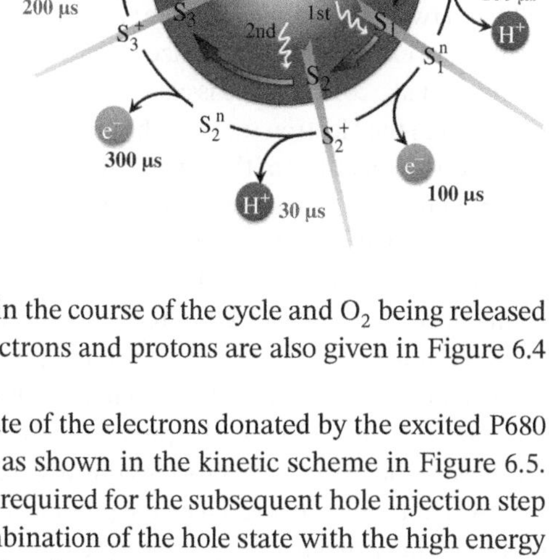

Figure 6.4 A schematic diagram of the Kok cycle for oxidation of water by the OEC shows the sequence and timing of four oxidation steps and proton transfers driven by excitation of P680. Reproduced with permission from Zaharieva and Dau (2019)/Frontiers Media/Public Domain CC BY 4.0.

into the OEC pushes it to the next S state, with four protons being released in the course of the cycle and O_2 being released in the end step. Timing aspects of this catalysis for the transfer of these electrons and protons are also given in Figure 6.4 (Ananyev et al., 2019; Kern et al., 2018; Zaharieva & Dau, 2019).

The positive charge in the OEC at any S state is directly coupled to the fate of the electrons donated by the excited P680 to the electron transfer chain because of the possibility of recombination, as shown in the kinetic scheme in Figure 6.5. It is evident for this catalyst coupled to the P680 photosystem that the time required for the subsequent hole injection step from tyr^+ to the OEC in the Kok cycle has to be competitive with the recombination of the hole state with the high energy electron being transferred down the PSII electron transfer chain through the pheophytin and two quinones, Q_A and Q_B. The recombination time is about 1 ms for the reduced primary quinone Q_{A-} with the oxidized tyrosine (Renger & Renger, 2008). Neglecting side reactions, any recombination of the first reduction of Q_A would be a geminate process. Of the five S

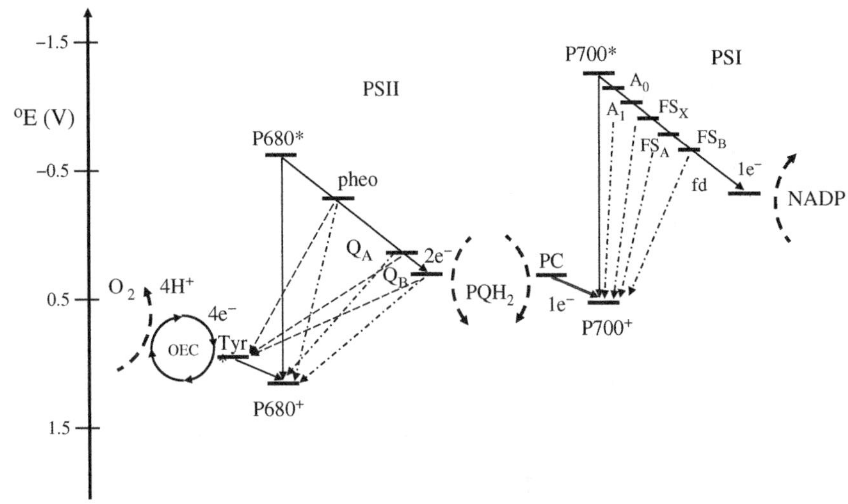

Figure 6.5 A schematic diagram of the kinetics of the coupled operation of PSII and the OEC. Possible back reactions are given as dotted arrows. Dashed lines represent diffusing species in solution. Coupled with PSII is the corresponding diagram of the kinetics of the indirect operation of PSI with possible back reactions to P700⁺ given as dotted arrows. Energetics are taken from Blankenship (2010). M. T. Spitler.

states depicted in Figure 6.4, only the S_1 state is stable in the dark, whereas another hole is required within 100 ms in order to push the OEC from S_2 and S_3 toward S_4. Otherwise, the system can revert to the S_1 state (Ananyev et al., 2019; Kern et al., 2018; Zaharieva & Dau, 2019).

This example implies that it is an important requirement for a photosystem to provide redox equivalents to a multi-electron, proton-coupled electron transfer (PCET) catalyst for a desired reaction on a time schedule determined by that catalyst. Here it is a relatively undemanding requirement on proton transfers given that the OEC reaction only sheds protons and does not consume them, as does a CO_2 fixation reaction. When protons are required in a reaction, the situation becomes more complex, as is known from the Calvin–Benson cycle. This concept of a timing sequence will be carried forward to the discussion on the multielectron PCET water oxidation and CO_2 reduction reactions at semiconductor electrodes or through molecular catalysts.

6.2.3 PSI Coupling to PSII and RuBisCO

PSI has a two-fold symmetry in the structure of its core, so the initial excitation of the P700 reaction center in PSI appears to result in an excited state delocalized over two or more reaction center chlorophylls until the initial charge separation occurs (Caffarri et al., 2014). There is also a two-fold duplication of the chlorophylls and quinones (A_0 and A_1) down two parallel pathways in the electron transfer chain until they join at the first of a sequence of three iron-sulfur (FS) clusters. An electron can travel down either of these two pathways to the initial cluster, the path preference being dependent upon the organism. Any recombination of electron with hole would be a geminate process. An electron reaching the third iron–sulfur cluster in PSI located at the edge of the thylakoid membrane with the stroma can reduce a soluble ferredoxin (fd), which in turn provides an electron to $NADP^+$ reductase to enable NADPH formation. At the input, P700 is regenerated from $P700^+$ in plants through a single electron reduction by a plastocyanin (PC) from the lumen. A kinetic scheme for electron transfers in PSI is shown in Figure 6.5 that includes possible electron–hole recombination pathways.

The redox chemistries at the output and input of PSI both involve just one electron and one proton (Caffarri et al., 2014; Fromme & Mathis, 2004). The production of NADPH at the output of PSI, along with the formation of ATP from ADP synthase at the stroma side provides the driving force for CO_2 fixation by RuBisCO. Both of these reactions involve a proton, with ATP formation being fed by the protons released into the lumen by the action of the OEC. The diffusive concentration profile of these two reactants projecting from the thylakoid membrane into the stroma eventually feeds the series of dark reactions in the Calvin–Benson cycle that enable CO_2 fixation by the RuBisCO. This diffusion process constitutes an indirect coupling of the output side of PSI with the distant site of carbon fixation by RuBisCO. Diffusion is a physical process that is well understood and will be seen as important for particulate man-made systems for energy transduction because diffusion profiles of reagents in solution are a function of the size and shape of a planar or point source. In PSI, the regeneration of P700 from a PC input reflects an indirect coupling of PSI with PSII. PC is created by the cofactor cytochrome b_6f from the plastoquinone QH_2 and is able to dock into PSI close to $P700^+$ to regenerate P700. The kinetic scheme of Figure 6.5 shows a diffusion-controlled process controlling both the input and output of PSI (Grotjohann & Fromme, 2013).

Of note here is the important role of protons in this step, implying that proton control and proton-coupled electron transfer will also be essential parts of a man-made system for CO_2 reduction. In contrast to the sensitivity of a photosynthetic system to the oxidizing power of P680, it is stable in the presence of the reductants produced by PSI, owing perhaps to the enhanced complexity inherent in a PCET oxidation reaction of an NADPH.

6.2.4 LHC Coupling

Another important coupling element of interest is the energy transfer from the LHC to the reaction centers in which many chromophores in the antenna absorb the incoming photons and funnel their excited state energy to the P680 or P700 entities with 90–100% efficiency (Croce & van Amerongen, 2020). P680 and P700 can themselves absorb a photon, but the probability is far higher that neighboring chromophores or those in a LHC absorb light and their excitation is transferred through various energy transfer steps to the reaction centers to initiate the reaction.

In higher plants, the LHC complexes are associated with the thylakoid membrane, where they are located around the core elements of the reaction centers. They increase light absorption in both PSII and PSI with the presence of several hundred chlorophyll a, chlorophyll b, and various carotene molecules in an effective concentration greater than 0.1 M. The composition of the LHCs will vary from PSI to PSII and is organism-dependent, but they form a dynamic system where different forms of a LHC can transfer energy to PSII or PSI in order to balance light absorption between the two reaction centers (Blankenship, 2015; Şener et al., 2011).

The energy transfer process can occur through a Forster resonance mechanism between two molecules or through delocalized quantum mechanical states that effectively search out an acceptor entity (Calhoun et al., 2009). However, the exceptional element of these LHCs is the lack of traps where the excitation would be quenched outside of the reaction center. A billion years of evolution in various organisms have produced a number of forms of this highly efficient structure of high-concentration chromophores (Blankenship, 2010). In man-made systems, however, it is very difficult to make an assembly of high-concentration chromophores for light harvesting without also creating deleterious quenching traps, although it has been done before in photographic science in the past century (Eachus et al., 1999; James, 1977; Tani, 1995). In that technology, arrays of dye aggregates self-assemble on silver halide crystal surfaces and funnel excitation energy to the furthest red-absorbing species with near unity yield to initiate sensitized electron transfer to the silver halide. It took, however, the better part of a century for photographic R&D labs to perfect this capability.

Aside from the degree of perfection required to make an artificial light-harvesting system, the natural photosynthetic systems reveal another vulnerability of any such molecular constructs. For example, if the reaction center chromophore is damaged or the dyes near it that couple it to the LHC are damaged, the entire associated LHC decouples and becomes nonproductive. The loss of hundreds of chromophores from the deactivation of a few would be very expensive in a man-made product and could only be tolerated when the device is long past its payback time. The implication for artificial systems is that light-harvesting systems may best be composed of solids, which are inherently more stable, through the positioning of high bandgap semiconductors within energy transfer distance of lower bandgap materials. This could be done in a straightforward manner with nanometer-sized thin layers, spheres, or rods.

6.2.5 Indirect Coupling to RuBisCo

The ATP and NADPH produced by PSI must diffuse a distance from the thylakoid membrane through the stroma to the location of the RuBisCO enzyme and participate in the Calvin–Benson Cycle to drive the formation of the precursor to CO_2 fixation. The role of NADPH in this cycle reveals the significant aspect of proton presence and delivery. With it, ATP may drive the Calvin–Benson cycle to allow RuBisCO to fix CO_2, but RuBisCO is the essential player and has been found to persist in natural systems for several billion years (Bracher et al., 2017). RuBisCO is a multifaceted, hexameric enzyme formed by the complexation of a number of cofactors that can vary in detail with species. It fixes CO_2 through a carboxylation of a 5-carbon sugar RuBP to produce a three-carbon, 3-phosphoglycerate (3PGA), which is converted in the cycle to glyceraldehyde-3-phosphate to serve as feedstock for the organism. The enzyme evolved before the oxygenation of the

Earth's atmosphere, under conditions under which there was no need for RuBisCO to discriminate between O_2 and CO_2, and it still does not discriminate well between these two (Erb & Zarzycki, 2018). RuBisCO is slow and inefficient in contrast to other biological enzymes and requires the energy-wasting side reaction of respiration to recover from any reduction of O_2.

The key player in any photosynthetic CO_2 reduction system is the catalyst, and the RuBisCO is not a very good one (Yokota, 2017). It is complicated to activate and is slow in turn over frequency (TOF), about 2–6 s^{-1}, which is about 1% of the velocity of normal enzymes. Its discrimination in preferring CO_2 relative to O_2 in a ratio of 40–90:1 depending upon the organism (Yokota, 2017). Biological enzymes have been observed to be either fast or selective, but not both, and so photosynthesis evolved to improve selectivity at the sacrifice of speed. With such a slow catalyst, nature uses an engineering solution to the problem of a high flux of reductant from the thylakoid electron transfer chain by increasing the number of RuBisCo catalysts in the stroma to the extent that their mass constitutes 30–50% of the leaf protein (Vitlin Gruber & Feiz, 2018). What is clear from the low TOF for RuBisCO is that it cannot be coupled directly to a photosystem with a short electron–hole recombination time and that an indirect coupling involving a slow diffusion process is necessary. Why this coupling with RuBisCO is indirect and that of PSII is direct coupling appears straightforward to understand. The oxidizing power of $P680^+$ is sufficient to cause significant damage to the organic infrastructure of the organism if allowed to diffuse freely. It must be carefully and securely controlled.

Certain photosynthetic systems have evolved physical structures for concentrating CO_2 about RuBisCO as a pathway to increased efficacy. Subcellular hollow structures known as carboxysomes in cyanobacteria (Bracher et al., 2017; Kerfeld et al., 2018; Turmo et al., 2017) and pyrenoids in green algae (Meyer et al., 2020) are known to sequester RuBisCO in their interior, exclude O_2, and allow carbonic anhydrases as a source of CO_2 to enter in order to concentrate the CO_2. Carbonic acid concentrations of up to 0.1 M can be attained while O_2 can be blocked from entering. Such a carboxysome function is depicted schematically in Figure 6.6 as a structural arrangement that allows the form of RuBisCO within cyanobacteria to be a faster and less selective catalyst, but still maintain an effectively higher percentage of CO_2 fixation relative to O_2 reduction. Although the kinetics of the chemistry for this concentration may be too slow for scale-up to practical application, it is evident that a CO_2 concentration mechanism would benefit the operation of man-made systems as well.

6.2.6 Photostability

The charge flow of protons and electrons through the photosystem inside the thylakoid membranes can become unbalanced and disrupted at high photon fluxes whenever the reaction centers overdrive the electron transfer chains (Derks et al., 2015). Electrons can accumulate at the end quinone QB in the PSII electron transfer chain, or they can build up at the endpoint iron–sulfur complex in PSI if there is too much NADPH in the stroma. Only at solar irradiances below about 20–30% of AM 1.5, depending upon the organism, can photosynthetic systems manage an effective and efficient electron flow.

Above this threshold, the organism can suffer chemical and structural damage from side reactions. During a high flux of excitation energy through the LHCs to the P680 in PSII, triplet-state chlorophyll can be produced and a damaging level of singlet oxygen can be created. Such damage is minimized or prevented through a number of photoprotection mechanisms

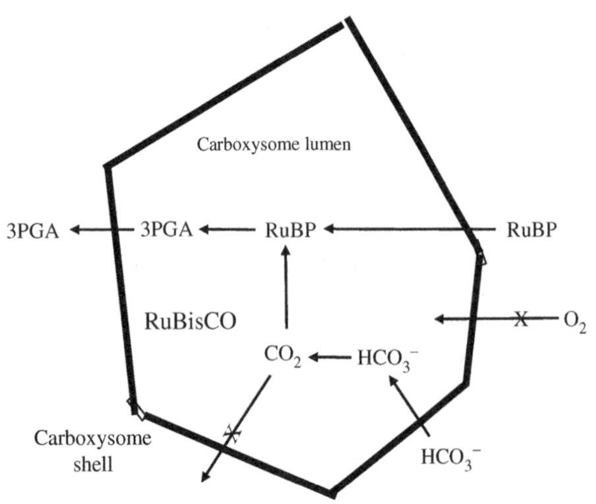

Figure 6.6 A scheme is given of the multi-faceted carboxysome enclosure and how it excludes O_2 with a heightened CO_2 concentration to increase the efficacy of CO_2 fixation by RuBisCO. Transport into the carboxysome occurs through gates at the apices and sides of the membrane structure. M. T. Spitler.

that have been embedded into the components of the PS apparatus. For example, the electron transfer chain driven by high light flux can create a pH imbalance across the thylakoid membrane that can cause a dissociative decoupling of the LHCs from the reaction center core, thereby shutting the reaction center down (Derks et al., 2015). When these photoprotection events do not suffice, the inherent repair system in the cell nucleus is stimulated to restore the system back to functionality.

These photoprotection events are allowed by plants to become significant above 20–30% of full solar irradiance because the organism has more than enough energy by then to sustain its physiological needs for growth and reproduction. In contrast, artificial systems must operate as if they were light-starved at AM 1.5 and operate at high efficiencies at this solar luminance. In contrast, it is at low-level, light-starved conditions where the photosynthetic process is called upon to be most efficient in order for the organism to survive. Given the various coupling mismatches in photosynthesis at the high irradiance of every day sunlight and the damage that can exceed the inherent photoprotection capability, photosynthetic organisms endure primarily because of their capability of self-repair.

There are designs and chemistries for self-repair in man-made systems, but these are few (Kanan & Nocera, 2008), and in their absence, it is clear that artificial ones must be designed and constructed to be stable at full solar exposure. Any cumulative damage from light-induced, deleterious side reactions, can only become significant long after the payback time of the device. The absence of inherent self-repair functions in system design and chemistry imposes severe demands and is the stimulus for many system design choices and materials research in solar photoconversion.

6.3 Water Oxidation

Photosynthetic systems use water as a sacrificial electron donor, and man-made systems presently under research for solar photoconversion and CO_2 reduction seek to do the same. This is a difficult problem that has been researched for a half-century since Fujishima and Honda in 1972 proposed it for semiconductor interfaces, although in perspective this is a short time compared to the billion or more years that passed before photosynthesis achieved this goal. As in nature, the artificial systems must deal with the byproducts of the oxidation of water, which are molecular oxygen and protons. O_2 being an atmospheric constituent, the tasks for man-made systems are then to arrange for the exit of O_2 to the outside world as well as to attain control over proton reactivity and transport since they need to be removed from the site of O_2 evolution to be consumed at the reduction reaction at the other end of the photosystem, be it for molecular hydrogen production or CO_2 reduction.

Research on oxygen evolution for artificial photosynthesis will be analyzed here within the perspective of the design elements of the photosynthetic systems identified above: indirect coupling and direct coupling. It will be seen how the system design and its component coupling interfaces dictate the TOF performance required of the catalyst. First will come an examination of the reactivity and structure of molecular catalysts in solution activated through an indirect pathway through addition of a strong oxidizing agent in solution such as Ce^{+4}, which has been the primary means of their study and development over the years (Parent et al., 2013). Next, the molecular catalysts will be examined that are directly coupled to a photosystem like the OEC of PSII is to P680. Many of the catalysts to be discussed have been selected from the high-performing molecular catalysts from solution research. A follow-up examination of photoelectrochemical (PEC) electrodes and particulate photoelectrolysis will find examples controlled by the same two design parameters.

The path of catalyst research and development for an indirect coupling through redox agents in solution has differed from those in directly coupled systems. In an indirect coupling context, the longevity of the catalyst as quantified by the turnover number (TON) assumes a greater role than TOF, because it serves as important guidance for the direction of structural improvements to the catalyst. In this situation, the TOF is considered secondarily and only as a threshold metric in comparison with the arrival frequency of solar photons.

In the direct coupling arrangement for water oxidation in PSII, the time required for the first hole transfer from the photosystem to the catalyst must be shorter than the recombination lifetime of electrons and holes in that photosystem. As will be seen, the same holds for man-made photosystems directly coupled to a catalyst. Subsequent hole transfers in any multi-electron catalysis must also be competitive with electron–hole recombination. Given this requirement, the velocity of the reaction between catalyst and hole and the catalyst TOF become controlling aspects of system performance once satisfactory TONs have been attained.

6.3.1 Molecular Water Oxidation

Molecular water oxidation catalysts will be examined first since a number of these molecular catalysts have been employed in later work on heterogeneous constructs for water splitting. This discussion will proceed within the context of the indirect coupling scheme described above since the initiation of molecular catalysis in solution usually involves a reaction with strongly oxidizing reagents such as Ce^{+4}, $HClO_4^-$, HSO_5^-, or IO_4^- in solution, usually, as with Ce^{+4}, at pH 1 (Parent et al., 2013). This is fundamental research with the goal of learning how to make and improve molecular catalysts with TON serving as a guiding figure of merit since the catalyst must survive long enough to study. In principle, these oxidizing equivalents in solution could also be generated at a photosystem elsewhere and, in an indirect coupling, be allowed to diffuse to the location of the catalyst.

It is useful to have some estimate of the minimum TON required for a successful catalyst in a device. For this estimate, a finance-based algorithm is presented here using dollars and financial conditions for the U.S. market. It considers the cost of the production of solar-driven fuel as an investment in competition in the marketplace with other investments for financial return. Order-of-magnitude estimates are made for each parameter, which allows them to be easily adjusted to specific situations. If a physical plant is built that will produce energy at \$4/gal equivalent of fuel over a 20-year period, then the present value of that \$4 over a 20-year time span with a competitive investment return of 4% p.a. is at onset \$1.83, (this percentage is for a distributed, roof-top consumer product where it competes with the dividend on a utility bond, whereas a corporation may require a 12% return for a large-scale facility), which is the capital available to build the portion of the structure to produce the gallon equivalent. Taking 10% of this capital as costs for the consumables and attributing all of this cost to the catalyst, \$.183 is available, assuming that one of either the reduction or oxidation catalysts dominates in cost. With the experience of estimates for the scaled-up price of a specialty chemical such as this catalyst to be \$100/kg, the \$.183 will buy 1.83 g and if the molecular weight of the catalyst is 1000 g/mol, then 1.83×10^{-3} mol of catalyst are available to produce the gallon equivalent over 20 years. Whether one produces 1 kg/gal equivalent hydrogen gas over 20 years, or pure octane, within a factor of 2 the moles of electrons passed in its solar production are 10^3 and with four electrons involved in the oxygen evolution cycle, the number of moles of oxygen produced in these cycles is 250. The quotient of 250 over 1.83×10^{-3} results in a minimum TON of 1.4×10^5 and if one wants the device to be operating at 90% of its original capability at the end of its lifetime, then the TON required of the catalyst is 1×10^6. Using these figures, the target TON range is then 10^5–10^6, with the higher end being a more secure goal. Some of the organometallic Ru catalysts featured below approach this particular estimate (Creus et al., 2016).

The history of the discovery and synthesis of molecular catalysts for water oxidation reveals the sheer difficulty in creating a specialized catalyst de novo as well as the determination of the researchers who have created them (Matheu et al., 2019; Zhang & Sun, 2019). Given the multi-electron nature of water splitting, transition metal complexes become the focus of research owing to the multivalent redox character of many transition metals.

Through electrochemical analyses (Azcarate et al., 2016) the characterization of many molecular catalysts in solution at metal electrodes have correlated the overpotential η for product evolution with currents, Faradaic efficiencies, and an assessment of durability in terms of TONs. The Ru-based "blue dimer" as a first example appeared in 1982 with a TON of 4–5 and a TOF of $4 \times 10^{-3}\ s^{-1}$ (Gilbert et al., 1985). The next generation of water oxidation catalysts appeared only years later with the development of mononuclear Ru catalysts such as the general structures in Figure 6.7a–c (Zong & Thummel, 2005), where a TON of 580 and a TOF of $3 \times 10^{-4}\ s^{-1}$ were reported for the structure in Figure 6.7a. This has been followed by an ever-increasing list of mononuclear candidates for water splitting (Wasylenko et al., 2011), which includes consideration of manganese, cobalt, and iron-based molecules (Brimblecombe et al., 2009).

This experience presages a challenging road to molecular catalysts for CO_2 reduction, but the extensive synthetic experience learned in creation of water oxidation catalysts can be applied to the CO_2 catalysis effort and accelerate it. These include synthetic approaches, theoretical developments, and analytical techniques that have accompanied the progress in molecular water oxidation catalysis (Matheu et al., 2019; Meyer et al., 2017). There are even "guides" now on how to make such catalysts (Garrido-Barros et al., 2017).

Here a summary is made of the most prominent examples from solution research in organometallic Ru and Ir complexes, followed by efforts in first row, "earth-abundant" Fe, Co, Mn, and Cu organometallic catalysis for water splitting. The suitability of a molecular catalyst for a role in either a direct or indirect coupling system is given in terms of their experimentally measured TOF and TON numbers.

An example of an expanded class of Ru-based catalysts is the $[(bda)(L)_2Ru(II)]$ structure given in Figure 6.7b, where bda is 2,2′-bipyridiene-6,6′-dicarboxylate and L is a methylpyridyl function (Gao et al., 2013). It is a monomeric Ru complex

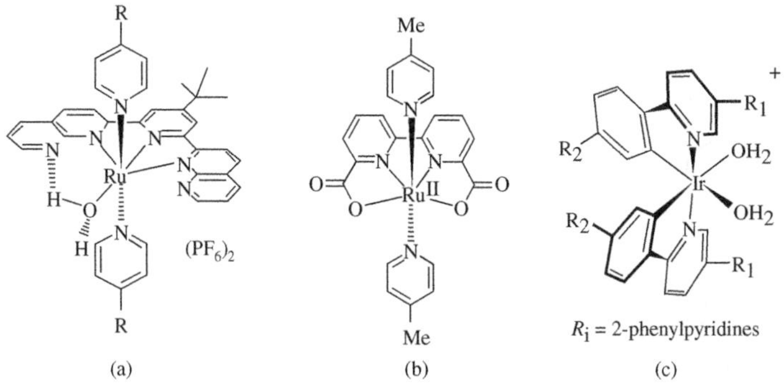

Figure 6.7 Examples are shown for the single center water oxidation catalysts used in early experimentation that are discussed in the text. (a) An initial development of mononuclear Ru catalysts is given here (Zong and Thummel 2005). (b) A member of an expanded class of more effective Ru catalysts is shown (Gao et al. 2013). (c) An example of an Ir complex that has been extensively studied (McDaniel et al. 2008). M. T. Spitler.

and is thereby synthetically easier in preparation than the Ru-dimers of the first generation and contains ligands that are resistant to oxidation. This flexible molecule is one of many with carboxylate ligands in the equatorial plane that are not coordinated to the central ruthenium and can effect proton transfer during O—O bond formation. Variants within this class have exhibited a TON > 50,000 and TOFs up to 1,000 s^{-1} (Duan et al., 2012).

Ir-based catalysts share a stability challenge with the first-row transition elements. Extensive studies have been made of Ir structures for water splitting (Blakemore et al. 2010) with a TON of 2,000 being reported and a TOF of 1.5 s^{-1} being reported for the structure in Figure 6.7c (McDaniel et al., 2008). However, much evidence has been gathered to show that many of these initial structures are pre-catalysts for the formation of iridium oxide solid catalysts or μ-dimers of reacted forms of the original structure (Lv et al., 2012). Complexes with this element do not yet compete well with the performance of Ru catalysts and a control of their active form is not guaranteed.

The absence of an effective earth-abundant catalyst does not preclude the present development of device designs and optimization that employ Ru. As pointed out above, the initial niche product markets for CO_2 reduction will bear market prices far above those derived from standard fossil fuels and so these devices can bear the expense of rare elements. The widespread use of costly metals in consumer products is already well known in the marketplace, in particular in catalytic converters for automobiles where Pt is used at the gram level for some billion automobiles on the planet. The key to the use of these expensive elements is a frugal usage of the element and an effective recycling system. However, a general and widespread marketing of solar devices for CO_2 reduction in the cost-driven fuel harvesting market will one day necessitate a transition to more abundant transition metals.

Even though Mn is employed by nature in the OEC, it is much more challenging to use such first-row transition metal compounds in man-made structures for water oxidation since they can be more labile in ligand substitution and the retention of their structure in any application will be controlled by environmental pH and the ligand composition (Dalle et al., 2019). Doubts have been raised about claims for catalysis that have been made in the literature for these first-row complexes since they are prone to decomposition into metal oxides, which can themselves be water oxidation catalysts, and rigorous analytical work is required to confirm the true nature of the active catalytic species (Folkman et al., 2018; Lv et al., 2012; Wu et al., 2015). There are few reports of performance comparable with the Ru compounds, but given the decades it has taken to advance the science of the Ru structures, much patience is in order in this first-row work because the development of a cheap element catalyst will see a payoff in the final product.

Notable examples of interest from the past decade are provided here for structures with Fe, Co, Mn, and Cu at their core, since many have been employed in heterogeneous, directly coupled systems. A binuclear example for Fe is depicted in Figure 6.8a, which has been evaluated for O_2 evolution when immersed in a Ce^{+4} solution and exhibits a TOF of 2.2 s^{-1} (Wickramasinghe et al., 2015). Numerous other Fe complexes have been examined as water oxidation catalysts (Blakemore et al., 2015; Singh & Spiccia, 2013; Zhang & Sun, 2019) with TONs of up to 2,000, but always with a lower TOF.

Organometallic Co forms have been synthesized, an initial effort being given in Figure 6.8b (Wasylenko et al., 2011) with a TOF of 80 s^{-1} as well as the Co cubane structure of Figure 6.8c (McCool et al., 2011) reported to exhibit TONs of 10^3. Mechanistically, however, it is not clear how the cubanes function (Ullman et al., 2014). Co-based catalysts have also been extensively studied in various other forms, including porphyrins (Nakazono et al., 2013; Wang & Groves, 2013) and salophens (Pizzolato et al., 2013). Co polyoxometallates have also been examined (Lv et al., 2012) because they are

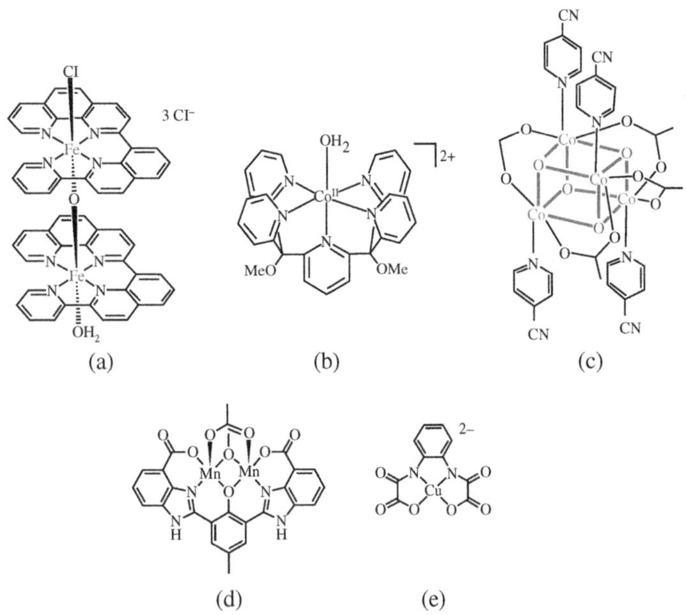

Figure 6.8 Initial research efforts in first row transition metal catalysts for water oxidation included these compounds, which are discussed in the text. (a) A binuclear example for Fe is depicted (Wickramasinghe et al. 2015). (b) Organometallic Co forms have been synthesized (Wasylenko et al. 2011). (c) A more effective Co cubane structure has been reported (McCool et al. 2011). (d) A structure emulating the multi-Mn based OEC has been tested (Karlson et al. 2011). (e) Organometallic copper complexes have also been studied (Lee et al. 2020). M. T. Spitler.

large enough to accumulate all four oxidizing equivalents for water splitting, with reported TOFs below $1\,s^{-1}$ and TONs approaching 100. However, solid CoO_2 is an effective agent for water oxidation and even a small decomposition of these Co catalysts to produce CoO_2 particles will confound the evaluation of the organometallic species. With Mn being present in the OEC, it is to be expected that there would be considerable exploration of Mn as a man-made catalyst, as reviews show (Blakemore et al., 2015). Structures emulating the multi-Mn-based OEC have been synthesized (Karlsson et al., 2011), although their effectiveness in water splitting is small (Brimblecombe et al., 2009; Kanady et al., 2014). Organometallic complexes have been made with Mn that evolve oxygen through oxidants in solution, this time a one-electron oxidant $Ru(bpy)_3^{3+}$, an example of which is given in Figure 6.8d. Electrochemical studies of organometallic copper complexes such as that given in Figure 6.8e have found them to be effective at water oxidation (Coggins et al., 2014; Lee et al., 2020), with TONs approaching 100 and high Faradaic efficiencies, although with overpotentials exceeding 400 mV.

6.3.2 Dye-Sensitized Photoelectrosynthesis Cell (DSPEC)

A good illustration of direct coupling in an artificial system for water oxidation can be seen in the dye-sensitized photoelectrosynthesis cell (DSPEC), which was introduced by T. J. Meyer (Brennaman et al., 2016; Swierk & Mallouk, 2013; Wang, Eberhart, et al., 2018). A form is depicted in Figure 6.9a where a linear dyad of a sensitizer coupled with a molecular catalyst is attached to mesoporous TiO_2 and produces photocurrent of $185\,\mu A/cm^2$ under illumination for the plateau photocurrent at a 0.2 V bias vs. NHE (Ashford et al., 2015).

Here, the photooxidation of a Ru dye produces a hole that is captured by an attached water oxidation catalyst. The sensitized electron transfer to the TiO_2 conduction band provides electrons that are drawn upon by a Pt electrode for hydrogen evolution.

The steady-state concentration of electrons in the conduction band is controlled not only by the back reaction of the electron with the oxidized dye and the catalyst but also by the rate of the fuel production at the cathode, which draws electrons away from the semiconductor/sensitizer interface. Subsequent hole transfers in any multielectron step in catalysis must also be competitive with recombination. Given this requirement, the velocity of the reaction between catalyst and hole in coupled designs becomes a controlling aspect of system performance. This is the point where the path of catalyst research and development for these coupled systems has differed from those in an indirect coupling through redox agents in solution.

This DSPEC system is analogous to PSII, where the excited reaction center dye is directly coupled to the OEC and pushes charge down the electron transfer chain to reduce plastoquinone. There are differences, however, from the design of the artificial system, and these can be seen through the scheme of Figure 6.9b. In a DSPEC, light-initiated excitation of the dye results in electron transfer to the TiO_2 conduction band with a rate constant k_{inj}. Subsequently, the electron can either be transported to the cathode at k_{trans} to then produce the target fuel with constant k_{red}, or be able to undergo a back reaction with oxidized dye at rate k_{dbr} or with oxidized catalyst at rate k_{cbr}. Oxidized dye can be regenerated through a hole transfer to the catalyst at a rate governed by k_{ht} or to a neighboring dye through a self-exchange reaction at k_{lht}.

One difference from PSII is found in the direct coupling of the photocatalytic half reaction in the artificial system of Figure 6.9a with its cathodic reaction, molecular hydrogen production, represented in Figure 6.9b as k_{red}. If $k_{red} < k_{cbr} + k_{dbr}$, then the back reactions at the anode can control the lifetime of electrons in the TiO_2 conduction band. Whereas the potential drop between that of the P680* donor level and the reduced plastoquinone is ~−780 mV, the potential drop between the TiO_2 conduction and SHE is only −200 mV, which amplifies the impact of the coupling of the TiO_2 to a reduced hydrogen fuel by the corresponding Boltzmann factor of the difference. On the other hand, if $k_{red} > k_{cbr} + k_{dbr}$ then the operational current would be at the maximum plateau current of a three-electrode photocurrent–voltage curve for the mesoporous TiO_2 electrode, which has not yet been observed in a complete operational system.

A more significant difference is the dominant role of the non-geminate recombination of electrons injected into the TiO_2 by sensitizing dyes with the hole states in the dye or catalyst, given as k_{dbr} and k_{cbr} in Figure 6.9b. Instead of electrons being directed in a downhill single electron transfer chain away from the site of $P680^+$ in PSII, this artificial photosystem creates a pool of electrons in the TiO_2 conduction band following dye photooxidation. The electrons diffuse through the TiO_2 particles in a random walk along the surface at k_{trans} to reach the cathode, where they can recombine along the entire pathlength. The hole state on the dye is also mobile on the surface in a lateral transport k_{lht} through self-exchange reactions with neighboring dyes (Hu & Meyer, 2015; Moia et al., 2016). This is in competition with hole transfer to the catalyst k_{ht} and enhances the role of non-geminate recombination of electrons with holes relative to a simple geminate recombination.

This recombination between an electron moving along a surface in parallel with a lateral hole conduction in the dye layer is an inherent characteristic of the double layer of semiconductor and dye. Detailed measurements and modeling of current–voltage curves obtained at well-defined surfaces of single-crystal TiO_2 electrodes sensitized with N_3 have revealed that an anodic bias of 400–600 mV is required to overcome this kind of recombination and to attain the maximum plateau yield of photocurrent (Watkins et al., 2015).

This is shown in Figure 6.10a, where the fitting of the experimental current–voltage curves with the model requires the use of the same stretched exponential rate constants that have been derived for recombination at dye-sensitized mesoporous TiO_2 electrodes. An example of the time dependence of recombination of TiO_2 conduction band electrons with oxidized dye is given in Figure 6.10b (Hu et al., 2020) where it is seen that recombination of injected electrons with oxidized dye can

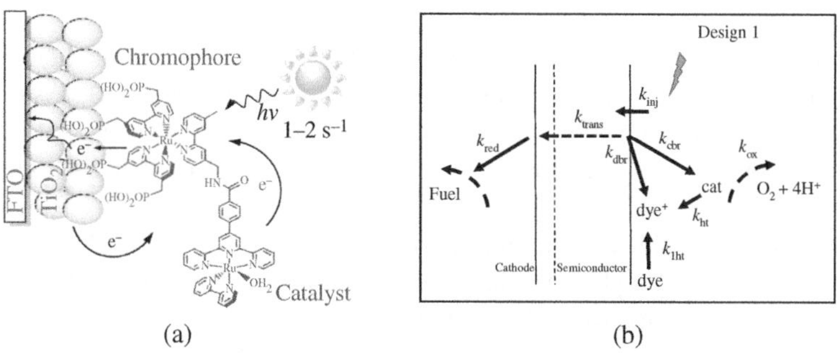

Figure 6.9 (a) A schematic of a dye-sensitized photoelectrosynthesis cell (DSPEC) with a Ru-based sensitizer attached to a nano-TiO_2 electrode. The hole in the oxidized sensitizer can be passed to an attached molecular catalyst for water oxidation. Reproduced from Brennaman et al. (2016)/with permission of American Chemical Society. (b) A kinetic scheme for the DSPEC operation is given with the solid arrows showing the direction of electron flow and the dashed arrows showing the diffusion of products in solution. k_{inj} is the rate of electron injection by the excited sensitizing dye. k_{dbr} is the back reaction of this electron with the oxidized dye. k_{lht} is the rate for lateral hole transfer at the surface. k_{ht} is the rate of hole transfer from the oxidized dye to the catalyst. k_{ox} is the rate of water oxidation at that catalyst. k_{trans} is the rate of transport of the electron through the solid to reach the cathode, where it is consumed through the rate k_{red} of reduced species or fuel. M. T. Spitler.

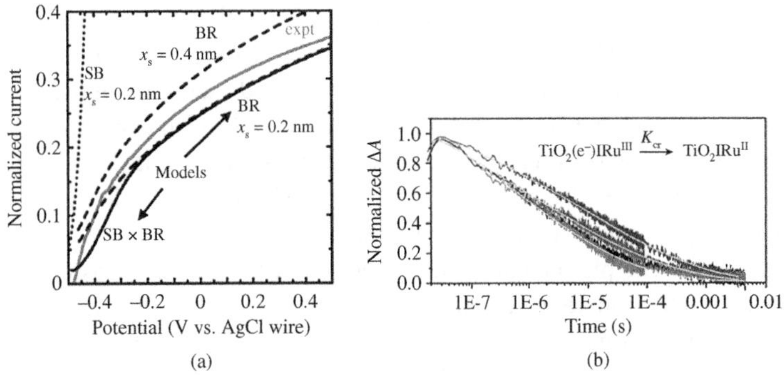

Figure 6.10 (a) Current–voltage curves for dye "N_3" at a $TiO_2(110)$ surface single-crystal electrode require 400 mV of positive bias to overcome recombination and shift more positively with the greater number of oxidized dyes caused by a higher light intensity. This behavior can be fitted to models SB × BR and BR only by using a range of recombination rates for the conduction band electron with the oxidized dye. Adapted from (Watkins et al., 2015). (b) The recombination of conduction band electrons from TiO_2 with oxidized dye is seen through the absorbance of the oxidized species for four different Ru sensitizers, with a range of recombination rates from 10^{-7} to 10^{-3} s. This decay correlates with the decline of the electric field in the double layer. Reproduced from Hu et al. (2020)/with permission of American Chemical Society.

range from 10^{-3} to 10^{-7} s and it is with the faster rates that k_{ht} and k_{ox} in surface catalysis must compete. This is the aspect of the system design that determines the required TOF for the catalyst.

One can reduce the probability of non-geminate recombination by consuming the hole states in a faster catalytic cycle. Whereas photosynthesis is dependent on the single OEC structure, man-made single metal organometallic catalysts have been developed to be ever faster in their TOF (Matheu et al., 2019). Furthermore, there is an available pool of hole states in the dye layer that can be gathered from a distance to feed into the catalyst during its cycle; the catalyst need not wait for the next photon to be absorbed, as is required in the geminate scheme of Figure 6.5. In variations of the DSPEC designs of Figure 6.9, this reduction in recombination has been accomplished with a $Ru(dpa)L_2$ catalyst from the group of Figure 6.7b that is measured to have a TOF of 10^4 s^{-1} and photocurrents of >1 mA (Gao et al., 2013), which is far greater than the 180 μA observed for the DSPEC design of Figure 6.9a, where the catalyst TON is 10^3 (Ashford et al., 2015).

Molecular spacers inserted between the dye and surface to increase their separation have also been used to attenuate overall recombination and to increase currents several fold (Ashford et al., 2015; Wang, Eberhart, et al., 2018), revealing that the velocity at which the hole state is removed from the surface is also important. Another conductive spacer entity inserted between the sensitizer dye and a nanoparticle WO_3 water oxidation catalyst at TiO_2 has been shown to increase photocurrents by 60% (Swierk et al., 2014; Zhao et al., 2012). Faradaic efficiencies for oxygen evolution in these and similar DSPEC systems can range from 10% to 70%, but milliampere current densities give one confidence that the system is operating according to its design.

Recognition of the differences between the design of this DSPEC system and the operation of PSII that reduce its efficiency has guided system structural changes for DSPEC improvement or evaluation. A variant on this design in Figure 6.11a is to use a core-shell structure for the solid where a TiO_2 shell is overlaid on SnO_2 core (Sherman et al., 2015; Song et al., 2013), the latter having a more positive energy for the band edge than TiO_2, thereby slowing down k_{cbr} and k_{dbr} (Alibabaei et al., 2013).

More work to accelerate catalytic currents features a side-by-side attachment of both the dye and catalyst to the solid surface, as is shown in Figure 6.11b for a Ru dye-sensitized TiO_2 and a coadsorbed Ru-based molecular catalyst or WO_3 catalyst. The arrangement may reduce the fraction of the solid surface available for light absorption, especially if the catalyst is a solid such as WO_3, but with appropriate photon management (Sprafke & Wehrspohn, 2015), this drawback can be overcome. As expected, this proximity of the catalyst to the surface can increase a back reaction between the injected electron and the hole states in the catalyst (Kirner & Finke, 2017). However, given the multielectron nature of the water oxidation, this increased contact between dyes and catalyst appears to allow a faster provision of holes for the follow-up steps in the reaction. This mobility of the hole state creates a vulnerability because of the hole's high oxidizing power, sufficient to cause damage to any unprotected component of the system. It is evident that the control of the location, direction, and rates of lateral hole hopping among molecules within the double layer at the TiO_2 surface is not just of scientific interest, but is a necessary part of technology development for DSPEC systems.

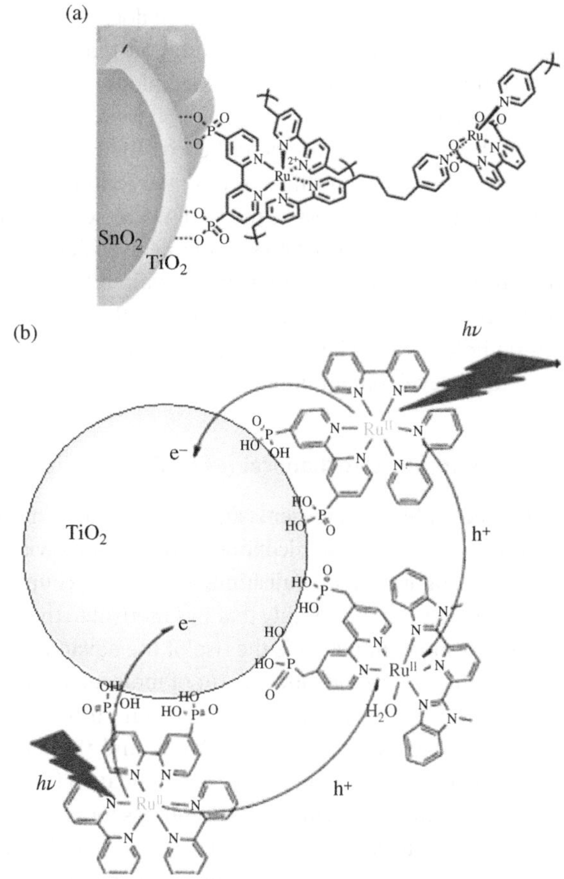

Figure 6.11 (a) An example of a core-shell DSPEC with a linear dyad attached to a thin layer of TiO_2 surrounding SnO_2 to accelerate transport and reduce recombination. Adapted from Swierk et al. (2014). (b) Another variation of a DSPEC is the side-by-side placement of sensitizer with catalyst where lateral hole transport in the dye layer increases the catalysis rate. Adapted from Sherman et al. (2015).

Although the DSPEC systems discussed above use a Ru-based sensitizer, it is also possible to use organic sensitizers such as porphyrin, phthalocyanines, and perylenes in a water splitting scheme (Kirner & Finke, 2017; Swierk et al., 2015). These are attempts to move away from Ru-based sensitizers to those with an earth-abundant atomic composition, and the dyes have been employed in the same set of system designs that have been highlighted above for Ru-based sensitizers. They feature a number of side-by-side arrangements of sensitizer and catalyst that have employed porphyrins, Zn Porphyrins, and perylene dyes adjacent to a number of catalysts, including molecular iridium complexes, IrO_2, the Ru family of Figure 6.7b described above, and cobalt oxides, all of which were deposited on TiO_2, SnO_2, or WO_3. Few of these have reported photocurrents above 100 μA/cm^2 (Kirner & Finke, 2017), although the faradaic efficiency of some has exceeded 60%. Linear dyadic arrangements of the dye and sensitizer have also been employed after the model of Figure 6.9a, and several have been modified by ALD overlayers to reduce back reactions and increase stability (Kamire et al., 2017).

The chromophores of these systems have high extinction coefficients and are planar in structure, where modifications of the periphery are often made to minimize aggregation, which is thought to degrade the performance of sensitizers. Aggregation is not bad per se, as the highly aggregated and highly efficient sensitizing dyes in silver halide photography have shown (Eachus et al., 1999; James, 1977; Tani, 1995). In photographic products, aggregation is used as a self-assembly process to control the wavelength of dye absorption and to effect a near-unity yield for electron transfer to the silver halide conduction band. It is poorly aggregated dye that is to be avoided because it creates traps for energy transfer quenching and electron transfer recombination, especially for organic chromophores with radiative lifetimes of nanoseconds instead of the μs times of the Ru-based sensitizers. This is hard to avoid with a dye attachment functionality to the surface that controls the ordering of the dyes with respect to one another (Choi et al., 2013) and interferes with the dye–dye, van der Waals interactions involved in good aggregate formation. Organic sensitizer layers also exhibit a lateral hole hopping comparable to Ru-based compounds (Moia et al., 2016), an aspect, which must also be controlled for a better yield for catalysis since these holes have been observed to initiate dye decomposition.

There are a great number of other formats and structures for chromophore-initiated water oxidation at solid surfaces, and these can involve polymers, carbon nanotubes, or graphitic carbon (Bai et al., 2019; Kirner & Finke, 2017;

Kumar et al., 2018). Their performance does not exceed the examples given above. It is evident that it is straightforward to assemble a system for water oxidation that functions poorly, but the value of each of these many designs is in the probability that they are the first step along a path to development of a truly efficient system. Such follow-up work on improvement is rare.

All of these systems are described by the rate scheme design given in Figure 6.9b. They feature the same advantages and disadvantages but differ in how they minimize the latter. The stability of the DSPEC systems discussed above will depend on the robustness of the designed assembly, the subcomponents of the design, or some combination of both. The assembly can fall apart when sensitizing dyes desorb from the solid surface or when dye and/or catalyst can decompose in the presence of the oxidizing power required for oxygen evolution. In an effort to preclude dye detachment, for example, ALD has been used to coat and surround the attachment function of the dye, with the consequence that it can eliminate lateral hole hopping between dyes (Kamire et al., 2017). The stability of the catalyst itself is a challenge and a matter for long-term development in molecular catalysts.

6.3.3 Photoelectrochemical (PEC) Water Splitting

PEC water splitting represents another approach that will split water to provide electrons for a solar-initiated CO_2 reduction. The search for a semiconductor that absorbs visible radiation and will be stable during the photoelectrolysis of water has been ongoing since Fujishima and Honda (Fujishima & Honda, 1972), but without success. It has been that lack of semiconductor photostability that has motivated the work on DSPECS above. A natural follow-up to these discovery efforts for novel semiconductors is the use of the advances in molecular water oxidation catalysts to assist reactivity of the bare semiconductor through attachment of molecular catalysts to their surface.

Much has been written on the research on materials and fundamentals of visible light absorbing semiconductors for water splitting (Archer & Nozik, 2008), and the recent in-depth examination of $BiVO_4$ reveals the analytical capability that may now be applied to possible candidates (Lee et al., 2019). In terms of new semiconductor materials, combinatorial approaches are at present the main sources for novel semiconductors, mostly metal oxides in nature, and these efforts have been elevated to consider ternary compounds and greater (Woodhouse & Parkinson, 2008; Zhou et al., 2020). The scheme for the operation of semiconductor water splitting is given in Figure 6.12a, and it is seen that this is once again a directly coupled system where the rate of oxygen evolution is connected to the reduction reaction at a counter electrode to produce a fuel. In Figure 6.12a, an n-type semiconductor electrode is illuminated with bandgap radiation to lower the quasi-fermi level for holes at the interface with an increase the concentration of valence band hole states.

These holes provide the oxidizing power to evolve oxygen from water. If they escape recombination with the holes, the electrons produced by the excitation are used in the cathodic reaction at the counterelectrode. Buried p/n junctions below the aqueous interface can serve to minimize electron–hole recombination and maximize the yield of O_2 evolution, which would be a useful device design for fuel harvesting if they could be made very cheaply.

Although the examples given thus far have considered only solid/liquid interfaces, the vapor phase supply of reactants for water splitting should not be ignored (Ali et al., 2019; Kistler et al., 2019; Modestino et al., 2015; Xiang et al., 2013). At the low solar-induced current flux of 20 mA/cm^2, ambient humidity suffices as a source of water for reaction. However, the kinetic

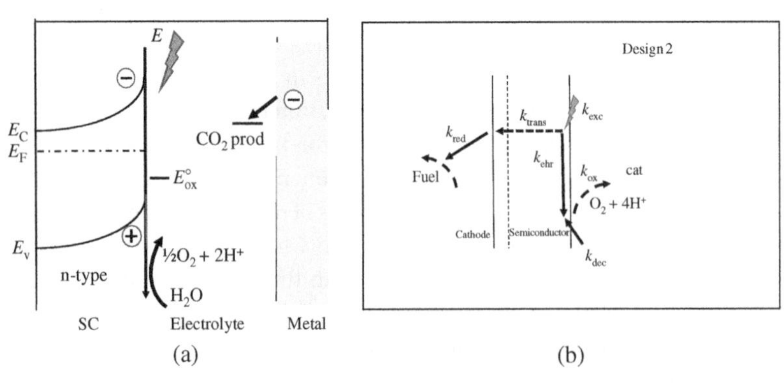

Figure 6.12 (a) The energetic scheme for the photooxidation of water at a semiconductor to produce electrons for the reduction of CO_2. (b) A kinetic representation of this process shows how the valence band hole oxidizes water at a rate k_{ox} and competes with electron–hole recombination at k_{ehr} and the decomposition of the solid at k_{dec}. M. T. Spitler.

pathways shown in Figure 6.12b are still applicable, and the same chemical and chemical engineering challenges and problems are encountered in the assembly of device prototypes. Membranes are required to reduce crossover of product gases to the opposing electrodes. The resistance to ion transport can be high, even in microfluidic systems. A distinction to be expected in this case is found in the differing catalytic pathways at the surface for fuel production between condensed-phase and gas-phase environments. One may be better than the other, but only experimentation will tell.

Figure 6.12a can be misleading in that it shows the holes at potential E_v in relation to the $E°_{ox}$ for an overall four electron, four proton oxidation of water. Oxidation of water from hole transfer will occur one electron at a time between the donor intermediate species in solution and the hole. These transfers will happen, of course, in coordination with assocated proton transfers, but the rate constant for each electron transfer reaction will be governed by a Marcus-Gerischer model that depends upon the unit free energies of the hole state, that is, E_v, and the $E°$ of an intermediate in the water oxidation process. This is similar to the operation of the Kok cycle in Figure 6.4, where the oxidizing equivalents are fed into the OEC all from the same potential of the oxidized tyrosine, but to intermediates, S_1–S_4 that have different oxidation potentials. This rate constant for hole transfer to intermediate is independent of light intensity and so is the ability of a hole to oxidize an intermediate for water splitting. However, as in the case of the OEC in photosynthesis, any hole transfer must be followed by another hole transfer within a set period of time, depending upon the catalyst, and this sequence is light-dependent. Otherwise, the reaction will reverse or undergo a side reaction, either of which will reduce the efficiency of the system. The concentration of holes at the surface available for reaction is often given by the shorthand metric of the quasi-fermi level for holes, which is the statistical measure for the number of holes on the semiconductor side of the interface. This is a steady-state figure that is dependent upon the rates of hole transfer and electron–hole recombination but is rarely ever measured (Tan et al., 1994).

PEC water splitting is the simplest of all systems in concept, but it has a design that is the most demanding on the semiconductor for electrolysis—it should absorb visible light and it should not decompose. Much effort has been invested in the design and engineering of devices to utilize this concept (Haussener et al., 2012; Nielander et al., 2015). A rate scheme for these systems is given in Figure 6.12b, which differs slightly from the DSPEC in Figure 6.9b. Excitation of the semiconductor at a rate k_{exc} results in electrons that can drive the reduction reaction at the cathode or recombine with rate constant k_{ehr} with the created valence band holes. In principle, a semiconductor could be found with a potential for the conduction band edge that is sufficiently negative in potential that it drives the cathodic reaction with such a velocity in k_{red} that the back reaction k_{ehr} at the anode is noncompetitive. Here the interdisciplinary nature of this research field comes to fore because the control of the electron–hole recombination rate in the bulk of the solid is within solid-state materials research whereas the velocity of the catalyst is a parameter varied by chemists through the synthesis of catalysts. Figure 6.9b also shows a branching point for the hole state at the semiconductor surface, where it can either oxidize water (k_{cat}) or decompose the semiconductor (k_{dec}). The rate constant of the latter is not directly known for any semiconductor, even after 50 years of research, so it is not known how fast the water oxidation reaction must proceed to suppress decomposition (Rajeshwar, 2008). Given the tendency of many semiconductors to decompose through formation of an inert surface layer, k_{dec} may be too difficult to determine. As can be the case in chemical product development during a search for a material with a specific combination of properties, Nature says no, and the researcher must move on to a new design or variation of design.

Such a variation is sought in the combination of a light-absorbing semiconductor with attached molecular water oxidation catalyst, as shown in Figure 6.13a for an Ir catalyst at TiO_2 (Liu et al., 2020; Parent et al., 2013; Wang et al., 2017). The rate constant scheme is the one presented in Figure 6.13b with the catalyst providing an additional pathway for water oxidation that competes with k_{dec}. An advantage of this approach is to make the water oxidation step independent of the surface chemistry of the light-absorbing semiconductor. It should also change the intensity dependence of water oxidation. Another approach substitutes thin-film organic structures from organic PVs for the light-absorbing medium (Kirner & Finke, 2017). There, k_{trans} can once again become a factor, although the reducing power of these organic thin films at the counter electrode for fuels production may be optimized more easily than with inorganic semiconductors.

A primary concern of all of this work has been the mode of attachment and contact of the catalysts with the semiconductor surface and its associated stability under light exposure. Work in the past on PEC PV solar cells, which is now textbook material (Gerischer, 1994; Memming, 2015), sought to produce electricity with light-absorbing semiconductors immersed in an electrolyte of fast outer sphere redox couples. Even the use of these high-velocity electron transfer reactions failed in time to outrun the decomposition reactions of the semiconductor. One should not expect these water oxidation catalysts in the scheme of Figure 6.13b to perform any better, but in the product sense, they need only last until the appropriate payback time.

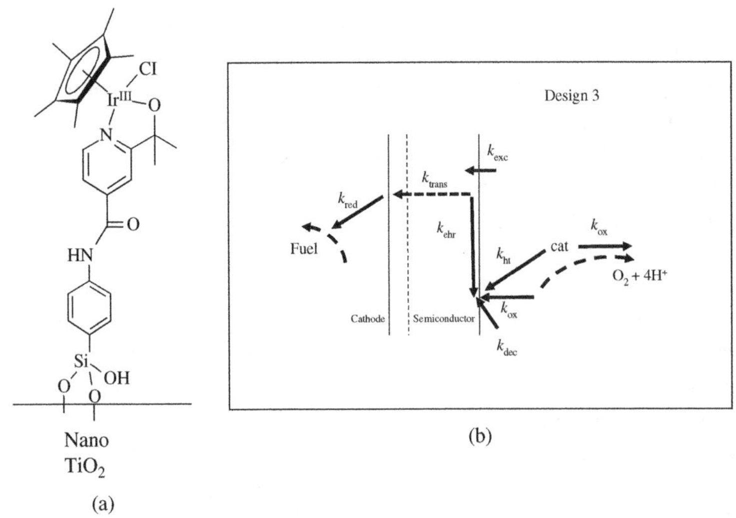

Figure 6.13 (a) The attachment of a molecular catalyst to a nano TiO_2 electrode surface for water oxidation. Adapted from Alibabaei et al. (2013). (b) The kinetic scheme with the presence of this catalyst introduces its water oxidation rate k_{ht} in competition with k_{ox} and k_{dec}. M. T. Spitler.

There are an array of specific catalysts and solid substrates that have been employed in the design of Figure 6.13b (Liu et al., 2020). Given the high extinction of the absorbing semiconductors and their good transport properties, photocurrents on the order of mA/cm^2 appear common. For the most part, the systems have been composed of known solids and catalysts selected from the molecular catalyst research described above that exhibit exceptional properties. They include the Ru and Ir catalysts as well as those containing Rh, Co, or Cu. Organometallic structures are common, but catalysis in the form of cubanes, porphyrins, and phthalocyanines has also been demonstrated. Almost all of this work retains the directly coupled design structure of Figure 6.13b, with occasional use of separation of the catalyst from the surface through a stable intervening transport species for conveyance of the oxidizing hole equivalents away from the surface. With the right structural components, this could effect a design change, where, in effect, a Nernstian oxidizing potential couple would be projected along this transport path away from the surface and an indirect coupling would be created in the system.

6.3.4 Particles

There has been a longstanding interest in particles for solar water splitting owing to the opportunity they present to make low-cost, high-surface-area substrates for reaction. Research with particulate and nanometer-sized semiconductors suspended in solution has paralleled the work on planar semiconductor PEC electrodes over several decades (Ola & Maroto-Valer, 2015; Ran et al., 2018; Wang & Domen, 2020), but in the limit of systems for very small particles, design features emerge that emulate the geminate recombination character of photosynthetic systems. Although there have been modeling efforts (Garcia-Esparza & Takanabe, 2016), a materials focus of this small particle solar photoconversion research has obscured these unique characteristics.

Schematically, the two primary designs for water oxidation and fuels reduction at particles, which have been pursued are shown in Figure 6.14a,b (Wang & Domen, 2020; Zhang & Sun, 2019). There is a single semiconductor particle for light absorption where a catalyst has been deposited to effect water oxidation, and a second deposition of a different catalyst for the reduction reaction. In Figure 6.14a example, H_2 is also shown as the fuel, but in the focus of this work, a reduced form of CO_2 such as CH_4 would be produced. Usually, these catalytic deposits on the semiconducting particle are metals or metal oxides in nature. The reaction rate scheme of Figure 6.14a is one where the particle is directly coupled to both catalytic reactions with the overall yield of product being determined by the slower catalytic reaction. The thermodynamic requirements for these reactions are shown in Figure 6.14c where it is seen that the valence band of the semiconductor where O_2 is evolved must rest at a more positive potential than the 1.23 V vs. SHE for water oxidation and the conduction band of a semiconductor must be more negative in potential than the fuel producing reaction desired, whether it be 0.0 V vs. SHE for H_2 evolution or otherwise for the many possible CO_2 reduction products. The particle arrangement of Figure 6.14b

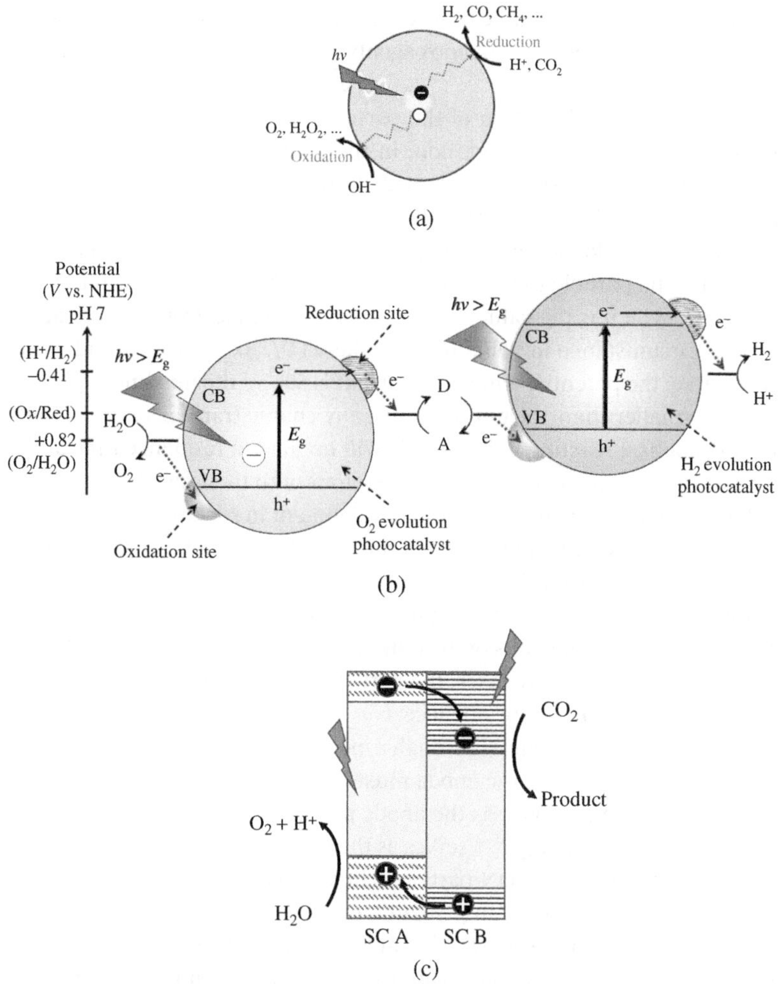

Figure 6.14 (a) Single-particle photooxidation of water to effect H_2 or CO_2 reduction. (b) A Z-scheme arrangement with two particles, one for water photooxidation and a second for its photoreduction connected through a mediating redox couple. Adapted permission from Maeda (2013). (c) A Z-scheme arrangement where the two halves are contacted electrically on either a microscopic or macroscopic scale. Adapted from Stolarczyk et al. (2018).

shows a Z-scheme arrangement where oxidation occurs at one particle and reduction at another with an indirect coupling through a redox agent to carry charge between them (Maeda, 2013; Wang, Suzuki, et al., 2018). Each photosystem is directly coupled to a catalytic process with the extraneous redox equivalents being projected into solution through the mediating redox couple. In Figure 6.14c the configuration where the two Z-scheme solids are connected is given, with the appropriate relation of valence and conduction bands to separate the electron–hole pair. The Z-schemes of Figure 6.14 are actually not analogs of photosynthetic Z-schemes because the man-made ones have both photosystems directly coupled to a catalyst, whereas the natural system of Figure 6.2 does not. It has only one.

Particle research for the photoelectrolysis of water has been exploratory and highly pragmatic in nature in that the experimental results are reported in terms of gas evolution rates and totals. Given the many practical advantages of the use of particles in a device, there have been many reports of examples of the photoelectrolysis of water at semiconductor particles (Wang & Domen, 2020), where most concern catalytic reactions at metal oxide surfaces that are directly coupled to each other through Figure 6.14a design. Metal oxides and ternary semiconductors have been of greatest interest because the highly positive character of their valence bands allows oxidation of water as a sacrificial electron donor. The oxidation is enabled or accelerated through loading of the surface with metal oxide catalysts for water oxidation, such as NiO_x or RuO_2. System efficiencies are several percent and their durability is usually less than 24 hr of illumination. Though few involve solids that absorb in the visible region, that has always been a long-term goal. Molecular catalysts would be of interest in this highly results-oriented field once they reach the desired TON in durability.

There is a dividing line in particle size that determines the nature of charge transfer reactions in reactive particle systems. Below this critical size, diffusive external charge transfer of the particles with redox agents may be neglected, and one needs only consider the reactivity of adsorbed species.

This determination of critical size can be made through a consideration of the particles as nano/microelectrodes for charge transfer where 3D diffusion is involved, along with any electromigration, in the voltammetry at these surfaces (Ciszkowska & Stojek, 1999; Murray, 2008; Zoski et al., 1991). For a spherical-shaped electrode, the diffusion velocity k_{dif} of a redox agent in cm/s to and from the particle is given by $\sim D/r$ where D is the redox ion diffusion constant and r is the radius of the particle. For a 100 nm-radius particle and a redox ion with a large D of 1×10^{-5} cm^2/s, k_{dif} is calculated to be 1 cm/s. Thus, for an electron transfer reaction with the particle to be competitive with diffusion away from the surface it must have an equal reaction velocity k_{et} of 1 cm/s, which is the value of a fast outer sphere electron transfer agent and exceeds the velocity of most redox reactions. This circumstance in voltammetry, where $(D/r)/k_{et} = 1$, is found at the half-wave potential of the steady-state voltammetric curve, the potential where the current is 50% of the maximum plateau value (Bard & Faulkner, 2001). Therefore, for particles smaller than 100 nm in radius, any charge transfer reaction must be from adsorbed species because even a fast diffusing and fast reacting redox couple will not have a residence time at the surface long enough for an effective yield of electron transfer to the particle to occur. Protons with their fast diffusion may be excepted. This phenomenon is responsible for the characteristic behavior of film development in silver halide photography (Spitler, 2006). These considerations will hold as long as the diffusion profile of the particle does not overlap with the profile of its neighbor, which occurs when the interparticle distance is less than twice the particle diameter. In that case, the effective particle radius is the size of the overlapping diffusion layers of the single particles.

For larger particles above this dividing line, one is dealing with small bits of the bulk solid, and macroscopic models may be used to describe the semiconductor half of the double layer. In many cases, the particles are small single crystals with crystallographic faces that enable the deposition of catalytic centers on specific facets.

In the Z-scheme of Figure 6.14b, in its indirect coupling of the two particles through a mediating redox agent that controls the operation of the scheme, the rate of reduction of the redox agent at the anode must balance with its rate of oxidation at the cathode. An example (Qi et al., 2018) of this design involves $BiVO_4$ as the anode particle for water oxidation and a ZrO_2-modified TaON particle as the cathode for H_2 production. $Fe(CN)_6^{-3/-4}$ serves as the mediating agent, being reduced at Au nanoparticles loaded onto the $BiVO_4$ and oxidized at the ZrO_2/TaON particle. Thus far, such systems have modest solar-to-hydrogen efficiencies of less than 1% (Wang & Domen, 2020; Wang, Suzuki, et al., 2018). Given that these are nano- or microelectrodes, a fast, 0.1 cm/s velocity redox couple such as $Fe(CN)_6^{-3/-4}$ at Pt (Winkler, 1995) is required as the intermediary to match the oxidation rate at the nanometer-sized metal deposit with the reduction rate at the micron-sized semiconductor particle, although it is not clear this consideration was made in the system design. If a large number of metal deposits are made on the particle, the full area of their overlapping diffusion layers defines their effective radius in spherical diffusion. The design in Figure 6.14b is a good demonstration of two photosystems being indirectly coupled through a redox couple, but since both oxidized and reduced forms have free access to the anode and cathode for catalysis, the overall efficiency is bound to be reduced.

Most examples of research on quantum structures for solar fuels have employed them with the same device designs as smaller forms of larger particles, with a similarly poor outcome (Lu et al., 2020). There are several aspects of smaller particulate systems, however, that are of note. One is the well-known quantum size effect upon the bandgap, which can increase as the size of the particle decreases. This moves the potential of the semiconductor conduction band to a more negative potential and that of the valence band to a more positive potential, depending upon the effective masses of electrons and holes. The primary benefit of this size effect is to increase the reducing power of electrons in the semiconductor conduction band, which will be of interest in the CO_2 reduction discussion (Holmes et al., 2012; Zhao et al., 2013). A difficulty with small particles is the time interval between absorption of photons, which can be long (Memming, 2015). In a repeat of this earlier estimate of the dark period, a calculation is made here of the frequency at which a photon is absorbed by a 6 nm radius CdSe particle under solar luminance. Taking the extinction coefficient of the particle to be 1×10^5 cm^{-1} and the solar flux over its absorbance profile to be 1×10^{16}/cm^2-s, an excitation of the particle will occur once every 2 ms. Taking the radiative lifetime of 20 ns for this particle (Jones et al., 2009), which can be manipulated by ligand composition and compound stoichiometry to be several times higher (Omogo et al., 2013), any separated and trapped charge carriers will be in the dark for 2 ms before a second excitation is scheduled to occur. Any recombination reaction should therefore be geminate in nature. Given the minimal exchange of charge carriers occurring between this small particle and solution in the dark and a directly coupled photoconversion design following Figure 6.14a, the catalyst would have to tolerate this time spacing in its reaction kinetics through some manner of charge trapping. A modification of the particle would have to be made to store and separate the charges until the four equivalents were acquired for the water oxidation. Core-shell structures can be made to effect this charge separation (Burke et al., 2021). In CdS nanorods, acceptor CdSe spheres have been included to trap holes (Amirav & Alivisatos, 2010). Another example is the introduction of p/n junctions in a structured nanowire that precludes electron–hole recombination (Hill et al., 2018). Without these structures, the quantum particle system will suffer a photochemical low-intensity reciprocity failure for O_2 evolution.

There are forms of nanostructures that are not just smaller forms of larger particles. Nanotubes, hollow nanospheres, and nanolayers can possibly offer new design opportunities. It has been discussed how the close coupling of the OEC with P680 is necessary because of the potent oxidizing equivalents involved. While organic species may be susceptible to oxidation, the inner surfaces of man-made hollow nanospheres or nanotubes of some metal oxides can well tolerate the presence of such oxidizing agents and could represent a volume into which a photosystem could project a Nernstian concentration of oxidants. This would enable an indirect coupling between the photosystem and a distant water oxidation site with a design similar to Figure 6.5 for PSI.

6.4 CO_2 Reduction

This section will deal with applications for the electrochemical and molecular catalysis of CO_2 reduction, beginning with the designs for large recycling markets and moving on to the atmospheric limit involved in the fuel harvesting market. In doing so, the electrochemical reduction of CO_2 at metal and semiconductor electrodes will be addressed first, which requires an inspection of the thermodynamics of the many complex species that arise in the reduction of CO_2. As the discussion moves from recycling to fuel harvesting, system designs are seen to play a larger role in defining the role and characteristics of the system catalyst, which can be heterogeneous, homogeneous, or molecular in character.

6.4.1 Recycling Applications

As has been mentioned, the scope of a recycling function of CO_2 reduction at a permanent location, such as at a fossil fuel power plant, may justify a significant investment in a physical plant. An example taken here is that of a farm of solar PV cells with an output that drives electrolyzers of a metallic character, where CO_2 emissions from the power plant may be reduced to a useful fuel or chemical product. As pictured in Figure 6.15, the output of the solar cells is seen to be decoupled from the electrolyzer system through electronic and electrical power conversion intermediaries (Bullock et al., 2017; Esiner et al., 2020; Huan et al., 2019; Reis Machado & Nunes da Ponte, 2018; Schreier et al., 2015). As current is passed through the electrodes where water is oxidized and CO_2 is reduced, it will be seen in the following discussion that control of the potential of the CO_2 electrode determines the product selectivity and faradaic efficiency. Different sets of membranes would be required for differing anolyte and catholyte composition and for current density operation higher than solar luminance.

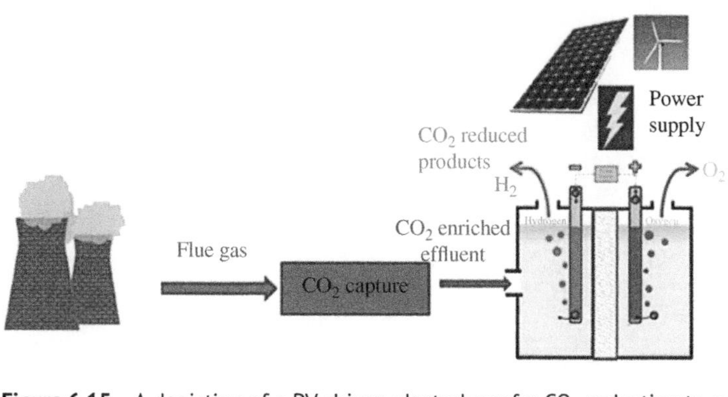

Figure 6.15 A depiction of a PV-driven electrolyzer for CO_2 reduction to useful chemicals in a recycling application. Intermediate between the output of the solar cells and the electrolyzer are power converter electronics to maintain operational control of the electrodes. Reproduced from Reis Machado and Nunes da Ponte (2018)/with permission of Elsevier.

6.4.2 Metals as Catalysts

Given this recycling form of a product for CO_2 reduction, it is useful to explore the catalytic properties for this reaction at metal electrodes owing to the extensive research that has been done in this area and the analyses that have been made (Hori, 2008; Nitopi et al., 2019). An array of metals have been found to be effective in reducing CO_2. Some produce C1 products, such as CH_4, CO, or $HCOO^-$, but only a few, such as Cu or various Cu-containing bimetallics, produce C2 products or higher, such as ethanol, ethylene, or propanol. In all cases, these reductions are in competition with H_2 production, which must be minimized.

Each of these possible reduction products for CO_2 has an electrochemical potential $E°$ and most fall within several hundred mV of each other, as is presented in Table 6.1 for a number of C1 and C2 molecular species at pH 7, a pH for minimum H_2 evolution. These $E°$ values are seen not to be much different than the $E°$ for reduction of H_2O. However, the activation energy for the initial steps of the CO_2 reduction reaction can be high, and an applied overpotential η of a volt or more can be required to obtain the cathodic current to make reduced carbon products. These reductions follow a host of possible mechanistic pathways, which are dependent upon the particular metal, the applied potential, and the specific crystallographic face of the metal. Some surfaces are a dead end for a particular intermediate and others facilitate its further reaction.

Some electrochemical terminology is defined here in order to provide a context for the role of η as a measure of an activation barrier (Gerischer, 1994). At a metal electrode/electrolyte interface at equilibrium, the Galvani potential drop $\Delta\Phi^0$ between the bulk of the metal and the solution is given by $\Delta\Phi^0 = E_{f,\,\text{metal}} - E_{f,\,\text{redox}}$ where $E_{f,\,\text{metal}}$ is the fermi level for electrons in the bulk and $E_{f,\,\text{redox}}$ is the potential of the redox couple in solution, in this case, the redox reactions for CO_2 reduction listed in Table 6.1. This potential drop across the Helmholtz double layer can be changed to a different value $\Delta\Phi_\text{H}$ through the application of a potential that causes a deviation η from $\Delta\Phi^0$:

$$\Delta\Phi_H = \Delta\Phi^0 + \eta$$

These expressions reveal that at metals, η relates to a charging of the double layer and a change of $\Delta\Phi$.

With Cu having shown the most promise of the metals studied thus far and having merited the deepest theoretical analyses of the kinetics of CO_2 reduction physical models, Cu catalysis will be discussed here first because a vocabulary has been developed for understanding the promise and limitations of its electrocatalysis. Salient aspects of this work on Cu, which is extensive, are presented here as a launching point for subsequent discussions of the use of metal catalysts in light-driven systems.

In Figure 6.16 a flow diagram is shown that represents possible mechanistic intermediate structures and pathways for several CO_2 reduction reactions that can occur at Cu surfaces, where the depicted structures and flow of reactions are taken from experimentation and compilations of experimental studies (Nitopi et al., 2019; Peterson et al., 2010). This abbreviated version of complex pathway diagrams illustrates several aspects of the catalytic reduction of CO_2 at metal surfaces significant to this discussion.

Inspection of Figure 6.16 reveals that the first electron transfer to CO_2 at the interface has been modeled to result in an array of a half dozen different forms of adsorbed species, with different oxidation states of carbon and oxygen having

Table 6.1 Thermodynamic $E°$ Values for CO_2 Reductions in Solution at pH 7.

	CO_2 reduction potentials
V vs. NHE	**Reaction (pH 7)**
−0.610	$CO_2 + 2H^+ + 2e^- \rightarrow HCOOH$
−0.530	$CO_2 + 2H^+ + 2e^- \rightarrow CO + H_2O$
−0.480	$CO_2 + 4H^+ + 4e^- \rightarrow H_2CO + H_2O$
−0.380	$CO_2 + 6H^+ + 6e^- \rightarrow CH_3OH + H_2O$
−0.240	$CO_2 + 8H^+ + 8e^- \rightarrow CH_4 + H_2O$
−0.349	$2CO_2 + 12H^+ + 12e^- \rightarrow C_2H_4 + 4H_2O$
−0.329	$2CO_2 + 12H^+ + 12e^- \rightarrow C_2H_5OH + H_2O$

Adapted from Evans et al. (2019).

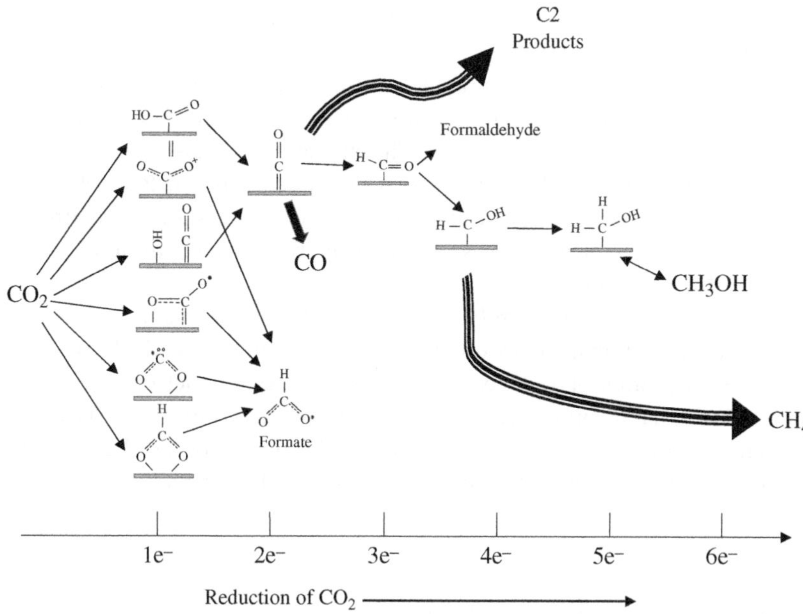

Figure 6.16 Depiction of proposed intermediates and kinetic paths for product evolution on a Cu surface as CO_2 is reduced. The structures and pathways are taken from Nitopi et al. (2019) and Peterson et al. (2010) where arrows represent the appropriate electron, proton, or PCET reactions. The dominant pathway to C2 products is proposed to originate in an adsorbed CO species and schematically follow the large arrow comprised of many possible surface intermediates. Similarly, the large arrow for CH_4 evolution passes through many surface-adsorbed forms. M. T. Spitler.

a specific binding energy ΔE with the Cu surface. These components can also have different degrees of interaction with neighboring C, O, and H atoms. The ΔE between the metal surface and the attached C, H, or O atoms will depend upon the particular crystallographic face, or metal facet, and the charge on the electrode as determined by $\Delta\Phi_H$. This ΔE for the C, H, and O atoms to the surface will control whether the reactants will proceed to the next step in the evolution of fuels and whether they will form single-carbon or multi-carbon products. It is a balance of weak vs. strong ΔE as a function of time for the particular intermediates.

The models for the first electron reduction of CO_2 at the surface show not only how CO_2 might interact with the Cu but also how added protons can be involved. A specific sequence of steps is given in Figure 6.16 to highlight how these parameters control the evolution of CH_3OH at Cu (Peterson et al., 2010). From a particular form of an adsorbed COOH on the top where the carbon is bound to the surface, OH may be eliminated in a second electron transfer step to form a Cu–CO complex that may release CO into the solution. This Cu–CO structure could also be reduced in three further PCETs going downwards in Figure 6.16 to create a surface complex that can be released into solution as CH_3OH.

Aside from the very clear message that the system must provide protons for the reactions to create methanol, this analysis also presents how the ΔE of the Cu–O and Cu–C interactions play a controlling role in product selectivity. With selected Cu surfaces and the appropriate overpotential, kinetic pathways can be created at the top of Figure 6.16 for the evolution of C2 products ethanol and ethylene (Hori, 2008). For example, at a 5 mA/cm^2 current density through Cu, a resultant η of -0.9 V will produce CO. At an η of -1.2 V CH_4 is evolved with a higher η producing C2 structures. It is also possible that O from H_2O in solution can end up in the product. There is obviously a great complexity in the CO_2 reduction reaction at metal electrodes.

Many pure metal electrodes other than Cu have been examined for their ability to reduce CO_2, but none produce multi-carbon species (Hori, 2008). Their behavior falls into groups: those where detection of CO_2 reduction is difficult given the evolution of H_2 at negative overpotentials, such as Ni, Fe, Pt, and Ti; those that produce only formate, Pb, Hg, Tl, In, Sn, Cd, and Bi; and those that produce almost exclusively carbon monoxide, Au, Ag, Zn, Pd, and Ga (Hori, 2008). Multi-metal mixtures of Cu with these metals have also been examined, but have not been observed to perform better than pure Cu in producing C2 products. In terms of the models of Figure 6.16, these metals do not bind C and O in a manner that allows them to evolve into the flow into C2 structures but produce CO instead. The different selectivity for CO_2 reduction at these various metals is determined by the ΔEs of specific intermediate metal-adsorbates (M-A) in the rate-determining steps for product formation where this ΔE is controlled by the specific M-A interaction, the charge on the surface, and the η that is required to attain that charge.

Most of these experimental reports of overpotentials were determined at a current flow through the electrodes, set a 1 mA/cm^2 or higher, where the resultant potential on the working electrode was used to determine the η for the reaction relative to the thermodynamic $E°$ for the reaction (Hori, 2008; Nitopi et al., 2019). The smallest η is −0.69 V for CO evolution at Au electrodes and increases to −1.1 to −1.2 V for other metals where one or two CO_2 reduction products are observed in complement to H_2 evolution. It is only for Cu that an array of products are observed at lower η values, such as methane, carbon monoxide, formate, and ethylene, among others. At less negative bias potentials the relative composition of these products varies, but loses in Faradaic efficiency to hydrogen evolution.

All of these systems are found to be sensitive to the nature of the electrolyte and the identity of the anions and cations within it (Liu & Guo, 2020; Nitopi et al., 2019). Given the depiction in Figure 6.16 of the various forms of adsorbed structures for reduced intermediates and the conclusions that the magnitude of ΔE for each of these structures and each of the different metals determines the outcome, it is implicit that any component of the electrolyte that interacts with the surface and disrupts these structures can influence the rate and specificity of the catalysis. This includes the pH of the electrolyte in controlling the H atom presence on the surface, the nature of the ions in solution, and in particular their size and charge.

6.4.3 PV-Driven CO_2 Reduction

With this background, a straightforward evaluation can be made of the solar-driven electrolysis system in Figure 6.15. In order to obtain a high Faradaic efficiency for CO_2 reduction, the electrolyzer will have to be run at a constant current mode at high current flux, at a high η, where product selectivity may well have to be sacrificed. In addition, the metal surface under such a current flux is dynamic and will reconstruct, thereby limiting the exploitation of a particular crystallographic face to focus selectivity (Nitopi et al., 2019). The metal to be employed and the current flux will have to be selected based on the value of the product mixture evolved under operating conditions.

The concept of using solar PV to drive electrolyzers for CO_2 reduction has been explored in the literature (Bullock et al., 2017; Esiner et al., 2020; Huan et al., 2019; Jeon et al., 2015; Schreier et al., 2015) and features many of the constraints and limitations evident from catalysis research on both water oxidation and CO_2 reduction to C1 and C3 compounds. The basic structural elements for a prototype in these works are those of Figure 6.15 and consist of PV devices connected in series or parallel to electrodes where the anode oxidizes water and the cathode reduces CO_2 to a desirable product. The anodes in these works are known dark water oxidation catalysts such as IrO_2, Co_3O_4, RuO_2, and CuO. They operate in an electrolyte near neutral pH, mostly in a HCO_3^- solution, and are separated from the cathode by membranes, which can be bipolar in nature, or nafion or ion exchange resins. At currents of 5–20 mA/cm^2, the demands on this membrane are low compared with the A/cm^2 requirements for a membrane in fuel cell operation. The cathodes for CO_2 reduction range in form from nanoparticles of Cu, bimetallic compositions of nanosized Cu and Ag, to polycrystalline Au or CuO and even include organic compounds known to reduce CO_2. As has been discussed above, the overpotential applied to the cathode determines the selectivity of the product evolved. As a result, there has been much analytical work done in this cell development on the electrochemistry of these cathode materials in order to determine the optimum overpotential and associated current for the particular cathode structure involved. The composition of the electrolyte and the pH are also variables in this optimization where control of the latter minimizes H_2 evolution.

This work determines the size of the PV cell arrangement and the power that it must deliver under solar illumination because the operating point of the electrolyzer with its optimum current flux must be close to the maximum power point of the associated PV cells (Gerischer, 1994; Winkler et al., 2013). Sometimes power electronics are inserted between the PV cells and the electrolyzer to match the two, but in these early efforts, the particular PV devices used, be they Si, perovskites, a-Si, or CIGS, are connected directly in a way that ensures the desired matching of electrolyzer with the voltage of its solar power source.

Most of these PV-driven electrolyzers evolve C1 products such as CO, $HCOO^-$, or CH_4, but fewer are able to create C2 ethylene and ethane. An 80% Faradaic efficiency has been measured for the triple junction perovskite-driven electrolysis of CO_2 to CO with an IrO_2 anode (Schreier et al., 2015). The production of CH_4 from CO_2 at a Au electrode at a Faradaic efficiency >80% at 7 mA/cm^2 has been reported for a PIN perovskite-driven cell and a RuO_2 anode (Esiner et al., 2020). Evolution of ethylene and ethane at Cu and CuO and bimetallic Au–Cu cathodes has been controlled at a selectivity that ranged from 30% to 70% corresponding to currents of 65 down to 10 mA/cm^2 (Bullock et al., 2017; Huan et al., 2019; Ren et al., 2015).

It is evident that with the wide latitude of design and materials options within this device design, it will be possible to construct one for recycling for fossil fuel power plants, assuming the CO_2 concentration and delivery problems can be solved and that it is overall cost-effective.

The electrolytic reduction of CO_2 reduction in the gas phase through a flow stream of CO_2 by the electrode may offer different mechanistic pathways than those in Figure 6.16 and a different set of reaction products. The electrostatics of the interface would differ from a solution double layer and there would be a greater facility for surface attachment of CO_2 and a quicker desorption of intermediates that would not have the residence time at the surface to react with other intermediate species. At the present, these differences can only be established experimentally.

6.4.4 Solar Fuel Harvesting

At the other end of the marketplace from recycling is the fuel harvesting concept, where the capital costs of the PV and electrolyzers are envisioned to be greatly reduced through their integration into a small-scale, composite device for CO_2 reduction. The photoabsorber is integrated with, or adjacent to, the location of catalysts for the chemical transformation of CO_2 into a useful chemical or fuel. It will require years of development and production to determine if the PV-electrolyzer in Figure 6.15 can reach the low-cost levels required for this application. The remediation market mentioned above and suitable technology for its application will be intermediate between recycling and fuel harvesting and will probably contain aspects of both. For example, it could contain a portable CO_2 capture system and a PV-driven catalysis at a solar farm located at a removed site.

Solar fuel harvesting from CO_2 is an area of research that has been actively pursued and many reviews have covered the topic in detail. The following examination of the science and technology for this application will cover a number of forms of integrated light-driven systems involving semiconductors and catalysts from the literature. They will be analyzed to identify their operational design features as has been done for photosynthetic and artificial water splitting systems. It will be found that there are few new design features to add to the discussion.

6.4.5 Semiconductor Photoanode-Driven Reduction of CO_2 at Metals

Given the background on metal electrocatalysis provided above and on semiconductor anodes involved in water splitting, the first semiconductor systems to be discussed include those following the design of Figure 6.12a, where the anodic reaction is the oxidation of water at a semiconductor and the cathodic reaction is the reduction of CO_2 at a metal electrode or other suitably modified cathode. The limiting factor in this concept is the long-standing problem of the photostability of the semiconductor (Ogura, 1987; Ogura & Takagi, 1986). Concerning the cathode, an analysis of the literature on this topic (Ogura, 1987) concludes that the ability to reduce CO_2 depends upon the potential E_c of the anode semiconductor since this represents the most negative potential that may be translated to the metallic cathode (White et al., 2015). For example, with bandgap radiation a GaN anode with an E_c of −0.89 vs. SHE produces CO_2 reduction products at a Cu metal cathode whereas a TiO_2 anode at pH 7 with an E_c of −0.52 V vs. SHE only drives production of H_2. C1 compounds have been detected as well as C_2H_4 as well when the Cu cathode is modified with the electrodeposition of Ag in small amounts. This is expected given the overpotentials of nearly −1.0 V required for electrocatalysis at Cu that have been outlined above. However, the magnitude of the current flow system is still directly coupled to the recombination rate of electrons with holes within the photoanode.

The use of a DSPEC approach to current generation for CO_2 reduction through Figure 6.9a has often been mentioned (Meyer et al., 2020), but not well explored. For example, n-type GaN with its more negative E_c than TiO_2 could be sensitized in the visible region with dyes (Chen et al., 2010) to avoid electrode decomposition from bandgap radiation and should be better than TiO_2 at generating current for CO_2 reduction at a cathode.

6.4.6 Semiconductor Electrodes

From a design perspective in the recycling application, it is clear that CO_2 reduction at metal electrodes in a two-electrode system is a dynamically controlled process with an applied voltage that must supply the overpotentials in a steady state at both the anode and cathode in order to attain a desired selectivity and faradaic efficiency. It will be seen that CO_2 reduction at plain semiconductor surfaces is not governed by applied potential in the same manner.

There are a number of semiconductors that have been researched for CO_2 reduction including many that are not suitable for water oxidation (White et al., 2015). These include Si, III–V, and II–VI compounds since their durability in a reductive environment is greater than under the oxidizing conditions of oxygen evolution from water. These CO_2-reducing solids have been explored in two main forms that parallel designs in the photooxidation of water at semiconductor surfaces described above. One employs planar semiconductor electrodes as light absorbers or current conductors, and a second exploits particles of these semiconductors as dispersed particles in solution. The planar macroscopic semiconductors will be addressed first, allowing the pertinent fundamentals for semiconductor electrodes to be presented, most of which are applicable to semiconductor particles in solution.

At semiconductor electrode surfaces, the electrochemical terminology for η as a measure of activation barrier is defined differently than at metals (Gerischer, 1994). At a semiconductor electrode/electrolyte interface at equilibrium, the Galvani potential drop $\Delta\Phi^0$ between the bulk of the semiconductor and the solution is given by $\Delta\Phi^0 = E_{f,\mathrm{sc}} - E_{f,\mathrm{redox}}$ where $E_{f,\mathrm{sc}}$ is the fermi level for electrons in the bulk of the semiconductor and $E_{f,\mathrm{redox}}$ falls within the semiconductor bandgap. At equilibrium, a potential drop $\Delta\Phi^0_{\mathrm{sc}}$ in the form of a space charge region forms across the semiconductor at the interface as well as the $\Delta\Phi^0_{\mathrm{H}}$ across the double layer:

$$\Delta\Phi^0 = \Delta\Phi^0_{\mathrm{sc}} + \Delta\Phi^0_{\mathrm{H}}$$

Through the application of a potential that causes a deviation η from $\Delta\Phi^0$, this drop can be changed to a different value $\Delta\Phi$ where, under conditions for a Schottky barrier, $\Delta\Phi^0_{\mathrm{H}}$ remains constant and the change occurs across the space charge region of the semiconductor, as expressed

$$\Delta\Phi_{\mathrm{sc}} = \Delta\Phi^0_{\mathrm{sc}} + \eta$$

This constancy of $\Delta\Phi^0_{\mathrm{H}}$ is a distinguishing feature of semiconductor electrochemistry. It is an accurate assumption when semiconductor electrodes are biased in the depletion region of the semiconductor electrode, but less so when inversion or accumulation layers begin to form. Its impact is seen through a simple mass action expression for cathodic current flow i_{c} for electrons from the conduction band of a p-type electrode to an acceptor at the surface $i_{\mathrm{c}} = k_{\mathrm{c}} n_{\mathrm{s}} N$. Here, k_{c} is the rate constant for electron transfer from the conduction band at E_{c} to the acceptor, n_{s} is the surface concentration of electrons in the semiconductor conduction band, and N is the number of acceptors activated to receive an electron from the conduction band. The only potential dependent term is n_{s} and this dependence for a dark current can be written in the formal expression:

$$n_{\mathrm{s}} = n_{\mathrm{b}} e^{-(\Delta\phi^0_{\mathrm{sc}} + \eta)/kT}$$

where n_{b} is the small, bulk concentration of electrons in the conduction band of this p-type material. This assumes charge equilibrium between bulk and surface. A diagram for this circumstance is given in Figure 6.17a.

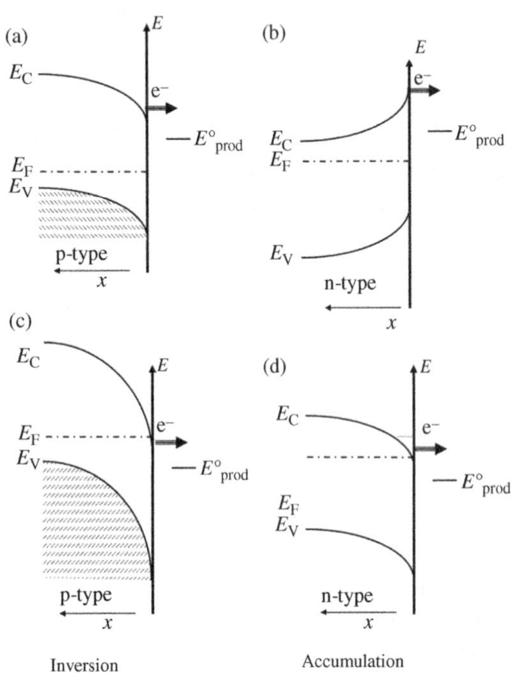

Figure 6.17 (a) Energy band diagram for a p-type semiconductor under negative bias. (b) Diagram for an n-type semiconductor under positive bias. (c) At extreme negative bias, a p-type semiconductor can form an inversion layer. (d) With a sufficient negative bias, an n-type semiconductor can form an accumulation layer. M. T. Spitler.

A similar expression can be written for a cathodic current over the conduction band at an n-type electrode:

$$n_s = n_D e^{-(\Delta\phi^0_{sc}+\eta)/kT}$$

where n_D is the donor doping density in the bulk of the n-type semiconductor. An energy level diagram for an n-type electrode is shown in Figure 6.17b.

In contrast to a metal electrode this states that when an overpotential is applied to the electrode only the concentration of electrons at the surface changes and not their potential. At electrode biases where a depletion region exists, $\Delta\Phi_H$ at a semiconductor surface does not change with η nor does any ΔE change for the binding of a reactant to its surface. At equilibrium, there is a ΔE for adsorption of CO_2 to a solid, as there is for any intermediate in CO_2 reduction, which is going to depend upon the crystal facet of the solid, the electrolyte, and the inherent and constant electric field across the Helmholtz layer. At a metal surface ΔE can change for the various intermediates as a result of the charging of the surface through changes in η and this controls the formation and evolution of the many structures of adsorbed species in Figure 6.16. This should not be expected at a semiconductor electrode. An exception is found in vapor-phase devices, where the Helmholtz double layer is ill-defined compared to its solution counterpart. For example, the C1 products at excited TiO_2 electrodes in solution are methanol, formaldehyde, and formic acid, whereas in the vapor phase only CH_4 is found (Olivo et al., 2017).

When the p-type electrode of Figure 6.17a is biased sufficiently negative, however, an inversion layer can form at the surface to turn the interface metallic in character, as depicted in Figure 6.17c. In this case, $\Delta\Phi_H$ can change with applied potential, but not necessarily in a 1:1 ratio with the applied bias. Similarly, an n-type electrode can be biased sufficiently negative to create a metal-like accumulation layer at the surface, as shown in Figure 6.17d. In addition, fermi-level pinning at a semiconductor by a surface state can fix $\Delta\Phi_{sc}$, allowing $\Delta\Phi_H$ to vary linearly with η. The models show that any η reported for CO_2 reduction at semiconductor electrodes reflects an interplay of band bending and Helmholtz layer charging and is difficult to interpret or compare with results from metals or other semiconductors.

The transfer of electrons from the conduction band to the intermediates in CO_2 reduction occurs at the same potential E_c for all electrons, as has been discussed above for the analogous case of hole transfer from the valence band in the water oxidation reaction. Any η applied to the electrode can change n_s significantly, and this concentration effect only influences how fast the solid can provide electrons to drive the multi-electron catalytic process at the surface reaction and does not change ΔE nor activation of any surface species. This parallels the discussion on the multielectron oxidation of water at semiconductor interfaces above and is analogous to PSII operation in Figure 6.2 where P680 feeds oxidizing equivalents into the OEC one at a time at the same potential in a timely manner to the various S states, otherwise, the reaction reverses. The timing sequence for these electron transfers will depend upon the particular surface, the catalyst that may be used, the composition of the electrolyte, and the involvement of proton or hydride transfers. Here, less is known about mechanisms for CO_2 reduction at semiconductor surfaces than at metal surfaces. Conceptually, however, the η dependence of a reaction at a semiconductor electrode could reveal timing information on some steps of the catalytic reaction mechanism.

6.4.7 Reduction of CO_2 at Semiconductor Surfaces

There are two main configurations in which semiconductor electrodes have been explored for the reduction of CO_2, and each is of use in a different system design. One uses photoexcited p-type electrodes for the cathodic reaction with CO_2, which can have application in a Z-scheme tandem design where the semiconductor anode for water splitting is also illuminated. This has been an active area of research. Another is a lesser pursued research topic in which dark current over an n-type semiconductor conduction band reduces CO_2 adsorbed from solution.

The ability of electrons in a semiconductor to reduce CO_2 will depend upon two factors. One is the chemistry of the CO_2 reduction intermediates at the surface of that particular semiconductor, which can be strongly influenced by a reactive electrolyte composition. Another is the potential E_c of the conduction band edge at the surface with respect to the $E°$ for the particular product, which is listed in Table 6.1 for pH 7 to account for the thermodynamics of proton involvement in the reaction. While the determination of the surface chemistry of reactants requires extensive experimental and theoretical analysis, as is clear from the examination of Cu catalysis above (Nitopi et al., 2019), tabulations of relative E_c values for semiconductors at a specific pH can be made. A representative sample is given in Figure 6.18. It is seen that these $E°$ values are more positive than the E_c of many semiconductors and that reduction should be possible at these surfaces. However, given the kinetic control of the reaction, the $E°$ values for each of the reaction intermediates involved, if it is possible to know them, are the determining parameters in the series of one-electron transfer reactions from E_c involved in the overall CO_2 reduction. The timing sequence of these electron transfers will be interactive with the involvement of protons and

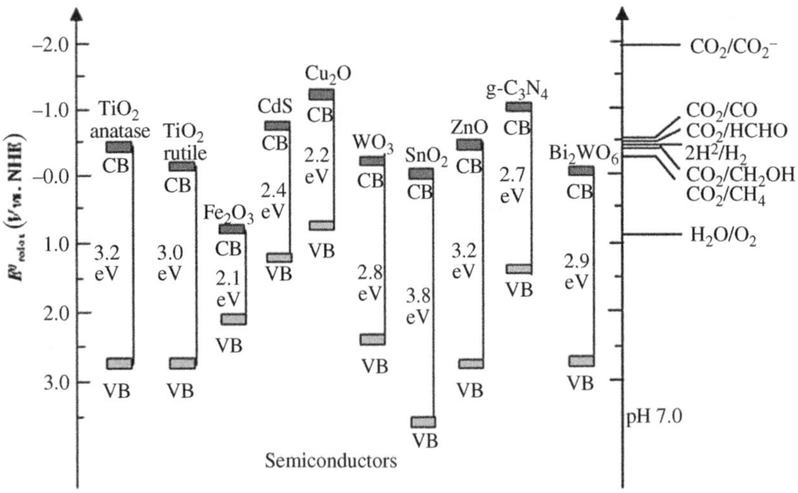

Figure 6.18 The energy band positions of a selection of semiconductors are given at pH 7 in relation to the $E°$ values for CO_2 reduction to various products. Reproduced with permission from Ola and Maroto-Valer (2015)/Elsevier/Public Domain CC BY 4.0.

hydride or proton transfers before or after any electron transfer reactions, as is implied in the proposed reaction schemes of Figure 6.16.

The bulk of experimentation with electrodes has been done on Si, II–VI, and III–V compounds (White et al., 2015; Zhang & Sun, 2019), both n- and p-type, and a few examples with aqueous electrolytes will be mentioned here to help define the system design features for solid electrodes. The greater part of the research on TiO_2 and other oxides pertains to particulate systems to be discussed later. Of greatest interest here are those semiconductors with an E_c sufficiently negative in potential for CO_2 reduction and a bandgap that can absorb a significant portion of visible light. Their surfaces are usually modified through the deposition of metals or metallic particles because bare surfaces of semiconductors for CO_2 reduction in the dark or under illumination are unexceptional as catalysts and evolve C1 and C2 products very slowly in comparison with metals (White et al., 2015; Zhang & Sun, 2019). The E_c values of most semiconductors in solution are also not sufficiently negative in potential to drive any surface-deposited Cu to the very high overpotentials required to make C2 complexes. In many experimental cases at semiconductors, the η required for CO_2 reduction to the C1 products has been sufficient to create an inversion layer, to cause shifts in $\Delta\Phi_H$ across the double layer, and to move the surface intermediates into a new regime of ΔE.

Photocathodes of p-type semiconductors provide an easy experimental system for the study of CO_2 reduction because dark reduction currents are small compared with photocurrents. Work with Group IV semiconductors has included p-Si, SiC, and even diamond (Hinogami et al., 1998; Hori, 2008). Considering low overpotential studies, work with Ni deposition on p-Si revealed the evolution of C1 products such as methanol and methane, along with small amounts of ethylene and ethane. The size of the metal particles from 5 to 20 nm and their character play a role in the η required to avoid H_2 evolution and to obtain high faradaic efficiencies. SiC has a more negative E_c and has also been shown as a dispersed powder to evolve these products under illumination when the surface is modified with metals (Cook et al., 1988).

Patches of metal deposited electrochemically on the surface of the semiconductor form metal-semiconductor junctions with a distance dependence of the junction potential that extends into the interior until "pinched off" by the band bending of the surrounding semiconductor-electrolyte Schottky barrier (Hinogami et al., 1998; Laskowski et al., 2020). The surface is heterogeneous, with metallic regions being scattered over a surface of bare semiconductor and serving as traps for electrons from the conduction band. Polarization of the semiconductor can result in different kinetics for CO_2 reduction in these neighboring metal/semiconductor regions since the double layer character of the Helmholtz layer at these materials two regions is different. This could be exploited in much the same manner as the Ag–Cu mixed metal electrode systems, where an intermediate can be produced at Ag that can further react at Cu. It would take advantage of the differing dependence of the catalytic kinetics at metal and semiconductor on this surface on η to provide a different means of control of staged, sequential reactions. Although this pinch-off effect has only been analyzed and quantified at Si electrodes, it should be possible to form on many metal-decorated semiconductor surfaces where the work function of the metal is within the bandgap of the semiconductor.

In a manner similar to Si photoelectrodes, GaP, GaAs, and InP as III–V solids have been investigated for their ability to photoreduce CO_2 in both their bare state and with metal deposits (Kumar et al., 2012). The negative E_c potentials of these compounds provide favorable energetics for the reaction and their bandgaps make them suitable for absorption of solar radiation. Depending upon the E_c, electrolyte composition, and the crystallographic face, the nature of the evolved C1 products, CO or formic acid, and their faradaic efficiencies can vary, but all require a bias potential. These semiconductors are well characterized as electronic materials and their crystal structure and crystallographic surface structures are well known so that theoretical DFT calculations can be done for the mechanism of product formation, as has been done for the (111) and (110) faces of GaP in solution (Muñoz-García & Carter, 2012; Xu & Carter, 2019, 2021). Among II–VI compounds CdTe and InP have been studied in both nonaqueous as well as aqueous electrolytes. CO production has been observed in both, but in the latter, early work with electrodes revealed that some electrolytes induce formic acid formation as well (Yoneyama et al., 1988).

Photocathodes can play a role in the Z-scheme design of Figure 6.14c, where they are directly coupled to the surface catalyst and contacted to a photoanode for water splitting (Halmann, 1978). Electron and hole photocurrents recombine at this junction. As has been discussed for the photoanode in Figure 6.12a, the anodic current produced is directly coupled to the rate of the electron–hole recombination with respect to the electron transfer rates in the catalytic reaction. The cathodic hole current behaves in a similar manner, but when the anode and cathode are connected electrically, the lower current controls the overall reaction. The faster catalysis chemistry is slowed down by the slower catalysis chemistry.

The system design changes for the dark reduction of CO_2 at n-type semiconductor electrodes over the conduction band. Little recent work is evident in this area, but the early work serves to illustrate the design concepts well. Here the semiconductor is driven in the dark as the cathode by a second semiconductor as the photoanode and operates as in Figure 6.12a. In such situations, reduction occurs at a potential E_c with a mixed surface reactivity if metal particles are present. Examinations of this electrode arrangement revealed that GaAs in a $KHCO_3$ electrolyte could evolve CH_3OH selectively at overpotentials greater than −1.0 V (Frese & Canfield, 1984) whereas illumination resulted in the additional evolution of CH_2O and $HCOO^-$ (Zafrir et al., 1983).

6.4.8 Molecular Catalysts

In much the same way that molecular catalysts have been evaluated electrochemically in solution for their ability to oxidize water to produce O_2, many molecular and organometallic catalysts have been examined and developed for their ability to reduce CO_2 (Dalle et al., 2019; Morris et al., 2009; Takeda et al., 2017; White et al., 2015; Yamazaki et al., 2015). The advantage of molecular catalysts for CO_2 reduction is that their inherent capability to be adjusted and modified synthetically acquires greater importance. Their structure could possibly be tuned to select one of the many desired CO_2 reduction products, modify its production rate, or increase its durability with a versatility that exceeds that of a bare metal cathode. Initial research in this area employed second and third-row transition metal elements, but more recent advances have been made with first-row metal complexes, which include porphyrins. There are also rare examples of purely organic catalysts (Barton et al., 2008; Dalle et al., 2019).

As with the molecular catalysts for water oxidation, electrochemical analyses at metals are used to characterize catalysts for CO_2 reduction in solution to correlate η for product evolution with currents, Faradaic efficiencies, and an assessment of durability in terms of TONs. This experimentation has been partitioned between nonaqueous and aqueous electrolytes with an emphasis on the former. It has been found that the production of C1 products dominates, with rare mentions of C2 evolution. Similar results have been found for the fewer studies on semiconductor surfaces (White et al., 2015). In water splitting success is quantified through the measure of a single species, O_2, whereas the electrocatalysis of CO_2 reduction can involve the many pathways and products represented in Figure 6.16. The evolution of C2 products from a single catalyst requires the reaction of two intermediate carbon structures, as the model structures in Figure 6.16 indicate, but research appears not to have advanced to that stage. Given this heightened complexity, progress in this area may be more difficult.

The pace of research into molecular CO_2 reduction catalysts has become intense within the last half dozen years (He & Janáky, 2020; Ola & Maroto-Valer, 2015). Promising opportunities have appeared, but the present TON, TOF, and faradaic efficiency levels do not yet compare with the water oxidation catalysts discussed above. Furthermore, it has been observed there are no good examples of molecular catalysts attached to semiconductor electrode surfaces (Dalle et al., 2019).

Initial research on molecular electrocatalysis featured complexes of Re, Ru, and other precious metals in solution and progress has moved the focus from the inner ligand coordination and structure to the secondary ligand form and modification (Francke et al., 2018; Kuramochi et al., 2018). In search of earth-abundant materials, molecular catalyst

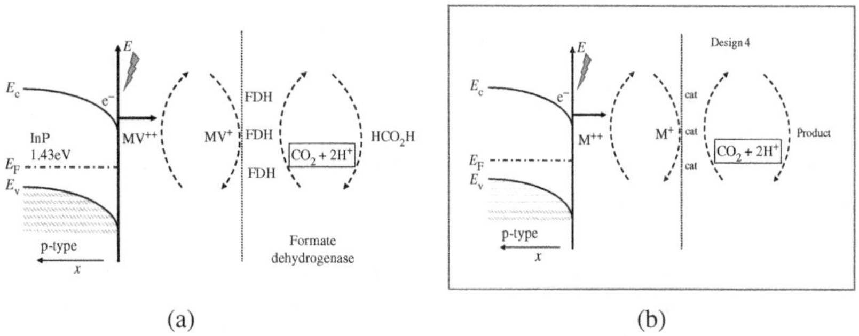

Figure 6.19 (a) A rare example in Parkinson and Weaver (1984) of indirect coupling for CO_2 reduction is depicted here for an illuminated InP electrode reducing an enzyme FDH through methyl viologen as mediator. (b) The kinetic design is given for (a). M. T. Spitler.

research has broadened out to first-row elements Mn, Fe, Cu, Ni, and Co, often in an aqueous environment. In analogy to early work with Re, Mn compounds have received extensive electrocatalytic exploration following the observation that a *fac*-Mn[(bpy)$(CO)_3$] complex evolves CO at >90% Faradaic efficiency in aqueous environments (Grills et al., 2018; Smieja et al., 2013). Fe and Cu porphyrins have been found to be efficient at CO production, with efficiencies of 90% for CO production but with TONs of about 100 (Manbeck & Fujita, 2015). Typical η values for these reductions were −1.0 V vs. SCE or higher. Ni cyclam complexes in aqueous solution reduced CO_2 at Hg with a TON of 8000 being reported for CO production at a 100% efficiency (Kuehnel et al., 2018). Fe porphyrins have exhibited the ability to produce CO from CO_2 with a TOF of 10^6 s^{-1} (Costentin et al., 2015), which would make this solution catalysis by metal electrodes competitive with electron–hole recombination when used as the cathode in the illuminated electrode configuration of Figure 6.13a. There are many fewer examples at semiconductor cathodes. At p-Si in acetonitrile, Ni and Co macrocycles at the surface were found to produce CO from CO_2 (Bradley & Tysak, 1982). Re complexes were also found to catalyze formation of CO at p-Si (Kumar et al., 2012). GaP and GaAs under illumination in water evolved CO with a faradaic efficiency of 80% but at different bias potentials owing to their differing E_c potentials (Petit et al., 1989). Pyridine in the electrolyte can play a strong catalytic role in directing the evolution of CH_4 at p-GaP with >90% Faradaic efficiency (Barton et al., 2008). It also influences the product evolution at $CuInS_2$, CdTe, and FeO_2 surfaces and other such additives are being sought (Yuan et al., 2016; Yuan & Hao, 2013). The observation that such a simple and widely available catalyst can function at defined crystallographic faces of these semiconductors has allowed a coordination of fundamental experimental and theoretical research similar to that done for Cu electrodes (Xu & Carter, 2019, 2021). In order to reach C2 and higher products, catalysis in staged steps of intermediates is envisioned (Benson et al., 2009), as was intended with the mixed Ag–Au CO_2 reductions at electrodes.

Enzymes have been used as model systems for study in fuel production since they possess a known ability to catalyze complex chemical reactions, including H_2 or CO_2 reduction (Evans et al., 2019; White et al., 2015). Most work uses surface adsorbed or attached enzymes on semiconductors or metals, with results that do not differ much from those of man-made catalysts. However, there is one report (Parkinson & Weaver, 1984) from 1984 of an indirectly coupled photosystem for formate production from CO_2 that is relevant to the present discussion.

This is shown in Figure 6.19, where an illuminated p-type InP cathode, potentiostatically controlled, reduces methyl viologen in solution, which then diffuses to a separate compartment. There the enzyme formate dehydrogenase (FDH) is located and is reduced. FDH then produces formate ions in solution with a 70–80% efficiency for a TON of 2×10^4 before, it is presumed, the enzyme is deactivated. This indirect coupling emulates the PSI function in Figure 6.2 and is known and exploited in fuel cell research (Anson & Stahl, 2020), but is not explored in the solar fuels field.

Electrocatalysis studies such as these on semiconductors serve an important role in catalyst development. However, with the exception of the Figure 6.19b design, any practical application of these catalysts in solution will fall in the cathodic reaction of Figure 6.9b, Figure 6.12b, or in the Z-scheme of Figure 6.14c. In the Z-scheme application, the molecular catalyst would replace any deposited metal in Figure 6.14c in order to provide the benefit of suppressing H_2 formation.

6.4.9 Particles for CO_2 Reduction

The practicality arguments made for the use of semiconductor particles in water splitting for H_2 production also apply to their use in CO_2 reduction. The role of photoactive particles in reduction of CO_2 is that played by the single particle system of

Figure 6.14a and the cathodes of the designs in Figure 6.14b. It is an area that has been extensively researched and reviewed (Ola & Maroto-Valer, 2015; Stolarczyk et al., 2018; Strange et al., 2020; Tan & Maroto-Valer, 2019; White et al., 2015). The discussion here will be limited to example systems that best illustrate the design features of particle-based catalysis. In the research on particles as cathodes, all of the techniques and materials for interfacial catalysis at planar semiconductors have been exploited. In addition, particle research allows a greater latitude of opportunity than solid electrodes owing to the ease of variation of particle forms and arrangements. There are fused particles of different spectral or catalytic character, nanoribbons, nanorods, nanospheres, nanobelts, and nanosheets selected to increase light absorption, charge separation, surface area, as well as to change E_c at the interface.

The use of a single particle to both split water and reduce CO_2 according to the design of Figure 6.14a imposes the most demands upon the semiconductor. It can vary in crystallographic morphology or dimensional aspect ratio, but it must have the stability during water oxidation and be able to reduce CO_2. The former requirement limits this area to oxide compounds, and the latter requires these oxides to be modified with inorganic or molecular cocatalysts. Of these oxides, TiO_2 has been the most researched and the results have been unimpressive, mostly attributable to the low visible light absorption (Ola & Maroto-Valer, 2015). The many variations on TiO_2 as the central element within these systems have sought to increase the catalytic rate and charge separation in order to minimize electron–hole recombination. Durability has been of lesser concern. A primary approach to CO_2 reduction at these particle surfaces is to modify the oxide with photo-deposited metal patches or nanoparticles of these metals, as has been discussed for semiconductor electrodes. This leads to the same collection of products as for pure metals. With the intent to accelerate the reaction, Co cyclam and porphyrin catalysts have also been immobilized in polymerized layers at surfaces. There are rare examples of performance that exceed 1%, but there has been little follow-up work on these opportunities. It is to be expected that the slower catalytic reaction of the two, water oxidation or CO_2 reduction, will limit the overall yield of the system. Yet, there is little evidence that the rates of either have been measured separately at these particles. Metal oxides other than TiO_2 have been studied and some have performed far better (He & Janáky, 2020; Tan & Maroto-Valer, 2019).

As surfaces for reduction of CO_2 in the Z-schemes of Figure 6.14b,c, both oxide and non-oxide semiconductors have been examined (Maeda, 2013). The latter have played a larger role because their instability against water oxidation is avoided and as Figure 6.18 shows, their E_c values can be far more negative than TiO_2. At these semiconductor surfaces, the dominant products have been C1 compounds, formic acid, formate, and methane. Overall, the catalysis reaction is still competing poorly with electron–hole recombination.

The primary use of quantum dots and nanoparticles in CO_2 reduction research is in the cathodic half of the Z-scheme of Figure 6.14c (Lu et al., 2020; Stolarczyk et al., 2018). The opportunity with nanoparticles in this role is the inherent ability of the researcher to control the crystallographic form of the solid, its morphology as sphere, rod, or sheet as well as the placement of cocatalyst on the surface. As pointed out above in the discussion on 3-D diffusion, with the nanometer size of these particles, only adsorbed intermediates or attached catalysts should play a role in a multi-electron catalytic reaction. The diffusion layer in solution for these various particles is limited to a distance equal to the size of the electroactive surface and in the presence of an electrolyte, electromigration effects can be ignored. In this cathodic half Z-scheme absorption of light by nanoparticles produces holes that are scavenged by a sacrificial donor and electrons that are captured by the CO_2 reduction catalyst. The behavior is influenced by the presence and nature of a capping ligand in which molecular catalysts can be imbedded. Prime examples are the bare surfaces of $CsPbBr_3$, which have been found to evolve CO from CO_2 upon illumination with 99+% selectivity (Burke et al., 2021; Hou et al., 2017; Zhang & Sun, 2019). II–VI quantum dots and ternary colloids have been investigated (Wu et al., 2019). With an Fe porphyrin derivative as a catalyst, $CuInS_2/ZnS$ structures can produce CO from CO_2 with a 99% efficiency. TONs of >400 have been reported (Lian et al., 2018). The performance of these systems is still low compared with others described above. The mismatch between the timing of photon absorption by the particle and of the sequential electron transfers of catalysis, which has been mentioned in the use of small particles for photooxidation, may play a role here as well. In this regard, however, the system may be reconfigured to be effective. For example, work on H_2 production from CdSe quantum particles in various environments showed high efficiency and TONs in excess of 10^6 with a Ni catalyst located not on the surface but imbedded in the dihydrolipoic acid (DHLA) capping layer as Ni(DHLA) (Burke et al., 2021).

There are other roles for nanostructures. Nanoparticles for water oxidation have been imbedded in a conductive graphite nanosheet along with nanoparticles for CO_2 reduction so that the graphite conducts charge within the two, this being a microscopic form of the Z-scheme of Figure 6.14c (Strange et al., 2020). There can also be a direct interaction of the internal photophysics of the quantum particle with the solution redox chemistry in solution, as has been discovered in Auger-assisted electron transfer (Zhu et al., 2014). In addition, hollow nanospheres of oxide semiconductors are well known

(Xiao et al., 2019), but have not been exploited as isolation chambers for catalytic reactions, an opportunity that nature has already exploited with the carboxysome structures described above where localized concentrations of CO_2 can be greatly elevated.

6.5 Conclusions

An examination of the kinetic design of the experimental work for CO_2 reduction in the fuel harvesting mode reveals that almost all involve directly coupled systems, both for the water oxidation and CO_2 halves. This is most clearly evident in the man-made Z-schemes of Figure 6.14b,c and stands in contrast to the photosynthetic structure of Figure 6.2, where only PSII is directly coupled to a catalyst and PSI is only indirectly coupled with carbon fixation. The electrolyzer scheme of Figure 6.15 for a recycling market stands distinct from these considerations where its electrodes are decoupled from the solar power source through the intervening electronics.

A straightforward strength of directly coupled designs is their ease of assembly. Semiconductor electrodes can simply be immersed in solution for water splitting or CO_2 reduction. In order to obtain reasonable yields of the product and a greater degree of durability, surface modification with cocatalysts is often necessary. Less easy are the DSPEC and the molecular catalytic schemes of Figures 6.9a and 6.13a, where surface attachment functions need to be developed. Particles and nanostructures modified in any of the ways found for their macroscopic counterparts can be dispersed in solution to provide a large surface area. Selectivity of the catalytic surface appears attainable for specific C1 products.

The present strategy in these directly coupled systems has been to make design changes to minimize the effect of the competition of the kinetics of the steps in catalysis with excitation quenching in the light absorber or the system's back reaction kinetics. Here the multidisciplinary nature of this research area comes to fore. Whereas a materials researcher may design a solid to vary the electron–hole recombination rate in an excited solid, a synthetic chemist in catalysis is needed to control the rates of molecular processes at the surface. The insertion of a p/n junction buried beneath the surface can remove one of the charge carriers from the surface and reduce electron/hole recombination through a mass action effect. For particles, core-shell structures can serve the same role. In a molecular DSPEC system, yields are increased by making the back reaction between the electron in the conduction band and the oxidized sensitizing dye or catalyst instead of the valence band hole. This creates a back reaction rate that can be controlled by inserting intermediate conductive connector species (Godin et al., 2017) or working with columnar nanorod arrays (Shan et al., 2019).

Catalyst development is the least mature element of these CO_2 reduction systems. Yet, these catalysts are inserted in directly coupled situations where demands on the catalyst performance are highest. In these cases, the system design determines what the kinetic TOF of the catalyst must be. A reverse strategy could also be employed, where an adequate and durable catalyst is developed and subsequently used in a system designed around it. An effective catalyst that is not competitive with the recombination reactions of photosystems, whether heterogeneous or homogeneous, could be developed for high-value C2 evolution and then featured in a system custom designed to optimize this capability. One need only consider the poor catalytic performance of RuBisCO to know that this is possible.

The kinetic schemes of these man-made systems contrast with those of the photosynthetic systems of Figure 6.2, where both indirect and direct coupling are employed. As discussed, the direct coupling of PSII to the OEC appears necessary because of the potent oxidizing power of the $P680^+$ species, which can damage or destroy the organic structures and components of the organism unless it is carefully controlled. Nature may have been forced to do this because it does not have access to the many robust materials available to modern researchers in the form of inorganic solids and compounds that can be made resistant to oxidation. The much-researched TiO_2 or $BiVO_4$ are examples of these materials and they can be made in many shapes as well in size from macro- to nano-scopic. They could possibly be used to enclose an environment where an indirect coupling could be used to oxidize water much like carboxysomes create a confined and favorable environment for RuBisCO function. It may be that man-made systems need not employ a directly coupled design to split water.

Most man-made systems ignore the opportunities that are presented by an indirect coupling of the catalytic action, in analogy to the relation between PSI and RuBisCO. In such a case, the criteria for a catalyst acquire a different emphasis. The TOF would probably drop to the insolation-determined frequency of $2\,s^{-1}$. TON would remain important, but selectivity would be highly valued, especially for C2 products. The efficiency and TONs of the preliminary experiment in Figure 6.19 show the inherent promise. In addition, there are a number of electrodes that can be used in the dark reduction of CO_2 to C1 and C2 products, but they require overpotentials of $-1.0\,V$ or more for metals and semiconductor E_c values of $-0.5\,V$ or higher. The creation of redox couples with this reducing power would require the use of PSI-type artificial photosystems to elevate the redox potential to a much more negative value.

In conclusion, this perspective on the kinetic designs for CO_2 photoreduction systems has revealed that the present thrust of scientific research is on catalysts to be directly coupled to a photosystem. This approach ignores the opportunities presented by an indirect coupling with the catalytic action as well as other aspects of photosynthetic systems presented above. It is clear that even with the immense amount of work that is in progress on solar-driven CO_2 reduction, there are still wide areas open for exploration.

References

Ali, S., Flores, M. C., Razzaq, A., Sorcar, S., Hiragond, C. B., Kim, H. R., Park, Y. H., Hwang, Y., Kim, H. S., Kim, H., Gong, E. H., Lee, J., & In, S.-I. (2019). Gas phase photocatalytic CO_2 reduction, "A brief overview for benchmarking". *Catalysts*, *9*(9), 90727.

Alibabaei, L., Luo, H., House, R. L., Hoertz, P. G., Lopez, R., & Meyer, T. J. (2013). Applications of metal oxide materials in dye sensitized photoelectrosynthesis cells for making solar fuels: Let the molecules do the work. *Journal of Materials Chemistry A*, *1*(13), 4133–4145.

Amirav, L., & Alivisatos, A. P. (2010). Photocatalytic hydrogen production with tunable nanorod heterostructures. *The Journal of Physical Chemistry Letters*, *1*(7), 1051–1054.

Ananyev, G., Roy-Chowdhury, S., Gates, C., Fromme, P., & Dismukes, G. C. (2019). The catalytic cycle of water oxidation in crystallized photosystem II complexes: Performance and requirements for formation of intermediates. *ACS Catalysis*, *9*(2), 1396–1407.

Anson, C. W., & Stahl, S. S. (2020). Mediated fuel cells: Soluble redox mediators and their applications to electrochemical reduction of O_2 and oxidation of H_2, alcohols, biomass, and complex fuels. *Chemical Reviews*, *120*(8), 3749–3786.

Archer, M. D., & Nozik, A. J.. (Eds.) (2008). *Nanostructured and photoelectrochemical systems for solar photon conversion*. Imperial College Press.

Ashford, D. L., Gish, M. K., Vannucci, A. K., Brennaman, M. K., Templeton, J. L., Papanikolas, J. M., & Meyer, T. J. (2015). Molecular chromophore–catalyst assemblies for solar fuel applications. *Chemical Reviews*, *115*(23), 13006–13049.

Azcarate, I., Costentin, C., Robert, M., & Savéant, J.-M. (2016). Through-space charge interaction substituent effects in molecular catalysis leading to the design of the most efficient catalyst of CO_2-to-CO electrochemical conversion. *Journal of the American Chemical Society*, *138*(51), 16639–16644.

Bai, Y., Wilbraham, L., Slater, B. J., Zwijnenburg, M. A., Sprick, R. S., & Cooper, A. I. (2019). Accelerated discovery of organic polymer photocatalysts for hydrogen evolution from water through the integration of experiment and theory. *Journal of the American Chemical Society*, *141*(22), 9063–9071.

Bard, A. J., & Faulkner, L. R. (2001). *Electrochemical methods: Fundamentals and applications*. Wiley.

Barton, E. E., Rampulla, D. M., & Bocarsly, A. B. (2008). Selective solar-driven reduction of CO_2 to methanol using a catalyzed p-GaP based photoelectrochemical cell. *Journal of the American Chemical Society*, *130*(20), 6342–6344.

Benson, E. E., Kubiak, C. P., Sathrum, A. J., & Smieja, J. M. (2009). Electrocatalytic and homogeneous approaches to conversion of CO_2 to liquid fuels. *Chemical Society Reviews*, *38*(1), 89–99.

Blakemore, J. D., Crabtree, R. H., & Brudvig, G. W. (2015). Molecular catalysts for water oxidation. *Chemical Reviews*, *115*(23), 12974–13005.

Blakemore, J. D., Schley, N. D., Balcells, D., Hull, J. F., Olack, G. W., Incarvito, C. D., Eisenstein, O., Brudvig, G. W., & Crabtree, R. H. (2010). Half-sandwich iridium complexes for homogeneous water-oxidation catalysis. *Journal of the American Chemical Society*, *132*(45), 16017–16029.

Blankenship, R. E. (2010). Early evolution of photosynthesis. *Plant Physiology*, *154*(2), 434–438.

Blankenship, R. E. (2014). *Molecular mechanisms of photosynthesis*. Blackwell Science.

Blankenship, R. E. (2015). Structural and functional dynamics of photosynthetic antenna complexes. *Proceedings of the National Academy of Sciences*, *112*(45), 13751–13752.

Bracher, A., Whitney, S. M., Hartl, F. U., & Hayer-Hartl, M. (2017). Biogenesis and metabolic maintenance of rubisco. *Annual Review of Plant Biology*, *68*(1), 29–60.

Bradley, M. G., & Tysak, T. (1982). p-Type silicon based photoelectrochemical cells for optical energy conversion: Electrochemistry of tetra-azomacrocyclic metal complexes at illuminated. *Journal of Electroanalytical Chemistry and Interfacial Electrochemistry*, *135*(1), 153–157.

Brennaman, M. K., Dillon, R. J., Alibabaei, L., Gish, M. K., Dares, C. J., Ashford, D. L., House, R. L., Meyer, G. J., Papanikolas, J. M., & Meyer, T. J. (2016). Finding the way to solar fuels with dye-sensitized photoelectrosynthesis cells. *Journal of the American Chemical Society, 138*(40), 13085–13102.

Brimblecombe, R., Dismukes, G. C., Swiegers, G. F., & Spiccia, L. (2009). Molecular water-oxidation catalysts for photoelectrochemical cells. *Dalton Transactions, 43*, 9374–9384.

Gurudayal, Bullock, J., Srankó, D. F., Towle, C. M., Lum, Y., Hettick, M., Scott, M. C., Javey, A., & Ager, J. (2017). Efficient solar-driven electrochemical CO_2 reduction to hydrocarbons and oxygenates. *Energy & Environmental Science, 10*(10), 2222–2230.

Burke, R., Bren, K. L., & Krauss, T. D. (2021). Semiconductor nanocrystal photocatalysis for the production of solar fuels. *The Journal of Chemical Physics, 154*(3), 030901.

Caffarri, S., Tibiletti, T., Jennings, R. C., & Santabarbara, S. (2014). A comparison between plant photosystem I and photosystem II architecture and functioning. *Current Protein and Peptide Science, 15*, 296–331.

Calhoun, T. R., Ginsberg, N. S., Schlau-Cohen, G. S., Cheng, Y.-C., Ballottari, M., Bassi, R., & Fleming, G. R. (2009). Quantum coherence enabled determination of the energy landscape in light-harvesting complex II. *The Journal of Physical Chemistry B, 113*(51), 16291–16295.

Chen, X. Y., Yip, C. T., Fung, M. K., Djurišić, A. B., & Chan, W. K. (2010). GaN-nanowire-based dye-sensitized solar cells. *Applied Physics A, 100*(1), 15–19.

Choi, D., Rowley, J. G., Spitler, M., & Parkinson, B. A. (2013). Dye sensitization of four low index TiO_2 single crystal photoelectrodes with a series of dicarboxylated cyanine dyes. *Langmuir, 29*(30), 9410–9419.

Ciszkowska, M., & Stojek, Z. (1999). Voltammetry in solutions of low ionic strength. Electrochemical and analytical aspects. *Journal of Electroanalytical Chemistry, 466*(2), 129–143.

Coggins, M. K., Zhang, M.-T., Chen, Z., Song, N., & Meyer, T. J. (2014). Single-site copper(II) water oxidation electrocatalysis: Rate enhancements with HPO_4^{2-} as a proton acceptor at pH 8. *Angewandte Chemie International Edition, 53*(45), 12226–12230.

Cook, R. L., MacDuff, R. C., & Sammells, A. F. (1988). Photoelectrochemical carbon dioxide reduction to hydrocarbons at ambient temperature and pressure. *Journal of the Electrochemical Society, 135*(12), 3069–3070.

Costentin, C., Robert, M., Savéant, J.-M., & Tatin, A. (2015). Efficient and selective molecular catalyst for the CO_2-to-CO electrochemical conversion in water. *Proceedings of the National Academy of Sciences, 112*(22), 6882–6886.

Creus, J., Matheu, R., Peñafiel, I., Moonshiram, D., Blondeau, P., Benet-Buchholz, J., García-Antón, J., Sala, X., Godard, C., & Llobet, A. (2016). A million turnover molecular anode for catalytic water oxidation. *Angewandte Chemie International Edition, 55*(49), 15382–15386.

Croce, R., & van Amerongen, H. (2020). Light harvesting in oxygenic photosynthesis: Structural biology meets spectroscopy. *Science, 369*(6506), eaay2058.

Dalle, K. E., Warnan, J., Leung, J. J., Reuillard, B., Karmel, I. S., & Reisner, E. (2019). Electro- and solar-driven fuel synthesis with first row transition metal complexes. *Chemical Reviews, 119*(4), 2752–2875.

Derks, A., Schaven, K., & Bruce, D. (2015). Diverse mechanisms for photoprotection in photosynthesis. Dynamic regulation of photosystem II excitation in response to rapid environmental change. *Biochimica Et Biophysica Acta (BBA)-Bioenergetics, 1847*(4), 468–485.

Duan, L., Araujo, C. M., Ahlquist, M. S. G., & Sun, L. (2012). Highly efficient and robust molecular ruthenium catalysts for water oxidation. *Proceedings of the National Academy of Sciences, 109*(39), 15584–15588.

Eachus, R. S., Marchetti, A. P., & Muenter, A. A. (1999). The photophysics of silver halide imaging materials. *Annual Review of Physical Chemistry, 50*(1), 117–144.

Erb, T. J., & Zarzycki, J. (2018). A short history of RubisCO: The rise and fall (?) of nature's predominant CO_2 fixing enzyme. *Current Opinion in Biotechnology, 49*, 100–107.

Esiner, S., Wang, J., & Janssen, R. A. J. (2020). Light-driven electrochemical carbon dioxide reduction to carbon monoxide and methane using perovskite photovoltaics. *Cell Reports Physical Science, 1*(5), 100058.

Evans, R. M., Siritanaratkul, B., Megarity, C. F., Pandey, K., Esterle, T. F., Badiani, S., & Armstrong, F. A. (2019). The value of enzymes in solar fuels research – Efficient electrocatalysts through evolution. *Chemical Society Reviews, 48*(7), 2039–2052.

Fan, D.-Y., Hope, A. B., Smith, P. J., Jia, H., Pace, R. J., Anderson, J. M., & Chow, W. S. (2007). The stoichiometry of the two photosystems in higher plants revisited. *Biochimica Et Biophysica Acta (BBA)-Bioenergetics, 1767*(8), 1064–1072.

Folkman, S. J., Soriano-Lopez, J., Galán-Mascarós, J. R., & Finke, R. G. (2018). Electrochemically driven water-oxidation catalysis beginning with six exemplary cobalt polyoxometalates: Is it molecular, homogeneous catalysis or electrode-bound, heterogeneous CoO_x catalysis? *Journal of the American Chemical Society, 140*(38), 12040–12055.

Francke, R., Schille, B., & Roemelt, M. (2018). Homogeneously catalyzed electroreduction of carbon dioxide—Methods, mechanisms, and catalysts. *Chemical Reviews, 118*(9), 4631–4701.

Frese, K. W., & Canfield, D. (1984). Reduction of CO_2 on n-GaAs electrodes and selective methanol synthesis. *Journal of the Electrochemical Society, 131*(11), 2518–2522.

Fromme, P., & Mathis, P. (2004). Unraveling the photosystem I reaction center: A history, or the sum of many efforts. *Photosynthesis Research, 80*, 109–124.

Fujishima, A., & Honda, K. (1972). Electrochemical photolysis of water at a semiconductor electrode. *Nature, 238*(5358), 37–38.

Gao, Y., Ding, X., Liu, J., Wang, L., Lu, Z., Li, L., & Sun, L. (2013). Visible light driven water splitting in a molecular device with unprecedentedly high photocurrent density. *Journal of the American Chemical Society, 135*(11), 4219–4222.

Garcia-Esparza, A. T., & Takanabe, K. (2016). A simplified theoretical guideline for overall water splitting using photocatalyst particles. *Journal of Materials Chemistry A, 4*(8), 2894–2908.

Garrido-Barros, P., Gimbert-Suriñach, C., Matheu, R., Sala, X., & Llobet, A. (2017). How to make an efficient and robust molecular catalyst for water oxidation. *Chemical Society Reviews, 46*(20), 6088–6098.

Gerischer, H. (1994). Principles of electrochemistry. In P. J. Gellings & H. J. Bouwmeester (Eds.), *CRC handbook of solid state electrochemistry* (pp. 9–73). CRC Press.

Gilbert, J. A., Eggleston, D. S., Murphy, W. R., Geselowitz, D. A., Gersten, S. W., Hodgson, D. J., & Meyer, T. J. (1985). Structure and redox properties of the water-oxidation catalyst $[(bpy)_2(OH_2)RuORu(OH_2)(bpy)_2]^{4+}$. *Journal of the American Chemical Society, 107*(13), 3855–3864.

Godin, R., Kafizas, A., & Durrant, J. R. (2017). Electron transfer dynamics in fuel producing photosystems. *Current Opinion in Electrochemistry, 2*(1), 136–143.

Goeppert, A., Czaun, M., Surya Prakash, G. K., & Olah, G. A. (2012). Air as the renewable carbon source of the future: An overview of CO_2 capture from the atmosphere. *Energy & Environmental Science, 5*(7), 7833–7853.

Grills, D. C., Ertem, M. Z., McKinnon, M., Ngo, K. T., & Rochford, J. (2018). Mechanistic aspects of CO_2 reduction catalysis with manganese-based molecular catalysts. *Coordination Chemistry Reviews, 374*, 173–217.

Grotjohann, I., & Fromme, P. (2013). Photosystem I. In W. J. Lennarz & M. D. Lane (Eds.), *Encyclopedia of biological chemistry* (2nd ed., pp. 503–507). Academic Press.

Halmann, M. (1978). Photoelectrochemical reduction of aqueous carbon dioxide on p-type gallium phosphide in liquid junction solar cells. *Nature, 275*(5676), 115–116.

Haussener, S., Xiang, C., Spurgeon, J. M., Ardo, S., Lewis, N. S., & Weber, A. Z. (2012). Modeling, simulation, and design criteria for photoelectrochemical water-splitting systems. *Energy & Environmental Science, 5*(12), 9922–9935.

Hayer-Hartl, M., & Hartl, F. U. (2020). Chaperone machineries of rubisco – The most abundant enzyme. *Trends in Biochemical Sciences, 45*(9), 748–763.

He, J., & Janáky, C. (2020). Recent advances in solar-driven carbon dioxide conversion: Expectations versus reality. *ACS Energy Letters, 5*(6), 1996–2014.

Hill, D. J., Teitsworth, T. S., Ritchie, E. T., Atkin, J. M., & Cahoon, J. F. (2018). Interplay of surface recombination and diode geometry for the performance of axial p–i–n nanowire solar cells. *ACS Nano, 12*(10), 10554–10563.

Hinogami, R., Nakamura, Y., Yae, S., & Nakato, Y. (1998). An approach to ideal semiconductor electrodes for efficient photoelectrochemical reduction of carbon dioxide by modification with small metal particles. *The Journal of Physical Chemistry B, 102*(6), 974–980.

Holmes, M. A., Townsend, T. K., & Osterloh, F. E. (2012). Quantum confinement controlled photocatalytic water splitting by suspended CdSe nanocrystals. *Chemical Communications, 48*(3), 371–373.

Hori, Y. (2008). Electrochemical CO_2 reduction on metal electrodes. In C. G. Vayenas, R. E. White, & M. E. Gamboa-Aldeco (Eds.), *Modern aspects of electrochemistry* (pp. 89–189). Springer New York.

Hou, J., Cao, S., Wu, Y., Gao, Z., Liang, F., Sun, Y., Lin, Z., & Sun, L. (2017). Inorganic colloidal perovskite quantum dots for robust solar CO_2 reduction. *Chemistry, 23*(40), 9481–9485.

Hu, K., & Meyer, G. J. (2015). Lateral intermolecular self-exchange reactions for hole and energy transport on mesoporous metal oxide thin films. *Langmuir, 31*(41), 11164–11178.

Hu, K., Sampaio, N., Schneider, J., Troian-Gautier, L., & Meyer, G. J. (2020). Perspectives on Dye Sensitization of Nanocrystalline Mesoporous Thin Films. *Journal of the American Chemical Society, 142*(38), 16099–16116.

Huan, T. N., Dalla Corte, D. A., Lamaison, S., Karapinar, D., Lutz, L., Menguy, N., Foldyna, M., Turren-Cruz, S. H., Hagfeldt, A., Bella, F., Fontecave, M., & Mougel, V. (2019). Low-cost high-efficiency system for solar-driven conversion of CO_2 to hydrocarbons. *Proceedings of the National Academy of Sciences, 116*(20), 9735–9740.

James, T. H. (1977). *The theory of the photographic process* (4th ed.). Macmillan.

Jeon, H. S., Koh, J. H., Park, S. J., Jee, M. S., Ko, D.-H., Hwang, Y. J., & Min, B. K. (2015). A monolithic and standalone solar-fuel device having comparable efficiency to photosynthesis in nature. *Journal of Materials Chemistry A*, *3*(11), 5835–5842.

Joliet, P., Barbieri, G., & Chabaud, R. (1969). Un noveau modele des centres photochimiques du systeme II. *Photochemistry and Photobiology*, *10*, 309–329.

Jones, M., Lo, S. S., & Scholes, G. D. (2009). Quantitative modeling of the role of surface traps in CdSe/CdS/ZnS nanocrystal photoluminescence decay dynamics. *Proceedings of the National Academy of Sciences*, *106*(9), 3011–3016.

Kamire, R. J., Materna, K. L., Hoffeditz, W. L., Phelan, B. T., Thomsen, J. M., Farha, O. K., Hupp, J. T., Brudvig, G. W., & Wasielewski, M. R. (2017). Photodriven oxidation of surface-bound iridium-based molecular water-oxidation catalysts on perylene-3,4-dicarboximide-sensitized TiO_2 electrodes protected by an Al_2O_3 layer. *The Journal of Physical Chemistry C*, *121*(7), 3752–3764.

Kanady, J. S., Lin, P.-H., Carsch, K. M., Nielsen, R. J., Takase, M. K., Goddard, W. A., & Agapie, T. (2014). Toward models for the full oxygen-evolving complex of photosystem II by ligand coordination to lower the symmetry of the Mn_3CaO_4 cubane: Demonstration that electronic effects facilitate binding of a fifth metal. *Journal of the American Chemical Society*, *136*(41), 14373–14376.

Kanan, M. W., & Nocera, D. G. (2008). In situ formation of an oxygen-evolving catalyst in neutral water containing phosphate and CO_2. *Science*, *321*(5892), 1072–1075.

Karlsson, E. A., Lee, B.-L., Åkermark, T., Johnston, E. V., Kärkäs, M. D., Sun, J., Hansson, Ö., Bäckvall, J. E., & Åkermark, B. (2011). Photosensitized water oxidation by use of a bioinspired manganese catalyst. *Angewandte Chemie International Edition*, *50*(49), 11715–11718.

Kerfeld, C. A., Aussignargues, C., Zarzycki, J., Cai, F., & Sutter, M. (2018). Bacterial microcompartments. *Nature Reviews Microbiology*, *16*(5), 277–290.

Kern J, Chatterjee R, Young ID, Fuller FD, Lassalle L, Ibrahim M, Gul, S., Fransson, T., Brewster, A.S., Alonso-Mori, R., Hussein, R., M. Zhang, L. Douthit, C. de Lichtenberg, M. H. Cheah, D. Shevela, J. Wersig, D. Sokaras, …, V. K. Yachandra Structures of the intermediates of Kok's photosynthetic water oxidation clock. *Nature*. 2018;*563*(7731):421–425.

Kirner, J. T., & Finke, R. G. (2017). Water-oxidation photoanodes using organic light-harvesting materials: A review. *Journal of Materials Chemistry A*, *5*(37), 19560–19592.

Kistler, T. A., Larson, D., Walczak, K., Agbo, P., Sharp, I. D., Weber, A. Z., & Danilovic, N. (2019). Integrated membrane-electrode-assembly photoelectrochemical cell under various feed conditions for solar water splitting. *Journal of the Electrochemical Society*, *166*, H3020–H3028.

Kok, B., Forbush, B., & McGloin, M. (1970). Cooperation of charges in photosynthetic O_2 evolution – I. A linear four step mechanism. *Photochemistry and Photobiology*, *11*(6), 457–475.

Kuehnel, M. F., Sahm, C. D., Neri, G., Lee, J. R., Orchard, K. L., Cowan, A. J., & Reisner, E. (2018). ZnSe quantum dots modified with a Ni(cyclam) catalyst for efficient visible-light driven CO_2 reduction in water. *Chemical Science*, *9*(9), 2501–2509.

Kumar, B., Llorente, M., Froehlich, J., Dang, T., Sathrum, A., & Kubiak, C. P. (2012). Photochemical and photoelectrochemical reduction of CO_2. *Annual Review of Physical Chemistry*, *63*(1), 541–569.

Kumar, S., Karthikeyan, S., & Lee, A. F. (2018). g-C_3N_4-based nanomaterials for visible light-driven photocatalysis. *Catalysts*, *8*(2), 20074.

Kuramochi, Y., Ishitani, O., & Ishida, H. (2018). Reaction mechanisms of catalytic photochemical CO_2 reduction using Re(I) and Ru(II) complexes. *Coordination Chemistry Reviews*, *373*, 333–356.

Laskowski, F. A. L., Oener, S. Z., Nellist, M. R., Gordon, A. M., Bain, D. C., Fehrs, J. L., & Boettcher, S. W. (2020). Nanoscale semiconductor/catalyst interfaces in photoelectrochemistry. *Nature Materials*, *19*(1), 69–76.

Lee, D. K., Lee, D., Lumley, M. A., & Choi, K.-S. (2019). Progress on ternary oxide-based photoanodes for use in photoelectrochemical cells for solar water splitting. *Chemical Society Reviews*, *48*(7), 2126–2157.

Lee, H., Wu, X., & Sun, L. (2020). Copper-based homogeneous and heterogeneous catalysts for electrochemical water oxidation. *Nanoscale*, *12*(7), 4187–4218.

Lian, S., Kodaimati, M. S., & Weiss, E. A. (2018). Photocatalytically active superstructures of quantum dots and iron porphyrins for reduction of CO_2 to Co in water. *ACS Nano*, *12*(1), 568–575.

Liu, H.-Y., Cody, C. C., Jacob-Dolan, J. A., Crabtree, R. H., & Brudvig, G. W. (2020). Surface-attached molecular catalysts on visible-light-absorbing semiconductors: Opportunities and challenges for a stable hybrid water-splitting photoanode. *ACS Energy Letters*, *5*(10), 3195–3202.

Liu, Y., & Guo, L. (2020). On factors limiting the performance of photoelectrochemical CO_2 reduction. *The Journal of Chemical Physics, 152*(10), 100901.

Lu, H., Huang, Z., Martinez, M. S., Johnson, J. C., Luther, J. M., & Beard, M. C. (2020). Transforming energy using quantum dots. *Energy & Environmental Science, 13*(5), 1347–1376.

Lv, H., Geletii, Y. V., Zhao, C., Vickers, J. W., Zhu, G., Luo, Z., Song, J., Lian, T., Musaev, D. G., & Hill, C. L. (2012). Polyoxometalate water oxidation catalysts and the production of green fuel. *Chemical Society Reviews, 41*(22), 7572–7589.

Maeda, K. (2013). Z-Scheme water splitting using two different semiconductor photocatalysts. *ACS Catalysis, 3*(7), 1486–1503.

Manbeck, G. F., & Fujita, E. (2015). A review of iron and cobalt porphyrins, phthalocyanines and related complexes for electrochemical and photochemical reduction of carbon dioxide. *Journal of Porphyrins and Phthalocyanines, 19*(1–3), 45–64.

Matheu, R., Garrido-Barros, P., Gil-Sepulcre, M., Ertem, M. Z., Sala, X., Gimbert-Suriñach, C., & Llobet, A. (2019). The development of molecular water oxidation catalysts. *Nature Reviews Chemistry, 3*(5), 331–341.

McCool, N. S., Robinson, D. M., Sheats, J. E., & Dismukes, G. C. (2011). A Co_4O_4 "cubane" water oxidation catalyst inspired by photosynthesis. *Journal of the American Chemical Society, 133*(30), 11446–11449.

McDaniel, N. D., Coughlin, F. J., Tinker, L. L., & Bernhard, S. (2008). Cyclometalated iridium(III) aquo complexes: Efficient and tunable catalysts for the homogeneous oxidation of water. *Journal of the American Chemical Society, 130*(1), 210–217.

Memming, R. (2015). *Semiconductor electrochemistry* (2nd ed.). Verlag GmBH.

Meyer, M. T., Itakura, A. K., Patena, W., Wang, L., He, S., Emrich-Mills, T., Lau, C. S., Yates, G., Mackinder, L. C., & Jonikas, M. C. (2020). Assembly of the algal CO_2-fixing organelle, the pyrenoid, is guided by a Rubisco-binding motif. *Science Advances, 6*(46), eabd2408.

Meyer, T. J., Sheridan, M. V., & Sherman, B. D. (2017). Mechanisms of molecular water oxidation in solution and on oxide surfaces. *Chemical Society Reviews, 46*(20), 6148–6169.

Modestino, M. A., Dumortier, M., Hosseini Hashemi, S. M., Haussener, S., Moser, C., & Psaltis, D. (2015). Vapor-fed microfluidic hydrogen generator. *Lab on a Chip, 15*(10), 2287–2296.

Moia, D., Szumska, A., Vaissier, V., Planells, M., Robertson, N., O'Regan, B. C., Nelson, J., & Barnes, P. R. (2016). Interdye hole transport accelerates recombination in dye sensitized mesoporous films. *Journal of the American Chemical Society, 138*(40), 13197–131206.

Morris, A. J., Meyer, G. J., & Fujita, E. (2009). Molecular approaches to the photocatalytic reduction of carbon dioxide for solar fuels. *Accounts of Chemical Research, 42*(12), 1983–1994.

Muñoz-García, A. B., & Carter, E. A. (2012). Non-innocent dissociation of H_2O on GaP(110): Implications for electrochemical reduction of CO_2. *Journal of the American Chemical Society, 134*(33), 13600–13603.

Murray, R. W. (2008). Nanoelectrochemistry: Metal nanoparticles, nanoelectrodes, and nanopores. *Chemical Reviews, 108*(7), 2688–2720.

Nakazono, T., Parent, A. R., & Sakai, K. (2013). Cobalt porphyrins as homogeneous catalysts for water oxidation. *Chemical Communications, 49*(56), 6325–6327.

Nielander, A. C., Shaner, M. R., Papadantonakis, K. M., Francis, S. A., & Lewis, N. S. (2015). A taxonomy for solar fuels generators. *Energy & Environmental Science, 8*(1), 16–25.

Nitopi, S., Bertheussen, E., Scott, S. B., Liu, X., Engstfeld, A. K., Horch, S., Seger, B., Stephens, I. E., Chan, K., Hahn, C., Nørskov, J. K., Jaramillo, T. F., & Chorkendorff, I. (2019). Progress and perspectives of electrochemical CO_2 reduction on copper in aqueous electrolyte. *Chemical Reviews, 119*(12), 7610–7672.

Ogura, K. (1987). Catalytic conversion of carbon monoxide and carbon dioxide into methanol with photocells. *Journal of the Electrochemical Society, 134*(11), 2749–2754.

Ogura, K., & Takagi, M. (1986). Electrocatalytic reduction of carbon dioxide to methanol part III. Use of an electrochemical photocell. *Journal of Electroanalytical Chemistry and Interfacial Electrochemistry, 201*(2), 359–365.

Ola, O., & Maroto-Valer, M. M. (2015). Review of material design and reactor engineering on TiO_2 photocatalysis for CO_2 reduction. *Journal of Photochemistry and Photobiology C: Photochemistry Reviews, 24*, 16–42.

Olivo, A., Ghedini, E., Signoretto, M., Compagnoni, M., & Rossetti, I. (2017). Liquid vs. gas phase CO_2 photoreduction process: Which is the effect of the reaction medium? *Energies, 10*(9), 91394.

Omogo, B., Aldana, J. F., & Heyes, C. D. (2013). Radiative and nonradiative lifetime engineering of quantum dots in multiple solvents by surface atom stoichiometry and ligands. *The Journal of Physical Chemistry C, 117*(5), 2317–2327.

Parent, A. R., Crabtree, R. H., & Brudvig, G. W. (2013). Comparison of primary oxidants for water-oxidation catalysis. *Chemical Society Reviews, 42*(6), 2247–2252.

Parkinson, B. A., & Weaver, P. F. (1984). Photoelectrochemical pumping of enzymatic CO_2 reduction. *Nature*, *309*(5964), 148–149.

Peterson, A. A., Abild-Pedersen, F., Studt, F., Rossmeisl, J., & Nørskov, J. K. (2010). How copper catalyzes the electroreduction of carbon dioxide into hydrocarbon fuels. *Energy & Environmental Science*, *3*(9), 1311–1315.

Petit, J.-P., Chartier, P., Beley, M., & Deville, J.-P. (1989). Molecular catalysts in photoelectrochemical cells: Study of an efficient system for the selective photoelectroreduction of CO_2: p-GaP or p-GaAs/Ni(cyclam)$^{2+}$, aqueous medium. *Journal of Electroanalytical Chemistry and Interfacial Electrochemistry*, *269*(2), 267–281.

Pizzolato, E., Natali, M., Posocco, B., Montellano López, A., Bazzan, I., Di Valentin, M., Galloni, P., Conte, V., Bonchio, M., Scandola, F., & Sartorel, A. (2013). Light driven water oxidation by a single site cobalt salophen catalyst. *Chemical Communications*, *49*(85), 9941–9943.

Qi, Y., Zhao, Y., Gao, Y., Li, D., Li, Z., Zhang, F., & Li, C. (2018). Redox-based visible-light-driven Z-scheme overall water splitting with apparent quantum efficiency exceeding 10%. *Joule*, *2*(11), 2393–2402.

Rajeshwar, K. (2008). Hydrogen generation from irradiated semiconductor-liquid interfaces. In K. Rajeshwar, R. McConnell, & S. Licht (Eds.), *Solar hydrogen generation: Toward a renewable energy future* (pp. 167–228). Springer New York.

Ran, J., Jaroniec, M., & Qiao, S.-Z. (2018). Cocatalysts in semiconductor-based photocatalytic CO_2 reduction: Achievements, challenges, and opportunities. *Advanced Materials*, *30*(7), 1704649.

Reis Machado, A. S., & Nunes da Ponte, M. (2018). CO_2 capture and electrochemical conversion. Current opinion in green and sustainable. *Chemistry*, *11*, 86–90.

Ren, D., Deng, Y., Handoko, A. D., Chen, C. S., Malkhandi, S., & Yeo, B. S. (2015). Selective electrochemical reduction of carbon dioxide to ethylene and ethanol on copper(I) oxide catalysts. *ACS Catalysis*, *5*(5), 2814–2821.

Renger, G., & Renger, T. (2008). Photosystem II: The machinery of photosynthetic water splitting. *Photosynthesis Research*, *98*(1), 53–80.

Sampaio, R. N., Müller, A. V., Polo, A. S., & Meyer, G. J. (2017). Correlation between charge recombination and lateral hole-hopping kinetics in a series of *cis*-Ru(phen′)(dcb)$(NCS)_2$ dye-sensitized solar cells. *ACS Applied Materials & Interfaces*, *9*(39), 33446–33454.

Schreier, M., Curvat, L., Giordano, F., Steier, L., Abate, A., Zakeeruddin, S. M., Luo, J., Mayer, M. T., & Grätzel, M. (2015). Efficient photosynthesis of carbon monoxide from CO_2 using perovskite photovoltaics. *Nature Communications*, *6*, 7326.

Şener, M., Strümpfer, J., Hsin, J., Chandler, D., Scheuring, S., Hunter, C. N., & Schulten, K. (2011). Förster energy transfer Theory as reflected in the structures of photosynthetic light-harvesting systems. *ChemPhysChem*, *12*(3), 518–531.

Shan, B., Brennaman, M. K., Troian-Gautier, L., Liu, Y., Nayak, A., Klug, C. M., Li, T. T., Bullock, R. M., & Meyer, T. J. (2019). A silicon-based heterojunction integrated with a molecular excited state in a water-splitting tandem cell. *Journal of the American Chemical Society*, *141*(26), 10390–10398.

Sherman, B. D., Ashford, D. L., Lapides, A. M., Sheridan, M. V., Wee, K.-R., & Meyer, T. J. (2015). Light-driven water splitting with a molecular electroassembly-based core/shell photoanode. *The Journal of Physical Chemistry Letters*, *6*(16), 3213–3217.

Singh, A., & Spiccia, L. (2013). Water oxidation catalysts based on abundant 1st row transition metals. *Coordination Chemistry Reviews*, *257*(17), 2607–2622.

Smieja, J. M., Sampson, M. D., Grice, K. A., Benson, E. E., Froehlich, J. D., & Kubiak, C. P. (2013). Manganese as a substitute for rhenium in CO_2 reduction catalysts: The importance of acids. *Inorganic Chemistry*, *52*(5), 2484–2491.

Song, W., Ito, A., Binstead, R. A., Hanson, K., Luo, H., Brennaman, M. K., Concepcion, J. J., & Meyer, T. J. (2013). Accumulation of multiple oxidative equivalents at a single site by cross-surface electron transfer on TiO_2. *Journal of the American Chemical Society*, *135*(31), 11587–11594.

Spitler, M. T. (2006). In M. T. Spitler & F. Willig (Eds.), *Physical constraints on the redox growth of nanoparticles in solution* (Vol. 6325, pp. 632508). SPIE.

Spitler, M. T., Modestino, M. A., Deutsch, T. G., Xiang, C. X., Durrant, J. R., Esposito, D. V., Haussener, S., Maldonado, S., Sharp, I. D., Parkinson, B. A., Ginley, D. S., Houle, F. A., Hannappel, T., Neale, N. R., Nocera, D. G., & McIntyre, P. C. (2020). Practical challenges in the development of photoelectrochemical solar fuels production. *Sustainable Energy & Fuels*, *4*(3), 985–995.

Sprafke, A. N., & Wehrspohn, R. B. (2015). Current concepts for optical path enhancement in solar cells. *Photon Management in Solar Cells*, 1–20.

Stolarczyk, J. K., Bhattacharyya, S., Polavarapu, L., & Feldmann, J. (2018). Challenges and prospects in solar water splitting and CO_2 reduction with inorganic and hybrid nanostructures. *ACS Catalysis*, *8*(4), 3602–3635.

Strange, L. E., Yadav, J., Li, X., & Pan, S. (2020). Editors' choice—Review—Creating electrocatalytic heterojunctions for efficient photoelectrochemical CO_2 reduction to chemical fuels. *Journal of the Electrochemical Society*, *167*(14), 146518.

Swierk, J. R., & Mallouk, T. E. (2013). Design and development of photoanodes for water-splitting dye-sensitized photoelectrochemical cells. *Chemical Society Reviews*, *42*(6), 2357–2387.

Swierk, J. R., McCool, N. S., Saunders, T. P., Barber, G. D., Strayer, M. E., Vargas-Barbosa, N. M., & Mallouk, T. E. (2014). Photovoltage effects of sintered IrO_2 nanoparticle catalysts in water-splitting dye-sensitized photoelectrochemical cells. *The Journal of Physical Chemistry C*, *118*(30), 17046–17053.

Swierk, J. R., Méndez-Hernández, D. D., McCool, N. S., Liddell, P., Terazono, Y., Pahk, I., Tomlin, J. J., Oster, N. V., Moore, T. A., Moore, A. L., Gust, D., & Mallouk, T. E. (2015). Metal-free organic sensitizers for use in water-splitting dye-sensitized photoelectrochemical cells. *Proceedings of the National Academy of Sciences*, *112*(6), 1681–1686.

Takeda, H., Cometto, C., Ishitani, O., & Robert, M. (2017). Electrons, photons, protons and earth-abundant metal complexes for molecular catalysis of CO_2 reduction. *ACS Catalysis*, *7*(1), 70–88.

Tan, J. Z. Y., & Maroto-Valer, M. M. (2019). A review of nanostructured non-titania photocatalysts and hole scavenging agents for CO_2 photoreduction processes. *Journal of Materials Chemistry A*, *7*(16), 9368–9385.

Tan, M. X., Kenyon, C. N., & Lewis, N. S. (1994). Experimental measurement of quasi-Fermi levels at an illuminated semiconductor/liquid contact. *The Journal of Physical Chemistry*, *98*(19), 4959–4962.

Tani, T. (1995). *Photographic sensitivity*. Oxford University Press.

Turmo, A., Gonzalez-Esquer, C. R., & Kerfeld, C. A. (2017). Carboxysomes: Metabolic modules for CO_2 fixation. *FEMS Microbiology Letters*, *364*(18), fnx176.

Ullman, A. M., Liu, Y., Huynh, M., Bediako, D. K., Wang, H., Anderson, B. L., Powers, D. C., Breen, J. J., Abruña, H. D., & Nocera, D. G. (2014). Water oxidation catalysis by Co(II) impurities in $Co(III)_4O_4$ cubanes. *Journal of the American Chemical Society*, *136*(50), 17681–17688.

Vitlin Gruber, A., & Feiz, L. (2018). Rubisco assembly in the chloroplast. *Frontiers in Molecular Biosciences*, *5*(24), 00024.

Wang, D., Eberhart, M. S., Sheridan, M. V., Hu, K., Sherman, B. D., Nayak, A., Wang, Y., Marquard, S. L., Dares, C. J., & Meyer, T. J. (2018). Stabilized photoanodes for water oxidation by integration of organic dyes, water oxidation catalysts, and electron-transfer mediators. *Proceedings of the National Academy of Sciences*, *115*(34), 8523.

Wang, D., & Groves, J. T. (2013). Efficient water oxidation catalyzed by homogeneous cationic cobalt porphyrins with critical roles for the buffer base. *Proceedings of the National Academy of Sciences*, *110*(39), 15579–15584.

Wang, M., Yang, Y., Shen, J., Jiang, J., & Sun, L. (2017). Visible-light-absorbing semiconductor/molecular catalyst hybrid photoelectrodes for H_2 or O_2 evolution: Recent advances and challenges. *Sustainable Energy & Fuels*, *1*(8), 1641–1663.

Wang, Q., & Domen, K. (2020). Particulate photocatalysts for light-driven water splitting: Mechanisms, challenges, and design strategies. *Chemical Reviews*, *120*(2), 919–985.

Wang, Y., Suzuki, H., Xie, J., Tomita, O., Martin, D. J., Higashi, M., Kong, D., Abe, R., & Tang, J. (2018). Mimicking natural photosynthesis: Solar to renewable H_2 fuel synthesis by Z-scheme water splitting systems. *Chemical Reviews*, *118*(10), 5201–5241.

Wasylenko, D. J., Ganesamoorthy, C., Borau-Garcia, J., & Berlinguette, C. P. (2011). Electrochemical evidence for catalyticwater oxidation mediated by a high-valent cobalt complex. *Chemical Communications*, *47*(14), 4249–4251.

Watkins, K. J., Parkinson, B. A., & Spitler, M. T. (2015). Physical models for charge transfer at single crystal oxide semiconductor surfaces as revealed by the doping density dependence of the collection efficiency of dye sensitized photocurrents. *The Journal of Physical Chemistry B*, *119*(24), 7579–7588.

White, J. L., Baruch, M. F., Pander, J. E., Hu, Y., Fortmeyer, I. C., Park, J. E., Zhang, T., Liao, K., Gu, J., Yan, Y., Shaw, T. W., Abelev, E., & Bocarsly, A. B. (2015). Light-driven heterogeneous reduction of carbon dioxide: Photocatalysts and photoelectrodes. *Chemical Reviews*, *115*(23), 12888–12935.

Wickramasinghe, L. D., Zhou, R., Zong, R., Vo, P., Gagnon, K. J., & Thummel, R. P. (2015). Iron complexes of square planar tetradentate polypyridyl-type ligands as catalysts for water oxidation. *Journal of the American Chemical Society*, *137*(41), 13260–13263.

de Wijn, R., & van Gorkom, H. J. (2001). Kinetics of electron transfer from QA to QB in photosystem II. *Biochemistry*, *40*(39), 11912–11922.

Winkler, K. (1995). The kinetics of electron transfer in $Fe(CN)_6^{3-/4-}$ redox system on platinum standard-size and ultramicroelectrodes. *Journal of Electroanalytical Chemistry*, *388*(1), 151–159.

Winkler, M. T., Cox, C. R., Nocera, D. G., & Buonassisi, T. (2013). Modeling integrated photovoltaic–electrochemical devices using steady-state equivalent circuits. *Proceedings of the National Academy of Sciences*, *110*(12), E1076–E1082.

Woodhouse, M., & Parkinson, B. A. (2008). Combinatorial discovery and optimization of a complex oxide with water photoelectrolysis activity. *Chemistry of Materials*, *20*(7), 2495–2502.

Wu, H.-L., Li, X.-B., Tung, C.-H., & Wu, L.-Z. (2019). Semiconductor quantum dots: An emerging candidate for CO_2 photoreduction. *Advanced Materials*, *31*(36), 1900709.

Wu, X., Li, F., Zhang, B., & Sun, L. (2015). Molecular complexes in water oxidation: Pre-catalysts or real catalysts. *Journal of Photochemistry and Photobiology C: Photochemistry Reviews*, *25*, 71–89.

Xiang, C., Chen, Y., & Lewis, N. S. (2013). Modeling an integrated photoelectrolysis system sustained by water vapor. *Energy & Environmental Science*, *6*(12), 3713–3721.

Xiao, M., Wang, Z., Lyu, M., Luo, B., Wang, S., Liu, G., Cheng, H. M., & Wang, L. (2019). Hollow nanostructures for photocatalysis: Advantages and challenges. *Advanced Materials*, *31*(38), 1801369.

Xu, S., & Carter, E. A. (2019). Theoretical insights into heterogeneous (photo)electrochemical CO_2 reduction. *Chemical Reviews*, *119*(11), 6631–6669.

Xu, S., & Carter, E. A. (2021). CO_2 photoelectrochemical reduction catalyzed by a GaP(001) photoelectrode. *ACS Catalysis*, *11*(3), 1233–1241.

Yamazaki, Y., Takeda, H., & Ishitani, O. (2015). Photocatalytic reduction of CO_2 using metal complexes. *Journal of Photochemistry and Photobiology C: Photochemistry Reviews*, *25*, 106–137.

Yokota, A. (2017). Revisiting RuBisCO. *Bioscience, Biotechnology, and Biochemistry*, *81*(11), 2039–2049.

Yoneyama, H., Sugimura, K., & Kuwabata, S. (1988). Effects of electrolytes on the photoelectrochemical reduction of carbon dioxide at illuminated p-type cadmium telluride and p-type indium phosphide electrodes in aqueous solutions. *Journal of Electroanalytical Chemistry and Interfacial Electrochemistry*, *249*(1), 143–153.

Yuan, J., & Hao, C. (2013). Solar-driven photoelectrochemical reduction of carbon dioxide to methanol at $CuInS_2$ thin film photocathode. *Solar Energy Materials and Solar Cells*, *108*, 170–174.

Yuan, J., Wang, P., Hao, C., & Yu, G. (2016). Photoelectrochemical reduction of carbon dioxide at $CuInS_2$/graphene hybrid thin film electrode. *Electrochimica Acta*, *193*, 1–6.

Zafrir, M., Ulman, M., Zuckerman, Y., & Halmann, M. (1983). Photoelectrochemical reduction of carbon dioxide to formic acid, formaldehyde and methanol on p-gallium arsenide in an aqueous V(II)–V(III) chloride redox system. *Journal of Electroanalytical Chemistry and Interfacial Electrochemistry*, *159*(2), 373–389.

Zaharieva, I., & Dau, H. (2019). Energetics and kinetics of S-state transitions monitored by delayed chlorophyll fluorescence. *Frontiers in Plant Science*, *10*(386), 00386.

Zhang, B., & Sun, L. (2019). Artificial photosynthesis: Opportunities and challenges of molecular catalysts. *Chemical Society Reviews*, *48*(7), 2216–22164.

Zhao, J., Holmes, M. A., & Osterloh, F. E. (2013). Quantum confinement controls photocatalysis: A free energy analysis for photocatalytic proton reduction at CdSe nanocrystals. *ACS Nano*, *7*(5), 4316–4325.

Zhao, Y., Swierk, J. R., Megiatto, J. D., Sherman, B., Youngblood, W. J., Qin, D., Lentz, D. M., Moore, A. L., Moore, T. A., Gust, D., & Mallouk, T. E. (2012). Improving the efficiency of water splitting in dye-sensitized solar cells by using a biomimetic electron transfer mediator. *Proceedings of the National Academy of Sciences*, *109*(39), 15612.

Zhou, L., Shinde, A., Guevarra, D., Richter, M. H., Stein, H. S., Wang, Y., Newhouse, P. F., Persson, K. A., & Gregoire, J. M. (2020). Combinatorial screening yields discovery of 29 metal oxide photoanodes for solar fuel generation. *Journal of Materials Chemistry A*, *8*(8), 4239–4243.

Zhu, H., Yang, Y., Hyeon-Deuk, K., Califano, M., Song, N., Wang, Y., Zhang, W., Prezhdo, O. V., & Lian, T. (2014). Auger-assisted electron transfer from photoexcited semiconductor quantum dots. *Nano Letters*, *14*(3), 1263–1269.

Zong, R., & Thummel, R. P. (2005). A new family of Ru complexes for water oxidation. *Journal of the American Chemical Society*, *127*(37), 12802–12803.

Zoski, C. G., Oldham, K. B., Mahon, P. J., Henderson, T. L. E., & Bond, A. M. (1991). Global kinetic analysis of cyclic voltammograms at a spherical electrode. *Journal of Electroanalytical Chemistry and Interfacial Electrochemistry*, *297*(1), 1–17.

7

Utilizing the Band Diagram Framework to Interpret the Operation of Photoelectrochemical Cells

Kirk H. Bevan, Botong Miao, and Asif Iqbal

Division of Materials Engineering, Faculty of Engineering, McGill University, H3A 0C5, QC, 3610 Rue University, Montreal, Canada

In this chapter, we explore the fundamental semiconductor concepts, which underpin the operation of photoelectrochemical (PEC) devices. The purpose of this chapter is to demonstrate how the current–voltage characteristics of PEC devices can be understood by utilizing the "band diagram" methodology commonly applied in the semiconductor device community (Memming, 2015; Pierret, 1996, 2002). This is accomplished through the application of both analytical and numerical modeling methods. To this end, we begin by overviewing foundational semiconductor concepts in Section 7.1, covering topics such as carrier generation and recombination. Subsequently, in Section 7.2 we review how one can understand the operational current–voltage characteristics of PEC devices in the dark through the use of band diagrams. In this "dark current" context, we also overview the underlying assumptions present in the band diagram framework. A short summary of reference electrodes is also provided. This discussion is then extended to illuminated PEC devices through an exploration of two analytical models in Section 7.3 (Gartner, 1959; Peter, 2013; Peter and Wijayantha, 2014). An emphasis is placed on understanding the phenomenological factors that impact upon the observable photocurrent as a function of the applied bias, such as photon absorption, carrier recombination, and hole transfer. Lastly, in Section 7.4, we explore how the operation of PEC devices can be further understood through the use of semiclassical device modeling. Concepts pertaining to the photovoltage and current–voltage characteristics are explained with the aid of numerically calculated band diagrams (Iqbal, 2018; Iqbal and Bevan, 2018a,b; Miao et al., 2019). To illustrate all of these concepts clearly, an emphasis is placed on analyzing photoanodes (Peter, 2013; Peter and Wijayantha, 2014; Xu et al., 2019). However, the approach is general and may also be extended to photocathodes with appropriate modifications (Bagal et al., 2019).

7.1 Semiconductor Concepts

Before exploring the behavior of semiconductor–liquid junctions, it is instructive to first overview basic semiconductor concepts. At finite temperatures a semiconductor consists of a valence band that is mostly filled with electrons and a nearly empty conduction band separated by a band gap E_G as shown in Figure 7.1a (Memming, 2015; Pierret, 1996). In a perfect "intrinsic" semiconductor there are no states that an electron can occupy within the band gap. At room temperature, due to thermal excitation there are always free electrons (charge carriers) in the conduction band above its minimum (E_C) and free holes in the valence band below its maximum (E_V). These electrons and holes move about freely within their respective bands, with holes behaving as positive charge carriers (Pierret, 1996, 2002). Both electrons and holes contribute to the overall conduction properties of a semiconductor. Under equilibrium conditions, the product of holes (p) and electrons (n) inside a semiconductor is generally given by

$$n_i^2 = np, \tag{7.1}$$

where n_i refers to the intrinsic carrier concentration. The relative concentration of each is thus

$$n = N_C e^{-(E_C - E_F)/k_B T} \quad \text{with} \quad N_C = 2\left(\frac{2\pi m_n^* k_B T}{h^2}\right)^{3/2} \tag{7.2}$$

Conversion of Water and CO_2 to Fuels using Solar Energy: Science, Technology and Materials, First Edition.
Edited by Oomman K. Varghese and Flavio L. Souza.

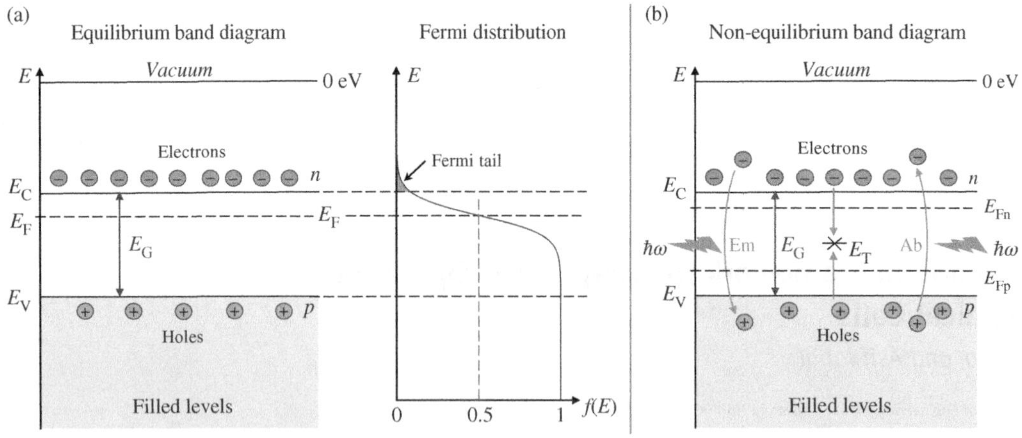

Figure 7.1 (a) Equilibrium energy band diagram of an n-type semiconductor juxtaposed against the Fermi–Dirac distribution $f(E)$. The conduction band minimum and valence band maximum are denoted as E_C and E_V, respectively. The Fermi level (E_F) is situated inside the band gap $E_G = E_C - E_V$. Below E_V the valence band is largely filled by electrons (shaded region), while above E_C only conducting electrons are present. All energies are taken as negative below the reference energy of vacuum. (b) Illuminated nonequilibrium energy band diagram for an isolated n-type semiconductor. Photons with an energy greater than the band gap ($\hbar\omega > E_G$) are absorbed (Ab) and excite electron–hole pairs. Electrons are denoted as negative circles, while holes are denoted as positive circles. Electron–hole pairs may recombine through radiative emission (Em) or via a trap energy level located at E_T. Quasi-Fermi levels E_{Fn} and E_{Fp} are utilized to describe the nonequilibrium electron and hole distributions, respectively.

and

$$p = N_V e^{-(E_F - E_V)/k_B T} \quad \text{with} \quad N_V = 2\left(\frac{2\pi m_p^* k_B T}{h^2}\right)^{3/2}. \tag{7.3}$$

Here, E_F is the Fermi level (or electrochemical potential) of electrons inside the semiconductor, k_B is Boltzmann's constant, h is Planck's constant, and T is the temperature. Electrons and holes move about their bands with "effective masses" of m_n^* and m_p^*, respectively, which further relate to the "effective density of states" for both electrons and holes given by N_C and N_V (Pierret, 1996). Note, here $k_B T$ is ascribed units of electron volts (eV). These expressions arise from assuming Boltzmann statistics for the electrons in the conduction band and holes in the valence band (Pierret, 1996, 2002). That is, they assume that only the "tails" of the Fermi–Dirac distribution $f(E) = 1/[1 + \exp((E - E_F)/k_B T)]$ and $[1 - f(E)]$, sketched in Figure 7.1a, respectively, overlap with the bottom of the conduction band and top of the valence band (Pierret, 2002). The effective density of states expressions for electrons (N_C) and holes (N_V) are parameterized by fitting the known bulk density of states near the band edges via the effective mass terms m_n^* and m_p^* (Pierret, 1996, 2002). Finally, it is worth mentioning that the notion of free charge carriers can become more limited in oxides when strong carrier–lattice interactions arise, which can complicate the application of such semiconductor concepts to PEC devices employing oxides (Iqbal et al., 2018).

As mentioned, a semiconductor without impurities/dopants is referred to as "intrinsic." A semiconductor with a concentration of impurities that donate electrons (N_D) to the conduction band is referred to as "n-type" and has the relative equilibrium carrier concentrations given by

$$\text{n-type: } n \approx N_D \Rightarrow p \approx n_i^2/N_D. \tag{7.4}$$

Such "donor" impurities create a situation where electrons are the majority carriers and holes are the minority carrier (Pierret, 2002). Likewise, a semiconductor with a concentration of impurities that accept electrons (N_A) from the conduction band is referred to as "p-type" and has relative equilibrium carrier concentrations given by

$$\text{p-type: } p \approx N_A \Rightarrow n \approx n_i^2/N_A. \tag{7.5}$$

Likewise, "acceptor" impurities create a situation where holes become the majority carrier and electrons the minority carrier (Pierret, 2002). To reflect that there is a change in the carrier concentration during doping, under equilibrium conditions the Fermi level rises toward the conduction band when donors are added and lowers to the valence band when acceptors are added. This is dictated by Equations 7.2 and 7.3, that are only strictly valid outside the "heavily doped regime" which is usually quantified by $(E_C - E_F) > 3k_B T$ and $(E_F - E_V) > 3k_B T$. When a semiconductor is "heavily doped" the Boltzmann approximations, from which Equations 7.2 and 7.3 are developed, no longer strictly apply since the Fermi distribution begins to encroach substantially on the semiconductor band edges. The equilibrium band diagram for a non-degenerate n-type semiconductor is shown in Figure 7.1a. The primary assumption in doping, to ensure that Equations 7.4 and 7.5 are obeyed, is that the dopant atoms have defect levels within $\sim k_B T$ from the band edges and are all therefore ionized at room temperature. This means that acceptor dopant levels will lie $\sim k_B T$ above E_V in order to accept electrons from the valence band and create holes, while donor dopant levels will lie $\sim k_B T$ below E_C in order to donate electrons to the conduction band. Due to this close proximity to the band edges, we do not typically draw dopant levels in a band diagram to avoid any confusion with defect trap states and recombination centers (Pierret, 2002).

Light absorption (Ab) by a semiconductor occurs primarily by promoting electrons across the band gap (see Figure 7.1b). Each photon absorbed thus produces one electron–hole pair. Photons with energies below the band gap are not absorbed, if one ignores defect assisted absorption, while those with an energy greater than or equal to the band gap can be absorbed as given by

$$\hbar\omega \geq E_G, \tag{7.6}$$

where ω is the angular frequency of the photon and $\hbar = h/2\pi$ (see Figure 7.1). This is a critical efficiency consideration in the operation of any semiconductor device seeking to utilize photoexcited electron–hole pairs generated to "do work" (Nelson, 2003; Pierret, 1996). Examples of such devices include solar cells and PEC cells (Nelson, 2003; Peter, 2013; Peter and Wijayantha, 2014; Pierret, 1996), the latter being the focus of this discussion.

In order to have photogenerated electron–hole pairs "do work," they must be separated spatially by an internal electric field. The process of forming internal electric fields in a PEC device will be discussed shortly, but for the remainder of this overview, let us focus on the mechanisms by which electron–hole pairs can recombine without "doing work" (Pierret, 2002). First, there is radiative recombination whereby electron–hole pairs rejoin and emit (Em) light in the process (see Figure 7.1b). Second, there is non-radiative recombination, whereby electron–hole pairs rejoin and emit heat in the process (also shown in Figure 7.1b). It is important to touch upon these two concepts, as they significantly impact the performance of PEC devices significantly. Auger recombination may also be present in a semiconductor (Pierret, 2002), but is believed to be a minor consideration in PEC cells and shall not be reviewed here.

Band-to-band radiative recombination is the primary mechanism by which electron–hole pairs lose their energy via light emission, as illustrated in Figure 7.1b (Pierret, 1996; Vasileska and Goodnick, 2006). In this process, electrons "jump down" from the conduction band to the valence band across the band gap (E_G) to recombine with holes. The energy lost through recombination must go somewhere (to maintain energy conservation) and in this mechanism it is released as a photon of light equal to the semiconductor band gap $\hbar\omega = E_G$. The amount of radiative recombination depends on both the density of electrons (n) and holes (p), such that the overall band-to-band (b) radiative recombination rate is given by

$$R_b = B(np - n_i^2), \tag{7.7}$$

where B is the bimolecular recombination constant, which is material dependent (Pierret, 1996, 2002; Vasileska and Goodnick, 2006). A close inspection of Equations 7.1 and 7.7 reveals that there is no net radiative emission in equilibrium (apart from blackbody radiation). Net radiative emission through electron–hole recombination is attained when a system is operating under nonequilibrium conditions (e.g., participating in current flow and/or light absorption). Formally speaking, local nonequilibrium may be defined by $np > n_i^2$. More generally, we define the nonequilibrium electron and hole concentrations as

$$n = N_C e^{-(E_C - E_{Fn})/k_B T} \tag{7.8}$$

and

$$p = N_V e^{-(E_{Fp} - E_V)/k_B T}. \tag{7.9}$$

To arrive at these definitions, we have introduced both electron and hole quasi-Fermi levels defined as E_{Fn} and E_{Fp}, respectively, which are also plotted in Figure 7.1b. The presence of distinct quasi-Fermi (for electrons and holes) levels is a key defining feature of a semiconductor operating under nonequilibrium conditions and is essential to understanding device performance (Pierret, 1996). The quasi-Fermi levels present in Figure 7.1b are induced through light absorption within the semiconductor shown. Excellent explanatory discussions on quasi-Fermi levels can be found in Pierret (1996) and Pierret (2002).

Generally, trap-assisted recombination is the most dominant non-radiative recombination mechanism in semiconductors with a high density of defects, such as those typically utilized in PEC cells (Iqbal et al., 2016; Li et al., 2021b; Miao et al., 2020). Often it is referred to as Shockley–Read–Hall (SRH) recombination. In this heat releasing mechanism (via phonons), electrons and holes recombine non-radiatively at a trap site with an energy level E_T within the band gap as shown in Figure 7.1b. A detailed derivation of the SRH recombination rate can be found in Pierret (2002), for the sake of brevity we provide it as

$$R_{SRH} = \frac{np - n_i^2}{\tau_n(n + n_i) + \tau_p(p + p_i)}. \tag{7.10}$$

Here, τ_n and τ_p are the electron and hole recombination lifetimes. Moreover, this particular formulation arises from the assumption that any trap states are located mid-gap close to the intrinsic Fermi level (E_i) (Pierret, 2002). The intrinsic Fermi level is denoted by $E_i = (E_C + E_V)/2 + k_B T \ln(\sqrt{N_C/N_V})/2$. Both non-radiative and radiative recombination mechanisms are crucial to developing an understanding of how PEC cells operate.

7.2 Semiconductor–Liquid Junctions in the Dark

Before forming a PEC device, we must first decide on methodology for referencing the electrochemical potentials (and energy levels) in the isolated semiconductor and liquid regions prior to forming a junction. Let us consider the n-type semiconductor and an aqueous solution before contact as shown in Figure 7.2a. In the semiconductor, we align all of the relevant energies at negative values below a vacuum reference set at 0 eV, including: E_C, E_V, and E_F. This is the “band diagram” convention in the solid-state physics community and is key to understanding the current–voltage properties of semiconductor devices (Pierret, 1996). The negative values reflect the fact that these are bound states, particularly for E_C and E_V, to which

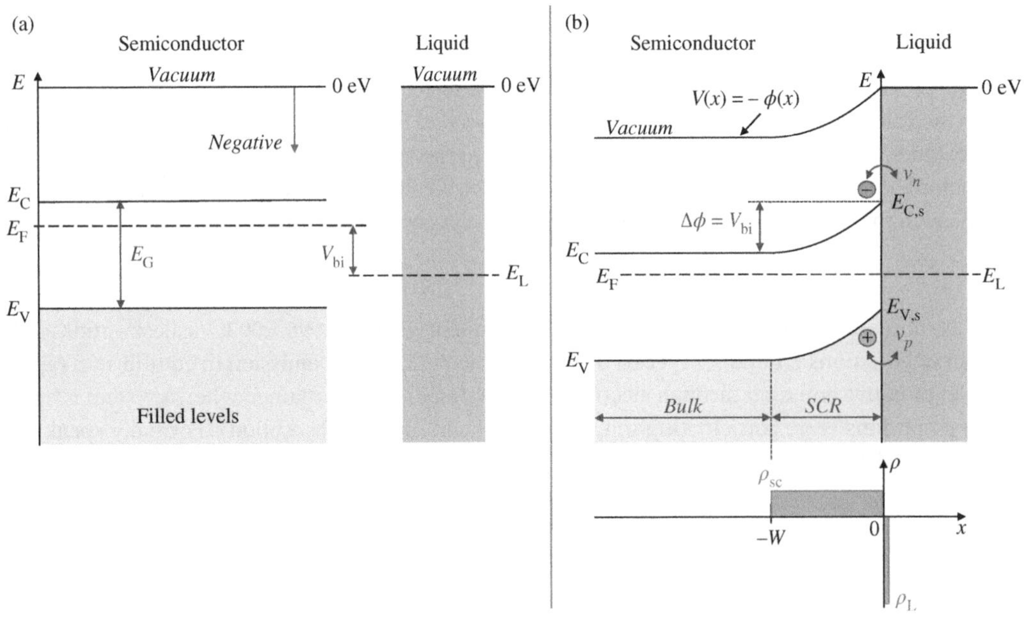

Figure 7.2 (a) Isolated juxtaposed equilibrium band diagrams for the semiconductor and the liquid region prior to contact. Prior to contact the semiconductor Fermi level (E_F) and liquid electrochemical potential (E_L) are separated by what will become the built-in voltage upon contact (V_{bi}). (b) Charge transfer occurs to facilitate equilibrium ($E_F = E_L$) when the semiconductor and liquid regions form a contact, facilitated by interfacial electron/hole transfer velocities $v_{n,p}$. This results in equal and opposite charge densities (ρ) on the semiconductor (ρ_{sc}) and liquid (ρ_L) sides as a function of position x through the junction. Since the semiconductor region typically has a much longer screening length than the liquid region much of the electrostatic potential drop (ϕ) created by these counter charges occurs in the portion of the semiconductor termed the “space charge region” (SCR) of width W. Under equilibrium conditions the absolute potential drop ($\Delta\phi$) across the interface is equal to the built-in voltage ($\Delta\phi = V_{bi}$). The band edges at the surface ($E_{C,s}$ and $E_{V,s}$) are shifted relative to those in the bulk by $\Delta\phi$.

an electron can be placed if "lowered" from a position in "vacuum" far away from the semiconductor. This vacuum reference picture (set at 0 eV), common in the device physics community, is discussed at length in Pierret (1996). Beyond defining bound states, to arrive at a band diagram description of the semiconductor–liquid interface, we must also define electrochemical potentials for each region. The semiconductor electrochemical potential (E_F) can be determined through occupation statistics (Pierret, 2002), following the Fermi–Dirac distribution discussed earlier (see Figure 7.1). The liquid electrochemical potential (E_L, as shown in Figure 7.2a) is a more subtle and complex subject that we shall not delve into. Interested readers are encouraged to consult Memming (2015), Smith and Stevenson (2007), and van de Krol (2012a,b). As an example, one might assume that the liquid is a very basic aqueous solution with pH $\approx$ 14, in which case E_L can be approximately aligned with the high concentration OH^-/O_2 redox couple (Iqbal, 2018; Memming, 2015; van de Krol, 2012a,b). Additionally, we shall not worry about the multitude of bound energy levels in the liquid (ion and solvent species bound states), as this will also greatly expand the discussion (Memming, 2015). This is not to imply that the topic is not important and deserving of a detailed analysis.

7.2.1 Charge Equilibration in the Dark

The key takeaway is that both the liquid and semiconductor region have electrochemical potentials, referenced with respect to vacuum at 0 eV, that must align ($E_F = E_L$) upon contact for "equilibrium" to be established as illustrated in Figure 7.2b (Bevan et al., 2021). This alignment is accomplished through charge transfer of electrons and holes across the semiconductor–liquid interface as determined by the velocity terms v_n and v_p, respectively (see Figure 7.2b). We utilize velocity terms, rather than reaction rate constants, because this turns out to more convenient for the purposes of computing current–voltage characteristics in a device model (Iqbal and Bevan, 2018a,b). Nevertheless, if one assumes that charge transfer occurs over a surface density of carriers with an interfacial region of width l (on the order of a few lattice constants or so) then the corresponding reaction rate terms become $k_n = v_n/l$ and $k_p = v_p/l$. In the end, this comes down to a choice of convention. It should be mentioned, however, that the surface semiconductor band edges ($E_{C,s}$ and $E_{V,s}$ in Figure 7.2b) should straddle the redox potentials of water (not shown) when the purpose of such a cell is to drive water splitting. A detailed discussion on this topic can be found in van de Krol (2012a).

The transfer of charge between the semiconductor and liquid regions, associated with establishing equilibrium ($E_F = E_L$), is physically present in the form of charged regions on opposite sides of the junction (see Figure 7.2b, bottom). In the n-type semiconductor example chosen, the donor atoms (with a density of N_D) lose some of their screening electrons to form a space charge region (SCR). This is plotted as a semiconductor (*sc*) charge density ρ_{sc} in Figure 7.2b. A region of precisely equal and opposite charge builds up on the liquid (*L*) side as required by charge conservation (ρ_L charge density in Figure 7.2b). These charges are responsible for forming the built-in voltage drop (V_{bi}) that aligns the electrochemical potentials of each region in Figure 7.2b. In general, the absolute potential drop in the semiconductor region can be defined as $\Delta\phi$ as shown in Figure 7.2b. From $\Delta\phi$ the width (W) of the SCR shown in Figure 7.2b can be computed as Memming (2015)

$$W = \sqrt{\frac{2\epsilon_o \epsilon_r \Delta\phi}{qN_D}}. \tag{7.11}$$

In this expression ϵ_o is the permittivity of free space, ϵ_r is the relative dielectric constant of the semiconductor, and N_D is its doping concentration (Gartner, 1959; Pierret, 1996). Later, we shall explore how this SCR width (W) is modulated under active device operation.

Note, however, that the overall electrostatic potential drop $V = -\phi$ due to the charges in the semiconductor and the liquid regions in Figure 7.2b (as summarized by the potential profile of the vacuum reference) occurs almost entirely across the semiconductor region. This arises because we have assumed that the concentration of supporting electrolyte ions, such as Na^+ and Cl^-, is much greater than that of dopants in the semiconductor (Memming, 2015; Schmickler, 1996). This means that the screening length of the semiconductor region will be much longer than that of the liquid, and, consequently, the semiconductor will absorb nearly all of the potential drop arising from charge transfer. Of course, our description of the screening process in Figure 7.2b simplifies the complex interfacial dynamics in the liquid that includes both inner Helmholtz plane and outer Helmholtz plane contributions (denoted as IHP and OHP, respectively, in Figure 7.3 (Brown et al., 2016)). Adsorption of H^+ and OH^- species in the IHP can also lead to a fixed bias independent (though pH dependent) potential drop at the semiconductor liquid interface. For the sake of brevity, we shall not discuss such adsorption phenomena here but instead refer the reader to the informative discussions in van de Krol (2012a,b).

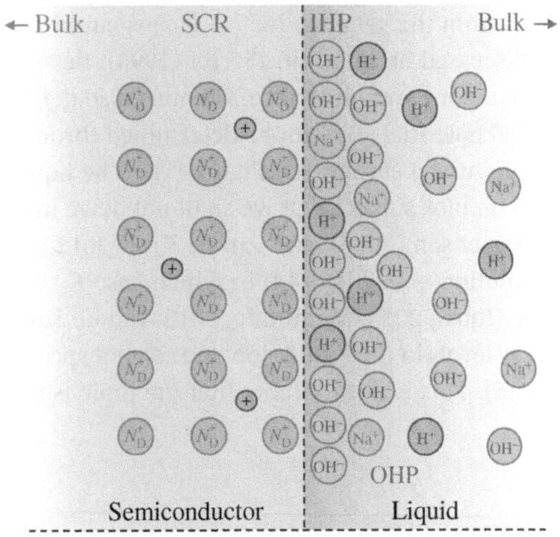

Figure 7.3 Microscopic view on charges at the semiconductor–liquid interface. Within an n-type anode the SCR is largely comprised of ionized donor atoms (N_D^+ circles). While screening in the liquid region is formed though the accumulation of electrolyte ions within both the inner and outer Helmholtz planes (IHP and OHP, respectively). Note charge screening within the liquid region is a complex process that is only phenomenologically illustrated here (Brown et al., 2016; Memming, 2015).

7.2.2 Semiconductor–Liquid Junctions Under Bias in the Dark

Under an applied bias the same conditions persist, with the screening length of the semiconductor being much longer than that of the liquid region. Thus, when a bias is applied (V_A) from an external source the vast majority of the potential drop change occurs across the semiconductor region. Consider first the action of a reverse applied bias ($V_A < 0$) with the respect to the equilibrium band alignment in Figure 7.4a. Here, in Figure 7.4b a reverse applied ($V_A < 0$) is shown to lift the semiconductor band edges in bulk closer to the liquid vacuum level. Conversely, in Figure 7.4c a forward applied bias ($V_A > 0$) pulls down the bulk semiconductor band edges further from the liquid vacuum. In both bias directions, the majority of the voltage drop occurs in the semiconductor region. Here, we do not discuss the relative bias fractions dropping across the working (semiconductor) and counter electrodes, the V_A we refer to is the potential, which drops across the semiconductor electrode (Memming, 2015). For a detailed discussion on working and counter electrodes see Memming (2015). The region over which the potential drops is a crucial point that distinguishes semiconductor–liquid junctions from metal–liquid junctions, where the latter are usually investigated in electrocatalysis (Memming, 2015). In a metal–liquid junction, we assume that all of the bias drop at the solid–liquid junction occurs in the IHP and OHP (see Figure 7.3). This assumption is justified because the screening length of elemental metals (e.g., Pt) is so short as to be around an atomic layer (Schmickler, 1996). However, in semiconductor–liquid junctions the screening length of the semiconductor is "generally" much longer than that of the liquid such that the great majority of V_A drops across the semiconductor region. In this context, we use the term "generally" because the screening length of a semiconductor can be shortened through extremely heavy doping and/or by biasing the semiconductor such that its surface carrier concentrations reach the accumulation or inversion regimes (Pierret, 1996)—surface states may also impact on screening dynamics (Iqbal et al., 2016; Miao et al., 2020). We will not consider these regimes, as they complicate the essence of our discussion but are important to be aware of nevertheless (Gao and Hamann, 2017; Iqbal et al., 2016; Li et al., 2021b; Miao et al., 2020).

These screening assumptions have a dramatic impact on how we understand charge transfer across semiconductor–liquid interfaces. Since the potential drop in the liquid region is largely independent of the applied bias, one may further assume that the interfacial reaction rate constants k_p and k_n are also largely independent of the applied bias (Memming, 2015; Peter, 2013; Peter and Wijayantha, 2014). *Thus, the current across the semiconductor–liquid junction is assumed to only depend on the concentration of electrons and holes that are available to react in the form of a mass action relationship* (Voit et al., 2015). The total current under an applied bias is defined as the sum of contributions from the independent electron and hole concentrations

$$J = J_p + J_n, \tag{7.12}$$

$$J_p = qv_p(p_s - p_{s0}), \tag{7.13}$$

$$J_n = -qv_n(n_s - n_{s0}), \tag{7.14}$$

where q is the elementary charge. Regarding v_p and v_n see the earlier discussion in Section 7.2.1. Here, n_s and p_s are, respectively, the electron and hole concentrations at the surface defined by

$$n_s = N_C e^{-(E_{Cs}-E_{Fn})/k_B T} \tag{7.15}$$

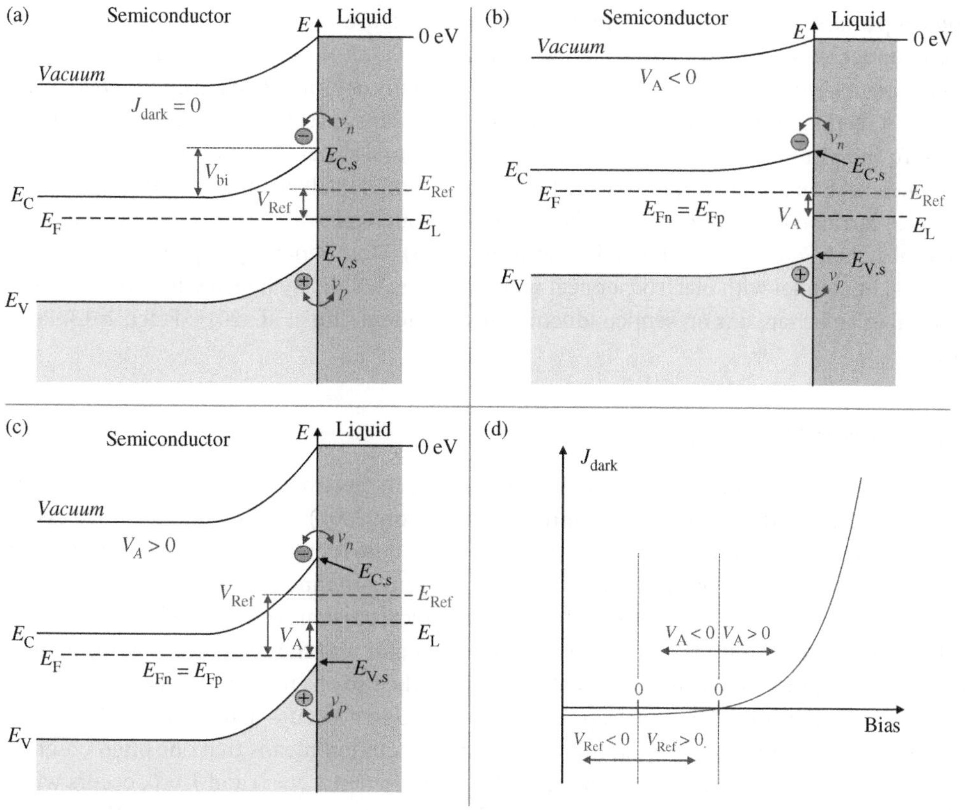

Figure 7.4 Band diagrams of the semiconductor–liquid interface: (a) in equilibrium when $J_{dark} = 0$; (b) when reverse biased such that $V_A < 0$ and $J_{dark} < 0$; and (c) when forward biased such that $V_A > 0$ and $J_{dark} > 0$. Under forward bias ($V_A > 0$) the bulk semiconductor bands are pulled further away from the liquid vacuum potential at 0 eV (increasing band bending). Whereas, under reverse bias the ($V_A < 0$) the bulk semiconductor bands are lifted closer to the liquid vacuum potential at 0 eV (reduced band bending). The hole transfer velocity is assumed to be much greater than the electron transfer velocity ($\upsilon_p \gg \upsilon_n$), such that the transport of holes (minority carrier) across the interface dominates J_{dark} leading to the current–voltage trend in (d). When the semiconductor Fermi level (E_F) is offset from the reference electrode electrochemical potential (E_{Ref}) in equilibrium, one obtains a nonzero current when $V_{Ref} = 0$.

and

$$p_s = N_V e^{-(E_{Fp}-E_{Vs})/k_B T}. \tag{7.16}$$

We define E_{Cs} and E_{Vs} as the conduction and valence band edge positions at the semiconductor–liquid interface as shown in Figures 7.2 and 7.4. Under equilibrium conditions when $E_F = E_{Fn} = E_{Fp} = E_L$ we have no net current across the interface ($J = 0$), such that electron and hole current surface concentrations are defined by n_{s0} and p_{s0}, respectively. Corrosion issues are not considered in this analysis but can be important (Su, 2016).

Taking a closer look at our example system, we can see that an n-type photoanode has been sketched in Figure 7.4—though it has not yet been subject to illumination and is being biased in the dark. To arrive at proper current–voltage description that is indicative of such photoanodes in the dark (Memming, 2015; Miao et al., 2019), we further assume that $\upsilon_p \gg \upsilon_n$ such that

$$J_{dark} \approx J_p = q\upsilon_p(p_s - p_{s0}). \tag{7.17}$$

In the parlance of the semiconductor community, such an anode would be described as a minority carrier device since it is majority n-type in bulk (see Figures 7.2 and 7.3), but instead conducts current via the transfer of minority hole carriers at interface (Pierret, 1996). Plotting Equation 7.17, one arrives at the current–voltage relationship provided in Figure 7.4d. Here, we can see that the dark current (J_{dark}) rises exponentially with respect to a forward applied bias ($V_A > 0$), because the surface hole concentration changes exponentially with linear variations in the hole quasi-Fermi level (E_{Fp}) position at the surface relative to (E_{Vs}) as dictated by Equation 7.16. Under dark conditions, $E_{Fp} = E_F$ is moved directly by the amount V_A

and thus the exponential relationship arises. Importantly, the exponential trend in Figure 7.4 does not arise from exponential changes in the transfer rate k_p under an applied bias, as would be assumed in a typical metal–liquid junction. *Rather, the exponential trend arises because the interfacial carrier concentration is exponentially dependent on the applied bias.* This, again, follows from the important screening arguments raised earlier, with necessary caveats in the accumulation and inversion regimes (Pierret, 1996). To support these conclusions in a given experimental system one needs to conduct measurements of the surface carrier concentration under an applied bias (Iqbal and Bevan, 2018a,b; Iqbal et al., 2016; Peter, 2013; Peter and Wijayantha, 2014). Surface carrier concentrations might be extracted through microwave reflectometry measurements and other methods (Cass et al., 2002; Peter, 2013; Peter and Wijayantha, 2014). More generally, applying such surface carrier concentration measurements in parallel with electrochemical measurements will likely help aid interpretations of both photocatalysis and electrocatalysis mechanisms on semiconducting metal oxides (Kim et al., 2018; Peter, 2013; Peter and Wijayantha, 2014; Su, 2016).

7.2.3 Biasing with Respect to Reference Electrodes

Though vacuum is taken as the "reference" in band diagram plots (Pierret, 1996), in electrochemical measurements one instead uses reference electrodes (Bard, 2000; Memming, 2015; Smith and Stevenson, 2007). A reference electrode essentially provides a "fixed" electrochemical potential located at some position below vacuum (denoted as E_{Ref} in Figure 7.2a–c). In the case of the standard hydrogen electrode E_{Ref} is situated −4.44 eV below vacuum as defined in the liquid region (Bard, 2000; Memming, 2015; Smith and Stevenson, 2007; Trasatti, 1986). In electrochemical current–voltage plots, one measures current as a function of the offset between the bulk Fermi level of the semiconductor and the reference electrochemical potential ($V_{\text{Ref}} = E_{\text{Ref}} - E_{\text{F}}$) as shown in Figure 7.2d. From a practical perspective, the use of a reference electrode potential (V_{Ref}) allows one to clearly delineate the portion of the total applied bias change that occurs across the working electrode (the semiconductor electrode in this example). However, this measurement notation means that one often observes nonzero current ($|J| > 0$) when $V_{\text{Ref}} = 0$. This is because the alignment of $E_{\text{F}} = E_{\text{L}}$, when $V_{\text{A}} = 0$ and $J = 0$, occurs when the bulk semiconductor Fermi level (E_{F}) is offset from the reference electrochemical potential (E_{L}) as shown in Figure 7.4a. In general, when E_{F} is below E_{Ref} we have $V_{\text{Ref}} > 0$ and refer to this bias region as "anodic bias." Conversely, when E_{F} is above E_{Ref} we have $V_{\text{Ref}} < 0$ and refer to this region as "cathodic bias." Both are plotted in Figure 7.4d. It is also important to note that the offset between the semiconductor band edges and reference electrode potential can be pH dependent, depending on the reference electrode chosen, as discussed in van de Krol (2012a). This pH dependence arises from H^+ and OH^- species adsorption as was briefly discussed in Section 7.2.1. The use of reference potentials (V_{Ref}) is helpful in the exploration of illuminated semiconductor–liquid interfaces as we shall discuss in Section 7.3.

7.3 Illuminated Semiconductor–Liquid Junctions

The illuminated operation of PEC cells is a complex process, involving many coupled nonequilibrium quantities. In this section, we shall analytically explore the basic current–voltage properties of PEC photoanodes. First, in Section 7.3.1, we shall undertake an operational analysis through a consideration of light absorption and carrier transport properties in the form of Gartner's model (which is able to capture basic current–voltage characteristics). Subsequently, in Section 7.3.2, we shall explore how the further inclusion of carrier recombination (via Peter's model) impacts upon both the performance and design of PEC photoanodes. A deeper exploration of the underlying carrier dynamics driving the operational behavior of PEC cells can be attained through semiclassical device modeling which is left to Section 7.4.

7.3.1 Gartner's Model

Just as was presumed in the dark-current model (see Section 7.2), Gartner's model begins with the assumption that the interfacial current in a photoanode arising from illumination is dominated by hole transport such that

$$J_{\text{light}} \approx J_p = J_{p,\text{drift}} + J_{p,\text{diff}}. \tag{7.18}$$

In this expression $J_{p,\text{drift}}$ and $J_{p,\text{diff}}$ are the hole drift and diffusion current contributions (see Figure 7.5a). In Gartner's model the drift current ($J_{p,\text{drift}}$) arises from photogenerated holes driven to the interface by electric field in the SCR. The SCR is shown in the lower portion of Figure 7.2b and its width (W, also marked in Figure 7.5) is determined by the potential drop

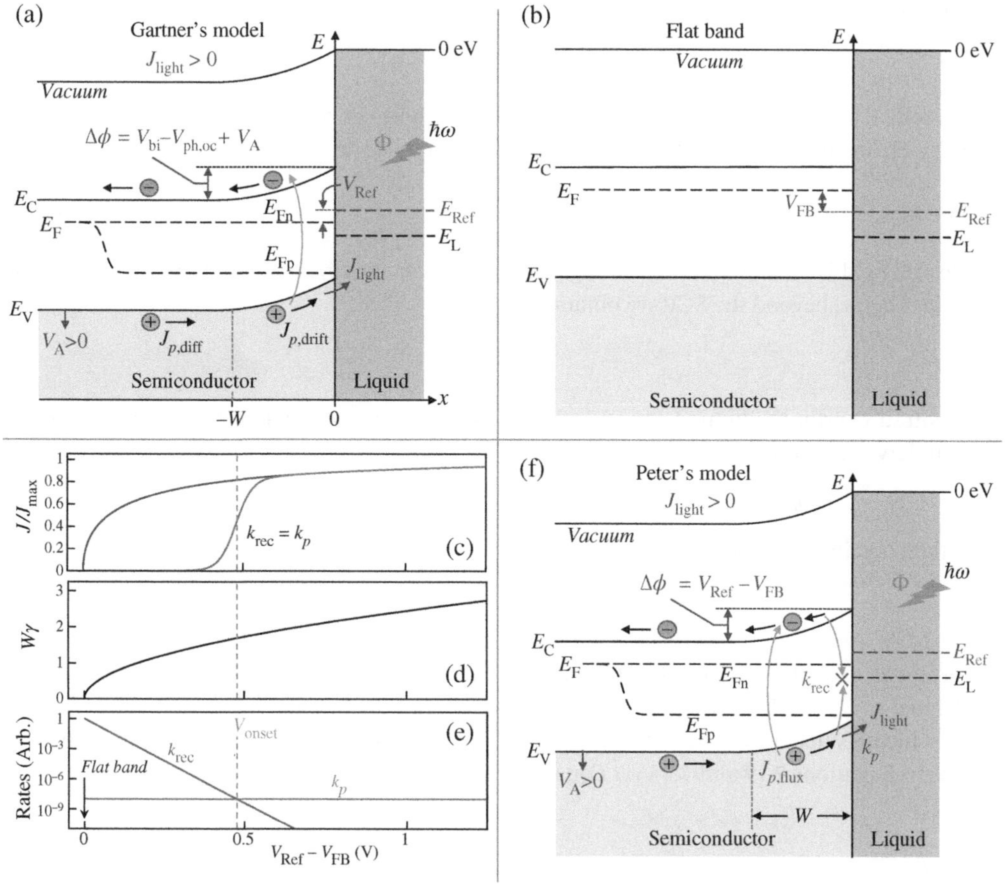

Figure 7.5 (a) Operation of a semiconductor photoanode within the Gartner model. (b) Flat band configuration for a photoanode, with the corresponding flat band reference electrode bias V_{FB}. (c) The current–voltage properties produced through Gartner's model (dark gray) juxtaposed against that produced by Peter's model (light gray). (d) The product of the depletion region width (W) under a given applied bias, multiplied by the semiconductor absorption coefficient (γ). (e) Competition between the interfacial recombination rate (k_{rec}) and hole transfer rate (k_p) at a given applied bias within Peter's model. The voltage scale in (c)–(e) is computed as $V_{Ref} - V_{FB}$, which is the current reference electrode potential V_{Ref} at a given applied bias V_A less the reference electrode potential attained in the flat band configuration (V_{FB}). (f) Operation of a semiconductor photoanode within Peter's model, where the additional impact of k_{rec} is considered. Illumination also induces quasi-Fermi level splitting ($E_{Fn} \neq E_{Fp}$) and the formation of an open-circuit photovoltage ($V_{ph,OC}$) that are not explicitly considered by either the Gartner or Peter models. Parameters utilized in (c)–(e) are provided at the end of Section 7.3.2.

($\Delta\phi$) in the SCR as given by Equation 7.11. If we define V_{FB} as the reference electrode bias at which the semiconductor bands are completely flattened (as shown in Figure 7.5b), then this potential drop may be measured via $\Delta\phi = V_{Ref} - V_{FB}$ (compare Figure 7.5a,b). Hence, an estimate of the space charge width (W) can be obtained directly from reference electrode measurements (Memming, 2015; Miao et al., 2020). This is valid so long as the semiconductor is not biased into the inversion or accumulation regimes as discussed in Section 7.2.

If one assumes that all of the holes generated by incident light within the SCR are rapidly swept to the interface and out of the semiconductor, then the drift current can be approximated as

$$J_{drift} \approx q \int_{-W}^{0} G(x)dx = q\Phi(1 - e^{-\gamma W}), \tag{7.19}$$

where Φ is the photon flux and γ is the monochromatic photon absorption coefficient. Note, here we define positive current flow in the $+x$-direction and are orienting the semiconductor region to be located in the region $x < 0$ (see Figure 7.5a). The generation term through which electron–hole pairs are produced upon illumination has the form (Gartner, 1959)

$$G(x) = \Phi\gamma e^{-|x|\gamma}. \tag{7.20}$$

While the hole diffusion current can be expressed as

$$J_{p,\text{diff}} = -qD_p\frac{dp}{dx}, \tag{7.21}$$

where D_p is the hole diffusion coefficient. To solve for $J_{p,\text{diff}}$, one needs to extract hole concentration p. In Gartner's model this is accomplished by starting with an approximate hole continuity equation of the form

$$\frac{dp}{dt} \approx -\frac{1}{q}\frac{dJ_{p,\text{diff}}}{dx} + G(x) - R(x), \tag{7.22}$$

beyond the SCR where the electric field is zero ($x < -W$). Noting that in steady-state $dp/dt = 0$ and approximating the recombination term by $R(x) \approx (p - p_0)/\tau$, beyond the SCR we obtain

$$\frac{p - p_0}{\tau} = D_p\frac{d^2p}{dx^2} + \Phi\gamma e^{-|x|\gamma}, \tag{7.23}$$

where p_0 is the bulk hole concentration prior to illumination and τ is the lifetime of excess carriers. Solving for p through this equation subject to the boundary conditions

$$[p(x)]_{x=-W} = p_0 \quad \text{and} \quad [p(x)]_{x=-\infty} = p_0 \tag{7.24}$$

following Gartner (1959), one then obtains via Equation 7.21 the result

$$J_{p,\text{diff}} = q\Phi\frac{\gamma L_p}{1 + \gamma L_p}e^{-\gamma W}, \tag{7.25}$$

which is the hole diffusion current at $x = -W$, between the SCR and the bulk (see Figure 7.5a). Here, we have made use of the hole diffusion length defined as $L_p = \sqrt{D_p\tau}$. Note that Gartner assumes that the boundary condition at $x = -W$ is $[p(x)]_{x=-W} = 0$ (Gartner, 1959), which results in a further additive qp_0D_p/L_p term in Equation 7.25. Further combining the drift and diffusion hole currents in Equations 7.19 and 7.25 via Equations 7.18 and 7.11 one arrives at

$$\begin{aligned} J_{\text{light}} &= q\Phi\left(1 - \frac{e^{-\gamma W}}{1 + \gamma L_p}\right) \\ &= q\Phi\left[1 - \exp\left(-\frac{\sqrt{2\gamma^2\epsilon_0\epsilon_r(V_{\text{Ref}} - V_{\text{FB}})}}{\sqrt{qN_D}}\right)\frac{1}{1 + \gamma L_p}\right], \end{aligned} \tag{7.26}$$

which is arrived at by utilizing Equation 7.11 and the relation $\Delta\phi = V_{\text{Ref}} - V_{\text{FB}}$ discussed earlier. A typical current–voltage plot obtained from the Gartner model is provided in Figure 7.5c. Note that the photocurrent is plotted with respect to $V_{\text{Ref}} - V_{\text{FB}}$, as this allows consistent comparison against Peter's model and numerical drift-diffusion calculations that will be discussed shortly. Here, we can see that the photocurrent rises and then saturates. Saturation occurs once the SCR increases to a sufficiently large width to encompass all of the incoming photons, as dictated by Equation 7.11 and the relation $\Delta\phi = V_{\text{Ref}} - V_{\text{FB}} = V_{\text{bi}} - V_{\text{ph,OC}} + V_A$ (see Figure 7.5). Hence, the maximum achievable photocurrent is defined by

$$J_{\max} = q\Phi. \tag{7.27}$$

Note, here we have introduced the concept of an open-circuit photovoltage ($V_{\text{ph,OC}}$), a critical and nontrivial concept in the operation of PEC devices that is the focus of a deeper discussion in Section 7.4.

Qualitatively, Gartner's model captures the basic behavior of photoanodes. However, it overestimates the photocurrent "onset" point (positioned at $V_{\text{Ref}} - V_{\text{FB}} = 0$ in Figure 7.5c). This overestimation arises from the assumption that all photogenerated holes in the SCR are rapidly swept to the semiconductor–liquid interface and "immediately" consumed in a photocatalytic reaction. However, Gartner's model correctly demonstrates that the depletion region width W must be greater than the inverse absorption coefficient γ^{-1} (i.e., $W > \gamma^{-1}$) for many of the generated holes to be swept to the semiconductor–liquid interface. The inverse absorption coefficient γ^{-1} can also be viewed as a measure of the light penetration depth. Indeed, the maximum photocurrent is achieved when $W\gamma > 2$ as determined by the exponential term in Equation 7.26. The term $W\gamma$ is plotted as a function of $V_{\text{Ref}} - V_{\text{FB}}$ in Figure 7.5d for the model parameters adopted. The practical consequence of this is that heavily doped photoanodes (with small depletion widths) possessing a comparatively small absorption coefficient will not be able to produce a wide enough depletion region to drive all of the holes produced by incident photons to the surface where they can then participate in photocatalytic reactions. Consequently, the maximum achievable conversion efficiency can be dramatically reduced when the depletion width and inverse absorption coefficient (light penetration depth) are not comparable in length scale. However, this analysis only holds for simplified planar electrodes. Light absorption characteristics can also be engineered through nano-structuring and other surface engineering techniques (Khan et al., 2002; Nelson, 2003; Xu et al., 2019).

7.3.2 Peter's Model

Peter's model builds on Gartner's model by directly considering electron–hole recombination in the SCR as shown in Figure 7.5f (Peter, 2013; Peter and Wijayantha, 2014). In Peter's model, one considers a hole flux ($J_{p,\mathrm{flux}}$) that arrives at the semiconductor–liquid interface region where electron transfer occurs (see Figure 7.5f). The hole flux term in Peter's model follows directly from Gartner's model taking the form

$$J_{p,\mathrm{flux}} = q\Phi\left(1 - \frac{e^{-\gamma W}}{1+\gamma L_p}\right). \tag{7.28}$$

However, in Peter's model the flux of holes from the semiconductor "interior" is not entirely successful at transferring to the liquid (e.g., to drive the oxygen evolution reaction in water) due to sluggish kinetics as determined by the hole transfer rate k_p. The portion of the hole flux ($J_{p,\mathrm{flux}}$) that is not able to transfer will be consumed by electron–hole recombination in the SCR, as determined by the recombination rate term k_{rec} marked near the semiconductor–liquid interface in Figure 7.5f. This leads to the charge flux conservation relation

$$J_{p,\mathrm{flux}} = q(k_p + k_{\mathrm{rec}})\delta p_{\mathrm{surf}}, \tag{7.29}$$

where $\delta p_{\mathrm{surf}} = (p_{\mathrm{s}} - p_{\mathrm{s0}})l$ represents the excess nonequilibrium density of holes per unit area at the surface contributed by the nonequilibrium flux—regarding l see also the earlier discussion in Section 7.2.1. Since Equations 7.28 and 7.29 represent the same property we obtain

$$\delta p_{\mathrm{surf}} = \frac{J_{p,\mathrm{flux}}}{q(k_p + k_{\mathrm{rec}})} = \Phi\left(1 - \frac{e^{-\gamma W}}{1+\gamma L_p}\right)\frac{1}{k_p + k_{\mathrm{rec}}}. \tag{7.30}$$

Those holes that do not recombine drive the net observable photocurrent leading to the expression

$$J_{\mathrm{light}} = qk_p\delta p_{\mathrm{surf}} = q\Phi\left(1 - \frac{e^{-\gamma W}}{1+\gamma L_p}\right)\frac{k_p}{k_p + k_{\mathrm{rec}}}. \tag{7.31}$$

Since Equation 7.31 is based on the nonequilibrium hole density near the semiconductor–liquid interface (δp_{surf}), it is primarily concerned with further capturing SRH recombination processes that would arise from defects on a surface (again, see the earlier discussion on recombination in Section 7.1). Nevertheless, to some degree bulk recombination is phenomenologically incorporated in Equation 7.28 via the hole diffusion length (L_p). Importantly, just as with the dark current model discussed in Section 7.2, here we assume that the hole transfer rate (k_p) is constant with respect to an applied bias. However, the hole recombination rate changes with the applied bias and to first-order can be approximated as

$$k_{\mathrm{rec}} = \sigma_p v_{\mathrm{th}} n_{\mathrm{s}}. \tag{7.32}$$

Within this expression σ_p represents the hole capture cross section and v_{th} is the thermal recombination velocity (Peter, 2013; Peter and Wijayantha, 2014; Pierret, 2002). The surface electron density in Equation 7.32 was defined previously in Equation 7.15 but can also be approximated in terms of the doping density (N_{D}) as

$$n_{\mathrm{s}} \approx N_{\mathrm{D}}e^{-\Delta\phi/k_{\mathrm{B}}T} \tag{7.33}$$

assuming an n-type semiconductor and a flat electron quasi-Fermi level (E_{Fn}). Importantly, both σ_p and v_{th} can be approximated as constants, thus as the band bending is altered via an applied bias (V_{A}) in the form of

$$\Delta\phi = V_{\mathrm{Ref}} - V_{\mathrm{FB}} = V_{\mathrm{bi}} - V_{\mathrm{ph,OC}} + V_{\mathrm{A}} \tag{7.34}$$

we end up with an exponential decrease in $k_{\mathrm{rec}} = \sigma_p v_{\mathrm{th}} N_{\mathrm{D}} \exp(-\Delta\phi/k_{\mathrm{B}}T)$ with respect to a linear increase in V_{A} (or "$V_{\mathrm{Ref}} - V_{\mathrm{FB}}$" when viewed in terms of a reference electrode). This subtle, but crucial insight, explains why the onset of the photocurrent can be so dramatically impacted upon by recombination processes within an photoelectrode. The maximum photocurrent as dictated by Equation 7.31 can only be achieved once $k_{\mathrm{rec}} \ll k_p$ as shown in Figure 7.5e, which is attained by suppressing k_{rec} though exponential decreases in n_{s} as band bending ($\Delta\phi$) is increased via application of a forward bias ($V_A > 0$). This means that by decreasing the recombination processes in the semiconductor (or near interfacial region) one can theoretically shift the onset for the photocurrent and thereby improve overall PEC performance (Iqbal and Bevan, 2018a,b; Iqbal et al., 2016; Peter, 2013; Peter and Wijayantha, 2014).

Peter's model also provides a dramatic but subtle insight into reaction kinetics that is often overlooked. The point at which the recombination rate equals the hole transfer rate ($k_p = k_{rec}$) is marked as a dashed gray vertical line in Figure 7.5c–e. From Figure 7.5c, we can see that the onset of the photocurrent (V_{onset}) in Peter's model occurs approximately at this rate intersection (plotted in Figure 7.5e) (Iqbal and Bevan, 2018a; Peter, 2013). Equating the two rates at this onset voltage (V_{onset}) one obtains

$$k_p = \sigma_p v_{th} N_D e^{-V_{onset}/k_B T} \Rightarrow V_{onset} = -k_B T \ln(k_p) + k_B T \ln(\sigma_p v_{th} N_D), \tag{7.35}$$

which shows that one obtains a $2.3k_BT$ shift in the onset point for every order of magnitude (×10) increase in the hole transfer rate (k_p)—note, we utilize units of electron volts for k_B. This means that the onset voltage (V_{onset}) is shifted linearly with exponential increases in the hole kinetics (captured by k_p). At room temperature (300 K) this shift in V_{onset} for each order of magnitude increase in k_p is a mere 59.5 meV—assuming k_{rec} remains fixed. Thus, focusing only on engineering novel kinetic catalysts (which determine k_p) to improve the onset voltage for unassisted photocatalytic water splitting is a daunting challenge (Iqbal and Bevan, 2018a,b; Iqbal et al., 2016; Miao et al., 2019; Peter, 2013). Clearly, one must focus on simultaneously engineering competing semiconductor recombination processes (via k_{rec}) to improve performance (Iqbal and Bevan, 2018a; Miao et al., 2019; Peter, 2013). These crucial insights underscore that integrated nature of photocatalysis design, which rests upon engineering both the semiconductor region and catalytic reactions. Of course, higher order effects may be present in the overall bias dependent behavior during photocatalysis, which are missing in the simplified model presented here, but still crucial challenges remain in developing an integrated design approach (Mesa et al., 2020; Peter, 2013; Peter and Wijayantha, 2014). The need for an integrated design approach is compactly and elegantly emphasized in Peter's model.

Both the Peter and Gartner model results assume a Ta_3N_5 photoanode with a band gap of 2.2 eV, a doping density of $N_D = 5 \times 10^{17}$ cm^{-3}, a relative dielectric constant $\epsilon_r = 17$, and a hole diffusion length of $L_p = 3$ nm (Nurlaela et al., 2016; Ziani et al., 2015). In Figure 7.5c, the assumed anode was illuminated at a wavelength of ∼400 nm corresponding to an absorption coefficient of $\gamma = 4 \times 10^5$ cm^{-1} (Nurlaela et al., 2016).

7.4 The Role of Numerical Modeling

Modeling can offer a deeper level of insight into the operational characteristics of photocatalytic systems than what can be achieved through purely analytical methods (such as those presented in Sections 7.2 and 7.3). To this end, in Section 7.4.1 we briefly outline a versatile semiclassical methodology for modeling semiconductor–liquid PEC junctions. Subsequently, in Section 7.4.2 some impactful insights into the operation of photocatalytic anodes that can be obtained through numerical modeling are presented and discussed.

7.4.1 Semiclassical Approach

First, let us consider how one may more rigorously model the semiconductor region situated at $x < 0$ as illustrated in Figures 7.5a and 7.6a. In the semiclassical framework, one begins with the electron (n) and hole (p) continuity equations as described by Pierret (2002)

$$\frac{dn}{dt} = \frac{1}{q}\frac{dJ_n}{dx} + G_n - R_n. \tag{7.36}$$

$$\frac{dp}{dt} = -\frac{1}{q}\frac{dJ_p}{dx} + G_p - R_p. \tag{7.37}$$

Earlier in Section 7.3.1, we considered an approximate form of Equation 7.37. The generation and recombination terms (G_n, R_n, G_p, and R_n) are usually arrived at via Equation 7.20 for light absorption, Equation 7.7 for radiative recombination, and Equation 7.10 for non-radiative recombination. A detailed discussion of generation and recombination processes were provided earlier in Sections 7.1 and 7.3.1. The electron and hole current densities found in Equations 7.36 and 7.37 are generally expressed as

$$J_n = qn\mu_n\mathcal{E} + qk_BT\mu_n\frac{dn}{dx} \tag{7.38}$$

$$J_p = qp\mu_p\mathcal{E} - qk_BT\mu_p\frac{dp}{dx} \tag{7.39}$$

with electron and hole mobilities given by μ_n and μ_p, respectively (Pierret, 1996, 2002). Note, we are still utilizing electron volts for k_BT as discussed earlier. The electric field in the J_n and J_p drift terms may be expressed as $\mathcal{E} = -d\phi/dx$, where $V = -\phi$ is the band bending potential seen by electrons such that electrons are pulled down to regions where V is lower (see Figure 7.2b). Alternately, the electrostatic potential may also be computed with an opposite sign as per the convention chosen (Martin, 2004; Pierret, 1996). Since discretization is performed in only one dimension (see Figure 7.6a), we have dropped vector notation from the electric field ($\mathcal{E}$) in this discussion. To obtain the steady-state response one needs to solve Equations 7.36 and 7.37 subject to $dn/dt = 0$ and $dp/dt = 0$ resulting in the form

$$\frac{dJ_n}{dx} = -q(G_n - R_n). \tag{7.40}$$

$$\frac{dJ_p}{dx} = q(G_p - R_p). \tag{7.41}$$

Further inserting Equations 7.38 and 7.39 into Equations 7.40 and 7.41, respectively, provides solvable equations for the electron (n) and hole (p) densities subject to their respective generation and recombination terms. When implemented numerically, Equations 7.40 and 7.41 are discretized with a variable grid spacing (Δ_i) as illustrated in Figure 7.6a. Discretization results in a system of matrix equations for n and p that can be solved through lower-upper (LU) decomposition, or a related numerical approach, as detailed in Iqbal (2018), Iqbal and Bevan (2018a,b), and Iqbal et al. (2016).

To solve Equations 7.40 and 7.41 numerically one must also impose boundary conditions on the electron (n) and hole (p) concentrations. Far away from the interface, at some distance much greater than the depletion region width ($x \ll -W$), the simulated region is truncated on a numerical grid with the boundary conditions $[n]_{x\ll -W} = N_D$ and $[p]_{x\ll -W} = n_i^2/N_D$ as shown in Figure 7.6a (Iqbal, 2018; Iqbal and Bevan, 2018a,b; Iqbal et al., 2016). In this bulk boundary limit ($x \ll -W$) the electron and hole quasi-Fermi levels converge ($E_{Fn} = E_{Fp}$) as illustrated in Figure 7.5a. Note, here we are assuming boundary conditions suitable for a photoanode (see Equation 7.4). However, these bulk boundary conditions take the form of Equation 7.5 when a photocathode is simulated (Bagal et al., 2019). To obtain the electron and hole concentrations at

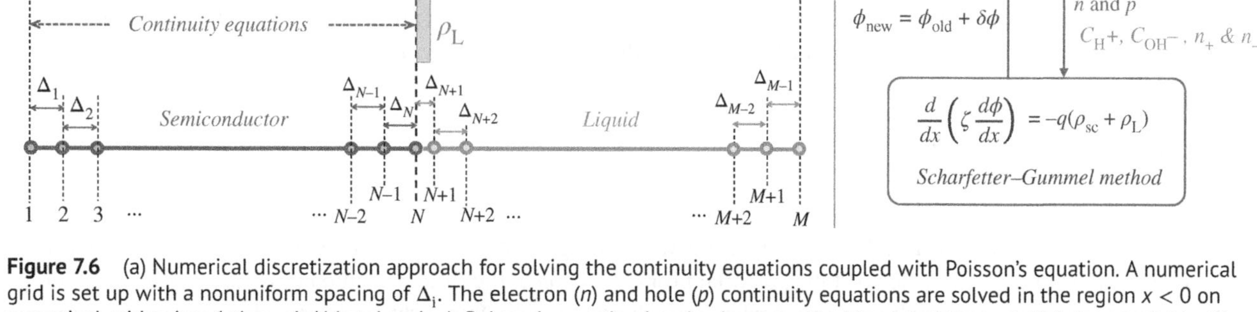

Figure 7.6 (a) Numerical discretization approach for solving the continuity equations coupled with Poisson's equation. A numerical grid is set up with a nonuniform spacing of Δ_i. The electron (n) and hole (p) continuity equations are solved in the region $x < 0$ on numerical grid points 1 through N (total region). Poisson's equation is solved across all grid points 1 through M (left region). Liquid charges are solved in the domain ($x > 0$) on grid points $N + 1$ through M (right region). When solving the continuity equations Dirichlet boundary conditions are set in the semiconductor bulk, while floating Neumann boundary conditions are applied at the semiconductor–liquid interface ($x = 0$). When solving Poisson's equation the Dirichlet boundary condition of $\phi = 0$ is set in the liquid bulk, while a floating Neumann boundary condition of $d\phi/dx = 0$ is set in the semiconductor bulk. The net semiconductor (ρ_{sc}) and liquid (ρ_L) charge densities must sum to zero. (b) A self-consistent solution is needed when solving for the interdependent electrostatic potential ϕ and the variable system charge densities n, p, c_{OH^-}, c_{H^+}, n_-, and n_+. The Scharfetter–Gummel method (Vasileska and Goodnick, 2006) can be applied in combination with the continuity equations and a Gouy–Chapman description of liquid screening.

the semiconductor–liquid interface free floating Neumann boundary conditions must be employed (shown in Figure 7.6a). This can be accomplished by incorporating the interfacial current expressions given by Equations 7.13 and 7.14 at $x = 0$ in Equations 7.40 and 7.41 as discussed in Iqbal (2018). Hence, the region over which the n and p concentrations are solved extends from the semiconductor bulk to $x = 0$ as illustrated in Figure 7.6a. Crucially, the use of Neumann boundary conditions for n and p at the semiconductor–liquid interface allows one to directly compute the interfacial electron and hole quasi-Fermi level (E_{Fn} and E_{Fp}) splitting under illumination via Equations 7.15 and 7.16. The relative positions of quasi-Fermi levels in an illuminated photoanode are extremely important to understanding its operation, as we shall discuss shortly in Section 7.4.2.

Beyond the semiconductor region, one must also solve for the charge concentrations in the liquid region to arrive at a comprehensive description of semiconductor–liquid junctions (as illustrated in Figures 7.3 and 7.6b). Through a first-order description of electrostatics, charges in the liquid can be captured by a Gouy–Chapman model of the form

$$c_{H^+} = c^0_{H^+} e^{[-\phi/k_B T]}, \tag{7.42}$$

$$c_{OH^-} = c^0_{OH^-} e^{[\phi/k_B T]}, \tag{7.43}$$

$$n_+ = c^+_{sup}\, e^{[-z\phi/k_B T]}, \tag{7.44}$$

$$n_- = c^-_{sup}\, e^{[z\phi/k_B T]}, \tag{7.45}$$

which is applied in the domain $x > 0$ as shown in Figure 7.6a. This expression includes the concentration of H^+ and OH^- species in water given by c_{H^+} and c_{OH^-}, relative to their respective bulk concentrations $c^0_{H^+}$ and $c^0_{OH^-}$, as determined by the electrostatic potential drop in the liquid ϕ. Note that we have assumed a band bending reference of $V = -\phi = 0$ in the liquid bulk region (as shown for a photoanode in Figure 7.2b). To ensure a short screening length in the liquid, as discussed in Section 7.2.1, there is often a high bulk concentration of supporting electrolyte ions c^+_{sup} and c^-_{sup} ions with a net ionic charge of z. The concentration of positively and negatively charged supporting electrolyte ions near the interface is given by n_+ and n_-, respectively. As also mentioned in Section 7.2.1, this Gouy–Chapman model excludes a detailed description of charge ordering. More advanced modeling should take into account pH dependent adsorption, as well as structuring of IHP and OHP as discussed earlier in the context of Figure 7.3. However, this model works sufficiently well when the semiconductor screening length is much longer than the liquid screening length. Under these conditions, as discussed in Section 7.2.1, the interfacial voltage drop occurs almost entirely across the semiconductor. Hence, the errors resulting from utilizing a Gouy–Chapman description of the liquid can be assumed to minimally impact on the physical trends inferred. However, specific adsorption by H^+ and OH^- species may require consideration to align the liquid and semiconductor electrochemical potentials (Miao et al., 2020; van de Krol, 2012a).

While the electron (n) and hole (p) concentrations are solved via Equations 7.38–7.41, and the liquid charge concentrations by Equations 7.42–7.45, both systems of equations are dependent upon the electrostatic potential drop ϕ. It can be solved via Poisson's equation

$$\frac{d}{dx}\left(\epsilon \frac{d\phi}{dx}\right) = -q(\rho_{sc} + \rho_L) \tag{7.46}$$

across both the semiconductor ($x < 0$) and liquid ($x > 0$) regions as depicted in Figure 7.6a. The dielectric constant $\epsilon = \epsilon_0 \epsilon_r$ is allowed to vary across both regions. The charge density in the semiconductor region is given by

$$\rho_{sc} = p - n + N_D^+ - N_A^-, \tag{7.47}$$

while that in the liquid is given by

$$\rho_L = c_{H^+} - c_{OH^-} + n_+ - n_-. \tag{7.48}$$

In photoanodes, we typically assume $N_A^- = 0$. Importantly, the charge sign on the right-hand side of Equation 7.46 is chosen to arrive at the band diagram electrostatic potential profile $V = -\phi$ of the form presented in Figure 7.2b—which drops in the positively charged region and rises in the negatively charged region. Similarly, the charge density sign assignments in Equations 7.47 and 7.48 are chosen to be consistent with Figures 7.2b and 7.6a. When combined together, the semiconductor (ρ_{sc}) and liquid (ρ_L) charge densities must sum/integrate to a net charge of zero to satisfy the condition of charge neutrality for the entire junction. This, in turn, allows one to establish a fixed electrostatic potential of $\phi = 0$ in the bulk liquid region ($x \gg 0$) and a Neumann boundary condition of $d\phi/dx = 0$ in the semiconductor bulk region ($x \ll -W$) as depicted in Figure 7.6a. The establishment of a floating electrostatic boundary condition in the semiconductor bulk allows one to solve for the photovoltage in PEC cells—a concept we shall explore shortly in Section 7.4.2. It is important to recognize that Equation 7.46 is solved over both the liquid and semiconductor regions (encompassing $x < 0$ and $x > 0$ as shown in Figure 7.6a). However, the semiconductor charge density is solved via Equations 7.36–7.39 in the domain $x < 0$ (see Figure 7.6a). Similarly, the liquid charges are only extracted via Equations 7.42–7.45 in the domain $x > 0$ (see Figure 7.6a). A full discussion of the variable grid implementation and discretization procedures concerned with solving these equations can be found in Iqbal and Bevan (2018a,b) and Iqbal (2018), which we shall not detail here for the sake of brevity.

Through a closer examination of Equations 7.38–7.48, one can discern that the electrostatic potential ϕ and charge densities $\rho_{sc} + \rho_L$ are interdependent quantities. One needs to know ϕ to solve for n, p, c_{OH^-}, c_{H^+}, n_-, and n_+ and vice versa. Computationally, this is addressed through the self-consistent solution of these interdependent quantities as shown in Figure 7.6b. Additionally, self-consistency also arises via n and p between the recombination-generation terms and the current flow expression in the continuity equations (see Equations 7.36–7.39). For the semiclassical approach adopted herein, where each charge density adopts a Boltzmann distribution, the Scharfetter–Gummel method provides a direct methodology for arriving at self-consistent representations of ϕ, n, p, c_{OH^-}, c_{H^+}, n_-, and n_+ (Gummel, 1964; Scharfetter and Gummel, 1969). A comprehensive explanation of the Scharfetter–Gummel method can be found in Vasileska and Goodnick (2006). In short, one begins with an initial guess for ϕ and through the Scharfetter–Gummel method obtains a perturbative correction $\delta\phi$ that aims to bring about self-consistency. From the perturbative correction one obtains a new potential guess $\phi_{new} = \phi_{old} + \delta\phi$. Smooth and rapid convergence often requires the use specialized self-consistent techniques to estimate $\delta\phi$ (Srivastava, 1984). From ϕ_{new}, one then obtains new guesses for the charge densities (n, p, c_{OH^-}, c_{H^+}, n_-, and n_+) which can then be reinserted into the perturbative form of Poisson's equation as arrived at via the Scharfetter–Gummel method (Vasileska and Goodnick, 2006). This then produces the next perturbative guess and the whole process is repeated over and over until self-consistent convergence between ϕ and $\rho_{sc} + \rho_L$ is obtained to within the desired computational tolerance (see Figure 7.6b). As with all self-consistent methods, the procedure for achieving convergence is somewhere between an art and a science (Martin, 2004). Once self-consistency is satisfactorily achieved to a defined tolerance, one may then compute the interfacial current density via Equations 7.12–7.16. Device operation is computed in this manner both under illumination and in the dark.

7.4.2 Insights from Semiclassical Modeling

The semiclassical approach discussed in Section 7.4.1 provides access to a great variety of operational PEC information (Cendula et al., 2014; Gaudy and Haussener, 2016). However, exploring the full range of device properties that can be captured via semiclassical modeling, or through higher order methodologies, still remains very much an active area of research investigation in the PEC literature (at the date of composition) (Gaudy and Haussener, 2016). For this reason, let us focus on two nontrivial but extremely important semiclassical insights: the photovoltage and the PEC voltage. Crucially, both of these properties follow from simultaneously solving both the continuity equations and electrostatics governing PEC device performance (as outlined in Section 7.4.1). These nontrivial quantities can also be readily extracted from experimental measurements, which makes their discussion all the more important.

The current–voltage (J–V) characteristics of a tantalum nitride (Ta_3N_5) photoanode (computed via Equations 7.13 and 7.14) are shown in Figure 7.7. The surface hole (p_s) and electron (n_s) concentrations in this system, required to solve Equations 7.13 and 7.14, are obtained by solving Equations 7.38–7.48 as a function of the applied bias (V_A). The application of V_A in both the light and the dark was discussed earlier in the context of Figures 7.4 and 7.5, respectively. In Figure 7.7a, the numerically computed current–voltage characteristics for both the illuminated and dark photoanode are shown. The same information is provided on a logarithmic scale in Figure 7.7b. In both plots the current is provided with respect to the reversible hydrogen electrode (RHE) on the lower axis. The flat band potential V_{FB}, measured with respect to RHE, is indicated in Figure 7.7 and may be compared to the $V_{Ref} - V_{FB}$ scale applied in Figure 7.5c–e. Additionally, in Figure 7.7 current flow (J) is normalized with respect to J_{max} as per Equation 7.27 and Figure 7.5c. Important operational current–voltage points are marked in Figure 7.7 as: O_D, O_L, α, and β (Miao et al., 2019). We shall discuss each of the points and their corresponding band diagrams, which are also computed by numerically solving Equations 7.38–7.48.

First, let us begin with the open-circuit dark (O_D) and illuminated (O_L) current–voltage points where $V_A = 0$. The open-circuit band diagram for our Ta_3N_5 anode in the dark (corresponding to O_D in Figure 7.7a) can be found in Figure 7.8a. Here, it can be seen that we have adopted an electrolyte with $pH \approx 14$, such that the liquid electrochemical potential E_L is situated about the OH^-/O_2 redox level (Iqbal and Bevan, 2018a,b; Iqbal et al., 2016). Equilibration between the semiconductor (Ta_3N_5) doped at 5×10^{17} cm^{-3} and the liquid results in a built-in potential of $V_{bi} = 1.18$ V (Ziani et al.,

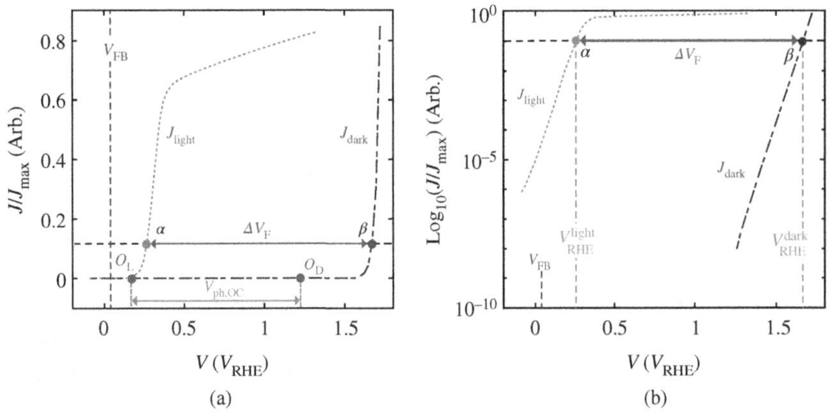

Figure 7.7 Current–voltage characteristics of a tantalum nitride photoanode in the dark (J_{dark}) and under illumination (J_{light}). Equivalent linear and logarithmic scale plots with respect to RHE are provided in (a) and (b), respectively. When plotting current–voltage characteristics with respect to a reference electrode, the offset between two equivalent current points in the light and the dark (α and β) provides the PEC voltage (ΔV_F). The offset between the reference electrode potentials in the dark (O_D) and under illumination (O_L) in the open-circuit condition ($V_A = 0$) provides the photovoltage $V_{ph,OC}$. The potential at which flat band is achieved is indicated by V_{FB}. All current–voltage characteristics are computed by self-consistently solving Equations 7.38–7.48. Parameters utilized are provided at the end of Section 7.4.2.

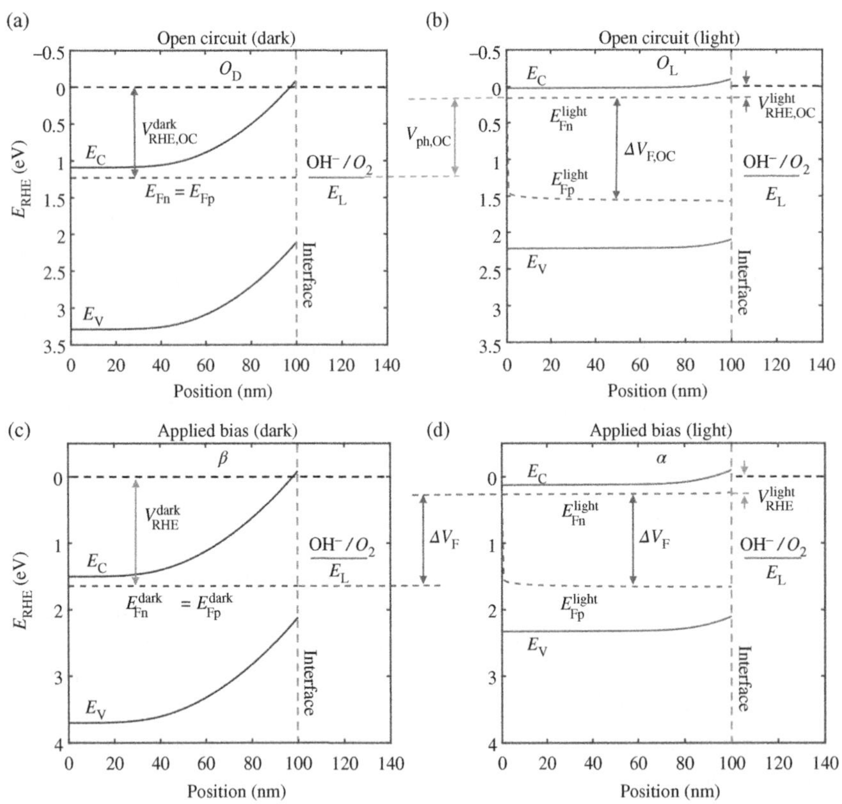

Figure 7.8 Operational band diagrams of a tantalum nitride photoanode in the dark and under illumination. (a) Band diagram of a dark photoanode when $V_A = 0$. (b) Band diagram of an illuminated photoanode in the open-circuit configuration ($V_A = 0$). (c) Biased band diagram of a photoanode biased at 0.413 V in the dark. (d) Band diagram of a photoanode under illumination biased at 0.104 V producing the same current as the dark configuration in (c). All band diagrams are computed by self-consistently solving Equations 7.38–7.48. Parameters utilized are provided at the end of Section 7.4.2.

2015). In all respects the numerical results of Figure 7.8a are phenomenologically identical to the dark anode properties discussed earlier in the context of Figures 7.2b and 7.4a.

Next, consider the open-circuit ($V_A = 0$) band diagram of our Ta_3N_5 anode under illumination as provided in Figure 7.8b—denoted as point O_L in Figure 7.7a. The generation of electron–hole pairs via illumination (see Equation 7.20) results in the formation of electron (E_{Fn}) and hole (E_{Fp}) quasi-Fermi levels in Figure 7.8b, just as was discussed early on in the context of Figure 7.1b for an isolated semiconductor. However, due to the presence of the built-in voltage (V_{bi} shown in Figure 7.8a), the generated electron–hole pairs can be efficiently separated by the interfacial electric field over the band bending region (see also Figure 7.5a). In the open-circuit configuration, the excess generated electrons are swept toward the bulk (where they have nowhere to go) and the excess holes are swept toward the interface (where only a small trickle can leak out via redox reactions, since $V_A = 0$). Hence, a nonequilibrium accumulation of opposing electron and hole charges builds-up on opposite sides of the space-charge region (see also Figure 7.2b). Under steady-state illumination this results in the formation of an electrostatic potential that counters V_{bi}, and is termed the open-circuit photovoltage $V_{ph,OC}$ as indicated in Figure 7.8 (Miao et al., 2019). This steady-state photovoltage is characterized by a balance between the electron–hole pair generation and recombination rates (see Equations 7.36 and 7.37), where further carrier recombination is facilitated by a flattening of the photoanode bands by $V_{ph,OC}$ (see Figure 7.8b). Flattening of the photoanodes bands lessens the separation of electron hole pairs in the anode SCR and thereby enhances recombination (for an illustrative depiction see Figure 7.5f).

The key takeaway here is that the open-circuit photovoltage $V_{ph,OC}$ can be directly measured. One merely needs to record the reference electrode potential in the dark, as well as the corresponding open-circuit reference electrode potential under illumination. These are indicated as $V^{dark}_{RHE,OC}$ and $V^{light}_{RHE,OC}$, respectively, in Figure 7.8a,b. The difference between these two is the open-circuit photovoltage, given as $V_{ph,OC} = V^{dark}_{RHE,OC} - V^{light}_{RHE,OC}$ in Figures 7.7a and 7.8. However, often the photovoltage is frequently confused with the nonequilibrium splitting of electron (E_{Fn}) and hole (E_{Fp}) quasi-Fermi levels shown in Figure 7.8b. The splitting of E_{Fn} and E_{Fp} under illumination can be termed the PEC potential ΔV_F (indicated in Figure 7.8b); it is a distinct concept from that of the photovoltage (Miao et al., 2019). If one returns to the earlier discussion in the context of Figure 7.1b, it can be seen that quasi-Fermi level splitting ($E_{Fn} \neq E_{Fp}$) can be obtained even within an isolated semiconductor under illumination without the formation of any electrostatic potential (i.e., voltage) whatsoever. Similarly, in the context of photoanodes the separation of E_{Fn} and E_{Fp} does not necessarily correspond to the electrostatic potential generated upon illumination (i.e., the photovoltage).

Nevertheless, the splitting between electron and hole quasi-Fermi levels can be ascribed physical meaning that we shall now explore. Let us begin by specifically quantifying the PEC voltage (ΔV_F) as the separation between the bulk electron quasi-Fermi level (E_{Fn}) and the surface hole quasi-Fermi (E_{Fp}) position. It can be defined either under open circuit conditions ($V_A = 0$) or at a given applied bias ($V_A > 0$), marked as $\Delta V_{F,OC}$ and ΔV_F in Figure 7.8b,d, respectively. To understand the meaning of ΔV_F and its physical manifestation, let us turn to examining the dark and illuminated current–voltage curves at the same operation current value marked as α and β in Figure 7.7. Here, it is important to recognize that following Equation 7.13, the current magnitude will only be equal (points α and β in Figure 7.7) when the surface hole concentration (p_s) is the same in both the dark and under illumination! Importantly, this is the same as saying that the hole quasi-Fermi levels are in the same position ($E^{dark}_{Fp} = E^{light}_{Fp}$, relative to the reference electrode) when the dark and illuminated currents are the same (see Equation 7.16). A direct calculation of the band diagrams for the equivalent current points α and β in Figure 7.8c,d reveals this to be the case. However, the current at points α and β in Figure 7.7 are measured with respect to the reference electrode potential (RHE for this study). This bias scale represents the offset between the reference electrode electrochemical potential and the bulk Fermi-level of the semiconductor electrode (as discussed earlier in the context of Figure 7.4). In the dark one must apply a bias of V^{dark}_{RHE} to reach the current magnitude β as indicated in Figure 7.7 and illustrated in Figure 7.8c. Yet, under illumination one can apply a smaller bias of V^{light}_{RHE} to achieve the same current magnitude at α in Figure 7.7. This is because the surface quasi-Fermi level position E^{light}_{Fp} is shifted by an amount ΔV_F from the bulk electron quasi-Fermi level (to which the reference electrode is measuring the offset) under illumination—compare Figure 7.8c,d. Thus, the offset between the illuminated and dark reference electrode potentials at which the same current magnitude is achieved are related by the PEC voltage in the manner $\Delta V_F = V^{dark}_{RHE} - V^{light}_{RHE}$. This correspondence can be discerned through a comparison of Figures 7.7 and 7.8c,d.

The central point of this discussion is to show that one is able to obtain clear insight into the physical manifestation of the photovoltage ($V_{ph,OC} = V^{dark}_{RHE,OC} - V^{light}_{RHE,OC}$, extracted when $V_A = 0$) and PEC voltage ($\Delta V_F = V^{dark}_{RHE} - V^{light}_{RHE}$, extracted when $J_{light} = J_{dark}$) by numerically solving the semiconductor continuity equations and Poisson's equation in self-consistent manner as detailed in Section 7.4.1. Through this approach, we are able to concurrently calculate the current–voltage characteristics and corresponding band diagrams at each bias point. As a result one can discern that the photovoltage ($V_{ph,OC}$) is an electrostatic potential countering the built-in voltage (V_{bi}) induced by electron–hole separation in the SCR (at $V_A = 0$). Similarly, the PEC voltage (ΔV_F) is physically manifested in the offset between the illuminated and dark reference electrode potentials at equivalent drive currents ($J_{light} = J_{dark}$). Of course further insights can be gleaned by solving Equations 7.38–7.48 and variations thereof. Recent work includes validation of the competition between the interfacial recombination and hole transfer rates as captured by Peter's model in Section 7.3.2 (Iqbal and Bevan, 2018a), as well as the impact of trap states and surface states on the performance of photoanodes (Iqbal et al., 2018; Miao et al., 2020).

In this numerical study a slightly different Ta_3N_5 illumination wavelength of 440 nm was chosen corresponding to an absorption coefficient of $\gamma = 8 \times 10^4$ cm^{-1} (Nurlaela et al., 2016). All other material parameters were set the same as those listed in Section 7.3.2. Additional numerical parameters (not required by the Gartner and Peter models) include: the electron affinity set at 3.74 eV; the carrier mobility set at 1.3 cm$^2 \cdot$ V$^{-1} \cdot$ s^{-1}; the bulk carrier lifetime set at 10^{-11} s; the supporting electrolyte concentration set at 1 M; and the liquid pH set at 14 to align the water electrochemical potential (E_L) with the OH^-/O_2 redox level (Hisatomi et al., 2014; Nurlaela et al., 2016; Ziani et al., 2015). The relative dielectric constant for water was set at $\epsilon_r = 80$.

7.5 Outlook

In this chapter, we have explored the manner in which band diagrams can be applied to understand the illuminated and dark current characteristics of photoanodes. However, the approach is general and can also be applied to photocathodes. Viewed more broadly, the band diagram approach provides a versatile and compact framework for interpreting the operation of PEC devices. Yet, as in any field of research, much more remains to be done. A major outstanding issue is the rigorous justification of the interfacial hole/electron transfer rate parameters (Mesa et al., 2020). It would also be interesting to understand PEC charge trapping and transfer dynamics more rigorously through comparison with transient spectroscopy experiments (Iqbal et al., 2021). As well, further effort is needed to incorporate more accurate descriptions of interfacial screening in the electrolyte region of the solid–liquid interface (Schmickler, 1996). Photocorrosion also needs be further tackled both experimentally and theoretically (Li et al., 2021a; Su, 2016). Compact models that go beyond the Gartner–Peter approach and yet provide faster computational insights than can be achieved with fully self-consistent calculations should also be of great utility. Additionally, our research community needs to work toward a methodology for integrating atomistic insights (e.g., via density functional theory) and experimental observations within PEC device level calculations (Iqbal et al., 2021, 2018; Mesa et al., 2020; Steiger et al., 2011). This remains an ongoing effort on all fronts, with plenty of room for exciting new insights and extensions to the present day understanding of PEC devices. Though the combined theoretical and experimental design of PEC devices new records in efficiency and performance will, hopefully, be achievable in the near term.

References

Bagal, I. V., Chodankar, N. R., Hassan, M. A., Waseem, A., Johar, M. A., Kim, D.-H., & Ryu, S.-W. (2019). Cu_2O as an emerging photocathode for solar water splitting - a status review. *International Journal of Hydrogen Energy*, *44*, 21351–21378.

Bard, A. J., & Faulkner, L. R. (2000). *Electrochemical methods: fundamentals and applications.* John Wiley & Sons, 2nd edition.

Bevan, K. H., Foong, Y. W., Shirani, J., Yuan, S., & Farraj, S. A. (2021). Physics applied to electrochemistry: tunneling reactions. *Journal of Applied Physics*, *129*, 090901.

Brown, M. A., Goel, A., & Abbas, Z. (2016). Effect of electrolyte concentration on the stern layer thickness at a charged interface. *Angewandte Chemie International Edition*, *55*, 3790–3794.

Cass, M. J., Duffy, N. W., Kirah, K., Peter, L. M., Pennock, S. R., Ushiroda, S., & Walker, A. B. (2002). Applications of microwave reflectance methods to the study of p-Si in fluoride solutions. *Journal of Electroanalytical Chemistry*, *538–539*, 191–203.

Cendula, P., Tilley, S. D., Gimenez, S., Bisquert, J., Schmid, M., Gratzel, M., & Schumacher, J. O. (2014). Calculation of the energy band diagram of a photoelectrochemical water splitting cell. *Journal of Physical Chemistry C*, *118*, 29599–29607.

Gao, Y., & Hamann, T. W. (2017). Elucidation of $CuWO_4$ surface states during photoelectrochemical water oxidation. *Journal of Physical Chemistry Letters*, *8*, 2700–2704.

Gartner, W. W. (1959). Depletion-layer photoeffects in semiconductors. *Physical Review*, *116*, 84–86.

Gaudy, Y. K., & Haussener, S. (2016). Utilizing modeling, experiments, and statistics for the analysis of water-splitting photoelectrodes. *Journal of Materials Chemistry A, 4*, 3100–3114.

Gummel, H. K. (1964). A self-consistent iterative scheme for one-dimensional steady state transistor calculations. *IEEE Transactions on Electron Devices, 11*, 455–465.

Hisatomi, T., Kubota, J., & Domen, K. (2014). Recent advances in semiconductors for photocatalytic and photoelectrochemical water splitting. *Chemical Society Reviews, 43*, 7520–7535.

Iqbal, A. (2018). *Modeling of carrier transport phenomena at photocatalytic semiconductor-liquid interfaces*. McGill University.

Iqbal, A., & Bevan, K. H. (2018a). The impact of boundary conditions on calculated photovoltages and photocurrents at photocatalytic interfaces. *MRS Communications, 8*, 466–473.

Iqbal, A., & Bevan, K. H. (2018b). Simultaneously solving the photovoltage and photocurrent at semiconductor-liquid interfaces. *Journal of Physical Chemistry C, 122*, 30–43.

Iqbal, A., Hossain, M. S., & Bevan, K. H. (2016). The role of relative rate constants in determining surface state phenomena at semiconductor–liquid interfaces. *Physical Chemistry Chemical Physics, 18*, 29466–29477.

Iqbal, A., Kafizas, A., Sotelo-Vazquez, C., Wilson, R., Ling, M., Taylor, A., Blackman, C., Bevan, K., Parkin, I., & Quesada-Cabrera, R. (2021). Charge transport phenomena in heterojunction photocatalysts: the WO_3/TiO_2 system as an archetypical model. *ACS Applied Materials & Interfaces, 13*, 9781–9793.

Iqbal, A., Yuan, S., Wang, Z., & Bevan, K. H. (2018). Impact of bulk trapping phenomena on the maximum attainable photovoltage of semiconductor–liquid interfaces. *Journal of Physical Chemistry C, 122*, 23878–23889.

Khan, S. U. M., Al-Shahry, M., & Ingler Jr, W. B. (2002). Efficient photochemical water splitting by a chemically modified n-TiO_2. *Science, 297*, 2243–2244.

Kim, J. S., Kim, B., Kim, H., & Kang, K. (2018). Recent progress on multimetal oxide catalysts for the oxygen evolution reaction. *Advanced Energy Materials, 8*, 1702774.

Li, K., Miao, B., Fa, W., Chen, R., Jin, J., Bevan, K., & Wang, D. (2021a). Evolution of surface oxidation on Ta_3N_5 as probed by a photoelectrochemical method. *ACS Applied Materials & Interfaces, 13*, 17420–17428.

Li, J., Wan, W., Triana, C. A., Chen, H., Zhao, Y., Mavrokefalos, C. K., & Patzke, G. R. (2021b). Reaction kinetics and interplay of two different surface states on hematite photoanodes for water oxidation. *Nature Communications, 12*, 255.

Martin, R. M. (2004). *Electronic structure: basic theory and practical methods*. Cambridge University Press.

Memming, R. (2015). *Semiconductor electrochemistry*. Wiley-VCH Verlag GmbH & Co. KGaA, 2nd Ed.

Mesa, C. A., Francàs, L., Yang, K. R., Garrido-Barros, P., Pastor, E., Ma, Y., Kafizas, A., Rosser, T. E., Mayer, M. T., Reisner, E., Grätzel, M., Batista, V. S., & Durrant, J. R. (2020). Multihole water oxidation catalysis on haematite photoanodes revealed by operando spectroelectrochemistry and DFT. *Nature Chemistry, 12*, 82–89.

Miao, B., Iqbal, A., & Bevan, K. H. (2019). Utilizing band diagrams to interpret the photovoltage and photocurrent in photoanodes: a semiclassical device modeling study. *Journal of Physical Chemistry C, 123*, 28593–28603.

Miao, B., Sangare, K., Iqbal, A., Marsan, B., & Bevan, K. H. (2020). Interpreting interfacial semiconductor–liquid capacitive characteristics impacted by surface states: a theoretical and experimental study of $CuGaS_2$. *Physical Chemistry Chemical Physics, 22*, 19631–19642.

Nelson, J. A. (2003). *The physics of solar cells*. Imperial College Press.

Nurlaela, E., Ziani, A., & Takanabe, K. (2016). Tantalum nitride for photocatalytic water splitting: concept and applications. *Materials for Renewable and Sustainable Energy, 5*, 18.

Peter, L. M. (2013). Energetics and kinetics of light-driven oxygen evolution at semiconductor electrodes: the example of hematite. *Journal of Solid State Electrochemistry, 17*, 315–326.

Peter, L. M., & Wijayantha, K. G. U. (2014). Photoelectrochemical water splitting at semiconductor electrodes: fundamental problems and new perspectives. *ChemPhysChem, 15*, 1983–1995.

Pierret, R. F. (1996). *Semiconductor device fundamentals*. Pearson Education.

Pierret, R. F. (2002). *Advanced semiconductor fundamentals*. Pearson, 2nd Ed.

Scharfetter, D. L., & Gummel, H. K. (1969). Large-signal analysis of a silicon read diode oscillator. *IEEE Transactions on Electron Devices, 16*, 64–77.

Schmickler, W. (1996). *Interfacial electrochemistry*. Oxford University Press, 1st Ed.

Smith, T. J., & Stevenson, K. J. (2007). Reference electrodes. In C. G. Zoski (Ed.), *Handbook of electrochemistry*, Chapter 4. Elsevier.

Srivastava, G. P. (1984). Broyden's method for self-consistent field convergence acceleration. *Journal of Physics A: Mathematical and General, 17*, L317–L321.

Steiger, S., Povolotskyi, M., Park, H., Kubis, T., & Klimeck, G. (2011). NEMO5: a parallel multiscale nanoelectronics modeling tool. *IEEE Transactions on Nanotechnology*, *10*, 1464–1474.

Su, L., & Vayssieres, J. (2016). A place in the sun for artificial photosynthesis. *ACS Energy Letters*, *1*, 121–135.

Trasatti, S. (1986). The absolute electrode potential: an explanatory note (Recommendations 1986). *Pure and Applied Chemistry*, *58*, 955–966.

van de Krol, R. (2012a). Principles of photoelectrochemical cells. In R. van de Krol, & M. Gratzel (Eds.), *Photoelectrochemical hydrogen production*, Chapter 2. Springer.

van de Krol, R. (2012b). Photoelectrochemical measurements. In R. van de Krol, & M. Gratzel (Eds.), *Photoelectrochemical hydrogen production*, Chapter 3. Springer.

Vasileska, D., & Goodnick, S. M. (2006). *Computational electronics*. Morgan & Claypool.

Voit, E. O., Martens, H. A., & Omholt, S. W. (2015). 150 years of the mass action law. *PLOS Computational Biology*, *11*, e1004012.

Xu, C., Anusuyadevi, P. R., Aymonier, C., Luque, R., & Marre, S. (2019). Nanostructured materials for photocatalysis. *Chemical Society Reviews*, *48*, 3868–3902.

Ziani, A., Nurlaela, E., Dhawale, D. S., Silva, D. A., Alarousu, E., Mohammed, O. F., & Takanabe, K. (2015). Carrier dynamics of a visible-light-responsive Ta_3N_5 photoanode for water oxidation. *Physical Chemistry Chemical Physics*, *17*, 2670–2677.

Section 2

Materials for Solar Fuel Generation

8

Materials Used for Solar Thermal/Thermochemical Processes for CO_2/H_2O Dissociation/Conversion

Heng Pan, Youjun Lu, and Bingchan Hu

State Key Laboratory of Multiphase Flow in Power Engineering (SKLMF), School of Energy and Power Engineering, Xi'an Jiaotong University, Xi'an, Shaanxi, China

8.1 Introduction

Development of commercially viable renewable energy conversion processes has become one of the major goals of the global research community. Solar energy is clean, sustainable, accessible, and abundant. However, the energy density of solar irradiation is low, unstable, and discontinuous and these bring technical challenges to its utilization (Lewis & Nocera, 2006; Steinfeld, 2005). Many technologies involving electrochemical (Bareiss et al., 2019; Dincer & Acar, 2015), thermochemical (Abanades et al., 2006; Kodama, 2003; Yadav & Banerjee, 2016), and photochemical (Colón, 2016; Ismail & Bahnemann, 2014) processes have emerged to convert and store solar energy as fuels for transportation and long-term utilization. Compared to other approaches, thermochemical processes integrated with concentrated solar irradiation could utilize the full solar spectrum and facilitate conversion of solar energy to chemical energy with high energy density and low losses.

H_2 is one of the most promising fuels for future energy needs due to its high energy density and carbon-free composition. It can be used directly as a fuel or converted to other fuels/chemicals through existing industrial conversion processes. In addition, the carbon element in an organic fuel can come from carbon dioxide captured in the existing industrial system, thus realizing the resource utilization of carbon dioxide. The conversion of carbon dioxide can be achieved by hydrogenation reactions with the produced hydrogen, or by direct reduction to carbon monoxide using focused solar energy to form synthesis gas with hydrogen, which can be converted to liquid fuels using other processes.

Solar thermochemical processes integrated with concentrated solar irradiation for CO_2/H_2O dissociation/conversion offer an attractive option for converting solar energy to chemical energy in gaseous fuels (CO/H_2) (Abanades et al., 2006; Kodama, 2003; Wang, 2020; Yadav & Banerjee, 2016). These processes include solar thermolysis, solar thermochemical, solar reforming, and solar gasification. In order to deepen the knowledge and help the advancement of research in the area, catalytic materials for these processes are introduced and explained in this chapter. The single-step thermal dissociation of CO_2/H_2O (solar thermolysis) is discussed first. Then the research progress in the two-step thermochemical cycles is summarized and materials are elaborated. Toward the end, the materials used for sulfur–iodine (S–I) cycle and metal-halide cycles to produce hydrogen, and the catalysts for solar gasification and reforming are discussed.

8.2 Solar Thermolysis of H_2O or CO_2

The single-step thermal dissociation of H_2O or CO_2 can be given as,

$$H_2O \rightarrow H_2 + \frac{1}{2}O_2 \tag{8.1a}$$

or

$$CO_2 \rightarrow CO + \frac{1}{2}O_2 \tag{8.1b}$$

Conversion of Water and CO_2 to Fuels using Solar Energy: Science, Technology and Materials, First Edition.
Edited by Oomman K. Varghese and Flavio L. Souza.

Direct thermal decomposition of H_2O/CO_2 needs very high temperatures, for example, H_2O decomposition is barely discernible at 2,000 K, while 25% of water vapor dissociates at 2,500 K at a pressure of 0.05 bar. The high temperature poses challenges in hydrogen–oxygen high-temperature separation (Fletcher & Moen, 1977). Effusion and electrolytic separations are two ways to separate hydrogen from the products. In addition, the extremely high temperatures required for these processes pose a serious challenge to the materials. Many materials have been proposed as candidates for this application for decades, but even zirconium-based ceramics cannot withstand the severe thermal shocks that often occur when operating at high fluxes of solar radiation. Therefore, it is more feasible to decompose CO_2/H_2O by a multi-step chemical reaction scheme, the so-called thermochemical cycle (the net reaction remains the decomposition of CO_2/H_2O), with the help of other reactants/products (Fletcher, 1999; Fletcher & Moen, 1977; Kogan, 1998; Traynor & Jensen, 2002).

8.3 Redox Pairs for Two-Step Thermochemical Cycles

The thermochemical cycle may consist of two or more steps. In most cases, the required maximum temperature decreases when more steps are employed, but the energy efficiency also decreases because of the energy losses associated with each step (Agrafiotis et al., 2015; Chaubey et al., 2013; Kodama & Gokon, 2007; Muhich et al., 2016; Perkins & Weimer, 2004; Perkins & Weimer, 2009; Xiao et al., 2012). Consequently, considerable efforts were put into the research on two-step cycles. A two-step thermochemical cycle can be expressed as,

Reduction:

$$MO_x \rightarrow MO_{x-\delta} + \frac{\delta}{2}O_2 \tag{8.2}$$

Oxidation:

$$MO_{x-\delta} + \delta CO_2 \rightarrow MO_x + \delta CO \tag{8.3a}$$

or

$$MO_{x-\delta} + \delta H_2O \rightarrow MO_x + \delta H_2 \tag{8.3b}$$

As shown in Figure 8.1, the two-step thermochemical cycle for H_2O/CO_2 splitting consists of, first, the reduction of metal oxides (MO_x) using concentrated solar radiation as the energy source for the high-temperature process as shown in Equation 8.2, and second, the oxidation of the reduced oxides ($MO_{x-\delta}$) with CO_2/H_2O as shown in Equation 4.3. The δ indicates that they are partial molar volumes. Metal oxides used as the CO_2/H_2O splitting medium are recycled and no corrosive or poisonous substance is produced.

The thermodynamic properties of the oxide redox pair ($MO_x/MO_{x-\delta}$) determine the conditions required for the reaction and the fuel output of a single cycle. The equilibrium thermodynamic expressions are shown as,

$$\Delta G_{\text{red}} = \Delta H_{\text{red}} - T_{\text{red}}\Delta S_{\text{red}} + \frac{1}{2}RT_{\text{red}} \ln\left(\frac{p_{O_2}}{p^0}\right) < 0 \tag{8.4}$$

$$\Delta G_{\text{ox}} = (\Delta H_{CO_2} - \Delta H_{\text{red}}) - T_{\text{ox}}(\Delta S_{CO_2} - \Delta S_{\text{red}}) + RT_{\text{ox}} \ln\left(\frac{p_{CO}}{p_{CO_2}}\right) < 0 \tag{8.5}$$

Here ΔH and ΔS are the standard reduction enthalpy and entropy, respectively. The subscripts red and ox indicate the reduction and oxidation, respectively. From this criterion, it can be concluded that materials with a lower reduction enthalpy and a higher reduction entropy would be easier to reduce. Also, an atmosphere with a lower oxygen partial pressure could

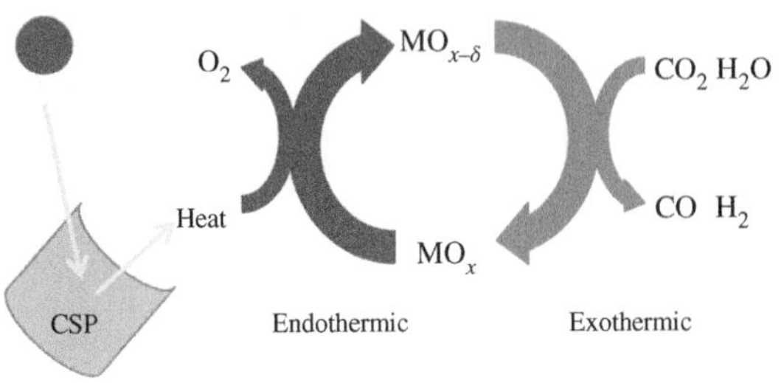

Figure 8.1 A schematic of a generic two-step solar thermal H_2O/CO_2 splitting cycle.

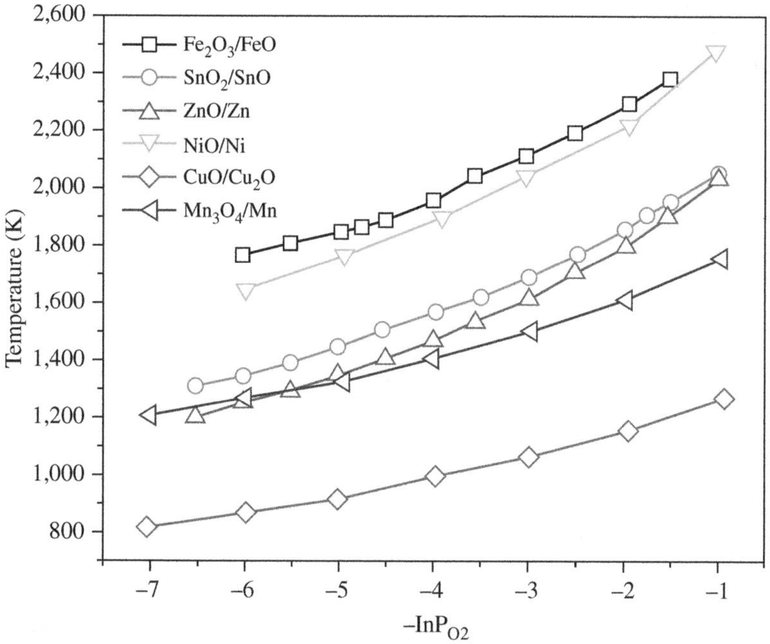

Figure 8.2 Contours of $\Delta G = 0$ of Equation (8.4) on a temperature vs. pressure graph with data from Bulfin et al. (2017).

enhance the reduction processes. These trends point out the way for material modification and system design, respectively. Figure 8.2 shows the conditions required for reduction/oxidation of different redox pairs.

Since the temperatures required for the thermal decomposition reactions are high, they often exceed the boiling temperatures of the reduced species. Thus, in general, redox pairs can be classified as volatile redox pairs (such as ZnO/Zn and SnO_2/SnO) and nonvolatile redox pairs (such as CeO_2/Ce_2O_3 and Fe_2O_3/FeO), depending on whether the redox material remains in the condensed state during the entire process or not. In addition, they can also be divided into stoichiometric (or phase-change) redox pairs and nonstoichiometric (partial reduction or oxygen-vacancy) redox pairs according to the reduction mechanism. We classify redox pairs as volatile and nonvolatile pairs to introduce them in this part.

8.3.1 Volatile Redox Pairs

As mentioned above, gaseous metal oxides or metal elements are produced in the reduction reactions of volatile redox pairs cycles. The solid-gas phase transition will result in a significant entropy increase, which is thermodynamically helpful for this step. As shown in Figure 8.3, the volatile pairs (ZnO/Zn, SnO_2/SnO) theoretically have higher fuel yields than the nonvolatile pairs (CeO_2/Ce_2O_3, Fe_2O_3/FeO). However, the gaseous products can easily recombine with oxygen during cooling, leading to a significant decrease in metal oxide conversion rate and fuel yield. Therefore, high-temperature gas separation devices are required in the cycles involving volatile redox pairs. The high-temperature gas separation is also a major obstacle to the large-scale application of these cycles.

8.3.1.1 ZnO/Zn Pair

The production of hydrogen from water using solar energy via ZnO/Zn cycle was first considered by Steinfeld et al. (Steinfeld, 2002). The ZnO/Zn cycle has received extensive attention in research for a long period due to its high fuel yields and energy efficiency (Abanades, 2012; Abanades et al., 2007; Bhosale, 2021; Jarrett et al., 2016; Loutzenhiser et al., 2010; Palumbo et al., 1998; Wegner et al., 2006; Weidenkaff et al., 2000). The decomposition reaction of ZnO in the first step is a high-temperature endothermic reaction (1,700–2,000 °C) and the produced zinc r exists in gaseous state. However, the produced zinc vapor can easily recombine with oxygen, which decreases the decomposition rate of zinc oxide. Fortunately, a tube furnace with temperature gradient has recently been designed as a quenching device (Abanades & Chambon, 2010; Müller & Steinfeld, 2008). It avoids the recombination of zinc vapor and O_2 produced by the solar thermal dissociation of ZnO. The decomposition rate of ZnO is directly proportional to the temperature and the mass flow rate of the inert gas, and inversely proportional to the oxygen partial pressure. When the Ar/Zn dilution is 530 and the cooling rate is 10^5 K/s,

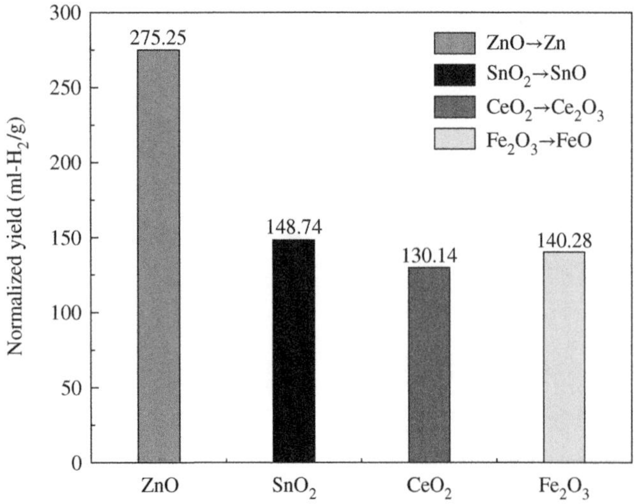

Figure 8.3 Calculated normalized yield of various metal oxides under standard conditions based on redox limit. Reproduced from Mao et al. (2020)/with permission of Elsevier.

the zinc yield can reach 94% (Gstoehl et al., 2008). In addition, other effective techniques for direct in-situ separation of zinc vapor/oxygen at high temperatures can also be used to avoid recombination.

In the second step, the obtained Zn is cooled to about 500 °C so that it can participate in the H_2O/CO_2 decomposition reaction. However, in this step, zinc oxide layers may form on the surface of Zn and these oxide layers would prevent effective contact between the fresh reactants. Thus, the use of Zn particles with high specific surface area can increase the reaction area and, thus, the H_2O/CO_2 conversation rate. In addition, the oxidation temperature and H_2O/CO_2 concentration also have important effects on the reaction processes. As shown in Figure 8.4, increasing the oxidation temperature and the concentration of H_2O can increase the oxidation reaction rate.

The energy conversion efficiency of this cycle can reach up to 29% (Steinfeld, 2002). However, there are still some problems with this process that hinder its commercialization. In 2016, a techno-economic analysis of a 110 MW ZnO/Zn solar

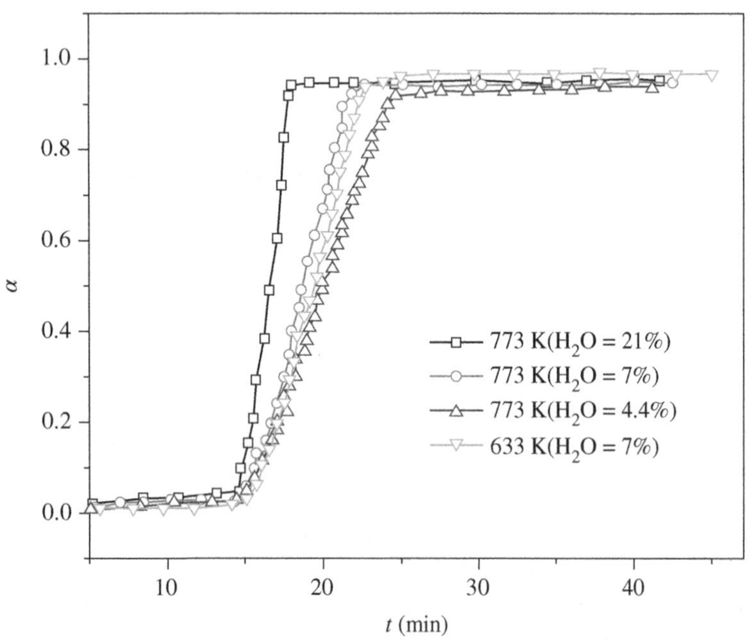

Figure 8.4 The conversion of Zn vs. time for Zn hydrolysis during isothermal thermogravimetry at different temperatures and steam mole fractions with data from Abanades (2012). Where α indicates the degree of reaction proceeding, $\alpha = 0$ means the reaction has not started, while $\alpha = 1$ means the reaction is complete.

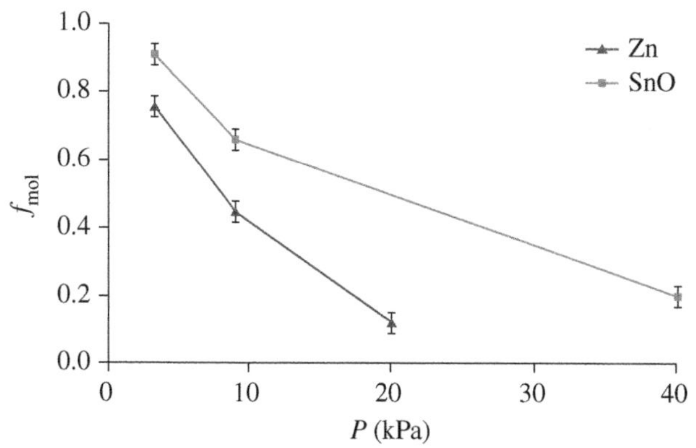

Figure 8.5 Comparison of the mole fraction of reduced species in the powders obtained from ZnO and SnO_2 reductions. Reproduced from Mao et al. (2020)/with permission of Elsevier.

hydrogen production facility showed a predicted hydrogen price of $8/kg without gas quenching and over $50/kg with gas quenching, which is not competitive with other hydrogen production technology pathways using nonrenewable energy sources (Koepf et al., 2017).

8.3.1.2 SnO_2/SnO Pair

Two-step thermochemical cycle based on SnO_2/SnO pair has been proven to be a promising cycle for hydrogen production via solar water splitting. It has many characteristics similar to the ZnO/Zn system (Abanades, 2012; Abanades, 2019; Abanades et al., 2008; Charvin et al., 2008; Leveque et al., 2014). SnO_2 undergoes thermal reduction at approximately 1,600 °C at atmospheric pressure. The advantages of this pair over ZnO/Zn pair are that the required reduction temperature can be reduced and no passivation occurs during the oxidation process (Chambon et al., 2010; Musgrave & Weimer, 2015; Palumbo et al., 1998). However, as shown in Figure 8.5, compared with ZnO/Zn cycle, the yield of fuels of SnO_2/SnO cycle is lower.

Recently, other volatile redox pairs, such as CdO/Cd (Sibieude et al., 1982) and GeO_2/GeO (Kang et al., 2009), were also investigated. These studies found that the volatile pairs have the advantage of higher theoretical fuel yield compared to nonvolatile pairs. However, the volatile pair cycles have the inherent technical drawback that the gases produced by reduction reaction would easily recombine with oxygen during cooling, resulting in a significantly lower reaction rate and thus lower fuel yield (Mao et al., 2020; Xiao et al., 2012). Despite considerable progress in volatile cycle reactor research over the past few decades, the major research groups still believe that commercialization of volatile pairs is highly unlikely (Bhosale et al., 2022; Mao et al., 2020).

8.3.2 Nonvolatile Redox Pairs

Considering the technical difficulties inherent in volatile cycling, researchers have turned their attention to the development of nonvolatile redox pairs in recent years. The nonvolatile redox pairs remain solid throughout the cycle, effectively avoiding the problem of product recombination in volatile cycles.

8.3.2.1 Fe_3O_4/FeO Pair

The Fe_3O_4/FeO cycle has long been considered promising because of its higher fuel yield and lower reduction temperature than that of ZnO/Zn cycle (Abanades & Villafan-Vidales, 2011; Charvin et al., 2007). Figure 8.6 shows the equilibrium composition of the Fe/O system (100 mol of N_2, 1 mol of Fe_2O_3). The decomposition reaction temperature of Fe_3O_4 exceeds 2,500 K in air. The lower the partial pressure of oxygen, the lower the temperature required. Fe_3O_4 can be reduced to FeO at 1,400 °C under the conditions of low oxygen partial pressure generated by inert gas. However, the disadvantage of this cycle is that the melting temperature of iron oxide is relatively low (about 1,350 °C), which exposes the material to sintering and even liquefaction during the actual process. Sintering greatly reduces the gas-solid contact area, thus causing the material to exhibit significant deactivation in the oxidation reaction. There were efforts to replace ferric ions with other transition metal ions in order to reduce the reduction temperature of materials and alleviate the sintering/liquefaction problem (Bush et al., 2019; Goikoetxea et al., 2019; Huang et al., 2018; Huang et al., 2019; Tamaura et al., 1999; Zhai et al., 2020). Ferrite materials,

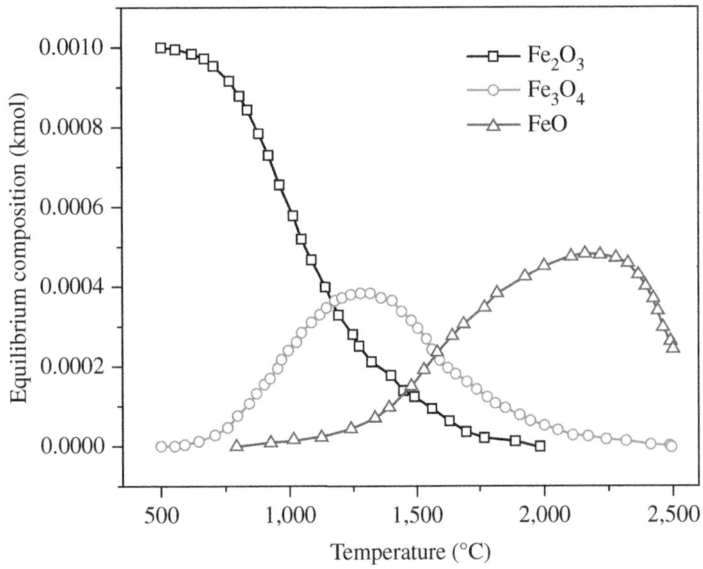

Figure 8.6 Equilibrium composition of the Fe/O system (100 mol of N_2, 1 mol of Fe_2O_3) as a function of temperature with data from Charvin et al. (2007).

MFe_2O_4, where M is Ni, Co, Zn, Mn, etc., that are commonly used in the electronics industry were thus introduced into this cycle. In addition, researchers tried to use support materials such as zirconia to keep the ferrites active at high temperatures. However, sintering-induced activity reduction is still an unavoidable problem in this cycle. Furthermore, the introduction of support material greatly reduces the fuel yield of the material (Gokon et al., 2009; Huang et al., 2019; Ishihara et al., 2008; Kodama et al., 2005; Kodama et al., 2006; Kodama et al., 2008).

8.3.2.2 CeO_2/$CeO_{2-\delta}$ Pairs

Ceria has been widely investigated as a catalyst and an oxygen storage material (Abanades & Flamant, 2006; Chueh et al., 2010; Mogensen et al., 2000; Otsuka et al., 1983; Otsuka et al., 1985; Yokokawa et al., 2006). It has a much higher melting point than iron oxides, which makes it very stable in high-temperature processes. Ceria could be partially reduced without major changes in lattice structure, favoring a high oxidation rate. Lattice cerium ions are partially reduced from Ce^{4+} to Ce^{3+}, with the reduction extent depending on the reduction conditions. In the 1980s, the thermochemical cycle for H_2O splitting using ceria as the oxygen carrier was first reported (Otsuka et al., 1983; Otsuka et al., 1985). At that time, the reduction method adopted was not thermal reduction, but H_2-assisted reduction. Complete reduction of CeO_2 to Ce_2O_3 (the fuel yield is 2.91 mol/kg) was achieved at a reduction temperature of 2,000 °C (Abanades & Flamant, 2006). However, at this temperature, ceria would enter a molten state and thus be partially lost through sublimation. Subsequently, a partial reduction of CeO_2 to $CeO_{2-\delta}$ was investigated to lower the reduction temperature, where δ is the stoichiometric number corresponding to the oxygen vacancies in the cerium lattice (Bhosale et al., 2019; Chueh et al., 2010). Figure 8.7 shows isothermal log δ vs. p_{O_2} for nonstoichiometric ceria. In contrast to the complete reduction process, partial reduction involves the formation of oxygen vacancy in the crystal lattice. The oxides keep their original crystal phase unchanged within a certain range of vacancy concentration. In 2010, researchers from the California Institute of Technology (Caltech) and ETH reported that a H_2O/CO_2 decomposition system based on partial reduction of cerium oxide maintained a stable fuel yield over 500 cycles (Chueh et al., 2010). Since then, almost all research groups in this field got involved in the study of the two-step cycle of ceria due to its rapid reaction rate and high stability. For the development of other redox materials, a criterion used was that the reduction should happen at a temperature below the material's melting point.

Doping is generally considered as an effective method to improve the extent of lattice reduction or reduce the operating temperature for the reduction reaction. Many dopants were studied with varying degrees of success. Both valence and ionic radius were supposed to be important factors determining the performance (Haeussler et al., 2019; Lee et al., 2013; Lu et al., 2019). However, despite the improvement in product yield, both Hf and Zr doped ceria showed an obvious drawback in oxidation kinetics (Haeussler et al., 2019; Kang et al., 2013; Zhao, Huang, et al., 2016; Zhu et al., 2018). This could also be understood through thermodynamics. The decrease in reduction enthalpy lowers the driving force for the oxidation

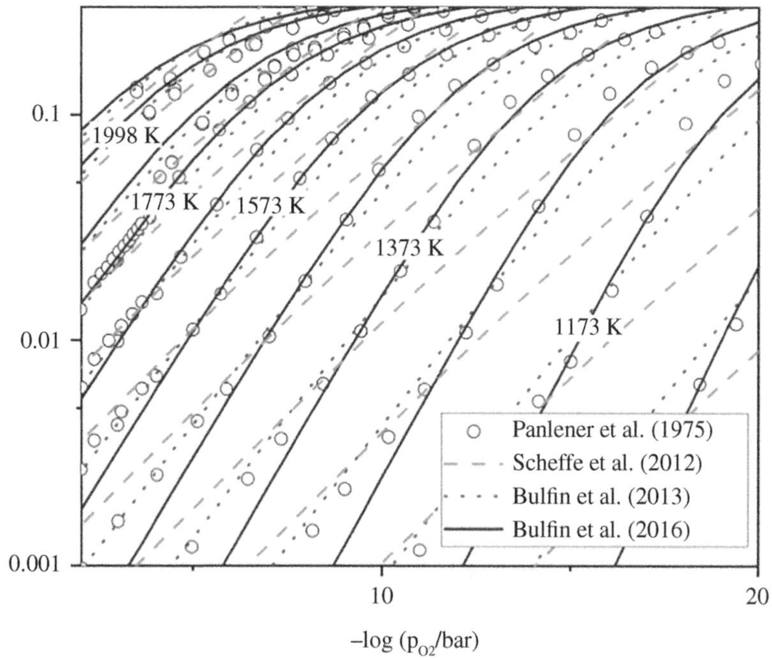

Figure 8.7 Isothermal log δ vs. p_{O_2} for nonstoichiometric ceria. Reproduced from Lu et al. (2019)/with permission of Elsevier.

reaction, leading to a slow oxidation rate. There is an inherent contradiction in the terms "easy reduction" and "rapid oxidation." This is because a lower reduction enthalpy favors reduction but does not favor oxidation.

8.3.2.3 $CoFe_2O_4/FeAl_2O_4$ Pairs

In 2010, a team from the University of Colorado reported a cobalt–iron–iron aluminate ($CoFe_2O_4/FeAl_2O_4$) thermochemical cycle, which is considered to be a complete reduction cycle,

$$CoFe_2O_4 + 3Al_2O_3 \rightarrow CoAl_2O_4 + 2FeAl_2O_4 + \frac{1}{2}O_2 \tag{8.6}$$

$$CoAl_2O_4 + 2FeAl_2O_4 + H_2O \rightarrow CoFe_2O_4 + 3Al_2O_3 + H_2 \tag{8.7}$$

It was believed that cobalt ferrite supported on alumina substrate could be reduced at 1,200 °C. Compared to simple cobalt ferrite, its reduction temperature could be lowered by 150–200 °C (Arifin et al., 2012; Scheffe et al., 2010). In addition, this cycle could be carried out in isothermal mode (the oxidation temperature is equal to the reduction temperature), which could effectively lower the sensible heat loss of solids in a non-isothermal cycle (Carrillo & Scheffe, 2019; Hoskins et al., 2019; Li et al., 2019; Muhich et al., 2013). However, the reaction process corresponding to Equation 8.6 was not realized in the subsequent studies. In addition, first principles studies showed that cobalt aluminate and ferric aluminate do not have the ability to decompose water. The water decomposition effect obtained in the experiment of this cycle is based on the partial reduction of cobalt ferrites, i.e., the oxygen vacancy mechanism (Muhich et al., 2015). Similarly, other ferrites have also been shown to undergo partial reduction at lower reduction temperatures. During the redox process, redox pairs of materials can switch between their oxygen-rich and oxygen-deficient states by forming or filling oxygen vacancies (Agrafiotis et al., 2012; Agrafiotis et al., 2013; Dimitrakis et al., 2016).

8.3.2.4 Perovskites

Considering that some perovskites have an oxygen vacancy formation mechanism similar to that of ceria, some scholars have studied the application of perovskites in two-step thermochemical cycles (Barcellos et al., 2018; Cooper et al., 2015; Demont & Abanades, 2014; Demont et al., 2014; Haeussler et al., 2018; McDaniel et al., 2013; McDaniel et al., 2014; Nair & Abanades, 2018; Sastre et al., 2017; Scheffe et al., 2013). Perovskites represent a large group of materials with an AMO_3 structure, where A is usually an alkali or rare earth metal and M is usually a transition metal that exhibits redox properties during the thermochemical cycle. Figure 8.8 shows the measured CO production rates vs. time during the oxidation of

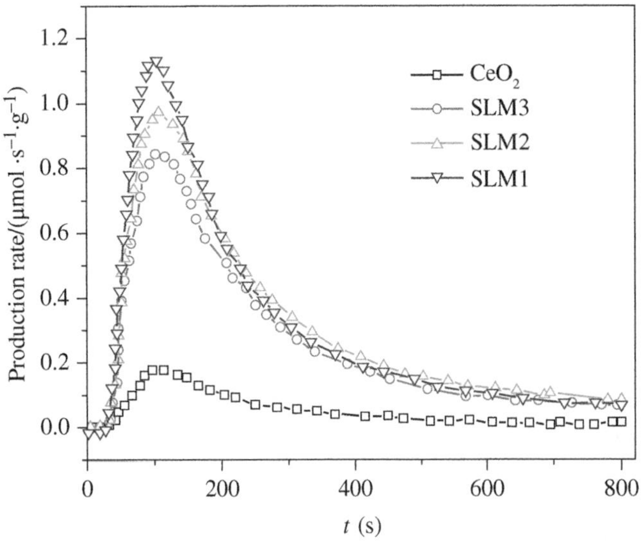

Figure 8.8 CO production rates as a function of time measured during oxidation in 40 vol% CO_2 at 1,000 °C with data from McDaniel et al. (2013).

Mn^{4+}/Mn^{3+}-based perovskites and CeO_2 in 40 vol% CO_2 at 1,000 °C. The Mn^{4+}/Mn^{3+}-based perovskite cycle shows lower reduction temperature and higher reduction degree than the ceria cycle, and is considered to be a viable alternative oxygen carrier for ceria (McDaniel et al., 2013, 2014). However, its oxidation performance was found to be poor. The oxidation reaction process was slow and the H_2O/CO_2 consumption was high, causing a significant energy loss in the oxidation process, thus, it is difficult to show better cycling performance than ceria (Ezbiri et al., 2017; McDaniel et al., 2013). However, in contrast to ceria, perovskite represents a large class of materials with a very wide scope for adjustment of its composition. The thermodynamic properties can be controlled by adjusting its composition, and it is very possible to find perovskite materials with lower reduction temperature and higher system efficiency than ceria. Aiming at this goal, scholars have extensively investigated the redox properties of perovskite materials and found that perovskites with Co, Fe, Mn, and Ti as redox ions have increasing reduction resistance in descending order (Bulfin, Vieten, Agrafiotis, Roeb, & Sattler, 2017; Bulfin, Vieten, Starr, Azarpira, et al., 2017; Vieten et al., 2016; Vieten et al., 2017). The reduction difficulty of ceria is between that of Mn and Ti. Therefore, an oxygen carrier with a better performance than ceria may be obtained by adjusting the ratio of Mn and Ti ions. In addition, scholars from Northwestern University have screened all kinds of possible perovskite structures based on the redox properties obtained from flux calculations (Emery et al., 2016). However, its practicality needs further experimental verification.

8.3.3 Redox Pairs: New Discoveries

As shown in Table 8.1, different redox pairs of two-step thermochemical cycles are compared in this section. Energy efficiency is particularly critical in this area since it has a decisive influence on the commercial prospects of this cycle. It is defined as the ratio of the calorific value of the produced gaseous fuel to the total solar energy input. The theoretical efficiency of the system under different reaction conditions can be obtained by combining the thermodynamic data of the redox material with some assumptions on the energy-consuming terms. Many thermodynamic models have been proposed to predict the energy conversion efficiency of the whole system. The results and conclusions are different due to different assumptions/parameters/approaches employed on the energy losses. Thus, it may not be meaningful to directly compare the results or build a new model without experimental support. Therefore, in this section, we only clarify the effect of different reaction modes (reaction conditions and technological approaches) on the overall efficiency. Generally, there are two problems limiting the practical application of a two-step cycle: reduction at high temperatures, and re-oxidation completely with excess H_2O/CO_2 at high oxidation temperatures. Ceria, for example, could achieve a high oxidation rate; however, to be reduced, it requires a high reduction temperature.

Recently, a few studies have shown that hybrid cycle techniques might overcome the challenges of a two-step thermochemical cycle (Gao et al., 2020; Pan et al., 2022; Zhai et al., 2018; Zhang, Xu, et al., 2015). In 2018, Zhai et al.

Table 8.1 The Redox Pairs Investigated for a Two-Step Thermochemical Cycle.

Material	Reduction temperature (K)	Oxidation temperature (K)	Operating conditions	Efficiency	References
ZnO/Zn	2,300	700	Quench or in situ electrolytic separation	29%	Steinfeld (2002)
SnO_2/SnO	1,873	773–873	Quench or in situ electrolytic separation	29.8%	Bhosale, Kumar, and Sutar (2017)
Fe_3O_4/FeO	1,900	798–978	Milling or granulation	20.4–25.1%	Abanades and Villafan-Vidales (2011), Bush et al. (2019), Charvin et al. (2007), Goikoetxea et al. (2019), Gokon et al. (2009), Huang et al. (2018), Huang et al. (2019), Ishihara et al. (2008), Kodama et al. (2005), Kodama et al. (2006); Kodama et al. (2008); Tamaura et al. (1999), and Zhai et al. (2020)
$CeO_2/CeO_{2-\delta}$	1,400–2,000	100–500	No	10–22%	Abanades and Flamant, (2006), Bhosale et al. (2019), Chueh et al. (2010), Haeussler et al. (2019), Kang et al., (2013), Lee et al. (2013), Lu et al. (2019), Mogensen et al. (2000), Otsuka et al. (1983, 1985), Yokokawa et al. (2006), Zhao, Huang, et al. (2016), and Zhu et al. (2018)
$(La, Sr)Mn_xAl_yO_{3-\delta}$	1,623	1223	No	>20%	McDaniel et al. (2013, 2014) and Scheffe et al. (2013)
GeO_2/GeO	1,673–2,073	973	Quench	34.6%	Kang et al. (2009)
MgO/Mg	1,723–1,823	623–823	Need reducing agents	No	Galvez et al. (2008)

reported a two-step thermochemical cycle based on Fe, Mg, Co, and Ni multi-ion mixed oxides. They could carry out H_2 fuel production at a reduction temperature of 1,100 °C. Their material showed a better performance than Mn-based perovskites (Zhai et al., 2018). In 2020, Gao et al. synthesized mixed oxides with the same composition and loaded them on the surface of SiC (Gao et al., 2020). They used microwave heating to study their reaction performance in a two-step thermochemical cycle. The above studies initially demonstrated the superior performance of high-entropy oxides in the two-step thermochemical cycle. However, the current research on high-entropy oxides for two-step thermochemical cycle is very limited. In addition, Zhang et al. reported a photo-thermochemical cycle that combines photochemistry and thermochemistry to avoid the use of high temperatures. However, the fuel productivity remains very low (~100 μmol/m^2, calculated based on light coverage) since the oxygen vacancies created by photoexcitation are limited.

There were also studies on the two-step thermochemical cycle based on metal sulfate, such as $CdSO_4/CdO$ (Bhosale, 2020), $ZnSO_4/ZnO$ (Bhosale, Kumar, AlMomani, & Gupta, 2017), and $CaSO_4/CaO$ (Bhosale, 2019). The principle of these cycles is to decompose sulfate at high temperature to produce metal oxide, sulfur dioxide, and oxygen, and then the metal oxide, sulfur dioxide, and water react to release hydrogen and complete the regeneration of sulfate. However, the current research on this cycle is mainly focused on thermodynamic calculations.

8.4 Materials for Sulfur–Iodine (S–I) Cycle

General Atomics (GA) initially introduced the S–I cycle in 1974 on an experimental scale (Norman et al., 1981). The S–I cycle is one of the most promising large-scale hydrogen production thermochemical cycles due to its high energy efficiency (close to 56%), good economy, and low environmental impact (Hammache & Bilgen, 1992; Khan et al., 2021; Kim et al., 2008; Kurata et al., 2001; Park et al., 2013; Park et al., 2019; Park et al., 2021; Vitart et al., 2006; Ying et al., 2020). As shown

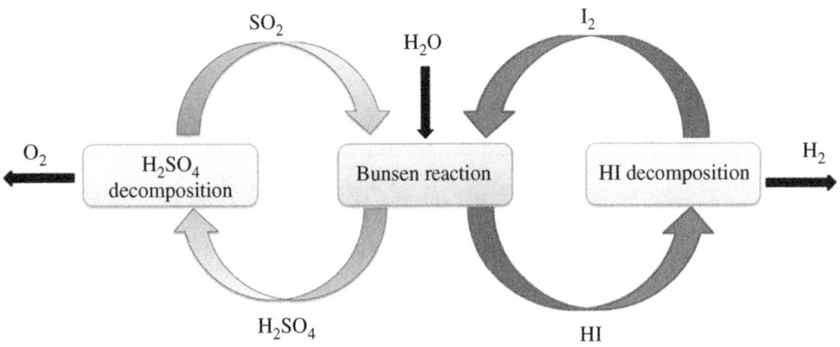

Figure 8.9 A schematic diagram of the S–I cycle.

in Figure 8.9, the S–I cycle can be performed in three reaction steps: Bunsen reaction (Equation 8.8), HI decomposition (Equation 8.9), and H_2SO_4 decomposition (Equation 8.10).

$$I_2(l) + SO_2 + 2H_2O \rightarrow H_2SO_4 + 2HI \tag{8.8}$$

$$2HI \rightarrow I_2 + H_2 \tag{8.9}$$

$$H_2SO_4 \rightarrow H_2O + SO_2 + 1/2O_2 \tag{8.10}$$

The Bunsen reaction could proceed quickly without any catalyst. However, catalysts are often required for the sustained formation of I_2 and SO_2 to increase the decomposition rate of HI and H_2SO_4. The extent and rate of sulfuric acid decomposition can be improved at very high temperatures (over 1,000 °C), which can be provided by solar concentrators. In addition, the circulating material contains highly corrosive substances such as sulfuric acid and sulfur dioxide. These highly corrosive substances can corrode the reactors at different temperatures and pressures. This section will describe the corrosion-resistant materials and catalysts explored for the decomposition of H_2SO_4 and HI.

8.4.1 Corrosion-Resistant Materials

Many studies are currently focusing on the selection and development of corrosion-resistant materials for S–I cycles. Corrosion ability of solutions increases with temperature, so in this section, we discuss corrosion-resistant materials according to different temperatures and corrosive substances. For extremely corrosive solution systems, four corrosion-resistant materials were screened: refractory metals, reactive metals, superalloys, and ceramics. Refractory metals and ceramics based on tantalum and niobium have shown excellent corrosion resistance (Choi et al., 2014; Kasahara et al., 2007; Kubo et al., 2013; Liberatore et al., 2019; Maatouk et al., 2009; Sakurai et al., 1999; Wong et al., 2007). In addition, the results of pressure corrosion resistance tests under evaporation and vaporization conditions of H_2SO_4 solutions showed that ceramic materials containing Si have higher corrosion resistance (Maatouk et al., 2009). Candidates for corrosion-resistant materials for different reaction processes are given in Figure 8.10. For the low-temperature Bunsen reaction part, ceramics (SiC, Si_3N_4, Al_2O_3), glass, fluorocarbons, nickel alloys, and rare metals are candidates. Zirconium and tantalum are the most suitable metals, but tantalum is susceptible to localized corrosion in specific reaction solutions (10 wt% H_2SO_4, 10 wt% HI, and 70 wt% I_2). SiC, Ta, and Nb can be used in the low-temperature process part of the HI concentration section, and the material of the HI decomposition part can be chosen from nickel-based alloys, SiC, Si_3N_4, or Mo. The H_2SO_4 concentration part can be made of SiC, Si_3N_4, FeSi, or Au (Sakurai et al., 1999), and the material for the H_2SO_4 decomposition part can be chosen from Incoloy 800, Hastelloy C276, Inconel 625, or SiC.

8.4.2 The Catalysts of HI Decomposition

HI decomposition is critical for hydrogen production in the S–I cycle. At 773 K, the conversion rate of HI in the homogeneous decomposition reaction is only about 1–2%, which is not sufficient for an industrial application. Therefore, catalysts must be used to increase the HI conversion rate (Choi et al., 2014; Kubo et al., 2013; Liberatore et al., 2019; Singhania & Bhaskarwar, 2017; Singhania & Bhaskarwar, 2018; Singhania & Bhaskarwar, 2018; Wong et al., 2007; Zhang et al., 2009; Zhang, Fu, et al., 2015; Zhang, Wang, Zhou, et al., 2008; Zhang, Zhou, Chen, et al., 2008; Zhang, Zhou, Wang, et al., 2008).

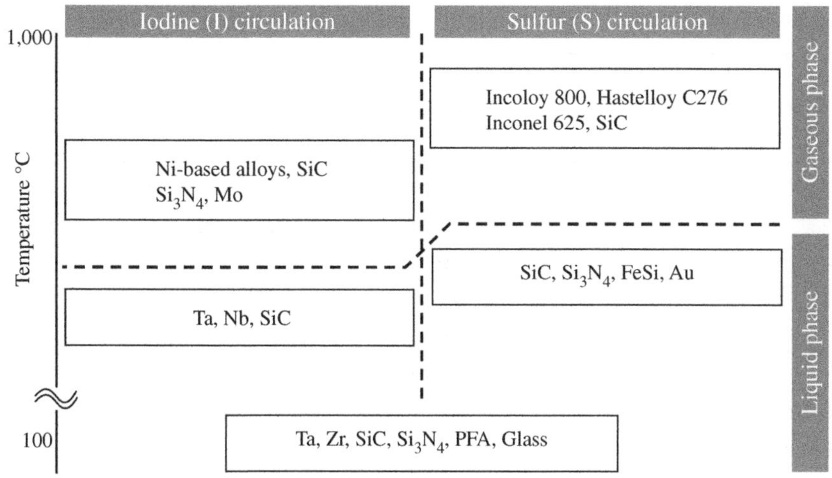

Figure 8.10 Candidate materials for prototypical environments of a sulfur–iodine process. Kubo et al. (2013)/with permission of Elsevier.

There are three main types of HI decomposition catalysts: (a) pure metals; (b) supported catalysts, such as carbon materials, CeO_2, and γ-Al_2O_3; and (c) metal-supported catalysts, such as Pt/AC, Ni/AC, Pt/γ-Al_2O_3, and Ni/γ-Al_2O_3.

Although various pure metals (Au, Mo, Ni, Ir, Rh, Pd, and Pt) were reported to catalyze this reaction, platinum (Pt) was found to be superior to all other metal catalysts in terms of activity (Liberatore et al., 2019; Wong et al., 2007). However, precious metal catalysts are expensive, so the development of inexpensive catalysts containing non-precious metals has become a research priority.

Supported catalysts can increase the overall specific surface area of the catalyst and ensure effective heat and mass transfer, which facilitate the adsorption and activation of reactants on the catalyst surface. The CeO_2 catalysts, especially gel CeO_2 300, strongly enhance the rate of decomposition of HI into H_2 (Zhang et al., 2009; Zhang, Fu, et al., 2015; Zhang, Zhou, Chen, et al., 2008; Zhang, Zhou, Wang, et al., 2008). In addition, activated carbon (AC) is a good catalyst due to its low cost, good activity, and stability. Petkovic et al. and Wang et al. investigated the catalytic decomposition of HI by different kinds of AC and explored the effect of surface properties and components of the catalysts on catalytic activity (Ginosar et al., 2011; Peikovic et al., 2009; Wang, Chen, et al., 2011). Table 8.2 shows the BET surface area, total pore volume, and micropore volume of some supported catalysts. The AC catalyst has a large specific surface area (about 1,000 m^2/g or even above 1,600 m^2/g with special production methods), while the specific surface area of the CeO_2 catalysts is only 100 m^2/g. In addition, the AC raw materials and preparation methods may play a significant role in the development of surface characteristics that eventually dictate catalyst activity and stability.

Metal-supported catalysts are catalysts made by loading metals onto the support material, such as Pt/AC and Ni/γ-Al_2O_3. As shown in Figure 8.11, Wang et al. compared AC, CMS, CNT, and GR with γ-$A1_2O_3$, and the results showed that the catalytic activity sequence of these materials was CMS > AC > GR > CNT > γ-Al_2O_3 (Lin et al., 2013; Wang et al., 2012). GA Company used the supported single-component Pt, Pd, Ni, and Ru catalysts, such as Pt/AC, Ni/AC, Pt/γ-Al_2O_3, and Ni/γ-Al_2O_3, for the catalytic decomposition of HI (O'keefe DR, Norman JH, Williamson DG., 1980; Oosawa et al., 1980; Oosawa et al., 1981). Table 8.3 shows the activation energy of decomposition of HI with different catalysts. Pt-containing catalysts, especially AC-supported Pt catalysts show high catalytic activity and stability. The performance of Pd-containing catalysts is better than that of Ni-containing catalysts. In addition, the support material also affects the activation energy. Although Pt/Pd-containing catalysts have good catalytic activity, such precious metal catalysts are costly and have poor stability. Therefore, cheaper Ni-containing catalysts were investigated. Favuzza et al. investigated the catalytic performance of different Ni precursors prepared by impregnation–roasting and coprecipitation methods, respectively. It was found that the activity of the catalysts decreased when the temperature of the Ni catalysts was lower than 650 °C, while the activity could be recovered when the temperature was brought above 650 °C. Therefore, Ni-containing catalysts could be used for HI decomposition reactions at temperatures below 650 °C (Favuzza et al., 2009).

It can be concluded that the catalytic activity of the catalysts containing the noble metal Pt is the highest when the same support material is used, and the catalytic activity of the catalysts with AC as the support material is the highest when the same loading metal is used. In addition, considering the economics of the catalysts, nickel, a transition metal, is a suitable

Table 8.2 Surface Areas and Pore Volume of Supported Catalysts.

Catalyst	BET surface area (m^2/g)	Total pore volume (ml/g)	Micropore volume (ml/g)	References
Commerce CeO_2	9.24	–	3.14×10^{-2}	Zhang et al. (2009)
Precipitation CeO_2	30.92	–	1.01×10^{-1}	Bhosale, Kumar, and Sutar (2017)
gel CeO_2 300	80.94	–	1.68×10^{-1}	
gel CeO_2 500	35.75	–	1.06×10^{-1}	
gel CeO_2 700	11.93	–	4.89×10^{-2}	
gel CeO_2 900	4.11	–	1.57×10^{-2}	
AC produced by steam activation	808.8472	0.547991	–	Wang, Chen, et al., (2011)
AC produced by $ZnCl_2$ activation	1453.6679	1.329389	–	
AC produced by H_3PO_4 activation	1675.0134	1.293954	–	
AC-COAL[a)]	954	0.52	–	Lin et al., (2013)
COAL-CLF[b)]	1032	0.63	–	
COAL-30HN[c)]	968	0.52	–	
COAL-60HN[d)]	981	0.54	–	
COAL-90HN[e)]	825	0.45	–	
AC	1298.7	0.18	–	Wang et al., (2012)
CMS	821.7	0.08	–	
CNT	174.2	0.59	–	
Graphite	40.0	0.02	–	
Al_2O_3	279.5	0.73	–	

a) Original samples.
b) Both 5M hydrochloric acid and 25 wt% hydrofluoric acid treatment on AC-COAL at 60 °C for 3 hr.
c) 2M nitric acid treatment on COAL-CLF at 30 °C for 3 hr.
d) 2M nitric acid treatment on COAL-CLF at 60 °C for 3 hr.
e) 2M nitric acid treatment on COAL-CLF at 90 °C for 3 hr.

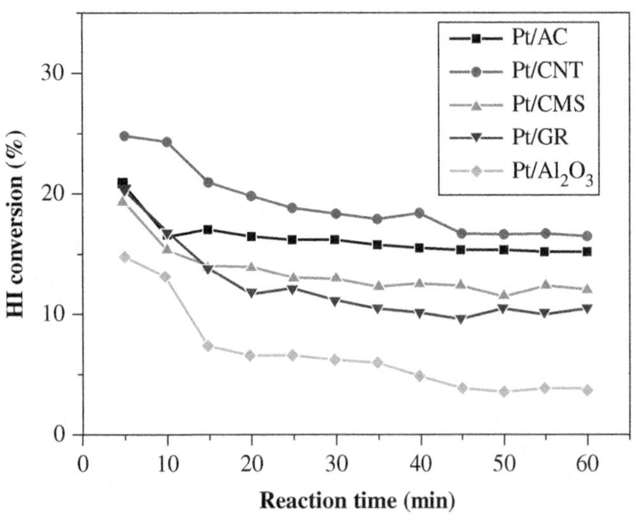

Figure 8.11 HI conversion over different Pt-supported catalysts at 500 °C with the reaction time. Reproduced from Wang et al. (2012)/with permission of Elsevier.

Table 8.3 Measured Activation Energies for Supported Metal Catalysts Used to Decompose Gaseous HI.

Catalyst	Activation energy (kcal/mol)	References
$PdCl_2$-Al_2O_3	13.0	O'keefe DR, Norman JH, Williamson DG., (1980)
$PtCl_2$-Al_2O_3	13.0	
NiO-Al_2O_3	22.0	
MoO_3-Al_2O_3	18.0	
Pt/γ-Al_2O_3	12.0	Oosawa et al. (1981)
Pt/H-Y-Zeolite	11.0	
Pt/H-Mordenite	9.4	
Pt/Teflon	7.9	
Pt/Tenax G.C.	7.9	
Pt/AC	11.0	
Pd/γ-Al_2O_3	13.0	
Pd/SiO_2-Al_2O_3	12.0	
Pd/H-Y-Zeolite	11.0	
Pd/H-Mordenite	13.0	
Pd/SiO_2	12.0	
Pd/Teflon	8.3	
Pd/Tenax G.C.	14.0	
Pd/AC	13.0	
Ni/γ-Al_2O_3	24.0	
Ni/H-Y-Zeolite	13.0	
Ni/H-Mordenite	9.9	
Ni/AC	8.3	

loading metal. However, the current catalysts used for decomposing HI are not sufficiently strong and are prone to breakage and destruction during prolonged operation, which leads to an increase in reaction resistance and a decrease in heat transfer performance. Therefore, the cycle is limited to the laboratory scale only. It is necessary to prepare high-performance catalysts in the laboratory and to study the molding process in order to apply them to industrial production.

8.4.3 The Catalysts for H_2SO_4 Decomposition

H_2SO_4 decomposition is common in sulfur-containing systems. The decomposition of H_2SO_4 can be divided into two steps (O'keefe DR, Norman JH, Williamson DG., 1980):

$$H_2SO_4 \rightarrow H_2O + SO_3 \tag{8.11}$$

$$SO_3 \rightarrow 1/2O_2 + SO_2 \tag{8.12}$$

H_2SO_4 is rapidly converted to SO_3 and H_2O in the temperature range of 350–500 °C, but the direct decomposition of SO_3 into SO_2 and O_2 requires very high temperatures (above 1,000 °C). Thus, SO_3 decomposition has been the focus of research in the area of H_2SO_4 decomposition (Corgnale et al., 2020; Khan et al., 2019; Nadar et al., 2018). Catalysts for the decomposition of SO_3 mainly include noble metals, transition metal oxides, and composite metal oxides.

The research on precious metal catalysts has been primarily focused on Pt-supported catalysts. The support materials include SiO_2, Al_2O_3, ZrO_2, TiO_2, $BaSO_4$, and SiC (Besenbruch et al., 1980; Brown & Revankar, 2012). Table 8.4 compares the catalyst surface area of typical Pt-containing catalysts. Pt/Al_2O_3 and Pt/ZrO_2 with high specific surface areas have higher

Table 8.4 Catalyst Surface Areas Before and After Reaction.

Catalyst	Fresh (BET SA, m^2/g)	Reaction at 800 °C (6 hr) (BET SA, m^2/g)	Reaction at 850 °C (6 hr) (BET SA, m^2/g)	Reaction at 800 °C (>200 hr) (BET SA, m^2/g)	References
Pt/Al_2O_3	179	77	63	–	
Pt/ZrO_2	51	25	23	–	Besenbruch et al. (1980), Brown and Revankar (2012), and Ginosar et al. (2007)
Pt/TiO_2	2.0	2.8	2.4	1.8	

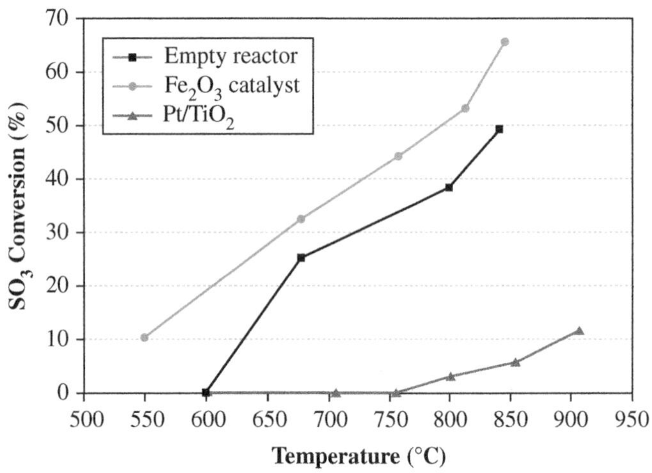

Figure 8.12 SO_3 conversion with direct resistive heating method. Reproduced from Li et al. (2012)/with permission of Elsevier.

catalytic activity, but the deactivation rate is faster and the stability is poor; while Pt/TiO_2 with lower specific surface area has better catalytic activity over a long time (Ginosar et al., 2007). In order to overcome the instability of the Pt/Al_2O_3 catalyst during long-term operation, Pt was loaded on SiC, and the catalytic activity and lifetime of Pt/SiC were tested (Zhang et al., 2011). The results showed that it could exhibit high stability.

In view of the high cost of Pt catalysts and the difficulties in applying them in practical industrial applications, the transition metal oxide catalysts such as Fe_2O_3, CuO, Cr_2O_3, TiO_2, Al_2O_3, ZnO, NiO, CoO, MnO_2, V_2O_5, and SiO_2, were developed (Dokiya et al., 1977; O'keefe DR, Norman JH, Williamson DG., 1980). Fe_2O_3 is currently found to be a catalyst with high catalytic activity and relatively low cost at high temperatures. Nevertheless, the catalytic activity of Fe_2O_3 is lower than that of the Pt-supported catalyst shown in Figure 8.12. In addition, CuO and Cr_2O_3 are highly active in monometallic oxide catalysts (Li et al., 2012). However, CuO and Cr_2O_3 are unstable at high temperatures, and especially Cr_2O_3 tends to volatilize at high temperatures, causing catalyst deactivation (Tagawa & Endo, 1989).

Low-cost composite metal oxide catalysts were also investigated to reduce the reaction temperature of SO_3 decomposition and further improve the thermal efficiency of the system. The composite oxide catalysts, $CuCr_2O_4$ and $CuFe_2O_4$, showed high activity at high temperatures, but the stability was not as good as Pt/SiC. These catalysts have low activity due to material sintering and easy decomposition. More research efforts are needed to develop cheap, highly reactive, and stable composite metal oxides.

8.5 Other Multi-Step Thermochemical Cycles

In addition to two-step thermochemical cycle and S–I cycle, there are a number of other multi-step thermochemical cycles developed for H_2O/CO_2 splitting/conversion. Table 8.5 shows a comparison of the main multi-step thermochemical cycles. It is worth noting that the heat driving each of these cycles can come from concentrated solar energy. We briefly discuss

Table 8.5 Comparison of Major Multi-step Thermochemical Cycles.

Cycles	Maximum temperature (°C)	List of elements	Number of chemical steps	References
S–I cycle	900–1,000	S, I	3	Choi et al., (2014), Kasahara et al. (2007), Kubo et al. (2013), Liberatore et al. (2019), Maatouk et al. (2009), Sakurai et al. (1999), and Wong et al. (2007)
Westinghouse cycle	827	S	2	Brecher et al. (1977), Farbman (1979), and Smitkova et al. (2011)
ISPRA Mark 13 cycle	120	Br, S	2	Van et al. (1980)
ISPRA Mark 9 cycle	644	Fe, Cl	3	Utgikar and Ward (2006)
ISPRA Mark 10 cycle	867	S, I, Na	5	Rosen (2008)
Cu-Cl cycle	700	Cu, Cl	3	Naterer et al. (2009) and Ozbilen et al. (2011)
UT-3 cycle	650	Ca, Br, Fe	4	Sakurai et al. 1996 and Aihara et al. (1988), Kameyama et al., (1989), Kameyama et al. (1992), and Sakurai et al. (1992)

several classical metal halide cycles in this section. The metal halide cycle is characterized by a series of reactions between the metal halide and water to split the water to produce hydrogen, and the European Joint Research Center (JRC) has proposed as many as 25 different thermochemical metal-halide system cycles since the late 1960s (De & Marchetti, 1970). The Mark 1 cycle is,

$$CaBr_2 + 2H_2O \rightarrow Ca(OH)_2 + HBr \tag{8.13}$$

$$HBr + Hg \rightarrow HgBr_2 + H_2 \tag{8.14}$$

$$HgBr_2 + Ca(OH)_2 \rightarrow CaBr_2 + HgO + H_2O \tag{8.15}$$

$$HgO \rightarrow Hg + 0.5O_2 \tag{8.16}$$

This cycle is not suitable for large-scale industrial production due to the highly toxic properties of mercury. Other hydrogen production cycles, such as manganese, vanadium, and chromium cycles, have been proposed in order to achieve a mercury-free process. Among them, Fe–Cl cycle is used as a substitute for other metal-halogen cycles because of its low price,

$$3FeCl_2 + 4H_2O \rightarrow Fe_3O_4 + 6HCl + H_2 \tag{8.17}$$

$$Fe_3O_4 + 8HCl \rightarrow FeCl_2 + 2FeCl_3 + 4H_2O \tag{8.18}$$

$$2FeCl_3 \rightarrow 2FeCl_2 + Cl_2 \tag{8.19}$$

$$Cl_2 + H_2O \rightarrow 2HCl + 0.5O_2 \tag{8.20}$$

The problem of Fe–Cl cycle is that it has low thermal efficiency, and the products HCl and Cl_2 are highly corrosive (Beghi, 1986; Knoche et al., 1978). The most famous metal halide cycle is the UT-3 cycle (Aihara et al., 1988). The reaction temperature of the UT-3 cycle is low, so it is easy to find the matching heat source. The UT-3 cycle has aroused a great interest among researchers due to its low cost, simple operation, and high hydrogen production efficiency,

$$CaBr_2 + H_2O \rightarrow CaO + 2HBr \tag{8.21}$$

$$CaO + Br_2 \rightarrow CaBr_2 + \frac{1}{2}O_2 \tag{8.22}$$

$$Fe_3O_4 + 8HBr \rightarrow 3FeBr_2 + 4H_2O + Br_2 \tag{8.23}$$

$$3FeBr_2 + 4H_2O \rightarrow Fe_3O_4 + 6HBr + H_2 \tag{8.24}$$

However, the UT-3 cycle requires materials that are resistant to HBr corrosion at 750 °C, and the cycle process is relatively long, so it is difficult to achieve industrial applications (Sakurai et al. 1996; Kameyama et al., 1989; Kameyama et al., 1992; Sakurai et al., 1992).

8.6 Catalysts for Solar Gasification and Reforming

8.6.1 Catalysts for Solar Gasification

The process of converting solid coal, biomass, and other fossil fuels into gaseous syngas through a gasification reaction driven by concentrated solar energy is known as "solar gasification" (Fang et al., 2021; Piatkowski et al., 2010; Suárez-Almeida et al., 2021). The advantages of these processes are: (a) the use of solar energy rather than burning fossil fuels to provide heat; (b) the reduction of pollution and environmental impact caused by the direct burning of fossil fuels; and (c) the fact that solar energy can be converted into a form that can be both stored and transported. The product syngas has greater calorific value, ideally higher than initial coal by 44–45%. The solar gasification processes are very complex and can be divided into two sequential processes: pyrolysis and gasification. The pyrolysis process is,

$$\text{coal/biomass} \rightarrow \text{char(carbon)} + CO + CO_2 + H_2 + CH_4 + \text{tars} \tag{8.25}$$

Then, the produced char is gasified with steam or CO_2. There are highly endothermic reactions as evident from the enthalpy change in the reactions (8.26)–(8.28), so the catalysts are needed for effectively catalyzing steam or CO_2 gasification,

$$C + H_2O \rightarrow CO + H_2 \qquad \Delta H = 131.3\ \text{kJ/mol} \tag{8.26}$$

$$C + CO_2 \rightarrow 2CO \qquad \Delta H = 172.4\ \text{kJ/mol} \tag{8.27}$$

$$CH_4 + H_2O \rightarrow CO + 3H_2 \qquad \Delta H = 206\ \text{kJ/mol} \tag{8.28}$$

Many catalysts were studied for gasification of coal (Johnson, 1976; McKee, 2021). The alkali-metal, alkaline-earth metal, and some transition-metal compounds could effectively catalyze steam or CO_2 gasification of coal (Alam & Debroy, 1984; Cazorla-Amorós et al., 1991; Furimsky & Sears, 1988; Guzman & Wolf, 1982; Kapteijn et al., 1986; Li & Cheng, 1995; Ohtsuka & Asami, 1991; Pérez-Florindo et al., 1993; Sams & Shadman, 1983; Song & Kim, 1993). The active catalysts may improve chemical conversion and reduce the operating temperature requirements for the solar coal gasification process. However, the use of the catalyst depends on the feasibility of recovering the catalyst from the residual coal ash. In_2O_3 and ZnO are proposed to be separated from remaining ash by In_2O and Zn evaporation at high temperatures (Chuayboon & Abanades, 2021; Chuayboon et al., 2018). Steinfeld et al. used solar furnaces to study similar processes for Zn and syngas production by combining ZnO reduction and CH_4 reforming (Steinfeld et al., 1998). They reported Zn yields of close to 90% in 5 kW volumetric gas-particle receiver-reactors at temperatures ranging from 1,200K to 1,600K, and the recovered metal Zn could be easily oxidized to zinc oxide in air. However, there are some additional economic and energy losses due to the energy required to carry out the catalyst separation process at high temperatures. At this point, it is difficult to assess with certainty technical and economic feasibility and efficiency of the proposed solar processes, including catalyst separation processes.

8.6.2 Catalysts for Solar Reforming of Methane

The steam or carbon dioxide reforming of methane has high endothermic properties and is the basis for increasing the calorific value of hydrocarbon feedstocks. The methane molecule has a highly symmetric tetrahedral structure and has a C—H bond energy of up to 438.8 kJ/mol. The heat required for such an endothermic reaction can be supplied by concentrating solar energy, thereby storing solar energy into fuels (Chen, Wang, Han, et al., 2018; Chen, Wang, Yan, et al., 2018; Li et al., 2020; Zhang et al., 2020; Zhao et al., 2021). The methane reforming reactions are,

$$CH_4 + H_2O \rightarrow CO + 3H_2 \tag{8.29}$$

$$CH_4 + CO_2 \rightarrow 2CO + 2H_2 \tag{8.30}$$

Although the theoretical conversion rate of the reforming reaction is very high, the actual conversion rate is usually much lower due to the kinetic limitations of the reaction. Therefore, the methane reforming reaction generally requires the use of a catalyst to enhance the reaction rate. According to the metal active components, the catalysts for methane reforming can be divided into precious metals and non-precious metals. The catalysts are generally attached to Al_2O_3, ZrO_2, Ce-ZrO_2, and other supports.

The main indicators for evaluating methane reforming catalysts are catalytic activity and carbon deposition resistance. Precious metal catalysts, such as Pt, Pd, Ru, and Rh, have high activity, high stability, and excellent carbon deposition resistance in the methane reforming reaction. Among them, the noble metal Pt is considered to be the catalyst with the best performance in activating methane (Gallucci et al., 2008; Graf et al., 2007; Özkara-Aydınoğlu et al., 2009). In addition, the effect of different supports on the performance of the methane dry reforming reaction was also investigated (Nagaoka et al., 2001). The experimental results showed that ZrO_2 support was the most suitable support for the methane dry-reforming reaction.

Despite the high catalytic activity and resistance to carbon deposition of precious metal catalysts, the large-scale use of these materials is limited by their high price, which severely limits their industrial application prospects. Non-precious metals such as Ni, Co, Fe, and other transition metals also show high catalytic activity in the methane reforming reaction (Boukha et al., 2007; Nagaoka et al., 2003; Wang, Chu, et al., 2011). Among these transition metal catalysts, metal Ni is widely used in large-scale industrial methane reforming research because the price is relatively low and catalytic activity is high. However, Ni-based catalysts have two major issues: (a) the catalyst is prone to deactivation due to carbon contamination during the reaction process, as the carbon particles generated in the process could cover the catalytically active sites of metal Ni; and (b) the catalyst can be deactivated under high-temperature conditions because the metal Ni microparticles dispersed on the surface of the carrier are easy to agglomerate to form larger metal Ni particles, which directly reduces the surface area of the metal Ni distributed on the surface of the carrier. The activity and stability of the Ni-based catalyst may be improved by rare earth-based promoters or alkaline-earth metal ions (Han et al., 2020; Han et al., 2021; Safavinia et al., 2020; Zhang, Xia, et al., 2015).

There have been some efforts to apply perovskites in methane reforming reactions (Fang et al., 2013; Sang & Sang, 2009; Su et al., 2014; Zhao, He, et al., 2016). The specific crystal structure of perovskite offers a wider range of elemental modulation (i.e., the ability of A and B cations to be substituted or partially substituted by different metal ions) and catalytic potential, which helps to facilitate the reaction by optimizing the composition ratio. Zheng et al. prepared $LaFeO_3$-supported CeO_2 for the study of methane reforming (Zheng et al., 2017). The experimental results show that the coexistence of Ce and Fe^{2+} increases the oxygen vacancies of the catalyst, thus improving the catalytic activity. The catalytic activity of the catalyst was highest when the $CeO_2/LaFeO_3$ ratio was 10. No carbon deposition was observed on the catalyst surface after the reaction. Zhao et al. prepared the $La_{2-x}Sr_xFeCoO_3$ by microemulsion method for methane reforming and experimentally verified the influence of the ratio of the elements at the A site on the catalytic performance. They found that the performance of the catalyst was the best when the ratio of La/Sr was 3:2 (Zhao et al., 2017).

8.6.3 Catalysts for Solar Reforming of Methanol

The conventional steam or carbon dioxide reforming of methanol to produce syngas uses fossil fuel as the source of the process heat. Alternatively, the heat required for such processes can be supplied by concentrated solar energy, thereby storing solar energy (Cheng et al., 2019; Hong et al., 2009; Liu et al., 2009; Wang et al., 2017). In 2006, Hong et al. proposed a comprehensive energy system of moderate-temperature solar thermal energy and methanol complementation (Hong et al., 2006). The system uses a trough collector to absorb the solar heat energy of 200 to 300°C to drive the decomposition reaction of methanol fuel and generate syngas.

At present, the most widely used catalysts in the industry are copper-based catalysts with high activity and selectivity for solar reforming of methanol (Herdem et al., 2019; Nakajima et al., 2016). Cu or CuO are mainly dispersed on metal oxides, such as ZnO, ZrO_2, Al_2O_3, and SiO_2. These catalysts are cheap and easy to obtain and have relatively high catalytic activity (Milani et al., 2015; Wang et al., 2015). With these catalysts, the temperature for complete conversion to methanol is usually high (above 300 °C), which can easily cause sintering of the catalyst. In addition, as the temperature increases, the by-product CO produced by the reaction gradually increases, leading to a decrease in the hydrogen produced (Sá et al., 2010). In recent years, In_2O_3 has attracted more attention due to its abundant oxygen vacancy on the surface (Sun et al., 2015). It was found that the In_2O_3 catalyst could still maintain very high selectivity and formation rate of methanol at high temperatures. The perovskite-type metal oxides have also been studied for reforming methanol with steam or CO_2 (Longo & Raccah, 1973; Parravano, 1952; Tejuca et al., 1989). However, perovskites require pretreatment in a hydrogen atmosphere. The pretreatment temperature of these catalysts is high, usually above 500 °C, which increases the energy consumption in the pretreatment stage. The optimal composition and doping ratio of the perovskites have not yet been fully identified for the methanol steam reforming reaction, and therefore, the application of perovskite in the process does not look possible in the near-term.

8.7 Summary and Outlook

The solar thermochemical processes integrated with concentrated solar irradiation for CO_2/H_2O dissociation/conversion are one of the most promising ways to utilize solar energy because of their potential for full-spectrum utilization and high solar–chemical efficiency. In this chapter, the materials for such processes are discussed. First, the single-step pyrolysis of CO_2/H_2O was introduced. Then, the recent contributions of redox pairs in two-step thermochemical cycle were reviewed. The operation method of two-step thermochemical cycle is simple, and hydrogen and oxygen are produced in different steps, which avoids the problem of explosive hydrogen–oxygen mixtures. The key to two-step thermochemical cycle development lies in the search for oxygen carriers with good redox properties at lower temperatures. Fortunately, perovskite was found to be a promising oxygen carrier due to the fact that its thermodynamic properties can be controlled by adjusting its composition.

The S–I cycle and metal halide cycle used for hydrogen production were also discussed from a material point of view. We reviewed the relevant catalyst materials as well as corrosion-resistant materials at high temperatures. Multi-step cycles can achieve medium-temperature thermochemical processes for CO_2/H_2O dissociation/conversion if suitable catalysts and corrosion-resistant materials are found. Finally, the catalysts for solar gasification and reforming were described. Solar gasification and reforming are currently the most viable solar thermochemical technologies with excellent prospects for industrialization.

Acknowledgment

This work is financially supported by the National Natural Science Foundation of China through contracts No.51925602 and No.51888103.

Conflict of Interest

The authors declare that they have no conflict of interest.

References

Abanades, S. (2012). Thermogravimetry analysis of CO_2 and H_2O reduction from solar nanosized Zn powder for thermochemical fuel production. *Industrial & Engineering Chemistry Research, 51*(2), 741–750.

Abanades, S. (2012). CO_2 and H_2O reduction by solar thermochemical looping using SnO_2/SnO redox reactions: Thermogravimetric analysis. *International Journal of Hydrogen Energy, 37*(10), 8223–8231.

Abanades, S. (2019). Metal oxides applied to thermochemical water-splitting for hydrogen production using concentrated solar energy. *ChemEngineering, 3*(3), 63.

Abanades, S., & Chambon, M. (2010). CO_2 dissociation and upgrading from two-step solar thermochemical processes based on ZnO/Zn and SnO_2/SnO redox pairs. *Energy & Fuels, 24*, 6667–6674.

Abanades, S., Charvin, P., & Flamant, G. (2007). Design and simulation of a solar chemical reactor for the thermal reduction of metal oxides: Case study of zinc oxide dissociation. *Chemical Engineering Science, 62*(22), 6323–6333.

Abanades, S., Charvin, P., Flamant, G., & Neveu, P. (2006). Screening of water-splitting thermochemical cycles potentially attractive for hydrogen production by concentrated solar energy. *Energy, 31*(14), 2805–2822.

Abanades, S., Charvin, P., Lemont, F., & Flamant, G. (2008). Novel two-step SnO_2/SnO water-splitting cycle for solar thermochemical production of hydrogen. *International Journal of Hydrogen Energy, 33*(21), 6021–6030.

Abanades, S., & Flamant, G. (2006). Thermochemical hydrogen production from a two-step solar-driven water-splitting cycle based on cerium oxides. *Solar Energy*, *80*(12), 1611–1623.

Abanades, S., & Villafan-Vidales, H. I. (2011). CO_2 and H_2O conversion to solar fuels via two-step solar thermochemical looping using iron oxide redox pair. *Chemical Engineering Journal*, *175*, 368–375.

Agrafiotis, C., Roeb, M., & Sattler, C. (2015). A review on solar thermal syngas production via redox pair-based water/carbon dioxide splitting thermochemical cycles. *Renewable and Sustainable Energy Reviews*, *42*, 254–285.

Agrafiotis, C., Zygogianni, A., Pagkoura, C., Kostoglou, M., & Konstandopoulos, A. G. (2013). Hydrogen production via solar-aided water splitting thermochemical cycles with nickel ferrite: Experiments and modeling. *AICHE Journal*, *59*(4), 1213–1225.

Agrafiotis, C. C., Pagkoura, C., Zygogianni, A., Karagiannakis, G., Kostoglou, M., & Konstandopoulos, A. G. (2012). Hydrogen production via solar-aided water splitting thermochemical cycles: Combustion synthesis and preliminary evaluation of spinel redox-pair materials. *International Journal of Hydrogen Energy*, *37*(11), 8964–8980.

Aihara, M., Umida, H., Yamaguchi, K., Tsutsumi, A., & Yoshida, K. (1988). Thermochemical hydrogen production by the UT-3 cycle consisting of Br-Ca-Fe compounds. Reaction of Ca compounds. *Nippon Kagaku Kaishi*, *12*, 2031–2036.

Alam, M., & Debroy, T. (1984). The effects of CO and CO_2 on the rate of Na_2CO_3 catalyzed Boudouard reaction. *Metallurgical Transactions B*, *15*, 400–403.

Arifin, D., Aston, V. J., Liang, X., McDaniel, A. H., & Weimer, A. W. (2012). $CoFe_2O_4$ on a porous Al_2O_3 nanostructure for solar thermochemical CO_2 splitting. *Energy & Environmental Science*, *5*(11), 9438–9443.

Barcellos, D. R., Sanders, M. D., Tong, J., McDaniel, A. H., & O'Hayre, R. P. (2018). $BaCe_{0.25}\ Mn_{0.75}\ O_{3-\delta}$—a promising perovskite-type oxide for solar thermochemical hydrogen production. *Energy & Environmental Science*, *11*(11), 3256–3265.

Bareiss, K., Cristina, D., Moeckl, M., & Hamacher, T. (2019). Life cycle assessment of hydrogen from proton exchange membrane water electrolysis in future energy systems. *Applied Energy*, *37*, 862–872.

Beghi, G. E. (1986). A decade of research on thermochemical hydrogen at the Joint Research Centre, Ispra. *Hydrogen Systems*, *11*(12), 761–771.

GE Besenbruch, O'Keefe DR, Norman JH. Thermochemical water splitting for hydrogen production. 1980.

Bhosale, R. R. (2019). Hydrogen production via thermochemical H_2O splitting using $CaSO_4$-CaO redox reactions. *International Journal of Hydrogen Energy*, *45*(5).

Bhosale, R. R. (2020). Solar driven $CdSO_4$/CdO thermochemical water splitting cycle for hydrogen generation. *International Journal of Hydrogen Energy*, *45*(10), 5829–5839.

Bhosale, R. R. (2021). Solar hydrogen production via ZnO/Zn based thermochemical water splitting cycle: Effect of partial reduction of ZnO. *International Journal of Hydrogen Energy*, *46*(6), 4739–4748.

Bhosale, R. R., Gupta, R. B., & Shende, R. V. (2022). Thermodynamic evaluation of solar assisted ZnO/Zn thermochemical CO_2 splitting cycle. *Environmental Research*, 113266.

Bhosale, R. R., Kumar, A., AlMomani, F., & Gupta, R. B. (2017). Solar thermochemical ZnO/$ZnSO_4$ water splitting cycle for hydrogen production. *International Journal of Hydrogen Energy*, *42*(37), 54.

Bhosale, R. R., Kumar, A., & Sutar, P. (2017). Thermodynamic analysis of solar driven SnO_2/SnO based thermochemical water splitting cycle. *Energy Conversion and Management*, *135*, 226–235.

Bhosale, R. R., Takalkar, G., Sutar, P., Kumar, A., AlMomani, F., & Khraisheh, M. (2019). A decade of ceria based solar thermochemical H_2O/CO_2 splitting cycle. *International Journal of Hydrogen Energy*, *44*(1), 34–60.

Boukha, Z., Kacimi, M., Pereira, M. F. R., Faria, J. L., Figueiredo, J. L., & Ziyad, M. (2007). Methane dry reforming on Ni loaded hydroxyapatite and fluoroapatite. *Applied Catalysis A: General*, *317*(2), 299–309.

Brecher, L. E., Spewock, S., & Warde, C. J. (1977). The Westinghouse sulfur cycle for the thermochemical decomposition of water. *International Journal of Hydrogen Energy*, *2*(1), 7–15.

Brown, N. R., & Revankar, S. T. (2012). A review of catalytic sulfur (VI) oxide decomposition experiments. *International Journal of Hydrogen Energy*, *37*(3), 2685–2698.

Bulfin, B., Hoffmann, L., de Oliveira, L., et al. (2016). Statistical thermodynamics of non-stoichiometric ceria and ceria zirconia solid solutions. *Physical Chemistry Chemical Physics*, *18*(33), 23147–23154.

Bulfin, B., Lowe, A. J., Keogh, K. A., et al. (2013). Analytical model of CeO_2 oxidation and reduction. *The Journal of Physical Chemistry C*, *117*, 24129–24137.

Bulfin, B., Vieten, J., Agrafiotis, C., Roeb, M., & Sattler, C. (2017). Applications and limitations of two step metal oxide thermochemical redox cycles: A review. *Journal of Materials Chemistry A*, *5*(36), 18951–18966.

Bulfin, B., Vieten, J., Starr, D. E., Azarpira, A., Zachäus, C., Hävecker, M., Skorupska, K., Schmücker, M., Roeb, M., & Sattler, C. (2017). Redox chemistry of $CaMnO_3$ and $Ca_{0.8}Sr_{0.2}MnO_3$ oxygen storage perovskites. *Journal of Materials Chemistry A*, *5*(17), 7912–7919.

Bush, H. E., Datta, R., & Loutzenhiser, P. G. (2019). Aluminum-doped strontium ferrites for a two-step solar thermochemical air separation cycle: Thermodynamic characterization and cycle analysis. *Solar Energy*, *188*, 775–786.

Carrillo, R. J., & Scheffe, J. R. (2019). Beyond ceria: theoretical investigation of isothermal and near-isothermal redox cycling of perovskites for solar thermochemical fuel production. *Energy & Fuels*, *33*(12), 12871–12884.

Cazorla-Amorós, D., Linares-Solano, A., Marcilla Gomis, A. F., & de Lecea, C. S.-M. (1991). Calcium catalytic active sites in carbon-gas reactions. Determination of the specific activity[J]. *Energy & Fuels*, *5*(6), 796–802.

Chambon, M., Abanades, S., & Flamant, G. (2010). Solar thermal reduction of ZnO and SnO_2: Characterization of the recombination reaction with O_2. *Chemical Engineering Science*, *65*(11), 3671–3680.

Charvin, P., Abanades, S., Flamant, G., & Lemort, F. (2007). Two-step water splitting thermochemical cycle based on iron oxide redox pair for solar hydrogen production. *Energy*, *32*(7), 1124–1133.

Charvin, P., Abanades, S., Lemont, F., & Flamant, G. (2008). Experimental study of SnO_2/SnO/Sn thermochemical systems for solar production of hydrogen. *AICHE Journal*, *54*(10), 2759–2767.

Chaubey, R., Sahu, S., James, O. O., & Maity, S. (2013). A review on development of industrial processes and emerging techniques for production of hydrogen from renewable and sustainable sources. *Renewable and Sustainable Energy Reviews*, *23*, 443–462.

Chen, X., Wang, F., Han, Y., Yu, R., & Cheng, Z. (2018). Thermochemical storage analysis of the dry reforming of methane in foam solar reactor. *Energy Conversion and Management*, *158*, 489–498.

Chen, X., Wang, F., Yan, X., Han, Y., Cheng, Z., & Jie, Z. (2018). Thermochemical performance of solar driven CO_2 reforming of methane in volumetric reactor with gradual foam structure. *Energy*, *151*, 545–555.

Cheng, Z. D., Men, J. J., Zhao, X. R., He, Y. L., & Tao, Y. B. (2019). A comprehensive study on parabolic trough solar receiver-reactors of methanol-steam reforming reaction for hydrogen production. *Energy Conversion and Management*, *186*, 278–292.

Choi, J. Y., Kim, Y. S., Sah, I., No, H. C., & Jang, C. (2014). Corrosion resistances of alloys in high temperature hydrogen iodide gas environment for sulfur–iodine thermochemical cycle. *International Journal of Hydrogen Energy*, *39*(27), 14557–14564.

Chuayboon, S., & Abanades, S. (2021). Thermodynamic and experimental investigation of solar-driven biomass pyro-gasification using H_2O, CO_2, or ZnO oxidants for clean syngas and metallurgical Zn production. *Processes*, *9*(4), 687.

Chuayboon, S., Abanades, S., & Rodat, S. (2018). Solar chemical looping gasification of biomass with the ZnO/Zn redox system for syngas and zinc production in a continuously-fed solar reactor. *Fuel*, *215*, 66–79.

Chueh, W. C., Falter, C., Abbott, M., Scipio, D., Furler, P., Haile, S. M., & Steinfeld, A. (2010). High-flux solar-driven thermochemical dissociation of CO_2 and H_2O using nonstoichiometric ceria. *Science*, *330*(6012), 1797–1801.

Colón, G. (2016). Towards the hydrogen production by photocatalysis. *Applied Catalysis A: General*, *518*, 48–59.

Cooper, T., Scheffe, J. R., Galvez, M. E., Jacot, R., Patzke, G., & Steinfeld, A. (2015). Lanthanum manganite perovskites with Ca/Sr A-site and Al B-site doping as effective oxygen exchange materials for solar thermochemical fuel production. *Energy Technology*, *3*(11), 1130–1142.

Corgnale, C., Gorensek, M. B., & Summers, W. A. (2020). Review of sulfuric acid decomposition processes for sulfur-based thermochemical hydrogen production cycles. *Processes*, *8*(11), 1383.

De, B. G., & Marchetti, C. (1970). Hydrogen, key to the energy market. *Eurospectra*, *9*(2), 46.

Demont, A., & Abanades, S. (2014). High redox activity of Sr-substituted lanthanum manganite perovskites for twostep thermochemical dissociation of CO_2. *RSC Advances*, *4*(97), 54885–54891.

Demont, A., Abanades, S., & Beche, E. (2014). Investigation of perovskite structures as oxygen-exchange redox materials for hydrogen production from thermochemical two-step water-splitting cycles. *The Journal of Physical Chemistry C*, *118*(24), 12682–12692.

Dimitrakis, D. A., Tsongidis, N. I., & Konstandopoulos, A. G. (2016). Reduction enthalpy and charge distribution of substituted ferrites and doped ceria for thermochemical water and carbon dioxide splitting with DFT+U. *Physical Chemistry Chemical Physics*, *18*(34), 23587–23595.

Dincer, I., & Acar, C. (2015). Review and evaluation of hydrogen production methods for better sustainability. *International Journal of Hydrogen Energy*, *40*(34), 11094–11111.

Dokiya, M., Kameyama, T., Fukuda, K., & Kotera, Y. (1977). The study of thermochemical hydrogen preparation. III. An oxygen-evolving step through the thermal splitting of sulfuric acid. *Bulletin of the Chemical Society of Japan*, *50*(10), 2657–2660.

Emery, A. A., Saal, J. E., Kirklin, S., Hegde, V. I., & Wolverton, C. (2016). High-throughput computational screening of perovskites for thermochemical water splitting application. *Chemistry of Materials*. acs.chemmater.6b01182

Ezbiri, M., Takacs, M., Stolz, B., Lungthok, J., Steinfeld, A., & Michalsky, R. (2017). Design principles of perovskites for solar-driven thermochemical splitting of CO_2. *Journal of Materials Chemistry A*, *5*(29), 15105–15115.

Fang, H., Li, X., Zhao, K., Huang, Z., Wei, G., & Li, H. (2013). The use of $La_{1-x}Sr_xFeO_3$ perovskite-type oxides as oxygen carriers in chemical-looping reforming of methane. *Fuel*, *108*, 465–473.

Fang, Y., Paul, M. C., Varjani, S., Li, X., Park, Y.-K., & You, S. (2021). Concentrated solar thermochemical gasification of biomass: Principles, applications, and development. *Renewable and Sustainable Energy Reviews*, *150*, 111484.

Farbman, G. H. (1979). Hydrogen production by the Westinghouse sulfur cycle process: program status. *International Journal of Hydrogen Energy*, *4*(2), 111–122.

Favuzza, P., Felici, C., Lanchi, M., Liberatore, R., Mazzocchia, C. V., Spadoni, A., Tarquini, P., & Tito, A. C. (2009). Decomposition of hydrogen iodide in the S-I thermochemical cycle over Ni catalyst systems. *International Journal of Hydrogen Energy*, *34*(9), 4049–4056.

Fletcher, E. A. (1999). Solar thermal and solar quasi-electrolytic processing and separations: Zinc from zinc oxide as an example. *Industrial & Engineering Chemistry Research*, *38*(6), 2275–2282.

Fletcher, E. A., & Moen, R. L. (1977). Hydrogen and oxygen from water. *Science*, *197*(4308), 1050–1056.

Furimsky, E., & Sears, P. (1988). Iron-catalyzed gasification of char in CO_2. *Energy & Fuels*, *2*, 634–639.

Gallucci, F., Tosti, S., & Basile, A. (2008). Pd-Ag tubular membrane reactors for methane dry reforming: A reactive method for CO_2 consumption and H_2 production. *Journal of Membrane Science*, *317*(1–2), 96–105.

Galvez, M. E., Frei, A., Albisetti, G., Lunardi, G., & Steinfeld, A. (2008). Solar hydrogen production via a two-step thermochemical process based on MgO/Mg redox reactions-Thermodynamic and kinetic analyses. *International Journal of Hydrogen Energy*, *33*(12), 2880–28902.

Gao, Y., Mao, Y., Song, Z., Zhao, X., Sun, J., Wang, W., Chen, G., & Chen, S. (2020). Efficient generation of hydrogen by two-step thermochemical cycles: Successive thermal reduction and water splitting reactions using equal-power microwave irradiation and a high entropy material. *Applied Energy*, *279*, 115777.

Ginosar, D. M., Petkovic, L. M., & Burch, K. C. (2011). Commercial activated carbon for the catalytic production of hydrogen via the sulfur–iodine thermochemical water splitting cycle. *International Journal of Hydrogen Energy*, *36*(15), 8908–8914.

Ginosar, D. M., Petkovic, L. M., Glenn, A. W., & Burch, K. C. (2007). Stability of supported platinum sulfuric acid decomposition catalysts for use in thermochemical water splitting cycles. *International Journal of Hydrogen Energy*, *32*(4), 482–488.

Goikoetxea, N. B., Gómez-Mancebo, M. B., Fernández-Saavedra, R., Borlaf, F., García-Pérez, F., Jiménez, J. A., Llorente, I., Rucandio, I., & Quejido, A. J. (2019). Understanding water-splitting thermochemical cycles based on nickel and cobalt ferrites for hydrogen production. *International Journal of Hydrogen Energy*, *44*(33), 17578–17585.

Gokon, N., Murayama, H., Umeda, J., Hatamachi, T., & Kodama, T. (2009). Monoclinic zirconia-supported Fe_3O_4 for the two-step water splitting thermochemical cycle at high thermal reduction temperatures of 1400–1600°C. *International Journal of Hydrogen Energy*, *34*(3), 1208–1217.

Graf, P. O., Mojet, B. L., Ommen, J., & Lefferts, L. (2007). Comparative study of steam reforming of methane, ethane and ethylene on Pt, Rh and Pd supported on yttrium-stabilized zirconia. *Applied Catalysis A: General*, *332*(2), 310–317.

Gstoehl, D., Brambilla, A., Schunk, L. O., & Steinfeld, A. (2008). A quenching apparatus for the gaseous products of the solar thermal dissociation of ZnO. *Journal of Materials Science*, *43*(14), 4729–4736.

Guzman, G. L., & Wolf, E. E. (1982). Kinetics of the potassium carbonate-catalyzed steam gasification of carbon and coal. *Industrial and Engineering Chemistry Process Design and Development*, *21*(1), 25–29.

Haeussler, A., Abanades, S., Jouannaux, J., Drobek, M., Ayral, A., & Julbe, A. (2019). Recent progress on ceria doping and shaping strategies for solar thermochemical water and CO_2 splitting cycles. *AIMS Materials Science*, *6*(5), 657–684.

Haeussler, A., Abanades, S., Jouannaux, J., & Julbe, A. (2018). Non-stoichiometric redox active perovskite materials for solar thermochemical fuel production: A review. *Catalysts*, *8*(12), 611.

Hammache, A., & Bilgen, E. (1992). Nuclear hydrogen production based on sulfuric acid decomposition process. *Journal of Energy Resources Technology*, *114*(3), 227–234.

Han, B., Zhao, L., Wang, F., Xu, L., Yu, H., Cui, Y., Zhang, J., & Shi, W. (2020). Effect of calcination temperature on the performance of the Ni@ SiO_2 catalyst in methane dry reforming. *Industrial & Engineering Chemistry Research*, *59*(30), 13370–13379.

Han, K., Yu, W., Xu, L., Deng, Z., Yu, H., & Wang, F. (2021). Reducing carbon deposition and enhancing reaction stability by ceria for methane dry reforming over Ni@SiO_2@CeO_2 catalyst. *Fuel*, *291*, 120182.

Herdem, M. S., Sinaki, M. Y., Farhad, S., & Hamdullahpur, F. (2019). An overview of the methanol reforming process: Comparison of fuels, catalysts, reformers, and systems. *International Journal of Energy Research*, *43*(10), 5076–5105.

Hong, H., Jin, H., & Liu, B. (2006). A Novel Solar-hybrid gas turbine combined cycle with inherent CO_2 separation using chemical-looping combustion by solar heat source. *ASME Trans, The Journal of Solar Energy Engineering*, *28*, 275–284.

Hong, H., Liu, Q., Jin, H., Hong, H., Liu, Q., & Jin, H. (2009). Solar hydrogen production integrating low-grade solar thermal energy and methanol steam reforming. *Journal of Energy Resources Technology*, *131*(1).

Hoskins, A. L., Millican, S. L., Czernik, C. E., Alshankiti, I., Netter, J. C., Wendelin, T. J., Musgrave, C. B., & Weimer, A. W. (2019). Continuous on-sun solar thermochemical hydrogen production via an isothermal redox cycle. *Applied Energy*, *249*, 368–376.

Huang, J., Fu, Y., Li, S., Kong, W., Zhang, J., & Sun, Y. (2019). Cobalt-based ferrites as efficient redox materials for thermochemical two-step CO_2-splitting: Enhanced performance due to cation diffusion. *Sustainable Energy & Fuels*, *3*(4), 975–984.

Huang, J., Fu, Y., Li, S., Kong, W., Zhang, J., & Sun, Y. (2018). Enhanced activity of MgFeO ferrites for two-step thermochemical CO_2 splitting. *Journal of CO_2 Utilization*, *27*, 450–458.

Ishihara, H., Kaneko, H., Hasegawa, N., & Tamaura, Y. (2008). Two-step water-splitting at 1273-1623 K using yttria-stabilized zirconia-iron oxide solid solution via co-precipitation and solid-state reaction. *Energy*, *33*(12), 1788–1793.

Ismail, A. A., & Bahnemann, D. W. (2014). Photochemical splitting of water for hydrogen production by photocatalysis: A review. *Solar Energy Materials & Solar Cells*, *128*, 85–101.

Jarrett, C., Chueh, W., Yuan, C., Kawajiri, Y., Sandhage, K. H., & Henry, A. (2016). Critical limitations on the efficiency of two-step thermochemical cycles. *Solar Energy*, *123*, 57–73.

Johnson, J. L. (1976). The use of catalysts in coal gasification[J]. *Catalysis Reviews: Science and Engineering*, *14*(1), 131–152.

Kameyama, H., Sato, T., Amir, R., Yoshida, K., Aihara, M., Sakurai, M., Tadokoro, Y., Kajiyama, T., Yamaguchi, T., & Sakai, N. (1992). Cycle simulation of the UT-3 thermochemical hydrogen production process. *International Journal of Hydrogen Energy*, *17*(10), 789–794.

Kameyama, H., Tomino, Y., Sato, T., Amir, R., Orihara, A., Aihara, M., & Yoshida, K. (1989). Process simulation of "MASCOT" plant using the UT-3 thermochemical cycle for hydrogen production. *International Journal of Hydrogen Energy*, *14*(5), 323–330.

Kang, K. S., Kim, C. H., Cho, W. C., Bae, K.-K., Kim, S.-H., & Park, C.-S. (2009). Novel two-step thermochemical cycle for hydrogen production from water using germanium oxide: KIER 4 thermochemical cycle. *International Journal of Hydrogen Energy*, *34*(10), 4283–4290.

Kang, M., Zhang, J., Wang, C., Wang, F., Zhao, N., Xiao, F., Wei, W., & Sun, Y. (2013). CO_2 splitting via two step thermochemical reactions over doped ceria/zirconia solid solutions. *RSC Advances*, *3*(41), 18878–18885.

Kapteijn, F., Porre, H., & Moulijn, J. (1986). CO_2 gasification of activated carbon catalyzed by earth alkaline elements. *AICHE Journal*, *32*(4), 691–695.

Kasahara, S., Kubo, S., Hino, R., Onuki, K., Nomura, M., & Nakao, S. I. (2007). Flowsheet study of the thermochemical water-splitting iodine–sulfur process for effective hydrogen production. *International Journal of Hydrogen Energy*, *32*(4), 489–496.

Khan, H. A., Iqbal, M. I., Jaleel, A., Abbas, I., Abbas, S. A., & Deog-Jung, K. (2019). Pt encapsulated hollow mesoporous SiO_2 sphere catalyst for sulfuric acid decomposition reaction in SI cycle. *International Journal of Hydrogen Energy*, *44*(4), 2312–2322.

Khan, H. A., JaleelA, M. E., Ahmed, S., Bhatti, U. H., Bilal, M., & Hussain (2021). Development of catalysts for sulfuric acid decomposition in the sulfur–iodine cycle: A review. *Catalysis Reviews*, 1–36.

Kim, C. S., Hong, S. D., Kim, Y. W., Kim, J. H., Lee, W. J., & Chang, J. (2008). Thermal design of a laboratory—scale SO_3 decomposer for nuclear hydrogen production. *International Journal of Hydrogen Energy*, *33*(14), 3688–3699.

Knoche, K. F., Cremer, H., Breywisch, D., Hegels, S., Steinborn, G., & Wüster, G. (1978). Experimental and theoretical investigation of thermochemical hydrogen production. *International Journal of Hydrogen Energy*, *3*(2), 209–216.

Kodama, T. (2003). High-temperature solar chemistry for converting solar heat to chemical fuels. *Progress in Energy and Combustion Science*, *29*, 567–597.

Kodama, T., & Gokon, N. (2007). Thermochemical cycles for high-temperature solar hydrogen production. *Chemical Reviews*, *107*(10), 4048–4077.

Kodama, T., Gokon, N., & Yamamoto, R. (2008). Thermochemical two-step water splitting by ZrO_2-supported $Ni_xFe_{3-x}O_4$ for solar hydrogen production. *Solar Energy*, *82*(1), 73–79.

Kodama, T., Kondoh, Y., Yamamoto, R., Andou, & Satou, N. (2005). Thermochemical hydrogen production by a redox system of ZrO_2-supported Co (II)-ferrite. *Solar Energy*, *78*(5), 623–631.

Kodama, T., Nakamuro, Y., & Mizuno, T. (2006). A two-step thermochemical water splitting by iron-oxide on stabilized zirconia. *Journal of Solar Energy Engineering*, *128*(1), 3–7.

Koepf, E., Alxneit, I., Wieckert, C., & Meier, A. (2017). A review of high temperature solar driven reactor technology: 25 years of experience in research and development at the Paul Scherrer Institute. *Applied Energy*, *188*, 620–651.

Kogan, A. (1998). Direct solar thermal splitting of water and on-site separation of the products—II. Experimental feasibility study. *International Journal of Hydrogen Energy*, *23*(2), 89–98.

Kubo, S., Futakawa, M., Ioka, I., Onuki, K., & Yamaguchi, A. (2013). Corrosion resistance of structural materials in high-temperature aqueous sulfuric acids in thermochemical water-splitting iodine–sulfur process. *International Journal of Hydrogen Energy*, *38*(16), 6577–6585.

Kurata, Y., Tachibana, K., & Suzuki, T. (2001). High temperature tensile properties of metallic materials exposed to a sulfuric acid decomposition gas environment. *Journal of the Japan Institute of Metals and Materials*.

Lee, C., Meng, Q. L., Kaneko, H., & Tamaura, Y. (2013). Solar hydrogen productivity of ceria–scandia solid solution using two-step water-splitting cycle. *Journal of Solar Energy Engineering*, *135*(1), 011002.

Leveque, G., Abanades, S., Jumas, J. C., & Olivier-Fourcade, J. (2014). Characterization of two-step tin-based redox system for thermochemical fuel production from solar-driven CO_2 and H_2O splitting cycle. *Industrial & Engineering Chemistry Research*, *53*(14), 5668–5677.

Lewis, N. S., & Nocera, D. G. (2006). Powering the planet: Chemical challenges in Sol. Energy utilization. *Proceedings of the National Academy of Sciences of the United States of America*, *103*, 15729–15735.

Li, H., Geng, T., Zhang, W., & Suppiah, S. (2012). Development SO_3 decomposition in the S–I cycle for hydrogen production. *Applied Energy*, *93*, 59–64.

Li, S., & Cheng, Y. (1995). Catalytic gasification of gas–coal char in CO_2. *Fuel*, *74*(3), 456–458.

Li, S., Kreider, P. B., Wheeler, V. M., & Lipiński, W. (2019). Thermodynamic analyses of fuel production via solar-driven ceria-based nonstoichiometric redox cycling: A case study of the isothermal membrane reactor system. *Journal of Solar Energy Engineering*, *141*(2).

Li, Z., Lin, Q., Li, M., Cao, J., Liu, F., Pan, H., Wang, Z., & Kawi, S. (2020). Recent advances in process and catalyst for CO_2 reforming of methane. *Renewable and Sustainable Energy Reviews*, *134*, 110312.

Liberatore, R., Favuzza, P., Felici, C., Tarquini, P., Narducci, G., & Borruto, A. (2019). Materials resistance to corrosion by I_2–HI–H_2O mixtures for the realization of a sulfur-iodine plant. *International Journal of Hydrogen Energy*, *44*(49), 26816–26834.

Lin, X., Zhang, Y., Wang, R., Zhou, J., & Cen, K. (2013). Influence of the structural and surface characteristics of activated carbon on the catalytic decomposition of hydrogen iodide in the sulfur–iodine cycle for hydrogen production. *International Journal of Hydrogen Energy*, *38*(35), 15003–15011.

Liu, Q., Hong, H., Yuan, J., Jin, H., & Cai, R. (2009). Experimental investigation of hydrogen production integrated methanol steam reforming with middle-temperature solar thermal energy. *Applied Energy*, *86*(2), 155–162.

Longo, J. M., & Raccah, P. M. (1973). The structure of La_2CuO_4 and $LaSrVO_4$. *Journal of Solid State Chemistry*, *6*(4), 526–531.

Loutzenhiser, P. G., Meier, A., & Steinfeld, A. (2010). Review of the two-step H_2O/CO_2-splitting solar thermochemical cycle based on Zn/ZnO redox reactions. *Materials*, *3*, 4922–4938.

Lu, Y., Zhu, L., Agrafiotis, C., Vieten, J., Roeb, M., & Sattler, C. (2019). Solar fuels production: Two-step thermochemical cycles with cerium-based oxides. *Progress in Energy and Combustion Science*, *75*, 100785.

Maatouk, S. C., Mokrani, N. B., Tabarant, M., Fleche, J. L., & Carles, P. (2009). Study of the miscibility gap in $H_2SO_4/HI/I_2/H_2O$ mixtures produced by the Bunsen reaction – Part I: Preliminary resultsat 308 K. *International Journal of Hydrogen Energy*, *34*(17), 7155–7161.

Mao, Y., Gao, Y., Dong, W., Wu, H., Song, Z., Zhao, X., Sun, J., & Wang, W. (2020). Hydrogen production via a two-step water splitting thermochemical cycle based on metal oxide – A review. *Applied Energy*, *267*, 114860.

McDaniel, A. H., Ambrosini, A., Coker, E. N., Miller, J. E., Chueh, W. C., O'Hayre, R., & Tong, J. (2014). Nonstoichiometric perovskite oxides for solar thermochemical H_2 and CO production. *Energy Procedia*, *49*, 2009–2018.

McDaniel, A. H., Miller, E. C., Arifin, D., Ambrosini, A., Coker, E. N., O'Hayre, R., Chueh, W. C., & Tong, J. (2013). Sr-and Mn-doped $LaAlO_{3-\delta}$ for solar thermochemical H_2 and CO production. *Energy & Environmental Science*, *6*(8), 2424–2428.

McKee, D. W. (2021). *The catalyzed gasification reactions of carbon[M]//Chemistry and physics of carbon* (pp. 1–118). CRC Press.

Milani, D., Khalilpour, R., Zahedi, G., & Abbas, A. (2015). A model-based analysis of CO_2 utilization in methanol synthesis plant. *Journal of CO_2 Utilization, 10*, 12–22.

Mogensen, M., Sammes, N. M., & Tompsett, G. A. (2000). Physical, chemical and electrochemical properties of pure and doped ceria. *Solid State Ionics, 129*(1-4), 63–94.

Muhich, C. L., Ehrhart, B. D., Al-Shankiti, I., Ward, B. J., Musgrave, C. B., & Weimer, A. W. (2016). A review and perspective of efficient hydrogen generation via solar thermal water splitting. *Wiley Interdisciplinary Reviews: Energy and Environment, 5*(3), 261–287.

Muhich, C. L., Ehrhart, B. D., Witte, V. A., Miller, S. L., Coker, E. N., Musgrave, C. B., & Weimer, A. W. (2015). Predicting the solar thermochemical water splitting ability and reaction mechanism of metal oxides: A case study of the hercynite family of water splitting cycles. *Energy & Environmental Science, 8*(12), 3687–3699.

Muhich, C. L., Evanko, B. W., Weston, K. C., Lichty, P., Liang, X., Martinek, J., Musgrave, C. B., & Weimer, A. W. (2013). Efficient generation of H_2 by splitting water with an isothermal redox cycle. *Science, 341*(6145), 540–542.

Müller, R., & Steinfeld, A. (2008). H_2O-splitting thermochemical cycle based on ZnO/Zn-redox: Quenching the effluents from the ZnO dissociation. *Chemical Engineering Science, 63*(1), 217–227.

Musgrave, C., & Weimer, A. (2015). A review and perspective of efficient hydrogen generation via solar thermal water splitting.

Nadar, A., Banerjee, A. M., Pai, M. R., Meena, S. S., Tewari, R., & Tripathi, A. K. (2018). Catalytic properties of dispersed iron oxides Fe_2O_3/MO_2 (M= Zr, Ce, Ti and Si) for sulfuric acid decomposition reaction: Role of support. *International Journal of Hydrogen Energy, 43*(1), 37–52.

Nagaoka, K., Seshan, K., Aika, K. I., & Lercher, J. A. (2001). Carbon deposition during carbon dioxide reforming of methane-comparison between Pt/Al_2O_3 and Pt/ZrO_2. *Journal of Catalysis, 197*(1), 34–42.

Nagaoka, K., Takanabe, K., & Aika, K. I. (2003). Influence of the reduction temperature on catalytic activity of Co/TiO_2 (anatase-type) for high pressure dry reforming of methane. *Applied Catalysis A: General, 255*(1), 13–21.

Nair, M. M., & Abanades, S. (2018). Experimental screening of perovskite oxides as efficient redox materials for solar thermochemical CO_2 conversion. *Sustainable Energy & Fuels, 2*(4), 843–854.

Nakajima, H., Lee, D., Lee, M. T., & Grigoropoulos, C. P. (2016). Hydrogen production with CuO/ZnO nanowire catalyst for a nanocatalytic solar thermal steam-methanol reformer. *International Journal of Hydrogen Energy, 41*(38), 16927–16931.

Naterer, G., Suppiah, S., Lewis, M., Gabriel, K., Dincer, I., Rosen, M. A., Fowler, M., Rizvi, G., Easton, E. B., Ikeda, B. M., & Kaye, M. H. (2009). Recent Canadian advances in nuclear-based hydrogen production and the thermochemical Cu–Cl cycle. *International Journal of Hydrogen Energy, 34*(7), 2901–2917.

Norman, J. H., Besenbruch, G. E., Brown, L. C., O'keefe, D. R., and Allen, C. L. (1981). Thermochemical water-splitting cycle, bench-scale investigations, and process engineering. Final report, February 1977-December 31, 1981.

Ohtsuka, Y., & Asami, K. (1991). Steam gasification of low-rank coals with a chlorine-free iron catalyst from ferric chloride. *Industrial & Engineering Chemistry Research, 30*, 1921–1926.

O'keefe DR, Norman JH, Williamson DG. (1980). Catalysis research in thermochemical water-splitting processes. *Catalysis Reviews: Science and Engineering, 22*(3), 325–369.

Oosawa, Y., Kumagai, T., Mizuta, S., Kondo, W., Takemori, Y., & Fujii, K. (1981). Kinetics of the catalytic decomposition of hydrogen iodide in the magnesium-iodine thermochemical cycle. *Bulletin of the Chemical Society of Japan, 54*(3), 742–748.

Oosawa, Y., Takemori, Y., & Fujii, K. (1980). Catalytic decomposition of hydrogen iodide in the magnesium-iodine thermochemical cycle – catalyst search. *Nippon Kagaku Kaishi, 7*, 1081–1087.

Otsuka, K., Hatano, M., & Morikawa, A. (1983). Hydrogen from water by reduced cerium oxide. *Journal of Catalysis*; (United States), *79*(2).

Otsuka, K., Hatano, M., & Morikawa, A. (1985). Decomposition of water by cerium oxide of δ-phase. *Inorganica Chimica Acta, 109*(3), 193–197.

Ozbilen, A., Dincer, I., & Rosen, M. A. (2011). A comparative life cycle analysis of hydrogen production via thermochemical water splitting using a Cu–Cl cycle. *International Journal of Hydrogen Energy, 36*(17), 11321–11327.

Özkara-Aydınoğlu, Ş., Özensoy, E., & Aksoylu, A. E. (2009). The effect of impregnation strategy on methane dry reforming activity of Ce promoted Pt/ZrO_2. *International Journal of Hydrogen Energy, 34*(24), 9711–9722.

Palumbo, R., Lédé, J., Boutin, O., Ricart, E. E., Steinfeld, A., Möller, S., Weidenkaff, A., Fletcher, E. A., & Bielicki, J. (1998). The production of Zn from ZnO in a high-temperature solar decomposition quench process—I. The scientific framework for the process[J]. *Chemical Engineering Science, 53*(14), 2503–2517.

Pan, H., Li, Y., Zhu, L., & Lu, Y. (2022). Solar-driven H_2O/CO_2 conversion to fuels via two-step electro-thermochemical cycle in a solid oxide electrochemical cell. *Energy Conversion and Management, 259*, 115578.

Panlener, R. J., Blumenthal, R. N., & Garnier, J. E. (1975). A thermodynamic study of nonstoichiometric cerium dioxide. *Journal of Physics and Chemistry of Solids*, *36*, 1213–1222.

Park, J., Cho, J. H., Jung, H., Jung, K.-D., Kumar, S., & Moon, I. (2013). Simulation and experimental study on the sulfuric acid decomposition process of SI cycle for hydrogen production. *International Journal of Hydrogen Energy*, *38*(14), 5507–5516.

Park, J. K., Lee, J., Jeon, J., Kim, J., Yoon, Y. S., & Um, W. (2021). Process optimization and safety assessment on a pilot-scale Bunsen process in sulfur–iodine cycle. *International Journal of Hydrogen Energy*, *46*(67), 33616–33634.

Park, J. K., Lee, S. Y., & Lee, I. B. (2019). Study of alternative reactor–separator network in Bunsen process of sulfur-iodine cycle for hydrogen production. *Journal of Chemical Engineering of Japan*, *52*(7), 638–649.

Parravano, G. (1952). Ferroelectric transitions and heterogenous catalysis. *Journal of Chemical Physics*, *20*(2), 342–343.

Peikovic, L. M., Ginosar, D. M., Rollins, H. W., Burch, K. C., Deiana, C., Silva, H. S., Sardella, M. F., & Granados, D. (2009). Activated carbon catalysts for the production of hydrogen via the sulfur–iodine thermochemical water splitting cycle. *International Journal of Hydrogen Energy*, *34*(9), 4057–4064.

Pérez-Florindo, A., Cazorla-Amorós, D., & Linares-Solano, A. (1993). CO_2-Carbon gasification catalyzed by alkaline-earths: Comparative study of the metal-carbon interaction and of the specific activity[J]. *Carbon*, *31*(3), 493–500.

Perkins, C., & Weimer, A. W. (2004). Likely near-term solar-thermal water splitting technologies. *International Journal of Hydrogen Energy*, *29*(15), 1587–1599.

Perkins, C., & Weimer, A. W. (2009). Solar-thermal production of renewable hydrogen. *AICHE Journal*, *55*(2), 286–293.

Piatkowski, N., Wieckert, C., Weimer, A. W., & Steinfeld, A. (2010). Solar-driven gasification of carbonaceous feedstock-a review. *Energy & Environmental Science*, *4*(1), 73–82.

Rosen, M. A. (2008). Exergy analysis of hydrogen production by thermochemical water decomposition using the Ispra Mark-10 Cycle. *International Journal of Hydrogen Energy*, *33*(23), 6921–6933.

Sá, S., Silva, H., Brandão, L., Sousa, J. M., & Mendes, A. (2010). Catalysts for methanol steam reforming—A review. *Applied Catalysis B: Environmental*, *99*(1-2), 43–57.

Safavinia, B., Wang, Y., Jiang, C., Wang, Y., Jiang, C., Roman, C., Darapaneni, P., Larriviere, J., Cullen, D. A., Dooley, K. M., & Dorman, J. A. (2020). Enhancing $Ce_xZr_{1-x}O_2$ activity for methane dry reforming using subsurface Ni dopants. *ACS Catalysis*, *10*(7), 4070–4079.

Sakurai, M., Aihara, M., Miyake, N., Tsutsumi, A., & Yoshida, K. (1992). Test of one-loop flow scheme for the UT-3 thermochemical hydrogen production process. *International Journal of Hydrogen Energy*, *17*(8), 587–592.

Sakurai, M., Bilgen, E., Tsutsumi, A., et al. (1996). Solar UT-3 thermochemical cycle for hydrogen production[J]. *Solar Energy*, *57*(1), 51–58.

Sakurai, M., Nakajima, H., Onuki, K., Ikenoya, K., & Shimizu, S. (1999). Preliminary process analysis for the closed cycle operation of the iodine-sulfur thermochemical hydrogen production process. *International Journal of Hydrogen Energy*, *24*(7), 603–612.

Sams, D., & Shadman, F. (1983). Catalytic effect of potassium on the rate of char-CO_2 gasification. *Fuel*, *62*, 880–882.

Sang, O. C., & Sang, H. M. (2009). Performance of $La_{1-x}Ce_xFe_{0.7}Ni_{0.3}O_3$ perovskite catalysts for methane steam reforming. *Catalysis Today*, *146*(1-2), 148–153.

Sastre, D., Carrillo, A. J., Serrano, D. P., Pizarro, P., & Coronado, J. M. (2017). Exploring the redox behavior of $La_{0.6}Sr_{0.4}Mn_{1-x}Al_xO_3$ perovskites for CO_2-splitting in thermochemical cycle. *Topics in Catalysis*, *60*(15–16), 1108–1118.

Scheffe, J. R., Li, J., & Weimer, A. W. (2010). A spinel ferrite/hercynite water-splitting redox cycle. *International Journal of Hydrogen Energy*, *35*(8), 3333–3340.

Scheffe, J. R., & Steinfeld, A. (2012). Thermodynamic analysis of cerium-based oxides for solar thermochemical fuel production. *Energy and Fuel*, *26*, 1928–1936.

Scheffe, J. R., Weibel, D., & Steinfeld, A. (2013). Lanthanum-strontium-manganese perovskites as redox materials for solar thermochemical splitting of H_2O and CO_2. *Energy & Fuels*, *27*(8), 4250–4257.

Sibieude, F., Ducarroir, M., Tofighi, A., & Ambriz, J. (1982). High temperature experiments with a solar furnace: The decomposition of Fe_3O_4, Mn_3O_4,CdO. *International Journal of Hydrogen Energy*, *7*(1), 79–88.

Singhania, A., & Bhaskarwar, A. N. (2017). Development of catalysts for hydrogen production from hydrogen iodide decomposition in thermo-chemical water-splitting sulfur-iodine cycle: A review. *Catalysis Reviews*.

Singhania, A., & Bhaskarwar, A. N. (2018). Catalytic performance of carbon nanotubes supported palladium catalyst for hydrogen production from hydrogen iodide decomposition in thermochemical sulfur iodine cycle. *Renewable Energy*, *127*, 509–513.

Singhania, A., & Bhaskarwar, A. N. (2018). TiO_2 as a catalyst for hydrogen production from hydrogen-iodide in thermo-chemical water-splitting sulfur-iodine cycle. *Fuel, 221*, 393–398.

Smitkova, M., Janíček, F., & Riccardi, J. (2011). Life cycle analysis of processes for hydrogen production. *International Journal of Hydrogen Energy, 36*(13), 7844–7851.

Song, B., & Kim, S. (1993). Catalytic activity of alkali and iron salt mixtures for steam–char gasification. *Fuel, 72*(6), 797–803.

Steinfeld, A. (2002). Solar hydrogen production via a two-step water-splitting thermochemical cycle based on Zn/ZnO redox reactions. *International Journal of Hydrogen Energy, 27*, 611–619.

Steinfeld, A. (2005). Solar thermochemical production of hydrogen: A review. *Solar Energy, 78*, 603–615.

Steinfeld, A., Brack, M., Meier, A., Weidenkaff, A., & Wuillemin, D. (1998). A solar chemical reactor for co-production of zinc and synthesis gas. *Energy, 23*(10), 803–814.

Su, Y. J., Pan, K. L., & Chang, M. B. (2014). Modifying perovskite-type oxide catalyst $LaNiO_3$ with Ce for carbon dioxide reforming of methane. *International Journal of Hydrogen Energy, 39*(10), 4917–4925.

Suárez-Almeida, M., Gómez-Barea, A., Ghoniem, A. F., & Pfeifer, C. (2021). Solar gasification of biomass in a dual fluidized bed. *Chemical Engineering Journal, 406*, 126665.

Sun, K., Fan, Z., Ye, J., Yan, J., Ge, Q., Li, Y., He, W., Yang, W., & Liu, C.-j. (2015). Hydrogenation of CO_2 to methanol over In_2O_3 catalyst. *Journal of CO_2 Utilization, 12*, 1–6.

Tagawa, H., & Endo, T. (1989). Catalytic decomposition of sulfuric acid using metal oxides as the oxygen generating reaction in thermochemical water splitting process. *International Journal of Hydrogen Energy, 14*(1), 11–17.

Tamaura, Y., Ueda, Y., Matsunami, J., Hasegawa, N., Nezuka, M., Sano, T., & Tsuji, M. (1999). Solar hydrogen production by using ferrites. *Solar Energy, 65*(1), 55–57.

Tejuca, L. G., Fierro, J., & Tascón, J. (1989). Structure and reactivity of perovskite-type oxides. *Advances in Catalysis, 36*(6), 237–328.

Traynor, A. J., & Jensen, R. J. (2002). Direct solar reduction of CO_2 to fuel: First prototype results. *Industrial & Engineering Chemistry Research, 41*(8), 1935–1939.

Utgikar, V., & Ward, B. (2006). Life cycle assessment of ISPRA Mark 9 thermochemical cycle for nuclear hydrogen production. *Journal of Chemical Technology & Biotechnology: International Research in Process, Environmental & Clean Technology, 81*(11), 1753–1759.

Van, V. D., Langenkamp, H., Schütz, G., Lalonde, D., Flamm, J., & Fiebelmann, P. (1980). Development and design of a continuous laboratory-scale plant for hydrogen production by the Mark-13 cycle. *International Journal of Hydrogen Energy, 5*(2), 131–139.

Vieten, J., Bulfin, B., Call, F., Lange, M., Schmücker, M., Francke, A., Roeb, M., & Sattler, C. (2016). Perovskite oxides for application in thermochemical air separation and oxygen storage. *Journal of Materials Chemistry A, 4*(35).

Vieten, J., Bulfin, B., Senholdt, M., Roeb, M., Sattler, C., & Schmücker, M. (2017). Redox thermodynamics and phase composition in the system $SrFeO_{3-\delta}$—$SrMnO_{3-\delta}$. *Solid State Ionics, 308*, 149–155.

Vitart, X., Duigou, A. L., & Carles, P. (2006). Hydrogen production using the sulfur-iodine cycle coupled to a VHTR: An overview. *Energy Conversion and Management, 47*(17), 2740–2747.

Wang, H., Gao, P., Zhao, T., Wei, W., & Sun, Y. (2015). Recent advances in the catalytic conversion of CO_2 to value added compounds. *SCIENCE CHINA Chemistry, 58*(1), 79–92.

Wang, H. S. (2020). *Solar thermochemical fuel generation*. Wind Solar Hybrid Renewable Energy System.

Wang, J., Wu, J., Xu, Z., & Li, M. (2017). Thermodynamic performance analysis of a fuel cell trigeneration system integrated with solar-assisted methanol reforming. *Energy Conversion and Management, 150*, 81–89.

Wang, L., Li, D., Ping, Z., Chen, S., & Xu, J. (2012). The HI catalytic decomposition for the lab-scale H_2 producing apparatus of the iodine–sulfur thermochemical cycle. *International Journal of Hydrogen Energy, 37*(8), 6415–6421.

Wang, N., Chu, W., Zhang, T., & Zhao, X.-S. (2011). Manganese promoting effects on the Co–Ce–Zr–Ox nano catalysts for methane dry reforming with carbon dioxide to hydrogen and carbon monoxide. *Chemical Engineering Journal, 170*(2–3), 457–463.

Wang, Z., Chen, Y., Zhou, C., Whiddon, R., Zhang, Y., Zhou, J., & Cen, K. (2011). Decomposition of hydrogen iodide via wood-based activated carbon catalysts for hydrogen production. *International Journal of Hydrogen Energy, 36*(1), 216–223.

Wegner, K., Ly, H. C., Weiss, R. J., Pratsinis, S. E., & Steinfeld, A. (2006). In situ formation and hydrolysis of Zn nanoparticles for H_2 production by the 2-step ZnO/Zn water-splitting thermochemical cycle. *International Journal of Hydrogen Energy, 31*(1), 55–61.

Weidenkaff, A., Reller, A. W., Wokaun, A., & Steinfeld, A. (2000). Thermogravimetric analysis of the ZnO/Zn water splitting cycle. *Thermochimica Acta, 359*(1), 69–75.

Wong, B., Buckingham, R. T., Brown, L. C., Russ, B. E., Besenbruch, G. E., Kaiparambil, A., Santhanakrishnan, R., & Roy, A. (2007). Construction materials development in sulfur–iodine thermochemical water-splitting process for hydrogen production. *International Journal of Hydrogen Energy, 32*(4), 497–504.

Xiao, L., Wu, S. Y., & Li, Y. R. (2012). Advances in solar hydrogen production via two-step water-splitting thermochemical cycles based on metal redox reactions. *Renewable Energy, 41*, 1–12.

Yadav, D., & Banerjee, R. (2016). A review of solar thermochemical processes. *Renewable and Sustainable Energy Reviews, 54*, 497–532.

Ying, Z., Wang, Y., Zheng, X., Geng, Z., Dou, B., & Cui, G. (2020). Experimental study and development of an improved sulfur–iodine cycle integrated with HI electrolysis for hydrogen production. *International Journal of Hydrogen Energy, 45*(24), 13176–13188.

Yokokawa, H., Horita, T., Sakai, N., Yamaji, K., Brito, M. E., Xiong, Y.-P., & Kishimoto, H. (2006). Ceria: Relation among thermodynamic, electronic hole and proton properties. *Solid State Ionics, 177*(19–25), 1705–1714.

Zhai, S., Rojas, J., Ahlborg, N., Cheng, C. H. M., Xie, C., Toney, M. F., Jung, I.-H., Chueh, W. C., & Majumdar, A. (2020). High-capacity thermochemical CO_2 dissociation using iron-poor ferrites. *Energy & Environmental Science, 13*(2), 592–600.

Zhai, S., Rojas, J., Ahlborg, N., Lim, K., Toney, M. F., Jin, H., Chueh, W. C., & Majumdar, A. (2018). The use of poly-cation oxides to lower the temperature of two-step thermochemical water splitting. *Energy & Environmental Science, 11*(8), 2172–2178.

Zhang, H., Shuai, Y., Lougou, B. G., Jiang, B., Wang, F., Cheng, Z., & Tan, H. (2020). Effects of multilayer porous ceramics on thermochemical energy conversion and storage efficiency in solar dry reforming of methane reactor. *Applied Energy, 265*, 114799.

Zhang, P., Su, T., Chen, Q. H., Wang, L. J., Chen, S. Z., & Xu, J. M. (2011). Catalytic decomposition of sulfuric acid on composite oxides and Pt/SiC. *International Journal of Hydrogen Energy, 37*(1), 760–764.

Zhang, R. J., Xia, G. F., Li, M. F., Yu, W. U., Hong, N. I. E., & Li, D.-D. (2015). Effect of support on the performance of Ni-based catalyst in methane dry reforming. *Journal of Fuel Chemistry and Technology, 43*(11), 1359–1365.

Zhang, Y., Fu, G., Wang, Z., Kuang, M., Whiddon, R., & Cen, K. (2015). HI Decomposition over carbon-based and Ni-impregnated catalysts of the sulfur–iodine cycle for hydrogen production. *Industrial & Engineering Chemistry Research, 54*(5), 1498–1504.

Zhang, Y., Wang, Z., Zhou, J., & Cen, K. (2009). Ceria as a catalyst for hydrogen iodide decomposition in sulfur–iodine cycle for hydrogen production. *International Journal of Hydrogen Energy, 34*(4), 1688–1695.

Zhang, Y., Wang, Z., Zhou, J., Liu, J., & Cen, K. (2008). Effect of preparation method on platinum–ceria catalysts for hydrogen iodide decomposition in sulfur–iodine cycle. *International Journal of Hydrogen Energy, 33*(2), 602–607.

Zhang, Y., Xu, C., Chen, J., Zhang, X., Wang, Z., Zhou, J., & Cen, K. (2015). A novel photo-thermochemical cycle for the dissociation of CO_2 using solar energy. *Applied Energy, 156*, 223–229.

Zhang, Y., Zhou, J., Chen, Y., Wang, Z., Liu, J., & Cen, K. (2008). Hydrogen iodide decomposition over nickel–ceria catalysts for hydrogen production in the sulfur–iodine cycle. *International Journal of Hydrogen Energy, 33*(20), 5477–5483.

Zhang, Y., Zhou, J., Wang, Z., Liu, J., & Cen, K. (2008). Influence of the oxidative/reductive treatments on Pt/CeO_2 catalyst for hydrogen iodide decomposition in sulfur–iodine cycle. *International Journal of Hydrogen Energy, 33*(9), 2211–2217.

Zhao, B., Huang, C., Ran, R., Wu, X., & Weng, D. (2016). Two-step thermochemical looping using modified ceria-based materials for splitting CO_2. *Journal of Materials Science, 51*(5), 2299–2306.

Zhao, K., He, F., Huang, Z., Wei, G., Zheng, A., Li, H., & Zhao, Z. (2016). Perovskite-type oxides $LaFe_{1-x}Co_xO_3$ for chemical looping steam methane reforming to syngas and hydrogen co-production. *Applied Energy, 168*(apr.15), 193–203.

Zhao, K., Li, L., Zheng, A., Huang, Z., He, F., Shen, Y., Wei, G., Li, H., & Zhao, Z. (2017). Synergistic improvements in stability and performance of the double perovskite-type oxides $La_{2-x}Sr_xFeCoO_6$ for chemical looping steam methane reforming. *Applied Energy, 197*, 393–404.

Zhao, Q., Su, B., Wang, H., He, A., He, R., Kong, H., & Hu, X. (2021). Mid/low-temperature solar hydrogen generation via dry reforming of methane enhanced in a membrane reactor. *Energy Conversion and Management, 240*, 114254.

Zheng, Y., Li, K., Wang, H., Tian, D., Wang, Y., Zhu, X., Wei, Y., Zheng, M., & Luo, Y. (2017). Designed oxygen carriers from macroporous $LaFeO_3$ supported CeO_2 for chemical-looping reforming of methane. *Applied Catalysis B: Environmental, 202*, 51–63.

Zhu, L., Lu, Y., & Li, F. (2018). Reactivity of Ni, Cr and Zr doped ceria in CO_2 splitting for CO production via two-step thermochemical cycle. *International Journal of Hydrogen Energy, 43*(30), 13754–13763.

9

Electrocatalytic Reduction of CO_2 to Value-Added Chemicals and Fuels

Qian Sun, Kamran Dastafkan, and Chuan Zhao

School of Chemistry, The University of New South Wales, Faculty of Science, Sydney, NSW, Australia

9.1 Introduction

Massive consumption of fossil fuels (e.g., coal, natural gas, and oil) has generated excessive carbon dioxide (CO_2) in Earth's atmosphere, triggering serious climate change (Chu & Majumdar, 2012; Feldman et al., 2015). Various strategies have been developed for CO_2 capture and utilization. These include chemical reforming (Graciani et al., 2014), biological transformation (Schuchmann & Müller, 2013), electrochemical reduction (Zhang, Zhao, & Gong, 2017), and photochemical conversion (Ouyang et al., 2017). Among these approaches, CO_2 reduction driven by electric power is effective in converting CO_2 into fuels.

The energy efficiency (EE) for CO_2-to-fuels conversion is defined as the ratio of "chemical energy stored in the products generated" to "input electric energy" (He, Ni, et al., 2018), which is calculated based on the cathodic electrochemical CO_2 reduction reaction (CO_2RR) coupled with anodic water oxidation reaction ($O_2 + 4H^+ + 4e^- \leftrightarrow 2H_2O$; 1.23 V vs. RHE). For example, the electrical power-to-carbon monoxide (CO) conversion efficiency can be calculated using the equation:

$$EE = \frac{E_{ox^0} - E_{red^0}}{E_{ox} - E_{red}} \times FE_{CO}$$

where $E_{ox}{}^0$ and $E_{red}{}^0$ are the standard thermodynamic potentials for water oxidation and CO_2RR to CO, respectively. E_{ox} and E_{red} are the applied potentials at anode and cathode, respectively. The anodic reaction is assumed to occur with $E_{ox} = 1.23$ V for calculation of half-cell cathodic energy efficiency (CEE). FE_{CO} is the Faradaic efficiency for CO generation, which can be used to evaluate the CO_2 reduction product selectivity (Tang & Xiao, 2022):

$$FE_{CO} = \frac{\alpha \times n_{CO} \times F}{Q}$$

where α and n_{CO} (mol) are the numbers of electrons and moles for CO formation, respectively. F and Q (C/mol) are the Faraday constant and the total passed charge, respectively. The full-cell EE can be calculated using the equation:

$$EE = \frac{E^0}{E_{\text{full cell}}} \times FE_{CO}$$

where $E_{\text{full cell}}$ is the voltage applied between anode and cathode and $E^0 = E_{ox}{}^0 - E_{red}{}^0$.

Promising half-cell efficiencies of 79%, 50%, and 44%, respectively, for CO_2 to formate (He, Ni, et al., 2018), C_{2+} (Kibria et al., 2018), and ethylene (Dinh et al., 2018) conversion, as well as overall efficiencies of 57% and 23%, respectively, toward CO (Ren et al., 2022), and ethanol (Li, Li, et al., 2020) production were reported.

When the electric power for CO_2RR is supplied by a solar cell, the solar-to-chemical conversion efficiency is defined as the ratio of "chemical power output" to "solar power input". For CO, the conversion efficiency is calculated by the equation (Schreier et al., 2017):

$$EE_{CO} = \frac{P_{CO}}{P_{solar}} = \frac{E^0_{CO/CO_2} J_{solar} FE_{CO}}{P_{solar}}$$

Conversion of Water and CO_2 to Fuels using Solar Energy: Science, Technology and Materials, First Edition.
Edited by Oomman K. Varghese and Flavio L. Souza.

where P_{CO} is the power drawn from the solar cell for CO_2 reduction to CO, P_{solar} is the intensity of sunlight incident on the solar cell, $E_{CO/CO2}{}^0 = 1.34$ V is the equilibrium potential for $CO_2 + 2e^- \rightarrow CO + 1/2O_2$ and J_{solar} is the current per illuminated area of the solar cell.

There are two systems used for solar CO_2RR: photovoltaic-activated electrochemical reactors (PV-EC) (Park et al., 2019) and photoelectrochemical (PEC) CO_2 reduction systems (Pan, Kim, et al., 2018). The PV-EC uses an external photovoltaic (PV) array to power an electrolyzer consisting of metallic electrodes. The coupling of PV and EC enables storage of solar energy as valuable chemicals generated electrochemically from CO_2. Examples of solar-to-fuel conversion efficiency obtained using PV-EC systems are 19.1% (Cheng et al., 2020), 3.9% (Wu et al., 2016), and 5.6% (Wu et al., 2016) for CO, hydrocarbons, and oxygenates production, respectively. A PEC system (Pan, Kim, et al., 2018) is also promising for artificial photosynthesis, where light is absorbed by a photoactive electrode (a semiconductor) to generate excited charge carriers, which drive catalytic reactions at the anode and the cathode. Examples of solar-to-fuel efficiency obtained from PEC systems are respectively 13.3%, 13.8%, and 19.4% for formate (Wulan et al., 2022), C_1 (Wulan et al., 2022), and CH_4 (Ren et al., 2017).

Electrochemical CO_2RR is one of the most promising approaches to convert CO_2 into value-added chemicals and fuels because applied potentials/currents can be controlled, the reaction device is simple, involves no net CO_2 emission, and the technology is scalable (Agarwal et al., 2011; Jia, Dastafkan, et al., 2019; Whipple & Kenis, 2010). CO_2RR not only lowers the CO_2 level in the air, but also reduces the strong dependence on traditional fossil fuels via CO_2 transformation into valuable chemicals (Jiang et al., 2018; Zhu et al., 2016). However, CO_2RR faces several challenges, including low CO_2 solubility in aqueous electrolytes (Jiang et al., 2018), sluggish kinetics of CO_2 reduction (Zhu et al., 2016), competitive hydrogen evolution reaction (HER) (Zhang, Sethuraman, et al., 2014), high enthalpy of C=O bond (806 kJ/mol) in CO_2, making the molecule highly stable, high overpotential, and slow reaction rates due to the large reorganizational energy (−1.9 V vs. standard hydrogen electrode [SHE]) for the transformation of linear CO_2 into bent $*CO_2^-$ (Yang, Dastafkan, et al., 2018). For example, the overpotential (η) for CO_2-to-CO conversion is calculated by the following equation:

$$\eta = E_{applied} - E^0_{CO}$$

where $E_{applied}$ is the applied potential, E^0_{CO} is the standard potential of CO_2 reduction to CO ($CO_2 + 2H^+ + 2e^- \rightarrow CO + H_2O$). Furthermore, the selectivity of CO_2RR is limited owing to the production of multiple products. Hence, there is a need to develop CO_2RR electrocatalysts that are active and product-selective.

To date, many kinds of catalysts have been developed for CO_2RR, and these include noble and transition metals (Mistry et al., 2017; Todorova et al., 2020), OD-Ms (Zhao, Liu, et al., 2017), metal alloys (Kim et al., 2017; Zhao et al., 2015), metal oxides (Li, Chen, et al., 2017), transition metal chalcogenides (TMCs) (Gao et al., 2020), single-atom catalysts (SACs) (Zhao, Dai, et al., 2017), and molecular catalysts (Nam, De Luna, et al., 2020). Earlier studies classified metal-based CO_2RR catalysts according to the product's selectivity (Hori et al., 1994). Sn, Hg, Pb, and In are promising for CO_2-to-HCOOH conversion; Au, Ag, Zn, and Pd favor CO production. Cu and Cu-based catalysts are effective in converting CO_2 to hydrocarbons and alcohols, while Ni, Co, and Fe show no significant activity to CO_2RR. Recent studies have also shown promising CO_2RR performances using metal oxides and TMCs (Gao et al., 2020; Gao, Jiao, et al., 2016; Zeng et al., 2018), although the number of reported studies devoted to metal selenides and tellurides is relatively rare. For example, copper selenides have been reported as promising materials for CO_2RR because of their nonstoichiometric ($Cu_{2-x}Se$) (Zhang et al., 2016) and stoichiometric (CuSe, Cu_2Se, and Cu_2Se_3) compositions (Li et al., 2010), structure stability, low cost, and high electrical conductivity (Li et al., 2010).

Carbon-based electrocatalysts for CO_2RR have gained special interest owing to their low cost and eco-friendly properties. Besides, carbonaceous materials such as carbon nanotubes (CNTs), graphene, and graphite exhibit large surface area, excellent conductivity, as well as remarkable chemical and mechanical stability (Jia, Dastafkan, et al., 2019). Moreover, the porous carbon materials facilitate electron and ion transfer, which promote CO_2 electroreduction by acting as catalyst supports or catalysts.

Over the past few decades, homogeneous catalysts, which activate the electrochemical reactions with the reactants, products, and catalyst existing in the same phase, most often the liquid phase, have been applied for CO_2RR (Benson et al., 2009). These catalysts exhibit good catalytic performances with FE almost unity for CO or HCOOH production and other low-carbon fuels (Qiao et al., 2014). However, homogeneous catalysts usually require relatively mild conditions (Wang et al., 2019). In contrast, heterogeneous catalysts, which promote electrochemical reactions with the phase of the catalyst differing from that of the reactants or products, are usually preferred in industrial applications (Wang et al., 2019), due to their capability of operation at relatively harsh environments (high pressures and temperatures) (Astruc et al., 2005), direct separation from the reactants and products afterward (Astruc et al., 2005), and reduced mass-transport limitations through the elimination of catalyst diffusion to the electrode (Corbin et al., 2019). Apart from these heterogeneous catalysts, molecular catalysts such as organometallic complexes, covalent-organic frameworks (COFs), metal-organic frameworks (MOFs), metal-free and polymer catalysts were also investigated as effective materials for CO_2RR. However, they are restricted by problems about overpotentials, selectivity, and undesired HER (Jiao et al., 2019).

Recently, SACs with isolated metal atoms distributing on solid supports have emerged as effective candidates for CO_2RR due to low coordination number, high metal utilization, and strong atom-support interactions. For example, Zhao et al. (Zhao, Dai, et al., 2017) reported a Ni SAC on N-doped porous carbon derived from zeolitic imidazolate framework for CO generation from CO_2 reduction. This Ni SAC exhibited an excellent turnover frequency (5273 h^{-1}), a high CO selectivity of over 71.9%, and a current density of 10.48 mA/cm^2 at 0.89 V overpotential, which were attributed to the low-coordinated surface $Ni–N_3$ active sites that facilitated the electron transfer and the formation of the key intermediate, $CO_2^{\cdot -}$. Following this pioneering study, a range of metal SACs such as Ru (Wang, Wang, Dares, & Meyer, 2017), Pd (Back & Jung, 2017), Pt (Zhao et al., 2019), Fe (Cheng et al., 2018), Co (Pan, Lin, et al., 2018), Cu (Abdel-Mageed et al., 2019), Zn (Yang, Song, et al., 2018), and Sn SAC (Zhao et al., 2018) were developed as promising materials for CO_2RR.

Developing highly efficient and selective electrocatalysts for CO_2RR is the key to making the technology commercially viable. Many book chapters (Hori, 2008; Sullivan et al., 2012; Sun, Jia, & Zhao, 2022) and reviews (Sun, Jia, Zhao, & Zhao, 2022; Varela, Ju, & Strasser, 2018; Yang, Dastafkan, et al., 2018) focus on the development and applications of effective catalysts, electrolysis cells, electrolytes, and characterizations for CO_2RR. Our book chapter is different because we summarize here the recent progress in *in situ* characterizations and detailed understanding of the CO_2RR mechanism. Another uniqueness of this chapter is that we systematically summarized the advantages and disadvantages of various electrocatalysts for CO_2 reduction, giving general information for the development of catalysts and paving the way to design advanced materials. In this chapter, after introducing the topic, we elucidate the reaction mechanisms and the present models of electrochemical cells and electrolytes for CO_2RR. Then, we discuss various types of materials developed as electrocatalysts for the purpose. This is followed by a discussion of the advanced *in situ* characterization techniques for CO_2RR electrocatalysts. Finally, we express our views on addressing outstanding challenges in the field and potential research directions.

9.2 Fundamentals of CO_2 Electroreduction (CO_2RR)

9.2.1 Reaction Mechanism of CO_2RR

CO_2 possesses a stable linear structure with strong C=O bond (806 kJ/mol) (Diercks et al., 2018), challenging CO_2 electroreduction. Therefore, a large activation barrier exists for CO_2RR, which in turn results in significant overpotentials (Verma et al., 2016). The electrochemical reduction of CO_2 molecules through different multi-electron/proton transfer pathways generates various products such as carbon monoxide (CO), methane (CH_4), methanol (CH_3OH), formic acid (HCOOH), ethylene (C_2H_4), ethanol (C_2H_5OH), acetate (CH_3COOH), and propanol (C_3H_7OH). Table 9.1 summarizes the CO_2RR pathways and HER, together with their standard potentials E vs. SHE (at pH = 7), which were converted from the potential values vs. RHE (SHE = RHE − 0.059*pH) in alkaline and acidic solutions (Handoko et al., 2018; Kortlever et al., 2015; Kuhl et al., 2012). Generally, CO_2 is first adsorbed and reduced to $^*CO_2^-$ on the catalyst surface during a CO_2RR process, followed by stepwise electron/proton transfer to various intermediates and products happening almost instantaneously, and finally, the desorption of products. Different reaction pathways exhibit similar equilibrium potentials, leading to a wide range of CO_2RR products. Besides, HER emerges as a significantly competitive side reaction in aqueous electrolytes.

The detailed reduction pathways that have been discovered until today for the formation of C_1 and C_2 products are discussed below. The mechanism of CO_2 reduction to many other products is largely unclear. The products exist in small amounts and can be detected only by nuclear magnetic resonance (NMR) (Kortlever et al., 2015). Generally, CO_2 is converted to C_1 and C_{2+} products via the CO pathway (Nie et al., 2014) and CO dimerization pathway (Garza et al., 2018), respectively. The production of CO via CO_2 reduction displayed the highest selectivity nearly 100% (Zheng, Jiang, et al., 2019). The reaction proceeds only through a two-electron/proton transfer process. The first step consists of CO_2 activation by the catalyst via a proton-coupled electron transfer. This is followed by the rate determining step (RDS) of *COOH formation, another proton–electron transfer to give *CO, and finally, the CO desorption (Hou et al., 2020). Formate is another CO_2RR product with similar high selectivity to CO (Han et al., 2018). In the catalytic pathway to generate formate, the protonation of C or O atom is induced by CO_2 reduction with a proton-coupled electron transfer. The C atom is protonated to OCHO* intermediate, which is an endothermic (+0.49 eV) process. This intermediate is further protonated to $HCOO^-$ via

Table 9.1 Half-Cell Reactions of Main CO_2RR Products and Their Standard Potentials E^0 in Base and Acid (V vs. SHE).

Products	Base		Acid	
	Half-cell reaction	E^0 (V vs. SHE)	Half-cell reaction	E^0 (V vs. SHE)
Hydrogen (H_2)	$2H_2O + 2e^- \rightarrow H_2 + 2OH^-$	−0.241	$2H^+ + 2e^- \rightarrow H_2$	−0.413
Carbon monoxide (CO)	$CO_2 + H_2O + 2e^- \rightarrow CO + 2OH^-$	−1.345	$CO_2 + 2H^+ + 2e^- \rightarrow CO + H_2O$	−1.517
Methane (CH_4)	$CO_2 + 6H_2O + 8e^- \rightarrow CH_4 + 8OH^-$	−1.072	$CO_2 + 8H^+ + 8e^- \rightarrow CH_4 + 2H_2O$	−0.244
Methanol (CH_3OH)	$CO_2 + 5H_2O + 6e^- \rightarrow CH_3OH + 6OH^-$	−1.225	$CO_2 + 6H^+ + 6e^- \rightarrow CH_3OH + H_2O$	−0.397
Formate (HCOOH)	$CO_2 + H_2O + 2e^- \rightarrow HCOO^- + OH^-$	−1.052	$CO_2 + 2H^+ + 2e^- \rightarrow HCOOH$	−0.584
Formate (HCOOH, HCO_3^- route)	$HCO_3^- + H_2O + 2e^- \rightarrow HCOO^- + 2OH^-$	−1.234	$HCO_3^- + 3H^+ + 2e^- \rightarrow HCOOH + H_2O$	−0.352
Oxalate ($H_2C_2O_4$)	$2CO_2 + 2e^- \rightarrow C_2O_4^{2-}$	−1.008	$2CO_2 + 2H^+ + 2e^- \rightarrow H_2C_2O_4$	−1.008
Formaldehyde (HCHO)	$CO_2 + 3H_2O + 4e^- \rightarrow HCHO + 4OH^-$	−1.383	$CO_2 + 4H^+ + 4e^- \rightarrow HCHO + H_2O$	−0.555
Carbon (C)	$CO_2 + 2H_2O + 4e^- \rightarrow C + 4OH^-$	−1.034	$CO_2 + 4H^+ + 4e^- \rightarrow C + 2H_2O$	−0.206
Glyoxal (CHOCHO)	$2CO_2 + 4H_2O + 6e^- \rightarrow CHOCHO + 6OH^-$	−1.382	$2CO_2 + 6H^+ + 6e^- \rightarrow CHOCHO + 2H_2O$	−0.554
Glycolaldehyde (CH_2OHCHO)	$2CO_2 + 6H_2O + 8e^- \rightarrow CH_2OHCHO + 8OH^-$	−1.298	$2CO_2 + 8H^+ + 8e^- \rightarrow CH_2OHCHO + 2H_2O$	−0.470
Ethylene (C_2H_4)	$2CO_2 + 8H_2O + 12e^- \rightarrow C_2H_4 + 12OH^-$	−1.156	$2CO_2 + 12H^+ + 12e^- \rightarrow C_2H_4 + 4H_2O$	−0.328
Ethane (C_2H_6)	$2CO_2 + 10H_2O + 14e^- \rightarrow C_2H_6 + 14OH^-$	−1.098	$2CO_2 + 14H^+ + 14e^- \rightarrow C_2H_6 + 4H_2O$	−0.296
Ethanol (C_2H_5OH)	$2CO_2 + 9H_2O + 12e^- \rightarrow C_2H_5OH + 12OH^-$	−1.157	$2CO_2 + 12H^+ + 12e^- \rightarrow C_2H_5OH + 3H_2O$	−0.329
Ethylene glycol (CH_2OHCH_2OH)	$2CO_2 + 8H_2O + 10e^- \rightarrow CH_2OHCH_2OH + 10OH^-$	−1.631	$2CO_2 + 10H^+ + 10e^- \rightarrow CH_2OHCH_2OH + 2H_2O$	−0.803
Acetaldehyde (CH_3CHO)	$2CO_2 + 7H_2O + 10e^- \rightarrow CH_3CHO + 10OH^-$	−1.188	$2CO_2 + 10H^+ + 10e^- \rightarrow CH_3CHO + 3H_2O$	−0.360
Acetate (CH_3COOH)	$2CO_2 + 5H_2O + 8e^- \rightarrow CH_3COO^- + 7OH^-$	−1.066	$2CO_2 + 8H^+ + 8e^- \rightarrow CH_3COOH + 2H_2O$	−0.315
Hydroxyacetone (CH_3COCH_2OH)	$3CO_2 + 10H_2O + 14e^- \rightarrow CH_3COCH_2OH + 14OH^-$	−1.199	$3CO_2 + 14H^+ + 14e^- \rightarrow CH_3COCH_2OH + 4H_2O$	−0.371
Propionaldehyde (CH_3CH_2CHO)	$3CO_2 + 11H_2O + 16e^- \rightarrow CH_3CH_2CHO + 16OH^-$	−1.147	$3CO_2 + 16H^+ + 16e^- \rightarrow CH_3CH_2CHO + 5H_2O$	−0.319
Allyl alcohol (CH_2CHCH_2OH)	$3CO_2 + 11H_2O + 16e^- \rightarrow CH_2CHCH_2OH + 16OH^-$	−1.191	$3CO_2 + 16H^+ + 16e^- \rightarrow CH_2CHCH_2OH + 5H_2O$	−0.363
Acetone ($(CH_3)_2CO$)	$3CO_2 + 11H_2O + 16e^- \rightarrow (CH_3)_2CO + 16OH^-$	−1.139	$3CO_2 + 16H^+ + 16e^- \rightarrow (CH_3)_2CO + 5H_2O$	−0.311
Propylene (CH_2CHCH_3)	$3CO_2 + 12H_2O + 18e^- \rightarrow CH_2CHCH_3 + 18OH^-$	−1.139	$3CO_2 + 18H^+ + 18e^- \rightarrow CH_2CHCH_3 + 6H_2O$	−0.311
Propane ($CH_3CH_2CH_3$)	$3CO_2 + 14H_2O + 20e^- \rightarrow CH_3CH_2CH_3 + 20OH^-$	−1.146	$3CO_2 + 20H^+ + 20e^- \rightarrow CH_3CH_2CH_3 + 6H_2O$	−0.275
n-Propanol (C_3H_7OH)	$3CO_2 + 13H_2O + 18e^- \rightarrow C_3H_7OH + 18OH^-$	−1.146	$3CO_2 + 18H^+ + 18e^- \rightarrow C_3H_7OH + 5H_2O$	−0.318

a second exothermic (−0.17 eV) proton-coupled electron transfer process. Finally, the formate is spontaneously released. An earlier study discovered that the pathway to formate must go through CO, separating from the hydrocarbon one (Hori et al., 1989). Carbene species ($^*CH_2$) derived from *CO were commonly proposed as an intermediate for methane and ethylene production. Methane is generated via double proton–electron transfer to $^*CH_2$, while ethylene is produced by $^*CH_2$ dimerization or CO insertion in a Fischer–Tropsch-like step, which is also the suggested pathway for alcohol generation (Hori et al., 1997).

In contrast with the "carbene" mechanism, the second C—O bond was broken in the process for CO_2-to-CH_4 conversion on Cu (Kumar et al., 2017) (Peterson et al., 2010). A possible pathway for CH_4 formation is given as the following:

$$CO_2 \rightarrow {}^*COOH \rightarrow {}^*CO \rightarrow {}^*CHO \rightarrow {}^*CH_2O \rightarrow {}^*CH_3O \rightarrow {}^*CH_4 + {}^*O \rightarrow {}^*CH_4 + {}^*OH \rightarrow CH_4 + H_2O.$$

where the charge transfer from *CO to *CHO is the RDS for CH_4 formation (Peterson et al., 2010).

The H_xCO species can be dimerized and subsequently deoxygenated to ethylene (Montoya et al., 2013). Notably, the intermediates $^*H_2CO$ and $^*H_3CO$ in this mechanism cannot elucidate the formaldehyde reduction to methanol and the methanol reduction to methane. Nie et al. (Nie et al., 2013) declared that CO reduction was the RDS for CO_2RR on Cu(111). The dominant pathway proceeded through CO reduction to COH and further to CH_x species, finally to methane and ethylene via $^*CH_2$ moieties coupling.

A comprehensive mechanism for CO_2RR on Cu (Figure 9.1) proposed by Kortlever et al. (Kortlever et al., 2015), similar to the work by Nie et al. (Nie et al., 2014), indicated distinct pathways to methane (C_1 pathway) and ethylene (C_2 pathway). Per the C_1 pathway, methane is formed via the CO intermediate reduction to *CHO or *COH. It is assumed that the second C—O bond is broken at an early stage; however, this is different from the late breaking of the C—O bond, which is more thermodynamically favorable according to the theoretical calculations (Peterson et al., 2010). Notably, in this C_1 pathway, the intermediates are dimerized to ethylene at high overpotentials. This process could occur on Cu(111) and Cu(100) with a current density up to 10 mA/cm^2.

Per the C_2 pathway, CO is dimerized to $^*C_2O_2^-$ intermediate via electron transfer at low overpotentials, followed by proton transfer, which is the RDS for CO reduction (RDS2 in Figure 9.1). Cu(100) favors ethylene formation at low potentials, because of the fact that the negatively charged CO dimer is most stable on square arrangements of four surface atoms. The CO dimerization via C—C coupling is considered as the key step to the formation of multi-carbon products. For example, the CO_2-to-C_2H_5OH conversion follows the pathway:

$$CO_2 \rightarrow {}^*CO \rightarrow {}^*CO - CO \rightarrow {}^*CH_2CHO \rightarrow C_2H_5OH.$$

where the *CO–CO dimerization to $^*CH_2CHO$ is the RDS for C_2H_5OH (Fan, Xia, Yang, et al., 2020).

Because of the complex reaction pathways, the selectivity toward a particular product or a family of products is a big challenge. Among all possible products of CO_2RR, CO, and HCOOH are generated with the highest Faradaic efficiency, above 99% (Han et al., 2018; Jiang et al., 2018). This is significantly higher than that of CH_4 (>55%) (Li, Cui, et al., 2017). Usually, C_{2+} products exhibit lower selectivity than the C_1 species, attributed to slower hydrogenation rate of C–C formation than that in the low-carbon intermediates. For example, C_2 species such as C_2H_4 and C_2H_5OH can achieve Faradaic efficiency of ~70% (Dinh et al., 2018) and ~30% (Zhuang et al., 2018), respectively, while the C_3 species such as propanol is limited to a low Faradaic efficiency of <30% (Zhuang et al., 2018). Therefore, CO is the most practical product, which can be utilized to manufacture bulk chemicals and fabricate synthetic fuels (Huan et al., 2017). C_2H_4 and C_2H_5OH are more favorable than their C_1 counterparts (CH_4) due to their higher energy storage and commercial value (Todorova et al., 2020). To improve the FE of a desired product, often rational designs of highly active and robust electrocatalysts are applied.

9.2.2 Electrochemical Cells

Several types of electrochemical cells have been developed for CO_2RR, including H-cell, flow cell, membrane electrode assembly (MEA), high-temperature molten salt cell, and solid oxide cell (Table 9.2).

9.2.2.1 H-Cell

To date, H-cells are the most renowned lab-scale CO_2RR reactors. Many studies on CO_2RR have been conducted in H-cells, mainly owing to easy operation, adaptability to various electrode materials and structures, and simple reactor set-up, leading to quick measurement of catalyst activity. Typically, a H-cell consists of a working cathode and a Pt anode separated by an ion exchange membrane, a reference electrode such as saturated calomel electrode (SCE) or Ag/AgCl electrode or

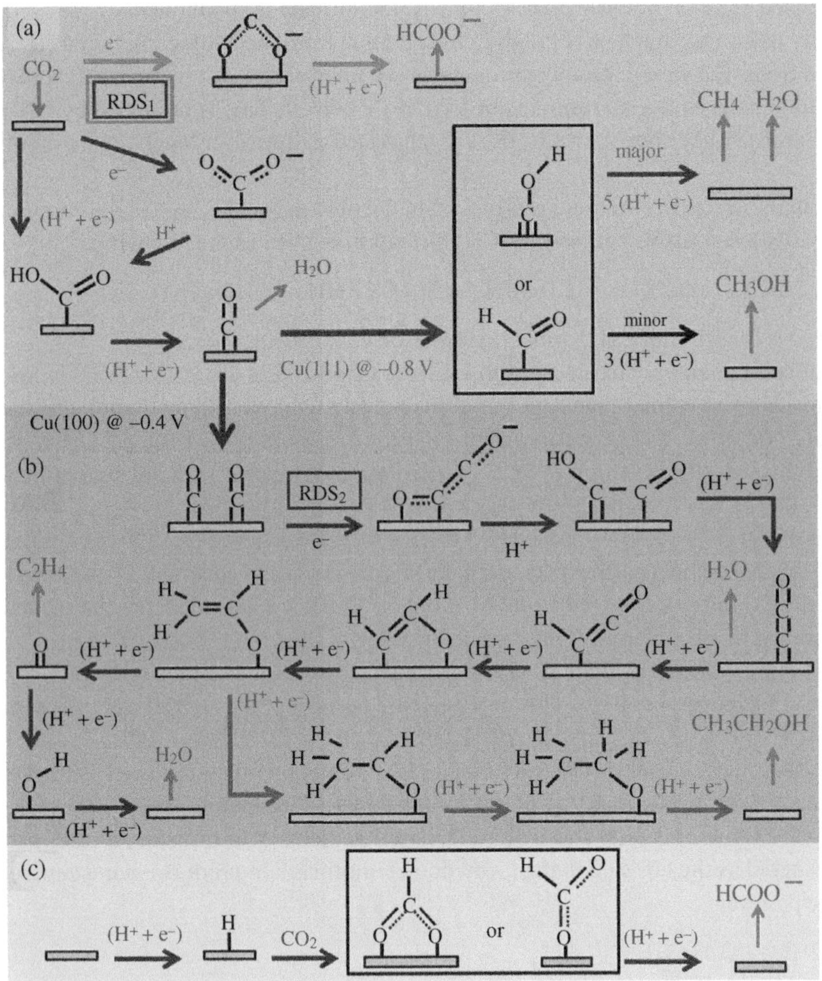

Figure 9.1 Possible reaction pathways for the electrocatalytic reduction of CO_2 to products on transition metals and molecular catalysts: (a) pathways from CO_2 to CO, CH_4, CH_3OH, and $HCOO^-$; (b) pathways from CO_2 to ethylene and ethanol; (c) pathway of CO_2 insertion into a metal–H bond yielding formate. Potentials are reported vs. RHE, while RDS indicates rate-determining steps and ($H^+ + e^-$) indicates steps in which either concerted or separated proton–electron transfer takes place. Source: Reproduced from Kortlever et al., 2015, with permission. Copyright 2015, American Chemical Society/American Chemical Society.

Hg/HgO electrode, and a liquid electrolyte such as $KHCO_3$ (Figure 9.2a) (Tan et al., 2019). In the anodic chamber, the anode usually processes water oxidization via the oxygen evolution reaction to balance the charge and complete the electrical circuit. While in the cathodic chamber, CO_2 dissolves into the electrolyte and transfers to the cathode surface, where it is electrochemically reduced to various products such as CO, CH_4, HCOOH, CH_3OH, and C_2H_5OH (Mariano et al., 2017).

Several issues in H-cell should be considered. First, the cell should be gastight for precise quantification of gas products. For the liquid product determination, the electrolysis time and the cell volume should be carefully considered to reach the limits of nuclear magnetic resonance (NMR). This is attributed to the low FE of most liquid products, especially for that of Cu-based catalysts (Kuhl et al., 2012). In this case, smaller electrode compartment volume and longer electrolysis duration are recommended for low-FE liquid product detection. In addition, although the desired CO_2RR products with high FE have been achieved using efficient catalysts, CO_2 electroreduction delivers relatively low current densities ($< 100\ mA/cm^2$) due to low CO_2 solubility in aqueous electrolytes, and the competitive HER half-reaction, hindering the feasibility of the technology (Jiang et al., 2016). To improve the CO_2 concentration and mass transfer, ionic liquid (IL) electrolytes with high CO_2 solubility (Ren et al., 2020) can be applied to CO_2RR, which activate CO_2, work as cocatalyst, and stabilize the key intermediates simultaneously, thus boosting CO_2 reduction. Another issue comes from the application of the cation exchange membrane such as Nafion membrane in H-cell. During the CO_2 electrolysis, K^+ migrates from anode to cathode across the Nafion membrane, resulting in anolyte pH decrease and catholyte pH increase, which further influences the

Table 9.2 A Comparison of the Features of Different CO_2 Electrolysis Cells.

			MEA				
Cell types	**H-cell**	**Flow cell**	**Cation exchange**	**Anion exchange**	**Bipolar**	**High-temperature molten salt cell**	**Solid oxide cell**
Electrolyte	Liquid ($KHCO_3$, ionic liquids, etc.)	Liquid ($KHCO_3$, KOH, ionic liquids, etc.)	Solid polymer (CEM)	Solid polymer (AEM)	Solid polymer (BPM)	Molten salt	Solid oxide
Cathode	Metal	Metal	Metal			Fe, alloy	Ni, perovskite
Anode	Pt	IrO_2, Ni foam	IrO_2			Ni, SnO_2	Perovskite
Temperature	<100 °C	< 100°C	< 100°C			400–800 °C	600–1000 °C
Maximum current	<100 mA/cm^2	>1 A/cm^2	100 mA/cm^2	Hundreds of mA/cm^2	200 mA/cm^2	>1 A/cm^2	>1 A/cm^2
Advantages	Rich products; Simple setup; High faradaic efficiency; Convenient to study half-cell reaction.	Rich products; High current densities; High faradaic efficiency; Controllable reaction environment.	Easier scale-up; Low ohmic losses.	Low ohmic losses; Stability.	Stability; pH balance; Less liquid product loss.	High current densities; High faradaic efficiency.	High current densities; High faradaic efficiency; Stability (~500 h); Decreased overpotentials; Increased charge transfer.
Drawbacks	Low current density; Insufficient stability; Limited mass transport and electrolytes; Not scalable; Membrane resistance.	Insufficient stability; Ohmic losses; Flooding; Complex setup (pumps); Carbonate salt generation.	Competitive HER; Membrane hydration; Cathode acidification.	Membrane and gas stream hydration (catholyte-free system); Carbonate ion generation.	Membrane and gas stream hydration; Cathode acidification; Ohmic losses.	Simple product; Insufficient stability; Poor energy efficiency.	Simple products (single carbon species); Extreme conditions require; Electrode passivation by impurities.

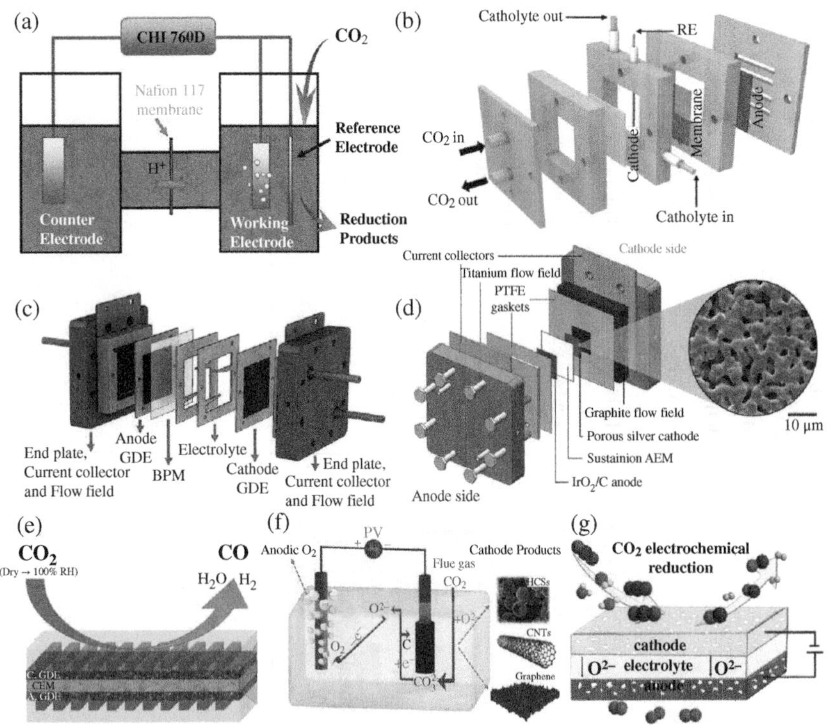

Figure 9.2 Electrochemical cells for CO_2RR. (a) H-cell. Source: Reproduced from Tan et al., 2019, with permission. Copyright 2019, Springer Nature/Springer Nature. (b) Flow cell. Source: Reproduced from Gong et al., 2019, with permission. Copyright 2019, Springer Nature/Springer Nature. (c) Bipolar-MEA. Source: Reproduced from Chen et al., 2020. Copyright 2018, Elsevier B.V. and Science/Elsevier. (d) Anion-MEA. Source: Larrazábal et al., 2019/Reproduced with permission from American Chemical Society. (e) Cation-MEA. Source: Reproduced from Shafaque et al., 2020, with permission. Copyright 2020, American Chemical Society/American Chemical Society. (f) High temperature molten salt cell. Source: Weng et al., 2019/Reproduced with permission from Elsevier. (g) Solid oxide cell. Source: Reproduced from Zhang, Hu, et al., 2017, with permission. Copyright 2017, Elsevier B.V. and Science/Elsevier.

FE, stability, and current density (Varela et al., 2016). Generally, H-cell is suitable for assessing and screening the CO_2RR catalysts in lab scale. But for industrial applications, efficient reactors with smaller resistance and higher mass transfer are recommended. Besides, highly efficient electrocatalysts should be developed to address the problems in H-cells and improve the CO_2RR performances.

9.2.2.2 Flow Cell

To meet the high reaction rates for practical CO_2 conversion, flow cells are designed. Usually, a flow cell utilizes the electrodes with gas diffusion layer (~50 nm) coated by catalysts (Ma et al., 2014). This configuration allows CO_2 to transfer directly through the gas diffusion layer (GDL) to the catalyst surface and react at the solid-liquid-gas triple phase boundary, which mitigates the problem of low CO_2 solubility in electrolytes and surmounts the mass transfer limitations in H-cells, reaching an efficient mass transfer efficiency over 100 mA/cm^2 (Li et al., 2018; Liang et al., 2020). The electrolyte in flow cell is circulated by peristaltic pumps to buffer the pH variation and maintain a stable concentration (Figure 9.2b) (Gong et al., 2019). Besides, flow cells allow the application of alkaline electrolytes, which are effective in suppressing HER and lowering the energy barrier of CO_2RR activation. The alkaline electrolytes promote oxygen evolution reaction, thereby improving the overall EE of CO_2 electroreduction (Gong et al., 2019). Therefore, compared with traditional H-cells, flow cells provide fundamentally different CO_2RR thermodynamics and kinetics, making them more promising for large-scale applications. Although flow cell contributes significantly to improve the reaction rate of CO_2RR, this technology still faces critical challenges such as insufficient EE, flooding (permeability of water molecules into the electrodes and proton transfer through the membrane during the operation at high current densities), stability, salt accumulation, and crossover of CO_2 and liquid products (Zhang, Luo, & Züttel, 2020). In addition, the preparation of flow cell is complex. To realize excellent CO_2RR performances, several strategies should be considered. These include exploring effective catalysts, optimizing CO_2 humidity, flow rate, and pressure, updating the electrolytes periodically to avoid electrolyte degradation, introducing a solid-support electrolyte layer, and varying the wetting properties of the electrodes (Weekes et al., 2018).

9.2.2.3 MEA

The three-electrode configuration is only used in labs to study the half-cell reactions, while the commercial systems run in a two-electrode mode. Different from the three-electrode catalysis systems of H-cell and flow cell, MEA resembles to the two-electrode system used in fuel cells (Ren et al., 2022). Therefore, the CO_2RR in MEA is controlled by the cell voltage, rather than the working electrode potential as no reference electrode or cell current is employed. Various types of membranes, primarily bipolar, anion-exchange, and cation-exchange membranes (BPMs, AEMs, and CEMs), are applied in MEA. BPMs promote H_2O dissociation into H^+ and OH^-, which are driven to cathode and anode, respectively (Vargas-Barbosa et al., 2014). AEMs such as Sustainion (Liu, Yang, et al., 2018) mediate the flow of anions (e.g., OH^-, CO_3^{2-}) from the cathode to the anode, while CEMs such as Nafion (Vennekötter et al., 2019) facilitate the flow of cations from the anode to cathode. The MEA is accordingly classified as bipolar-MEA (Figure 9.2c) (Chen et al., 2020), anion-MEA (Figure 9.2d) (Larrazábal et al., 2019), and cation-MEA (Figure 9.2e) (Shafaque et al., 2020). MEA systems are developed to address the issues in flow cells where the catalyst layer on cathodic gas diffusion electrode directly contacts the polymer membranes in a zero-gap manner. By removing the catholyte between the cathode and membrane, MEAs present reduced ohmic losses and cathode catalyst contamination from the impurities in catholyte, thereby delivering better FE and stability. Besides, gaseous and liquid products exhibit little diffusion into catholyte through membrane, resulting in high-concentration liquid products. In addition, the application of GDL in MEA guarantees large surface area and efficient contact between catalyst and CO_2 molecules, thus uplifting the reaction rates. Despite these advantages, MEA suffers from problems of membrane hydration, gas stream hydration, cathode acidification, and carbonate formation in alkaline electrolytes. A self-cleaning CO_2 reduction strategy with short and periodic reductions at applied voltages (Xu et al., 2021) effectively avoided saturation and suppressed carbonate salt precipitation.

Recently, a bipolar-MEA was employed to realize scalable CO_2 conversion to HCOOH with high FE (Chen et al., 2020). In this bipolar-MEA, a catholyte layer with 1.27 mm thickness was added to separate the cation exchange layer of the BPM and the carbon-supported SnO_2 cathode, which effectively mitigated H^+ concentration and achieved a high FE_{HCOOH} of 90% at 500 mA/cm^2. This bipolar-MEA configuration facilitated a tunable catholyte flow layer and the application of gas diffusion electrodes at high current densities (>100 mA/cm^2) (Chen et al., 2020), which enable CO_2RR at large scales beyond lab-scale devices (25 cm^2). Additionally, bipolar-MEA was also confirmed to be promising for CO_2-to-CO conversion at practical current densities (200 mA/cm^2) for long periods of time (14 hr) (Li et al., 2016).

Using solid electrolytes is a novel approach to enrich the liquid product from CO_2 conversion in MEA. For example, Xia et al. (Xia et al., 2019) sandwiched solid electrolytes between AEM and CEM through which deionized water was passed. In the cell with solid electrolyte, cathode and anode were separated by AEM and CEM, respectively. On the cathode where CO_2 was reduced, humidified CO_2 was supplied to facilitate CO_2 mass transfer, while 0.5 M H_2SO_4 was circulated at the anode side for water oxidation. The porous styrene–divinylbenzene copolymer used as solid electrolyte in this work was composed of sulfonic acid and quaternary amino functional groups for H^+ and $HCOO^-$ conduction, respectively. The electrochemically formed ions (H^+, $HCOO^-$) were combined to generate pure liquid fuel. Using a Bi catalyst, this MEA produced pure HCOOH concentrations up to 12 M and presented remarkable stability by generating 0.1 M HCOOH over 100 hr, and a large efficiency (48.5% at 100 mA/cm^2). On the other hand, using a Cu catalyst, C_{2+}-oxygenated fuels such as acetic acid, ethanol, and *n*-propanol were generated. Following this study, an all-solid-state MEA system was designed to produce high-purity and high-concentration formic acid vapors and solutions (Fan, Xia, Zhu, et al., 2020). Using a porous solid electrolyte layer and inert gas flow, the HCOOH vapors were efficiently removed. Combined with a GBs-enriched Bi catalyst, the MEA system achieved pure formate production, with high formate partial current densities (~450 mA/cm^2 at −1.05 V, $j_{formate} = FE_{formate}\, j_{total}$, where $FE_{formate}$ and j_{total} are the formate Faradaic efficiency and total current density, respectively), maximum selectivity (~97%), durable stability (100 hr), and ultrahigh concentration (~100 wt%, relative purity of formate in the liquid CO_2RR products) for pure HCOOH solution.

Humidification has been proven effective in MEA cells, such as bipolar-MEA (Salvatore et al., 2017). A sharp increase in cell voltage in BPM-based cells supplied with non-humidified CO_2 (relative humidity of 15%) comes from BPM dehydration. Humidified CO_2 has also been shown to sustain the hydration of AEM (Liu, Yang, et al., 2018), which maintained a FE_{CO} of 98% and 3 V applied potential with only 3 μV/hr increase at 200 mA/cm^2 for five months. This slow voltage increase is probably due to the dehydration of the membrane. To achieve the excellent cell lifetime, apart from the extended three-phase boundaries in Ag cathode, it is crucial to maintain the membrane hydration all the time, which is closely related to the circulation of the diluted $KHCO_3$ in the anode. Similarly, improved CO_2 humidification was demonstrated to increase CEM hydration for CO_2 conversion to CO (Shafaque et al., 2020). Given 50% and 100% CO_2 humidification, operating current densities of 665 and 755 mA/cm^2 were respectively realized, and these values are respectively 16% and 30% higher than 575 mA/cm^2 delivered by non-humidified CO_2. Besides, compared with dry CO_2, both 50% and fully humidified CO_2 exhibited lower ohmic losses at a wide range of operating current densities (125–755 mA/cm^2). Especially, the 100% CO_2 humidification delivered the smallest ohmic losses with a striking fourfold decrease at 575 mA/cm^2.

9.2.2.4 High-Temperature Molten Salt Cell

The above discussed CO_2 electrolysis devices (H-cell, flow cell, MEA) operate at low temperatures (< 100 °C), even down to room temperature (Chen et al., 2020; Gong et al., 2019; Larrazábal et al., 2019; Mariano et al., 2017; Shafaque et al., 2020). High-temperature molten salt electrolysis cell is capable of reducing CO_2 at a temperature in the range of 400–800 °C. The cell uses carbonate or chloride molten salts or their combinations as electrolytes (Figure 9.2f) (Weng et al., 2019). Compared with aqueous electrolytes and room-temperature ILs, the high-temperature molten salts deliver the advantages of wide potential windows contributing to high FE (Xiao & Wang, 2014), high-temperature properties promoting electro-active ions diffusion and CO_2 conversion (Weng et al., 2019), high CO_2 solubility (Weng et al., 2019), and no need for complex or expensive catalysts (Yin et al., 2013). Besides, the free energy requirements and overpotential losses are reduced at high temperatures, which increase the mobility and kinetics and result in power losses in the electrolysis cell. Additionally, it enables storage of concentrated solar power. The renewable electricity (solar, hydroelectric) could be used to promote CO_2 reduction with a net CO_2 conversion efficiency of 70–98% (Jiang et al., 2019). The high-temperature process could outperform CO_2RR near room temperature with intrinsic advantages in terms of reaction thermodynamics, kinetics, and product selectivity. No extra energy input is needed to sustain the operating temperature of an industrial cell, as the system can be self-heated by the electrolysis process's generated Joule heat (Sun et al., 2011). In the high-temperature molten salt cells, CO_2 is electrochemically reduced to advanced carbon materials such as carbon spheres, CNTs, and graphene on cathode surface (Deng et al., 2017). Thus, it is necessary to interrupt the CO_2 electrolysis frequently and replace the cathodes in a timely manner. Although high-temperature molten salt cells exhibit advantages of high current densities and FE, they still endure insufficient stability and poor EE. In addition, there is a lack of knowledge about the molten salt composition optimization, molten salt structure, and the gas behavior in molten salt and electrode/electrolyte interface (Jiang et al., 2019). More efforts should be invested in developing simple electro-deposited products and robust and affordable inert anode materials for oxygen evolution for the further advancement of the technology of high-temperature CO_2 electroreduction using molten salt cells. Other engineering issues should be addressed properly to realize efficient CO_2 reduction. These issues are related to CO_2 feedstock pretreatment, stability and durability of the reactor, product separation, mass and energy balance, EE, system design, and power supply (Jiang et al., 2019).

9.2.2.5 Solid Oxide Cell

Another high-temperature CO_2 electrocatalysis device is solid oxide cell, which operates at a high temperature in the range 600–1000 °C and utilizes ion-conducting solid oxide electrolytes (Figure 9.2g) (Lu et al., 2018; Song et al., 2018; Zhang, Hu, et al., 2017). In a solid oxide cell, CO_2 diffuses directly to cathode and is reduced chemically to gas products exclusively (Ebbesen et al., 2014). CO_2 is reduced to CO and O^{2-} at cathode. CO is further transformed into valuable hydrocarbons, while O^{2-} transfers to anode and is evolved into pure gas O_2. The oxygen ions are used as charge carriers in solid oxide cells at high temperatures (>600 °C). This is different from the case of molten salt cells at moderate temperatures (400–600 °C), where the charge carriers are carbanions for CO_2 conversion into elemental carbon species (CNTs, spheres). It is also different from the CO_2RR at low temperatures (<200 °C) with alkaline, acidic, and polymer electrolytes to dissolve CO_2 gas and to involve proton or hydroxyl ions in the cathode reaction (Jiang et al., 2021). Solid oxide cells are promising for practical applications due to their operational stability, long life (~500 hr) (Yang et al., 2017), and low cost. It facilitates co-electrolysis of CO_2 and H_2O for syngas production (Ebbesen & Mogensen, 2009). The high operating temperature in solid oxide cells is effective in increasing the rate of the electrode reaction, maximizing the EE, and obtaining higher current densities than those achieved at the same potentials by other types of cells utilizing liquid electrolytes. However, solid oxide cells are limited to simple products of CO, which may be further reduced to solid carbon. These cells encounter problems such as metal particle oxidation, carbon deposition, grain coarsening of the electrode materials, impurities contamination, and difficulty in sealing. To address these problems and improve CO_2RR performances, efficient cathode and anode systems should be explored, beyond the very specialized electrodes (Ni-yttria stabilized zirconia (YSZ), lanthanum strontium manganite-YSZ), in terms of the durability, mechanical, and thermal stability. Another tough challenge for CO_2RR in solid oxide cells is the difficulty in economic viability assessment of industrial CO_2 reduction, as no techno-economic analysis (TEA) has been reported in this field. Therefore, flow cells and MEA have more potentials than solid oxide cells for industrial-scale CO_2RR applications (Sánchez et al., 2019).

9.2.3 Electrolytes

Most CO_2RR were performed in aqueous electrolytes such as $KHCO_3$, KOH, $NaHCO_3$, and $CsHCO_3$. Electrolytes play significant roles in the CO_2RR process. Their concentrations, pH, species (aqueous or IL), buffer capacity, etc. are important. Generally, CO_2RR in acid is challenging because of more facile HER than CO_2 reduction. For example, both Fe–N–C (Varela, Kroschel, et al., 2018) and Au (Ringe et al., 2020) are active for CO_2RR in neutral and alkaline media, but exhibit nearly 100% selectivity to HER in CO_2-saturated 0.1 M $HClO_4$ (pH = 1). The CO_2-to-CO conversion was feasible in mild acids (pH ≥ 3.0), such as K_2HPO_4 (Ringe et al., 2020) and $HClO_4/NaClO_4$ (Bondue et al., 2020), while the CO_2 reduction to multi-carbons on Cu was achieved in strong acid (pH 0.67) with mixed H_3PO_4/KCl (Huang et al., 2021). Both these indicated that alkali cations are essential for CO_2RR, similar to the work by Monteiro et al. (Monteiro et al., 2021), which confirmed the indispensable effect of alkali cations for the CO_2^- species stabilization. To understand how alkali cations suppress HER in acid, Gu et al. (Gu et al., 2022) demonstrated efficient CO_2 reduction to formic acid, CO, and hydrocarbons on carbon-supported SnO_2, Au, and Cu nanoparticles (NPs), respectively. Alkali cations in the cathode double layer shielded the electric field and suppressed the hydronium ions transfer to cathode, resulting in lower concentrated hydronium ions in the outer Helmholtz plane and thus suppressed HER. Besides, cations strengthened the Stern layer field and stabilized the key intermediates during CO_2RR.

Apart from the progress in acidic CO_2RR, many improvements have also been achieved in alkaline electrolytes (De Arquer et al., 2020; Dinh et al., 2018). Integrating flow-cell and highly alkaline electrolyte can significantly reduce the reaction overpotential and push the production rates for C_{2+} products. However, CO_2RR is not stable in alkaline solutions, because of the reaction between CO_2 and OH^- to form bicarbonate/carbonate (Jouny et al., 2019; Verma et al., 2017), which further induced significant reactant loss and large energy penalty for CO_2 recovery. This issue can be circumvented by the first step of CO_2 reduction to CO and subsequent electrochemical CO reduction (Du et al., 2022; Ji et al., 2022; Jouny et al., 2018; Li, Chang, et al., 2021; Liu et al., 2022; Ma et al., 2022; Overa et al., 2022; Rong et al., 2022; Shen et al., 2022; Wang et al., 2022; Xia et al., 2021; Zhou et al., 2022; Zhu, Xia, et al., 2021).

Normally, aqueous $KHCO_3$ outperformed $NaHCO_3$ electrolyte in the case of CO_2-to-CO conversion by ZnN_x/C SAC with high CO selectivity (Yang, Song, et al., 2018). This is attributed to more available protons provided by K^+ than Na^+, which buffer the pH increase near the cathode surface and improve the concentration of CO_2 dissolved in the electrolyte. Another study on the electrolyte concentration effects on CO_2-to-CH_4 conversion demonstrated increasing CH_4 partial current density (j_{CH4}) and yield rate with $KHCO_3$ concentration (Han et al., 2020), which could be explained by the electrolyte buffer capacity-local interfacial pH relationship. j_{CH4} is proportionally dependent on proton activity because of the 8 protons transferred from CO_2 to CH_4 formation. Higher $KHCO_3$ concentration led to larger bulk pH, while higher buffer capacity resulted in smaller interfacial pH near catalyst surface (pH_{local}). Thus, lower $KHCO_3$ concentration induced larger pH_{local}, due to the decreased protons local concentration. Regarding the effects of electrolyte pH and proton concentrations on CO_2RR by Fe–N–C SAC, CH_4 and H_2 production were discovered to be dependent on pH (Varela, Kroschel, et al., 2018), while CO formation was not, which suggested the rate-limiting step of decoupled electron–proton transfer for CO production. This work also achieved precise regulation of CO/H_2 ratio by controlling the pH values, which selectively affected the formation of CO or H_2.

Although aqueous electrolytes are intensively applied for CO_2RR, these electrolytes are limited by low CO_2 solubility (Jiang et al., 2018) and intense HER (Zhang, Sethuraman, et al., 2014). Compared to these electrolytes, ILs, a kind of organic salts existing in liquid phase at ambient temperatures, are more effective for CO_2RR (Chang et al., 2021; Chen et al., 2016; Faggion Jr et al., 2019; Feng et al., 2018), because of their high CO_2 solubility (Shin & Lee, 2008; Soriano et al., 2009; Zhang, Zhang, et al., 2017), improved CO_2 transfer (Asadi et al., 2014), enhanced selectivity (Rosen et al., 2011), inhibited HER, and cocatalyst effect (Ren et al., 2020). For example, Mn-C_3N_4/CNT SAC exhibited maximum j_{CO} of 29.7 mA/cm^2 at 0.62 V overpotential in CO_2-saturated 1-butyl-3-methylimidazolium tetrafluoroborate (Feng, Gao, Zheng, et al., 2020). This current density is higher than that in aqueous $KHCO_3$, due to the application of ILs with high CO_2 solubility for efficient mass transfer and the cocatalyst effect. However, bulk IL electrolytes with high viscosity and poor ionic conductivity normally induce issues of sluggish mass transfer and high energy consumption (Galan-Sanchez, 2008; Kelley et al., 2017; Zhang et al., 2008). In addition, ILs are expensive and are not economical for large-scale applications for practical CO_2RR. This could be overcome by introducing the nanoconfined ILs into porous Ni sites (Ren et al., 2020; Sun, Zhao, Ren, & Zhao, 2022), which would alter the CO_2–Ni site interactions and create a solid/liquid interface with high CO_2 concentration, thus boosting CO_2-to-CO conversion.

9.3 Electrocatalysts for CO_2RR

In this section, the development and applications of various electrocatalyst materials for CO_2RR are discussed. We focus on metals, metal compounds, SACs, and molecular catalysts. Table 9.3 compares the CO_2RR performances of these electrocatalysts.

Table 9.3 Summary of the performance of electrocatalysts for CO_2RRs

Materials	Electrolyte	E/V vs. RHE	Main product [FE/%]	j/mA/cm²	Stability/h	References
5.1 nm Pd nanosheets	0.1 M $KHCO_3$	−0.5	CO[94]	14 mA/cm² @ −1 V	8	(Zhu et al., 2018)
Pd NPs	0.5 M $NaHCO_3$	−0.05	HCOOH [>95]	2.4	3	(Min & Kanan, 2015)
h-Zn	0.1 M $KHCO_3$	−0.95	CO[85.4]	9.5 mA/cm² @ −0.85 V	30	(Won et al., 2016)
OD-Cu	0.5 M $NaHCO_3$	−0.9	C_2H_4+C_2H_6 [55]	/	/	(Dutta et al., 2016)
$Ag_{70}Hg_{30}$	0.1 M $KHCO_3$	−0.9	HCOOH [85]	5.1	40	(Yang, Chen, et al., 2019)
$Cu_{1.63}Se_{(1/3)}$	0.5 M H_2SO_4 (anolyte) [Bmim]PF_6–CH_3CN-H_2O (catholyte)	−2.1 V vs. Ag/Ag^+	CH_3OH [77.6]	41.5	/	(Yang, Zhu, et al., 2019)
Sn-pNWs	0.1 M $KHCO_3$	−0.8	HCOOH [80]	10 mA/cm² @ −1 V	15	(Kumar et al., 2017)
mSnO_2 NPs-350	0.5 M $KHCO_3$	−1.3	HCOOH [90]	8.3	12	(Wei et al., 2020)
		−1.2	C_1 [97.4]			
Cu-on-Cu_3N	0.1 M $KHCO_3$	−0.95	C_{2+} [64±2%]	14	30	(Liang et al., 2018)
Cu_3N NCs	0.1 M $KHCO_3$	−1.6	C_2H_4 [60]	30	20	(Yin et al., 2019)
$Ce(OH)_x$–doped-Cu	1 M KOH	−0.7	C_2H_5OH [43]	128	/	(Luo et al., 2019)
Ir_{AC}−1.7	0.5 M $KHCO_3$	−0.66	CO [97.6]	18.6	24	(Sun et al., 2019)
Ni–N-CNT	0.5 M $KHCO_3$	−0.8	CO [94]	28	8	(Sun et al., 2021)
Ni/Fe–N–C	0.5 M $KHCO_3$	−0.7	CO [98]	7.4	30	(Ren, Tan, et al., 2019)
Cu-CeO_2-4%	0.1 M $KHCO_3$	−1.8	CH_4 [~58]	~56	2.2	(Wang, Chen, et al., 2018)
Sb–N_4	0.5 M $KHCO_3$	−0.8	HCOOH [94]	/	10	(Jiang, Wang, Pei, et al., 2020)
$In^{\delta+}$–N_4	0.5 M $KHCO_3$	−0.65	HCOOH [96]	8.87	/	(Shang et al., 2020)
$VitB_{12}$@rGO	0.5 M $KHCO_3$	−0.8	CO [94.5]	14.36 mA/cm² @ −1 V	10	(Jia, Ching, et al., 2020)
Fe _MOF-525	1 M $TBAP_6$/CH_3CN	−1.3 V vs. NHE	CO [54]	2.3	4	(Hod et al., 2015)
COF-367-Co	0.5 M $KHCO_3$	−0.67	CO [90]	1.8	24	(Lin et al., 2015)
NBPC	0.5 M $KHCO_3$	−0.4	CO [83]	1.2 mA/cm² @ −0.8 V	20	(Jia, Ren, et al., 2020)
Polyethylenimine	0.1 M $KHCO_3$	−1.8 V vs. SCE	HCOOH [87]	9.5	24	(Zhang, Kang, et al., 2014)

9.3.1 Metals

Metallic catalysts have been extensively explored for CO_2RR over the past few years. In earlier studies, bulk metals were investigated as potential catalysts to promote CO_2 conversion due to their simple configuration, easy preparation, and robust structure. However, bulk metals present unsatisfactory catalytic activity because of their limited surface area. A range of strategies were developed to enhance the CO_2RR performances of the metals, noble metals, transition metals, OD-Ms, and metal alloys. These are discussed in this section.

9.3.1.1 Noble Metals

The noble metal, Pt, dominantly exhibits high activity for the competing HER and can be poisoned by CO, which therefore is normally not useful for CO_2RR (Guo et al., 2015; Vasileff et al., 2018). The existence of the intermediate *CO during CO_2RR process is especially crucial in tuning the HER activity on the Cu/Pt interface (Varela et al., 2013). The Pt-containing systems are not stable and lead to Pt segregation, attributed to the strong interaction between Pt and CO, which in turn promotes HER over CO_2RR.

Other noble metals such as Pd, Ag, and Au are highly selective for CO and HCOOH production due to the strong *COOH binding energy but weaker *CO binding energy on these metal surfaces. The binding energy of intermediates *CO and *COOH can be tuned by altering the catalyst surface structure (i.e., particle size, lattice strain, facets) and composition, thereby optimizing the selectivity and activity. For example, the Pd icosahedra/C catalyst achieved a maximum FE_{CO} of 91.1% at −0.8 V vs. RHE, 1.7-fold higher than that of Pd octahedra/C at −0.7 V vs. RHE. Besides, Pd icosahedra/C outperformed Pd octahedra/C with larger j_{CO} and mass activity in a wide potential range, from −0.6 to −1.1 V vs. RHE, as well as with greater stability (over 10 hr of continuous CO_2 electrolysis at −0.9 V vs. RHE). This improvement originated from the tensile strain on the icosahedra surface, which promoted catalytic activity by shifting up the d-band center, favoring COOH* adsorption (Huang et al., 2017). Such a shape-dependent performance also applies to triangular Ag nanoplates (Liu et al., 2017) and atomically ordered $AuCu_3$ NPs with *fct* Au (face-centered tetragonal Au) overlayer (epitaxial coating) on the surface (o-$AuCu_3$@*fct* Au NPs) (Yu et al., 2021). The triangular Ag nanoplates presented a 2.2-fold larger j_{CO} than the similarly sized Ag NPs, improved FE (96.8%) at −0.856 V vs. RHE, EE (61.7%), and a considerable durability (7 days). The o-$AuCu_3$@*fct* Au NPs outperformed those with epitaxial Au overlayer (face-centered cubic, *fcc* Au) with lower onset potential, much higher current density, larger FE_{CO} (94.5% vs. 72.2% at −0.8 V vs. RHE), 18.2-fold higher mass activity at −0.6 V vs. RHE, and long-term stability over 20 hr. In another study, the superiority of Pd(310) facet over the (100) and (110) facets for CO production was revealed. The FE_{CO} for the (310) facet was the highest (90.6%, Figure 9.3a). A large j_{CO} for over 6 hr of CO_2 electrolysis at −0.9 V vs. RHE was obtained. This performance was attributed to the concave cubic shape of the (310) facilitating COOH* formation and CO* desorption (Dong et al., 2019). Similarly, Au(110) and Au(211) were confirmed to exhibit at least 20-fold higher activity than Au(110) for CO production because of their under-coordinated sites (Mezzavilla et al., 2019).

Au (Zhu et al., 2013), Ag (Kim et al., 2015), and Pd NPs (Gao et al., 2015) exhibit size-dependent CO_2RR properties. The 8 nm Au NPs, 5 nm Ag NPs, and 3.7 nm Pd NPs were demonstrated to have optimal performance for CO production, superior to the corresponding NPs of other sizes. The improved performance of catalysts with an appropriate size was attributed to the presence of certain ratios of active sites, such as edge, corner, and terrace sites. Similarly, Au nanowires (NWs) (Zhu et al., 2014), Ag nanoplates (Liu et al., 2017), and Pd nanosheets (Zhu et al., 2018) also deliver highly size-dependent selectivity to CO, favored by the abundant edge sites. Recently, Zhao and coworkers reported a nanoporous Ag network that exhibited (Yang, Ma, et al., 2018) ligament size-dependent selectivity to CO. The catalysts with average ligament sizes of 21 and 87 nm (np-Ag (21 nm), np-Ag (87 nm)) were prepared by dealloying $Mg_{80}Ag_{20}$ (wt%) in 1 wt% citric acid and 5 wt% phosphoric acid, respectively. Compared with the np-Ag (78 nm), the np-Ag (21 nm) delivered a two-fold higher FE_{CO} (85%), more than 2.5-fold increase in j_{CO} at −0.8 V vs. RHE, and good stability for over 6 hr of CO_2 electrolysis at −0.9 V vs. RHE with a larger j_{CO}. This is because of the enhanced electrochemical surface area, higher local pH derived from ligament size effect, and more defect sites (i.e., GBs) in ligaments. According to these studies, the active sites of catalysts are structurally stable during CO_2RR process. Recently, a different surface reconstruction phenomenon was observed on defect-rich ultrathin Pd nanosheets (Pd NSs) (Zhao, Tan, et al., 2020). The Pd NSs with dominant (111) facets changed into crumpled sheet-like structures with prevalent (100) regions. This transformation increased the active site's density and lowered the CO intermediates binding energy, thus boosting CO_2 electro-conversion to CO with high Faradaic efficiency (93%) and a site-specific activity of 6.6 mA/cm^2 at a moderate overpotential (590 mV). CO intermediate was identified as the key driving factor for the structural transformation of the 2D metal nanosheets during CO_2 reduction. This work emphasized the instability of defective metal nanosheets and highlighted the significance of surface engineering in alternating electrocatalytic performances.

Beside the shape-dependent (Yu et al., 2021) and size-dependent activities (Mistry et al., 2014), noble metals such as Au exhibit grain boundary (GB)-dependent CO_2RR activity. Feng et al. (Feng et al., 2015) prepared Au NPs on carbon nanotubes (Au/CNT) via vapor deposition with a high GB density ($49.5 \pm 4.5\,\mu m^{-1}$), which is greater than 28.1 ± 2.4, 13.3 ± 1.0, and $4.2 \pm 0.3\,\mu m^{-1}$ obtained for Au/CNT samples annealed at 200, 300, and 400 °C, respectively. Au/CNT was the suitable catalyst to promote CO_2-to-CO conversion with a high FE_{CO} (~94%) and good stability of over 12 hr CO_2 electrolysis at −0.5 V vs. RHE. Nevertheless, the mass activity declined from 15 to 13 A/g. Recently, they discovered that GB surface terminations on Au electrode would selectively promote CO_2-to-CO conversion over a GB field (representing the local dislocation density) dependent length scale (Mariano et al., 2017). The surface-area-normalized activity of Au was in linear relationship with the surface density of GBs. This work demonstrated that GB engineering could be regarded as a powerful approach in heterogeneous catalysis.

Mesoporous structure is another factor influencing the catalyst activity and product selectivity. For example, a series of gold inverse opal (Au-IO) thin films (Figure 9.3b) were fabricated with different thicknesses (0.5, 1.6, 2.7 μm) by varying the electrodeposition time of Au on colloidal crystal templates (Hall et al., 2015). With increasing porous film thickness, the HER activity was suppressed dramatically (Figure 9.3c) because of the generation of diffusion gradients within the pores of the meso-structured Au-IO film. The CO production remained stable at modest overpotentials (Figure 9.3d) because the induced alkalinity within the porous network during CO_2 electrocatalysis had no effect on the CO_2 concentration due to its slow hydration kinetics. A high FE_{CO} of 99% was achieved at a low overpotential of −0.4 V by the electrodes with optimal mesoporosity. Additionally, the 0.5 and 1.6 μm Au-IO films presented identical j_{CO} normalized by electrochemically active surface area (ECSA) from −0.3 to −0.7 V vs. RHE, while the 2.7 μm Au-IO film displayed a two-fold decreased j_{CO}, because of the onset of transport limitations in the thickest sample. This work confirmed that the mesostructured electrodes would be promising for enhancing CO_2RR.

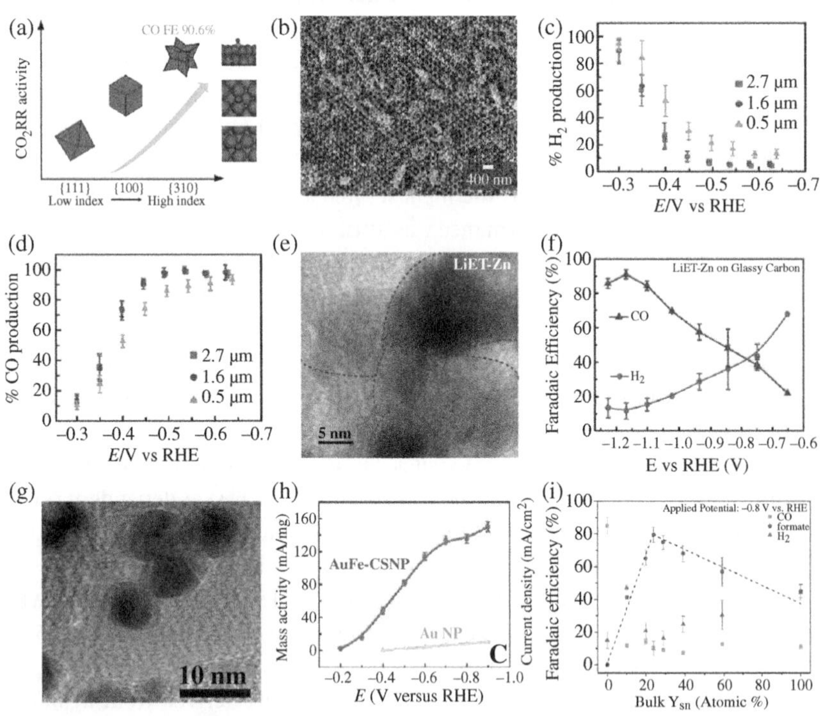

Figure 9.3 (a) Facet effect on Pd NCs with the same size for CO_2RR. Source: Reproduced from Dong et al., 2019, with permission. Copyright 2018, Elsevier Ltd/Elsevier. (b) Top-down SEM image of a Au-IO thin film. Faradaic efficiency for (c) H_2 evolution and (d) CO for 0.5, 1.6, and 2.7 μm thick Au-IO samples evaluated in CO_2-saturated 0.1 M $KHCO_3$ electrolyte, pH 6.7. Source: Reproduced from Hall et al., 2015/with permission. Copyright 2020, Wiley-VCH GmbH/John Wiley & Sons. (e) TEM images of LiET-Zn. (f) FEs of H_2 and CO of 0.2 mg/cm² loading LiET-Zn. Source: Jiang et al., 2017/Reproduced with permission from American Chemical Society. (g) TEM of AuFe-CSNP. (h) Mass activity (in mA/mg) for CO_2RR to CO exhibited by AuFe-CSNPs, Au-NPs, and Au foils. Source: Sun et al., 2017/Reproduced with permission from American Chemical Society. (i) CO_2 reduction Faradaic efficiencies of AgSn/SnO_x catalysts. Source: Reproduced from Luc et al., 2017, with permission. Copyright 2017, American Chemical Society/American Chemical Society.

9.3.1.2 Transition Metals

Transition metals are much cheaper than noble metals. To enhance their catalytic activity toward CO_2RR, the morphology and structure of bulk transition metals can be manipulated to expose active sites, orient facets, and induce surface strain. For example, a hierarchical hexagonal Zn catalyst (*h*-Zn) was prepared by electrodepositing $ZnCl_2$ on Zn foil as a highly CO-selective catalyst with a FE_{CO} of 85.4 % at −0.95 V vs. RHE, 60.3% higher than that of bare Zn foil. This activity enhancement was attributed to the hexagonal morphology with dominant Zn(101) facets, which more effectively stabilized the intermediate. COOH than the (002) facets, thus lowering the reduction potential for CO_2 conversion to CO (Won et al., 2016). Besides, the *h*-Zn exhibited an unprecedented stability for long-term operation (30 hr at −0.85 V vs. RHE) with stable FE_{CO} (80%) and j_{CO} (9.5 mA/cm^2), superior to that reported for Zn-based catalysts (Quan et al., 2015; Rosen et al., 2015) and noble metal catalysts (Chen et al., 2012; Rosen et al., 2011). In another study, Cu nanocubes were treated by controlled chemical etching to create high-energy Cu(111) facets, which showed significantly improved activity toward CO_2 reduction with respect to the unetched Cu nanocubes (Wang et al., 2016). Additionally, two Bi nanosheets exhibited impressive selectivity to HCOOH. The Bi nanosheets derived from bismuth oxyiodide transformation outperformed commercial Bi for CO_2-to-HCOOH conversion with a high FE_{HCOOH} of ~100% from −1.5 to −1.7 V vs. saturated calomel electrode (SCE), a maximum j_{HCOO^-} (24 mA/cm^2 at −1.74 V vs. SCE vs. <6 mA/cm^2), and an excellent durability for over 10 hr operation, due to the favorable stabilization of intermediate (OCHO*) on Bi (001) surface relative to COOH* or H* (Han et al., 2018). Bi nanosheets prepared via a wet chemical method also showed eight-fold larger j_{HCOO^-} (54 mA/cm^2 around −1 V vs. RHE) than the pure Bi, a high FE_{HCOOH} (99% at −0.58 V vs. RHE), a small onset overpotential (<90 mV) and a high durability (no performance decay after 75 hr of operation) (Yang, Elnabawy, et al., 2020). This is due to the unique compressive strain on the Bi (111) facet, which facilitated a weaker adsorption than the (011) Bi nanosheets, reducing the possibility of being poisoned by reaction intermediates and improving the activity of all active sites (step edge sites and terraces). Their studies indicated the promotion of CO_2RR activity by the oriented facets created through a morphology manipulation.

Porous structure and excellent mass transport properties contributed significantly to improving the CO_2RR performances, which were confirmed by the studies on three-dimensional electrodes based on porous Cu fiber (Kas et al., 2016) and Sn-based gas diffusion electrodes (Wang, Wang, Wu, et al., 2017). The porous fiber Cu electrode exhibited a maximum FE_{CO} of ~72% and a durability of 24 hr at −0.4 V vs. RHE. Additionally, Cu hollow fibers facilitated CO_2-to-CO conversion at −0.4 V vs. RHE with over 15–400 times larger j_{CO} than polycrystalline Cu and Cu NPs, and presented comparable performance at low potentials (−0.2 to −0.6 V vs. RHE) to noble metals (Au NPs (Chen et al., 2012) and nanoporous Ag (Lu et al., 2014)) in aqueous electrolytes. This was attributed to its defect-rich porous structure that provided excellent mass transport, large surface area, and three-phase boundary for gas-liquid reactions. Apart from this, the remarkable CO production at very low overpotentials was probably related to the property of the copper sites arising from the preparation of Cu hollow fiber via hydrogen-induced reduction of CuO. The defect-rich porous structure is responsible for the enhanced CO_2 reduction to CO (Kas et al., 2016). On the other hand, Sn-based gas diffusion electrode with 90 s electrodeposited Sn (ESGDE-90) was highly selective to HCOOH with a maximum FE of 73.01 ± 3.42%, a current density of 34.21 ± 1.14 mA/cm^2, and a production rate of 1772.81 ± 59.08 μmol·m^{-2}·s^{-1}, which were 59%, 72%, and 72% higher than the corresponding values obtained from ESGDE-30. This is because of the enhanced rate of transfer of reactants and the 3D reaction zone, which provided a large surface area to expose active sites and facilitated fast mass and electron transfer (Wang, Wang, Wu, et al., 2017).

In the subsequent study, Cu NPs with varying particle sizes (2–15 nm) were synthesized by inverse micelle encapsulation and were evaluated for CO_2RR (Reske et al., 2014). Smaller Cu NPs below 5 nm favored CO/H_2 formation over hydrocarbons (CH_4/C_2H_4), while larger NPs (5–15 nm) promoted CH_4/C_2H_4 production over CO/H_2. This activity variation came from a stronger chemisorption to intermediates *COOH and *CO_2– by the low-coordinated surface sites, triggering H—H bond formation via Heyrovsky mechanism for HER [H_2 is generated from the reaction between a solvated proton from water and one adsorbed surface hydrogen ($H_{ad} + H^+ + e^- \rightarrow H_2$)]. This was in agreement with another work on ZnNPs (Jeon et al., 2018) for CO_2-to-CO conversion, which showed an increased HER on small NPs (<3 nm) because of the enhanced content of low-coordinated sites. The NPs with 3–5 nm size exhibited high activity and selectivity to CO (~70%), while larger NPs (>5 nm) revealed a decline in activity and a stable selectivity. This selectivity trend was related to the increased content of low-coordinated sites on small NPs. Besides, the presence of residual cationic Zn species was revealed, as the small NPs favored stabilization of the oxidic Zn species, thus facilitating the stabilization of the zinc hydroxide species, which are playing a crucial role in the selectivity trends. A similar study of cube-shaped Cu nanocrystals (NCs) with different sizes of 24, 44, and 63 nm also exhibited size-dependent performance for CO_2RR (Loiudice et al., 2016). The 44 nm Cu cubes yielded the highest selectivity to CO_2RR with 80% FE_{CO2RR} and 41% FE_{C2H4}, as well as larger current density than Cu foil at −1.1 V vs. RHE, because of the edge sites that adsorbed and stabilized the intermediate specie COOH* and inhibited HER.

9.3.1.3 Oxide-Derived Metals

OD-Ms encompass high-quality NCs and the crystal planes connected by GBs, which have been proven effective in boosting CO_2RR (Chen et al., 2012; Dutta et al., 2016). The oxide-derived Au (OD-Au) fabricated via Au film anodization was expected to be highly selective to CO formation over 8 hr of CO_2 electrolysis with FEs above 98% and 96% at −0.4 and −0.35 V vs. RHE, respectively, which is higher than that yielded by other Au electrodes (Chen et al., 2012). The active sites at the OD-Au NPs surface were shown to dramatically increase the stabilization of $CO_2{\cdot}^-$ intermediate compared to the polycrystalline Au. Similarly, OD-Cu foam (Dutta et al., 2016) prepared by a template-assisted electrodeposition exposed abundant specific surface sites for C—C coupling (dominant (100) surface texture) and temporally trapped the gaseous intermediates inside the mesoporous structure, thereby reaching high $FE_{C2H4+C2H6}$ of 55%, and a larger j_{C2+} than j_{H2} over a broad potential range (−0.4 to −1.0 vs. RHE). To further clarify the origin of catalytic activity for producing certain C_{2+} molecules on OD-Cu, very recently, Cheng et al. (Cheng et al., 2021) investigated the OD-Cu catalysts derived from Cu_2O octahedrons dominated by (111) facets. The defect concentrations were controlled by various annealing temperatures. Step sites and some defects disappeared and transformed into flat (111) facet generally after annealing. Theoretical calculations revealed that the concentrations of planar-square (p-sq, the Σ3 GB derived from two flat (111) planes with 60° lattice orientation difference), step-square (s-sq, similar to Cu(S)-[n(111) × (100)] step sites with the introduction of (100) steps into (111) basal plane), and convex-square (cv-sq, created by rotation and then splicing of two (111) planes) sites decreased from 11.54% to 9.37%, 4.93%, and 1.67% after annealing at 450, 500, and 650 K, respectively. The planar-square sites and convex-square sites favored C_2H_4 formation, while step-square sites (n(111) × (100)) were found responsible for alcohol production, because of the acetaldehyde intermediates stabilization and Cu–O interactions destabilization by the geometric effect.

GBs are also found to impact the CO_2RR behaviors. For example, the OD-Cu catalyst composed of highly interconnected NCs linked by GBs, exhibited lower and higher specific activity for H^+ and CO_2 reduction, respectively, compared with other Cu materials. These characteristics resulted in high selectivity to CO (FE_{CO}: ~45%) and HCOOH (FE_{HCOOH}: 33%) at remarkably low overpotentials (0.19–0.39 V for CO, 0.25–0.35 V for HCOOH), while enabling C_2 hydrocarbons production at relatively negative potentials (< −0.6 V vs. RHE) on OD-Cu developed by reducing Cu_2O films (Li & Kanan, 2012). The catalyst also enabled >1 mA/cm^2 geometric current densities at <0.4 V overpotential and detectable current density at <0.1 V overpotential, which was inaccessible by metal electrodes under comparable conditions as reported previously. Besides, CO_2RR using the developed electrode was resistant to deactivation for at least several hours. The catalyst was indeed superior to the polycrystalline Cu that suffered rapid deactivation. Similarly, ZnO was controllably reduced using a Li electrochemical tuning (LiET) method to get interconnected ultrasmall Zn NPs with abundant GBs (LiET-Zn) for CO_2-to-CO conversion (Figure 9.3e) (Jiang et al., 2017). The LiET-Zn with freshly exposed GBs delivered a j_{CO} of ~23 mA/cm^2 at −948 mV overpotential, which was five times higher than that produced by a OD-Zn NPs with GBs eliminated during the *in situ* electro-reduction process. Besides, a maximum FE_{CO} of ~91.1% was obtained from LiET-Zn deposited on a glassy carbon substrate (Figure 9.3f). Good stability of LiET-Zn was detected. The FE_{CO} remained stable (>80%) while the total current density increased slightly from 26.5 to 30 mA/cm^2 during 4 hr continuous electrolysis at −1.17 V vs. RHE.

In subsequent study, a size-dependent CO_2RR performance was observed in Cu_2O derived from Cu NPs (Kas et al., 2014). The catalytic activity predominantly depended on Cu_2O film thickness, which was related to a local pH variation. A thinner film deposited with total charges of 3, 6, 7, and 9 C/cm^2 favored C_2H_4 formation besides CO and H_2 with a high C_2H_4/CH_4 ratio of ~8 to 12, while a thick film (11 C/cm^2) generated remarkable amounts of C_2H_6. Although the selectivity to hydrocarbons decreased upon using larger layer thickness, the overall current density increased, and thus, the partial current density for the production of hydrocarbons only slightly decreased from 15 to 12 mA/cm^2 at −1.1 V vs. RHE. In addition, the thinnest samples (3 C/cm^2) delivered a long-term stability of 5 hr of CO_2 electrolysis without hydrocarbon production deactivation, which outperformed the polycrystalline Cu with rapid deactivation within 40 min.

9.3.1.4 Metal Alloys

Alloy materials also exhibit outstanding CO_2RR performance. Composition and morphology engineering of alloys are effective approaches to enhance performance and decrease the catalyst preparation cost. To date, many studies have been conducted to understand the catalytic activity of alloy materials and engineer them for efficient CO_2RR. For example, Au–Fe core-shell NPs (Figure 9.3g) were developed via a solvothermal method as an exclusively selective catalyst for CO formation

(Sun et al., 2017). Compared with pure Au NPs, the Au–Fe alloy achieved 2.2 times higher FE_{CO} (97.6% vs. 30.5%) and ~100 times larger mass activity (48.2 vs. 0.5 mA/mg, where mass activity $= j_{CO}/\beta m_{cat}$, in which β is the Au or Au–Fe weight ratio in total carbon-supported NP catalysts, m_{cat} is the catalyst weight; Figure 9.3h) at −0.4 V vs. RHE. This is because of the Fe sites at subsurface and surface defects due to surface Fe leaching. Besides, the Au–Fe alloy was quite stable with constant FE_{CO} (>90%) and current density (19.7 mA/cm^2) for over 90 hr of CO_2 electrolysis. This improvement is substantial compared to Au NPs and Au foil that suffered rapid deactivation after 12 hr with respectively 55% (from 2.8 to 1.26 mA/cm^2) and 87% (from 1.69 to 0.22 mA/cm^2) loss in the current density. These outstanding performances were attributed to the more stable dimpled Au surface structure arising from the AuFe NP fabrication process followed by Fe leaching than the simple formation of Au NP and Au film, as the surface layer might have prevented the diffusion of Au atoms from interior. In another study, a series of metal alloys M_xIn_{1-x} (M = Zn, Cu, Ni, Co, and Fe) were investigated as electrocatalysts for CO_2-to-CO conversion (He et al., 2017). The catalyst activity for CO evolution followed the order Fe < Co < Zn < Ni < Cu, while the hydrogen evolution followed the order of Zn < Cu < Ni < Co < Fe. This was in good agreement with the magnitudes of heat of CO adsorption (ΔH_{CO}) on these metal surfaces. Typically, large ΔH_{CO} facilitates hydrogen and hydrocarbon evolution, while small ΔH_{CO} favors the HCOOH formation. The introduction of indium (In) was found effective in suppressing hydrogen evolution in these alloys, among which the Cu_xIn_{1-x} alloy ($x = 0.5 - 0.8$) presented a high CO FE (~90%). Smaller FE_{CO} and larger current density were detected at higher Cu:In ratios, with $Cu_{0.8}In_{0.2}$ exhibiting maximum FE_{CO} and high CO formation rate (3 mA/cm^2), while $Ni_{0.02}In_{0.98}$ showed the highest selectivity (60%) and CO formation rate (0.5 mA/cm^2). Interestingly, the Fe–In, Co–In, and Ni–In alloys were also highly selective to CO when the alloy was indium rich.

Other important alloys include bimetallic systems such as BiSn (Yang, Wu, et al., 2020), CuSn (Zheng, Ji, et al., 2019), Ag–Sn (Luc et al., 2017), Pb–Pd (Lu et al., 2019), and Pd–Sn (Bai et al., 2017). These alloys are regarded as effective catalysts to promote CO_2-to-HCOOH conversion. BiSn (Yang, Wu, et al., 2020) and CuSn (Zheng, Ji, et al., 2019) were reported to be favoring HCOOH production. BiSn outperformed Sn NPs with high FE_{HCOOH} (95.8% at −0.88 V vs. RHE), great j_{HCOOH} (74.6 mA/cm^2 at −1.38 V vs. RHE), and good stability for over 50 hr of operation. Nonetheless, CuSn presented the best $FE_{formate}$ (95%), highest j_{HCOOH} (31 mA/cm^2 at −0.5 V vs. RHE), and stability for over 50 hr of operation. These were due to the incorporation of Bi or Cu atoms into Sn species that suppressed H_2/CO formation and facilitated formate production. A bifunctional Pb–Pd alloy system was reported with a Pb-rich structure as active catalyst for CO_2-to-HCOOH conversion in an integrated CO_2 electrolyzer-formate fuel cell device (Lu et al., 2019) at cathodic conditions. At anodic conditions, the Pd rich alloy was active for HCOOH-to-CO_2 conversion. The alloy catalyst yielded a $FE_{formate}$ of above 85% with an $j_{formate}$ of 26.1 mA/cm^2 at −2 V vs. Ag/AgCl, surpassing the pure Pb and Pd. Besides, the integrated cell could operate stably for over 12 days with a high FE_{HCOOH} of ~80%. A similar FE_{HCOOH} (~80%) was achieved using a core-shell nanostructured Ag–Sn catalyst (Luc et al., 2017), consisting of a bimetallic core for high electronic conductivity and an ultrathin partially oxidized shell (SnO_x) with an optimal thickness of ~1.7 nm for CO_2 reduction. The SnO_x shell was stable for long-term operation, which was confirmed by the 24 hr of durable operation of the Ag–Sn catalyst. Compared with the state-of-the-art formate selective catalysts, Ag–Sn exhibited remarkable performance with high FE (~80%) and high $j_{formate}$ (~16 mA/cm^2 at −0.8 V vs. RHE). Interestingly, FE_{HCOOH} showed a volcano-like relationship with Sn concentration in the Ag–Sn catalyst (Figure 9.3i), which stemmed from the more favorable stabilization of OCHO* over COOH* on SnO (Weng et al., 2019) surface with oxygen vacancy sites. This competing effect was attributed to lattice expansion and the electrical conductivity loss upon increasing the thickness of SnO_x layer.

Tuning surface electronic structures of catalysts by varying the surface configurations has been proven to be a powerful approach to boost CO_2-to-HCOOH conversion. For example, a Pd–Sn alloy with the $PdSnO_2$ composition favored the intermediate HCOO* and formate product generation, delivering a nearly 100% FE_{HCOOH} with a complete suppression of H_2/CO at low overpotential of 0.26 V (Bai et al., 2017). Additionally, Pd–Sn was stable for over 5 hr of CO_2 electrolysis with negligible decay in current density at −0.43 V vs. RHE. Another AgHg alloy with a controllable amount of Ag–Hg bonding sites was fabricated using a one-step wet-chemistry method for CO_2-to-HCOOH/CO conversion (Yang, Chen, et al., 2019). The $Ag_{70}Hg_{30}$ alloy was highly selective to HCOOH with 85% production at −0.9 V vs. RHE, while the $Ag_{91}Hg_9$ alloy favored CO formation with FE_{CO} of 58% at −0.7 V vs. RHE. This switchable selectivity was ascribed to the competing effects of the preferred formation of HCOO* or *COOH intermediates at Ag–Hg and Ag–Ag bonding sites. Additionally, the $Ag_{70}Hg_{30}$ catalyst exhibited not only a superior $j_{formate}$ than $Ag_{91}Hg_9$ (5.1 vs. 2.1 mA/cm^2 at −0.9 V_{RHE}), but also an excellent stability with high FE_{HCOOH} (> 80%), constant FE_{CO+H2}, and constant current density for over 40 hr of electrolysis.

Alloy catalysts were also used for the electrocatalytic conversion of CO_2 to other low-carbon fuels, except HCOOH, as well. For example, alloy structures namely Cu-Pt (Guo et al., 2015), Ni-Ga (Torelli et al., 2016), Cu–Pd (Ma et al., 2017), CuAl (Zhong et al., 2020), and CuAg (Hoang et al., 2018) were developed to obtain hydrocarbon products such as CH_4,

C_2H_4, C_2H_6, and C_2H_5OH. Among these alloy systems, Ni–Ga (Torelli et al., 2016) and Cu–Pd (Ma et al., 2017) exhibited superior CO_2RR performances because of their multi-phase composition. The three Ni–Ga alloy films with different phases (NiGa, Ni_3Ga, and Ni_5Ga_3) were selective to CO_2 conversion into CH_4, C_2H_4, and C_2H_6, respectively (Torelli et al., 2016). All the three phases exhibited an onset potential of −0.48 V vs. RHE, about >250 mV more positive than that of polycrystalline Cu foil. Because of the stronger interaction of Ni_xGa_y phase with CO than Cu, the chemisorbed CO was further reduced to C_2 products and CH_4. In another study, a series of CuPd catalysts with ordered, disordered, and phase-separated atomic arrangements (Cu_{at}:Pd_{at} = 1 : 1), as well as disordered Cu_3Pd and $CuPd_3$, were investigated for CO_2 electroreduction (Ma et al., 2017). The ordered CuPd outperformed the disordered and phase-separated CuPd for the formation of C_1 products with high selectivity above 80%. The phase-separated CuPd and Cu_3Pd favored the formation of C_2 products with a FE_{C2} of more than 60%. The results indicated that the C_1 intermediates probably got dimerized on surfaces with neighboring Cu atoms. This work suggested that the geometric arrangements in alloy structures play an important role in CO_2RR activity and selectivity.

Precisely controlling the composition of catalysts is another promising approach to improve CO_2RR activity and selectivity. For example, a series of Cu–Pt catalysts with controllable atomic ratios were fabricated for CO_2-to-CH_4 conversion (Guo et al., 2015). The Cu–Pt with an optimized atomic molar ratio of 3 : 1 achieved the highest FE_{CH4} of 21%, lowest onset potential (ca. −0.972 V), and largest CO_2 reduction current density (ca. 0.598 mA/cm^2 at −1.3 V vs. SCE), because of more adsorbed CO* formation with the increased Cu content, thus improving the CH_4 production. However, too high Cu content resulted in higher density of adsorbed CO* but lower density of adsorbed protons. Both Cu–Pt in this work and Cu-Au in another study (Zhao et al., 2014) contributed to the best CO_2RR performance, but they had different optimal atomic ratios, Cu:Pt (3 : 1) and Au:Cu (3 : 1), because of the different proton affinities of Pt and Au.

Electrodeposited nanoporous CuAg alloy films were confirmed as efficient catalysts for CO_2-to-C_2H_4/C_2H_5OH conversion (Hoang et al., 2018). The alloy film with 6% Ag was the most active for CO_2RR and delivered 60% FE_{C2H4} and 25% FE_{C2H5OH} at −0.7 V vs. RHE and a total current density of ∼ −300 mA/cm^2 in flow cell, superior to the Cu-wire and Cu–Ag films. The high selectivity originated from the enhanced Cu_2O overlayer stabilization combined with the optimal CO intermediate availability provided by the incorporated Ag. Also, a high FE_{C2H4} approaching 80% at 400 mA/cm^2 during 50 hr of CO_2 electrolysis in 1 M KOH, a high FE_{C2H4} of 73 ± 4% and a CO_2-to-C_2H_4 half-cell power conversion efficiency of 55 ± 2% at 150 mA/cm^2 for over 50 hr of CO_2 electrolysis in 3 M KOH + 3 M KI were achieved in a flow cell using a Cu–Al alloy catalyst prepared by incorporating Al on Cu (Zhong et al., 2020), far surpassing the porous Cu and deposited Cu. Theoretical calculations indicated that Cu–Al alloys would provide abundant active sites and surface orientations with near-optimal CO binding energy for CO_2 reduction. Besides, *in situ* X-ray absorption measurements demonstrated that the C–C dimerization was enhanced in the alloy structure due to favorable Cu coordination environment afforded by the Cu–Al alloy.

9.3.2 Metal Compounds

Transition metal-based compounds involve a wide range of CO_2RR catalysts, and to date, various structural modifications have been applied to optimize the electrochemical performance. The catalysts of this type fall into the families of metal dichalcogenides, metal oxides, metal nitrides, or metal hydroxides. In this section, classification and development of various transition metal compounds for CO_2RR are discussed.

9.3.2.1 Metal Chalcogenides

Transition metal dichalcogenides (TMDs) have shown huge potential for CO_2RR. They are low-cost materials possessing high electrical conductivity, layered structures, large active surface, and active sites (defects, edges) in abundance (Asadi et al., 2016; Shao et al., 2021). Asadi et al. (Asadi et al., 2016) discovered that WSe_2 nanoflakes could carry out CO_2 reduction at an overpotential as low as −0.54 V vs. RHE with a current density of 18.95 mA/cm^2, higher than that of bulk Ag (0.19 mA/cm^2), Ag NPs (1.57 mA/cm^2) and bulk MoS_2 (3.4 mA/cm^2). Moreover, a FE_{CO} of 24%, a value 8 times higher than that of MoS_2, was achieved using WSe_2. Theoretical calculations indicated that TMDs ($MoSe_2$, MoS_2, WSe_2, WS_2) outperform Ag in stabilizing the intermediate CO* and have faster kinetics for CO* formation on TMDs, resulting in lower overpotentials. In addition, the calculated work functions of the TMD monolayers followed the order of $MoS_2 > WS_2 > MoSe_2 > WSe_2$, demonstrating that WSe_2 is the best TMD compound to activate CO_2 reduction. TMD-alloy composites usually exhibit enhanced CO_2RR performances with respect to pristine binary TMDs. A ternary TMDs (MoSeS alloy) with abundant surface-exposed atoms as single active sites was applied for CO_2-to-CO conversion, which delivered a FE_{CO} of 45.2% at −1.15 V vs. RHE, much higher than that of MoS_2 (16.6%) and $MoSe_2$ (30.5%). The d-band electronic structure of Mo atoms was tailored by the shortened Mo–S and lengthened Mo–Se bonds. The off-center charge around Mo facilitated COOH* intermediate stabilization and promoted the rate-limiting CO desorption step. In a following theoretical study, $MoSe_2$ and MoS_2 were proven as promising catalysts for CO_2RR (Chan et al., 2014). The bonding of COOH and CHO intermediates to the bridging S atoms and Se atoms at the edges was found to be more facile than the bonding of intermediate CO to the metal atom, and this indicates the effect of the active edge sites on CO_2 reduction.

Copper selenides are potential catalysts for CO_2RR due to their structural stability, existence in multi-oxidation states and compositions, as well as the presence of unsaturated Se atoms along edge sites, which improve the density of exposed active sites. However, only a few studies have been devoted to understanding the CO_2RR performance of copper selenides. Mi et al. (Mi et al., 2018) prepared $Cu_{1.8}Se$ NWs on Cu foam using chemical oxidization and subsequent selenization. These $Cu_{1.8}Se$ NWs were highly selective to C_2 species with FE_{C2H4} and FE_{C2H5OH} approaching 55% and 24%, respectively, at −1.1 V vs. RHE. The NWs also exhibited a stable current density (−14.6 mA/cm^2) and FE_{C2+} (79%) for over 25 hr of CO_2 electrolysis at −1.1 V vs. RHE. The superior activity and selectivity originated from the unique structure of $Cu_{1.8}Se$ NWs. The binder-free construction facilitated fast electron transfer and sufficient electrolyte-active material contact. The hierarchical structure ensured fast electron transfer to get a faster reaction kinetics. The high ECSA, as derived from the NP formation on $Cu_{1.8}Se$ NWs, offered abundant active sites. Furthermore, a large C_2/C_1 selectivity ratio of 36 was achieved. This was attributed to the high CO_2 pressure (10 atm) and the adsorbed Se species. Hence, the surface CO concentration and C_1 formation were decreased. Another nanocatalyst, $Cu_{1.63}Se_{(1/3)}$ fabricated by a solvothermal method, showed outstanding selectivity toward CH_3OH with a FE_{CH3OH} of 77.6% and a high current density of 41.5 mA/cm^2 at a low overpotential of 285 mV (Figure 9.4a) (Yang, Zhu, et al., 2019). This high activity and selectivity originated from improved CO_2 solubility and transport to the catalyst surface provided by IL (a type of organic salt existing in liquid phase at ambient temperatures) added to the electrolyte, the presence of appropriate Cu active sites aiding the adsorbed $CO_2\cdot^-$ to accept e^-/H^+ for CH_3OH

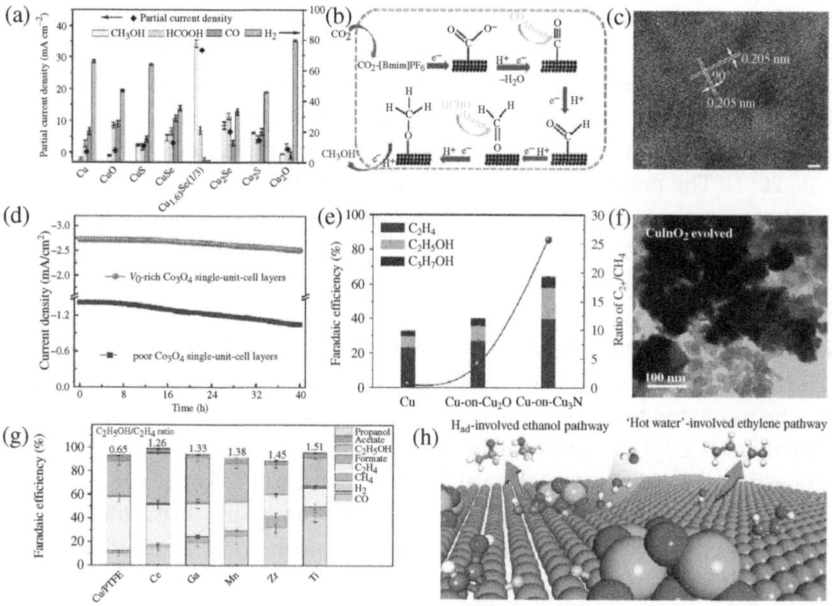

Figure 9.4 (a) Partial current density and FE of methanol over various Cu-based catalysts at −2.1 V vs. Ag/Ag$^+$. (b) Proposed mechanism on $Cu_{2-x}Se_{(y)}$ electrode for CO_2 reduction to methanol. Source: Reproduced from Yang, Zhu, et al., 2019, with permission. Copyright 2019, Springer Nature/Springer Nature. (c) HRTEM image of the *V*o-rich Co_3O_4 single-unit-cell layer, scale bar: 1 nm. (d) Chronoamperometry results at the potential of −0.87 V vs. SCE. Source: Gao, Sun, et al., 2017/Reproduced with permission from Springer Nature/CC BY 4.0. (e) Comparison of faradaic efficiency for C_{2+} and the ratio of C_{2+}/CH_4 at −0.95 V vs. RHE on Cu, Cu-on-Cu_2O, and Cu-on-Cu_3N. Source: Reproduced from Liang et al., 2018, with permission. Copyright 2018, Springer Nature/Springer Nature. (f) BF-STEM micrographs in the evolved $CuInO_2$, showing the core-shell-like structure of $In(OH)_3$ around Cu-rich regions. Source: Larrazábal et al., 2016/Reproduced with permission from American Chemical Society. (g) Product distribution of various hydroxides/oxides modified Cu/PTFE electrodes, along with corresponding C_2H_5OH/C_2H_4 ratio. (h) Density functional theory calculations on the ethylene and ethanol pathways. Source: Reproduced from Luo et al., 2019, with permission. Copyright 2019, Springer Nature (https://pubs.acs.org/doi/full/10.1021/acscatal.6b02067/Springer Nature.

generation (Figure 9.4b), as well as the optimal size and morphology of the Cu–Se catalyst for CO_2 reduction, which are dependent on the solvent's composition (volume ratio $V_{diethylenetriamine}/V_{H2O}$) used for catalyst fabrication.

In addition to TMDs, some other chalcogenide materials have also been found to be highly selective toward CO_2RR products; these include layer-stacked bulk MoS_2 with Mo-terminated edges (Asadi et al., 2014), CdS with S vacancy (Feng, Gao, Feng, et al., 2020), PbS NCs (Zhang, Liu, et al., 2019), SnS_2 monolayers (He, Liu, et al., 2018), CuS nanosheet arrays (Zhao, Peng, et al., 2017), and Mo–Bi bimetallic chalcogenide nanosheets (Sun et al., 2016).

9.3.2.2 Metal Oxides

Catalyst systems encompassing fully or partially oxidized metals have been proven promising in altering the active sites' properties, thereby enhancing the selectivity and activity to CO_2RR products. Tailoring GBs is effective in tuning the binding energies of reaction intermediates and accelerating CO_2RR. For example, SnO_2 quantum wires (QWs) consisting of individual quantum dots (QDs) and numerous GBs were prepared by a hydrothermal method and applied for CO_2-to-HCOOH conversion (Liu, Xiao, et al., 2019). Compared with SnO_2 NPs possessing larger ECSA, the ultrathin SnO_2 QWs with exposed GBs exhibited improved FE_{HCOOH} (>80%), FE_{C1} (90%), and EE (>50%) over a wide potential range, from −0.5 to −1.3 V vs. RHE, as well as maximum FE_{HCOOH} (87.3%) and EE (52.7%). Besides, a higher j_{HCOOH} than in the case of SnO_2 NPs was obtained at −1.156 V vs. RHE (13.7 vs. 9.1 mA/cm^2). For the stability of the ultrathin SnO_2 QWs, a roughly fluctuating current density (~9.9 mA/cm^2) was observed, which showed only slight increase at −0.956 V vs. RHE for over 25,000 s of CO_2 electrolysis. In another study, SnO_2 porous nanowire catalyst (Sn-pNWs) with a high density of GBs was prepared using a scalable plasma synthesis technology (Kumar et al., 2017). Compared with the crystalline SnO_2 nanowire (Sn NWs) without GBs, the Sn-pNWs activated CO_2-to-HCOOH conversion at lower overpotential of ca. 350 mV and reached 10 times higher FE_{HCCOH} of 80% at −0.8 V vs. RHE. The Sn-pNWs exhibited a larger j_{HCOOH} (4.8 mA/cm^2 at −0.8 V vs. RHE) than Sn-NPs and Sn-NWs. Additionally, Sn-pNWs displayed a relatively stable performance with high $FE_{HCOOH + CO}$ (92%) for over 15 hr of CO_2 electrolysis at −0.5 V vs. RHE. This performance was superior to the case of Sn-NWs (75%). The authors attributed the improved catalytic activity of Sn-pNWs to high extents of GBs, which introduced new active sites.

Oxygen vacancies in metal oxides tailor the electronic structures, charge transport, and surface properties and, therefore, promote CO_2 activation and dissociation. A theoretical study demonstrated that the reduced (Weng et al., 2019) surface of anatase TiO_2 with oxygen vacancies would favor CO_2 binding and charge transfer to CO_2, due to the formation of hybridized orbitals favoring such a transfer (He et al., 2010). To further investigate the relationship between oxygen vacancies and CO_2RR properties, oxygen vacancy-decorated Co_3O_4 single-unit-cell layers (Figure 9.4c) were fabricated and applied for CO_2-to-HCOOH conversion (Gao, Sun, et al., 2017). The presence of oxygen(II) vacancy as the main defect lowered the activation barrier of the rate-limiting proton transfer step by facilitating the stabilization of formate anion radical intermediate. Therefore, the vacancy-rich Co_3O_4 single-unit-cell layer exhibited a high FE_{HCOOH} of 85% and current density of 2.7 mA/cm^2 for over 40 hr of CO_2 electrocatalysis (Figure 9.4d) and large ECSA-corrected current density at −0.87 V vs. RHE. The performance of the stoichiometric structure was inferior.

Engineering the metal oxide surface was shown to be efficient in boosting CO_2RR activity and selectivity. A CuCl-derived Cu catalyst with reconstructed Cu(I) oxide surface, uniform morphology, and abundant surface active sites was prepared by a wet-oxidation method (Kibria et al., 2018). The reconstructed surface obtained a high C_{2+} selectivity (FE_{C2+} 73%, FE_{C2H4} 56%) and an overall current density of 17 mA/cm^2 in H-cell. The surface-engineered Cu catalyst also exhibited a record-high FE_{C2+} of ~80% and half-cell power conversion efficiency of 50% at a high partial current density of 336 mA/cm^2 at −0.68 V vs. RHE in flow cell, outperforming the evaporated Cu. In another study, the surface of mesoporous SnO_2 in the form of hierarchical nanosheets and nanospheres was manipulated in a controlled manner using fluorinated surfactants with different lengths of hydrophobic chains (Wei et al., 2020). The surfaces of both 0D SnO_2 nanospheres (average size 165 nm) and 2D SnO_2 nanosheets (thickness 22.5 nm) were decorated with ordered mesopore arrays (pore size 16 nm). After a subsequent calcination, fully crystallized hierarchical mesopore structures (5 nm pores for 0D SnO_2, 16 nm pores for 2D SnO_2) with enhanced surface area were formed. The optimal dimensionality and mesoscopic structure facilitated the enhancement in surface area, the exposure of active sites, the improvement in electrolyte accessibility, and the enhancement of charge transfer, as well as adsorption and desorption of the intermediates and products. Therefore, the 2D SnO_2 nanosheets presented high FE_{HCOOH} (90% at −1.3 V vs. RHE) and FE_{C1} (97.4% at −1.2 V vs. RHE) and stability for at least 12 hr of operation, as well as larger current density than SnO_2 NPs. Furthermore, the structural engineering and the formation of metal–metal oxide interfaces were explored to boost CO_2RR activity. For example, $CuSnO_2$ NPs (Wang, Qiao, et al., 2018), Cu_2O cubes (Arán-Ais et al., 2020), Cu-Cu_2O (Zhu et al., 2019), Au-CeO_x (Gao, Zhang, et al., 2017), Sn–Cu/SnO_x core/shell (Ye et al., 2020), and partially oxidized Co 4-atom-thick layers (Gao, Lin, et al., 2016) exhibited excellent catalytic activity because of their engineered metal–metal oxide interfaces, which lowered the energy barrier for the formation of the key intermediates during CO_2RR.

9.3.2.3 Metal Nitrides

Metal nitrides were confirmed as potential electrocatalysts for CO_2 conversion. For example, Liang et al. (Liang et al., 2018) deposited Cu on Cu_3N by the surface electroreduction of CuO-on-Cu_3N intermediate consisting of a layer of CuO formed at the surface of a Cu_3N NC. The intermediate was prepared by the ligand exchange in octadecyl amine-capped Cu_3N NCs using azide (N_3^-) and simultaneous surface oxidation in ambient air. The obtained Cu-on-Cu_3N showed reduced spacing between NCs after ligand exchange; Cu was distributed across the NP volume, and N was concentrated in the core with a lower concentration on the surface. This demonstrated the surface reconstruction during the catalyst preparation. Cu-on-Cu_3N exhibited a remarkable stability (>30 hr) and a high FE_{C2+} of $64 \pm 2\%$ at −0.95 V vs. RHE with $39 \pm 2\%$ C_2H_4, $19 \pm 1\%$ C_2H_5OH, and $6 \pm 1\%$ C_3H_7OH, respectively (Figure 9.4e). Cu-on-Cu_3N exhibited 40-fold and 2.2-fold enhancement in C_{2+}/CH_4 ratio and j_{C2+}, respectively, than the pure Cu. This performance was even superior to that exhibited by Cu-on-Cu_2O fabricated using a similar synthetic strategy (6.3-fold enhancement in C_{2+}/CH_4 ratio, 4.4-fold higher j_{C2+}). In this catalyst, Cu_3N as the inner support was effective in suppressing Cu^+ reduction to modify the electronic structure of surface metal. This facilitated CO adsorption and reduced the CO dimerization energy barrier, thereby promoting CO_2 electroreduction. The perovskite-type Cu_3N nanocubes (25 nm) exhibited outstanding size-dependent CO_2RR performance with a high FE_{C2H4} of 60%, a mass activity of 34 A/g, and a C_2H_4/CH_4 molar ratio >2000 at −1.6 V vs. RHE (Yin et al., 2019). The high selectivity stemmed from the stabilization of Cu(I) (100) by Cu_3N structure, which promoted the coupling of intermediates CO* and CHO* to form CO-CHO on the (100) Cu_3N surface, resulting in the formation of C_2H_4. In addition, Cu-on-Cu_3N showed much-enhanced stability compared to Cu nanocubes (Grosse et al., 2018; Roberts et al., 2015). While a FE_{C2H4} of 60% showed by the material in the first hour was reduced to a stabilized value of ~53%, the current density decreased from 135 mA/cm^2 in the first hour to 101 mA/cm^2 after 20 hr of CO_2 electrolysis at −1.6 V vs. RHE. This indicated more efficient Cu^+ stabilization by N^{3-} in the perovskite-type Cu_3N structure.

Probing the intrinsic active sites helps in understanding the structure-activity relationship and laying foundations for the rational design of CO_2RR catalysts. Zhang et al. (Zhang, Liang, et al., 2020) unveiled the intrinsic origin of InN nanosheets using in situ measurements. During CO_2 electrocatalysis, InN nanosheets were reconstructed to an In-rich surface, which resulted in a surface charge redistribution to improve HCOO* adsorption and further reduction to HCOOH. Therefore, the self-reconstructed InN nanosheets presented a maximum FE for the carbonaceous product (95%), 1.4-fold higher than that of the In nanosheets. Besides, InN nanosheets delivered a stable FE_{HCOOH} (~88%), and an increased j_{HCOOH} (from −30.0 to −40.7 mA/cm^2) during 3 hr of CO_2RR performed at −0.9 V vs. RHE. This suggested that the InN nanosheets underwent a reconstruction process during CO_2 electrolysis. Following this study, Cu_3PdN NCs were prepared via solution synthesis and applied for CO_2-to-HCOOH conversion. The material exhibited a FE_{HCOOH} of 43% at −1 V vs. RHE and a FE_{HCOOH} of 91% at −0.9 V vs. RHE, higher than those of Cu_3N and Cu_3Pd NCs fabricated using the same synthesis method (Jia, Hao, et al., 2020). The enhanced selectivity was attributed to the optimal synergistic effect of N and Pd modification which provided more active sites, avoided NCs agglomeration, and the role of RGO as support in improving the conductivity. Additionally, the Cu_3PdN NC stability study conducted using chronoamperometry method showed a total current density of −7 mA/cm^2 during 17 hr of CO_2 electrolysis at −1.4 V vs. RHE. This current density was significantly different from that determined using the cyclic voltammetric curve due to the mass transfer issue. The decrease in the initial current density could be attributed to the production of numerous bubbles on the electrode surface due to HER, which interfered the electrode equilibrium state. The Cu_3PdN NC maintained its original crystalline structure after 17 hr of CO_2RR operation.

Apart from the above-mentioned metal nitrides, Ni_3N supported on carbon (Ni_3N/C) was shown effective to promote the CO_2-to-CO conversion compared with Ni/C and NiO/C. Ni_3N/C exhibited high FE_{CO} (85.7% at −0.9 V vs. RHE) and j_{CO} (8.7 mA/cm^2 at −1 V vs. RHE) for aqueous CO_2RR, superior to Ni/C (16.8%, 0.73 mA/cm^2) and NiO/C (< 5%). The Ni_3N/C catalyst also displayed decent stability with high FE_{CO} (>84%) during 6 hr of CO_2 electrolysis. For the gas phase CO_2RR in MEA, a much higher FE_{CO} of 92.5% and an increased j_{CO} of 23.3 mA/cm^2 at a cell voltage of 2.8 V were obtained using Ni_3N/C. At 2.8 V, the cathode potential in the cell was more positive than that in the aqueous phase (−0.79 V vs. −1.12 V with respect to normal hydrogen electrode (NHE)), suggesting a reduced activation energy in the gas phase electrolysis, because of the direct CO_2RR at the CO_2-catalyst (gas-solid) interface without competitive water adsorption. A decent stability, with a high FE_{CO} (>90%), was detected at a cell voltage of 2.8 V, in a 70 hr CO_2RR experiment with Ni_3N/C. These excellent performances were due to the abundance of strong base sites in Ni_3N that enriched the catalyst surface with CO_2, and the utilization of the gas phase electrolysis, which synergistically improved the catalyst reactivity (Hou et al., 2018).

9.3.2.4 Metal Hydroxides

Metal hydroxides also have great potential to activate CO_2RR. For example, a heterogeneous nanostructure with Cu-rich cores and $In(OH)_3$ shell-like matrix (Figure 9.4f) evolved from the segregation of Cu and In in Cu–In nanoalloy through consecutive electrocatalytic cycles. The Cu–In nanoalloy was prepared by the in situ reduction of $CuInO_2$ and In_2O_3-supported Cu NPs (GnO et al., 2016). The $In(OH)_3$ phase at shell played a significant role in CO_2-to-CO conversion. By removing the active $In(OH)_3$ shell, the CO selectivity decreased dramatically but recovered with the reconstruction of $In(OH)_3$ phase upon catalyst re-equilibration during a subsequent CO_2 catalysis. In another study, hydroxide and oxide doping were employed to alter the adsorbed hydrogen on Cu (Luo et al., 2019). Among a series of metal hydroxide-interface-doped-Cu catalysts, $Ce(OH)_x$-doped-Cu delivered the best CO_2RR performance with a FE_{C2H5OH} of 43% and j_{C2H5OH} of 128 mA/cm^2 at −0.7 V vs. RHE (Figure 9.4g), surpassing the Cu control (29%, 87 mA/cm^2). $Ce(OH)_x$-doped-Cu also provided a stable operation for over 6 hr at 250 mA/cm^2. Theoretical calculations demonstrated that introducing $Ce(OH)_x$ phase accelerated water dissociation and optimized the hydrogen adsorption. Surface hydrogen (H_{ad}) adsorption on Cu favored the ethanol electroproduction over ethylene by attacking the Cu—C bond of the *HCCOH intermediate, which is a key intermediate in the branching of ethanol vs. ethylene (Figure 9.4h).

9.3.3 Single-Atom Catalysts

Compared with the bulk materials and NPs, SACs with isolated single atoms on supports, have attracted considerable attention as promising CO_2RR catalysts, due to their maximum metal utilization, strong atom-support interactions, suppressed HER, and outstanding catalytic activity (Sun, Jia, Zhao, & Zhao, 2022). For example, the enriched Ni atoms on porous N-doped carbon surfaces outperformed the Ni NPs with higher j_{CO}, FE_{CO}, and excellent stability and activity (Yang, Qiu, et al., 2018). Similar results were also detected for Pd SAC over Pd NPs (He et al., 2020), and Zn SAC over Zn NPs (Yang, Song, et al., 2018).

Many strategies have been developed to successfully prepare materials mainly covered by single atoms, and these include mass-selected soft loading (Abbet et al., 2000), physical methods (Sun et al., 2013) like atomic layer deposition (Sun et al., 2013), and wet-chemistry approaches (Pinna, 1998). Among these, the wet-chemistry methods are the most commonly applied. The approaches include defect engineering (Wang, Han, et al., 2018), spatial confinement (Liu, Li, et al., 2019), and coordination (Pan, Lin, et al., 2018). To be specific, the ligand-mediated method (Yang, Shang, et al., 2019), freeze-vacuum drying-calcination method (Zu et al., 2019), and electrospinning-pyrolysis (Yang, Wu, et al., 2019) are employed to fabricate high-metal loading SACs at a large-scale, laying foundations for industrial CO_2RR applications.

A wide range of metal SACs have been developed so far to promote CO_2 electroreduction, and these include Ni (Zhao, Dai, et al., 2017), Ru (Wang, Wang, Dares, & Meyer, 2017), Pd (Back & Jung, 2017), Pt (Zhao et al., 2019), Fe (Cheng et al., 2018), Co (Pan, Lin, et al., 2018), Cu (Abdel-Mageed et al., 2019), Zn (Yang, Song, et al., 2018), and Sn (Zhao et al., 2018). Among them, transition metal SACs based on Ni, Fe, and Co favor CO_2-to-CO conversion (Büchele et al., 2020; Chen, Zou, et al., 2019; Geng et al., 2019), while Cu SACs are effective in generating multi-carbon hydrocarbons (C_2H_4, C_2H_5OH, and C_3H_7OH) (Abdel-Mageed et al., 2019). In order to maintain the unique structure of single-atom active sites, different kinds of supporting materials, such as carbon supports (CNT (Wang, Jia, Song, et al., 2017), graphene oxide (Cheng et al., 2019), MOFs-derived nanocarbons (Han et al., 2019), and carbon nitride (Chen, 2018)), and metal complexes (CeO_2 (Wang, Chen, et al., 2018), TiO_2 (Wan et al., 2018), MoS_2 (Biroju et al., 2017), and TiC (Back & Jung, 2017)), have been explored for SACs preparation. In this section, applications of SACs for CO_2RR are discussed.

9.3.3.1 Noble Metal SACs

Noble metal-based SACs (i.e., Au, Ag, and Pd) not only solve the issue of high cost associated with the use of bulk noble metals but also deliver excellent selectivity and activity for CO electroproduction, as they bind *COOH stronger than *CO. Recently, Li et al. (Li, Chen, et al., 2020) employed a dual-atom Ag_2/graphene catalyst (Ag_2-G) with well-defined AgN_3–AgN_3 active sites for CO_2RR. Ag_2-G could catalyze CO_2 reduction at a high potential (−0.25 V vs. RHE) and deliver a larger j_{CO} than Ag_1-G SAC and the traditional Ag NPs over a wide potential range (−0.3 to −1.0 V vs. RHE). It exhibited a high FE_{CO} (93.4%), a current density of 11.87 mA/cm^2, and a durability for over 36 h of operation at −0.7 V vs. RHE. Theoretical calculations indicated that AgN_3–AgN_3 active sites in Ag_2-G outperformed AgN_4 active sites in Ag_1-G to produce CO, due to stronger *CO_2 adsorption and lower energy barrier for *COOH formation. In addition, AgN_3–AgN_3 active sites facilitated *CO_2 stabilization by concomitantly interacting with C and an O atom in CO_2 through two Ag atoms, thereby generating more *COOH and obtaining higher FE_{CO}. Following this study, a Pd SAC with Pd-N_4 sites on carbon support (Pd

NC) was developed for CO_2-to-CO conversion (He et al., 2020). The Pd-N_4 active sites were selective toward CO formation without palladium hydride formation, a common problem with Pd NP-based CO_2RR. Because of the *HOCO intermediate stabilization by Pd-N_4 sites, the Pd NC could generate CO even at potentials as low as −0.2 and −0.25 V vs. RHE with a FE_{CO} of 4.9% and 8.3%, respectively. Besides, the Pd NC exhibited a maximum FE_{CO} of 55% at −0.5 V vs. RHE that was 2.5 times higher than that of the commercially purchased carbon-supported Pd with 10 wt% Pd loading (Pd/C) (22% at −0.5 V vs. RHE). Pd NC outperformed Pd/C with 3.8–9.0 times higher j_{CO} (0.244 vs. 0.027 mA/cm^2 at −0.4 V vs. RHE) and 3.9-fold larger j_{CO} (2.2 vs. 0.57 mA/cm^2 −0.8 V vs. RHE). The Pd NC also exhibited distinguishable CO_2RR activity with far higher mass activity than its Pd NPs counterpart (373.0 vs. 28.5 mA/mg at −0.8 V vs. RHE). Both Pd NC and Pd/C were stable over 3 h of CO_2 electrolysis at −0.6 V vs. RHE. No reduction in FE_{CO} or current density was observed.

Using α-$Co(OH)_2$ as the supporting material, an Ir SAC was fabricated with uniform Ir atom distribution and 1.7 wt% Ir loading (Ir_{AC}-1.7), which was used for CO_2-to-CO conversion (Sun et al., 2019). While the Ir_{AC}-1.7 delivered a maximum FE_{CO} of 97.6% and exhibited stability for over 24 hr of electrolysis at −0.66 V vs. RHE, the Ir NPs (2 nm) on α-$Co(OH)_2$ showed no activity for CO_2RR. The superior electrochemical performance originated from the Ir sites favoring strong CO_2 adsorption and activation, stabilization of intermediate $CO_2^{\cdot-}$, as well as the synergistic effect of the catalyst with $Co(OH)_2$ support.

9.3.3.2 Transition Metal SACs

Introducing transition metals to carbon frameworks to form metal–nitrogen–carbon catalysts (M–N–C, M = Mn, Fe, Co, Ni, Cu) has been recognized as an effective strategy to improve CO_2RR performance. Ju et al. (Ju et al., 2017) distributed M–N sites on porous carbons (Figure 9.5a) and discovered that Ni–N–C would exhibit a maximum FE_{CO} of 85% at −0.78 V vs. RHE, Fe–N–C would show high activity for CO formation at low overpotentials, and Co–N–C would favor H_2 production (Figure 9.5b). Generally, the activity of M–N–C-based SACs toward CO formation follows the sequence: Ni–N–C > Fe–N–C > Mn–N–C > Cu–N–C > Co–N–C (Figure 9.5c). Similar conclusions were made by Hu et al. (Hu et al., 2018) and Pan et al. (Pan, Zhang, et al., 2018). They reported that M–N species would provide active sites and promote CO_2-to-CO conversion. Pan et al. (Pan, Zhang, et al., 2018) reported that the edge-hosted M–N_{2+2}–C_8 (M = Fe, Co) moieties

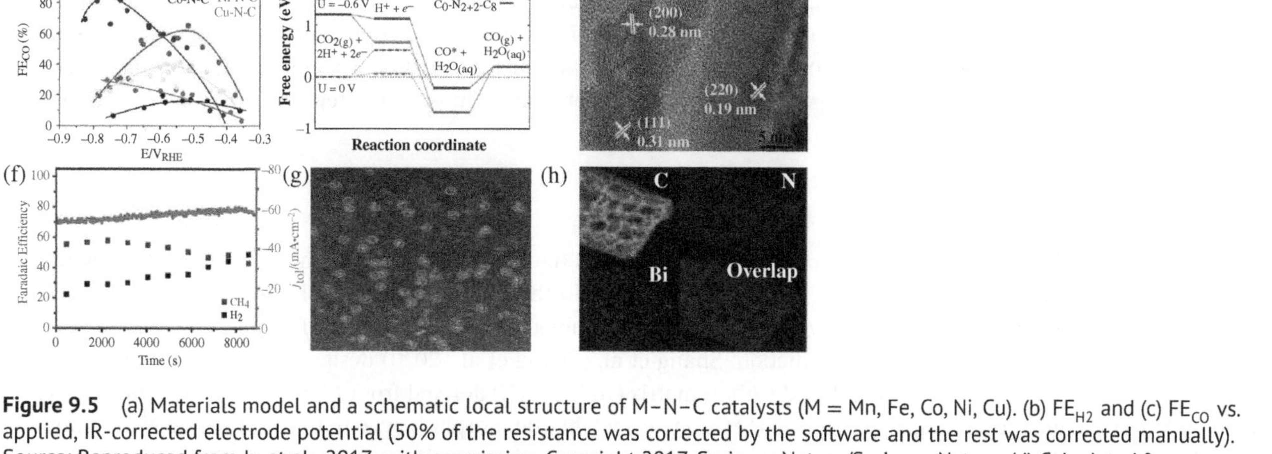

Figure 9.5 (a) Materials model and a schematic local structure of M–N–C catalysts (M = Mn, Fe, Co, Ni, Cu). (b) FE_{H2} and (c) FE_{CO} vs. applied, IR-corrected electrode potential (50% of the resistance was corrected by the software and the rest was corrected manually). Source: Reproduced from Ju et al., 2017, with permission. Copyright 2017, Springer Nature/Springer Nature. (d) Calculated free energy evolution of CO_2 reduction to CO on M–N_{2+2}–C_8 sites under an applied electrode potential (U) of 0 and −0.6 V. Source: Reproduced from Pan, Zhang, et al., 2018, with permission. Copyright 2018, American Chemical Society/American Chemical Society. (e) HRTEM image and (f) stability of Cu-CeO_2-4% at −1.8 V. Source: Wang, Chen, et al., 2018/Reproduced with permission from American Chemical Society. (g) Magnified HAADF-STEM image and (h) Energy dispersive X-ray spectroscopy (EDS) elemental mapping results of Bi SAs/NC. Source: Zhang, Wang, et al., 2019/Reproduced with permission from American Chemical Society.

would bridge two adjacent armchair-like graphitic layers, while the bulk-hosted M–N_4–C_{10} moieties would be embedded in a graphitic layer. As a result, the M–N–C phase at edge sites was the catalytically active site that promoted *COOH dissociation (Figure 9.5d). Fe–N_4 was more active than Co–N_4 for CO_2-to-CO conversion. It showed a larger j_{CO} in the range −0.3 to −0.9 V vs. RHE, a higher FE_{CO} (93% at 0.47 V overpotential vs. 45% at 0.48 V overpotential), and a better stability (over 20 hr of operation at −0.58 V vs. RHE). Very recently, our group designed a gram-scale (> 1 g) Ni-N-CNT SAC by distributing Ni atoms on CNTs via a direct high-temperature pyrolysis of nickel acetate tetrahydrate, phenanthroline, and CNTs (Sun et al., 2021). The Ni-N-CNT with a remarkable Ni loading (2 wt%) depicted a maximum j_{CO} of 28 mA/cm^2, high FE_{CO} of 94% and an excellent stability over 8 hr at −0.8 V vs. RHE. The isolated individual Ni atoms were the active centers for CO_2RR. The CNT support provided short pathways for electron and mass transport. By plotting the FE_{CO} of M-N-CNT-based catalysts (M = Ni, Co, Cu, Fe, Mn, Zn, Pt, or Ru) against the metal atomic number, a volcano-like trend with Ni-N-CNT as the catalyst with the highest activity for CO_2-to-CO conversion was obtained. This superior activity of Ni-N-CNT is related to the d-band center position of Ni^{2+} being the most ideal for CO_2RR. Our group also recently demonstrated a novel isolated diatomic Ni/Fe–N–C derived from zeolitic imidazolate framework (ZIF-8) for CO_2-to-CO conversion (Ren, Tan, et al., 2019). The Ni/Fe–N–C NPs exhibited uniform sizes of 50–100 nm and maintained rhombic dodecahedron shape with distinct edges. When applied for CO_2 electrocatalysis, the Ni/Fe–N–C not only outperformed Ni–N–C and Fe–N–C with a larger j_{CO} over a wide potential range from −0.4 to −1 V vs. RHE. The maximum j_{CO} was 19.7 mA/cm^2 at −1 V vs. RHE. It exhibited a high FE_{CO} of 98% and was stable for over 30 h of operation at −0.7 V vs. RHE. This outstanding selectivity originated from the dual Ni–Fe centers as active sites that favored COOH* formation and CO desorption, thereby promoting CO_2 reduction.

The enhanced CO_2RR performances can also be realized by anchoring single atoms onto other supports such as CeO_2 nanorods (Wang, Chen, et al., 2018), g-C_3N_4 (Chen, Zhao, et al., 2019), graphdiyne (Feng, Su, Ding, et al., 2020), TiC, and TiN (Back & Jung, 2017). For example, Wang et al. (Wang, Chen, et al., 2018) demonstrated an effective Cu SAC with isolated Cu atoms dispersed on mesoporous CeO_2 nanorods for CO_2RR (Figure 9.5e). Every Cu active site for CO_2-to-CH_4 conversion was surrounded by three oxygen vacancies. The Cu SAC catalyst with 4% Cu species (4% is the percentage of Cu/Ce in the composite characterized by inductively coupled plasma-atomic emission spectrometry) presented several-fold higher partial current density for deep reduction products (~40, 70, and 90 mA/cm^2 at −1.6, −1.8, and −2.0 V vs. RHE, respectively) than pure Cu and undoped CeO_2. It showed a high FE_{CH4} of 58% at −1.8 V vs. RHE and a stability for over 8000 s of operation (Figure 9.5f). This performance apparently originated from the synergistic activity of the Cu sites, the multiple oxygen vacancies, and the CeO_2 framework.

9.3.3.3 Other Metal SACs

In addition to noble and transitional metal SACs (M = Ni, Fe, Co, Cu), some studies have demonstrated that other M–N–C SACs such as Bi SAs/NC (Zhang, Wang, et al., 2019), Sn–N (Zhao et al., 2018), Sb–NC (Jia, Hong, et al., 2019) and $In^{\delta+}$–N_4 (Shang et al., 2020) are effective in boosting CO_2RR. Zhang et al. (Zhang, Wang, et al., 2019) distributed Bi–N_4 sites on porous carbon networks (Figure 9.5g,h) derived from MOF via pyrolysis of the mixed Bi-MOF precursor and dicyandiamide. The obtained Bi SAs/NC were highly selective to CO with a FE_{CO} of 97%, a current density of 3.9 mA/cm^2, and a high turnover frequency (TOF) of 5535 h^{-1} at a low overpotential (0.39 V), outperforming most of the reported catalysts for CO_2-to-CO conversion. The TOF was calculated using the following equation:

$$\text{TOF} = \frac{j_{\text{product}}/\text{nF}}{\text{m}_{\text{cat}}\omega/\text{M}_{Bi}}$$

j_{product} is the partial current density for specific product (A), n is the electron transfer number for product formation, F is the Faraday constant (96,485 C/mol), m_{cat} is the catalyst mass in the electrode (g), ω is the Bi loading in the catalyst, and M_{Bi} is the atomic mass of Bi (208.98 g/mol). Bi-N_4 sites promoted CO_2 activation and facilitated COOH* formation simultaneously, and thus boosting CO production. Shang et al. (Shang et al., 2020) designed an In SAC with exclusive $In^{\delta+}$-N_4 atomic interface moieties on N-doped carbon matrix (In SAs/NC) derived from MOFs using a wet-impregnation method and subsequent pyrolysis. The In SAs/NC exhibited a polyhedral morphology with uniform In and N distributions over the carbon architecture. Because of the active $In^{\delta+}$-N_4 species, the intermediate HCOO* formation was energetically more favorable than COOH*, which promoted CO_2-to-HCOOH conversion. In SAs/NC exhibited extremely large TOF, 12500 h^{-1} at −0.95 V vs. RHE, high FE_{formate} (above 85% over wide potential range from −0.65 to −0.95 V with maximum value of 96%), 40-times larger j_{formate} than In NPs/NC (8.52 vs. 0.21 mA/cm^2), and remarkable stability for over 60 hr of operation at a low potential (−0.65 V vs. RHE). Similarly, $Sb^{\delta+}$-N_4 ($0 < \delta < 3$) active sites (Jiang, Wang, Pei, et al., 2020) and

$Sn^{\delta+}$ sites on N-doped carbon (Zu et al., 2019) also contributed to formate formation with high production rate, indicating the promotion of CO_2RR by single metal atom sites.

An interesting Sn SAC was prepared by electrospinning-pyrolysis and demonstrated to have high selectivity to HCOOH before an acid etching and to CO after the etching (Zhao et al., 2018). The Sn single atoms loaded on carbon nanofiber (Sn-CF SAC) exhibited a high FE_{HCOOH} of 62% and j_{HCOOH} of 11.1 mA/cm^2 at a moderate overpotential of 0.69 V, along with a maximum FE_{CO} of 91% at a low overpotential of 0.49 V. A stability for over 24 hr of operation was demonstrated. The outstanding CO_2RR performance came from the active Sn sites facilitating HCOO* and COOH* formation, respectively, for HCOOH and CO production, as well as the porous carbon structure enabling fast mass transport.

9.3.4 Molecular Catalysts

Molecular catalysts such as organometallic complexes, COFs, and MOFs (Nam, De Luna, et al., 2020) possess M–N active centers similar to those in M–N–C SACs. Moreover, molecular catalysts also include metal-free and polymerized structures. Generally, these catalysts have advantages such as availability of well-defined active sites, tunable steric and electronic properties, and promising activity and selectivity for CO_2RR. While transition metal porphyrin or phthalocyanine derivatives such as cobalt and nickel phthalocyanines (CoPc, NiPc) and cobalt tetraphenylporphyrin (CoTPP) (Zhang, Wu, et al., 2017) are commonly used as molecular catalysts for selective CO formation. Cu-based molecular catalysts such as the copper rubeanate metal-organic framework (CR-MOF) favor hydrocarbon production, i.e., CO_2-to-HCOOH conversion (Hinogami et al., 2012). Additionally, metal-free and polymerized catalysts such as polydopamine (Coskun et al., 2017), polyethylenimine (PEI) (Zhang, Kang, et al., 2014), B-graphene (Sreekanth et al., 2015), and N-doped carbon nanofibers (N-CNFs) (Kumar et al., 2013) were also reported to show outstanding CO_2RR performance. In this section, various molecular catalysts for CO_2 electrocatalysis are discussed.

9.3.4.1 Organometallic Complexes

Immobilizing molecular systems on conductive supports is a promising strategy to prepare CO_2RR-active molecular catalysts. Recently, Jia et al. (Jia, Ching, et al., 2020), immobilized Vitamin B_{12} (containing Co(II)) (Figure 9.6a) on the surface of reduced graphene oxide (rGO) and obtained a hybrid $VitB_{12}$@rGO catalyst for CO_2 reduction. In the unique structure of $VitB_{12}$, Co–N_4 centers, and redox-active benzimidazole are two possible active sites that are covalently linked to Co atoms. As a result, the CO_2-to-CO conversion was promoted with the hybrid $VitB_{12}$@rGO catalyst, as evidenced by the fact that the catalyst yielded a high FE_{CO} of 94.5%, j_{CO} of 6.24 mA/cm^2, a TOF of up to 28.6 s^{-1} and stability for over 10 hr of operation at −0.8 V vs. RHE. The maximum j_{CO} was 13.6 mA/cm^2 at −1.0 V vs. RHE. The TOF was also higher (52.4 s^{-1}) at this potential. Theoretical calculations showed that the Co–N_4 catalytic site would exhibit a low energy barrier for COOH* formation (0.13 eV) at −0.5 V (Figure 9.6b), which was the rate-limiting step for CO production. Interestingly, COOH* was bonded too strongly (by ~2 eV) with the truncated $VitB_{12}$ structure without the benzimidazole unit, indicating that this chemical bond was favorable to facilitate the two-electron reduction pathway in the unique structure of $VitB_{12}$. However, due to the steric hindrance from the neighboring hydrocarbon chains, COOH* binding was restricted for the reduction pathway at benzimidazole site, resulting in unstable COOH* attachment. Therefore, the Co–N_4 unit was regarded as the primary active site for CO_2 reduction.

Recently, various molecular complexes were developed to enhance CO_2RR performances, including CoPc (Ren, Joulié, et al., 2019), cobalt aminopyridine (Chapovetsky et al., 2016), polyoxometalate–metalloporphyrin (Wang, Huang, et al., 2018), tannin–lead(II) (TA–Pb) complex (Shi et al., 2020), Cu porphyrin (Weng et al., 2016), Ru pyridyl carbene (Gonell et al., 2019), Mn^{III}-corrole complex (De et al., 2020), and so forth. Among these organometallic complexes, TA–Pb complex (Shi et al., 2020), Cu porphyrin (Weng et al., 2016), and Mn^{III}-corrole complex (De et al., 2020) showed high selectivity and activity to the formation of CO_2RR products such as HCOOH, CH_4, C_2H_4, and CH_3COOH. This property of the TA-Pb complex, Cu porphyrin, and Mn^{III}-corrole complex was attributed, respectively, to the active phase transformation of TA-Pb complex to $Pb_3(CO_3)_2(OH)_2$ via $PbCO_3$, the built-in hydroxyl groups in porphyrin and copper(I) metal center, and the electroactive Mn^{II} species reduced from Mn^{III}. These active centers could bind to CO_2 and stabilize the reduced intermediates, thereby promoting CO_2 conversion. Additionally, an interesting work on iron tetraphenylporphyrin (FeTPP) showed that the phenolic groups introduced in all ortho and ortho′ positions (in the six-carbon atom of the benzene ring, take a carbon atom with a substituent as the standard, and the position of the carbon atom adjacent to the standard is called ortho position) of the TPP phenyl groups could significantly accelerate CO_2RR (Costentin et al., 2012). Because of the phenolic hydroxyl substituents facilitating high local protons concentration, a remarkable 50 million turnovers after

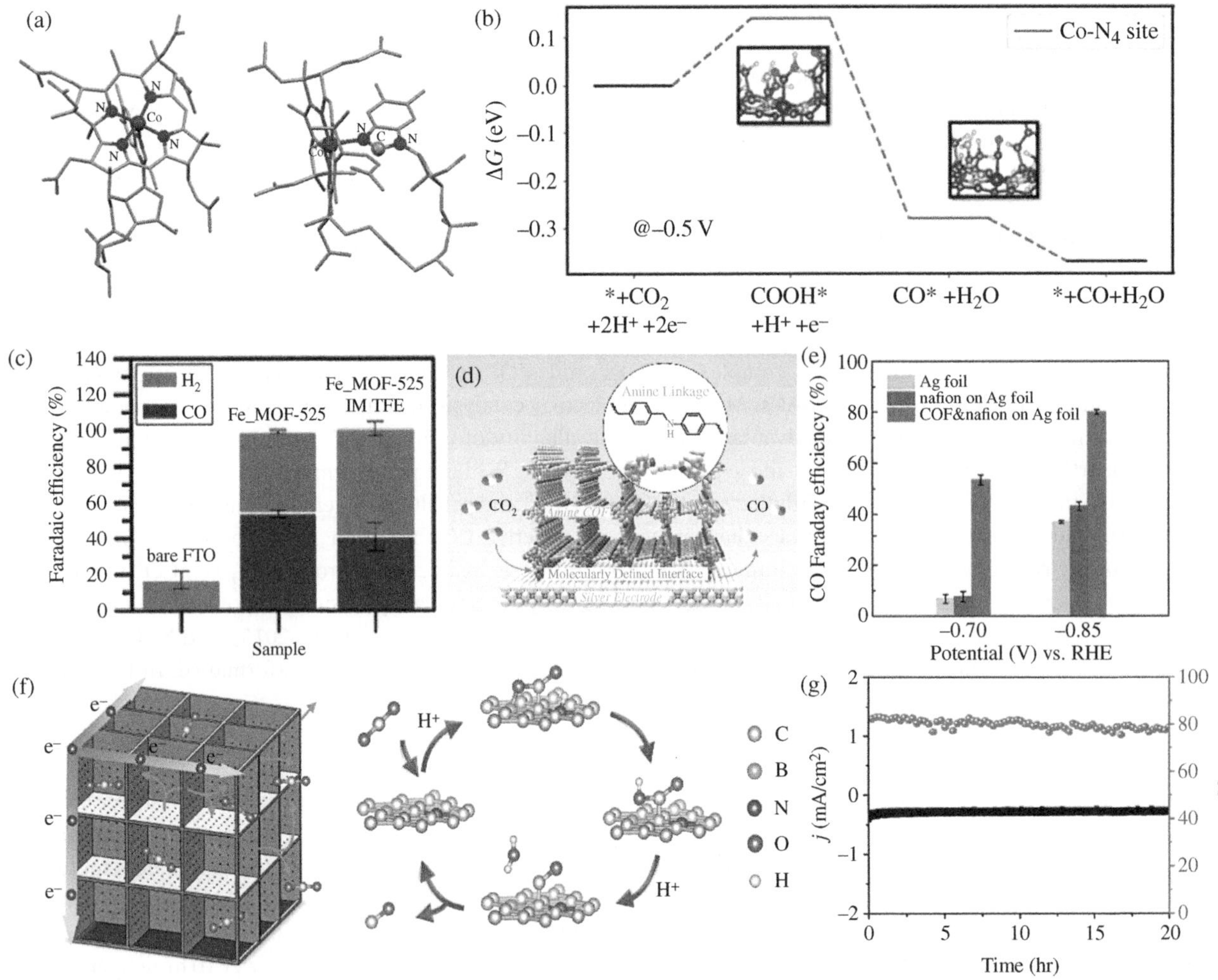

Figure 9.6 (a) Two different orientations of the relaxed structure of Vitamin B_{12} ($VitB_{12}$) (without a cyano unit), Co–N bonds from the central Co-corrin system and Co–N bonds between the Co-benzimidazole unit [(cobalt), (nitrogen), (oxygen), (phosphorus), and (carbon)]. Hydrogen atoms are removed for clarity. (b) Potential energy diagram of two-electron CO_2 reduction at the Co–N_4 site, showing low barrier to COOH* formation followed by a gradual downhill energy landscape favorably promoting CO formation. Source: Reproduced from Jia, Ching, et al., 2020, with permission. Copyright 2020, American Chemical Society/American Chemical Society. (c) Faradaic efficiency over approximately 4 h of electrolysis for Fe_MOF-525 without added TFE, Fe_MOF-525 with added TFE, and a bare FTO blank. Source: Reproduced from Hod et al., 2015, with permission. Copyright 2015, American Chemical Society/American Chemical Society. (d) Illustration of COF-300-AR with amine linkage for CO_2RR. (e) Faradic efficiency for CO production on the concerted electrode of COF-300-AR and Ag at various potentials. Bare silver foil and Nafion solution were used as controls. Source: Reproduced from Liu, Chu, et al., 2018, with permission. Copyright 2018, Elsevier. (f) Illustration of the CO_2RR process on NBPC catalysts. (g) Stability test of NBPC during a 20 hr electrolysis at −0.4 V vs. RHE. Source: Reproduced from Jia, Ren, et al., 2020, with permission. Copyright 2020, American Chemical Society/American Chemical Society.

4 hr of CO_2 catalysis, a high FE_{CO} of 94%, and an averaged current density of 0.31 mA/cm^2 at a low overpotential of 0.466 V were achieved by the electrogenerated iron(0) complex.

9.3.4.2 MOF and COF Catalysts

MOFs and COFs usually exhibit suitable structural properties such as high surface area and tunable porosity, which favor exposing abundant active sites, shortening the charge transfer distance, providing excellent physical and chemical stability, and improving their lifetime for applications in electrochemical reactions. These materials possess diverse metal and functional groups, exhibit confinement effects for the metal single atoms to prevent migration and aggregation, and have periodic metal arrangements such as the ordered cylindrical array of the π-conjugated system, which is conducive to electron delocalization and affords COFs' outstanding electron transport property (Diercks et al., 2018). Some MOFs and COFs display excellent CO_2RR performances due to the uniformly distributed metal active sites. A cobalt-porphyrin

MOF, $Al_2(OH)_2$TCPP-Co (TCPP-H_2 = 4,4′,4″,4‴-(porphyrin-5,10,15,20-tetrayl)tetrabenzoate) (Kornienko et al., 2015) was prepared by assembling the catalytic linker units cobalt-metalated TCPP into a porous MOF thin film. *In situ* spectra and electrochemical measurements indicated that Co(II) species were reduced to active Co(I) sites during CO_2 catalysis, which resulted in a high FE_{CO} of 76% and durability for over 7 hr of operation with a per-site turnover number (TON) of 1400 at −0.7 V vs. RHE. Another MOF catalyst, iron tetrakis(4-carboxyphenyl)porphyrin incorporated into MOF-525 (Fe_MOF-525), was also reported as a selective structure to CO formation (Hod et al., 2015). The MOF porous structure enabled facile accessibility to solvent, reactant, and electrolyte to the Fe catalytic sites. Moreover, the metalloporphyrinic linkers worked as electrocatalysts and redox-hopping-based conduits for delivering reducing equivalents to active sites that did not directly contact the underlying electrode. The electrode consisted of a Fe_MOF-525 film deposited on a fluorine-doped tin oxide (FTO) glass substrate. Consequently, Fe_MOF-525 added with a weak Brönsted acid (2,2,2-trifluoroethanol (TFE)) delivered a total FE of approximately 100% with 54 ± 2 and 45 ± 1% for CO and H_2, respectively. This material outperformed bare FTO glass electrode and Fe_MOF-525 containing no TFE (Figure 9.6c). Besides, the catalyst was stable enough to perform CO_2RR for over 6 hr with a constant current (~12 mA/cm^2).

Similar to MOFs, COFs with covalently connected structural linkages have also been recognized as active materials for CO_2-to-CO conversion. For example, a COF-367-Co COF (Lin et al., 2015) with cobalt porphyrin as building blocks exhibited a high FE_{CO} of 90% and TONs up to 290,000 at −0.67 V vs. RHE. This activity was 26-fold higher compared to the case of molecular cobalt complex. The material suffered no degradation when subjected to 24 hr of operation. The COF-367-Co was shown to influence the catalytic cobalt centers electronically and sterically, similar to the behavior of redox ligands in molecular systems (Eisenberg & Gray, 2011), thereby possessing high selectivity and activity. Another example is COF-300-AR COF structure containing amine linkage (Liu, Chu, et al., 2018). Compared with the bare Ag electrode, the COF-300-AR modified Ag electrode (Figure 9.6d) exhibited significantly increased FE_{CO} of 53% and 80% at −0.7 and −0.85 V vs. RHE (Figure 9.6e), respectively. This superior and selective activity promoting CO_2 reduction to CO came from the porous structure of COFs enabling fast molecular diffusion to the electrode surface and also from the amine groups facilitating carbamate intermediates formation.

9.3.4.3 Metal-Free and Polymerized Catalysts

Heteroatom doping is a promising strategy to modify carbon materials and enhance their CO_2RR catalytic performance. For example, a N and B co-doped porous carbon (NBPC), was prepared by the liquid nitrogen-assisted freeze drying of NaCl-glucose solution containing carbon, nitrogen, and boron precursors followed by a two-step solid pyrolysis process. The material exhibited a uniform 3D hierarchical structure and high doping level (N, 10.2 at.%; B, 8.86 at.%) (Jia, Ren, et al., 2020). The 3D porous structure provided abundant active sites with maximum exposure and facilitated ion transfer, the doped N atoms improved H* transfer and COOH* formation, and the doped B atoms favored formation of CO_2* and CO* intermediates (Figure 9.6f). Because of this superior structure and the synergistic activity of N and B atoms, NBPC showed a high FE_{CO} of 83% at −0.4 V vs. RHE and a durable CO_2 catalysis for over 20 hr (Figure 9.6g). The j_{CO} was far larger than that in the case of BPC and NPC over a wide potential range (from −0.3 to −0.8 V vs. RHE). Other atoms were also reported to be used as dopants in carbon structures to enhance CO_2RR significantly and these include F-C (Xie et al., 2018), N, F–C (Pan et al., 2019), and N, S-porous CNF (Yang, Wu, et al., 2018).

In addition to heteroatom-doped carbon catalysts, some polymers such as polypyrrole (Aydin & Köleli, 2004), polyaniline (Köleli et al., 2004), polydopamine (Coskun et al., 2017), fluoropolymer (Wicks et al., 2021), and PEI (Zhang, Kang, et al., 2014) have attracted much attention owing to their excellent activity and selectivity to CO_2RR. For example, polypyrrole was used to catalyze CO_2 reduction and produced HCHO, HCOOH, and CH_3COOH with maximum FEs of 1.8, 40.5%, and 62.2% at 20 bar, respectively (Aydin & Köleli, 2004). It was suggested that CO_2 adsorbed on the 0.3 μm polymer film via hydrogen bond facilitated the transfer of surface adsorbed H_{ad} to molecular CO_2, thus promoting CO_2 conversion. Polydopamine (Coskun et al., 2017) and PEI (Zhang, Kang, et al., 2014) were also reported to have remarkable performance for CO_2RR. Polydopamine displayed a $j_{CO+HCOOH}$ of 18 mA/cm^2 at 0.21 V overpotential for over 16 hr at high FE (>80%). This performance was originating from hydrogen-bonded catalytic sites repeating sequentially in amide-carbonyl groups. PEI (Zhang, Kang, et al., 2014) on N-doped carbon nanotube (PEI-NCNT) and that on N-doped graphenated carbon nanotubes (PEI-NGCNT) showed a high current density (7.2 and 9.5 mA/cm^2, respectively) and FE_{HCOOH} (85% and 87%, respectively), surpassing that on NCNT (3 mA/cm^2, 59%) and CNT (1.3 mA/cm^2, 5%). These suggested that PEI worked as cocatalyst to promote CO_2RR on NCNT. In addition, both PEI-NCNT and PEI-NGCNT were stable for over 24 hr of CO_2 electrolysis at −1.8 V vs. SCE with the latter one delivering a larger j_{total}, far surpassing the stability of NCNT and CNT. The improved performance of PEI was attributed to the H-bond interaction between NCNT-N-C(O)O·$^-$···H-N-PEI, which stabilized the CO_2·$^-$ intermediate, enriched CO_2 concentration and promoted CO_2-to-HCOOH conversion.

Following conclusions could be drawn from a comparison of different materials discussed in Table 9.3. Much progress has been made in the development of traditional catalysts (metals, metal compounds, MOF/COF, and their composites) in terms of low overpotential, high selectivity (close to unity), large current density, and long-term stability. However, the performance of these materials is challenged by inexplicit active sites present in the materials and limited atom utilization, which significantly hinder attaining an in-depth understanding of CO_2RR reaction process and developing high-performance catalysts. Compared with these materials, SACs are more intensively explored for CO_2RR. SACs exhibit remarkable electrochemical performance, due to their high catalytic activity, maximum metal atom utilization, well-defined and clearly identified active sites, inhibited HER, and favorable geometric and electronic effects. Some SACs facilitate remarkable CO_2RR performance, with 100% FE_{CO} (Ren et al., 2022), industry-compatible j_{CO} approaching 447.6 mA/cm^2 (Chen et al., 2021), high durability (over 200 hr) (Zu et al., 2019), as well as high selectivity (90.3%) to valuable chemicals such as acetate (Qiu et al., 2022).

The catalyst stability is an important parameter in the evaluation of CO_2RR performance. It is also the key to industrial-scale deployment. Metal electrodes such as noble metals normally provide great stability for CO_2 reduction at low overpotentials (Liu et al., 2017; Zhu et al., 2014). The low-cost metals such as Zn (Won et al., 2016), and Bi (Han et al., 2018) also deliver good stability to CO_2 reduction. Cu materials are unique for the production of multi-carbon products and hydrocarbons with good stability (Kas et al., 2014). Metal alloys (Au–Fe (Sun et al., 2017), BiSn (Yang, Wu, et al., 2020), Ag–Sn catalyst (Luc et al., 2017), Ag–Hg (Yang, Chen, et al., 2019), and CuAg (Hoang et al., 2018)) offer promising stability. Similar stability was detected for metal compounds (Mi et al., 2018). Although molecule catalysts suffer from poor electroconductivity and stability, these can be addressed by immobilizing the metal complexes onto conductive carbon supports (CNTs, carbon paper) and constructing MOFs/COFs that avoid segregation with the arrangement of metal complexes in a coplanar configuration and mitigate catalyst leaching. In most cases, these catalysts survive only for <100 hr, far below the practical requirements for technology commercialization. Intensive studies have been devoted to evaluating the long-time performance of SACs in CO_2RR. The superior stability of SACs originates from their strong metal-support interactions. The Ni SAC was stable over 120 hr of continuous CO_2 reduction with no decay in FE_{CO} and only slightly oscillating current density due to random perturbation (Liu et al., 2021). Another $Sn^{\delta+}$ single atom on N–GN displayed ultra-stability (over 200 hr of operation) without activity deactivation (Zu et al., 2019). The attention paid to the recyclability of SACs for CO_2RR is relatively low. The Cu–SA/NPC could be used only for five sequential CO_2RR experiments (Zhao, Nie, et al., 2020). Therefore, a balance between activity and stability of SACs should be achieved via rational design of supports, ligands, metal centers, etc.

9.4 *In Situ* Characterizations of Electrocatalysts for CO_2RR

The complex process in CO_2RR necessitates further exploration of reaction mechanisms along with the development of efficient CO_2RR electrocatalysts by investigating the catalyst structure, surface evolution, the key reaction intermediates, as well as active centers for CO_2 reduction. *In situ* and operando characterization techniques, such as *in situ* UV–vis, *in situ* Raman spectroscopy, *in situ* Fourier-transform infrared spectroscopy (FTIR), and operando X-ray absorption spectroscopy (XAS), have been developed to gain deep insights into the CO_2RR mechanism. In this section, *in situ* characterization techniques used for CO_2RR are introduced.

9.4.1 *In Situ* Raman

Raman spectroscopy provides valuable information about the structural fingerprints of surface-adsorbed species, which can be employed to investigate the chemical state of a catalyst surface during CO_2RR. Yang et al. (Yang, Zhang, et al., 2020) applied *in situ* Raman spectroscopy to investigate the origin of the excellent activity of multi-hollow Cu_2O catalyst and corresponding selectivity toward C_{2+} products. The *in situ* Raman spectroscopy was operated with the 532 nm laser excitation using a flow cell with 2 M KOH electrolyte at −0.61 V vs. RHE. The multi-hollow Cu_2O presented peaks at 223, 419, 524, and 624 cm^{-1} (Figure 9.7a), similar to those of solid and fragmental Cu_2O, which originated due to $2\Gamma^-{}_{12}$, $4\Gamma^-{}_{12}$, $\Gamma^+{}_{25}$, and $\Gamma^-{}_{12} + \Gamma^+{}_{25}$ phonon modes, respectively. In addition, the multi-hollow Cu_2O retained the characteristic $2\Gamma^-{}_{12}$ mode after 20 min of CO_2 catalysis, while the Raman signals for solid and fragmental Cu_2O (Figure 9.7b) disappeared rapidly within 2 min, indicating that Cu^+ species were protected against reduction by the confined geometry of multi-hollow Cu_2O.

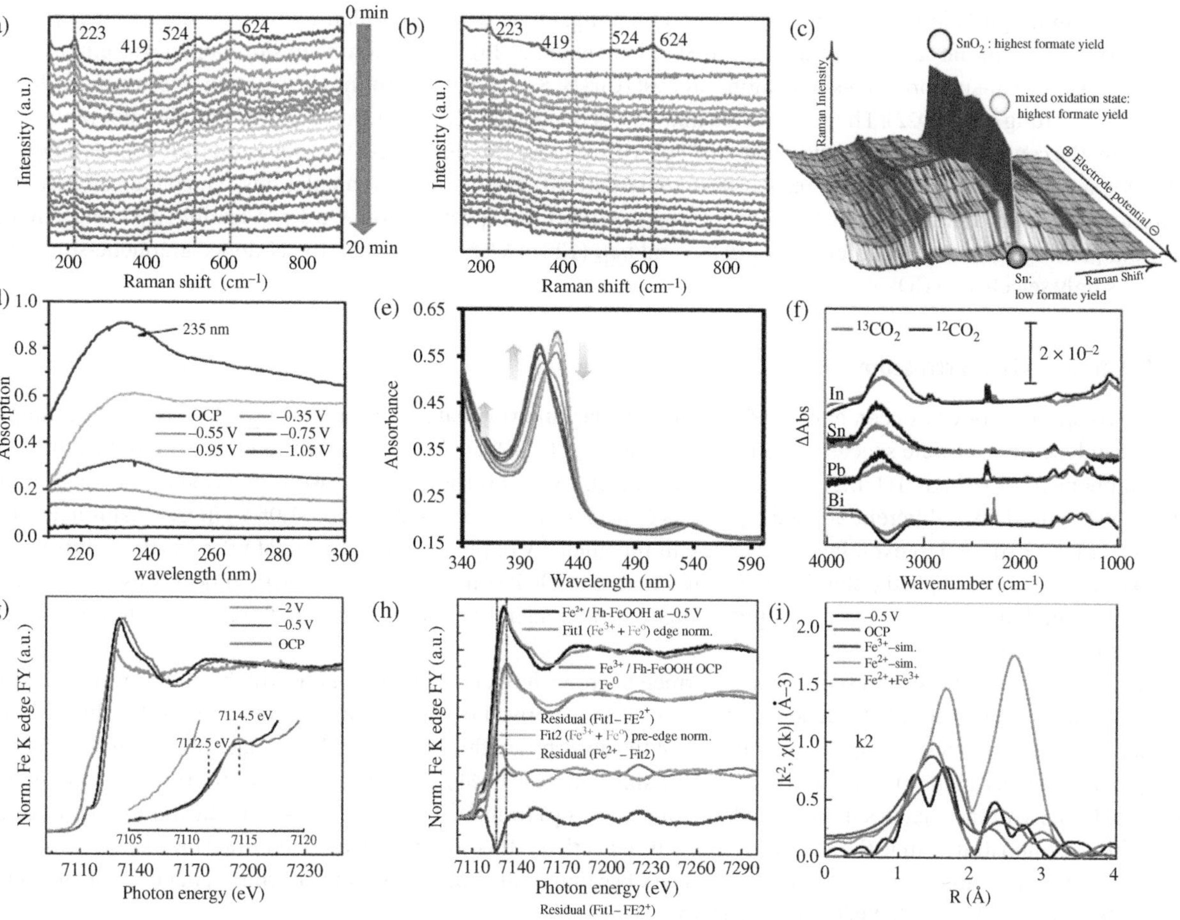

Figure 9.7 Operando Raman spectra of (a) multi-hollow and (b) fragmental Cu_2O as a function of reaction time at −0.61 V vs. RHE as a function of reaction time at −0.61 V vs. RHE. Source: Reproduced from Yang, Zhang, et al., 2020, with permission. Copyright 2020, American Chemical Society. (c) *In situ* Raman of SnO_2. Source: Reproduced from Dutta et al., 2015, with permission. Copyright 2015, American Chemical Society. (d) The *in situ* ultraviolet absorption spectroscopy during CO_2RR process for Bi-rGO. Source: Reproduced from Jiang, Wang, Xiao, et al., 2020, with permission. Copyright 2020, Elsevier. (e) *In situ* UV of the $[Al_2(OH)_2TCPPCo]$ MOF film under various potentials from 0.2 to −0.7 V vs. RHE. Source: Reproduced from Kornienko et al., 2015, with permission. Copyright 2015, American Chemical Society. (f) ATR-IR spectra of thin films of In, Sn, Pb and Bi on exposure to CO_2 under reducing conditions. Source: Reproduced from Pander III et al., 2016, with permission. Copyright 2016, American Chemical Society. (g) FY Fe K edge XANES spectra during cyclic voltammetry. (h) Normalized Fe K edge spectra at −0.5 V in 0.05 M $KHCO_3$. (i) Fourier transform EXAFS spectra. Source: Reproduced from Genovese et al., 2018, with permission. Copyright 2018, Springer Nature/CC BY 4.0.

In another work, potential- and time-dependent *in situ* Raman spectroscopy was employed to monitor the changes in the oxidation state of SnO_2 during CO_2-to-HCOOH conversion (Dutta et al., 2015). At moderately cathodic potentials, a high $FE_{formate}$ was exhibited by the catalyst due to the SnO_2 phase remaining at surface. The peaks were identified as E_g, A_{1g}, and B_{2g} bands at 482, 623, and 762 cm^{-1}, respectively. However, when the surface SnO_2 was reduced to metallic Sn at very negative potentials, $FE_{formate}$ decreased significantly (Figure 9.7c). Recently, Pan et al. (Pan et al., 2022) employed *in situ* Raman to correlate the stability and the mechanism of CO_2-to-formate conversion performed by the multi-layer SnO_x. During CO_2RR process, the Sn(IV) was first reduced to Sn(II), which served as the active site to promote CO_2 reduction through a proton-coupled electron transfer for formate production. The possible further conversion of Sn(II) species to metallic Sn was suppressed by the multi-layer structure, which helped the material overcome the oxide layer attenuation, thus improving the durable selectivity for formate. The spontaneous oxidation of Sn(II) to Sn(IV) species was observed at the end of electrolytic reaction. The effective reversibility of the Sn species by the multi-layer structure contributed to the excellent reproducibility of CO_2RR, achieving high $FE_{formate}$ (93.2%) and stability (>40 hr) at −1.15 V vs. RHE.

In situ Raman was effectively used for probing the actual valence states of the metal elements in the BiCu nanosheet single atom alloy catalyst during CO_2-to-HCOOH conversion (Luo et al., 2021), monitoring the state change of Ni in nickel(II)

2,9,16,23-tetra(amino)phthalocyanine during CO_2RR (Liu, Yang, et al., 2020), and identifying the binding energy of the CO_2RR intermediates on the catalyst surface (Nam, Shekhah, et al., 2020). Very recently, *in situ* Raman was employed to track the synthesis of Cu-supported Fe-single-atom catalyst (Cu–Fe SA) and to determine the reaction intermediates during CO_2 electrolysis (Hung et al., 2022). They discovered that Fe—N bond would break during CO_2RR. The phthalocyanine ring detached from the central Fe ions and converted to Cu-FeSA. Cu-FeSA exhibited strong *CO adsorption on Fe sites following the reduction reaction. The *in situ* Raman results were consistent with the theoretical calculations, which indicated that the Fe atoms attracted surrounding intermediates and promoted CO_2-to-CH_4 conversion. Cu-FeSA exhibited 32 times higher activity than pristine Cu, delivering high FE_{CH4} (64%) and j_{CH4} (128 mA/cm^2); however, the N-doped graphene-supported material was only selective to CO.

9.4.2 *In Situ* UV–vis Spectroscopy

In situ UV–vis spectroscopy has been employed to detect the key intermediates during the CO_2RR process. For example, the electrocatalytic intermediate species were identified during CO_2RR catalyzed by a hybrid composite of bismuth and reduced graphene oxide (Bi-rGO) (Jiang, Wang, Xiao, et al., 2020). The absorption signals were collected first at open circuit potential (OCP) and then at different applied potentials (−0.35, −055, −0.75, −0.95, and −1.05 V) in CO_2-saturated 0.1 M $KHCO_3$. The Bi-rGO(p) catalyst exhibited an increase in the intensity of the absorption peak at near 235 nm featuring the $CO_2^{\cdot-}$ intermediate under applied potentials compared to that at OCP condition. This absorption peak was also stronger for Bi-rGO(p) than that of Bi(p) at the same applied potential, demonstrating the existence of more $CO_2^{\cdot-}$ intermediates on Bi-rGO(p) (Figure 9.7d).

Very recently, *in situ* UV–vis spectroscopy was performed to probe the oxidation state changes of the multi-layer SnO_x after CO_2 electrocatalysis (Pan et al., 2022). Generally, the defects were derived from the reduction in metal oxide. The decrease in the defect concentration resulted in the attenuation of the absorption, thus significantly stronger peak between 350 and 800 nm was observed between 0.02 and −0.23 V. The blue shift from 450 to 440 nm indicated more absorption attenuation at 350–450 nm, because of the changes in intershell transitions (s → p) from $O^{2-} \rightarrow Sn^{II}$ to $O^{2-} \rightarrow Sn^{IV}$ ligand-to-metal charge transfer. These demonstrated that the O^{2-} reinsertion in defective lattice of SnO_x facilitated Sn(II) transformation into Sn(IV) species. At a potential of −0.28 V, the reverse peak became weaker and shifted to contrary tendency concurrently, because of the Sn(IV) reduction to Sn(II) species again.

In situ UV–vis spectroscopy is also promising in characterizing the properties of the redox species during CO_2RR process. For example, Co(II) was reduced to Co(I) during CO_2RR catalyzed by COF-366-Co (Lin et al., 2015), as confirmed by the weaker Co(II) Soret band at 422 nm accompanied with the Co(I) Soret band increase at 408 nm upon varying the reduction potentials. Similarly, in another study, the [$Al_2(OH)_2$TCPPCo] MOF film presented typical Soret ($S_0 \rightarrow S_2$) and Q ($S_0 \rightarrow S_1$) bands at 422 and 530 nm at OCP conditions, respectively (Kornienko et al., 2015). The intensity of Co(II) Soret band at 422 nm decreased with varying the potential from 0 to −0.7 V vs. RHE, while the Co(I) Soret band at 408 nm became stronger (Figure 9.7e), indicating the reduction of Co(II) to Co(I) during CO_2 electrolysis.

In situ UV–vis spectroscopy was powerful in elucidating the active sites of Fe(II) species to accommodate CO_2-to-CO conversion as Fe(II)-CO adducts, which were the predominant molecular species (Lu et al., 2021). It helped also in identifying the origin of the outstanding CH_4 selectivity (78%) catalyzed by the single-atom Cu-embedded carbon dots (Cu-CDs) (Cai et al., 2021). The single Cu sites in the Cu-CDs were active for CO_2-to-CH_4 conversion, as no shift in *in situ* UV–vis spectra compared with those of CDs was observed upon decreasing the potential from OCP (0.86 V vs. RHE) to −1.64 V vs. RHE. This agreed well with the DFT calculations, which demonstrated that the unique CuN_2O_2 active sites altered the electronic structure of Cu sites and improved the energy barrier of CH_4 and H_2.

9.4.3 *In Situ* FTIR Spectroscopy

In situ ATR-IR (attenuated total reflection-infrared) spectroscopy is generally sensitive to the molecular vibrations of surface-adsorbed species, which helps to understand the CO_2RR process. For example, Pander et al. (Pander III et al., 2016) applied *in situ* ATR-IR spectroscopy to probe the mechanism of CO_2RR by metal electrodes made of In, Sn, Pb, and Bi. All the metal electrodes exhibited gaseous CO_2 peaks at 2350 cm^{-1} in the *in situ* ATR-IR spectra. Specifically, surface-confined In carbonate and Sn carbonate species at 1500 and 1385 cm^{-1} were detected, respectively. However, no absorption peaks were observed for Pb carbonate (1410 cm^{-1}) and Bi carbonate (1389 cm^{-1}) (Figure 9.7f), probably due to the direct reduction of CO_2 at the metallic sites of Pb and Bi electrodes. Recently, Fang et al. (Fang et al., 2022) performed *in situ* FTIR (Fourier transform infrared) spectroscopy to investigate the CO_2RR progress on Zn nanosheets (NSs), Zn NPs, and Zn foil. The peaks at 1391 and 1421 cm^{-1} were assigned to the C–O stretch of COOH and COO–, respectively. The C=O stretch of COOH at 1613 cm^{-1} indicated the reduction of CO_2 and further hydrogenation to the COOH* intermediate. Peaks for the H–O–H bending of water molecules, surface-bonded Zn–CO, and the adsorbed CO_2 were detected at 1641, 1944, and 2048, and 2335 cm^{-1}, respectively. The intermediate peak became stronger with decreasing negative potential. Especially, Zn NSs displayed stronger COOH and COO^- peaks than Zn NPs and Zn foil, indicating their more effective ability to stabilize the intermediate COOH* and facilitate CO production. The inferior electrocatalytic performance of Zn NPs and Zn foil was attributed to the active site occupation by the adsorbed carbonate anions (1538 cm^{-1}). Upon further CO_2RR operation, Zn NSs exhibited intermediately strengthened COOH* signals, which demonstrated the CO_2 reduction on Zn NSs surface. Impressively, the coordinatively unsaturated pit sites derived from the in situ structural reconstruction of $Bi(OH)_3$ NS were identified by *in situ* FTIR (Yuan et al., 2022), which facilitated stabilizing the intermediate *OCHO to improve the reaction kinetics for formate production. Therefore, the reconstructed Bi NS outperformed intact Bi NS with improved catalytic activity, delivering a 2.6-fold larger current density, high FE_{HCOO-} (>95%), extraordinary turnover frequency (0.35 s^{-1} at −0.98 vs. RHE), and stability (over 10 hr) in H-cell, industrial-scale j_{HCOO-} (325 mA/cm^2) in flow cell, and ultrahigh durability (over 110 hr of operation at 200 mA/cm^2) in MEA.

In situ FTIR spectroscopy was reported to be powerful in identifying the CO_2 reduction pathways created by Ni/NCTs-50 SAC (Hou et al., 2020), PyNiPc/CNT (Ma et al., 2020), Fe–N–C SAC (Qin et al., 2019), ZrO_2@Ni-NC (Wang et al., 2021), $[Cu_4ZnCl_4(btdd)_3]$ (Zhu, Huang, et al., 2021), $[MnBr(2, 2'\text{-bipyridine})(CO)_3]$ (Reuillard et al., 2017), and COF_{bpyMn} (Dubed Bandomo et al., 2021). *In situ* ATR-IR spectroscopy was also employed to reveal the key intermediate during CO_2RR catalyzed by F-doped $Ni–N_4$ sites on carbon nanosheets (Han et al., 2021). The detected peaks at 1400 and 1630 cm^{-1} were ascribed to the C=O stretching of *COOH and the consumption of H_2O, respectively. The intensity of these peaks increased gradually by applying more negative potentials and reached the maximum at −0.97 V, demonstrating that the *COOH intermediate formation was the RDS for CO_2RR. Similarly, in another study, *in situ* ATR-IR spectroscopy was applied to probe the reaction intermediates of CO_2RR catalyzed by single atom $Sn^{\delta+}$ on N-doped graphene (Zu et al., 2019). The IR peaks of $CO_2{\cdot}–$ groups (1273 and 1694 cm^{-1}) and $HCOO^-$ (1538 cm^{-1}) were gradually strengthened with the electrolysis time. The stronger $HCOO^-$ IR peak was caused by the favorable proton-trapping capacity of $CO_2{\cdot}^-$ radicals. Besides, the peaks of monodentate carbonate (m-CO_3^{2-}) groups (1460 and 1520 cm^{-1}) and the asymmetric OCO stretches of b-CO_3^{2-} groups (1648 cm^{-1}) were detected. The HCO^{3-} band located at 1415 cm^{-1} was also intensified with CO_2 electrocatalysis time, due to its adsorption on catalyst surface in $KHCO_3$. Therefore, possible reduction mechanism on $Sn^{\delta+}$ SAC was proposed as the following: ($CO_2(g) + * \rightarrow CO_2*$, $CO_2* + e^- \rightarrow CO_2{\cdot}–*$, $CO_2{\cdot}–* + HCO^{3-} + e^- \rightarrow HCOO–* + CO_3^{2-}$, $HCOO–* \rightarrow HCOO^- + *$). Recently, *in situ* ATR-FTIR spectroscopy was applied to ascertain the C–C coupling mechanism for CO_2 conversion to multi-carbon products on nafion-coated functionalized multi-wall carbon nanotubes (Fe-n-f-CNTs) at −0.8 V vs. RHE in CO_2-saturated in 0.5 M $KHCO_3$ (Lakshmanan et al., 2022). Prominent electrolyte peaks were observed at 1633 and 1361 cm^{-1}, which were assigned to CO_3^{2-} and HCO_3^-, respectively. Besides, COO^- peak (1411 cm^{-1}) was detected, which was the key intermediate for *CO formation and further to C–C coupling.

9.4.4 Operando XAS

XAS, including X-ray absorption near-edge structure (XANES) and extended X-ray absorption fine structure (EXAFS), is capable of characterizing the electronic states and fine structures of the catalysts. The XANES probes chemical states and the electronic structure of target atoms, while EXAFS provides information about coordination configurations and chemical bond lengths of probing atoms. Therefore, XAS is widely used for the *in situ* characterizations of the CO_2RR electrocatalysts. For example, Yang et al. (Yang et al., 2022) developed operando electrochemical resonant soft X-ray scattering (EC-RSoXS) to characterize the chemical identity of active sites during the structural transformation of Cu NP ensembled through 1 μm thick electrolyte. Operando scattering-enhanced XAS was able to simultaneously study the size-dependent catalyst stability under beam exposure and monitor the potential-dependent surface structural changes. Small NPs (7 nm) exhibited an obvious oxidation to CuO induced by soft X-ray beam despite only sub-second X-ray exposure, while large NPs (18 nm) with stronger beam damage resistance allowed reliable detection of electroreduction of surface Cu_2O to metallic Cu. The particle–particle interactions of large ensembles of NPs were statistically probed using small-angle X-ray scattering (SAXS). The technique used in this work was crucial in elucidating the property of the active site via investigating the chemical environment and interparticle information in nanocatalysts.

Efforts were devoted to performing *in situ*/operando XAS on molecular catalysts to obtain an in-depth mechanistic understanding (Zhang, Wang, et al., 2020), determine the coordination environment and electronic structure (Liu, Deb, et al., 2020), and elucidate the working states of Fe-based molecular catalysts during CO_2RR process (Lu et al., 2021).

One important application of XAS experiments was on studying the stability of molecules during CO_2 reduction (Boutin et al., 2019). It was revealed that the active sites of Cu molecular species evolved into NPs under CO_2RR conditions. For instance, Cu(II) phthalocyanine underwent structural and oxidation state changes to form ~ 2 nm metallic Cu clusters, the active sites for CO_2-to-CH_4 conversion, which were reversibly converted to the original structure after CO_2 electrolysis, as detected by *in situ* XAS (Weng et al., 2018).

Operando XAS was found to be powerful to monitor the structural evolution of the NiN_4 sites (Yang, Hung, et al., 2018), Fe-N sites in Fe^{3+}–N–C SAC (Gu et al., 2019), Fe-N sites in Fe-PANI SAC (Leonard et al., 2018), and Sb single atom on N-doped carbon nanosheets (Jiang, Wang, Pei, et al., 2020) during CO_2RR. It was also useful for probing the active site in the Ni-TAPc SAC (Huang et al., 2019), which helped to understand the mechanism of CO_2 electroreduction. For example, Gu et al. (Gu et al., 2019) employed operdando XAS to track the structural evolution of iron species in Fe^{3+}–N–C SAC during CO_2RR process. The Fe K-edge XANES spectra exhibited no obvious shift between the in situ and the as-prepared Fe^{3+}–N–C samples at −0.4 V vs. RHE. The *in situ*-measured Fe^{3+}–N–C sample displayed a similar edge energy to that of Fe^{3+}TPPCl, suggesting that iron ions had invariant +3 oxidation state during the CO_2RR at −0.4 V vs. RHE. However, the Fe K-edge shifted to lower energies similar to that of FeO at a more negative potential (<−0.5 V vs. RHE), indicating the reduction of Fe^{3+} to Fe^{2+}. This reduction was accompanied by a change in the first shell coordination number from 4 to 3, as confirmed by the fitted EXAFS spectra. The coordinated Fe^{3+}-pyrrolic N_4 center facilitated fast CO_2 adsorption and COOH intermediate formation, thereby promoting CO_2-to-CO conversion with a high FE_{CO} of 90% at −0.45 V vs. RHE.

While in the above study, Fe^{3+} reduction to Fe^{2+} was detected by operando XAS, in a subsequent work the oxidation of Fe atoms during CO_2RR was probed (Lakshmanan et al., 2022). Very recently, *in situ* XAS was applied to reveal the dynamics of Fe single-atom active sites in Fe-n-f-CNTs. The oxidation state of Fe atoms in Fe-n-f-CNTs remained unchanged, as the XANES spectra of Fe K-edge exhibited no distinct energy shift up to −0.1 V vs. RHE, similar to Fe_2O_3. This indicated no CO_2RR engagement with Fe active sites and a stable local coordination environment at this initial potential. Given a more negative potential −0.2 V vs. RHE, the XANES spectra evolved close to FeO and remained no changes until −0.6 V vs. RHE, which confirmed the +2 oxidation state of Fe single atoms during CO_2RR. Upon applying a potential in the range −0.7 to −1.0 V vs. RHE, Fe-n-f-CNTs displayed a XANES spectrum close to Fe_2O_3, indicating that Fe atoms were oxidized to a higher state (>Fe^{+2}). These higher oxidation states of Fe SA, Fe–$(O)_3$, are more active for CO intermediates formation, which facilitated the generation of multi-carbon products. Therefore, the Fe SAC showed high $FE_{ethanol}$ (45% at −0.8 V vs. RHE) and yield rate (56.42 μmol/h) which was seven-fold higher than the metal-free samples (f-CNTs, n-f-CNTs). The material was found to be stable for over 10 hr of operation because of the Fe SA stabilization by carboxylic groups on -n-f-CNTs during the structural deformation at Fe–$(O)_3$ due to *H on SO_3. This work paved the way to design distinct single-atom coordination configurations for improving the performance of CO_2 conversion into multi-carbon products.

Using operando XAS, Genovese et al. (Genovese et al., 2018) investigated the relationship between the high $FE_{acetic\ acid + formic\ acid}$ (97.4% at −0.5 V vs. RHE) and the N-coordinated single atom or polyatomic Fe species. Fe K edge XANES of Fe/N–C showed that the pre-edge resonance decreased at 7114.5 eV by 2 eV at −0.5 V, which was fitted with the spectra of Fe(III) and Fe^0 components (Figure 9.7g). The negative residual intensity of the pre-edge was assigned to the down-shift in the Fe(II) region, while the positive residual demonstrated overestimated contribution by Fe^0 (Figure 9.7h). EXAFS spectra of Fe/N–C displayed a characteristic bimodal distribution of Fe-(O, OH) bonding lengths, attributed to the reduction of some Fe(III) species to Fe(II) (Figure 9.7i). This work indicated a high activity and selectivity by Fe/N–C catalyst originated from the synergistic effect of small *Fh*-FeOOH clusters or Fe single atoms and the carbon surface chemistry. The coordinated Fe(II)–N sites contributed to CO_2-to-acetic acid conversion with a high $FE_{acetic\ acid}$ of 60.9% at −0.5 V vs. RHE. Similarly, operando XAS was utilized to study the relationship between the activity (FE_{CO} 93% at −0.4 V vs. RHE) and the role of monovalent Fe(I) site in a single-Fe-atom catalyst (Li, Zeng, et al., 2021). It was also used to reveal the real active sites in a series of molecularly dispersed electrocatalysts (Zhang, Wang, et al., 2020), to help explore the origin of the excellent CO_2RR performance and design more effective catalysts.

9.5 Summary and Perspectives

Electroreduction of CO_2 is a promising approach to convert CO_2 to value-added chemicals and fuels, which could reduce the dependence on traditional fossil fuels and solve environmental issues. However, the industrial application of CO_2RR has not yet been completely realized, mainly due to the lack of highly efficient electrocatalysts and electrolyzers. Guided by the fundamentals of CO_2RR (i.e., reaction mechanism, various types of electrochemical cells, electrolytes), many efforts have been devoted to developing electrocatalysts for CO_2RR. In this chapter, we summarized recent progresses in developing effective CO_2RR catalysts, including metals, metal compounds, SACs, and molecular catalysts, along with a description of the advanced *in situ* characterization techniques (*in situ* Raman, *in situ* UV–vis, *in situ* FTIR, operando XAS) used to investigate the catalyst electronic structure evolution, active sites/species, and reaction mechanisms during CO_2RR process at atomic level.

9.5.1 Challenges for CO_2RR

Although considerable efforts have been invested so far to enhance the efficiency of CO_2RR, substantial challenges still remain. First, a CO_2 molecule is very inert and thermodynamically stable because of its strong C=O bond (806 kJ/mol). The high CO_2 activation barrier and the reaction pathways including multi-electron–proton transfers for generating various products (C_1, C_2, C_3 species) lead to sluggish kinetics, high overpotentials, and poor selectivity to desired products. Moreover, the formation of the key intermediate ($CO_2^{\cdot-}$) at the first electron transfer from CO_2 molecule ($CO_2 + e^- \rightarrow CO_2^{\cdot-}$) involves a large reorganizational energy (−1.9 V vs. SHE), which imposes high overpotential and slow reaction rates. Additionally, the competing side reaction HER generates H_2 as an important by-product competing with CO_2 reduction. Efficient CO_2RR is largely limited by insufficient intrinsically active and selective electrocatalysts. Most importantly, CO_2RR is far from commercialization, due to the low CO_2 solubility in aqueous electrolytes, lack of durable catalysts, insufficient study on CO_2 and electrolyte mass transfer, and other aspects including reaction volume, pressure, temperature, pH, products separation and collection. To boost commercial CO_2RR, several strategies are promising. These include designing highly selective, active, and durable catalysts, applying ILs with high CO_2 solubility to enrich CO_2 concentration, using suitable reactors (flow cells, MEA), and designing economic CO_2 reduction systems (Chang et al., 2021; Jouny et al., 2018; Orella et al., 2020; Verma et al., 2016). Many efforts have been devoted to achieving commercial CO_2RR. For instance, Wen et al. (Wen et al., 2022) designed a flow-through induced dynamics triple-phase boundaries (FTDT) cell using forced convection of CO_2-saturated aqueous electrolyte through porous electrode. The design improved CO_2, electrons, protons, and products transfer by the in situ formed CO_2(g)-liquid-catalyst interfaces, and yielded a high current density (3.37 A/cm^2). Large-scale CO_2RR was demonstrated with a stable CO yield in a 100 cm^2 FTDT cell (23.4 hr^{-1} for 50 hr), and also in a 4 × 100 cm^2 electrolyzer stack (90.6 ± 4.0 l hr^{-1} in the initial 2 hr).

9.5.2 Comparison with HER

To better understand the development of CO_2RR catalysts, a comparison of different types of catalyst materials and their applications for CO_2RR electrolysis and HER is given in Table 9.4. HER is a competitive side reaction to CO_2RR and a well-established strategy in water splitting. We highlight challenges that should be addressed to advance the electrocatalysts and realize practical CO_2RR in future.

Considering the importance of HER side reaction competing with CO_2RR at cathode as well as the OER process at anode, both catalyst development and understanding the mechanism of CO_2RR could be boosted by the development of efficient catalyst materials and electrolytic systems for electrochemical water splitting reactions. Almost all of the discussed categories of electrocatalysts in this section have promising activity toward HER and OER processes. This implies the importance of the rational design of catalyst structures, specifically transition metal nanostructures, heterostructures, alloy structures and multicomponent structures including molecular building blocks, to reach high intrinsic activity and product selectivity in different reactions related to CO_2RR at similar electrolytic conditions. To achieve efficient water splitting catalysts, the following aspects should be considered seriously: the impacts of size, shape, morphology, intermetallic synergy, hetero-phasic charge transfer, and surface properties in 2D transition metal-based nanostructures, 3D metal oxide-based nanostructures, alloy-based multiphase catalysts, and MOF-derived SACs. Furthermore, the cell design has a significant impact on the state of both cathodic CO_2RR and anodic OER reactions in an electrochemical CO_2RR cell. The advanced flow and stack cell designs promote the efficiency of water splitting process also. Hence, proper cell design could benefit both CO_2RR and water electrolysis technologies.

9.5.3 Perspectives for CO_2RR

Regarding the existing innate limitations and challenges for practical CO_2RR, herein, we suggest the following perspectives toward fulfilling the desired efficiency in CO_2RR:

Table 9.4 Comparison of the Properties of Catalysts for CO_2RR and HER

Materials	CO_2RR	HER
Noble metals	√ Strong *COOH binding energy but weaker *CO binding energy. √ Possibility of altering the catalyst surface structures (particle size, lattice strain, facets) and compositions. √ Size/shape/GB-dependent performance. × Expensive. × Unsatisfactory catalytic activity. × Limited surface area.	√ Excellent catalytic activity. × Expensive.
Transition metals	√ Cheaper than noble metals. √ Possibility of manipulating the morphology and structure. Easy fabrication. √ Size/shape/GB-dependent performances. √ Weak CO_2 adsorption. × Only Cu metals are specific for multi-carbon hydrocarbon production. × Unsatisfactory catalytic activity.	√ Cost-effective. √ Abundant. × Hard to develop both active and stable catalysts to meet the industrial demands of water electrolysis.
Metal alloys	√ Could engineer the composition and morphology. √ Reduced cost. √ Could manipulate the activity, selectivity, overpotential, and FE. √ Multiple binding sites for reduction intermediates. √ Close proximity and aggregation of different phases may cause variation in the activity. × Reduction in FE with decreasing potentials (e.g., CuSn) and unsatisfactory partial current density (e.g., NiGa).	√ Tunable physicochemical properties of the host metal or the main catalytic center. √ Synergistic effects (electronic, and strain effects) of different metallic phases. √ Elevated catalytic performance, particularly the stability in acidic environments. × Hard to balance the precious metal mass loading and catalytic performance.
TMDs	√ Low cost. √ Good electrical conductivity. √ Layered structures. √ Prominent active surface. √ High specific capacity. √ Defect-rich species. × Insufficient understanding of the effects of catalyst structure caused by S or Se. × Poor selectivity to other products (CH_4, CH_3OH, C_2H_4, EtOH) besides CO and HCOOH.	√ Superior catalytic performance. √ Unique electronic states and chemical properties. √ Large anions and suitable bandgap for metal selenides. × Insufficient understanding of the catalytic mechanism by TMDs with defects. × Relatively unstable because of the over-oxidization and severe corrosion under high applied potentials.
Metal oxides	√ Could alter the binding nature of active sites. √ Could change the properties of active sites (GBs, oxygen vacancies). √ Enhanced selectivity and activity at improved conditions such as low overpotential. √ Could engineer the oxide surface. × Poor electrical conductivity.	√ Low cost, abundant. √ Favorable oxidation states for redox reactions. × Limited surface area. × Poor conductivity. × Remote possibility for a large-scale application. × Poor activity and stability in acidic solutions. × The trade-off between activity and stability is difficult to balance.

(Continued)

Table 9.4 (Continued)

Materials	CO_2RR	HER
Metal nitrides	√ High selectivity to multi-carbon hydrocarbons (e.g., Cu_3N). √ Increased surface basicity (e.g., Ni_3N). × Generally unstable in aqueous electrolytes.	√ Precious-metal-like performance. √ Metallic conductivity. × A poor understanding of the modulation mechanism for enhanced activity. × Hard to establish large-scale and reproducible synthesis of metal nitrides with controlled morphology and precise composition.
Metal hydroxides	√ Selective to multi-carbon hydrocarbon production. × Low electrical conductivity. × Limited studies.	√ High intrinsic catalytic activity. √ Enhanced active surface area. √ Facile charge transfer. × Low electrical conductivity.
SACs	√ Outstanding catalytic performance. √ High metal utilization. √ Strong atom-support interactions. √ Low coordination number. √ Low-cost and diverse precursors (calcinated M–N–C SAC). √ Large specific surface area and 2D structure (graphene-supported SAC). √ Well-defined pore structure, controllable shape and size of MOF-derived carbon matrix, and availability of various metal nodes and organic linkers (MOF-derived SAC). × Migration, agglomeration, and sintering of particles/metals, low metal loading, and difficulty in large-scale production. × Less controllable (calcinated M–N–C SAC). × Low metal loading and time-intensive to graphene growth (graphene-supported SACs). × Complicated synthetic process (MOF-derived SAC).	√ Well-defined active sites. √ High activity. √ Uncoordinated environments. √ Maximized utilization of metal centers. × Difficulty in large-scale production (aggregation, migration).
Metal complexes	√ Superior activity and selectivity. √ Precisely tunable. √ Well-defined active sites. √ High elemental abundance. √ Various metal centers, ligands, and supports. × Insufficient conductivity. × Restricted mass transport due to catalyst aggregation.	√ High elemental abundance. √ High activity. × Hard to design a stable and efficient metal complex for water splitting that operates at high catalytic turnover number and frequency.
MOF and COF catalysts	√ High surface area. √ Tunable porosity. √ Diversity in metal and functional groups, confinement of metal atoms to prevent migration and aggregation. √ Periodic metal arrangements to afford superior electron transfer properties. × Complicated preparation process. × Poor conductivity.	√ Tunable porous structure and high internal specific surface area with abundant active sites. × Poor electrical conductivity and chemical stability.

(Continued)

Table 9.4 (Continued)

Materials	CO_2RR	HER
Metal-free catalysts	√ Enhanced performance by the introduced heteroatoms. √ Modified electronic structure and increased charge mobility by heteroatom doping. √ Synergistic effect between the polymer and the carbon support to promote CO_2RR. × Poor activity of g-C_3N_4 because of the abundance of HER active sites. × Hard to understand the effects of the amorphous polymer chemical compositions and morphology on the activity and concentration of active sites.	√ Enhanced activity due to the introduction of active sites or heteroatoms compared to the activity of the catalysts without heteroatoms. × Chemically inert. × Lower current density for acidic OER compared to that produced by the metal-based catalysts. × Limited density of active sites.

(i) Selection of suitable electrochemical cells for industrial CO_2RR applications. Among the available CO_2RR electrochemical cells, H-cells are applied extensively for lab-scale measurements, due to their easy operation, low cost, and facile assembly. However, the practical use of H-cells is restricted by the low solubility of CO_2 in aqueous electrolytes, limited electrode area, competing HER, and low selectivity to C_{2+} species. To overcome these problems, flow cells were developed. In these cells, CO_2 gas is directly applied to cathode and the electrolyte is circulated to buffer the pH variation, hence, the mass diffusion and production rate are improved, and the side reaction HER is suppressed. MEA consists of two electrodes and an ion-exchange membrane in between. This cell configuration significantly decreases the resistance due to mass/electron transfer, improves the FE, and is more suitable for practical applications. Both flow cells and MEA utilize working electrodes with catalysts on gas diffusion layers, providing hydrophobic channels to facilitate CO_2 diffusion and improving cell performances. These cells operate at room temperature and should be generally applied to evaluate the performance of the developed catalysts. Molten salt and solid oxide cells normally operate at high temperatures (400–1000 °C), but the output is limited to simple products. Proper selection of electrolyzer is important in making CO_2RR viable industrially, where large current densities, high FEs, and durable stability are required.

(ii) Rational design of efficient metals and metal compounds for CO_2RR. Bulk noble metals favor the formation of CO and HCOOH because of their strong *COOH binding energy. However, the use of noble metals imposes extravagant operation costs. Compared with noble metals, transition metals are much cheaper and widely applied for CO_2RR. For example, Zn, Sn, and Bi are selective to CO and HCOOH, while Cu and Cu-based materials facilitate CO_2 conversion to multi-carbon hydrocarbons. Normally, metals exhibit size-, shape-, and GB-dependent CO_2RR performance, attributed to their manipulated morphology and structure. However, bulk metallic electrodes suffer from weak CO_2 adsorption and hence unsatisfactory catalytic activity. To improve the selectivity, metal alloys and metal-based compounds (i.e., metal chalcogenides, metal oxides, metal nitrides, and metal hydroxides) are explored as promising catalysts for CO_2RR. Basically, the rational design of transition metal-based structures requires sufficient mechanistic understanding, which can be achieved with the assistance of advanced characterization techniques and theoretical calculations.

(iii) Large-scale production of SACs. SACs present superior CO_2RR performances owing to their uniformly distributed active sites, strong atom-support interactions, and maximum metal utilization. While some noble metals-(Au, Ag, Pd) and transition metal-(Ni, Fe, Co) based SACs are highly selective to CO or HCOOH production, Cu SACs favor the formation of multi-carbon hydrocarbons. Nonetheless, SACs currently suffer from particles/metals agglomeration, low metal loading, and difficulty in mass production. Therefore, developing SACs with high metal loadings at large scale is crucial to establishing CO_2RR as an industrial process. Novel fast and scalable synthesis of SACs should be developed to fully exploit the potentials of SACs.

(iv) Exploration of molecular catalysts. Molecular catalysts such as organometallic complexes, metal-free and polymerized catalysts, along with MOF and COF catalysts possess great potentials for CO_2RR, especially when the tunability of their steric and electronic properties around active sites is considered. Most studies on molecular catalysts are based on the intermediate-electrode surface binding energy and theoretical calculations. A further understanding of the electrode–electrolyte interface, mass transport of CO_2, HCO_3^-, CO_3^{2-}, and OH^-, as well as chemical states and organic structures is needed. Because of the complexity of molecular catalyst systems, detailed investigations on the lifetime, activity, selectivity, and identification of active species should be conducted. Besides, other issues such as catalyst aggregation, leaching, demetallation (e.g., the CoPc demetallation is induced by the electroreduction of Fe impurities, which are in an oxidized state, leading to severe degradation of catalytic activity), and poor conductivity also need to be addressed. Hybrid molecular catalysts immobilized onto a conducting electrode surface can combine the advantages of heterogeneous and homogeneous catalysis and present a very promising strategy.

(v) Understanding of the complex CO_2RR mechanism. The CO_2RR process involves multi electron–proton transfer steps with multiple intermediates or products, which make the mechanistic study complicated. Although significant progresses have been achieved in this regard, a further understanding of the mechanistic pathways in CO_2RR is needed. Theoretical calculations and advanced *in situ* characterizations are powerful to identify the active sites/species, intermediates, electronic structure evolution, and reaction pathways. A majority of the present studies focus on a certain material or a certain category, posing concerns for a universal application. A combination of multiple and systematic *in situ* approaches could provide the desired insights into complex CO_2RR mechanisms, and promote subsequently the development of efficient electrocatalysts.

References

Abbet, S., Sanchez, A., Heiz, U., Schneider, W. D., Ferrari, A. M., Pacchioni, G., & Rösch, N. (2000). Acetylene cyclotrimerization on supported size-selected Pd_n clusters ($1 \leq n \leq 30$): one atom is enough! *Journal of the American Chemical Society*, *122*(14), 3453–3457.

Abdel-Mageed, A. M., Rungtaweevoranit, B., Parlinska-Wojtan, M., Pei, X., Yaghi, O. M., & Behm, R. J. (2019). Highly active and stable single-atom Cu catalysts supported by a metal–organic framework. *Journal of the American Chemical Society*, *141*(13), 5201–5210.

Agarwal, A. S., Zhai, Y., Hill, D., & Sridhar, N. (2011). The electrochemical reduction of carbon dioxide to formate/formic acid: engineering and economic feasibility. *ChemSusChem*, *4*(9), 1301–1310.

Arán-Ais, R. M., Rizo, R., Grosse, P., Algara-Siller, G., Dembélé, K., Plodinec, M., Lunkenbein, T., Chee, S. W., & Cuenya, B. R. (2020). Imaging electrochemically synthesized Cu_2O cubes and their morphological evolution under conditions relevant to CO_2 electroreduction. *Nature Communications*, *11*(1), 1–8.

Asadi, M., Kim, K., Liu, C., Addepalli, A. V., Abbasi, P., Yasaei, P., Phillips, P., Behranginia, A., Cerrato, J. M., Haasch, R., Zapol, P., Kumar, B., Klie, R. F., Abidae, J., Curtiss, L. A., & Salehi-Khojin, A. (2016). Nanostructured transition metal dichalcogenide electrocatalysts for CO_2 reduction in ionic liquid. *Science*, *353*(6298), 467–470.

Asadi, M., Kumar, B., Behranginia, A., Rosen, B. A., Baskin, A., Repnin, N., Pisasale, D., Phillips, P., Zhu, W., Haasch, R., Klie, R. F., Král, P., Abiade, J., & Salehi-Khojin, A. (2014). Robust carbon dioxide reduction on molybdenum disulphide edges. *Nature Communications*, *5*(1), 1–8.

Astruc, D., Lu, F., & Aranzaes, J. R. (2005). Nanoparticles as recyclable catalysts: the frontier between homogeneous and heterogeneous catalysis. *Angewandte Chemie, International Edition*, *44*(48), 7852–7872.

Aydin, R., & Köleli, F. (2004). Electrocatalytic conversion of CO_2 on a polypyrrole electrode under high pressure in methanol. *Synthetic Metals*, *144*(1), 75–80.

Back, S., & Jung, Y. (2017). TiC-and TiN-supported single-atom catalysts for dramatic improvements in CO_2 electrochemical reduction to CH_4. *ACS Energy Letters*, *2*(5), 969–975.

Bai, X., Chen, W., Zhao, C., Li, S., Song, Y., Ge, R., Wei, S., & Sun, Y. (2017). Exclusive formation of formic acid from CO_2 electroreduction by a tunable Pd-Sn alloy. *Angewandte Chemie*, *129*(40), 12387–12391.

Benson, E. E., Kubiak, C. P., Sathrum, A. J., & Smieja, J. M. (2009). Electrocatalytic and homogeneous approaches to conversion of CO_2 to liquid fuels. *Chemical Society Reviews*, *38*(1), 89–99.

Biroju, R. K., Das, D., Sharma, R., Pal, S., Mawlong, L. P., Bhorkar, K., Giri, P. K., Singh, A. K., & Narayanan, T. N. (2017). Hydrogen evolution reaction activity of graphene–MoS_2 van der Waals heterostructures. *ACS Energy Letters*, *2*(6), 1355–1361.

Bondue, C. J., Graf, M., Goyal, A., & Koper, M. T. (2020). Suppression of hydrogen evolution in acidic electrolytes by electrochemical CO_2 reduction. *Journal of the American Chemical Society*, *143*(1), 279–285.

Boutin, E., Wang, M., Lin, J. C., Mesnage, M., Mendoza, D., Lassalle-Kaiser, B., Hahn, C., Jaramillo, T. F., & Robert, M. (2019). Aqueous electrochemical reduction of carbon dioxide and carbon monoxide into methanol with cobalt phthalocyanine. *Angewandte Chemie, International Edition*, *58*(45), 16172–16176.

Büchele, S., Martin, A. J., Mitchell, S., Krumeich, F., Collins, S. M., Xi, S., Borgna, A., & Pérez-Ramírez, J. (2020). Structure sensitivity and evolution of nickel-bearing nitrogen-doped carbons in the electrochemical reduction of CO_2. *ACS Catalysis, 10*(5), 3444–3454.

Cai, Y., Fu, J., Zhou, Y., Chang, Y., Min, Q., Zhu, J., Lin, Y., & Zhu, W. (2021). Insights on forming N, O-coordinated Cu single-atom catalysts for electrochemical reduction CO_2 to methane. *Nature Communications, 12*(1), 1–9.

Chan, K., Tsai, C., Hansen, H. A., & Nørskov, J. K. (2014). Molybdenum sulfides and selenides as possible electrocatalysts for CO_2 reduction. *ChemCatChem, 6*(7), 1899–1905.

Chang, F., Zhan, G., Wu, Z., Duan, Y., Shi, S., Zeng, S., Zhang, X., & Zhang, S. (2021). Technoeconomic analysis and process design for CO_2 electroreduction to CO in ionic liquid electrolyte. *ACS Sustainable Chemistry & Engineering, 9*(27), 9045–9052.

Chapovetsky, A., Do, T. H., Haiges, R., Takase, M. K., & Marinescu, S. C. (2016). Proton-assisted reduction of CO_2 by cobalt aminopyridine macrocycles. *Journal of the American Chemical Society, 138*(18), 5765–5768.

Chen, J. G. (2018). Electrochemical CO_2 reduction via low-valent nickel single-atom catalyst. *Joule, 2*(4), 587–589.

Chen, L., Guo, S. X., Li, F., Bentley, C., Horne, M., Bond, A. M., & Zhang, J. (2016). Electrochemical reduction of CO_2 at metal electrodes in a distillable ionic liquid. *ChemSusChem, 9*(11), 1271–1278.

Chen, Y., Li, C. W., & Kanan, M. W. (2012). Aqueous CO_2 reduction at very low overpotential on oxide-derived Au nanoparticles. *Journal of the American Chemical Society, 134*(49), 19969–19972.

Chen, Y., Vise, A., Klein, W. E., Cetinbas, F. C., Kariuki, N. N., Myers, D. J., Smith, W. A., Deutsch, T. G., & Neyerlin, K. C. (2020). A robust, scalable platform for the electrochemical conversion of CO_2 to formate; identifying pathways to higher energy efficiencies. *ACS Energy Letters, 5*(6), 1825–1833.

Chen, Y., Zou, L., Liu, H., Chen, C., Wang, Q., Gu, M., Yang, B., Zou, Z., Fang, J., & Yang, H. (2019). Fe and N Co-doped porous carbon nanospheres with high density of active sites for efficient CO_2 electroreduction. *The Journal of Physical Chemistry C, 123*(27), 16651–16659.

Chen, Z., Zhang, X., Liu, W., Jiao, M., Mou, K., Zhang, X., & Liu, L. (2021). Amination strategy to boost the CO_2 electroreduction current density of M–N/C single-atom catalysts to the industrial application level. *Energy & Environmental Science, 14*(4), 2349–2356.

Chen, Z., Zhao, J., Cabrera, C. R., & Chen, Z. (2019). Computational screening of efficient single-atom catalysts based on graphitic carbon nitride (g-C_3N_4) for nitrogen electroreduction. *Small Methods, 3*(6), 1800368.

Cheng, D., Zhao, Z., Zhang, G., Yang, P., Li, L., Gao, H., Liu, S., Chang, X., Chen, S., Wang, T., Ozin, G. A., Liu, Z., & Gong, J. (2021). The nature of active sites for carbon dioxide electroreduction over oxide-derived copper catalysts. *Nature Communications, 12*(1), 1–8.

Cheng, Q., Mao, K., Ma, L., Yang, L., Zou, L., Zou, Z., Hu, Z., & Yang, H. (2018). Encapsulation of iron nitride by Fe–N–C shell enabling highly efficient electroreduction of CO_2 to CO. *ACS Energy Letters, 3*(5), 1205–1211.

Cheng, W. H., Richter, M. H., Sullivan, I., Larson, D. M., Xiang, C., Brunschwig, B. S., & Atwater, H. A. (2020). CO_2 reduction to CO with 19% efficiency in a solar-driven gas diffusion electrode flow cell under outdoor solar illumination. *ACS Energy Letters, 5*(2), 470–476.

Cheng, Y., Zhao, S., Li, H., He, S., Veder, J. P., Johannessen, B., Xiao, J., Lu, S., Pan, J., Chisholm, M. F., Yang, S., Liu, C., Chen, J. G., & Jiang, S. (2019). Unsaturated edge-anchored Ni single atoms on porous microwave exfoliated graphene oxide for electrochemical CO_2. *Applied Catalysis B: Environmental, 243*, 294–303.

Chu, S., & Majumdar, A. (2012). Opportunities and challenges for a sustainable energy future. *Nature, 488*(7411), 294–303.

Corbin, N., Zeng, J., Williams, K., & Manthiram, K. (2019). Heterogeneous molecular catalysts for electrocatalytic CO_2 reduction. *Nano Research, 12*, 2093–2125.

Coskun, H., Aljabour, A., De Luna, P., Farka, D., Greunz, T., Stifter, D., Kus, M., Zhang, X., Liu, M., Hassel, A. W., Schofberger, W., & Sradler, P. (2017). Biofunctionalized conductive polymers enable efficient CO_2 electroreduction. *Science, 3*(8), e1700686.

Costentin, C., Drouet, S., Robert, M., & Savéant, J. M. (2012). A local proton source enhances CO_2 electroreduction to CO by a molecular Fe catalyst. *Science, 338*(6103), 90–94.

De Arquer, F. P. G., Dinh, C. T., Ozden, A., Wicks, J., McCallum, C., Kirmani, A. R., Nam, D. H., Gabardp, C., Seifitokaldani, A., Wang, X., Li, Y. C., Li, F., Edwards, J., Richter, L. J., Thorpe, S. J., Sinton, D., & Sargent, E. H. (2020). CO_2 electrolysis to multicarbon products at activities greater than 1 A cm^{-2}. *Science, 367*(6478), 661–666.

De, R., Gonglach, S., Paul, S., Haas, M., Sreejith, S., Gerschel, P., Apfel, U. P., Vuong, T. H., Rabeah, J., Roy, S., & Schöfberger, W. (2020). Electrocatalytic reduction of CO_2 to acetic acid by a molecular manganese corrole complex. *Angewandte Chemie, 132*(26), 10614–10621.

Deng, B., Mao, X., Xiao, W., & Wang, D. (2017). Microbubble effect-assisted electrolytic synthesis of hollow carbon spheres from CO_2. *Journal of Materials Chemistry A*, *5*(25), 12822–12827.

Diercks, C. S., Liu, Y., Cordova, K. E., & Yaghi, O. M. (2018). The role of reticular chemistry in the design of CO_2 reduction catalysts. *Nature Materials*, *17*(4), 301–307.

Dinh, C. T., Burdyny, T., Kibria, M. G., Seifitokaldani, A., Gabardo, C. M., De Arquer, F. P. G., Kiani, A., Edwards, J. P., De Luna, P., Bushuyev, O. S., Zou, C. Q., Quintero-Bermudez, R., Pang, Y. J., Sinton, D., & Sargent, E. H. (2018). CO_2 electroreduction to ethylene via hydroxide-mediated copper catalysis at an abrupt interface. *Science*, *360*(6390), 783–787.

Dong, H., Zhang, L., Yang, P., Chang, X., Zhu, W., Ren, X., Zhao, Z., & Gong, J. (2019). Facet design promotes electroreduction of carbon dioxide to carbon monoxide on palladium nanocrystals. *Chemical Engineering Science*, *194*, 29–35.

Du, H., Liu, L., Cai, Y., Wang, Y., Zhang, J., Min, Q., & Zhou, W. (2022). In situ formed N-containing copper nanoparticles: a high-performance catalyst toward carbon monoxide electroreduction to multicarbon products with high faradaic efficiency and current density. *Nanoscale*, *14*(19), 7262–7268.

Dubed Bandomo, G. C., Mondal, S. S., Franco, F., Bucci, A., Martin-Diaconescu, V., Ortuño, M. A., Langevelde, P. H., Shafir, A., López, N., & Lloret-Fillol, J. (2021). Mechanically constrained catalytic $Mn(CO)_3Br$ single sites in a two-dimensional covalent organic framework for CO_2 electroreduction in H_2O. *ACS Catalysis*, *11*, 7210–7222.

Dutta, A., Kuzume, A., Rahaman, M., Vesztergom, S., & Broekmann, P. (2015). Monitoring the chemical state of catalysts for CO_2 electroreduction: an in operando study. *ACS Catalysis*, *5*(12), 7498–7502.

Dutta, A., Rahaman, M., Luedi, N. C., Mohos, M., & Broekmann, P. (2016). Morphology matters: tuning the product distribution of CO_2 electroreduction on oxide-derived Cu foam catalysts. *ACS Catalysis*, *6*(6), 3804–3814.

Ebbesen, S. D., Jensen, S. H., Hauch, A., & Mogensen, M. B. (2014). High temperature electrolysis in alkaline cells, solid proton conducting cells, and solid oxide cells. *Chemical Reviews*, *114*(21), 10697–10734.

Ebbesen, S. D., & Mogensen, M. (2009). Electrolysis of carbon dioxide in solid oxide electrolysis cells. *Journal of Power Sources*, *193*(1), 349–358.

Eisenberg, R., & Gray, H. B. (2011). Noninnocence in metal complexes: A dithiolene dawn. *Inorganic Chemistry*, *50*(20), 9741–9751.

Faggion, D., Jr., Gonçalves, W. D., & Dupont, J. (2019). CO_2 electroreduction in ionic liquids. *Frontiers in Chemistry*, *7*, 102.

Fan, L., Xia, C., Yang, F., Wang, J., Wang, H., & Lu, Y. (2020). Strategies in catalysts and electrolyzer design for electrochemical CO_2 reduction toward C_{2+} products. Science. *Advances*, *6*(8), eaay3111.

Fan, L., Xia, C., Zhu, P., Lu, Y., & Wang, H. (2020). Electrochemical CO_2 reduction to high-concentration pure formic acid solutions in an all-solid-state reactor. *Nature Communications*, *11*(1), 1–9.

Fang, Y., Yu, Y., Zhao, W., Yang, P., Huang, T., & Gao, S. (2022). In situ electrochemical restructuring integrating corrosion engineering to fabricate Zn nanosheets for efficient CO_2 electroreduction. *Electrocatalysis*, 1–10.

Feldman, D. R., Collins, W. D., Gero, P. J., Torn, M. S., Mlawer, E. J., & Shippert, T. R. (2015). Observational determination of surface radiative forcing by CO_2 from 2000 to 2010. *Nature*, *519*(7543), 339–343.

Feng, J., Gao, H., Feng, J., Liu, L., Zeng, S., Dong, H., Bai, Y., Liu, L., & Zhang, X. (2020). Morphology modulation-engineered Flowerlike In_2S_3 via ionothermal method for efficient CO_2 electroreduction. *ChemCatChem*, *12*(3), 926–931.

Feng, J., Gao, H., Zheng, L., Chen, Z., Zeng, S., Jiang, C., Dong, H., Liu, L., Zhang, S., & Zhang, X. (2020). A $Mn-N_3$ single-atom catalyst embedded in graphitic carbon nitride for efficient CO_2 electroreduction. *Nature Communications*, *11*(1), 1–8.

Feng, J., Zeng, S., Feng, J., Dong, H., & Zhang, X. (2018). CO_2 electroreduction in ionic liquids: a review. *Chinese Journal of Chemistry*, *36*(10), 961–970.

Feng, X., Jiang, K., Fan, S., & Kanan, M. W. (2015). Grain-boundary-dependent CO_2 electroreduction activity. *Journal of the American Chemical Society*, *137*(14), 4606–4609.

Feng, Z., Su, G., Ding, H., Ma, Y., Li, Y., Tang, Y., & Dai, X. (2020). Atomic alkali metal anchoring on graphdiyne as single-atom catalysts for capture and conversion of CO_2 to HCOOH. *Molecular Catalysis*, *494*, 111142.

Galan-Sanchez LM. Functionalised ionic liquids, absorption solvents for CO_2 and olefin separation. PhD thesis, Technische Universiteit Eindhoven. 2008.

Gao, D., Zhang, Y., Zhou, Z., Cai, F., Zhao, X., Huang, W., Zhang, J., Wu, Z., & Han, B. (2017). Enhancing CO_2 electroreduction with the metal–oxide interface. *Journal of the American Chemical Society*, *139*(16), 5652–5655.

Gao, D., Zhou, H., Wang, J., Miao, S., Yang, F., Wang, G., Wang, J., & Bao, X. (2015). Size-dependent electrocatalytic reduction of CO_2 over Pd nanoparticles. *Journal of the American Chemical Society*, *137*(13), 4288–4291.

Gao, F. Y., Hu, S. J., Zhang, X. L., Zheng, Y. R., Wang, H. J., Niu, Z. Z., Yang, P. P., Bao, R. C., Ma, T., Dang, Z., Guan, Y., Zheng, X. S., Zheng, X., Zhu, J. H., Gao, M. R., & Yu, S. H. (2020). High-curvature transition-metal chalcogenide nanostructures with a pronounced proximity effect enable fast and selective CO_2 electroreduction. *Angewandte Chemie*, *132*(22), 8784–8790.

Gao, S., Jiao, X., Sun, Z., Zhang, W., Sun, Y., Zhang, W., Sun, Y., Wang, C., Hu, Q., Zu, X., Yang, F., Yang, S., Liang, L., Wu, J., & Xie, Y. (2016). Ultrathin Co_3O_4 layers realizing optimized CO_2 electroreduction to formate. *Angewandte Chemie, International Edition*, *55*(2), 698–702.

Gao, S., Lin, Y., Jiao, X., Sun, Y., Luo, Q., Zhang, W., Li, D., Yang, J., & Xie, Y. (2016). Partially oxidized atomic cobalt layers for carbon dioxide electroreduction to liquid fuel. *Nature*, *529*(7584), 68–71.

Gao, S., Sun, Z., Liu, W., Jiao, X., Zu, X., Hu, Q., Sun, Y., Yao, T., Zhang, W., Wei, S., & Xie, Y. (2017). Atomic layer confined vacancies for atomic-level insights into carbon dioxide electroreduction. *Nature Communications*, *8*(1), 1–9.

Garza, A. J., Bell, A. T., & Head-Gordon, M. (2018). Mechanism of CO_2 reduction at copper surfaces: pathways to C_2 products. *ACS Catalysis*, *8*(2), 1490–1499.

Geng, Z., Cao, Y., Chen, W., Kong, X., Liu, Y., Yao, T., & Lin, Y. (2019). Regulating the coordination environment of Co single atoms for achieving efficient electrocatalytic activity in CO_2 reduction. *Applied Catalysis B: Environmental*, *240*, 234–240.

Genovese, C., Schuster, M. E., Gibson, E. K., Gianolio, D., Posligua, V., Grau-Crespo, R., Cibin, G., Wells, P. P., Garai, D., Solokha, V., Calderon, S. K., Velasco-Velez, J. J., Ampelli, C., Perathoner, S., Held, G., Centi, G., & Arrigo, R. (2018). Operando spectroscopy study of the carbon dioxide electro-reduction by iron species on nitrogen-doped carbon. *Nature Communications*, *9*(1), 1–12.

GnO, L., Martín, A. J., Mitchell, S., Hauert, R., & Pérez-Ramírez, J. (2016). Enhanced reduction of CO_2 to CO over Cu–In electrocatalysts: catalyst evolution is the key. *ACS Catalysis*, *6*(9), 6265–6274.

Gonell, S., Massey, M. D., Moseley, I. P., Schauer, C. K., Muckerman, J. T., & Miller, A. J. M. (2019). The trans effect in electrocatalytic CO_2 reduction: mechanistic studies of asymmetric ruthenium pyridyl-carbene catalysts. *Journal of the American Chemical Society*, *141*(16), 6658–6671.

Gong, Q., Ding, P., Xu, M., Zhu, X., Wang, M., Deng, J., Ma, Q., Han, N., Zhu, Y., Lu, J., Feng, Z., Li, Y., Zhou, W., & Li, Y. (2019). Structural defects on converted bismuth oxide nanotubes enable highly active electrocatalysis of carbon dioxide reduction. *Nature Communications*, *10*(1), 1–10.

Graciani, J., Mudiyanselage, K., Xu, F., Baber, A. E., Evans, J., Senanayake, S. D., Stacchiola, D. J., Liu, P., Hrbek, J., Sanz, J. F., & Rodriguez, J. A. (2014). Highly active copper-ceria and copper-ceria-titania catalysts for methanol synthesis from CO_2. *Science*, *345*(6196), 546–550.

Grosse, P., Gao, D., Scholten, F., Sinev, I., Mistry, H., & Roldan, C. B. (2018). Dynamic changes in the structure, chemical state and catalytic selectivity of Cu nanocubes during CO_2 electroreduction: size and support effects. *Angewandte Chemie, International Edition*, *57*(21), 6192–6197.

Gu, J., Hsu, C., Bai, L., Chen, H., & Hu, X. (2019). Atomically dispersed Fe^{3+} sites catalyze efficient CO_2 electroreduction to CO. *Science*, *364*(6445), 1091–1094.

Gu, J., Liu, S., Ni, W., Ren, W., Haussener, S., & Hu, X. (2022). Modulating electric field distribution by alkali cations for CO_2 electroreduction in strongly acidic medium. *Nature Catalysis*, *5*(4), 268–276.

Guo, X., Zhang, Y., Deng, C., Li, X., Xue, Y., Yan, Y., & Sun, K. (2015). Composition dependent activity of Cu–Pt nanocrystals for electrochemical reduction of CO_2. *Chemical Communications*, *51*(7), 1345–1348.

Hall, A. S., Yoon, Y., Wuttig, A., & Surendranath, Y. (2015). Mesostructure-induced selectivity in CO_2 reduction catalysis. *Journal of the American Chemical Society*, *137*(47), 14834–14837.

Han, A., Wang, B., Kumar, A., Qin, Y., Jin, J., Wang, X., Yang, C., Dong, B., Jia, Y., Liu, J., & Sun, X. (2019). Recent advances for MOF-derived carbon-supported single-atom catalysts. *Small Methods*, *3*(9), 1800471.

Han, L., Song, S., Liu, M., Yao, S., Liang, Z., Cheng, H., Ren, Z., Liu, W., Lin, R., Qi, G., Liu, X., Wu, Q., Luo, J., & Xin, H. L. (2020). Stable and efficient single-atom Zn catalyst for CO_2 reduction to CH_4. *Journal of the American Chemical Society*, *142*(29), 12563–12567.

Han, N., Wang, Y., Yang, H., Deng, J., Wu, J., Li, Y., & Li, Y. (2018). Ultrathin bismuth nanosheets from in situ topotactic transformation for selective electrocatalytic CO_2 reduction to formate. *Nature Communications*, *9*(1), 1–8.

Han, S., Ma, D., Zhou, S., Zhang, K., Wei, W., Du, Y., Wu, X., Xu, Q., Zou, R., & Zhu, Q. (2021). Fluorine-tuned single-atom catalysts with dense surface $Ni\text{-}N_4$ sites on ultrathin carbon nanosheets for efficient CO_2 electroreduction. *Applied Catalysis B: Environmental*, *283*, 119591.

Handoko, A. D., Wei, F., Yeo, B. S., & Seh, Z. W. (2018). Understanding heterogeneous electrocatalytic carbon dioxide reduction through operando techniques. *Nature Catalysis*, *1*(12), 922–934.

He, H., Zapol, P., & Curtiss, L. A. (2010). A theoretical study of CO_2 anions on anatase (101) surface. *The Journal of Physical Chemistry C, 114*(49), 21474–21481.

He, J., Dettelbach, K. E., Salvatore, D. A., Li, T., & Berlinguette, C. P. (2017). High-throughput synthesis of mixed-metal electrocatalysts for CO_2 reduction. *Angewandte Chemie, 129*(22), 6164–6168.

He, J., Liu, X., Liu, H., Zhao, Z., Ding, Y., & Luo, J. (2018). Highly selective electrocatalytic reduction of CO_2 to formate over Tin (IV) sulfide monolayers. *Journal of Catalysis, 364*, 125–130.

He, Q., Lee, J. H., Liu, D., Liu, Y., Lin, Z., Xie, Z., Hwang, S., Kattel, S., Song, L., & Chen, J. G. (2020). Accelerating CO_2 electroreduction to CO over Pd single-atom catalyst. *Advanced Functional Materials, 30*(17), 2000407.

He, S., Ni, F., Ji, Y., Wang, L., Wen, Y., Bai, H., Liu, G. J., Zhang, Y., Li, Y., Zhang, B., & Peng, H. S. (2018). The p-orbital delocalization of main-group metals to boost CO_2 electroreduction. *Angewandte Chemie, 130*(49), 16346–16351.

Hinogami, R., Yotsuhashi, S., Deguchi, M., Zenitani, Y., Hashiba, H., & Yamada, Y. (2012). Electrochemical reduction of carbon dioxide using a copper rubeanate metal organic framework. *ECS Electrochemistry Letters, 1*(4), H17–H19.

Hoang, T. T., Verma, S., Ma, S., Fister, T. T., Timoshenko, J., Frenkel, A. I., Kenis, P. J. A., & Gewirth, A. A. (2018). Nanoporous copper–silver alloys by additive-controlled electrodeposition for the selective electroreduction of CO_2 to ethylene and ethanol. *Journal of the American Chemical Society, 140*(17), 5791–5797.

Hod, I., Sampson, M. D., Deria, P., Kubiak, C. P., Farha, O. K., & Hupp, J. T. (2015). Fe-porphyrin-based metal–organic framework films as high-surface concentration, heterogeneous catalysts for electrochemical reduction of CO_2. *ACS Catalysis, 5*(11), 6302–6309.

Hori, Y. (2008). Electrochemical CO_2 reduction on metal electrodes. Modern Aspects of. *Electrochemistry*, 89–189.

Hori, Y., Murata, A., & Takahashi, R. (1989). Formation of hydrocarbons in the electrochemical reduction of carbon dioxide at a copper electrode in aqueous solution. *Journal of the Chemical Society, Faraday Transactions 1: Physical Chemistry in Condensed Phases, 85*(8), 2309–2326.

Hori, Y., Takahashi, R., Yoshinami, Y., & Murata, A. (1997). Electrochemical reduction of CO at a copper electrode. *The Journal of Physical Chemistry B, 101*(36), 7075–7081.

Hori, Y., Wakebe, H., Tsukamoto, T., & Koga, O. (1994). Electrocatalytic process of CO selectivity in electrochemical reduction of CO_2 at metal electrodes in aqueous media. *Electrochimica Acta, 39*(11-12), 1833–1839.

Hou, P., Wang, X., Wang, Z., & Kang, P. (2018). Gas phase electrolysis of carbon dioxide to carbon monoxide using nickel nitride as the carbon enrichment catalyst. *ACS Applied Materials & Interfaces, 10*(44), 38024–38031.

Hou, Y., Liang, Y. L., Shi, P. C., & Cao, R. (2020). Atomically dispersed Ni species on N-doped carbon nanotubes for electroreduction of CO_2 with nearly 100% CO selectivity. *Applied Catalysis B: Environmental, 271*, 118929.

Hu, X.-M., Hval, H. H., Bjerglund, E. T., Dalgaard, K. J., Madsen, M. R., Pohl, M. M., Welter, E., Lamagni, P., Buhl, K. B., Bremholm, M., Beller, M., Pedersen, S. U., Skrydstrup, T., & Daasbjerg, K. (2018). Selective CO_2 reduction to CO in water using earth-abundant metal and nitrogen-doped carbon electrocatalysts. *ACS Catalysis, 8*(7), 6255–6264.

Huan, T. N., Ranjbar, N., Rousse, G., Sougrati, M., Zitolo, A., Mougel, V., Jaouen, F., & Fontecave, M. (2017). Electrochemical reduction of CO_2 catalyzed by Fe-NC materials: a structure–selectivity study. *ACS Catalysis, 7*(3), 1520–1525.

Huang, H., Jia, H., Liu, Z., Gao, P., Zhao, J., Luo, Z., Yang, J., & Zeng, J. (2017). Understanding of strain effects in the electrochemical reduction of CO_2: using Pd nanostructures as an ideal platform. *Angewandte Chemie, 129*(13), 3648–3652.

Huang, J. E., Li, F., Ozden, A., Sedighian Rasouli, A., García de Arquer, F. P., Liu, S., Zhang, S., Luo, M., Wang, X., Lum, Y., Xu, Y., Bertens, K., Miao, R., Dinh, C. T., Sinton, D., & Sargent, E. H. (2021). CO_2 electrolysis to multicarbon products in strong acid. *Science, 372*(6546), 1074–1078.

Huang, Y., Liu, S., Yang, H., Hung, S., Ding, J., Cai, W., Liu, L., Gao, J., Li, X., Zhao, T., Liu, B., Ren, X., & Kuang, Z. (2019). Electrifying model single-atom catalyst for elucidating the CO_2 reduction reaction. *Angewandte Chemie, International Edition, 59*, 798–803.

Hung, S. F., Xu, A., Wang, X., Li, F., Hsu, S. H., Li, Y., Wicks, J., Cervantes, E. G., Rasouli, A. S., Li, Y. C., Luo, M., Nam, D. H., Wang, N., Peng, T., Yan, Y., Lee, G., & Sargent, E. H. (2022). A metal-supported single-atom catalytic site enables carbon dioxide hydrogenation. *Nature Communications, 13*(1), 1–9.

Jeon, H. S., Sinev, I., Scholten, F., Divins, N. J., Zegkinoglou, I., Pielsticker, L., & Cuenya, B. R. (2018). Operando evolution of the structure and oxidation state of size-controlled Zn nanoparticles during CO_2 electroreduction. *Journal of the American Chemical Society, 140*(30), 9383–9386.

Ji, Y., Chen, Z., Yang, C., Wang, Y., Xu, J., Guan, A., Chen, J., Sham, T. K., Luo, J., Yang, Y., Xu, X., & Zheng, G. (2022). Selective CO-to-acetate electroproduction via intermediate adsorption tuning on ordered Cu-Pd sites. *Nature Catalysis, 5*, 251–258.

Jia, C., Ching, K., Kumar, P. V., Zhao, C., Kumar, N., Chen, X., & Das, B. (2020). Vitamin B_{12} on graphene for highly efficient CO_2 electroreduction. *ACS Applied Materials & Interfaces*, *12*(37), 41288–41293.

Jia, C., Dastafkan, K., Ren, W., Yang, W., & Zhao, C. (2019). Carbon-based catalysts for electrochemical CO_2 reduction. *Sustainable Energy & Fuels*, *3*(11), 2890–2906.

Jia, C., Ren, W., Chen, X., Yang, W., & Zhao, C. (2020). (N, B) dual heteroatom-doped hierarchical porous carbon framework for efficient electroreduction of carbon dioxide. *ACS Sustainable Chemistry & Engineering*, *8*(15), 6003–6010.

Jia, J., Hao, X., Chang, Y., Jia, M., & Wen, Z. (2020). Rational design of Cu_3PdN nanocrystals for selective electroreduction of carbon dioxide to formic acid. *Journal of Colloid and Interface Science*, *586*, 491–497.

Jia, M., Hong, S., Wu, T., Li, X., Soo, Y., & Sun, Z. (2019). Single Sb sites for efficient electrochemical CO_2 reduction. *Chemical Communications*, *55*(80), 12024–12027.

Jiang, K., Siahrostami, S., Zheng, T., Hu, Y., Hwang, S., Stavitski, E., Peng, Y., Dynes, J., Gangisetty, M., Su, D., Attenkofer, K., & Wang, H. (2018). Isolated Ni single atoms in graphene nanosheets for high-performance CO_2 reduction. *Energy & Environmental Science*, *11*(4), 893–903.

Jiang, K., Wang, H., Cai, W., & Wang, H. (2017). Li electrochemical tuning of metal oxide for highly selective CO_2 reduction. *ACS Nano*, *11*(6), 6451–6458.

Jiang, R., Gao, M., Mao, X., & Wang, D. (2019). Advancements and potentials of molten salt CO_2 capture and electrochemical transformation (MSCC-ET) process. *Current Opinion in Electrochemistry*, *17*, 38–46.

Jiang, X., Cai, F., Gao, D., Dong, J., Miao, S., Wang, G., & Bao, X. (2016). Electrocatalytic reduction of carbon dioxide over reduced nanoporous zinc oxide. *Electrochemistry Communications*, *68*, 67–70.

Jiang, X., Wang, Q., Xiao, X., Chen, J., Shen, Y., & Wang, M. (2020). Interfacial engineering of bismuth with reduced graphene oxide hybrid for improving CO_2 electroreduction performance. *Electrochimica Acta*, *357*, 136840.

Jiang, Y., Chen, F., & Xia, C. (2021). A review on cathode processes and materials for electro-reduction of carbon dioxide in solid oxide electrolysis cells. *Journal of Power Sources*, *493*, 229713.

Jiang, Z., Wang, T., Pei, J., Shang, H., Zhou, D., Li, H., Dong, J., Wang, Y., Cao, R., Zhuang, Z., Chen, W., Wang, D., Zhang, J., & Li, Y. (2020). Discovery of main group single Sb-N_4 active sites for CO_2 electroreduction to formate with high efficiency. *Energy & Environmental Science*, *13*(9), 2856–2863.

Jiao, J., Lin, R., Liu, S., Cheong, W. C., Zhang, C., Chen, Z., Pan, Y., Tang, J., Wu, K., Hung, S. F., Chen, H., Zheng, L., Lu, Q., Yang, X., Xu, B., Xiao, H., Li, J., Wang, D., Peng, Q., … Li, Y. (2019). Copper atom-pair catalyst anchored on alloy nanowires for selective and efficient electrochemical reduction of CO_2. *Nature Chemistry*, *11*(3), 222–228.

Jouny, M., Hutchings, G. S., & Jiao, F. (2019). Carbon monoxide electroreduction as an emerging platform for carbon utilization. *Nature Catalysis*, *2*(12), 1062–1070.

Jouny, M., Luc, W., & Jiao, F. (2018). High-rate electroreduction of carbon monoxide to multi-carbon products. *Nature Catalysis*, *1*(10), 748–755.

Jouny, M., Luc, W., & Jiao, F. (2018). General techno-economic analysis of CO_2 electrolysis systems. *Industrial and Engineering Chemistry Research*, *57*(6), 2165–2177.

Ju, W., Bagger, A., Hao, G.-P., Varela, A. S., Sinev, I., Bon, V., Cuenya, B. R., Kaskel, S., Rossmeisl, J., & Strasser, P. (2017). Understanding activity and selectivity of metal-nitrogen-doped carbon catalysts for electrochemical reduction of CO_2. *Nature Communications*, *8*(1), 1–9.

Kas, R., Hummadi, K. K., Kortlever, R., De Wit, P., Milbrat, A., Luiten-Olieman, M. W., Benes, N. E., Koper, M. T. M., & Mul, G. (2016). Three-dimensional porous hollow fibre copper electrodes for efficient and high-rate electrochemical carbon dioxide reduction. *Nature Communications*, *7*(1), 1–7.

Kas, R., Kortlever, R., Milbrat, A., Koper, M. T., Mul, G., & Baltrusaitis, J. (2014). Electrochemical CO_2 reduction on Cu_2O-derived copper nanoparticles: controlling the catalytic selectivity of hydrocarbons. *Physical Chemistry Chemical Physics*, *16*(24), 12194–12201.

Kelley, S. P., Flores, L. A., Shannon, M. S., Bara, J. E., & Rogers, R. D. (2017). Understanding carbon dioxide solubility in ionic liquids by exploring the link with liquid clathrate formation. Chemistry–A. *European Journal*, *23*(57), 14332–14337.

Kibria, M. G., Dinh, C. T., Seifitokaldani, A., De Luna, P., Burdyny, T., Quintero-Bermudez, R., Ross, M. B., Bushuyev, O. S., De Arquer, F. P. G., Yang, P. D., Sinton, D., & Sargent, E. H. (2018). A surface reconstruction route to high productivity and selectivity in CO_2 electroreduction toward C_{2+} hydrocarbons. *Advanced Materials*, *30*(49), 1804867.

Kim, C., Jeon, H. S., Eom, T., Jee, M. S., Kim, H., Friend, C. M., Min, B. K., & Hwang, Y. J. (2015). Achieving selective and efficient electrocatalytic activity for CO_2 reduction using immobilized silver nanoparticles. *Journal of the American Chemical Society*, *137*(43), 13844–13850.

Kim, D., Xie, C., Becknell, N., Yu, Y., Karamad, M., Chan, K., Crumlin, E. J., Nørskov, J. K., & Yang, P. (2017). Electrochemical activation of CO_2 through atomic ordering transformations of AuCu nanoparticles. *Journal of the American Chemical Society, 139*(24), 8329–8336.

Köleli, F., Röpke, T., & Hamann, C. H. (2004). The reduction of CO_2 on polyaniline electrode in a membrane cell. *Synthetic Metals, 140*(1), 65–68.

Kornienko, N., Zhao, Y., Kley, C. S., Zhu, C., Kim, D., Lin, S., Chang, C. J., Yaghi, O. M., & Yang, P. (2015). Metal–organic frameworks for electrocatalytic reduction of carbon dioxide. *Journal of the American Chemical Society, 137*(44), 14129–14135.

Kortlever, R., Shen, J., Schouten, K. J. P., Calle-Vallejo, F., & Koper, M. T. M. (2015). Catalysts and reaction pathways for the electrochemical reduction of carbon dioxide. *The Journal of Physical Chemistry Letters, 6*(20), 4073–4082.

Kuhl, K. P., Cave, E. R., Abram, D. N., & Jaramillo, T. F. (2012). New insights into the electrochemical reduction of carbon dioxide on metallic copper surfaces. *Energy & Environmental Science, 5*(5), 7050–7059.

Kumar, B., Asadi, M., Pisasale, D., Sinha-Ray, S., Rosen, B. A., Haasch, R., Abiade, J., Yarin, A. L., & Salehi-Khojin, A. (2013). Renewable and metal-free carbon nanofibre catalysts for carbon dioxide reduction. *Nature Communications, 4*(1), 1–8.

Kumar, B., Atla, V., Brian, J. P., Kumari, S., Nguyen, T. Q., Sunkara, M., & Spurgeon, J. M. (2017). Reduced SnO_2 porous nanowires with a high density of grain boundaries as catalysts for efficient electrochemical CO_2-into-HCOOH conversion. *Angewandte Chemie, International Edition, 56*(13), 3645–3649.

Lakshmanan, K., Huang, W. H., Chala, S. A., Taklu, B. W., Moges, E. A., Lee, J. F., Huang, P., Lee, Y. C., Tsai, M. C., Su, W. N., & Hwang, B. J. (2022). Highly active oxygen coordinated configuration of Fe single-atom catalyst toward electrochemical reduction of CO_2 into multi-carbon products. *Advanced Functional Materials, 2109310.*

Larrazábal, G. O., Strøm-Hansen, P., Heli, J. P., Zeiter, K., Therkildsen, K. T., Chorkendorff, I., & Seger, B. (2019). Analysis of mass flows and membrane cross-over in CO_2 reduction at high current densities in an MEA-type electrolyzer. *ACS Applied Materials & Interfaces, 11*(44), 41281–41288.

Leonard, N., Ju, W., Sinev, I., Steinberg, J., Luo, F., Varela, A. S., Cuenya, B. R., & Strasser, P. (2018). The chemical identity, state and structure of catalytically active centers during the electrochemical CO_2 reduction on porous Fe–nitrogen–carbon (Fe–N–C) materials. *Chemical Science, 9*(22), 5064–5073.

Li, C. W., & Kanan, M. W. (2012). CO_2 reduction at low overpotential on Cu electrodes resulting from the reduction of thick Cu_2O films. *Journal of the American Chemical Society, 134*(17), 7231–7234.

Li, D., Zheng, Z., Lei, Y., Ge, S., Zhang, Y., Zhang, Y., Wong, K. W., Yang, F., & Lau, W. M. (2010). Design and growth of dendritic $Cu_{2-x}Se$ and bunchy CuSe hierarchical crystalline aggregations. *CrystEngComm, 12*(6), 1856–1861.

Li, F., Chen, L., Knowles, G. P., MacFarlane, D. R., & Zhang, J. (2017). Hierarchical mesoporous SnO_2 nanosheets on carbon cloth: a robust and flexible electrocatalyst for CO_2 reduction with high efficiency and selectivity. *Angewandte Chemie, 129*(2), 520–524.

Li, F. W., Li, Y. C., Wang, Z., Li, J., Nam, D. H., Lum, Y., Luo, M. C., Wang, X., Ozden, A., Hung, S. F., Chen, B., Wang, Y. H., Wicks, J., Xu, Y., Li, Y. L., Gabardo, C. M., Dinh, C. T., Wang, Y., Zhuang, T. T., … Sargent, E. H. (2020). Cooperative CO_2-to-ethanol conversion via enriched intermediates at molecule–metal catalyst interfaces. *Nature Catalysis, 3*(1), 75–82.

Li, J., Chang, X., Zhang, H., Malkani, A. S., Cheng, M., Xu, B., & Lu, Q. (2021). Electrokinetic and in situ spectroscopic investigations of CO electrochemical reduction on copper. *Nature Communications, 12*(1), 1–11.

Li, J., Chen, G., Zhu, Y., Liang, Z., Pei, A., Wu, C., Wang, H., Lee, H., Liu, K., Chu, S., & Cui, Y. (2018). Efficient electrocatalytic CO_2 reduction on a three-phase interface. *Nature Catalysis, 1*(8), 592–600.

Li, X., Zeng, Y., Tung, C., Lu, Y., Baskaran, S., Hung, S., Wang, S., Xu, C., Wang, J., Chan, T., Chen, H., Jiang, J., Yu, Q., Huang, Y., Li, J., Zhang, T., & Liu, B. (2021). Unveiling the in situ generation of a monovalent Fe(I) site in the single-Fe-atom catalyst for electrochemical CO_2 reduction. *ACS Catalysis, 11*, 7292–7301.

Li, Y., Chen, C., Cao, R., Pan, Z., He, H., & Zhou, K. (2020). Dual-atom Ag_2/graphene catalyst for efficient electroreduction of CO_2 to CO. *Applied Catalysis B: Environmental, 268*, 118747.

Li, Y., Cui, F., Ross, M. B., Kim, D., Sun, Y., & Yang, P. (2017). Structure-sensitive CO_2 electroreduction to hydrocarbons on ultrathin 5-fold twinned copper nanowires. *Nano Letters, 17*(2), 1312–1317.

Li, Y. C., Zhou, D., Yan, Z., Gonçalves, R. H., Salvatore, D. A., Berlinguette, C. P., & Mallouk, T. E. (2016). Electrolysis of CO_2 to syngas in bipolar membrane-based electrochemical cells. *ACS Energy Letters, 1*(6), 1149–1153.

Liang, S., Altaf, N., Huang, L., Gao, Y., & Wang, Q. (2020). Electrolytic cell design for electrochemical CO_2 reduction. *Journal of CO_2 Utilization, 35*, 90–105.

Liang, Z., Zhuang, T., Seifitokaldani, A., Li, J., Huang, C., Tan, C., Li, Y., Luna, P. D., Dinh, C. T., Hu, Y., Xiao, Q., Hsieh, P., Wang, Y., Li, F., Quintero-Bermudez, R., Zhou, Y., Chen, P., Pang, Y., Lo, S. C., … Sargent, E. H. (2018). Copper-on-nitride enhances the stable electrosynthesis of multi-carbon products from CO_2. *Nature Communications*, *9*(1), 1–8.

Lin, S., Diercks, C. S., Zhang, Y., Kornienko, N., Nichols, E. M., Zhao, Y., Paris, A. R., Kmi, D., Yang, P., Yaghi, O. M., & Chang, C. J. (2015). Covalent organic frameworks comprising cobalt porphyrins for catalytic CO_2 reduction in water. *Science*, *349*(6253), 1208–1213.

Liu, H., Chu, J., Yin, Z., Cai, X., Zhuang, L., & Deng, H. (2018). Covalent organic frameworks linked by amine bonding for concerted electrochemical reduction of CO_2. *Chem*, *4*(7), 1696–1709.

Liu, J., You, F., He, B., Wu, Y., Wang, D., Zhou, W., Qian, C., Yang, G., Liu, G., Wang, H., Guo, Y., Gu, L., Feng, L., Li, S., & Zhao, Y. (2022). Directing the architecture of surface-clean Cu_2O for CO electroreduction. *Journal of the American Chemical Society*, *144*(27), 12410–12420.

Liu, S., Tao, H., Zeng, L., Liu, Q., Xu, Z., Liu, Q., & Luo, J. (2017). Shape-dependent electrocatalytic reduction of CO_2 to CO on triangular silver nanoplates. *Journal of the American Chemical Society*, *139*(6), 2160–2163.

Liu, S., Xiao, J., Lu, X. F., Wang, J., Wang, X., & Lou, X. (2019). Efficient electrochemical reduction of CO_2 to HCOOH over Sub-2 nm SnO_2 quantum wires with exposed grain boundaries. *Angewandte Chemie, International Edition*, *58*(25), 8499–8503.

Liu, S., Yang, H. B., Hung, S. F., Ding, J., Cai, W., Liu, L., Gao, J., Li, X., Ren, X., Kuang, Z., Huang, Y., Zhang, T., & Liu, B. (2020). Elucidating the electrocatalytic CO_2 reduction reaction over a model single-atom nickel catalyst. *Angewandte Chemie, International Edition*, *59*(2), 798–803.

Liu, W., Wei, S., Bai, P., Yang, C., & Xu, L. (2021). Robust coal matrix intensifies electron/substrate interaction of nickel-nitrogen (Ni-N) active sites for efficient CO_2 electroreduction at industrial current density. *Applied Catalysis B: Environmental*, *299*, 120661.

Liu, Y., Deb, A., Leung, K. Y., Nie, W., Dean, W. S., Penner-Hahn, J. E., & McCrory, C. C. L. (2020). Determining the coordination environment and electronic structure of polymer-encapsulated cobalt phthalocyanine under electrocatalytic CO_2 reduction conditions using in situ X-Ray absorption spectroscopy. *Dalton Transactions*, *49*(45), 16329–16339.

Liu, Y., Li, Z., Yu, Q., Chen, Y., Chai, Z., Zhao, G., Liu, S., Cheong, W. C., Pan, Y., Zhang, Q., Gu, L., Zheng, L., Wang, Y., Lu, Y., Wang, D., Chen, C., Peng, Q., Liu, Y., Liu, L., … Li, Y. (2019). A general strategy for fabricating isolated single metal atomic site catalysts in Y zeolite. *Journal of the American Chemical Society*, *141*(23), 9305–9311.

Liu, Z., Yang, H., Kutz, R., & Masel, R. I. (2018). CO_2 electrolysis to CO and O_2 at high selectivity, stability and efficiency using Sustainion membranes. *Journal of the Electrochemical Society*, *165*(15), J3371–J3377.

Loiudice, A., Lobaccaro, P., Kamali, E. A., Thao, T., Huang, B. H., Ager, J. W., & Buonsanti, R. (2016). Tailoring copper nanocrystals towards C_2 products in electrochemical CO_2 reduction. *Angewandte Chemie, International Edition*, *55*(19), 5789–5792.

Lu, J., Zhu, C., Pan, C., Lin, W., Lemmon, J. P., Chen, F., Li, C., & Xie, K. (2018). Highly efficient electrochemical reforming of CH_4/CO_2 in a solid oxide electrolyser. Science. *Advances*, *4*(3), eaar5100.

Lu, Q., Rosen, J., Zhou, Y., Hutchings, G. S., Kimmel, Y. C., Chen, J. G., & Jiao, F. (2014). A selective and efficient electrocatalyst for carbon dioxide reduction. *Nature Communications*, *5*(1), 1–6.

Lu, X., Ahsaine, H. A., Dereli, B., Garcia-Esparza, A. T., Reinhard, M., Shinagawa, T., Li, D., Adil, K., Tchalala, M. R., Kroll, T., Eddaoudi, M., Sokaras, D., Cavallo, L., & Takanabe, K. (2021). Operando elucidation on the working state of immobilized fluorinated iron porphyrin for selective aqueous electroreduction of CO_2 to CO. *ACS Catalysis*, *11*, 6499–6509.

Lu, X., Wu, Y., Yuan, X., & Wang, H. (2019). An integrated CO_2 electrolyzer and formate fuel cell enabled by a reversibly restructuring Pb–Pd bimetallic catalyst. *Angewandte Chemie, International Edition*, *58*(12), 4031–4035.

Luc, W., Collins, C., Wang, S., Xin, H., He, K., Kang, Y., & Jiao, F. (2017). Ag–Sn bimetallic catalyst with a core–shell structure for CO_2 reduction. *Journal of the American Chemical Society*, *139*(5), 1885–1893.

Luo, M., Wang, Z., Li, Y. C., Li, J., Li, F., Lum, Y., Nam, D. H., Chen, B., Wicks, J., Xu, A., Zhuang, T., Leow, W. R., Wang, X., Dinh, C. T., Wang, Y., Wang, Y., Sinton, D., & Sargent, E. H. (2019). Hydroxide promotes carbon dioxide electroreduction to ethanol on copper via tuning of adsorbed hydrogen. *Nature Communications*, *10*(1), 1–7.

Luo P, Wang H, Li L, Nan B, Li Y, Lu T. Boosting proton reduction via isolated Cu atoms in Bi lattice for efficient electroreduction of CO_2 to formate. 2021.

Ma, D., Han, S., Cao, C., Li, X., Wu, X., & Zhu, Q. (2020). Remarkable electrocatalytic CO_2 reduction with ultrahigh CO/H_2 ratio over single-molecularly immobilized pyrrolidinonyl nickel phthalocyanine. *Applied Catalysis B: Environmental*, *264*, 118530.

Ma, M., Deng, W., Xu, A., Hochfilzer, D., Qiao, Y., Chan, K., Chorkendorff, I., & Seger, B. (2022). Local reaction environment for selective electroreduction of carbon monoxide. *Energy & Environmental Science*, *15*, 2470–2478.

Ma, S., Luo, R., Moniri, S., Lan, Y., & Kenis, P. J. A. (2014). Efficient electrochemical flow system with improved anode for the conversion of CO_2 to CO. *Journal of the Electrochemical Society, 161*(10), F1124–F1130.

Ma, S., Sadakiyo, M., Heima, M., Luo, R., Haasch, R. T., Gold, J. I., Yamauchi, M., & Kenis, P. J. A. (2017). Electroreduction of carbon dioxide to hydrocarbons using bimetallic Cu–Pd catalysts with different mixing patterns. *Journal of the American Chemical Society, 139*(1), 47–50.

Mariano, R. G., McKelvey, K., White, H. S., & Kanan, M. W. (2017). Selective increase in CO_2 electroreduction activity at grain-boundary surface terminations. *Science, 358*(6367), 1187–1192.

Mezzavilla, S., Horch, S., Stephens, I. E., Seger, B., & Chorkendorff, I. (2019). Structure sensitivity in the electrocatalytic reduction of CO_2 with Gold catalysts. *Angewandte Chemie, International Edition, 58*(12), 3774–3778.

Mi, Y., Peng, X., Liu, X., & Luo, J. (2018). Selective formation of C_2 products from electrochemical CO_2 reduction over $Cu_{1.8}Se$ nanowires. *ACS Applied Energy Materials, 1*(10), 5119–5123.

Min, X., & Kanan, M. W. (2015). Pd-catalyzed electrohydrogenation of carbon dioxide to formate: high mass activity at low overpotential and identification of the deactivation pathway. *Journal of the American Chemical Society, 137*(14), 4701–4708.

Mistry, H., Choi, Y. W., Bagger, A., Scholten, F., Bonifacio, C. S., Sinev, I., Divins, N. J., Zegkinoglou, I., Jeon, H. S., Kisslinger, K., Stach, E. A., Yang, J. C., Rossmeisl, J., & Cuenya, B. R. (2017). Enhanced carbon dioxide electroreduction to carbon monoxide over defect-rich plasma-activated silver catalysts. *Angewandte Chemie, 129*(38), 11552–11556.

Mistry, H., Reske, R., Zeng, Z., Zhao, Z.-J., Greeley, J., Strasser, P., & Cuenya, B. R. (2014). Exceptional size-dependent activity enhancement in the electroreduction of CO_2 over Au nanoparticles. *Journal of the American Chemical Society, 136*(47), 16473–16476.

Monteiro, M. C., Dattila, F., Hagedoorn, B., García-Muelas, R., López, N., & Koper, M. T. M. (2021). Absence of CO_2 electroreduction on copper, gold and silver electrodes without metal cations in solution. *Nature Catalysis, 4*(8), 654–662.

Montoya, J. H., Peterson, A. A., & Nørskov, J. K. (2013). Insights into C-C coupling in CO_2 electroreduction on copper electrodes. *ChemCatChem, 5*(3), 737–742.

Nam, D.-H., De Luna, P., Rosas-Hernández, A., Thevenon, A., Li, F., Agapie, T., et al. (2020). Molecular enhancement of heterogeneous CO_2 reduction. *Nature Materials, 19*(3), 266–276.

Nam, D.-H., Shekhah, O., Lee, G., Mallick, A., Jiang, H., Li, F., Chen, B., Wicks, J., Eddaoudi, M., & Sargent, E. H. (2020). Intermediate binding control using metal–organic frameworks enhances electrochemical CO_2 reduction. *Journal of the American Chemical Society, 142*(51), 21513–21521.

Nie, X., Esopi, M. R., Janik, M. J., & Asthagiri, A. (2013). Selectivity of CO_2 reduction on copper electrodes: the role of the kinetics of elementary steps. *Angewandte Chemie, 125*(9), 2519–2522.

Nie, X., Luo, W., Janik, M. J., & Asthagiri, A. (2014). Reaction mechanisms of CO_2 electrochemical reduction on Cu(111) determined with density functional theory. *Journal of Catalysis, 312*, 108–122.

Orella, M. J., Brown, S. M., Leonard, M. E., Román-Leshkov, Y., & Brushett, F. R. (2020). A general technoeconomic model for evaluating emerging electrolytic processes. *Energy Technology, 8*(11), 1900994.

Ouyang, T., Huang, H. H., Wang, J. W., Zhong, D. C., & Lu, T. B. (2017). A dinuclear cobalt cryptate as a homogeneous photocatalyst for highly selective and efficient visible-light driven CO_2 reduction to CO in CH_3CN/H_2O solution. *Angewandte Chemie, 129*(3), 756–761.

Overa, S., Crandall, B., Shrimant, B., Tian, D., Ko, B. H., Shin, H., Bae, C., & Jiao, F. (2022). Enhancing acetate selectivity by coupling anodic oxidation in carbon monoxide electroreduction. *Nature Catalysis, 5*, 738–745.

Pan, F., Li, B., Xiang, X., Wang, G., & Li, Y. (2019). Efficient CO_2 Electroreduction by highly dense and active pyridinic nitrogen on holey carbon layers with fluorine engineering. *ACS Catalysis, 9*(3), 2124–2133.

Pan, F., Zhang, H., Liu, K., Cullen, D., More, K., Wang, M., Feng, Z., Wang, G., Wu, G., & Li, Y. (2018). Unveiling active sites of CO_2 reduction on nitrogen-coordinated and atomically dispersed iron and cobalt catalysts. *ACS Catalysis, 8*(4), 3116–3122.

Pan, H., Gong, J., & Zhang, Y. (2022). Enabling durable selectivity of CO_2 electroreduction to formate achieved by a multi-layer SnO_x structure. *Applied Surface Science, 579*, 151971.

Pan, L., Kim, J. H., Mayer, M. T., Son, M.-K., Ummadisingu, A., Lee, J. S., Hagfeldt, A., Luo, J. S., & GrätzeL, M. (2018). Boosting the performance of Cu_2O photocathodes for unassisted solar water splitting devices. *Nature Catalysis, 1*(6), 412–420.

Pan, Y., Lin, R., Chen, Y., Liu, S., Zhu, W., Cao, X., Chen, W., Wu, K., Cheong, W. C., Wang, Y., Zheng, L., Luo, J., Lin, Y., Liu, Y., Liu, C., Li, J., Lu, Q., Chen, X., Wang, D., … Li, Y. (2018). Design of single-atom Co–N_5 catalytic site: A robust electrocatalyst for CO_2 reduction with nearly 100% CO selectivity and remarkable stability. *Journal of the American Chemical Society, 140*(12), 4218–4221.

Pander, J. E., III, Baruch, M. F., & Bocarsly, A. B. (2016). Probing the mechanism of aqueous CO_2 reduction on post-transition-metal electrodes using ATR-IR spectroelectrochemistry. *ACS Catalysis*, *6*(11), 7824–7833.

Park, H., Park, I. J., Lee, M. G., Kwon, K. C., Hong, S. P., Kim, D. H., Lee, S. A., Lee, T. H., Kim, C., Moon, C. W., Son, D. Y., Jung, G. H., Yang, H. S., Lee, J. R., Lee, J., Park, N. G., Kim, S. Y., Kim, J. Y., & Jang, H. W. (2019). Water splitting exceeding 17% solar-to-hydrogen conversion efficiency using solution-processed Ni-based electrocatalysts and perovskite/Si tandem solar cell. *ACS Applied Materials & Interfaces*, *11*(37), 33835–33843.

Peterson, A. A., Abild-Pedersen, F., Studt, F., Rossmeisl, J., & Nørskov, J. K. (2010). How copper catalyzes the electroreduction of carbon dioxide into hydrocarbon fuels. *Energy & Environmental Science*, *3*(9), 1311–1315.

Pinna, F. (1998). Supported metal catalysts preparation. *Catalysis Today*, *41*(1-3), 129–137.

Qiao, J., Liu, Y., Hong, F., & Zhang, J. (2014). A review of catalysts for the electroreduction of carbon dioxide to produce low-carbon fuels. *Chemical Society Reviews*, *43*(2), 631–675.

Qin, X., Zhu, S., Xiao, F., Zhang, L., & Shao, M. (2019). Active sites on heterogeneous single-iron-atom electrocatalysts in CO_2 reduction reaction. *ACS Energy Letters*, *4*(7), 1778–1783.

Qiu, X. F., Huang, J. R., Yu, C., Zhao, Z. H., Zhu, H. L., Ke, Z., Liao, P., & Chen, X. (2022). A stable and conductive covalent organic framework with isolated active sites for highly selective electroreduction of carbon dioxide to acetate. *Angewandte Chemie*, *134*(36), e202206470.

Quan, F., Zhong, D., Song, H., Jia, F., & Zhang, L. (2015). A highly efficient zinc catalyst for selective electroreduction of carbon dioxide in aqueous NaCl solution. *Journal of Materials Chemistry A*, *3*(32), 16409–16413.

Ren, J., Ouyang, S., Xu, H., Meng, X., Wang, T., Wang, D., & Ye, J. (2017). Targeting activation of CO_2 and H_2 over Ru-loaded ultrathin layered double hydroxides to achieve efficient photothermal CO_2 methanation in flow-type system. *Advanced Energy Materials*, *7*(5), 1601657.

Ren, S., Joulié, D., Salvatore, D., Torbensen, K., Wang, M., Robert, M., & Berlinguette, C. P. (2019). Molecular electrocatalysts can mediate fast, selective CO_2 reduction in a flow cell. *Science*, *365*(6451), 367–369.

Ren, W., Tan, X., Chen, X., Zhang, G., Zhao, K., Yang, W., Zhao, Y., Smith, S. C., & Zhao, C. (2020). Confinement of ionic liquids at single-Ni-sites boost electroreduction of CO_2 in aqueous electrolytes. *ACS Catalysis*, *10*(22), 13171–13178.

Ren, W., Tan, X., Yang, W., Jia, C., Xu, S., Wang, K., Smith, S. C., & Zhao, C. (2019). Isolated diatomic Ni-Fe metal–nitrogen sites for synergistic electroreduction of CO_2. *Angewandte Chemie, International Edition*, *58*(21), 6972–6976.

Ren, W. H., Tan, X., Jia, C., Krammer, A., Sun, Q., Qu, J. T., Smith, S., Schüler, A., Hu, X. L., & Zhao, C. (2022). Electronic regulation of Ni single atom by confined Ni nanoparticles for fast and energy-efficient CO_2 electroreduction. *Angewandte Chemie*, *61*, e202203335.

Reske, R., Mistry, H., Behafarid, F., Roldan Cuenya, B., & Strasser, P. (2014). Particle size effects in the catalytic electroreduction of CO_2 on Cu nanoparticles. *Journal of the American Chemical Society*, *136*(19), 6978–6986.

Reuillard, B., Ly, K. H., Rosser, T. E., Kuehnel, M. F., Zebger, I., & Reisner, E. (2017). Tuning product selectivity for aqueous CO_2 reduction with a Mn (bipyridine)-pyrene catalyst immobilized on a carbon nanotube electrode. *Journal of the American Chemical Society*, *139*(41), 14425–14435.

Ringe, S., Morales-Guio, C. G., Chen, L. D., Fields, M., Jaramillo, T. F., Hahn, C., & Chan, K. (2020). Double layer charging driven carbon dioxide adsorption limits the rate of electrochemical carbon dioxide reduction on Gold. *Nature Communications*, *11*(1), 1–11.

Roberts, F. S., Kuhl, K. P., & Nilsson, A. (2015). High selectivity for ethylene from carbon dioxide reduction over copper nanocube electrocatalysts. *Angewandte Chemie*, *127*(17), 5268–5271.

Rong, W., Zou, H., Tan, S., Hu, E., Li, F., Tang, C., Dai, H., Wei, S., Ji, Y., & Duan, L. (2022). Few-atom copper catalyst for the electrochemical reduction of CO to acetate: Synergetic catalysis between neighboring Cu atoms. *CCS Chemistry*, 1–13.

Rosen, B. A., Salehi-Khojin, A., Thorson, M. R., Zhu, W., Whipple, D. T., Kenis, P. J., & Masel, R. I. (2011). Ionic liquid–mediated selective conversion of CO_2 to CO at low overpotentials. *Science*, *334*(6056), 643–644.

Rosen, J., Hutchings, G. S., Lu, Q., Forest, R. V., Moore, A., & Jiao, F. (2015). Electrodeposited Zn dendrites with enhanced CO selectivity for electrocatalytic CO_2 reduction. *ACS Catalysis*, *5*(8), 4586–4591.

Salvatore, D. A., Weekes, D. M., He, J., Dettelbach, K. E., Li, Y. C., Mallouk, T. E., & Berlinguette, C. P. (2017). Electrolysis of gaseous CO_2 to CO in a flow cell with a bipolar membrane. *ACS Energy Letters*, *3*(1), 149–154.

Sánchez, O. G., Birdja, Y. Y., Bulut, M., Vaes, J., Breugelmans, T., & Pant, D. (2019). Recent advances in industrial CO_2 electroreduction. *Current Opinion in Green and Sustainable Chemistry*, *16*, 47–56.

Schreier, M., Héroguel, F., Steier, L., Ahmad, S., Luterbacher, J. S., Mayer, M. T., Luo, J. S., & Grätzel, M. (2017). Solar conversion of CO_2 to CO using earth-abundant electrocatalysts prepared by atomic layer modification of CuO. *Nature Energy*, *2*(7), 1–9.

Schuchmann, K., & Müller, V. J. S. (2013). Direct and reversible hydrogenation of CO_2 to formate by a bacterial carbon dioxide reductase. *Science, 342*(6164), 1382–1385.

Shafaque, H. W., Lee, C., Fahy, K. F., Lee, J. K., LaManna, J. M., Baltic, E., Hussey, D. S., Jacobson, D. L., & Bazylak, A. (2020). Boosting membrane hydration for high current densities in membrane electrode assembly CO_2 electrolysis. *ACS Applied Materials & Interfaces, 12*(49), 54585–54595.

Shang, H., Wang, T., Pei, J., Jiang, Z., Zhou, D., Wang, Y., Wang, Y., Li, H., Dong, J., Zhuang, Z., Chen, W., Wang, D., Zhang, J., & Li, Y. (2020). Design of a single-atom indium$^{\delta+}$-N_4 interface for efficient electroreduction of CO_2 to formate. *Angewandte Chemie, International Edition, 59*(50), 22465–22469.

Shao, X., Zhang, X., Liu, Y., Qiao, J., Zhou, X., Xu, N., Malcombe, J. L., Yi, J., & Zhang, J. (2021). Metal chalcogenide-associated catalysts enabling CO_2 electroreduction to produce low-carbon fuels for energy storage and emission reduction: catalyst structure, morphology, performance, and mechanism. *Journal of Materials Chemistry A, 9*, 2526–2559.

Shen, H., Wang, Y., Chakraborty, T., Zhou, G., Wang, C., Fu, X., Zhang, J., Li, C., Xu, F., Cao, L., Mueller, T., & Wang, C. (2022). Asymmetrical C–C coupling for electroreduction of CO on bimetallic Cu–Pd catalysts. *ACS Catalysis, 12*(9), 5275–5283.

Shi, Y., Ji, Y., Long, J., Liang, Y., Liu, Y., Yu, Y., Xiao, J., & Zhang, B. (2020). Unveiling hydrocerussite as an electrochemically stable active phase for efficient carbon dioxide electroreduction to formate. *Nature Communications, 11*(1), 1–10.

Shin, E. K., & Lee, B. C. (2008). High-pressure phase behavior of carbon dioxide with ionic liquids: 1-alkyl-3-methylimidazolium trifluoromethanesulfonate. *Journal of Chemical & Engineering Data, 53*(12), 2728–2734.

Song, Y., Zhou, Z., Zhang, X., Zhou, Y., Gong, H., Lv, H., Liu, Q., Wang, G., & Bao, X. (2018). Pure CO_2 electrolysis over an Ni/YSZ cathode in a solid oxide electrolysis cell. *Journal of Materials Chemistry A, 6*(28), 13661–13667.

Soriano, A. N., Doma, B. T., Jr., & Li, M. (2009). Carbon dioxide solubility in some ionic liquids at moderate pressures. *Journal of the Taiwan Institute of Chemical Engineers, 40*(4), 387–393.

Sreekanth, N., Nazrulla, M. A., Vineesh, T. V., Sailaja, K., & Phani, K. L. (2015). Metal-free boron-doped graphene for selective electroreduction of carbon dioxide to formic acid/formate. *Chemical Communications, 51*(89), 16061–16064.

Sullivan, B. P., Krist, K., & Guard, H. (2012). *Electrochemical and electrocatalytic reactions of carbon dioxide*. Elsevier.

Sun, K., Cheng, T., Wu, L., Hu, Y., Zhou, J., Maclennan, A., Jiang, Z., Gao, Y., Goddard, W. A., & Wang, Z. (2017). Ultrahigh mass activity for carbon dioxide reduction enabled by gold–iron core–shell nanoparticles. *Journal of the American Chemical Society, 139*(44), 15608–15611.

Sun, Q., Jia, C., & Zhao, C. (2022). Ionic liquids for electrochemical CO_2 reduction. *Encyclopedia of Ionic Liquids*.

Sun, Q., Jia, C., Zhao, Y., & Zhao, C. (2022). Single atom-based catalysts for electrochemical CO_2 reduction. *Chinese Journal of Catalysis, 43*(7), 1547–1597.

Sun, Q., Ren, W., Zhao, Y., & Zhao, C. (2021). Gram-scale synthesis of single-atom metal-N-CNT catalysts for highly efficient CO_2 electroreduction. *Chemical Communications, 57*(12), 1514–1517.

Sun, Q., Zhao, Y., Ren, W., & Zhao, C. (2022). Electroreduction of low concentration CO_2 at atomically dispersed Ni-NC catalysts with nanoconfined ionic liquids. *Applied Catalysis B: Environmental, 304*, 120963.

Sun, S., Zhang, G., Gauquelin, N., Chen, N., Zhou, J., Yang, S., Chen, W., Meng, X., Geng, D., Banis, M. N., Li, R., Ye, S., Knights, S., Botton, G. A., Sham, T. K., & Sun, X. L. (2013). Single-atom catalysis using Pt/graphene achieved through atomic layer deposition. *Scientific Reports, 3*(1), 1–9.

Sun, X., Chen, C., Liu, S., Hong, S., Zhu, Q., Qian, Q., Han, B., Zhang, J., & Zheng, L. (2019). Aqueous CO_2 reduction with high efficiency using α-$Co(OH)_2$-supported atomic Ir electrocatalysts. *Angewandte Chemie, 131*(14), 4717–4721.

Sun, X., Zhu, Q., Kang, X., Liu, H., Qian, Q., Zhang, Z., & Han, B. (2016). Molybdenum–bismuth bimetallic chalcogenide nanosheets for highly efficient electrocatalytic reduction of carbon dioxide to methanol. *Angewandte Chemie, 128*(23), 6883–6887.

Sun, Z., Liu, C., Lu, G., Song, X., Sun, S., Sun, Y., & Yu, J. (2011). Coupled thermoelectric model and effects of current fluctuation on thermal balance in magnesium electrolysis cell. *Energy & Fuels, 25*(6), 2655–2663.

Tan, W., Cao, B., Xiao, W., Zhang, M., Wang, S., Xie, S., Xie, D., Cheng, F., Guo, Q., & Liu, P. (2019). Electrochemical reduction of CO_2 on hollow cubic Cu_2O@Au nanocomposites. *Nanoscale Research Letters, 14*(1), 1–7.

Tang, B., & Xiao, F. X. (2022). An overview of solar-driven photoelectrochemical CO_2 conversion to chemical fuels. *ACS Catalysis, 12*(15), 9023–9057.

Todorova, T. K., Schreiber, M. W., & Fontecave, M. (2020). Mechanistic Understanding of CO_2 reduction reaction (CO_2RR) towards multicarbon products by heterogeneous copper-based catalysts. *ACS Catalysis, 10*(3), 1754–1768.

Torelli, D. A., Francis, S. A., Crompton, J. C., Javier, A., Thompson, J. R., Brunschwig, B. S., Soriaga, M. P., & Lewis, N. S. (2016). Nickel–gallium-catalyzed electrochemical reduction of CO_2 to highly reduced products at low overpotentials. *ACS Catalysis*, *6*(3), 2100–2104.

Varela, A. S., Ju, W., & Strasser, P. (2018). Molecular nitrogen–carbon catalysts, solid metal organic framework catalysts, and solid metal/nitrogen-doped carbon (MNC) catalysts for the electrochemical CO_2 reduction. *Advanced Energy Materials*, *8*(30), 1703614.

Varela, A. S., Kroschel, M., Leonard, N. D., Ju, W., Steinberg, J., Bagger, A., Rossmeisl, J., & Strasser, P. (2018). ph effects on the selectivity of the electrocatalytic CO_2 reduction on graphene-embedded Fe–N–C motifs: bridging concepts between molecular homogeneous and solid-state heterogeneous catalysis. *ACS Energy Letters*, *3*(4), 812–817.

Varela, A. S., Kroschel, M., Reier, T., & Strasser, P. (2016). Controlling the selectivity of CO_2 electroreduction on copper: the effect of the electrolyte concentration and the importance of the local pH. *Catalysis Today*, *260*, 8–13.

Varela, A. S., Schlaup, C., Jovanov, Z. P., Malacrida, P., Horch, S., Stephens, I. E. L., & Chorkendorff, I. (2013). CO_2 electroreduction on well-defined bimetallic surfaces: Cu overlayers on Pt(111) and Pt(211). *The Journal of Physical Chemistry C*, *117*(40), 20500–20508.

Vargas-Barbosa, N. M., Geise, G. M., Hickner, M. A., & Mallouk, T. E. (2014). Assessing the utility of bipolar membranes for use in photoelectrochemical water-splitting cells. *ChemSusChem*, *7*(11), 3017–3020.

Vasileff, A., Xu, C., Jiao, Y., Zheng, Y., & Qiao, S. (2018). Surface and interface engineering in copper-based bimetallic materials for selective CO_2 electroreduction. *Chem*, *4*(8), 1809–1831.

Vennekötter, J.-B., Scheuermann, T., Sengpiel, R., & Wessling, M. (2019). The electrolyte matters: stable systems for high rate electrochemical CO_2 reduction. *Journal of CO_2 Utilization*, *32*, 202–213.

Verma, S., Hamasaki, Y., Kim, C., Huang, W., Lu, S., Jhong, H. R. M., Gewirth, A. A., Fujigaya, T., Nakashima, N., & Kenis, P. J. A. (2017). Insights into the low overpotential electroreduction of CO_2 to CO on a supported gold catalyst in an alkaline flow electrolyzer. *ACS Energy Letters*, *3*(1), 193–198.

Verma, S., Kim, B., Jhong, H. R. M., Ma, S., & Kenis, P. J. A. (2016). A gross-margin model for defining technoeconomic benchmarks in the electroreduction of CO_2. *ChemSusChem*, *9*(15), 1972–1979.

Wan, J., Chen, W., Jia, C., Zheng, L., Dong, J., Zheng, X., Wang, Y., Yan, W., Chen, C., Peng, Q., Wang, D., & Li, Y. (2018). Defect effects on TiO_2 nanosheets: stabilizing single atomic site Au and promoting catalytic properties. *Advanced Materials*, *30*(11), 1705369.

Wang, H., Jia, J., Song, P., Wang, Q., Li, D., Min, S., Qian, C., Wang, L., Li, Y., Ma, C., Wu, T., Yuan, J., Antonietti, M., & Ozin, G. A. (2017). Efficient electrocatalytic reduction of CO_2 by nitrogen-doped nanoporous carbon/carbon nanotube membranes: a step towards the electrochemical CO_2 refinery. *Angewandte Chemie*, *129*(27), 7955–7960.

Wang, L., Chen, W., Zhang, D., Du, Y., Amal, R., Qiao, S., Wu, J., & Yin, Z. (2019). Surface strategies for catalytic CO_2 reduction: from two-dimensional materials to nanoclusters to single atoms. *Chemical Society Reviews*, *48*(21), 5310–5349.

Wang, P., Qiao, M., Shao, Q., Pi, Y., Zhu, X., Li, Y., & Hunag, X. (2018). Phase and structure engineering of copper tin heterostructures for efficient electrochemical carbon dioxide reduction. *Nature Communications*, *9*(1), 1–10.

Wang, Q., Wang, X., Wu, C., Cheng, Y., Sun, Q., Dong, H., & Yu, H. (2017). Electrodeposition of tin on Nafion-bonded carbon black as an active catalyst layer for efficient electroreduction of CO_2 to formic acid. *Scientific Reports*, *7*(1), 1–11.

Wang, X., Feng, S., Lu, W., Zhao, Y., Zheng, S., Zheng, W., Sang, X., Zheng, L., Xie, Y., Li, Z., Yang, B., Lei, L., Wang, S., & Hou, Y. (2021). A new strategy for accelerating dynamic proton transfer of electrochemical CO_2 reduction at high current densities. *Advanced Functional Materials*, *31*(50), 2104243.

Wang, X., Ou, P., Ozden, A., Hung, S. F., Tam, J., Gabardo, C. M., Howe, J. Y., Sisler, J., Bertens, K., García de Arquer, F. P., Miao, R., O'Brien, C. P., Wang, Z., Abed, J., Rasouli, A. S., Sun, M., Ip, A. H., Sinton, D., & Sargent, E. H. (2022). Efficient electrosynthesis of n-propanol from carbon monoxide using a Ag–Ru–Cu catalyst. *Nature Energy*, *7*(2), 170–176.

Wang, Y., Chen, Z., Han, P., Du, Y., Gu, Z., Xu, X., & Zheng, G. (2018). Single-atomic Cu with multiple oxygen vacancies on ceria for electrocatalytic CO_2 reduction to CH_4. *ACS Catalysis*, *8*(8), 7113–7119.

Wang, Y., Han, P., Lv, X., Zhang, L., & Zheng, G. (2018). Defect and interface engineering for aqueous electrocatalytic CO_2 reduction. *Joule*, *2*(12), 2551–2582.

Wang, Y., Huang, Q., He, C., Chen, Y., Liu, J., Shen, F., & Lan, Y. (2018). Oriented electron transmission in polyoxometalate-metalloporphyrin organic framework for highly selective electroreduction of CO_2. *Nature Communications*, *9*(1), 1–8.

Wang, Y., Wang, D., Dares, C., & Meyer, T. J. (2017). Single-site, heterogeneous electrocatalytic reduction of CO_2 in water as the solvent. *ACS Energy Letters*, *2*(6), 1395–1399.

Wang, Z., Yang, G., Zhang, Z., Jin, M., & Yin, Y. (2016). Selectivity on etching: Creation of high-energy facets on copper nanocrystals for CO_2 electrochemical reduction. *ACS Nano*, *10*(4), 4559–4564.

Weekes, D. M., Salvatore, D. A., Reyes, A., Huang, A., & Berlinguette, C. P. (2018). Electrolytic CO_2 reduction in a flow cell. *Accounts of Chemical Research*, *51*(4), 910–918.

Wei, F., Wang, T., Jiang, X., Ai, Y., Cui, A., Cui, J., Fu, J., Cheng, J., Lei, L., Hou, Y., & Liu, S. (2020). Controllably engineering mesoporous surface and dimensionality of SnO_2 toward high-performance CO_2 electroreduction. *Advanced Functional Materials*, *30*(39), 2002092.

Wen, G., Ren, B., Wang, X., Luo, D., Dou, H., Zheng, Y., Gao, R., Gostick, J., Yu, A., & Chen, Z. (2022). Continuous CO_2 electrolysis using a CO_2 exsolution-induced flow cell. *Nature Energy*, *7*, 978–988.

Weng, W., Tang, L., & Xiao, W. (2019). Capture and electro-splitting of CO_2 in molten salts. *Journal of Energy Chemistry*, *28*, 128–143.

Weng, Z., Jiang, J., Wu, Y., Wu, Z., Guo, X., Materna, K. L., Liu, W., Batista, V. S., Brudvig, G. W., & Wang, H. (2016). Electrochemical CO_2 reduction to hydrocarbons on a heterogeneous molecular Cu catalyst in aqueous solution. *Journal of the American Chemical Society*, *138*(26), 8076–8079.

Weng, Z., Wu, Y., Wang, M., Jiang, J., Yang, K., Huo, S., Wang, X., Ma, Q., Brudvig, G. W., Batista, V. S., Liang, Y., Feng, Z., & Wang, H. (2018). Active sites of copper-complex catalytic materials for electrochemical carbon dioxide reduction. *Nature Communications*, *9*(1), 1–9.

Whipple, D. T., & Kenis, P. J. (2010). Prospects of CO_2 utilization via direct heterogeneous electrochemical reduction. *The Journal of Physical Chemistry Letters*, *1*(24), 3451–3458.

Wicks, J., Jue, M. L., Beck, V. A., Oakdale, J. S., Dudukovic, N. A., Clemens, A. L., Liang, S., Ellis, M. E., Lee, G., Baker, S. E., Duoss, E. B., & Sargent, E. H. (2021). 3D-printable fluoropolymer gas diffusion layers for CO_2 electroreduction. *Advanced Materials*, *33*(7), 2003855.

Won, D. H., Shin, H., Koh, J., Chung, J., Lee, H. S., Kim, H., & Woo, S. I. (2016). Highly efficient, selective, and stable CO_2 electroreduction on a hexagonal Zn catalyst. *Angewandte Chemie, International Edition*, *55*(32), 9297–9300.

Wu, J., Ma, S., Sun, J., Gold, J. I., Tiwary, C., Kim, B., Zhu, L. Y., Chopra, N., Odeh, I. N., Vajtai, R., Yu, A. Z., Luo, R., Lou, J., Ding, G. Q., Kenis, P. J. A., & Ajayan, P. M. (2016). A metal-free electrocatalyst for carbon dioxide reduction to multi-carbon hydrocarbons and oxygenates. *Nature Communications*, *7*(1), 1–6.

Wulan, B., Zhao, L., Tan, D., Cao, X., Ma, J., & Zhang, J. (2022). Electrochemically driven interfacial transformation for high-performing solar-to-fuel electrocatalytic conversion. *Advanced Energy Materials*, *12*, 2103960.

Xia, C., Zhu, P., Jiang, Q., Pan, Y., Liang, W., Stavitsk, E., Alshareef, H. N., & Wang, H. (2019). Continuous production of pure liquid fuel solutions via electrocatalytic CO_2 reduction using solid-electrolyte devices. *Nature Energy*, *4*(9), 776–785.

Xia, R., Lv, J., Ma, X., & Jiao, F. (2021). Enhanced multi-carbon selectivity via CO electroreduction approach. *Journal of Catalysis*, *398*, 185–191.

Xiao, W., & Wang, D. (2014). The electrochemical reduction processes of solid compounds in high temperature molten salts. *Chemical Society Reviews*, *43*(10), 3215–3228.

Xie, J., Zhao, X., Wu, M., Li, Q., Wang, Y., & Yao, J. (2018). Metal-free fluorine-doped carbon electrocatalyst for CO_2 reduction outcompeting hydrogen evolution. *Angewandte Chemie*, *130*(31), 9788–9792.

Xu, Y., Edwards, J. P., Liu, S., Miao, R. K., Huang, J. E., Gabardo, C. M., O'Brien, C. P., Li, J., Sargent, E. H., & Sinton, D. (2021). Self-cleaning CO_2 reduction systems: unsteady electrochemical forcing enables stability. *ACS Energy Letters*, *6*(2), 809–815.

Yang, D., Zhu, Q., Chen, C., Liu, H., Liu, Z., Zhao, Z., Zhang, X., Liu, S., & Han, B. (2019). Selective electroreduction of carbon dioxide to methanol on copper selenide nanocatalysts. *Nature Communications*, *10*(1), 1–9.

Yang, F., Elnabawy, A. O., Schimmenti, R., Song, P., Wang, J., Peng, Z., Yao, S., Deng, R., Song, S., Lin, Y., Mavrikakis, M., & Xu, W. (2020). Bismuthene for highly efficient carbon dioxide electroreduction reaction. *Nature Communications*, *11*(1), 1–8.

Yang, F., Song, P., Liu, X., Mei, B., Xing, W., Jiang, Z., Gu, L., & Xu, W. (2018). Highly efficient CO_2 electroreduction on ZnN_4-based single-atom catalyst. *Angewandte Chemie, International Edition*, *57*(38), 12303–12307.

Yang, H., Kaczur, J. J., Sajjad, S. D., & Masel, R. I. (2017). Electrochemical conversion of CO_2 to formic acid utilizing Sustainion™ membranes. *Journal of CO_2 Utilization*, *20*, 208–217.

Yang, H., Shang, L., Zhang, Q., Shi, R., Waterhouse, G. I., Gu, L., & Zhang, T. (2019). A universal ligand mediated method for large scale synthesis of transition metal single atom catalysts. *Nature Communications*, *10*(1), 1–9.

Yang, H., Wu, Y., Li, G., Lin, Q., Hu, Q., Zhang, Q., Liu, J., & He, C. (2019). Scalable production of efficient single-atom copper decorated carbon membranes for CO_2 electroreduction to methanol. *Journal of the American Chemical Society*, *141*(32), 12717–12723.

Yang, H., Wu, Y., Lin, Q., Fan, L., Chai, X., Zhang, Q., Liu, J., He, C., & Lin, Z. (2018). Composition tailoring via N and S co-doping and structure tuning by constructing hierarchical pores: metal-free catalysts for high-performance electrochemical reduction of CO_2. *Angewandte Chemie, 130*(47), 15702–15706.

Yang, H. B., Hung, S., Liu, S., Yuan, K., Miao, S., Zhang, L., Huang, X., Wang, H., Cai, W., Chen, R., Gao, J., Yang, X., Chen, W., Huang, Y., Chen, H., Li, C., Zhang, T., & Liu, B. (2018). Atomically dispersed Ni (I) as the active site for electrochemical CO_2 reduction. *Nature Energy, 3*(2), 140–147.

Yang, J., Qiu, Z., Zhao, C., Wei, W., Chen, W., Li, Z., Qu, Y., Dong, J., Luo, J., Li, Z., & Wu, Y. (2018). In situ thermal atomization to convert supported nickel nanoparticles into surface-bound nickel single-atom catalysts. *Angewandte Chemie, International Edition, 57*(43), 14095–14100.

Yang, P., Zhang, X., Gao, F., Zheng, Y., Niu, Z., Yu, X., Liu, R., Wu, Z., Qin, S., Chi, L., Duan, Y., Ma, T., Zheng, X., Zhu, J., Wang, H., Gao, M., & Yu, S. (2020). Protecting copper oxidation state via intermediate confinement for selective CO_2 electroreduction to C_{2+} Fuels. *Journal of the American Chemical Society, 142*(13), 6400–6408.

Yang, Q., Wu, Q., Liu, Y., Luo, S., Wu, X., Zhao, X., Zou, H., Long, B., Chen, W., Liao, Y., Li, L., Shen, P., Duan, L., & Quan, Z. (2020). Novel Bi-doped amorphous SnO_x nanoshells for efficient electrochemical CO_2 reduction into formate at low overpotentials. *Advanced Materials, 32*(36), 2002822.

Yang, W., Chen, S., Ren, W., Zhao, Y., Chen, X., Jia, C., Liu, J., & Zhao, C. (2019). Nanostructured amalgams with tuneable silver–mercury bonding sites for selective electroreduction of carbon dioxide into formate and carbon monoxide. *Journal of Materials Chemistry A, 7*(26), 15907–15912.

Yang, W., Dastafkan, K., Jia, C., & Zhao, C. (2018). Design of electrocatalysts and electrochemical cells for carbon dioxide reduction reactions. *Advanced Materials Technologies, 3*(9), 1700377.

Yang, W., Ma, W., Zhang, Z., & Zhao, C. (2018). Ligament size-dependent electrocatalytic activity of nanoporous Ag network for CO_2 reduction. *Faraday Discussions, 210*, 289–299.

Yang, Y., Roh, I., Louisia, S., Chen, C., Jin, J., Yu, S., Salmeron, M. B., Wang, C., & Yang, P. (2022). Operando resonant soft X-ray scattering studies of chemical environment and interparticle dynamics of Cu nanocatalysts for CO_2 electroreduction. *Journal of the American Chemical Society, 144*(20), 8927–8931.

Ye, K., Zhou, Z., Shao, J., Lin, L., Gao, D., Ta, N., Si, R., Wang, G., & Bao, X. (2020). In situ reconstruction of a hierarchical Sn-Cu/SnO_x core/shell catalyst for high-performance CO_2 electroreduction. *Angewandte Chemie, International Edition, 59*(12), 4814–4821.

Yin, H., Mao, X., Tang, D., Xiao, W., Xing, L., Zhu, H., Wang, D., & Sadoway, D. R. (2013). Capture and electrochemical conversion of CO_2 to value-added carbon and oxygen by molten salt electrolysis. *Energy & Environmental Science, 6*(5), 1538–1545.

Yin, Z., Yu, C., Zhao, Z., Guo, X., Shen, M., Li, N., Muzzio, M., Li, J., Liu, H., Lin, H., Yin, J., Lu, G., Su, D., & Sun, S. (2019). Cu_3N nanocubes for selective electrochemical reduction of CO_2 to ethylene. *Nano Letters, 19*(12), 8658–8663.

Yu, D., Gao, L., Sun, T., Guo, J., Yuan, Y., Zhang, J., Li, M., Li, X., Liu, M., Ma, C., Liu, Q., Pan, A., Yang, J., & Huang, H. (2021). Strain-stabilized metastable face-centered tetragonal gold overlayer for efficient CO_2 electroreduction. *Nano Letters, 21*(2), 1003–1010.

Yuan, Y., Wang, Q., Qiao, Y., Chen, X., Yang, Z., Lai, W., Chen, T., Zhang, G., Duan, H., Liu, M., & Huang, H. (2022). In situ structural reconstruction to generate the active sites for CO_2 electroreduction on bismuth ultrathin nanosheets. *Advanced Energy Materials, 12*(29), 2200970.

Zeng, L., Shi, J., Luo, J., & Chen, H. (2018). Silver sulfide anchored on reduced graphene oxide as a high-performance catalyst for CO_2 electroreduction. *Journal of Power Sources, 398*, 83–90.

Zhang, A., Liang, Y., Li, H., Zhang, B., Liu, Z., Chang, Q., Zhang, H., Zhu, C., Geng, Z., Zhu, W., & Zeng, J. (2020). In-situ surface reconstruction of InN nanosheets for efficient CO_2 electroreduction into formate. *Nano Letters, 20*(11), 8229–8235.

Zhang, E., Wang, T., Yu, K., Liu, J., Chen, W., Li, A., Rong, H., Lin, R., Ji, S., Zheng, X., Wang, Y., Zheng, L., Chen, C., Wang, D., Zhang, J., & Li, Y. (2019). Bismuth single atoms resulting from transformation of metal–organic frameworks and their use as electrocatalysts for CO_2 reduction. *Journal of the American Chemical Society, 141*(42), 16569–16573.

Zhang, J., Luo, W., & Züttel, A. (2020). Crossover of liquid products from electrochemical CO_2 reduction through gas diffusion electrode and anion exchange membrane. *Journal of Catalysis, 385*, 140–145.

Zhang, L., Hu, S., Zhu, X., & Yang, W. (2017). Electrochemical reduction of CO_2 in solid oxide electrolysis cells. *Journal of Energy Chemistry, 26*(4), 593–601.

Zhang, L., Zhao, Z. J., & Gong, J. L. (2017). Nanostructured materials for heterogeneous electrocatalytic CO_2 reduction and their related reaction mechanisms. *Angewandte Chemie, International Edition, 56*(38), 11326–11353.

Zhang, S., Kang, P., Ubnoske, S., Brennaman, M. K., Song, N., House, R. L., Glass, J. T., & Meyer, T. J. (2014). Polyethylenimine-enhanced electrocatalytic reduction of CO_2 to formate at nitrogen-doped carbon nanomaterials. *Journal of the American Chemical Society, 136*(22), 7845–7848.

Zhang, S., Sun, C., Zeng, J., Sun, Q., Wang, G., Wang, Y., Wu, Y., Dou, S., Gao, M., & Li, Z. (2016). Ambient aqueous synthesis of ultrasmall PEGylated $Cu_{2-x}Se$ nanoparticles as a multifunctional theranostic agent for multimodal imaging guided photothermal therapy of cancer. *Advanced Materials, 28*(40), 8927–8936.

Zhang, S., Zhang, J., Zhang, Y., & Deng, Y. (2017). Nanoconfined ionic liquids. *Chemical Reviews, 117*(10), 6755–6833.

Zhang, X., Liu, Z., & Wang, W. (2008). Screening of ionic liquids to capture CO_2 by COSMO-RS and experiments. *AICHE Journal, 54*(10), 2717–2728.

Zhang, X., Wang, Y., Gu, M., Wang, M., Zhang, Z., Pan, W., Jiang, Z., Zheng, H., Lucero, M., Wang, H., Sterbinsky, G. E., Ma, Q., Wang, Y., Feng, Z., Li, J., Dai, H., & Liang, Y. (2020). Molecular engineering of dispersed nickel phthalocyanines on carbon nanotubes for selective CO_2 reduction. *Nature Energy, 5*(9), 684–692.

Zhang, X., Wu, Z., Zhang, X., Li, L., Li, Y., Xu, H., Li, X., Yu, X., Zhang, Z., Liang, Y., & Wang, H. (2017). Highly selective and active CO_2 reduction electrocatalysts based on cobalt phthalocyanine/carbon nanotube hybrid structures. *Nature Communications, 8*(1), 1–8.

Zhang, Y. J., Sethuraman, V., Michalsky, R., & Peterson, A. A. (2014). Competition between CO_2 reduction and H_2 evolution on transition-metal electrocatalysts. *ACS Catalysis, 4*(10), 3742–3748.

Zhang, Z., Liu, C., Brosnahan, J. T., Zhou, H., Xu, W., & Zhang, S. (2019). Revealing structural evolution of PbS nanocrystal catalysts in electrochemical CO_2 reduction using in situ synchrotron radiation X-ray diffraction. *Journal of Materials Chemistry A, 7*(41), 23775–23780.

Zhao, C. M., Dai, X. Y., Yao, T., Chen, W. X., Wang, X. Q., Wang, J., Yang, J., Wei, S. Q., Wu, Y., & Li, Y. D. (2017). Ionic exchange of metal–organic frameworks to access single nickel sites for efficient electroreduction of CO_2. *Journal of the American Chemical Society, 139*(24), 8078–8081.

Zhao, D., Chen, Z., Yang, W., Liu, S., Zhang, X., Yu, Y., Cheong, W. C., Zheng, L., Ren, F., Ying, G., Cao, X., Wang, D., Peng, Q., Wang, G., & Chen, C. (2019). MXene (Ti_3C_2) vacancy-confined single-atom catalyst for efficient functionalization of CO_2. *Journal of the American Chemical Society, 141*(9), 4086–4093.

Zhao, K., Liu, Y., Quan, X., Chen, S., & Yu, H. (2017). CO_2 electroreduction at low overpotential on oxide-derived Cu/carbons fabricated from metal organic framework. *ACS Applied Materials & Interfaces, 9*(6), 5302–5311.

Zhao, K., Nie, X., Wang, H., Chen, S., Quan, X., Yu, H., Choi, W., Zhang, G., Kim, B., & Chen, J. G. (2020). Selective electroreduction of CO_2 to acetone by single copper atoms anchored on N-doped porous carbon. *Nature Communications, 11*(1), 1–10.

Zhao, W., Yang, L., Yin, Y., & Jin, M. (2014). Thermodynamic controlled synthesis of intermetallic Au_3Cu alloy nanocrystals from Cu microparticles. *Journal of Materials Chemistry A, 2*(4), 902–906.

Zhao, X., Luo, B., Long, R., Wang, C., & Xiong, Y. J. (2015). Composition-dependent activity of Cu–Pt alloy nanocubes for electrocatalytic CO_2 reduction. *Journal of Materials Chemistry A, 3*(8), 4134–4138.

Zhao, Y., Liang, J., Wang, C., Ma, J., & Wallace, G. G. (2018). Tunable and efficient tin modified nitrogen-doped carbon nanofibers for electrochemical reduction of aqueous carbon dioxide. *Advanced Energy Materials, 8*(10), 1702524.

Zhao, Y., Tan, X., Yang, W., Jia, C., Chen, X., Ren, W., Smith, S. C., & Zhao, C. (2020). Surface reconstruction of ultrathin palladium nanosheets during electrocatalytic CO_2 reduction. *Angewandte Chemie, 132*(48), 21677–21682.

Zhao, Z., Peng, X., Liu, X., Sun, X., Shi, J., Han, L., Li, G., & Luo, J. (2017). Efficient and stable electroreduction of CO_2 to CH_4 on CuS nanosheet arrays. *Journal of Materials Chemistry A, 5*(38), 20239–20243.

Zheng, T., Jiang, K., Ta, N., Hu, Y., Zeng, J., Liu, J., & Wang, H. (2019). Large-scale and highly selective CO_2 electrocatalytic reduction on nickel single-atom catalyst. *Joule, 3*(1), 265–278.

Zheng, X., Ji, Y., Tang, J., Wang, J., Liu, B., Steinrück, H. G., Lim, K., Li, Y., Toney, M. F., Chan, K., & Cui, Y. (2019). Theory-guided Sn/Cu alloying for efficient CO_2 electroreduction at low overpotentials. *Nature Catalysis, 2*(1), 55–61.

Zhong, M., Tran, K., Min, Y., Wang, C., Wang, Z., Dinh, C. T., Luna, P. D., Yu, Z., Rasouli, A. S., Brodersen, P., Sun, S., Voznyy, O., Tan, C. S., Askerka, M., Che, F., Liu, M., Seifitokaldani, A., Pang, Y., Lo, S. C., … Sargent, E. H. (2020). Accelerated discovery of CO_2 electrocatalysts using active machine learning. *Nature, 581*(7807), 178–183.

Zhou, Y., Ganganahalli, R., Verma, S., Tan, H. R., & Yeo, B. S. (2022). Production of C_3–C_6 acetate esters via CO electroreduction in a membrane electrode assembly cell. *Angewandte Chemie, International Edition, 61*, e202202859.

Zhu, D. D., Liu, J. L., & Qiao, S. Z. (2016). Recent advances in inorganic heterogeneous electrocatalysts for reduction of carbon dioxide. *Advanced Materials, 28*(18), 3423–3452.

Zhu, H., Huang, J., Zhang, X., Wang, C., Huang, N., Liao, P. Q., & Chen, X. (2021). Highly efficient electroconversion of CO_2 into CH_4 by a metal–organic framework with trigonal pyramidal Cu(I)N_3 active sites. *ACS Catalysis, 11*(18), 11786–11792.

Zhu, P., Xia, C., Liu, C., Jiang, K., Gao, G., Zhang, X., Xia, Y., Lei, Y., Alshareef, H. N., Senftle, T. P., & Wang, H. (2021). Direct and continuous generation of pure acetic acid solutions via electrocatalytic carbon monoxide reduction. *Proceedings of the National Academy of Sciences, 118*(2), e2010868118.

Zhu, Q., Sun, X., Yang, D., Ma, J., Kang, X., Zheng, L., et al. (2019). Carbon dioxide electroreduction to C_2 products over copper-cuprous oxide derived from electrosynthesized copper complex. *Nature Communications, 10*(1), 1–11.

Zhu, W., Michalsky, R., On, M., Lv, H., Guo, S., Wright, C. J., Sun, X., Peterson, A. A., & Sun, S. (2013). Monodisperse Au nanoparticles for selective electrocatalytic reduction of CO_2 to CO. *Journal of the American Chemical Society, 135*(45), 16833–16836.

Zhu, W., Zhang, L., Yang, P., Hu, C., Luo, Z., Chang, X., Zhao, Z., & Gong, J. (2018). Low-coordinated edge sites on ultrathin palladium nanosheets boost carbon dioxide electroreduction performance. *Angewandte Chemie, 130*(36), 11718–11722.

Zhu, W., Zhang, Y.-J., Zhang, H., Lv, H., Li, Q., Michalsky, R., Peterson, A. A., & Sun, S. (2014). Active and selective conversion of CO_2 to CO on ultrathin Au nanowires. *Journal of the American Chemical Society, 136*(46), 16132–16135.

Zhuang, T., Liang, Z., Seifitokaldani, A., Li, Y., De Luna, P., Burdyny, T., Che, F., Meng, F., Min, Y., Quintero-Bermudez, R., Dinh, C. T., Pang, Y., Zhong, M., Zhang, B., Li, J., Chen, P., Zheng, X., Liang, H., Ge, W., … Sargent, E. H. (2018). Steering post-C–C coupling selectivity enables high efficiency electroreduction of carbon dioxide to multi-carbon alcohols. *Nature Catalysis, 1*(6), 421–428.

Zu, X., Li, X., Liu, W., Sun, Y., Xu, J., Yao, T., Yan, W., Gao, S., Wang, C., Wei, S., & Xie, Y. (2019). Efficient and robust carbon dioxide electroreduction enabled by atomically dispersed $Sn^{\delta+}$ sites. *Advanced Materials, 31*(15), 1808135.

10

Ceramic Materials for Photocatalytic/Photoelectrochemical Fuel Generation

Appu V. Raghu[1] *and Takashi Tachikawa*[1,2]

[1] *Laser Molecular Photoscience Laboratory, Molecular Photoscience Research Center, Kobe University, Kobe, Japan*
[2] *Department of Chemistry, Graduate School of Science, Kobe University, Kobe, Japan*

10.1 Introduction

Fossil fuels account for more than 82% of the global energy supply (Biresselioglu & Yelkenci, 2016; Nguyen et al., 2020). Though the global energy demand is increasing, the world can no more rely on fossil fuels for energy due to the fact that the combustion of fossil fuels causes greenhouse gas accumulation in the atmosphere and associated climate change (Hunt et al., 2020; Kirkinen et al., 2008). Because solar energy is extensively distributed, abundant, and inexhaustible, it has become a promising alternative to the conventional fossil fuels (Bie et al., 2021; Jones & Olsson, 2017). The conversion of energy harvested from sun to fuels is attractive because fuels are storable, usable upon demand, and transportable efficiently. Sustainable development requires the establishment of a clean energy system to avoid environmental problems (He & Janáky, 2020). Hydrogen is the most effective and attractive solar fuel because it offers high gravimetric energy density and can burn without generating any pollutants. Solar-driven hydrogen production via water splitting has drawn extensive attention as a green energy technology because it is clean and renewable (Kim et al., 2019; Staffell et al., 2019). Fujishima and Honda discovered the splitting of water into hydrogen and oxygen by illuminating a TiO_2 electrode with ultraviolet (UV) light in a photoelectrochemical (PEC) cell in 1972 (Fujishima & Honda, 1972). TiO_2, a ceramic semiconductor, functioned as a photocatalyst that facilitated the conversion of solar energy to chemical energy in the bonds of hydrogen molecules, allowing long-term storage for meeting energy needs at a later time (Nakata & Fujishima, 2012). A visible light-active semiconductor photocatalyst in the form of suspended particles in a suitable electrolyte (referred as a photocatalytic system) or an electrode in a PEC cell can be used for both energy conversion and environmental remediation. Nevertheless, insufficient energy conversion efficiency and stability are problems with many photocatalytic systems, and this has limited the advancement of the photocatalyst-based solar fuel generation technologies (Kubacka et al., 2012; Ros et al., 2020). Research efforts have been intensified in recent years to solve problems with photocatalysts, especially ceramics, due to the promising nature of photocatalytic/PEC water splitting as a sustainable technology (Bak et al., 2010; Kumar et al., 2016; Lhermitte & Sivula, 2019; Peerakiatkhajohn et al., 2021; Wick & Tilley, 2016; Yang & Wang, 2018).

10.2 Photocatalytic/Photoelectrochemical Fuel Generation

The concept of solar to chemical energy conversion by a PEC process is based on using n-type semiconductors for oxidation and/or p-type semiconductors for reduction. A PEC reaction occurs under a light illumination with a photon energy larger than the bandgap energy of semiconductor. The key role in a PEC reaction is handled by the photogenerated minority carriers (holes for the n-type semiconductor and electrons for the p-type semiconductor). For an n-type semiconductor used as a photoanode or a p-type semiconductor used as a photocathode in a PEC cell, the minority carriers in the semiconductor can migrate to the interface between the semiconductor and electrolyte under the influence of the electric field in the depletion layer formed in the semiconductor. The minority carriers are responsible for the PEC reaction at the surface of the semiconductor, causing oxidation of the electrolyte species for n-type semiconductors and reduction for p-type semiconductors (Bisquert et al., 2016; Iqbal & Siddique, 2018; Luo et al., 2019; Zhang, Nagashima, & Tachikawa, 2020).

Conversion of Water and CO_2 to Fuels using Solar Energy: Science, Technology and Materials, First Edition.
Edited by Oomman K. Varghese and Flavio L. Souza.

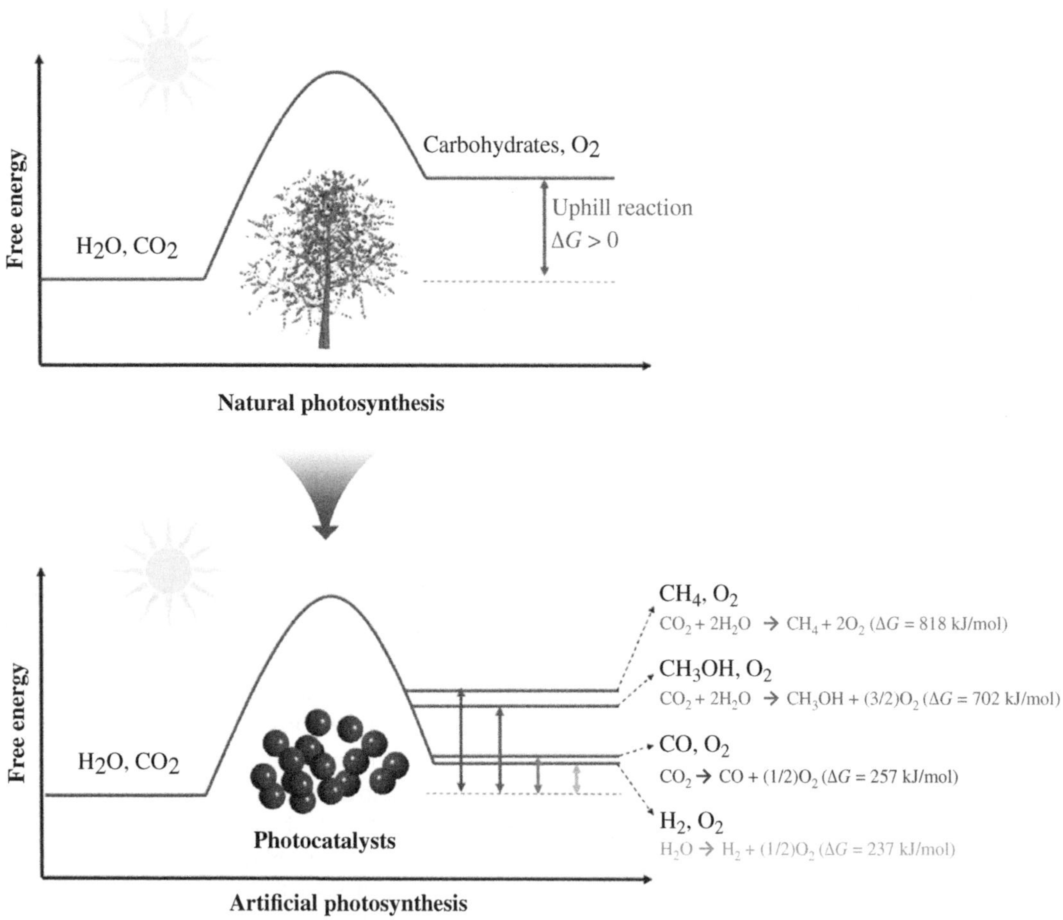

Figure 10.1 Schematic illustration of natural and artificial photosynthetic systems. Source: Reproduced with permission from Wang & Domen, 2020. Copyright 2020 American Chemical Society (Wang & Domen, 2020)/Springer Nature.

A PEC device can use solar energy directly to split water into hydrogen and oxygen, while the conversion is indirect for technologies such as photovoltaic (PV) cell-assisted water electrolysis. The use of solar energy enables a PEC cell to operate at a lower operating voltage as compared to a commercial electrolyzer, which is designed to operate with two orders of magnitude higher currents and significantly higher efficiencies.

The band bending inside the semiconductor can be changed through the application of an external electrical bias. The flat band condition, where the band edges are flat corresponds, to the bias at which the current density is zero. The energy band structure of the semiconductor and the characteristics of the redox species in the liquid determine the direction and magnitude of any flow of charges between the solid and liquid (Figure 10.1). The band positions of the semiconductors also play a key role in determining the PEC reactions occurring at the semiconductor–electrolyte interface. The conduction band (CB) edge defines the reducing power of the photogenerated electrons, while the valence band (VB) edge defines the oxidation power of the holes.

For generating fuels, a photocatalyst can be employed in two primary configurations: (a) as particulates dispersed in an appropriate solution and (b) as an electrode in a PEC cell as shown in Figure 10.2. In a particulate system (Figure 10.2a), the free-standing photocatalyst powders are either suspended in a solution or fixed at a reactor bed. When a semiconductor absorbs the photons with an energy greater than the bandgap energy (E_g), electrons are excited to the CB, leaving the holes in the VB. The separated electrons and holes then migrate through the semiconductor to the semiconductor–electrolyte interface, where redox reactions occur to generate fuels. Charge recombination is a most important loss of the excited charge carriers and an important factor that limits the efficiency of solar-to-chemical energy conversion process involving semiconductor photocatalysts. Sometimes the charge recombination can reach its maximum level (90%) within a 10 ns period after the generation.

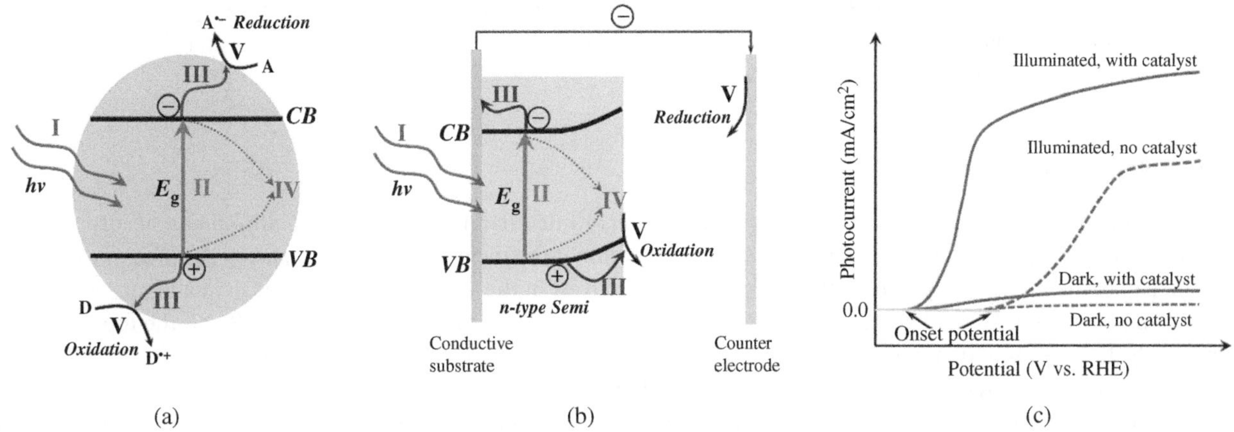

Figure 10.2 Schematic illustration of processes in (a) a particulate photocatalyst, (b) a PEC cell with an n-type photoelectrode under illumination. (c) Photocurrent density–voltage curves obtained from an n-type semiconductor photoanode. Note: (I) Light absorption, (II) Charge separation, (III) Charge migration, (IV) Charge recombination, and (V) redox reaction. CB = Conduction band, VB = Valence band, E_g = Bandgap, A = Acceptor, and D = Donor. Source: Reproduced from Li & Wu, 2015/Royal Society of Chemistry.

The apparent quantum efficiency (QE) is normally used to determine the photocatalyst performance (Li & Wu, 2015; Serpone et al., 1996),

$$QE = \frac{2 \times \text{number of } H_2 \text{ molecules}}{\text{number of incident photons}} \tag{10.1}$$

$$QE = \frac{4 \times \text{number of } O_2 \text{ molecules}}{\text{number of incident photons}} \tag{10.2}$$

In a PEC water splitting cell (Figure 10.2b) with an n-type semiconductor as the photoanode, hydrogen is generated at the cathode (e.g., Pt) while oxygen is evolved at the photoanode surface. In a PEC system, the Fermi level of the photoelectrode can be adjusted via a potentiostat, and therefore, the redox reactions can be controlled by choosing appropriate potentials so that reaction selectivity can be manipulated. Also, the solar-to-chemical energy conversion efficiency (η) can be measured for a two-electrode PEC cell, which needs an external bias to drive a redox reaction, as follows (Chen et al., 2013; Li & Wu, 2015),

$$\eta = \frac{J(V_{redox} - V_{app})}{P_{light}} \tag{10.3}$$

where J is the measured photocurrent density (mA/cm^2), V_{redox} is the redox potential of a reaction. V_{redox} = 1.23 V for electrochemical water splitting. V_{app} is the applied external bias (V). P_{light} is the measured intensity of the full-spectrum solar radiation (mW/cm^2). Moreover, the incident photon-to-current efficiency (IPCE) is defined as the ratio of photocurrent, which provides the rate of electron flow through the external circuit, vs. the rate of incident photons with a particular wavelength. IPCE, which is also called external quantum efficiency (EQE), of a two-electrode PEC is typically quantified by the relation (Chen et al., 2013; Li & Wu, 2015),

$$\text{IPCE}(\lambda) = \text{EQE}(\lambda) = \frac{1239.8(V \times \text{nm}).J_{ph}(\text{mA/cm}^2)}{\lambda(\text{nm}).P_\lambda(\text{mW/cm}^2)} = \frac{\text{electrons/cm}^2/\text{s}}{\text{photons/cm}^2/\text{s}} \tag{10.4}$$

where 1239.8 (V × nm) is a constant obtained by the multiplication of Planck's constant (h) with the ratio of speed of light (c) and electronic charge (e multiplied by 10^{-9}), λ is the wavelength of the incident light in nm, and P_λ is the measured irradiance at the specific wavelength. The electrochemical impedance spectroscopy can be used to identify the space charge capacitance near the semiconductor surface in an electrolyte. The donor density in a semiconductor photoelectrode and space charge capacitance can be correlated by using Mott–Schottky equation, which is expressed as (Chen et al., 2013; Li & Wu, 2015),

$$\frac{1}{C_{sc}^2} = \frac{2}{\varepsilon_0 \varepsilon_r e N_D A^2}\left(V_{app} - V_{fb} - \frac{kT}{e}\right) \tag{10.5}$$

The values of N_D and V_{fb} can be determined from the plot of $1/C^2$ vs. V_{app} (called Mott–Schottky plot). Here, ε_0 is the vacuum permittivity, ε_r is the relative permittivity, N_D is the donor density, A is the photoelectrode area, T is temperature, k is the Boltzmann constant, and V_{fb} is the flat band potential.

10.2.1 Photon Absorption

A semiconductor photocatalytic process to convert solar energy into chemical fuels begins with photon absorption. The irradiation of semiconductor particles results in optical excitation if the incident light has an energy equal to or more than the bandgap of the material. Upon photon absorption, the electrons in the VB are excited to the CB in a femtosecond time scale (Wang & Domen, 2020). The rapid relaxation of electrons and holes to the bottom of the CB and the top of the VB, respectively, follows excitation, on a similar time scale. Thus, upon irradiation by a photon of energy ($h\nu$) greater than E_g an electron is readily promoted from the VB to the CB, with minimal deviation in its momentum.

In a semiconductor material, the absorption of light can be specified based on the absorption coefficient of the material, $\alpha(\lambda)$. For a given photon energy, $E_2 - E_1 = h\nu$, this coefficient is summed over all possible band transitions from the VB to the CB, which is expressed as (Luque & Hegedus, 2003; Wang & Domen, 2020),

$$\alpha(h\nu) \propto P_{12}\, \mathrm{gv}E_1\, \mathrm{gc}(E_2) \tag{10.6}$$

where P_{12} is the probability of the transition of an electron from the initial state E_1 to the final state E_2, while $g_V(E_1)$ and $g_C(E_2)$ are the electron densities in the initial and final states, respectively. With respect to intrinsic absorption properties of the photocatalyst, like its bandgap and band positions, the value of $\alpha(\lambda)$ also changes.

The mechanism of the photocatalytic system for the water splitting is schematically represented in Figure 10.3. A Gibbs energy change of 237 kJ/mol[1] or 1.23 eV (therefore the bandgap energy of the photocatalyst should be over 1.23 eV [<1000 nm]) is needed for a successful water splitting reaction. But for the utilization of visible light, the bandgap of the semiconductor should be less than 3.0 eV (>400 m). The number of excited electrons at the photocatalyst–water interface (in the case of a p-type semiconductor) participating in the reduction of water determines the total amount of hydrogen evolved during the water splitting process. In the charge-generation process, the semiconductor photocatalyst should first have a low bandgap to absorb as much light as possible, and reflection or scattering of light by the photocatalyst should be minimized. Second, using the absorbed photons, the semiconductor photocatalyst should have a high efficiency in generating excited charges, instead of generating phonons or heat. The charge recombination and separation/migration processes after the successful excited charge creation are the main processes inside the semiconductor photocatalyst, which can largely affect the photocatalytic reaction efficiency for the water splitting. Moreover, the charge recombination decreases the excited charges by either releasing heat or emitting light. In certain materials (e.g., organic molecules), the photoexcited electron–hole pairs, called excitons, have high binding energy. The electrons and holes need to overcome this energy barrier to get separated and participate in the redox reactions.

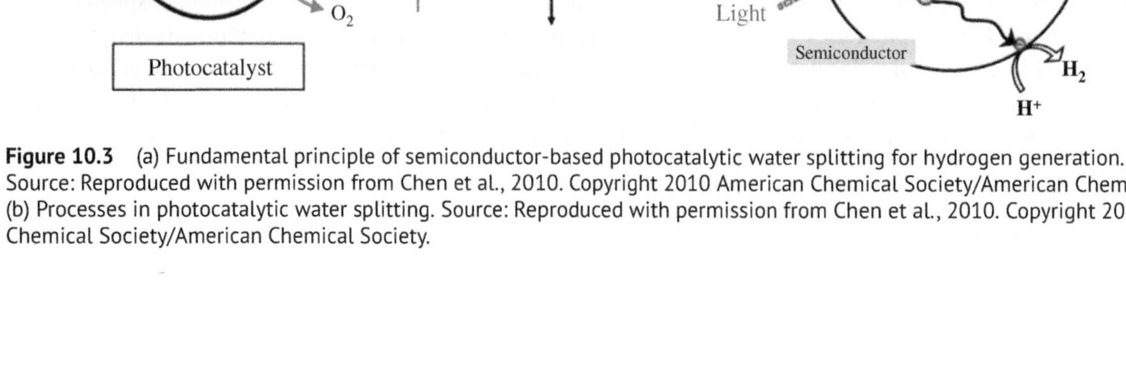

Figure 10.3 (a) Fundamental principle of semiconductor-based photocatalytic water splitting for hydrogen generation. Source: Reproduced with permission from Chen et al., 2010. Copyright 2010 American Chemical Society/American Chemical Society. (b) Processes in photocatalytic water splitting. Source: Reproduced with permission from Chen et al., 2010. Copyright 2010 American Chemical Society/American Chemical Society.

10.2.2 Requirements of Materials Useful as Photocatalysts

The materials used as photocatalysts (for photocatalytic or PEC processes) must be capable of causing light absorption, charge carrier generation and separation, and transfer to the electrolyte solution for redox reactions. As shown in Figure 10.4, semiconductor bandgap is a key factor in deciding the theoretical maximum of solar-to-hydrogen (STH) efficiency.

A PEC device involves only a few components to integrate the light-absorbing and electrocatalytic functions of materials for water splitting. A PEC cell consists of two electrodes, an anode for the oxidation reaction and a cathode for the reduction reaction, and a membrane to separate the products of the two half-cell reactions. In contrast with an electrochemical cell, a PEC cell has at least one electrode, a semiconductor that can absorb solar photons with energy greater than the bandgap energy. The photovoltage generated by the subsequent separation of photogenerated charges can directly drive electrochemical solar fuel production reaction if the photovoltage is sufficiently high. Although it is possible to envisage a PEC cell with one semiconductor photoelectrode and one standard metallic (dark) electrode, identifying a single semiconductor material that can both develop sufficient photovoltage for solar fuel production (approximately 1.5–1.8 V for water splitting) and harvest a significant portion of the solar spectrum is a major challenge (Chen et al., 2010). Therefore, the

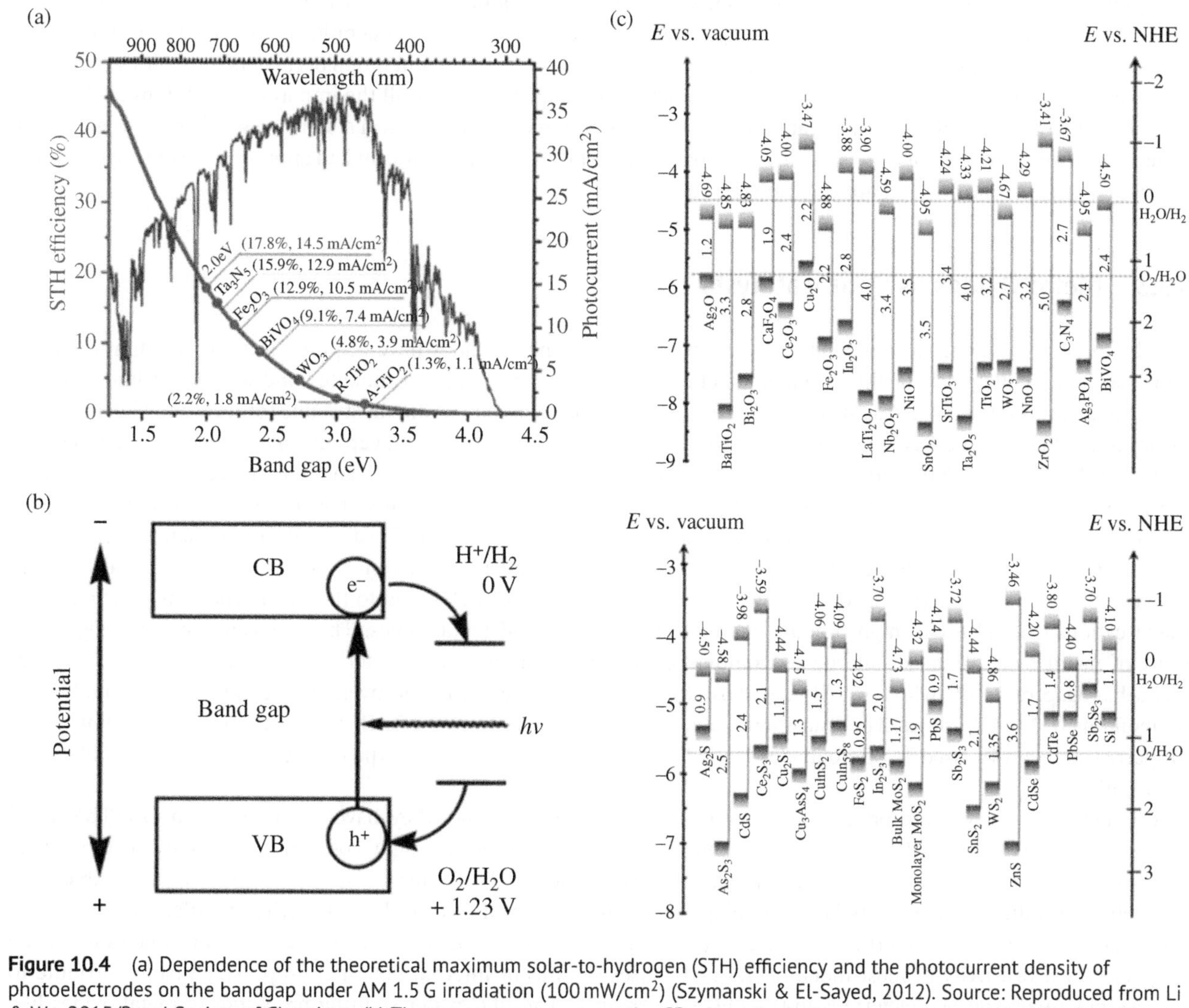

Figure 10.4 (a) Dependence of the theoretical maximum solar-to-hydrogen (STH) efficiency and the photocurrent density of photoelectrodes on the bandgap under AM 1.5 G irradiation (100 mW/cm^2) (Szymanski & El-Sayed, 2012). Source: Reproduced from Li & Wu, 2015/Royal Society of Chemistry. (b) The top squares represent the CB edges; the bottom squares present the VB edges. The top numbers show the exact CB level and the number between squares is the bandgap. The two dashed lines indicate the water redox reaction potentials (Kudo & Miseki, 2009). Source: Reproduced from Kudo & Miseki, 2009, with permission from Royal Society of Chemistry/Royal Society of Chemistry. (c) Band edge positions with respect to the vacuum level and the normal hydrogen electrode (NHE) for selected semiconductors at pH 0. Source: Reproduced from Li & Wu, 2015/Royal Society of Chemistry.

basic strategy is to find out photocatalytic surfaces have the ability to absorb light and produce electron–hole pairs. These electrons and holes can then proceed, respectively, for hydrogen evolution reaction (HER) and oxygen evolution reaction (OER) reaction on the catalyst surface. The CB edge of the catalyst should be more negative than the reduction potential of H^+ to H_2 (0.00 V vs. RHE), while the VB edge should be more positive than the oxidation potential of H_2O to O_2 (+1.23 V) (Li & Wu, 2015).

$$2H^+ + 2e^- \rightarrow H_{2,}\ E_0 = 0.00 \tag{10.7}$$

$$2H_2O \rightarrow O_2 + 4H^+ + 4e^-, E' = +1.23\ \text{V} \tag{10.8}$$

$$H_2O \rightarrow H_2 + \frac{1}{2}O_2, \Delta G_0 = +237.2\ \text{kJ/mol} \tag{10.9}$$

The main requirements to accomplish successful PEC water splitting are absorption of a broad region of the solar spectrum, minimal overpotential on performing the HER or the OER, efficient photogenerated electron–hole excitation and separation, long-term stable operation, cost-effectiveness, availability of scalable device fabrication techniques, use of earth-abundant materials and minimized carrier transport losses to the back contact and semiconductor–electrolyte surface (Kabir et al., 2018; Luo et al., 2019; Ong et al., 2016). In recent decades, researchers have devoted tremendous efforts to exploiting photocatalysts with broad-spectrum solar energy absorption (light harvesting and absorption), increasing the quantity and quality of active sites (photogenerated carrier (e^-–h^+ pairs) generation), and separating the charge efficiently in the bulk and interface of the photocatalyst system (separation and migration of carriers) by a series of designs, constructions, and modifications. Nevertheless, photoelectrode materials satisfy all the important conditions such as low cost of production, suitability for fabrication of electrodes with large areas, chemical stability for long-term operation in aqueous environments, strong absorption over a broad region of the solar spectrum, and an appropriate band edge position for reduction/oxidation of water are very rare to find.

10.3 Metal Oxides as Photocatalysts

Amidst the prevalently used semiconductors as photocatalysts, metal oxides have acquired a lot of attention owing to their exceptional chemical stability, suitable band edge positions, low cost, and tunable bandgaps (Marschall & Wang, 2014; Sivula & van de Krol, 2016; Szymanski & El-Sayed, 2012). Being one of the most stable forms of metal compounds in nature, oxides can be easily synthesized. Oxides have a broad range of bandgap and band edge positions (Walsh et al., 2009). The wide bandgaps of metal oxides such as TiO_2 (E_g = 3.0–3.2 eV) and ZnO (3.4 eV) are rendered by the high ionic character of O 2p and the metal s orbital. The narrower bandgaps of metal oxides are found in those containing transition metal cations with d^n electronic configurations, such as Fe_2O_3 (d^5 electronic configuration, bandgap 2.2 eV) and Cu_2O (d^{10} electronic configuration, bandgap 2.1 eV) and these materials possess visible light absorption. A narrow bandgap is observed in case of metal oxides such as PbO (2.1 eV), due to the involvement of post-transition metal cations (like Pb) with high binding energy s states, O 2p, and metal states. The bandgap and band edge positions of metal oxides can be tailored by combining two metal cations to form ternary oxides. For example, $BiVO_4$ has a bandgap of 2.3 eV with the CB edge close to 0 V vs. RHE, owing to the overlap of empty Bi 6p orbitals with antibonding V 3d-O 2p states. The transition metal oxides such as TiO_2, WO_3, ZnO, and Fe_2O_3 typically exhibit n-type properties and can be used as photoanodes while NiO, CuO, and Cu_2O have p-type characteristics and are suitable as photocathodes.

Although metal oxides are promising materials for PEC water splitting, they have inherent drawbacks. The metal oxides such as ZnO and TiO_2 have a wide bandgap and hence possess no light absorption capability in the visible and infra-red regions of the solar spectrum, whereas oxides such as Fe_2O_3 have a favorable bandgap for visible light absorption but, due to the slow polaron hopping and resultant limited electrical conductivity, it suffers from sluggish photogenerated carrier transport and extraction (Yang et al., 2017). A large amount of energy is consumed due to the large electrical resistivity of most metal oxides and causes significant internal potential drop, which can contribute to the overpotentials and be deleterious to the PEC performance. The PEC water splitting properties of metal oxides like TiO_2, ZnO, $BiVO_4$, Cu_2O, WO_3, and Fe_2O_3 have been extensively studied (Alves et al., 2018; Chae et al., 2017; Shan et al., 2015; Spagnoli et al., 2009; Sullivan et al., 2016; Walsh et al., 2009; Wang et al., 2011; Yan et al., 2019; Zhong et al., 2007; Zhu et al., 2016).

10.3.1 Doping and Surface Treatments

Poor electrical conductivity exhibited by the wide bandgap metal oxide semiconductors is one of the major challenges that cause a significant loss of electron–hole pairs due to recombination and associated reduction in their photoactivity (Xiao et al., 2019; Yang et al., 2017). Either addition of extrinsic dopants (or impurities) or creation of intrinsic defects can be used for enhancing the electrical and optical properties of metal oxides. The dopants, or intrinsic defects, can modify the electronic properties of the oxides and narrow the bandgap by introducing energy states within the band structure.

Introduction of impurity states (elemental doping or co-doping) within the bandgap and thus narrowing the bandgaps of metal oxides is one strategy. For doping in wide bandgap metal oxides, elements such as C, N, S, V, B, W, and Sn are used as common dopants to increase the light absorption range and photoactivity (Alves et al., 2018; Chae et al., 2017; Marschall & Wang, 2014; Sivula & van de Krol, 2016; Sullivan et al., 2016; Walsh et al., 2009; Yan et al., 2019; Yang et al., 2017). The nitrogen dopants can be incorporated into wide-bandgap ceramics such as TiO_2 and ZnO by annealing metal oxides in ammonia gas at elevated temperatures (Alves et al., 2018; Asahi et al., 2001; Yuan et al., 2006; Zhu et al., 2016). Wang et al. reported the synthesis of N-doped TiO_2 nanowire arrays by an ion implantation method (Li, Peng, et al., 2019). The IPCE of N-TiO_2 reached 17% at 450 nm with the help of OER cocatalysts. Yang et al. doped TiO_2 nanostructures with sulfur and oxygen vacancies by using aluminum reduction followed by annealing in H_2S atmosphere (Deng et al., 2017). A superior visible light absorption as compared to their mono-doped counterparts (S-doped TiO_2 and oxygen-deficient TiO_2) was exhibited by the co-doped TiO_2. This improvement was attributed to the synergistic effect between S and oxygen vacancies, by the simultaneous improvement of visible light absorption and charge collection efficiency. A simultaneous improvement of electrical properties and light absorption capability was achieved in TiO_2 samples co-doped with Ti^{3+} and S. Similar synergistic effect was also observed in TiO_2 nanowires co-doped with Ti^{3+} and N (Suzuki et al., 2016).

Much efforts have been directed toward the bandgap modification of metal oxides to improve their photoactivity (Hensel et al., 2010; Hoang et al., 2012; Wang et al., 2009; Wang, Ren, et al., 2015; Wang, Xiao, et al., 2015; Xu et al., 2012; Yang et al., 2009; Yang, Wang, et al., 2013; Yu et al., 2005). Yang et al. reported that nitrogen doping in ZnO nanowires would significantly increase donor densities, and hence, the photoactivity of the material (Yang et al., 2009). Similarly, Wang et al. reported a H_2 treatment method to introduce oxygen vacancies in TiO_2 nanowire arrays (Wang et al., 2009). Oxygen vacancies are the shallow donors here that can result in significantly superior carrier density of TiO_2 by several orders of magnitudes. Increasing the donor densities directly improved the charge separation efficiency and diminished the energy loss in TiO_2 nanowire-arrayed photoanodes. The photocurrent densities of the H_2-treated TiO_2 (H-TiO_2) nanowires at the temperature range from 300 to 450 °C were persistently higher than the untreated TiO_2 nanowires (Hensel et al., 2010; Hoang et al., 2012; Wang et al., 2009; Wang, Xiao, et al., 2015; Yang, Wang, et al., 2013). Later, Xie et al. generated oxygen vacancies in $SrTiO_3$ by thermal treatment in argon gas and synthesized highly photoactive $SrTiO_{3-x}$ (Xie et al., 2011). Notably, these oxygen-starved wide-bandgap metal oxides exhibited improved photoactivity even when oxygen vacancies were created via different methods. These investigations unambiguously show that introducing extrinsic dopant and/or intrinsic defects is an effective approach to modify the electronic properties and PEC activities of metal oxides for water splitting. It should be mentioned that higher carrier densities might lead to the photogenerated carrier recombination, thus resulting in lower photoactivity.

It was reported that $NaTaO_3$ photocatalysts possess high efficiency for water splitting. The material yielded an H_2 production rate of 1280 $\mu mol \cdot g^{-1} \cdot h^{-1}$ in the absence of sacrificial agents (Kato et al., 2003; Kato & Kudo, 1999). Subsequently, Kudo et al. discovered that Sr doping in the crystal structure of $NaTaO_3$ could enhance the charge segregation and change the particle size by varying the density of doping (Nakanishi et al., 2017). The resultant Sr-doped $NaTaO_3$ demonstrated a much higher H_2 production rate of 27,200 $\mu mol \cdot g^{-1} \cdot h^{-1}$ with the addition of 0.1 wt% NiO as the cocatalyst under UV irradiation (Iwase et al., 2009). Similarly, it was reported by Fang et al. that the introduction of copper into the $LaNiO_3$ would result in a distorted lattice structure and hence would influence the catalytic action of the oxides (Li et al., 2010). $SrTiO_3$ with a narrower bandgap (3.2 eV) than $NaTaO_3$ (4.0 eV) has continued to draw interest as a water splitting photocatalyst. The sluggish kinetics of water splitting properties can be augmented via surface adornment of cocatalysts. Recently, Takata et al. investigated water splitting activity of Al-doped $SrTiO_3$ loaded with Rh, Cr, and Co species as cocatalysts using sequential photodeposition (Takata et al., 2020). The loading of Rh (0.1 wt%), Cr_2O_3 (0.05 wt%), and CoOOH (0.05 wt%) into Al-doped $SrTiO_3$ significantly increased the water splitting activity. The Cr_2O_3 shell inhibited only the ORR by blocking the access of evolved O_2 to the surface of the Rh core, and the presence of Rh promotes both the HER and the ORR. The shell made of Cr species is expected to be hydrated in water to form $Cr(III)O_{1.5-m}(OH)_{2m} \cdot xH_2O$, thus the photodeposited oxyhydroxide

Co(III)OOH promoted the OER (Takata et al., 2020). Even though quantum efficiencies of H_2 evolution have reached almost unity in the UV region, it is still necessary to develop photocatalysts active under visible light irradiation.

10.3.2 Long-Term Stability

The long-term stability in PEC water splitting devices is one of the key factors and therefore the light absorbers must be chosen diligently to avoid photocorrosion. The wide bandgap of metal oxide water splitting photocatalysts, as in the case of La-doped $NaTaO_3$ (E_g = 4.1 eV) and Zn-doped Ga_2O_3 (E_g = 4.4 eV), prevents sunlight-triggered water splitting (Deng et al., 2017; Sudrajat et al., 2020). Photostability is another important property required for photoelectrodes. TiO_2 is one of the most stable photoanode materials for water oxidation among the wide bandgap metal oxides. The long-term stability of PEC performance for the doped TiO_2 has been a concern, albeit doping and defect engineering has proved to be effective in promoting the photoactivity efficacy of TiO_2. Recently, Yang et al. investigated the long-term photostability of pristine TiO_2 (rutile, anatase, and mixed-phased TiO_2) and hydrogen-treated TiO_2 photoanodes (Yang et al., 2015). The hydrogen-treated TiO_2 was equally stable as the pristine samples. Therefore, the controlled introduction of oxygen vacancies does not necessarily affect the photostability of TiO_2. Intriguingly, they found that all TiO_2 photoanodes indeed suffered from photo-hole-induced photocorrosion. Adding hole sacrificial reagent has been proven as an efficient approach to diminish the photocorrosion of ZnO photoanodes. However, most hole sacrificial reagents are important chemical fuels like methanol. Thus, it is undesirable to use sacrificial reagents for practical applications. Improving the PEC stability of metal oxide materials remains a technical challenge.

10.3.3 Heterostructures

In addition to homogeneous metal oxides, combining multiple metal oxides rationally into heterostructures has been evidenced to be competent in aiding the separation of photoinduced electron–hole pairs. For instance, the heterojunction formed by combining $BiVO_4$ and WO_3 facilitates both charge separation and broad-spectrum light absorption (Rao et al., 2014; Suzuki et al., 2016). Hou et al. developed the heterojunction between CuO and TiO_2 for the formation of an electric field at the interface within the hybrid, which could facilitate the separation of photogenerated electron and hole pairs (Hou et al., 2016). The prepared hybrid TiO_2/CuO/Cu mesoporous nanofibers exhibited robust stability and a high photocatalytic H_2 evolution rate of ~851.3 $\mu mol \cdot g^{-1} \cdot h^{-1}$, which was profoundly enhanced by more than 3.5 times with respect to those of the pristine TiO_2 counterparts and commercial P25. Figure 10.5a,b shows the schematic diagram of H_2 evolution mechanism and the energy band structure of hybrid TiO_2/CuO/Cu. The amounts of hydrogen generated from aqueous suspensions over the six photocatalyst samples determined using gas chromatography (GC) are shown in Figure 10.5c. These six samples consisted of 6.5 wt% polyvinylpyrrolidone (PVP), 30 wt% tetrabutyl titanate (TBOT), and $Cu(Ac)_2$ in 0, 2, 2, 1, and 3 wt% corresponding to labels A, B, C, D, and E. Sample B had no foaming agent (diisopropyl azodiformate (DIPA)) and other samples had 10 wt% DIPA. Trang et al. studied the photocatalyst with enhanced the photogenerated charge transfer properties by the formation of the p–n heterojunction between Cu_2O and TiO_2. They modified TiO_2 nanotubes with varying amounts of Cu for the H_2 production under sunlight irradiation. The TiO_2 nanotubes with 1.5 wt% Cu ($Cu_{1.5}$-TiO_2) modification resulted in highest photocatalytic activity for H_2 generation under sunlight (Figure 10.5e). Figure 10.5d shows the general mechanism of photocatalytic H_2 generation under sunlight by glycerol as hole scavenger for TiO_2 nanotube and $Cu_{1.5}$-TiO_2 nanotube photocatalysts. Their results showed that the efficiency of H_2 production would slightly decrease after four cycles of hydrogen production (Figure 10.5f) (Trang et al., 2019). Sharma and coworkers also demonstrated 25 times higher photocurrent density (2.52 mA/cm^2 at 0.8 V vs. SCE) in $Cu_2O/SrTiO_3$ heterojunction with an overall thickness of 343 nm as compared to that of pristine Cu_2O (0.10 mA/cm^2 at 0.8 V vs. SCE) (Sharma et al., 2014). The transfer of photoexcited electrons from CB of Cu_2O to that of $SrTiO_3$ (movement of holes from $SrTiO_3$ VB to Cu_2O VB) prevented photocorrosion of Cu_2O via the influence of the electric field formed at the p–n junction of the $Cu_2O/SrTiO_3$ heterostructure.

Another common perovskite oxide, barium titanate ($BaTiO_3$), with a bandgap energy of about 3.2 eV and its appropriate CB and VB positions compared with the redox potentials of H^+/H_2 and O_2/H_2O, can split water into hydrogen and oxygen. For an efficient photocatalytic reaction, the electrons with high reduction power in the CB and holes with oxidation ability in the VB are required. A high recombination rate of photoexcited holes and electrons is the main reason for low efficiency of the photocatalytic processes. Thus, approaches such as noble metal loading, ion doping, heterostructure formation, and size/morphology control have been explored to enhance the separation and migration of photogenerated

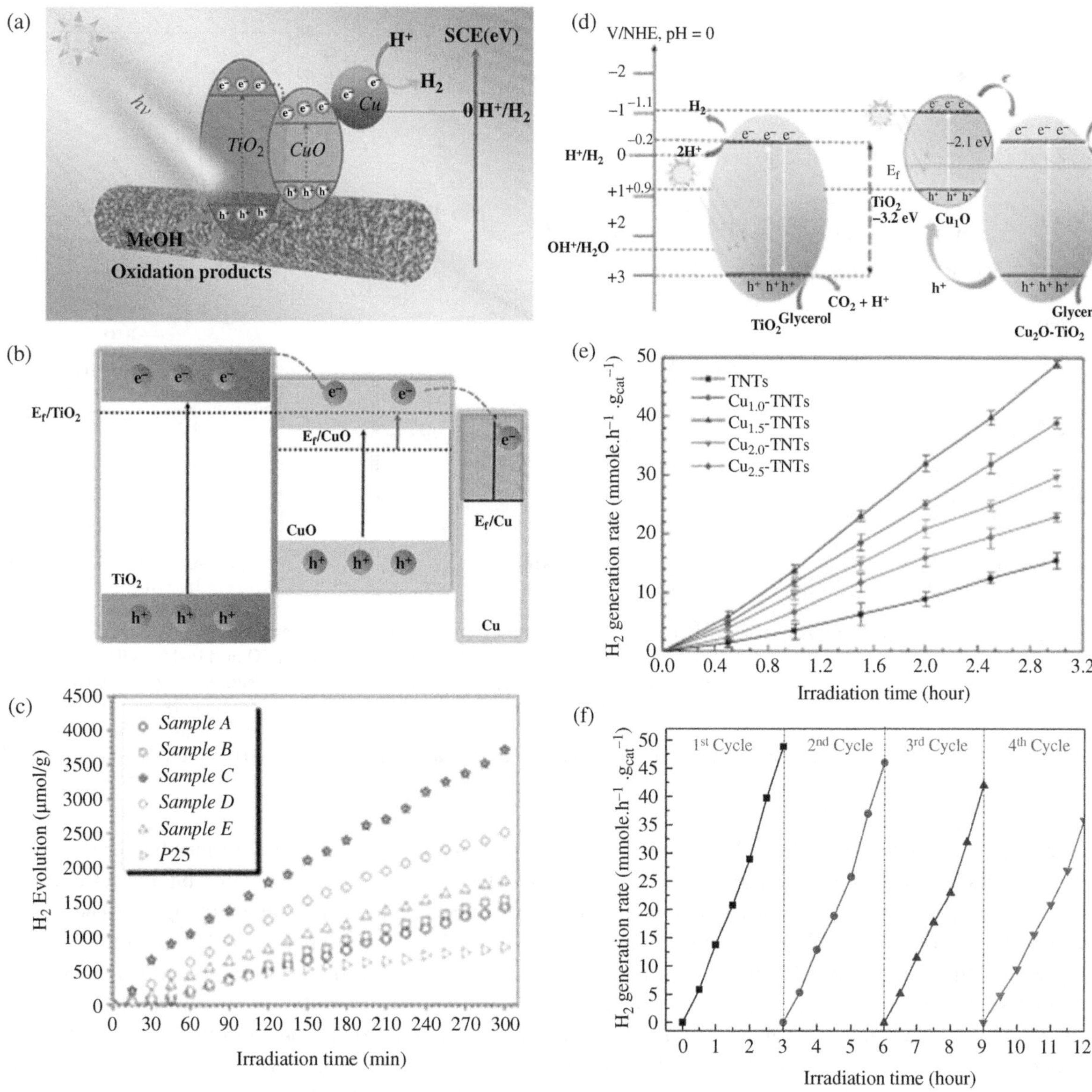

Figure 10.5 (a) Schematic of the photocatalytic mechanism of H_2 generation over ternary TiO_2/CuO/Cu mesoporous nanofibers under xenon lamp irradiation. (b) Schematic illustration of the energy band structures of TiO_2/CuO/Cu nanofibers. (c) H_2 evolution over the photocatalysts of samples A–E and P25 under different irradiation times (Hou et al., 2016). Source: Reproduced with permission from Hou et al., 2016. Copyright 2016 American Chemical Society/American Chemical Society. (d) Proposed reaction mechanism for the photocatalytic H_2 generation with glycerol water mixture solution under sunlight of (e) TiO_2 and (b) $Cu_{1.5}$-titanium dioxide nanotubes (TNTs) sample (a) H_2 generation of TNTs and Cu_x-TNTs samples in 5 vol% glycerol aqueous solution, and (f) recycling photocatalytic test for H_2 production of $Cu_{1.5}$-TNTs under sunlight (Trang et al., 2019). Source: Reproduced from Trang et al., 2019, with permission from Elsevier/Elsevier.

carriers. Pt-modified Mo-doped $BaTiO_3$ exhibited band-to-band visible-light absorption and remarkable improvement in hydrogen production efficiency. Xie et al. optimized Pt-modified Mo-doped $BaTiO_3$ by doping with 2 at% of Mo and 0.4 wt% Pt. The material exhibited a hydrogen evolution rate of 63, which was near two times more than that of pure $BaTiO_3$ (35 $\mu mol{\cdot}g^{-1}{\cdot}h^{-1}$) (Xie et al., 2019). A schematic representation of bandgap tuning in $BaTiO_3$ by Mo doping is given in Figure 10.6a. Also, the periodic photocatalytic hydrogen evolution in aqueous methyl alcohol solution with Pt/$BaTiO_3$ and Pt/Mo-$BaTiO_3$ photocatalysts under simulated sunlight irradiation is shown in Figure 10.6c. To observe the importance of performance difference of ferroelectric and non-ferroelectric heterojunction photoelectrodes, Li et al. developed a $BaTiO_3$@α-Fe_2O_3 heterojunction (Li, Fang, et al., 2019). The heterojunction of $BaTiO_3$ nanostructures with

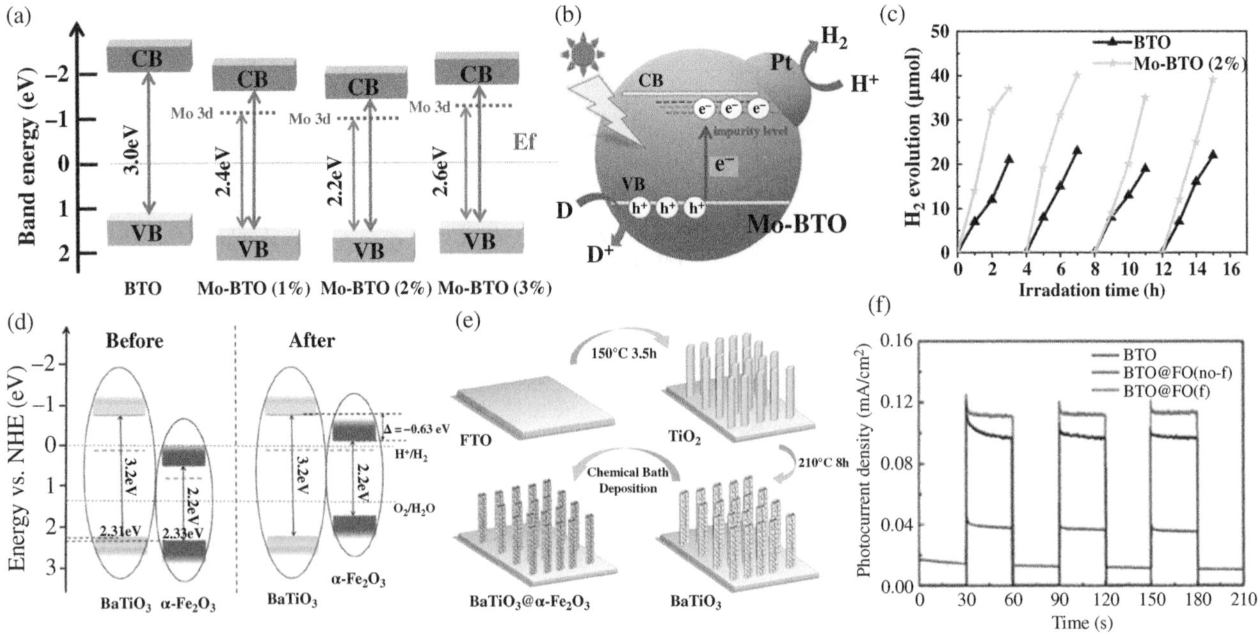

Figure 10.6 (a) Energy band diagram of the $BaTiO_3$ (BTO) and Mo-BTO samples. (b) Schematic representation of photocatalysis of Pt/Mo-BTO. (c) Periodic photocatalytic hydrogen evolution in aqueous methyl alcohol solution with Pt/BTO and Pt/Mo-BTO photocatalysts under simulated sunlight irradiation (Benko & Koffyberg, 1986). Source: Reproduced from Rao et al., 2014, with permission from Elsevier/Elsevier. (d) CB and VB potentials of $BaTiO_3$ and α-Fe_2O_3 before and after constructing heterojunction. (e) Schematic illustration of the growth procedure of $BaTiO_3$@α-Fe_2O_3 heterojunctions. (f) Measurements of transient photocurrent density (Li, Fang, et al., 2019). Source: Reproduced with permission from Li, Fang, et al., 2019. Copyright 2019 American Chemical Society/American Chemical Society.

α-Fe_2O_3 nanoparticles exhibited an enhanced photocurrent density 0.37 mA/cm^2, which was 3.7 times higher than that of pure $BaTiO_3$ nanostructures. The enhanced PEC performance of $BaTiO_3$@α-Fe_2O_3 heterojunction photoelectrode was mainly attributed to the effective charge transfer and separation. The detailed mechanism is shown in Figure 10.6d–f.

Rh-doped $BaTiO_3$ nanocrystals with photocatalytic H_2 evolution capabilities were reported by Nishioka and Maeda (Nishiok & Maeda, 2015). The photocatalytic H_2 evolution was observed in Rh-doped $BaTiO_3$ in an aqueous methanol solution under visible light irradiation. $BaTiO_3$ has a bandgap of 3.0 eV and is generally a wide-gap n-type semiconductor. New absorption bands in visible light region were observed after doping Rh species into the lattice of $BaTiO_3$. When Rh-doped $BaTiO_3$ was modified with the water reduction enhancer, nanoparticulate platinum, it was capable of producing H_2 from water containing an electron donor such as methanol and iodide, in the presence of visible light ($\lambda > 420$ nm) (Nishiok & Maeda, 2015). For hydrogen evolution, visible spectra-driven Z-scheme H_2O splitting was achieved using Pt-modified Rh-doped $BaTiO_3$ as building block in combination with PtO_x-loaded WO_3 as an O_2 evolution photocatalyst in the presence of an IO_3^-/I^- shuttle redox mediator. The cathodic photo-response of the porous Rh-doped $BaTiO_3$ electrode due to water reduction in a neutral aqueous Na_2SO_4 solution in the presence of visible light was elucidated from the PEC analysis. Rh-doped $BaTiO_3$ was also found to work as a potential photocathode that could generate H_2 from water in a PEC cell (Nishiok & Maeda, 2015). Gao et al. recently developed Zr-doped $CaTiO_3$ photocatalysts that possessed augmented photocatalytic action towards the H_2 production (Gao et al., 2021). The surface structures of $CaTiO_3$ could be regulated by the Zr doping. The resultant nanosteps with more exposed active sites accompanied by an increased surface area, oxygen vacancies, and reduction ability efficiently promoted the separation of photoexcited carriers, thus leading to an increase in the photocatalytic ability. Chen et al. reported the photocatalytic hydrogen evolution of $BaTiO_3$ materials under UV light (200–400 nm) illumination (Chen et al., 2018). In the presence of triethanolamine (TEOA) as a sacrificial reagent, the H_2 production rate was enhanced eight times that of commercial $BaTiO_3$, with the former having a hydrogen production rate of 789.7 $\mu mol \cdot g^{-1} \cdot h^{-1}$ and the latter having a hydrogen production rate of 98.5 $\mu mol \cdot g^{-1} \cdot h^{-1}$. The enhanced photocatalytic activity compared with the commercial $BaTiO_3$ was due to the small particle size below 10 nm that suppressed the recombination of photoinduced carriers.

Ferrite electrodes such as Fe_3O_4 (Büchler et al., 1998), p-type $CaFe_2O_4$ (Ida et al., 2010; Ida et al., 2012; Matsumoto et al., 1988; Sekizawa et al., 2014), $MgFe_2O_4$ (Benko & Koffyberg, 1986; Selvaraj et al., 2011), $Ti_xFe_{3-x}O_4$ (De Haart & Blasse, 1985), $Li_{0.5}Fe_{2.5}O_4$ (Archer et al., 1981), p- and n-type $Co_xFe_{3-x}O_4$ and $CoTi_xFe_{2-x}O_4$ (Yang, Mao, et al., 2013), p-type $CoFe_2O_4$ (Yang, Mao, et al., 2013), p- and n-type $NiFe_2O_4$ (Rekhila et al., 2013), n-type $ZnFe_2O_4$ (Matsumoto et al., 1985), $Zn_xTi_yFe_{3-x-y}O_4$ (Matsumoto et al., 1985), and $CdFe_2O_4$ (Kung et al., 1977) have also been explored for their PEC properties under visible light irradiation. $ZnFe_2O_4/Fe_2O_3$ layers on stainless steel were prepared by Borse and coworkers and the rates of HER at the $ZnFe_2O_4$ and the $ZnFe_2O_4/Fe_2O_3$ photoanodes were determined to be 46.3 and 99.0 $\mu mol{\cdot}cm^{-2}{\cdot}h^{-1}$, respectively (Dom et al., 2013). Lee and coworkers reported the synthesis and characterization of p-type $CaFe_2O_4$-modified TaON. They explored the p-type $CaFe_2O_4$ altered TaON and $BiVO_4$ to form p–n-heterojunction photoanodes for PEC application (Kim et al., 2013). MFe_2O_4 and $M^1{}_xM^2{}_yFe_{3-x-y}O_4$ type semiconducting ferrites were studied as potential water splitting photocatalysts owing to their ability to absorb visible light and their claimed resistance against photocorrosion (Dillert et al., 2015). The ability of Pt-loaded Fe_3O_4 to evolve H_2 from water–ethanol mixtures under visible light illumination was studied by Mangrulkar et al. (Mangrulkar et al., 2012). Under illumination, Fe_3O_4 surface showed an increase in HER rate (60 $\mu mol/h^1$ at 85–100 °C) with temperature of the ethanol–water suspension. Maximum performance was achieved at pH 10 with a quantum efficiency of 0.5×10^{-2} in the presence of the hole scavenger Na_2SO_3 (0.025 M). Visible light-driven H_2 and O_2 formation with a $CaFe_2O_4/PbBi_2Nb_{1.9}W_{0.1}O_9$ p–n-heterojunction loaded with platinum was explored by Kim et al. (Kim et al., 2005). A quantum yield of 38×10^{-2} for the O_2 production in the presence of $AgNO_3$ as the sacrificial reagent under illumination with $\lambda \geq 420$ nm was observed by the authors (Kim et al., 2005). The submicron-sized mesoporous $NiFe_2O_4$ spheres were synthesized by Hong et al. and the enhanced H_2 evolution rates exhibited by the amorphous $NiFe_2O_4$ ($\geq$235 m^2/g) were compared to the reference material (81 m^2/g) and the crystalline spinel $NiFe_2O_4$ (121 m^2/g) (Hong et al., 2014). Kim and coauthors elucidated the H_2 evolution capability of core–shell $NiFe_2O_4/TiO_2$ nanoparticles in H_2O–methanol mixtures under UV illumination (Kim et al., 2014).

10.4 Other Ceramic Materials

Transition metal carbides and nitrides are attractive materials for photoelectrode fabrication due to their favorable physical properties, such as chemical stability and high electrical conductivity. The early transition metal carbides, nitrides, and sulfides are considered as a substitute to platinum group metals because of their outstanding electronic and tremendous electrocatalytic properties. Metal carbides and nitrides show different electronic and electrocatalytic properties from those of the parent metals, but similar properties to those of platinum group metals in terms of d-band contraction and redistribution of the density of states near the Fermi levels. The formation of a metal-nitrogen bond causes a less amount of deficiency of the metal d-band occupation, leading the nitride surface to exhibit an electron-donating ability. Metal nitrides have near or greater electrocatalytic properties to those of noble metals for different chemical reactions, including HERs (Li, Yu, et al., 2018; Zhang, Ma, et al., 2020).

10.4.1 Nitrides

The transition metal nitrides such as W_2N have gained significant attention for their applications because of their exceptional properties such as high melting point (>2000 K), exceptional hardness (>10 GPa), chemical inertness in acidic and basic solutions, radiation resistance, high electrical conductivity, and durability (Chakrapani et al., 2009). Recently, Zhong et al. showed that the W_2N is a good electrocatalyst for oxygen reduction reaction (Hensel et al., 2010). Chakrapani et al. reported the tungsten oxide (WO_3) nanowire array samples were nitrided in a NH_3 atmosphere to get comprehensive conversion to tungsten nitride (W_2N) nanowires and the bandgap of WO_3 reduced from 2.9 to 2.2 eV after nitridation to W_2N. WO_3 nanowire arrays exhibited maximum incident photon-to-current conversion efficiency of 85% at 370 nm at 1.2 V vs. SCE (Chakrapani et al., 2009). The efficient light absorption and charge transport in nanowire architecture was attributed to the high quantum efficiency. The pure W_2N electrodes were unstable with respect to photocorrosion and the W_2N–WO_3 nanowires exhibited improvement in stability compared to pure W_2N nanowire arrays. Choi et al. have opted Si wire with the diameters of 1.6–14.6 µm arrays overlaid with an ultrathin titanium nitride film with thickness of 2–20 nm for an efficient and robust photocathode for hydrogen production (Choi et al., 2016). The Si wire overlaid titanium nitride films showed a photocurrent onset potential of +0.4 V vs. RHE, and a Faradaic efficiency of nearly 100% at 0 V vs. RHE over 20 hr of reaction. Such a remarkable performance was due to the fact that TiN performed the triple roles facilitating antireflection, protection of the Si surface, and electrocatalysis leading to hydrogen production.

Boron nitride (BN) has also attracted attention because it is nontoxic, low cost, and exhibits exceptional chemical and thermal stability. The stacked form of hexagonal lattice of 2-D BN sheets, known as 3-D layered structure of BN, could be associated with the layered graphitic carbon nitride/perylene diimide (g-C_3N_4/PDI) and reduced graphene oxide (rGO) to enhance h^+ separation efficiency (Hargreaves et al., 2020; Kofuji et al., 2018). Kofuji et al. prepared a metal-free g-C_3N_4/PDI-BN-rGO photocatalyst to efficiently produce H_2O_2 (Kofuji et al., 2018). The photoexcited g-C_3N_4/PDI moiety transfers the CB electrons to rGO, leading to selective synthesis of H_2O_2 by two-electron reduction of O_2 on the rGO surface (Kofuji et al., 2018). The VB holes formed on the g-C_3N_4/PDI move to BN for successful water oxidation. The g-C_3N_4/PDI-BN-rGO catalyst showed a 0.27% solar-to-chemical conversion (SCC) efficiency for H_2O_2 production. The rGO helped in forming endoperoxide via two-electron reduction of O_2. The separation of the h^+ in the VB of g-C_3N_4/PDI moiety could further suppress recombination of charge carriers and increase the photocatalytic activity (Hargreaves et al., 2020; Kofuji et al., 2018).

Ta_3N_5 is used as a photocatalyst for hydrogen generation because of its chemical stability and narrow bandgap of 2.0 eV. Importantly, nanosized Ta_3N_5 particles (30–50 nm) loaded with a Pt cocatalyst exhibited more photocatalytic activity for hydrogen evolution from an aqueous solution containing methanol as an electron donor under visible light conditions, than the bulk Ta_3N_5 particles (300–500 nm) (Maeda et al., 2009). The enhanced rate of water reduction is mainly due to the lower defect density in nanosized Ta_3N_5. Ordered mesoporous carbon nitride attached with nanosized Ta_3N_5 exhibited an enhanced H_2 evolution (Nurlaela et al., 2014). The pore size of 7 nm, small particle size, and high surface area also contributed to the higher activity. Ta_3N_5 structures having thin walls (thickness 2 nm) and pore size of 4 nm were effectively formed by nitriding an amorphous silica-coated mesoporous Ta_2O_5, followed by an alkaline treatment. This combination showed a three times higher activity than the bulk Ta_3N_5 particles. Other works related to increasing porosity by using colloidal silica and template of g-C_3N_4 were also reported (Yuliati et al., 2010). There are reports on the heterojunctions of Ta_3N_5 with TiO_2 and ZrO_2 used for enhancing the H_2 evolution properties (Maeda et al., 2010; Tabata et al., 2010).

10.4.2 Oxynitrides

Xu et al. explored an intriguing 2D charge transportation phenomenon in the $LaTa_2O_{6.77}N_{0.15}$ photocatalyst with a double perovskite structure, where the $LaTa_2O_{6.77}N_{0.15}$ was prepared by exfoliating a $RbLaTa_2O_{6.77}N_{0.15}$ precursor. It was found that the hybridization between O 2p and N 2p orbitals uplifted the VB maximum, which resulted in the intense light absorption in $LaTa_2O_{6.77}N_{0.15}$. The surface terminated $TaO(N)_6$ octahedrons provided instantaneous charge generation and absence of bulk charge diffusion steps caused photocatalytic hydrogen and oxygen production of almost four times and eight times higher than their bulk $LaTa_2O_{6.77}N_{0.15}$ system. The lower thickness of $LaTa_2O_{6.77}N_{0.15}$ significantly reduced bulk charge diffusion steps and extended the lifetime of photogenerated charges (Lv et al., 2017). Luo and coworkers modified the p-blocking centers through doping of A/B sites to alter the CB band and F-anion doping, which significantly shifted the band edge of the CB and enhanced the catalytic efficiency of the lattice oxygen for better water splitting (Hua et al., 2018).

10.4.3 Carbides

β-silicon carbide (β-SiC or cubic 3C-SiC) exhibits excellent physical and chemical properties such as low specific weight, chemical inertness, and insolubility in water, which are essential requirements for heterogeneous catalytic reactions. β-SiC has a characteristic bandgap of 2.2 eV. It is a p-type non-metal semiconductor (Mishra et al., 2015). Zhang et al. demonstrated a method to recycle crystalline SiC in the waste discharged from PV industry by reusing the material as a photocatalyst for hydrogen production (Zhang et al., 2017). The silicon carbide particles (some silicon oxycarbides that existed on the surface of the SiC) recovered from the PV industrial slicing waste consisted of 3C-SiC and hexagonal 6H-SiC. The hydrogen production from water was done in the presence of a sacrificial reagent (Na_2SO_3-Na_2S) to speed up the separation of photoinduced electron–hole pairs. The highest hydrogen production rate (191.8 $\mu mol \cdot g^{-1} \cdot h^{-1}$) was achieved in Pt/SiC obtained by impregnation and calcination method. This rate was higher than that obtained from the material fabricated using photoreduction method (Mishra et al., 2015).

Zhang et al. studied the pH effects of the addition of methanol and the loading of a Pt cocatalyst on the photocatalytic hydrogen generation efficiency of SiC nanoparticles (Zhang et al., 2014). They found that SiC nanoparticles would show high photocatalytic hydrogen generation efficiency in water under both ultraviolet and visible light irradiation. 3C-SiC with 2.36 eV bandgap showed a lower photocurrent (less than ~1 mA/cm^2) than the value (8 mA/cm^2) under AM 1.5 G (100 mW/cm^2) illumination. Zhou et al. proved that the photocatalytic performance of the SiC/CNTs composite could be efficiently enhanced through modifications of the morphology and interfaces of the SiC active layer (Zhou et al., 2015). The synthesized coaxial core–shell nanotube structure of C@SiCNTs showed a high photoactivity and good stability in water splitting. The synergistic effects of CNTs and SiC 1-D core–shell nanotube heterostructure improved light absorption, as well as the formation of tight Schottky-type heterojunctions between CNTs and SiC, which resulted in an enhanced photocatalytic activity by promoting charge separation at the interfaces of nanoheterostructures. The optimized nanoporous 3C-SiC(111) photoanode attained a maximum photocurrent density of 2.30 mA/cm^2 at 1.23 V vs. RHE under AM 1.5 G (100 mW/cm^2) illumination. This material considerably outperformed its planar counterpart (photocurrent density 0.69 mA/cm^2 at 1.23 V vs. RHE). The photoanode showed an improved light-harvesting efficiency (LHE) of over 93%, a high charge-separation efficiency (Φ_{sep}) of 38%, and a high charge-injection efficiency (Φ_{ox}) of 91% for water oxidation at 1.23 vs. RHE, which were significantly higher than those shown by its planar counterpart (LHE = 78%, Φ_{sep} = 28%, and Φ_{ox} = 53% at 1.23 V vs. RHE) (Jian et al., 2021).

Gao et al. reported that SiC would absorb visible light and splits water to form hydrogen, however, at a very low rate (24.9 μL·g^{-1}·h^{-1}) (Gao et al., 2007). The incorporation of sulfur would play a key role in the electronic structure modulations of TiO_2 in the interface and provide superior light absorption property. They modified the heterojunction compounds by adding different oxidation states of sulfur (S^{2-}, S^{4+}, and S^{6+}) in TiO_2. Another photocatalyst with superior photocatalytic performance was reported by Mishra et al. (Mishra et al., 2015). They synthesized sulfate-modified TiO_2/β-SiC heterojunction by varying amount of sulfate (5–20 wt%). The superior photocatalytic activity of the 15 wt% S-doped TiO_2/β-SiC catalysts (hydrogen energy production of 1254 μmol in 3 hr) was mainly due to the synergistic interaction of ternary species, and establishment of the heterojunction (Mishra et al., 2015). The charge recombination rate decreased, and surface charge transfer improved. A combination of SiC and TiO_2 with CdS was reported by Nagakawa et al. to overcome the problems of photocorrosion and fast recombination in CdS (Nagakawa et al., 2018). They reported that when CdS was subjected to visible light irradiation, the excited electrons in CdS moved to the CB of TiO_2 and the holes generated in CdS moved to the VB of SiC. The increased carrier separation efficiency reduced photocorrosion and recombination rate. The CdS/SiC/TiO_2 photocatalyst with a Pt cocatalyst showed a H_2 generation capacity of 1090.4 μmol·g^{-1}·h^{-1}.

Li et al. reported cubic silicon carbide (3C-SiC) nanoparticles with an amorphous silicon oxide (SiO_x) shell. This geometry brought plentiful surface states to trap the photogenerated carriers, thus enhancing the charge separation remarkably (Li & Sun, 2021). The thin SiO_x surface coating shell would help to prevent 3C-SiC from corrosion under visible light. The 3C-SiC nanoparticles could reduce CO_2 into CH_4 with a selectivity of over 90%. The surface states trapped the photogenerated electrons and suppressed the electron–hole recombination, while the SiO_x shell prevented silicon carbide from corrosion under visible light illumination. Hao et al. demonstrated that under visible light irradiation isolated Pd and Au nanoparticles on SiC could successfully catalyze the hydrogenation of nitrobenzene with molecular hydrogen to form aniline with high activity and selectivity at near-room temperature conditions (Hao et al., 2018). The localized surface plasmon resonance of the Au nanoparticles not only resulted in effective light absorption but also produced energetic electrons on the silicon carbide surface to activate adsorbed nitrobenzene molecules. The atomic hydrogen bonded on the SiC surface and thus reacted with activated nitrobenzene to produce aniline. The enhanced catalytic activity was mainly due to the combined effects from Pd and Au nanoparticles and SiC surface (Hao et al., 2018).

10.4.4 MXenes

MXene is a new family of 2D layered structures based on transition metal carbides and/or nitrides made from a bulk crystal called MAX, which are ternary carbides with the general formula $M_{n+1}AX_n$, where n = 1–3, M denotes a transition metal, A is an element such as aluminum or silicon, and X is either carbon or nitrogen (Eklund et al., 2017; Pei et al., 2021).

Recently, various nanostructures of $Ti_3C_2Ti_x$, where T_x is the surface functional group, such as F, O, and OH, have been explored to enhance the efficiency of hydrogen production. Su et al. prepared a series of $Ti_3C_2T_x$/TiO_2 composite photocatalysts with a monolayer and multilayers $Ti_3C_2T_x$ as the cocatalyst (Su et al., 2019). The PEC results showed that a monolayer $Ti_3C_2T_x$/TiO_2 composite under a 200 W Hg lamp embedded with a cut off filter of 285–325 nm exhibited an enhanced hydrogen production rate (2650 μmol·g^{-1}·h^{-1}), which was more than two-fold and nine-fold higher, compared to the case of their multilayer counterpart (H_2 production rate: 920 μmol·g^{-1}·h^{-1}) and pure TiO_2 (H_2 production rate: 290 μmol·g^{-1}·h^{-1}), respectively. The charge-carrier separation at the $Ti_3C_2T_x$/TiO_2 interface and superior electrical conductivity of the monolayer $Ti_3C_2T_x$ enhanced the PEC performance.

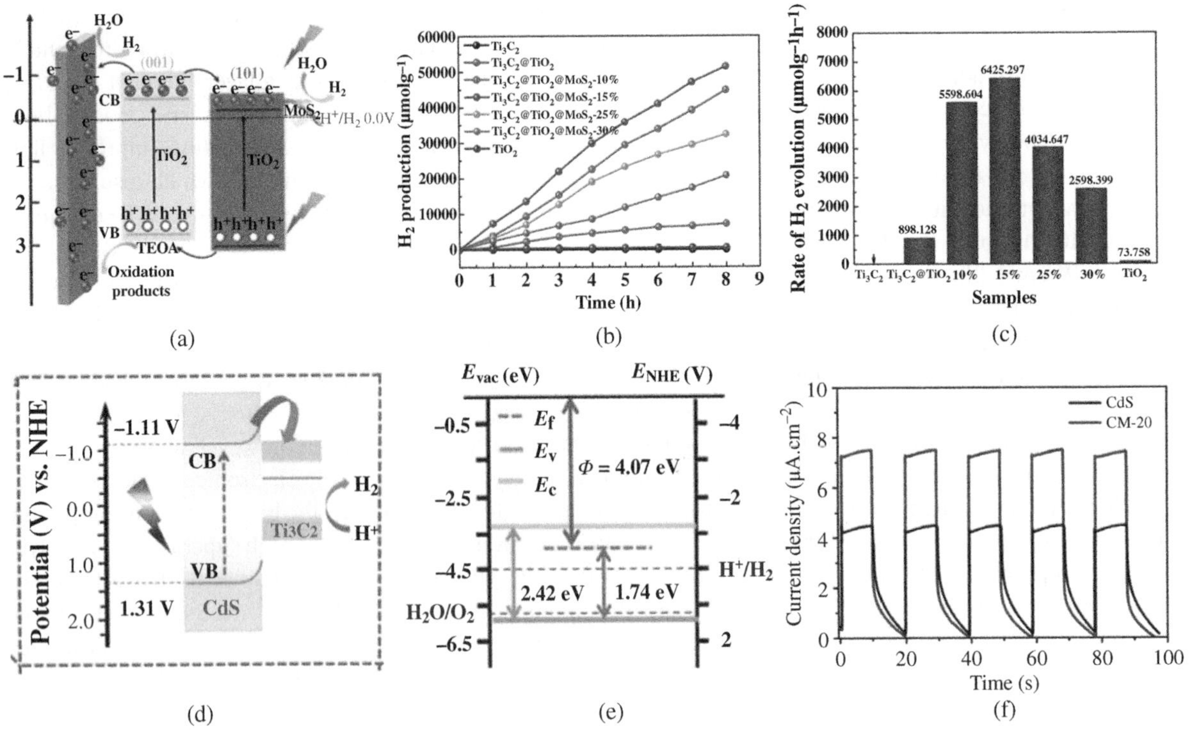

Figure 10.7 (a) Schematic representation of photocatalytic reaction mechanism for $Ti_3C_2@TiO_2@MoS_2$ composites under solar light irradiation (b) photocatalytic H_2 production, (c) the rate of photocatalytic H_2 production of different catalysts in aqueous acetone solution with TEOA within 8 h (Li, Yin, et al., 2019), Source: Reproduced from Li, Yin, et al., 2019, with permission from Elsevier/Elsevier. (d) Band energy diagram and (e) representative schematic Schottky heterojunction, and (f) photocurrent-density curves of 1D CdS/2D MXene for hydrogen production (Xiao et al., 2020). Source: Reproduced from Xiao et al., 2020, with permission from Elsevier/Elsevier.

Li et al. demonstrated 2D–2D–2D structure of $Ti_3C_2@TiO_2@MoS_2$ composite (Li, Yin, et al., 2019). They deposited MoS_2 (15 wt%) on TiO_2 nanosheet already attached on a highly conductive Ti_3C_2 MXene to fabricate the composite. The Ti_3C_2 MXene worked both as a source of titanium and a pathway for transferring photogenerated electrons (Figure 10.7a). The d surface embedded MoS_2 deposited on the (101) facets of TiO_2 could steal photogenerated electrons of (101) facets and act as reduction active sites. The $Ti_3C_2@TiO_2@MoS_2$ composite showed a hydrogen production rate of 6425.297 $\mu mol \cdot g^{-1} \cdot h^{-1}$ and very stable performance under the reaction condition (Figure 10.7b,c). This is mainly due to the in-situ growth of TiO_2 nanosheets making good interfacial contact with Ti_3C_2 having an extraordinary electronic conductivity, which facilitated the separation of carriers. The deposition of MoS_2 on electrons-rich (101) facets of TiO_2 nanosheets helped to decrease the charge recombination rate by capturing photogenerated electrons and making TiO_2 nanosheets to have high-efficiency photogeneration of electron–hole pairs. The co-exposed (101) and (001) facets of TiO_2 nanosheets also helped in enhancing charge transfer and separation (Li, Yin, et al., 2019). Xiao et al. have demonstrated 1D/2D Schottky heterojunction using one-dimensional (1D) CdS nanorod/2D Ti_3C_2 MXene nanosheets for accelerated charge separation and a lower Schottky barrier for solar-driven hydrogen evolution from water splitting (Figure 10.7a,b). They obtained an enhanced hydrogen evolution rate of 2407 $\mu mol \cdot g^{-1} \cdot h^{-1}$ under visible light irradiation (Figure 10.7f). 1D/2D Schottky heterojunction could enhance the rate of hydrogen evolution without using any noble metal-containing cocatalysts (Li, Yin, et al., 2019).

Li et al. designed $Ti_3C_2T_x/TiO_2$ nanoflowers that showed a hydrogen production rate of 526 $\mu mol \cdot g^{-1} \cdot h^{-1}$. This rate was about four-fold higher than that given by TiO_2 nanobelts. The enhanced PEC performance suggested that the noble metal-free $Ti_3C_2T_x$ is an alternative cocatalyst for the costly and precious noble metals, like Pt, Au, etc. To further enhance the H_2 production activity, Yuan et al. prepared $Ti_3C_2T_x$ nanofibers by hydrolyzation and selective etching of Ti_3AlC_2 MAX ceramics (Yuan et al., 2018). These $Ti_3C_2T_x$ nanofibers had more catalytic active sites due to increased surface area leading to enhanced hydrogen production, high cycling stability, and long-term viability. Recently, Li et al. had successfully designed $Ti_3C_2T_x$ quantum dots by a self-assembly method (Li, Ding, et al., 2019). This material showed a higher photocatalytic activity under artificial sunlight (300 W Xe arc lamp integrated with an AM-1.5 filter), which was nearly ten-fold higher than that exhibited by $g\text{-}C_3N_4/Ti_3C_2T_x$ sheets (524.3 $\mu mol \cdot g^{-1} \cdot h^{-1}$). Significantly, $Ti_3C_2T_x$ quantum dots (QDs) offered more abundant active edge sites, and exceptional electronic conductivity as compared to conventional $Ti_3C_2T_x$ sheets. The photogenerated carriers in g-C_3N_4/$Ti_3C_2T_x$ QDs composites can increase the hydrogen production activity. Therefore, due to exceptional physical properties, g-C_3N_4/$Ti_3C_2T_x$ quantum dot composites performed a remarkable enhancement in the photocatalytic H_2 production rate of 5111.8 $\mu mol \cdot g^{-1} \cdot h^{-1}$. Enhanced physicochemical properties of $Ti_3C_2T_x$ through surface termination groups were reported by Li et al. (Li, Ding, et al., 2019; Pang et al., 2019). These results agreed with the finding of Sun et al., who observed significant augmentation of hydrogen production (88 $\mu mol \cdot g^{-1} \cdot h^{-1}$) over O-$Ti_3C_2T_x$, compared to conventional samples (Sun et al., 2018). Yang et al. synthesized O-$Ti_3C_2T_x$/CdS composites by using radiofrequency oxygen plasma method. The material provided stable transfer channel for electrons to repress the recombination of e^-–h^+ pairs and sufficient g adsorption of water molecules and hydrogen ions on the surface of the catalyst (Yang et al., 2019). Xu et al. prepared layered g-C_3N_4/plasma-treated $Ti_3C_2T_x$ photocatalyst by plasma treatment and observed an increase of Ti—O bond concentration with a decrease in the concentration of Ti—C, Ti—F, and Ti—OH bonds (Xu et al., 2020). The plasma-treated $Ti_3C_2T_x$ photocatalyst worked as an excellent acceptor of photogenerated electrons, leading to substantially enhanced photocatalytic activity.

Tahir et al. decorated 2D hierarchical g-C_3N_4 (2D HCN) nanosheets on exfoliated TiC multilayers embedded with in situ-grown TiO_2 (Tahir, 2021). The material yielded a hydrogen evolution rate of 182.5 $\mu mol \cdot g^{-1} \cdot h^{-1}$. An H_2 production rate of 310 $\mu mol \cdot g^{-1} \cdot h^{-1}$ was obtained from the optimized 10TiC-48/HCN 2D/2D heterojunction. This high H_2 production was primarily because of the speedy electron transfer from g-C_3N_4 to TiC due to the synergistic effects of higher conductivity and creation of an interface between HCN and TiO_2 (Figure 10.8a–c). Peng and coworkers developed active TiO_2@$Ti_3C_2T_x$ photocatalysts, which efficiently split water to produce H_2 gas at 860 $\mu mol \cdot g^{-1} \cdot h^{-1}$ (Peng et al., 2018). The photocatalytic reaction followed a dual-carrier-separation mechanism, where the Cu acted as the electron mediator and reduction cocatalyst and MXene as the hole mediator to enhance the separation of photogenerated charges. Ramalingam et al. reported a $Ti_3C_2T_x$ MXene as a support to host S and N-coordinated ruthenium single atom (Ru_{SA}) catalyst for HER. The optimized Ru_{SA}-N-S-$Ti_3C_2T_x$ catalyst exhibited a low overpotential of 76 mV to achieve a current density of 10 mA/cm^2 (Peng et al., 2018). Moreover, the integration of the Ru_{SA}-N-S-$Ti_3C_2T_x$ catalyst on a Si-based photocathode allowed a PEC H_2 production with a significantly elevated photocurrent density of 37.6 mA/cm^2, which is superior to the noble metal catalysts like platinum coupled with Si photocathodes (Ramalingam et al., 2019). Li et al. confirmed the

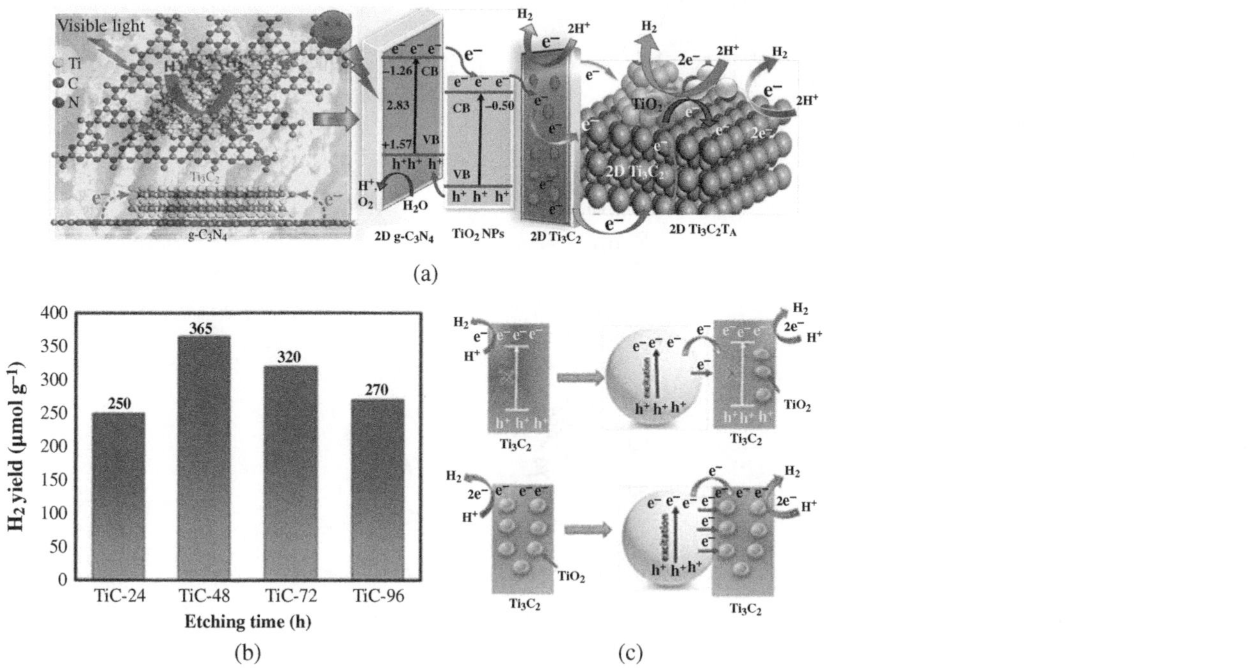

Figure 10.8 (a) Schematic illustration of the probable photocatalytic process for hydrogen production over the 2D/2D CN/$Ti_3C_2T_A$ heterojunction composite under visible light. (b) Effect of etchant time on growth of TiO_2 nanoparticles embedded over Ti_3C_2 nanosheets for photocatalytic hydrogen production. (c) Scheme on charge carrier production and separation over Ti_3C_2 MXene nanosheets (Tahir, 2021). Source: Reproduced with permission from Tahir, 2021. Copyright 2021 American Chemical Society/American Chemical Society.

interaction between the reactive metal and the support at the interface of transition metal carbide MXenes and platinum nanoparticles. Pt nanoparticles formed bonds with the MXene metal sites and created a bimetallic alloy and ordered intermetallic structures for H_2O and C–H activation, respectively (Li, Cui, et al., 2018; Li, Yu, et al., 2018).

Yan et al. investigated the properties of composite photoanode of $Ti_3C_2T_x/BiVO_4$ film grown on the FTO substrate (Yan et al., 2019). A photocurrent density in the range 2.1 to 2.95 mA/cm^2 was achieved by heat-treating bare $BiVO_4$/FTO (without $Ti_3C_2T_x$) photoanode in argon. The charge recombination at the interface between $BiVO_4$ and FTO was successfully suppressed by improving the contact between the $BiVO_4$ film and FTO. Coating of the thin $Ti_3C_2T_x$ flakes on $BiVO_4$ increased the photocurrent density further to 3.45 mA/cm^2 (at 1.23 V vs. RHE under 1 sun illumination), providing a 73% surface charge separation efficiency and 0.78% photoconversion efficiency. The $Ti_3C_2T_x$ flakes acted as cocatalysts enabling the facile transfer of the photogenerated carriers and decreasing the charge recombination on the surface of $BiVO_4$. The $Ti_3C_2T_x$ flakes also served as a sensitizer to improve the visible light absorption in the composite photoanode. For an improved PEC water splitting performance, TiO_2/Ti_3C_2 heterostructures were developed as photoanodes by Yu and coworkers (Yu et al., 2018). TiO_2/Ti_3C_2 hybrid structure exhibited significantly improved photoactivity for PEC water splitting. The applied bias photon-to-current efficiency of TiO_2/Ti_3C_2 photoanode was 6 times superior as compared to TiO_2. The effective extraction of electrons to the reaction interfaces occurred through the creation of interfacial Ti^{3+}, which was initiated from the interaction of both TiO_2 and defective Ti_3C_2. Additionally, the recombination of photogenerated holes with electrons in TiO_2 would be considerably decreased by the efficient transfer of holes to Ti_3C_2 hole reservoir system. This could enable the separation and transfer of charge carriers. Development of a combination of Ti_3C_2 nanoparticles with cadmium sulfide that exhibited an elevated visible-light photocatalytic H_2 production rate of 14,342 μmol·h^{-1}·g^{-1} (Figure 10.9a-b) and an apparent quantum efficiency of 40.1% at 420 nm was reported by Ran et al. (Ran et al., 2017). The Ti_3C_2 nanoparticles also served as an able cocatalyst on ZnS or $Zn_xCd_{1-x}S$ and this superior level of performance occurred due to favorable electrical conductivity, Fermi level position, and H_2 evolution capacity of Ti_3C_2 nanoparticles (Figure 10.9c,d) (Yu et al., 2018).

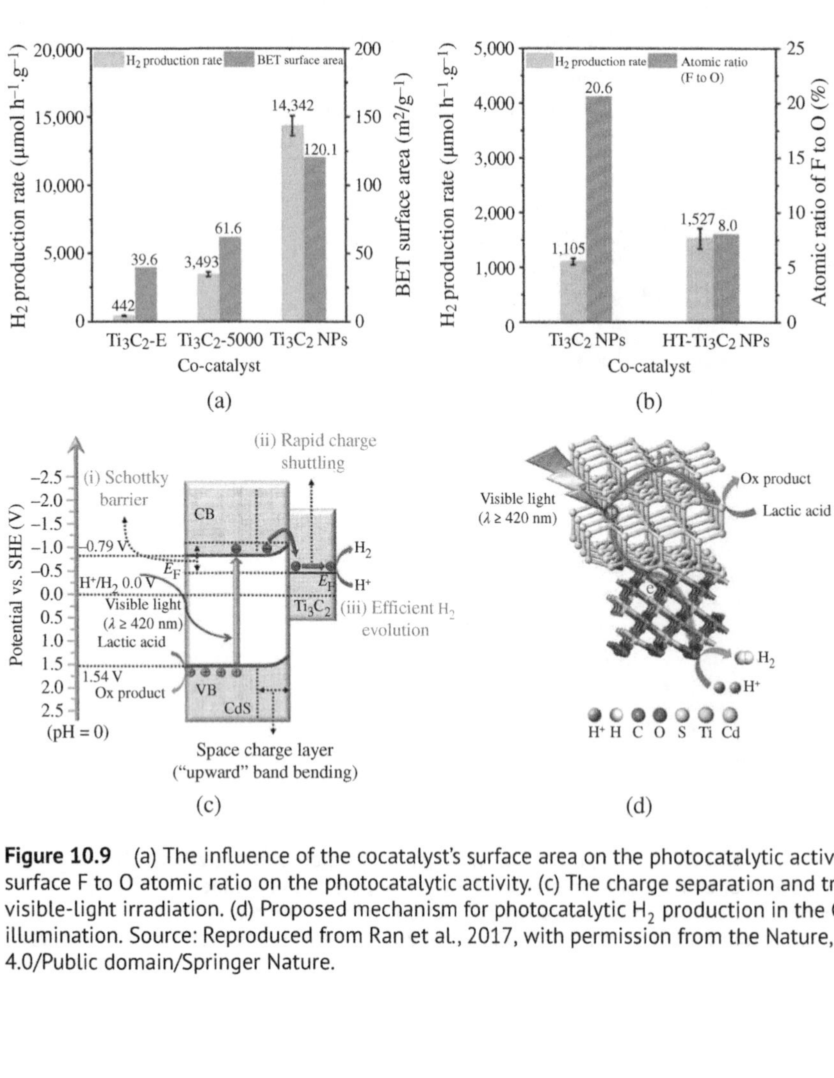

Figure 10.9 (a) The influence of the cocatalyst's surface area on the photocatalytic activity. (b) The influence of the cocatalyst's surface F to O atomic ratio on the photocatalytic activity. (c) The charge separation and transfer in the CdS/Ti_3C_2 system under visible-light irradiation. (d) Proposed mechanism for photocatalytic H_2 production in the CdS/Ti_3C_2 system under visible-light illumination. Source: Reproduced from Ran et al., 2017, with permission from the Nature, copyright 2017 (Ran et al., 2017)/CCBY 4.0/Public domain/Springer Nature.

10.5 Challenges

The development of practical solar to hydrogen PEC applications requires development of photoelectrodes with proper electronic structures and improved stability. So far, many metal oxides have been investigated for high-efficiency PEC, but still, several metal oxides are facing the challenges of partial light absorption, high recombination rates, and poor electrical conductivity. Although there have been efforts to improve optical and electronic properties via doping with different elements or crystal defects, the overall visible light photoactivity is still very low in the visible light region. New low-cost and visible-active semiconductor ceramic materials are still needed to reach the goal of achieving commercial production of hydrogen fuel. Currently, researchers are searching for new methods to enhance the stability, attain high solar to hydrogen efficiencies, reduce the cost, and increase durability. In order to improve the rate and efficiency of photocatalytic processes, creating active sites on the surface, enhancing light absorption rate, and accelerating the separation and transfer of charge carriers by forming robust interfaces between photocatalysts are required. The involvement of cocatalysts can enhance the performance of PEC technology. This strategy has the advantages such as decreasing the overpotential of photocatalytic reaction, extracting photogenerated charge carriers, and host active sites, as well as improving the stability of photocatalysts by inhibiting photocorrosion. The search for photocatalysts exhibiting excellent physicochemical stability in an environment containing oxidative and/or reductive photocorrosion processes is still continuing.

10.6 Conclusion

A range of ceramic materials used in photocatalytic or PEC hydrogen production is discussed in this chapter. Ceramic photocatalysts offer low-cost, photocorrosion resistive, and efficient hydrogen production performance. A detailed account of the development of doped and heterostructured ceramic materials and their modification with different types of materials to tune the structural and electronic properties for efficient light absorption and charge separation and transfer is given. Despite the limited success in the development of stable and efficient photoelectrodes, the cost and scalability remain as important issues for their large-scale production.

References

Alves, S. A., Goulart, L. A., & Mascaro, L. H. (2018). A novel WO_3/MoS_2 photocatalyst applied to the decolorization of the textile dye Reactive Blue 198. *Journal of Solid State Electrochemistry*, *22*, 1449–1458.

Archer, M. D., Morris, G. C., & Yim, G. K. (1981). Electrochemical approaches to solar energy conversion: A brief overview and preliminary results obtained with n-type cobalt ferrite. *Journal of Electroanalytical Chemistry and Interfacial Electrochemistry*, *118*, 89–100.

Asahi, R., Morikawa, T., Ohwaki, T., Aoki, K., & Taga, Y. (2001). Visible-light photocatalysis in nitrogen-doped titanium oxides. *Science*, *293*, 269–272.

Bak, T., Nowotny, J., Rekas, M., & Sorrell, C. C. (2010). Photo-electrochemical hydrogen generation from water using solar energy. *Materials-related aspects, International Journal of Hydrogen Energy*, *27*, 991–1022.

Benko, F. A., & Koffyberg, F. P. (1986). The effect of defects on some photoelectrochemical properties of semiconducting $MgFe_2O_4$. *Materials Research Bulletin*, *21*, 1183–1188.

Bie, J., Wang, S., Guan, Y., Zhang, X., Sun, Y., Sun, C., Tang, K., Huang, J., Cao, M., Ling, J., Tan, K., Shen, Y., & Wang, L. (2021). Close-spaced sublimation of CdZnTe: In films for solar energy water splitting. *Energy and Fuels*, *35*, 8234–8245.

Biresselioglu, M. E., & Yelkenci, T. (2016). Scrutinizing the causality relationships between prices, production and consumption of fossil fuels: A panel data approach. *Energy*, *102*, 44–53.

Büchler, M., Schmuki, P., Böhni, H., Stenberg, T., & Mäntylä, T. (1998). Comparison of the semiconductive properties of sputter-deposited iron oxides with the passive film on iron. *Journal of The Electrochemical Society*, *145*, 378–385.

Chae, S. Y., Lee, C. S., Jung, H., Joo, O.-S., Min, B. K., Kim, J. H., & Hwang, Y. J. (2017). Insight into charge separation in $WO_3/BiVO_4$ heterojunction for solar water splitting. *ACS Applied Materials & Interfaces*, *9*, 19780–19790.

Chakrapani, V., Thangala, J., & Sunkara, M. K. (2009). WO_3 and W_2N nanowire arrays for photoelectrochemical hydrogen production. *International Journal of Hydrogen Energy*, *34*, 9050–9059.

Chen, T., Meng, J., Wu, S., Pei, J., Lin, Q., Wei, X., Li, J., & Zhang, Z. (2018). Room temperature synthesized $BaTiO_3$ for photocatalytic hydrogen evolution. *Journal of Alloys and Compounds*, *754*, 184–189.

Chen, X., Shen, S., Guo, L., & Mao, S. S. (2010). Semiconductor-based photocatalytic hydrogen generation. *Chemical Reviews*, *110*, 6503–6570.

Chen, Z., Dinh, H. N., & Miller, E. (2013). *Photoelectrochemical Water Splitting Standards, Experimental Methods, and Protocols* (pp. 1–126). Springer-Verlag , ISBN 978-1-4614-8298-7.

Choi, S. K., Chae, W.-S., Song, B., Cho, C.-H., Choi, J., Han, D. S., Choi, W., & Park, H. (2016). Photoelectrochemical hydrogen production on silicon microwire arrays overlaid with ultrathin titanium nitride. *Journal of Materials Chemistry A*, *4*, 14008–14016.

De Haart, L. G. J., & Blasse, G. (1985). Photoelectrochemical properties of ferrites with the spinel structure. *Solid State Ionics*, *16*, 137–139.

Deng, Z., Zhou, J., Miao, L., Liu, C., Peng, Y., Sun, L., & Tanemura, S. (2017). The emergence of solar thermal utilization: Solar-driven steam generation. *Journal of Materials Chemistry A*, *5*, 7691–7709.

Dillert, R., Taffa, D. H., Wark, M., Bredow, T., & Bahnemann, D. W. (2015). Research Update: Photoelectrochemical water splitting and photocatalytic hydrogen production using ferrites (MFe_2O_4) under visible light irradiation. *APL Materials*, *3*, 104001.

Dom, R., Kumar, G. S., Hebalkar, N. Y., Joshi, S. V., & Borse, P. H. (2013). Eco-friendly ferrite nanocomposite photoelectrode for improved solar hydrogen generation. *RSC Advances*, *3*, 15217–15224.

Eklund, P., Rosen, J., & Persson, P.r. O. Å. (2017). Layered ternary $M_{n+1}AX_n$ phases and their 2D derivative MXene: an overview from a thin-film perspective. *Journal of Physics D: Applied Physics*, *50*, 113001.

Fujishima, A., & Honda, K. (1972). Electrochemical photolysis of water at a semiconductor electrode. *Nature*, *238*, 37–38.

Gao, Q., Meng, J., Yang, Y., Lin, Q., Lu, Y., Wei, X., Li, J., Han, G., & Zhang, Z. (2021). Zirconium doping in calcium titanate perovskite oxides with surface nanostep structure for promoting photocatalytic hydrogen evolution. *Applied Surface Science*, *542*, 148544.

Gao, Y., Wang, Y., & Wang, Y. (2007). Photocatalytic hydrogen evolution from water on SiC under visible light irradiation. *Reaction Kinetics and Catalysis Letters*, *91*, 13–19.

Bisquert, J., Giménez, S., Bertoluzzi, L., & Herraiz-Cardona, I. (2016). Analysis of photoelectrochemical systems by impedance spectroscopy. In S. Giménez & J. Bisquert (Eds.), *Principles, Photoelectrochemical Solar Fuel Production From Basic Principles to Advanced Devices* (pp. 281–321). Springer International Publishing, ISBN 978-3-319-29641-8.

Hao, C.-H., Guo, X.-N., Sankar, M., Yang, H., Ma, B., Zhang, Y.-F., Tong, X.-L., Jin, G.-Q., & Guo, X.-Y. (2018). Synergistic effect of segregated Pd and Au nanoparticles on semiconducting SiC for efficient photocatalytic hydrogenation of nitroarenes. *ACS Applied Materials & Interfaces*, *10*, 23029–23036.

Hargreaves, J. S. J., Chung, Y.-M., Ahn, W.-S., Hisatomi, T., Domen, K., Kung, M. C., & Kung, H. H. (2020). Minimizing energy demand and environmental impact for sustainable NH3 and H_2O_2 production—A perspective on contributions from thermal, electro-, and photo-catalysis. *Applied Catalysis A: General*, *594*, 117419.

He, J., & Janáky, C. (2020). Recent advances in solar-driven carbon. *ACS Energy Letters*, *5*, 1996–2014.

Hensel, J., Wang, G., Li, Y., & Zhang, J. Z. (2010). Synergistic effect of CdSe quantum dot sensitization and nitrogen doping of TiO_2 nanostructures for photoelectrochemical solar hydrogen generation. *Nano Letters*, *10*, 478–483.

Hoang, S., Berglund, S. P., Hahn, N. T., Bard, A. J., & Mullins, C. B. (2012). Enhancing visible light photo-oxidation of water with TiO_2 nanowire arrays via cotreatment with H_2 and NH_3: Synergistic effects between Ti^{3+} and N. *Journal of the American Chemical Society*, *134*, 3659–3662.

Hong, D., Yamada, Y., Sheehan, M., Shikano, S., Kuo, C.-H., Tian, M., Tsung, C.-K., & Fukuzumi, S. (2014). Mesoporous nickel ferrites with spinel structure prepared by an aerosol spray pyrolysis method for photocatalytic hydrogen evolution. *ACS Sustainable Chemistry & Engineering*, *2*, 2588–2594.

Hou, H., Shang, M., Gao, F., Wang, L., Liu, Q., Zheng, J., Yang, Z., & Yang, W. (2016). Highly efficient photocatalytic hydrogen evolution in ternary hybrid TiO_2/CuO/Cu thoroughly mesoporous nano fibers. *ACS Applied Materials & Interfaces*, *8*, 20128–20137.

Hua, B., Li, M., Pang, W., Tang, W., Zhao, S., Jin, Z., Zeng, Y., Amirkhiz, B. S., & Luo, J.-L. (2018). Activating *p*-blocking centers in perovskite for efficient water splitting. *Chem*, *4*, 2902–2916.

Hunt, N. D., Liebman, M., Thakrar, S. K., & Hill, J. D. (2020). Fossil energy use, climate change impacts, and air quality-related human health damages of conventional and diversified cropping systems in Iowa, USA. *Environmental Science & Technology*, *54*, 11002–11014.

Ida, S., Yamada, K., Matsuka, M., Hagiwara, H., & Ishihara, T. (2012). Photoelectrochemical hydrogen production from water using p-type and n-type oxide semiconductor electrodes. *Electrochimica Acta*, *82*, 397–401.

Ida, S., Yamada, K., Matsunaga, T., Hagiwara, H., Matsumoto, Y., & Ishihara, T. (2010). Preparation of P-Type $CaFe_2O_4$ photocathodes for producing hydrogen from water. *Journal of the American Chemical Society*, *132*, 17343–17345.

Iqbal, M. Z., & Siddique, S. (2018). Recent progress in efficiency of hydrogen evolution process based photoelectrochemical cell. *International Journal of Hydrogen Energy*, *43*, 21502–21523.

Iwase, A., Kato, H., & Kudo, A. (2009). The effect of alkaline earth metal ion dopants on photocatalytic water splitting by $NaTaO_3$ powder. *ChemSusChem*, *2*, 873–877.

Jian, J.-X., Jokubavicius, V., Syväjärvi, M., Yakimova, R., & Sun, J. (2021). Nanoporous cubic silicon carbide photoanodes for enhanced solar water splitting. *ACS Nano*, *15*, 5502–5512.

Jones, L. E., & Olsson, G. (2017). Solar photovoltaic and wind energy providing water. *Global Challenges*, *1*, 1600022.

Kabir, E., Kumar, P., Kumar, S., Adelodun, A. A., & Kim, K.-H. (2018). Solar energy: Potential and future prospects. *Renewable and Sustainable Energy Reviews*, *82*, 894–900.

Kato, H., Asakura, K., & Kudo, A. (2003). Highly efficient water splitting into H_2 and O_2 over lanthanum-doped $NaTaO_3$ photocatalysts with high crystallinity and surface nanostructure. *Journal of the American Chemical Society*, *125*, 3082–3089.

Kato, H., & Kudo, A. (1999). Highly efficient decomposition of pure water into H_2 and O_2 over $NaTaO_3$ photocatalysts. *Catalysis Letters*, *58*, 153–155.

Kim, E. S., Nishimura, N., Magesh, G., Kim, J. Y., Jang, J.-W., Jun, H., Kubota, J., Domen, K., & Lee, J. S. (2013). Fabrication of $CaFe_2O_4$/TaON heterojunction photoanode for photoelectrochemical water oxidation. *Journal of the American Chemical Society*, *135*, 5375–5383.

Kim, H. G., Borse, P. H., Choi, W., & Lee, J. S. (2005). Photocatalytic nanodiodes for visible-light photocatalysis. *Angewandte Chemie International Edition*, *44*, 4585–4589.

Kim, H. S., Kim, D., Kwak, B. S., Han, G. B., Um, M.-H., & Kang, M. (2014). Synthesis of magnetically separable core@shell structured $NiFe_2O_4$@TiO_2 nanomaterial and its use for photocatalytic hydrogen production by methanol/water splitting. *Chemical Engineering Journal*, *243*, 272–279.

Kim, J. H., Hansora, D., Sharma, P., Jang, J.-W., & Lee, J. S. (2019). Toward practical solar hydrogen production – An artificial photosynthetic leaf-to-farm challenge. *Chemical Society Reviews*, *48*, 1908–1971.

Kirkinen, J., Palosuo, T., Holmgren, K., & Savolainen, I. (2008). Greenhouse impact due to the use of combustible fuels: Life cycle viewpoint and relative radiative forcing commitment. *Environmental Management*, *42*, 458–469.

Kofuji, Y., Isobe, Y., Shiraishi, Y., Sakamoto, H., Ichikawa, S., Tanaka, S., & Hirai, T. (2018). Hydrogen peroxide production on a carbon nitride – boron nitride-reduced graphene oxide hybrid photocatalyst under visible light. *ChemCatChem*, *10*, 2070–2077.

Kubacka, A., Fernández-García, M., & Colón, G. (2012). Advanced nanoarchitectures for solar photocatalytic applications. *Chemical Reviews*, *112*, 1555–1614.

Kudo, A., & Miseki, Y. (2009). Heterogeneous photocatalyst materials for water splitting. *Chemical Society Reviews*, *38*, 253–278.

Kumar, D. P., Reddy, N. L., Karthik, M., Neppolian, B., Madhavan, J., & Shankar, M. V. (2016). Solar light sensitized p-Ag_2O/n-TiO_2 nanotubes heterojunction photocatalysts for enhanced hydrogen production in aqueous-glycerol solution. *Solar Energy Materials and Solar Cells*, *154*, 78–87.

Kung, H. H., Jarrett, H. S., Sleight, A. W., Ferretti, A., Kung, H. H., Jarrett, H. S., Sleight, A. W., & Ferretti, A. (1977). Semiconducting oxide anodes in photoassisted electrolysis of water. *Journal of Applied Physics*, *48*, 2463–2469.

Lhermitte, C. R., & Sivula, K. (2019). Alternative oxidation reactions for solar-driven fuel production. *ACS Catalysis*, *9*, 2007–2017.

Li, C., Fang, T., Hu, H., Wang, Y., Liu, X., Zhou, S., Fu, J., & Wang, W. (2019). Synthesis and SPEF-induced enhanced photoelectrochemical performance of the $BaTiO_3$@α-Fe_2O_3 ferroelectric heterojunction. *The Journal of Physical Chemistry C*, *123*, 22431–22438.

Li, H., & Sun, J. (2021). Highly selective photocatalytic CO_2 reduction to CH_4 by ball-milled cubic silicon carbide nanoparticles under visible-light irradiation. *ACS Applied Materials & Interfaces*, *13*, 5073–5078.

Li, J., & Wu, N. (2015). Semiconductor-based photocatalysts and photoelectrochemical cells for solar fuel generation: a review. *Catalysis Science & Technology*, *5*, 1360–1384.

Li, J., Zeng, J., Jia, L., & Fang, W. (2010). Investigations on the effect of Cu^{2+}/Cu_{1+} redox couples and oxygen vacancies on photocatalytic activity of treated $LaNi_{1-x}Cu_xO_3$ (x=0.1, 0.4, 0.5). *International Journal of Hydrogen Energy*, *35*, 12733–12740.

Li, Y., Ding, L., Guo, Y., Liang, Z., Cui, H., & Tian, J. (2019). Boosting the photocatalytic ability of g-C_3N_4 for hydrogen production by Ti_3C_2 MXene quantum dots. *ACS Applied Materials & Interfaces*, *11*, 41440–41447.

Li, Y., Peng, Y.-K., Hu, L., Zheng, J., Prabhakaran, D., Wu, S., Puchtler, T. J., Li, M., Wong, K.-Y., Taylor, R. A., & Tsang, S. C. E. (2019). Photocatalytic water splitting by N-TiO_2 on MgO (111) with exceptional quantum efficiencies at elevated temperatures. *Nature Communications*, *10*, 4421–4430.

Li, Y., Yin, Z., Ji, G., Liang, Z., Xue, Y., Guo, Y., Tian, J., Wang, X., & Cui, H. (2019). 2D/2D/2D heterojunction of Ti_3C_2 MXene/MoS_2 nanosheets/TiO_2 nanosheets with exposed (001) facets toward enhanced photocatalytic hydrogen production activity. *Applied Catalysis B: Environmental*, *246*, 12–20.

Li, Z., Cui, Y., Wu, Z., Milligan, C., Zhou, L., Mitchell, G., Xu, B., Shi, E., Miller, J. T., Ribeiro, F. H., & Wu, Y. (2018). Reactive metal–support interactions at moderate temperature in two-dimensional niobium-carbide-supported platinum catalysts. *Nature Catalysis*, *1*, 349–355.

Li, Z., Yu, L., Milligan, C., Ma, T., Zhou, L., Cui, Y., Qi, Z., Libretto, N., Xu, B., Luo, J., Shi, E., Wu, Z., Xin, H., Delgass, W. N., Miller, J. T., & Wu, Y. (2018). Two-dimensional transition metal carbides as supports for tuning the chemistry of catalytic nanoparticles. *Nature Communications*, *9*, 5258.

Luo, J., Zhang, S., Sun, M., Yang, L., Luo, S., & Crittenden, J. C. (2019). A critical review on energy conversion and environmental remediation of photocatalysts with remodeling crystal lattice, surface, and interface. *ACS Nano*, *13*, 9811–9840.

Luque, A., & Hegedus, S. (2003). *Handbook of Photovoltaic Science and Engineering* (pp. 1–43). Wiley, ISBN:9780470014004.

Lv, M., Sun, X., Wei, S., Shen, C., Mi, Y., & Xu, X. (2017). Ultrathin lanthanum tantalate perovskite nanosheets modified by nitrogen doping for efficient photocatalytic water splitting. *ACS Nano*, *11*, 11441–11448.

Maeda, K., Higashi, M., Lu, D., Abe, R., & Domen, K. (2010). Efficient nonsacrificial water splitting through two-step photoexcitation by visible light using a modified oxynitride as a hydrogen evolution photocatalyst. *Journal of the American Chemical Society*, *132*, 5858–5868.

Maeda, K., Nishimura, N., & Domen, K. (2009). A precursor route to prepare tantalum (V) nitride nanoparticles with enhanced photocatalytic activity for hydrogen evolution under visible light. *Applied Catalysis A: General*, *370*, 88–92.

Mangrulkar, P. A., Polshettiwar, V., Labhsetwar, N. K., Varma, R. S., & Rayalu, S. S. (2012). Nano-ferrites for water splitting: Unprecedented high photocatalytic hydrogen production under visible light. *Nanoscale*, *4*, 5202–5209.

Marschall, R., & Wang, L. (2014). Non-metal doping of transition metal oxides for visible-light photocatalysis. *Catalysis Today*, *225*, 111–135.

Matsumoto, Y., Omae, M., Watanabe, I., & Sato, E.-i. (1985). Photoelectrochernical properties of the Zn-Ti-Fe spinel oxides. *Journal of The Electrochemical Society*, *133*, 711–716.

Matsumoto, Y., Sugiyama, K., & Sato, E.-I. (1988). Improvement of $CaFe_2O_4$ photocathode by doping with Na and Mg. *Journal of Solid State Chemistry*, *74*, 117–125.

Mishra, G., Parida, K., & Singh, S. (2015). Facile fabrication of S-TiO_2/β-SiC nanocomposite photocatalyst for hydrogen evolution under visible light irradiation. *ACS Sustainable Chemistry & Engineering*, *3*, 245–253.

Nagakawa, H., Ochiai, T., Takekuma, Y., Konuma, S., & Nagata, M. (2018). Effective photocatalytic hydrogen evolution by cascadal carrier transfer in the reverse direction. *ACS Omega*, *3*, 12770–12777.

Nakanishi, H., Iizuka, K., Takayama, T., Iwase, A., & Kudo, A. (2017). Highly active $NaTaO_3$-based photocatalysts for CO_2 reduction to form CO using water as the electron donor. *ChemSusChem*, *10*, 112–118.

Nakata, K., & Fujishima, A. (2012). TiO_2 photocatalysis: Design and Applications. *Journal of Photochemistry and Photobiology C*, *13*, 169–189.

Nguyen, V.-H., Do, H. H., Nguyen, T. V., Singh, P., Raizada, P., Sharma, A., Sankar Sana, S., Grace, A. N., Shokouhimehr, M., Ahn, S. H., Xia, C., Kim, S. Y., & Le, Q. V. (2020). Perovskite oxide-based photocatalysts for solar-driven hydrogen production: Progress and perspectives. *Solar Energy*, *211*, 584–599.

Nishiok, S., & Maeda, K. (2015). Hydrothermal synthesis of rhodium-doped barium titanate nanocrystals for enhanced photocatalytic hydrogen evolution under visible light. *RSC Advances*, *5*, 100123–100128.

Nurlaela, E., Ould-Chikh, S., Harb, M., del Gobbo, S., Aouine, M., Puzenat, E., Sautet, P., Domen, K., Basse, J.-M., & Takanabe, K. (2014). Critical role of the semiconductor–electrolyte interface in photocatalytic performance for water-splitting reactions using Ta_3N_5 particles. *Chemistry of Materials*, *26*, 4812–4825.

Ong, W.-J., Tan, L.-L., Ng, Y. H., Yong, S.-T., & Chai, S.-P. (2016). Graphitic carbon nitride (g-C_3N_4)-based photocatalysts for artificial photosynthesis and environmental remediation: Are we a step closer to achieving sustainability? *Chemical Reviews*, *116*, 7159–7329.

Pang, J., Mendes, R. G., Bachmatiuk, A., Zhao, L., Ta, H. Q., Gemming, T., Liu, H., Liu, Z., & Rummeli, M. H. (2019). Applications of 2D MXenes in energy conversion and storage systems. *Chemical Society Reviews*, *48*, 72–133.

Peerakiatkhajohn, P., Yun, J.-H., Wang, S., & Wang, L. (2021). Review of recent progress in unassisted photoelectrochemical water splitting: from material modification to configuration design. *Journal of Photonics for Energy*, *7*, 0120067.

Pei, Y., Zhang, X., Hui, Z., Zhou, J., Huang, X., Sun, G., & Huang, W. (2021). $Ti_3C_2T_X$ MXene for sensing applications: Recent progress, design principles, and future perspectives. *ACS Nano*, *15*, 3996–4017.

Peng, C., Wei, P., Lia, X., Liu, Y., Cao, Y., Wang, H., Yu, H., Peng, F., Zhang, L., Zhang, B., & Lv, K. (2018). High efficiency photocatalytic hydrogen production over ternary $Cu/TiO_2@\ Ti_3C_2T_x$ enabled by low-work-function 2D titanium carbide. *Nano Energy*, *53*, 97–107.

Ramalingam, V., Varadhan, P., Fu, H.-C., Kim, H., Zhang, D., Chen, S., Song, L., Ma, D., Wang, Y., Alshareef, H. N., & He, J.-H. (2019). Heteroatom-mediated interactions between ruthenium single atoms and an MXene support for efficient hydrogen evolution. *Advanced Materials*, *31*, 1903841.

Ran, J., Gao, G., Li, F.-T., Ma, T.-Y., Du, A., & Qiao, S.-Z. (2017). Ti_3C_2 MXene co-catalyst on metal sulfide photo-absorbers for enhanced visible-light photocatalytic hydrogen production. *Nature Communications*, *8*, 13907.

Rao, P. M., Cai, L., Liu, C., Cho, I. S., Lee, C. H., Weisse, J. M., Yang, P., & Zheng, X. (2014). Simultaneously efficient light absorption and charge separation in $WO_3/BiVO_4$ core/shell nanowire photoanode for photoelectrochemical water oxidation. *Nano Letters*, *14*, 1099–1105.

Rekhila, G., Bessekhouad, Y., & Trari, M. (2013). Visible light hydrogen production on the novel ferrite $NiFe_2O_4$. *International Journal of Hydrogen Energy*, *38*, 6335–6343.

Ros, C., Andre, T., & Morante, J. R. (2020). Photoelectrochemical water splitting: A road from stable metal oxides to protected thin film solar cells. *Journal of Materials Chemistry A*, *8*, 10625–10669.

Sekizawa, K., Nonaka, T., Arai, T., & Morikawa, T. (2014). Structural improvement of $CaFe_2O_4$ by metal doping toward enhanced cathodic photocurrent. *ACS Applied Materials & Interfaces*, *6*, 10969–10973.

Selvaraj, S., Moon, H., & Kim, D.-H. (2011). Synthesis and photoelectrochemical properties of $Fe_2O_3/ZnFe_2O_4$ composite photoanodes for use in solar water oxidation. *Chemistry of Materials*, *23*, 4863–4869.

Serpone, N., Sauvé, G., Koch, R., Tahiri, H., Pichat, P., Piccinini, P., Pelizzetti, E., & Hidaka, H. (1996). Standardization protocol of process efficiencies and activation parameters in heterogeneous photocatalysis: Relative photonic efficiencies ζ_r. *Journal of Photochemistry and Photobiology A: Chemistry*, *94*, 191–203.

Shan, L.-w., Wang, G.-l., Suriyaprakash, J., Li, D., Liu, L.-z., & Dong, L.-m. (2015). Solar light driven pure water splitting of B-doped $BiVO_4$ synthesized via a sol–gel method. *Journal of Alloys and Compounds*, *636*, 131–137.

Sharma, D., Upadhyay, S., Satsangi, V. R., Shrivastav, R., Waghmare, U. V., & Dass, S. (2014). Improved photoelectrochemical water splitting performance of $Cu_2O/SrTiO_3$ heterojunction photoelectrode. *The Journal of Physical Chemistry C*, *118*, 25320–25329.

Sivula, K., & van de Krol, R. (2016). Semiconducting materials for photoelectrochemical energy conversion. *Nature Reviews Materials*, *1*, 15010.

Spagnoli, D., Gilbert, B., Waychunas, G. A., & Banfield, J. F. (2009). Prediction of the effects of size and morphology on the structure of water around hematite nanoparticles. *Geochimica et Cosmochimica Acta*, *73*, 4023–4033.

Staffell, I., Scamman, D., Abad, A. V., Balcombe, P., Dodds, P. E., Ekins, P., Shah, N., & Ward, K. R. (2019). The role of hydrogen and fuel cells in the global energy system. *Energy & Environmental Science*, *12*, 463–491.

Su, T., Hood, Z. D., Naguib, M., Bai, L., Luo, S., Rouleau, C. M., Ivanov, I. N., Ji, H., Qin, Z., & Wu, Z. (2019). Monolayer $Ti_3C_2T_x$ as an effective co-catalyst for enhanced photocatalytic hydrogen production over TiO_2. *ACS Applied Energy Materials*, *2*, 4640–4651.

Sudrajat, H., Kitta, M., Ito, R., Nagai, S., Yoshida, T., Katoh, R., Ohtani, B., Ichikuni, N., & Onishi, H. (2020). Water-splitting activity of La-doped $NaTaO_3$ photocatalysts sensitive to spatial distribution of dopants. *The Journal of Physical Chemistry C*, *124*, 15285–15294.

Sullivan, I., Zoellner, B., & Maggard, P. A. (2016). Copper(I)-based *p*-type oxides for photoelectrochemical and photovoltaic solar energy conversion. *Chemistry of Materials*, *28*, 5999–6016.

Sun, Y., Jin, D., Sun, Y., Meng, X., Gao, Y., & Agnese, D. (2018). $g\text{-}C_3N_4/Ti_3C_2T_x$ (MXenes) composite with oxidized surface groups for efficient photocatalytic hydrogen evolution. *Journal of Materials Chemistry A*, *6*, 9124–9131.

Suzuki, H., Tomita, O., Higashia, M., & Abe, R. (2016). Design of nitrogen-doped layered tantalates for non-sacrificial and selective hydrogen evolution from water under visible light. *Journal of Materials Chemistry A*, *4*, 14444–14452.

Szymanski, P., & El-Sayed, M. A. (2012). Some recent developments in photoelectrochemical water splitting using nanostructured TiO_2: A short review. *Theoretical Chemistry Accounts volume*, *131*, 1202.

Tabata, M., Maeda, K., Higashi, M., Lu, D., Takata, T., Abe, R., & Domen, K. (2010). Modified Ta_3N_5 powder as a photocatalyst for O_2 evolution in a two-step water splitting system with an iodate/iodide shuttle redox mediator under visible light. *Langmuir*, *26*, 9161–9165.

Tahir, M. (2021). Investigating the in fluential effect of etchant time in constructing 2D/2D HCN/MXene heterojunction with controlled growth of TiO_2 NPs for stimulating photocatalytic H_2 production. *Energy Fuels*, *35*, 6807–6822.

Takata, T., Jiang, J., Sakata, Y., Nakabayashi, M., Shibata, N., Nandal, V., Seki, K., Hisatomi, T., & Domen, K. (2020). Photocatalytic water splitting with a quantum efficiency of almost unity. *Nature*, *581*, 411–414.

Trang, T. N. Q., Tu, L. T. N., Man, T. V., Mathesh, M., Nam, N. D., & Thu, V. T. H. (2019). A high-efficiency photoelectrochemistry of Cu_2O/TiO_2 nanotubes based composite for hydrogen evolution under sunlight. *Composites Part B: Engineering*, *174*, 106969–106978.

Walsh, A., Yan, Y., Huda, M. N., Al-Jassim, M. M., & Wei, S.-H. (2009). Band edge electronic structure of $BiVO_4$: Elucidating the role of the Bi s and Vd orbitals. *Chemistry of Materials*, *21*, 547–551.

Wang, G., Wang, H., Ling, Y., Tang, Y., Yang, X., Fitzmorris, R. C., Wang, C., Zhang, J. Z., & Li, Y. (2011). Hydrogen-treated TiO_2 nanowire arrays for photoelectrochemical water splitting. *Nano Letters*, *11*, 3026–3033.

Wang, G., Xiao, X., Li, W., Lin, Z., Zhao, Z., Chen, C., Wang, C., Li, Y., Huang, X., Miao, L., Jiang, C., Huang, Y., & Duan, X. (2015). Significantly enhanced visible light photoelectrochemical activity in TiO_2 nanowire arrays by nitrogen implantation. *Nano Letters*, *15*, 4692–4698.

Wang, J., Tafen, D. N., Lewis, J. P., Hong, Z., Manivannan, A., Zhi, M., Li, M., & Wu, N. (2009). Origin of photocatalytic activity of nitrogen-doped TiO_2 nanobelts. *Journal of the American Chemical Society*, *131*, 12290–12297.

Wang, M., Ren, F., Zhou, J., Cai, G., Cai, L., Hu, Y., Wang, D., Liu, Y., Guo, L., & Shen, S. (2015). N doping to ZnO nanorods for photoelectrochemical water splitting under visible light: Engineered impurity distribution and terraced band structure. *Scientific Reports volume*, *5*, 1925.

Wang, Q., & Domen, K. (2020). Particulate photocatalysts for light-driven water splitting: Mechanisms, challenges and design strategies. *Chemical Reviews*, *120*, 919–985.

Wick, R., & Tilley, S. D. (2016). Photovoltaic and photoelectrochemical solar energy conversion with Cu_2O. *The Journal of Physical Chemistry C*, *119*, 26243–26257.

Xiao, M., Hao, M., Lyu, M., Moore, E. G., Zhang, C., Luo, B., Hou, J., Lipton-Duffin, J., & Wang, L. (2019). Surface ligands stabilized lead halide perovskite quantum dot photocatalyst for visible light-driven hydrogen generation. *Advanced Functional Materials*, *29*, 1905683.

Xiao, R., Zhao, C., Zou, Z., Chen, Z., Tian, L., Xu, H., Tang, H., Liu, Q., Lin, Z., & Yang, X. (2020). In situ fabrication of 1D CdS nanorod/2D Ti_3C_2 MXene nanosheet Schottky heterojunction toward enhanced photocatalytic hydrogen evolution. *Applied Catalysis B: Environmental*, *268*, 118382.

Xie, K., Umezawa, N., Zhang, N., Reunchan, P., Zhang, Y., & Ye, J. (2011). Self-doped $SrTiO_{3-\delta}$ photocatalyst with enhanced activity for artificial photosynthesis under visible light. *Energy & Environmental Science*, *4*, 4211–4219.

Xie, P., Yang, F., Li, R., Ai, C., Lin, C., & Lin, S. (2019). Improving hydrogen evolution activity of perovskite $BaTiO_3$ with Mo doping: Experiments and first-principles analysis. *International Journal of Hydrogen Energy*, *44*, 11695–11704.

Xu, F., Zhang, D., Liao, Y., Wang, G., Shi, X., Zhang, H., & Xiang, Q. (2020). Synthesis and photocatalytic H_2-production activity of plasma-treated $Ti_3C_2T_x$ MXene modified graphitic carbon nitride. *Journal of the American Ceramic Society*, *103*, 849–858.

Xu, M., Da, P., Wu, H., Zhao, D., & Zheng, G. (2012). Controlled Sn-doping in TiO_2 nanowire photoanodes with enhanced photoelectrochemical conversion. *Nano Letters*, *12*, 1503–1508.

Yan, D., Fu, X., Shang, Z., Liu, J., & Luo, H. (2019). A $BiVO_4$ film photoanode with re-annealing treatment and 2D thin $Ti_3C_2T_X$ flakes decoration for enhanced photoelectrochemical water oxidation. *Chemical Engineering Journal*, *361*, 853–861.

Yang, C., Wang, Z., Lin, T., Yin, H., Lü, X., Wan, D., Xu, T., Zheng, C., Lin, J., Huang, F., Xie, X., & Jiang, M. (2013). Core-shell nanostructured "Black" rutile titania as excellent catalyst for hydrogen production enhanced by sulfur doping. *Journal of the American Chemical Society*, *135*, 17831–17838.

Yang, H., Mao, Y., Li, M., Liu, P., & Tong, Y. (2013). Electrochemical synthesis of $CoFe_2O_4$ porous nanosheets for visible light driven photoelectrochemical applications. *New Journal of Chemistry*, *37*, 2965–2968.

Yang, X., & Wang, D. (2018). Photocatalysis: From fundamental principles to materials and applications. *ACS Applied Energy Materials*, *1*, 6657–6693.

Yang, X., Wolcott, A., Wang, G., Sobo, A., Fitzmorris, R. C., Qian, F., Zhang, J. Z., & Li, Y. (2009). Nitrogen-doped ZnO nanowire arrays for photoelectrochemical water splitting. *Nano Letters*, *9*, 2331–2336.

Yang, Y., Ling, Y., Wang, G., Liu, T., Wang, F., Zhai, T., Tong, Y., & Li, Y. (2015). Photohole induced corrosion of titanium dioxide: Mechanism and solutions. *Nano Letters*, *15*, 7051–7057.

Yang, Y., Niu, S., Han, D., Liu, T., Wang, G., & Li, Y. (2017). Progress in developing metal oxide nanomaterials for photoelectrochemical water splitting. *Advanced Energy Materials*, *7*, 1700555.

Yang, Y., Zhang, D., & Xiang, Q. (2019). Plasma-modified $Ti_3C_2T_x$/CdS hybrids with oxygen-containing groups for high-efficiency photocatalytic hydrogen production. *Nanoscale*, *11*, 18797–18805.

Yu, J. C., Ho, W., Yu, J., Yip, H., Wong, P. K., & Zhao, J. (2005). Efficient visible-light-induced photocatalytic disinfection on sulfur-doped nanocrystalline titania. *Environmental Science & Technology*, *39*, 1175–1179.

Yu, X., Wang, T., Yin, W., & Zhang, Y. (2018). Ti_3C_2 MXene nanoparticles modified metal oxide composites for enhanced photoelectrochemical water splitting. *International Journal of Hydrogen Energy*, *44*, 2704–2710.

Yuan, J., Chen, M., Shi, J., & Shangguan, W. (2006). Preparations and photocatalytic hydrogen evolution of N-doped TiO_2 from urea and titanium tetrachloride. *International Journal of Hydrogen Energy*, *31*, 1326–1331.

Yuan, W., Cheng, L., An, Y., Wu, H., Yao, N., Fan, X., & Guo, X. (2018). MXene nano fibers as highly active catalysts for hydrogen evolution reaction. *ACS Sustainable Chemistry & Engineering*, *6*, 8976–8982.

Yuliati, L., Yang, J.-H., Wang, X., Maeda, K., Takata, T., Antoniettic, M., & Domen, K. (2010). Highly active tantalum(v) nitridenanoparticles prepared from a mesoporous carbon nitride template for photocatalytic hydrogen evolution under visible light irradiation. *Journal of Materials Chemistry*, *20*, 4295–4298.

Zhang, C. (. J.)., Ma, Y., Zhang, X., Abdolhosseinzadeh, S., Sheng, H., Lan, W., Pakdel, A., Heier, J., & Nüesch, F. (2020). Two-dimensional transition metal carbides and nitrides (MXenes): Synthesis, properties, and electrochemical energy storage applications. *Energy & Environmental Materials*, *3*, 29–55.

Zhang, Y., Hu, Y., Zeng, H., Zhong, L., Liu, K., Cao, H., Li, W., & Yan, H. (2017). Silicon carbide recovered from photovoltaic industry waste as photocatalysts for hydrogen production. *Journal of Hazardous Materials*, *329*, 22–29.

Zhang, Y., Xia, T., Wallenmeyer, P., Harris, C. X., Peterson, A. A., Corsiglia, G. A., Murowchick, J., & Chen, X. (2014). Photocatalytic hydrogen generation from pure water using silicon carbide nanoparticles. *Energy Technology*, *2*, 183–187.

Zhang, Z., Nagashima, H., & Tachikawa, T. (2020). Ultra-narrow depletion layers in a hematite mesocrystal-based photoanode for boosting multihole water oxidation. *Angew. Chem. Int. Ed.*, *59*, 9047–9054.

Zhong, H., Zhang, H., Liang, Y., Zhang, J., Wang, M., & Wang, X. (2007). A novel non-noble electrocatalyst for oxygen reduction in proton exchange membrane fuel cells. *Journal of Power Sources*, *164*, 572–577.

Zhou, X., Li, X., Gao, Q., Yuan, J., Wen, J., Fang, Y., Liu, W., Zhang, S., & Liu, Y. (2015). Metal-free carbon nanotube–SiC nanowire heterostructures with enhanced photocatalytic H_2 evolution under visible light irradiation. *Catalysis Science & Technology*, *5*, 2798–2806.

Zhu, Z., Kao, C.-T., Tang, B.-H., Chang, W.-C., & Wu, R.-J. (2016). Efficient hydrogen production by photocatalytic water-splitting using Pt-doped TiO_2 hollow spheres under visible light. *Ceramics International*, *42*, 6749–6754.

11

Gallium Nitride-Based Artificial Photosynthesis Integrated Devices for Solar Hydrogen Generation and Carbon Dioxide Reduction

Baowen Zhou, Peng Zhou, Wanjae Dong, and Zetian Mi

Department of Electrical Engineering and Computer Science, University of Michigan, Ann Arbor, MI, USA

11.1 Introduction

Artificial photosynthesis generation of clean fuels and chemicals, for example, green hydrogen (H_2) and methanol (CH_3OH) from H_2O and CO_2, respectively, by using virtually unlimited solar energy presents a promising approach for recycling the gigatons of carbon dioxide annually emitted into the atmosphere, which is considered one of the grand challenges of the twenty-first century (Berardi et al., 2014; Kibria & Mi, 2016; Kim et al., 2015; Li, Mu, et al., 2015; Liu, Colon, et al., 2016; Tian et al., 2012, 2021; Xu et al., 2017; Zhang & Lin, 2014). Over the past decades, considerable efforts have been devoted to developing artificial photosynthesis devices derived from semiconductors like TiO_2 (Habisreutinger et al., 2013), $BiVO_4$ (Malathi et al., 2018), C_3N_4 (Li, Zhang, et al., 2015; Ong et al., 2016; Zhao et al., 2021), WO_3 (Jin et al., 2015; Sun et al., 2018), and CdS (Sun et al., 2018), by coupling with different cocatalysts (Kim et al., 2016; Smith et al., 2020; Tu et al., 2014; Zahran et al., 2019) but with limited success. Some of the critical challenges for artificial photosynthesis include very low efficiency, poor stability, and uncontrolled product selectivity, which are directly related to the inefficient light absorption and poor stability of the semiconductor light absorbers as well as sluggish kinetics of surface chemical reactions (Berardi et al., 2014; Chowdhury et al., 2018; Zhang & Lin, 2014; Zhang & Sun, 2019).

Artificial photosynthesis is a complex convolution of physical and chemical processes, which are highly associated with the complex behaviors of photons, charge carriers, as well as various chemical species involved in the reaction processes (Berardi et al., 2014; Frischmann et al., 2013; Gao et al., 2021; Liu, Inagaki, & Gong, 2016; Tran et al., 2012; Xu et al., 2017; Zahran et al., 2019; Zhang et al., 2021). Therefore, the design and construction of an artificial photosynthesis integrated device (APID) by coupling suitable semiconductor light absorbers with rationally designed cocatalysts is of scientific and practical significance (Berardi et al., 2014; Kim et al., 2015; Liu, Colon, et al., 2016; Maeda, 2013; Nocera, 2012; Reece et al., 2011; Wen & Li, 2013; Yang et al., 2014). In this context, epitaxial III-nitride nanostructures, for example, GaN, InN, AlN, BN, and their alloys, have emerged as the ideal building blocks of next-generation APIDs (Kibria et al., 2016; Kibria & Mi, 2016). The bandgap of III-nitride nanostructures can be tuned across nearly the entire solar spectrum while straddling a broad range of redox potentials under visible and near-infrared light irradiation (Kibria & Mi, 2016). Their electrical properties, including doping, p and n-type conduction, surface band bending, and surface polarity and polarization can also be precisely controlled utilizing state-of-the-art molecular beam epitaxy (MBE) technology (Kibria et al., 2016). The epitaxial III-nitride nanostructures exhibit excellent charge carrier mobility and efficient light absorption. Significantly, their abundant nonpolar surfaces can be tailored to be N-rich to protect against photo-corrosion and oxidation, resulting in unprecedented stability in harsh photocatalysis reactions (Guan, Chowdhury, Wang, et al., 2018; Vanka et al., 2018). Moreover, their unique N-rich surface holds great potential in interaction with CO_2 and other molecules, promising highly efficient surface chemical reactions by integrating with rational catalytic centers. Over the past decades, GaN-based epitaxial nanostructures have not only enabled the demonstration of some of the most efficient and stable APIDs capable of direct solar water splitting and hydrogen fuel generation (Kibria, Chowdhury, Zhao, AlOtaibi, et al., 2015; Kibria, Chowdhury, Zhao, Trudeau, et al., 2015; Wang et al., 2011), but have also realized some benchmarking solar-driven CO_2 reduction reactions (CO_2RR) toward fuels and chemicals (AlOtaibi et al., 2015, 2016).

In this chapter, the distinct properties and merits of epitaxial III-nitride nanostructures in the context of artificial photosynthesis are first presented. Recent advances in solar water splitting and green hydrogen fuel generation utilizing

Conversion of Water and CO_2 to Fuels using Solar Energy: Science, Technology and Materials, First Edition.
Edited by Oomman K. Varghese and Flavio L. Souza.

GaN-based photochemical (PC) and photoelectrochemical (PEC) APIDs are then discussed in detail. Recent progress on solar-driven CO_2 CO_2RR toward carbon fuels over GaN-based APIDs is then described. Subsequently, GaN-catalyzed organic transformations are briefly presented. In the end, the critical challenges and prospects for constructing practical GaN-based APIDs are discussed in the context of providing a scalable solution for sustainable and green fuels and chemicals production from sunlight, H_2O, and CO_2.

11.2 Merits of III-Nitride Nanostructures for Artificial Photosynthesis

To date, most of the semiconductors, for example, TiO_2 (Abidov et al., 2014; Habisreutinger et al., 2013; He et al., 2020; Neatu et al., 2014; She et al., 2019; Tu et al., 2013; Xu et al., 2018) and CdS (Bie et al., 2019; Jin et al., 2015; Low et al., 2019; Zong et al., 2008, 2011) for artificial photosynthesis were prepared by conventional approaches, for example, solvent thermal methods, calcination. Because of the limitations of the fabrication techniques, it is of great difficulty to precisely tailor the crystal structure, band structure (bandgap and/or surface band bending), as well as charge carrier mobility of the mentioned-above semiconductors, thus suffering from uncontrollable optical and electronic properties. Moreover, it is well known that the surface properties of the semiconductors play a vital role in artificial photosynthesis by influencing the geometric and electronic effects of catalytic sites and/or by directly involving the reactants/intermediates interaction. However, the modification of the semiconductors' surface properties by the conventional methods has still remained a grand challenge. Therefore, a revolutionary means for improving the structural, optoelectronic, and surface properties of the semiconductor materials is under urgent demand, especially if at atomic scale.

MBE was one of the most significant inventions in materials science in the second half of the twentieth century. It is a powerful approach for the controlled growth of high-quality III-nitrides at atomic scale (Izyumskaya et al., 2019; Jain et al., 2000; Morkoc, 2001; Yoshikawa et al., 2009; Zhao & Mi, 2017). As a result, the epitaxial III-nitride nanostructures have demonstrated tunable properties, providing a wide window for tuning the optoelectronic behavior (Georgakilas et al., 2004; Roul et al., 2015; Zhao & Mi, 2017). For example, the bandgap of III-nitrides can be adjusted in a wide range from 3.4 to 0.65 eV by varying the concentrations and types of the dopants, enabling a broad spectrum sunlight absorption (Figure 11.1a). In parallel, the conduction band (CB) and valence band (VB) edge positions can be tailored to straddle the redox potentials for water splitting and CO_2RR. Additionally, the structural properties of the epitaxial GaN can be finely tuned at atomic scale by MBE. Due to the efficient surface strain relaxation during the epitaxial process, the as-grown GaN with wurtzite structure is nearly dislocation-free. Together with the engineered band bending, the direction of the charge carrier flow

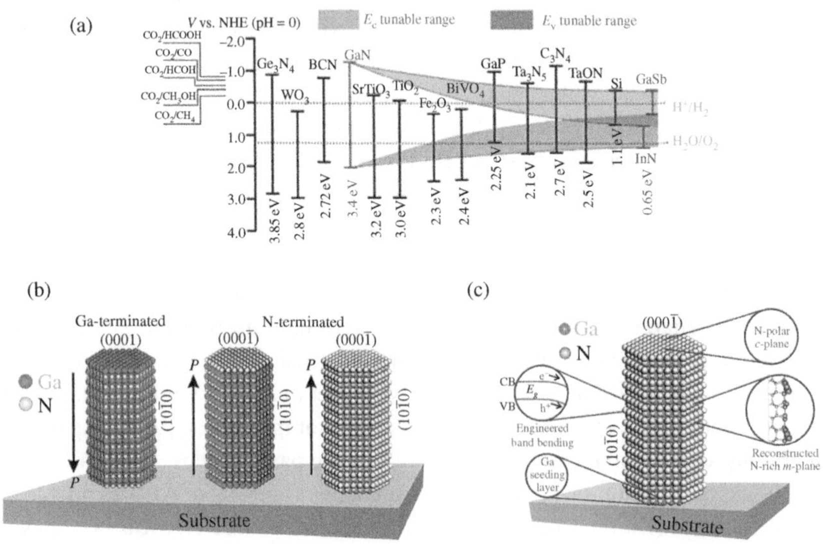

Figure 11.1 (a) Schematic of the tunable Ga(In,Sb)N band structure compared to other photocatalysts, including the redox potentials of water splitting and CO_2RR. Adapted from Kibria et al. (2016) and Kibria and Mi (2016). (b) Models for single crystal wurtzite GaN nanowires (NWs). (c) Schematic of the key atomic and electronic structures of the epitaxial GaN. Reproduced from Kibria and Mi (2016)/with permission of John Wiley & Sons.

can be facilely modulated (Okumura et al., 2016). It is worth mentioning that the surface properties of the epitaxial GaN could be controlled by terminating with various atoms (Figure 11.1b). The reconstructed N-rich *m*-plane of GaN, shown in Figure 11.1c, serves as an efficient protection layer that makes the material highly stable against photooxidation. This feature is substantially different from the existing systems and has enabled the material to demonstrate thousands of hours of stable operation for solar-driven water splitting and hydrogen fuel production (Vanka et al., 2019). More interestingly, partial oxygen substitution was discovered to show a self-improving effect via in situ formation of a gallium oxynitride layer in a silicon/gallium nitride photocathode used for solar-hydrogen evolution (Zeng et al., 2021).

For artificial photosynthesis, the grand challenge is the sluggish kinetics of the surface chemical reactions (Graf et al., 2021; Lei et al., 2021; Luo et al., 2018; Xu et al., 2021). Apart from the optoelectronic behavior, the surface properties of the semiconductors also play a critical role in the performance of artificial photosynthesis via interaction with reactants/intermediates (Dong et al., 2020; Kusior et al., 2019; Zheng & Shu, 2020). For the well-known four proton-coupled electron-transfer processes of water splitting, ab initio molecular dynamics theoretical calculations showed that the surface –NH sites of wurtzite GaN were thermodynamically more favored than –OH sites for localizing the photogenerated holes to dissociate water with reduced free energy (Ertem et al., 2015). In addition, it was discovered that the electronic properties of the adsorbed water and hydroxyl groups at GaN interface were substantially distinct from that on the surface of the well-studied TiO_2 (Wang et al., 2012), which could explain their performance difference for H_2 evolution. For CO_2RR, the surface of the epitaxial III-nitrides can potentially provide an electron-density-tunable platform for unique interaction with the inert CO_2 molecules and other species. Therefore, by coordinating with rationally designed cocatalysts, III-nitride semiconductors hold grand promise in promoting the sluggish kinetics of water splitting and CO_2RR, and selectively mediating the formation of high-value fuels and chemicals. It is also noted that the dopants of III-nitrides during the epitaxy process are found to affect the reaction remarkably, offering more options for optimizing the performance (AlOtaibi et al., 2016). As such, III-nitride nanostructures have emerged as a near-ideal building block for next-generation APIDs, promising breakthrough advances in artificial photosynthesis.

11.3 Recent Advances in III-Nitrides for Artificial Photosynthesis

Epitaxial III-nitrides have demonstrated several superior features for artificial photosynthesis (Chen, Ji, et al., 2020; Kibria & Mi, 2016). First, the band structure of GaN with 1-dimensional (1-D) morphology is highly tunable for photon absorption over a broad wavelength range without compromising the redox potentials while the nearly dislocation-free wurtzite structure, unique surface polarization, and high electron mobility of GaN facilitate electron–hole separation. Notably, the tunable surface of GaN exhibits unique interaction with the molecules/species of CO_2, water, hydroxyl, and so forth by incorporating with suitable catalytic centers. Moreover, the N-rich nonpolar surface is very efficient against photocorrosion, thus promising high efficiency and superior stability. These remarkable attributes enable III-nitrides as one of the most appropriate candidates for building new-generation APIDs. To date, several notable achievements have been made in both water splitting and carbon dioxide reduction for synthesizing fuels and chemicals over III-nitride-based artificial photosynthesis devices, which are discussed in the following sections.

11.3.1 Solar Water Splitting

Developing an efficient solar water splitting device is an important step to directly produce solar fuel by converting solar energy to chemicals without the external electricity input. The green hydrogen can be produced on-site, easily stored, and distributed to meet the ever-growing global energy demand. Solar water splitting is an endothermic reaction with Gibbs energy of 273.2 kJ/mol, corresponding to 1.23 eV (Walter et al., 2010). However, due to the energy loss caused by the charge carrier transport in semiconductors and overpotential for catalytic reactions, a photovoltage in the range of 1.7 eV is generally required to split water. Based on the configuration of the system, solar water splitting is classified into PEC water splitting and PC water splitting, each possessing different features and advantages. In the scheme of PEC solar water splitting, two electrodes consisting of a cathode and an anode are electrically connected by external circuits for hydrogen (H_2) and oxygen (O_2) generation, respectively, at cathode and anode. For PEC water splitting, one of the great virtues is the spatially separated production of H_2 and O_2 without the separation issue. On the contrary, PC water splitting benefits from the simple configuration without external circuit, despite the fact that H_2 and O_2 are simultaneously produced over a single photocatalyst with a separation issue. Later, GaN-based PEC water splitting will be first presented.

11.3.1.1 Photoelectrochemical Water Splitting

In 1972, unassisted solar water splitting was first demonstrated with a small but measurable solar-to-hydrogen (STH) efficiency using rutile TiO_2 photoanode (Fujishima & Honda, 1972). A few decades later, multi-junction tandem devices presented a promising pathway toward higher STH efficiency since the semiconductors with different bandgaps could absorb the broadband solar spectrum and increase the photovoltage (Khaselev & Turner, 1998). An ideal tandem photoelectrode should have excellent light absorption properties, suitable band edge positions, and defect-free material qualities to prevent bulk, interfacial, and surface recombination. To date, tandem III–V semiconductors have achieved a high STH efficiency >10% (May et al., 2015; Young et al., 2017). Moreover, the use of corrosion protection layers on III–V semiconductors led to improved photoelectrode stability >20 hr (Cheng et al., 2018; Verlage et al., 2015). Theoretical studies have shown that a double-junction PEC photoelectrode consisting of top and bottom cells with energy bandgaps of ~1.8 and ~1.1 eV, respectively, can achieve STH efficiency up to ~30% (Young et al., 2017), which has not been realized in tandem devices so far due to difficulty in epitaxial integration of top and bottom light absorbers. Recent works have shown that Si and InGaN are practically suitable for double-junction PEC photocathodes because bandgaps (1.1 eV for Si and tunable 0.6–3.4 eV for InGaN) near-perfectly match theoretical values and the CB of InGaN is approximately aligned with that of Si, resulting in efficient electron transport across the heterointerface (Vanka et al., 2018). To date, however, the growth of InGaN epilayers on Si substrates has been extremely difficult due to the large lattice mismatch and the formation of defects and dislocations. Moreover, a tunnel junction with good electrical conductivity, optical transparency, and low defect density is essential to realize the InGaN/Si tandem photoelectrode. To address these problems, a systematic study on the design and performance of III-nitride photoelectrodes is needed.

GaN Nanostructures on Si Photocathode An important component of a PEC water splitting is an efficient and durable photocathode. However, the most widely used Si photocathode generally suffers from inferior stability in aqueous solutions (Bae et al., 2017). In this regard, tremendous attention has been paid to exploring surface protection schemes. The surface protection layers require stability in aqueous media, efficient charge carrier transport, and optical transparency to allow light transmission to Si absorber. GaN, which is one of the most produced semiconductor materials in optoelectronic industries, can effectively protect the Si photocathode from photocorrosion since the surfaces of GaN can be terminated by nitrogen atoms in a nitrogen-rich atmosphere during the epitaxy process. GaN also has a large energy bandgap (~3.4 eV) and is optically transparent for the visible and infrared light. Therefore, the use of GaN nanowires (NWs) as a passivation scheme for Si photocathode offers robust protection and promotes the reaction kinetics and efficiency simultaneously (Figure 11.2) (Vanka et al., 2018). The CB edges of n-type doped GaN and Si are approximately aligned, which facilitates extraction of photogenerated electrons from Si substrate to GaN NWs and Pt cocatalysts without a significant energy barrier. Moreover, the large surface area of GaN NWs is favorable for exposing the reaction sites with high efficiency and lowering the charge carrier transfer resistance at the interface between semiconductor and electrolyte compared to that of the planar Si photocathode. As a result, Pt-decorated GaN/Si photocathodes show an excellent saturated photocurrent density of ~38 mA/cm^2 with an appreciable onset potential of ~0.5 V vs. reversible hydrogen electrode (V_{RHE}) in 0.5 M H_2SO_4 under AM 1.5G solar-simulated illumination (100 mW/cm^2). When an external potential is applied for water splitting, applied bias photon-to-current efficiency (ABPE) can describe the performance of a photocathode according to the following equation:

$$\text{ABPE}\,(\%) = \frac{J_{ph} \times \left[V_{RHE} - E^0_{rev}\right] \times 100}{P_{light}}$$

where P_{light} is the power of the incident light, J_{ph} is the cathodic photocurrent density, and V_{RHE} is the applied electrical bias versus reversible hydrogen electrode, and $E^0_{rev} = 0\,V_{RHE}$ for hydrogen evolution reaction. Although this ABPE determined for a photocathode can be different from the ABPE measured in a two-electrode configuration [= J_{ph} (1.23 – $V_{\text{bias between anode and cathoe}}$) × 100/$P_{light}$], which includes both HER and OER activities of the photoelectrodes (Chen et al., 2010), it is a useful parameter for comparing the performance of photocathodes evaluated in a three-electrode system. Pt-decorated GaN/Si photocathodes revealed a superior ABPE of 10.5%, indicating that III-nitride could grow directly on Si wafers and act as a promising scheme to achieve an efficient and durable photocathode for water splitting to produce hydrogen. The integration of several other cocatalysts on GaN NWs was also studied, which showed promising efficiency and stability for PEC water splitting (Zhou et al., 2018).

InGaN Photocathode and Photoanode The energy bandgap of GaN can be reduced from 3.4 to 0.6 eV by incorporating indium element for broadband light absorption (César et al., 2011). Uniquely, the VB and CB edges of InGaN can straddle with hydrogen evolution and water oxidation potentials for a bandgap possibly below 1.5 eV. This means that although the bandgap of InGaN is reduced from 3.4 to 1.5 eV, the CB minimum can still be located above the energy level of hydrogen

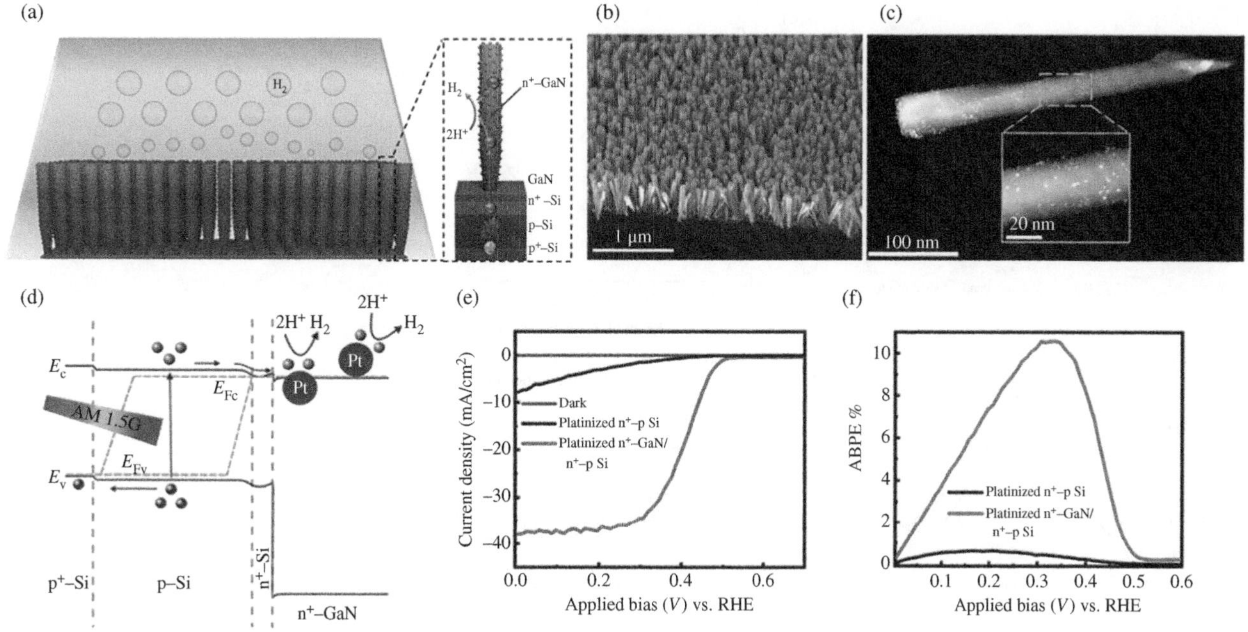

Figure 11.2 (a) Schematic of n^+-GaN NWs/n^+-p Si photocathode showing light absorption by the underlying Si wafer, electron transfer from Si wafer to GaN NWs, and proton reduction on platinized GaN NWs. (b) Scanning electron microscope (SEM) and (c) scanning transmission electron microscope-high angle annular dark field (STEM-HAADF) images of platinized GaN NWs. Pt nanoparticles (NPs) distributed on the GaN NWs. (d) Schematic of the energy band diagram of Pt-decorated n^+-GaN NWs/n^+-p Si photocathode under light illumination. (e) J–V curves and (f) ABPE of the photocathodes under AM 1.5G 1-sun illumination in 0.5 M H_2SO_4. Reproduced from Vanka et al. (2018)/with permission of American Chemical Society.

evolution reaction (HER) and the VB maximum can be positioned below the water oxidation level, resulting in efficient electron and hole transport (Chu, Vanka, et al., 2018). Therefore, investigation of the synthesis and PEC water splitting properties of p-type InGaN NWs on Si photocathode was performed (Figure 11.3) (Wang, Vanka, et al., 2019). p-type InGaN NWs and n-type Si wafers are connected through a tunnel junction, which is a low-resistivity connection and serves as an electron-blocking layer. The top p–InGaN NWs are the light absorbers and reaction sites for HER. Under light illumination, photoexcited holes are collected through tunnel junction, whereas photoexcited electrons migrate to the InGaN/electrolyte interface to participate in HER. By changing the growth conditions of InGaN NWs, photoluminescence (PL) emission wavelengths can be varied from ~500 to ~750 nm, corresponding to energy bandgap values of ~2.48 to ~1.65 eV, and In compositions of ~24% to ~51%, respectively. It is also shown that, as In composition increases, the PL intensity gets weaker and the width becomes broader, indicating the defect formation. The photocurrent density at 0 V_{RHE} shows an increasing trend with In composition from 24% to 42%, due to the reduced bandgap and more efficient light absorption. A significant reduction of photocurrent density, however, has been observed for In composition >42%, which is likely due to the increased number of defects and non-radiative recombination of In-rich InGaN NWs. It can be seen that as the In composition increases, there is a beneficial light absorption effect, but there is also disadvantage of the increased defect density. Growth of In-rich InGaN NWs with high-crystal quality is essential to achieve efficient PEC water splitting.

Double-Junction InGaN/Si Photoelectrode The InGaN/Si double-junction photocathode consists of a p-type InGaN NW top and a planar n^+–p Si bottom light absorber (Figure 11.4) (Navid et al., 2021). The top and bottom photon harvesters are connected through a p^{++}-InGaN/n^{++}-InGaN NW tunnel junction. Photogenerated electrons migrate from n^+–p Si junction to n^+-InGaN NWs, which then recombine with photogenerated holes from the top p^+-InGaN in the tunnel junction. On the other hand, photogenerated electrons in the top p^+-InGaN NWs transfer to the lateral surfaces and participate in proton reduction arising from the downward surface band bending. Theoretical calculations show that the photocurrent density of an $In_{0.46}Ga_{0.54}N$ top junction ($E_g \sim 1.8$ eV) can reach up to ~18 mA/cm^2 under AM 1.5G illumination. In addition, the built-in potential at two junctions of n^+–p Si and p^{++}-InGaN/n^{++}-InGaN can generate sufficient potential (>2 V) for unassisted solar water splitting.

To date, it is extremely difficult to grow high crystal-quality InGaN NWs with an ideal bandgap of ~1.8 eV (In composition ~46%) because of defect formation and phase separation. Moreover, InGaN/Si photocathode suffers from surface recombination induced by surface defects and the poor adhesion of cocatalysts. Taking these factors into account,

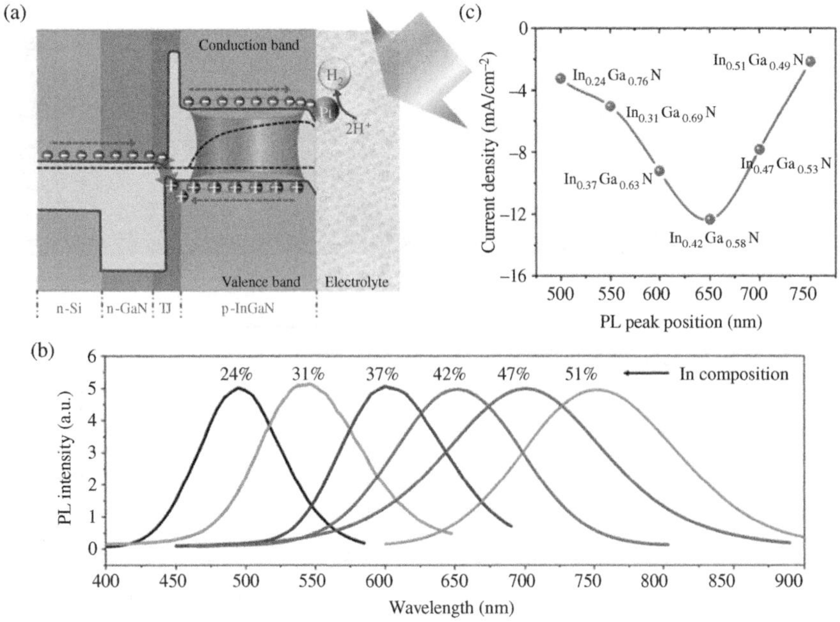

Figure 11.3 (a) Schematic energy band diagram of InGaN tunnel junction nanowire photocathode under illumination. (b) Room-temperature photoluminescence spectra of different InGaN NWs, showing a large variation of the emission peaks from 500 to 750 nm. (c) Variations of the photocurrent density (measured at 0 V_{RHE}) vs. indium composition of p-InGaN tunnel junction nanowire photocathodes. Reproduced from Wang, Vanka, et al. (2019)/with permission of Elsevier.

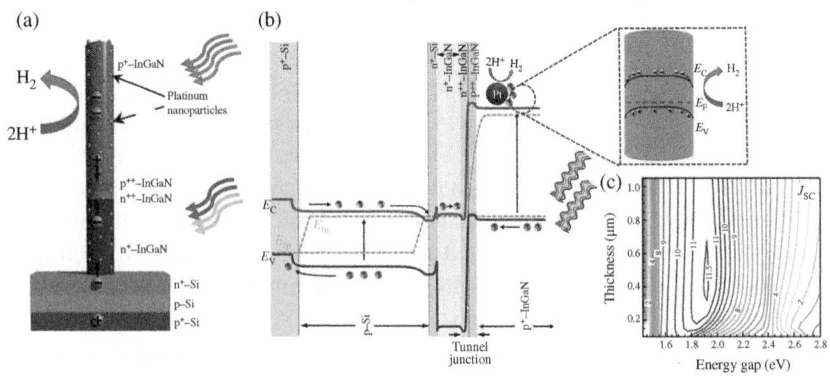

Figure 11.4 (a) Schematic of Pt-decorated p$^+$-InGaN/tunnel junction/n$^+$-InGaN nanowire on n$^+$-p Si wafer. (b) Energy band diagram under light illumination. (c) Top p$^+$-InGaN thickness and energy gap dependence of the theoretically calculated short circuit photocurrent density. Reproduced from Navid et al. (2021)/with permission of AIP Publishing.

InGaN/Si double-junction has been demonstrated as photocathode for unassisted solar water splitting by integrating p$^+$-InGaN top light absorber (bandgap energy of 2.2 eV) with the Si p–n junction bottom absorber through a nearly defect-free n$^+$-InGaN/p$^+$-InGaN NW tunnel junction (Figure 11.5) (Vanka et al., 2020). The tunnel junction is epitaxially grown on top of the bottom n$^+$-InGaN segment instead of Si. The tunnel junction in a NW segment possesses much fewer defects and dislocations than the one directly integrated into Si substrate. In addition, an ultrathin Al_2O_3 (1–2 nm) passivation layer is deposited on InGaN NWs by atomic layer deposition, which passivates the surface defects, thus dramatically inhibiting the surface recombination of electron–hole pairs and stabilizing Pt cocatalyst on the InGaN NWs. InGaN/Si double-junction photocathode with Pt and Al_2O_3 protective layer shows unassisted solar water splitting activity in two-electrode configuration. It exhibits $V_{on} \sim 0.7$ V vs. IrO_x counter electrode and photocurrent density of ~8.4 mA/cm^2 at 0 V vs. IrO_x with STH ~10.3% under AM 1.5G 1-sun illumination in 0.5 M H_2SO_4. This is the highest STH efficiency for Si and non-III–V-based photocathodes. Moreover, the high photocurrent density of ~8.2 mA/cm^2 at 0 V vs. IrO_x is stable for 100 hr under light illumination, which is the longest stability reported for any photoelectrode in two-electrode measurement, including III–V photoelectrodes (Table 11.1).

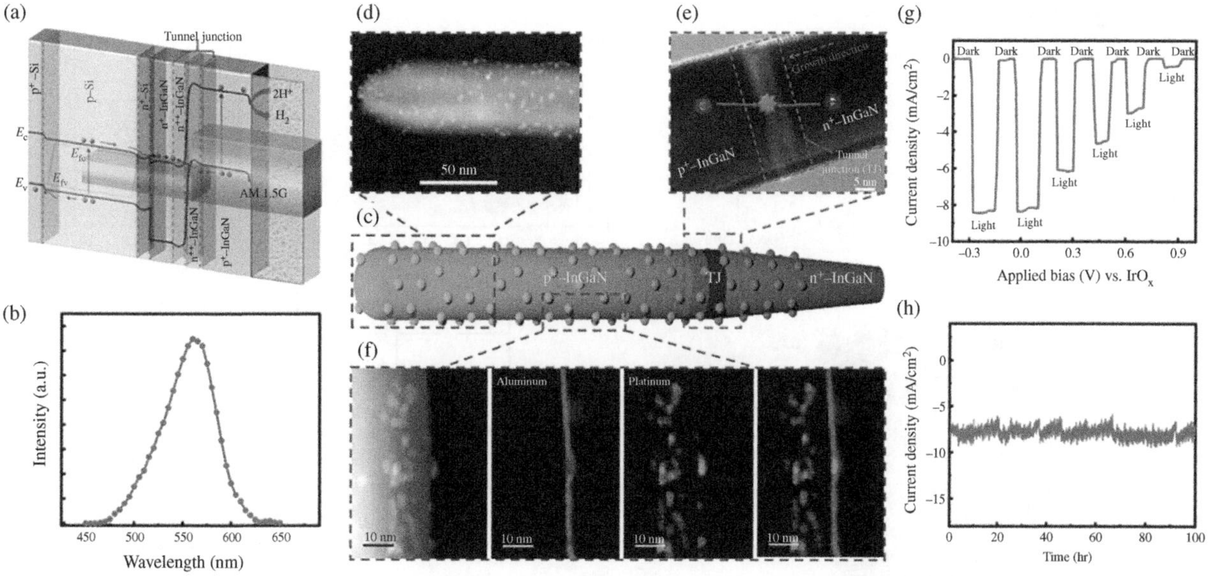

Figure 11.5 (a) Band-diagram of the p$^+$-InGaN/tunnel junction/n$^+$-p Si photocathode. The top p$^+$-InGaN absorbs photons with $h\nu \approx 2.2$ eV, or higher, and bottom Si absorbs photons passing through the top subcell but with $h\nu$ greater than the bandgap of Si. (b) Photoluminescence spectrum of p$^+$-InGaN NWs. (c) Schematic of single nanowire with Pt and Al_2O_3. (d) STEM-HAADF image of top p$^+$-InGaN segment. (e) Bright-field STEM image of tunnel junction region. The photogenerated holes from top p$^+$-InGaN and electrons from bottom n$^+$-InGaN/Si recombine in the tunnel junction. (f) STEM-HAADF and EELS elemental mapping showing Pt and Al_2O_3 on the nanowire sidewall. (g) Two-electrode chopped $J-V$ curves of InGaN/Si double-junction device. (h) Unassisted solar water splitting stability measurement at 0 V vs. IrO_x in 0.5 M H_2SO_4 under AM 1.5G 1-sun illumination. Reproduced from Vanka et al. (2020)/with permission of American Chemical Society.

Table 11.1 Summary of representative unassisted solar water splitting photoelectrodes using a 2-electrode configuration.

Photoelectrode	Counter electrode	Electrolyte	STH efficiency (%)	Stability (hr)	References
Al_2O_3–Pt–*InGaN/Si* double-junction photocathode	IrO_x	0.5 M H_2SO_4	10.3	100	Vanka et al. (2020)
Co–OEC/*triple-junction a-Si* photoanode	NiMoZn	1.5 M KNO_3	4.7	2	Reece et al. (2011)
Ni–TiO_2–*InGaP/GaAs* photoanode	Ni–Mo	1 M KOH	10.5	40	Verlage et al. (2015)
Rh–AlInP–*GaInP/GaInAs* photocathode	RuO_2	1 M $HClO_4$	14	2	May et al. (2015)
PtRu–*GaInP/GaInAs* photocathode	RuO_x	3 M H_2SO_4	15.4	1	Young et al. (2017)
Rh–TiO_2–*GaInP/GaInAs/GaAs* photocathode	RuO_2	1 M $HClO_4$	19.3	20	Cheng et al. (2018)
Pt-GaN–GaInP/GaAs/Ge phothocathode	Pt	0.1 M H_2SO_4	12.6	70	Wang, Schwartz, et al. (2019)

High-Efficiency Semiconductor Photocathodes Protected by Multifunctional GaN Nanostructures Various semiconductor photocathodes such as TiO_2/AlInP-protected GaInP/GaInAs double-junction (Cheng et al., 2018), inverted metamorphic GaInP/AlInP and GaInP/GaInAs double-junctions (Young et al., 2017), and metal-protected GaInP/GaInAs/Ge triple-junction (3J) have achieved relatively high STH efficiency >13% (May et al., 2015). However, these reported photoelectrodes usually have a poor lifetime due to the severe corrosion under harsh reaction conditions. To enhance the stability of high-efficiency photoelectrodes, *GaN-protected* $GaInP_2$/GaAs/Ge triple-junction (GaN/3J) photocathodes have been demonstrated (Figure 11.6) (Wang, Schwartz, et al., 2019). The use of GaN protection layer can result in enhanced light trapping due to the antireflective nature. The large surface area originating from the textured surface morphology leads to increase the number of active sites, thus improving the reactivity. When PEC measurements are carried out in a three-electrode configuration in 0.1 M H_2SO_4 electrolyte under AM 1.5 G 1-sun illumination, the GaN-protected 3J

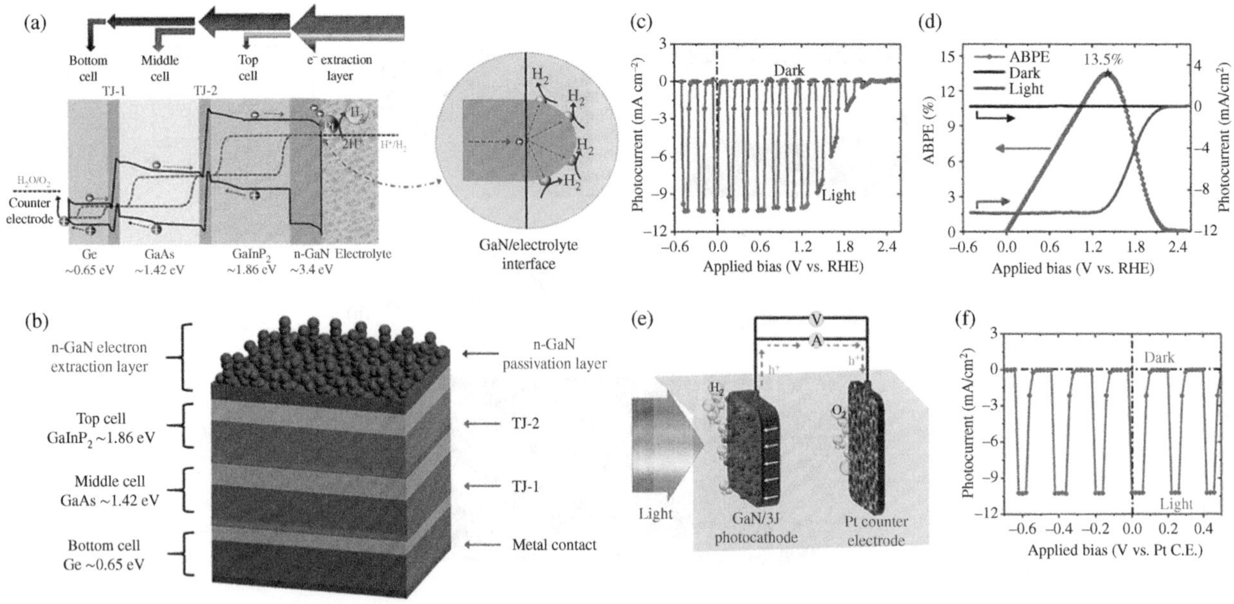

Figure 11.6 (a) Schematic energy band diagram showing the efficient extraction of photogenerated charge carriers (electrons) for proton reduction by the GaN nanostructures. (b) Schematic illustration of the $GaInP_2$/GaAs/Ge 3J structure protected by multifunctional GaN nanostructures. (c) *J–V* curves and (d) ABPE of the GaN/3J photocathode in three-electrode configuration under AM 1.5G 1-sun illumination. (e) Schematic illustration of the two-electrode measurement of the GaN/3J photocathode versus a Pt counter electrode. (f) *J–V* measurement of the GaN/3J photocathode under chopped illumination. Reproduced from Wang, Schwartz, et al. (2019)/with permission of American Chemical Society.

photocathode exhibits the saturated photocurrent density of ~10.3 mA/cm^2, large onset potential of ~2.2 V_{RHE}, and a maximum ABPE of ~13.5%. The GaN/3J photocathode also was tested in two-electrode configuration with the use of Pt as a counter electrode. The STH efficiency was ~12.6%. Stable solar water splitting was demonstrated for >50 hr in a two-electrode configuration. In short, GaN nanostructures were found to be grown on various semiconductor substrates and the fine control of physical structure and chemical composition of the material would help achieve performance improvement in PEC water splitting. In addition, benefitting from the exceptional structural, optical, and electronic properties, as discussed below, nanostructured III-nitrides have also shown notable advances in photocatalytic overall water splitting (OWS).

11.3.1.2 Photocatalytic Overall Water Splitting

Different from the bias-dependent PEC water splitting, the bias-free photocatalytic OWS can simultaneously produce hydrogen and oxygen with the stoichiometric ratio of 2:1 on single photocatalyst materials. This poses stringent thermodynamic and kinetic requirements for the photocatalyst materials. The photocatalyst materials for OWS not only need to satisfy the requirements for the potentials of the water reduction and oxidation but also contain the active surface catalytic sites for the water redox reaction. Though some photocatalyst oxides, such as TiO_2 and $SrTiO_2$, possess suitable band-edge potentials for water redox reactions, the excessively positive potential of O 2p states in the VB edge of photocatalyst oxides undesirably shifts the bandgap toward ultraviolet region. Thus, those photocatalyst oxides commonly show very low efficiency under solar light, which consists of <5% ultraviolet light and ~43% visible light. Hence, it is highly desirable to explore visible-light-responsive semiconducting materials for OWS. Compared to the oxides, the nitrides, such as III-nitrides, own the relatively less positive potential in the VB edge, mainly consisting of N 2p states, thus leading to a smaller bandgap. Hence, some III-nitrides, such as InGaN nanostructures, are developed in the field of photocatalytic OWS. Currently, some strategies, including multiband design, cocatalyst modification, and surface tuning, have been successfully used to enhance the photocatalytic activity of III-nitride photocatalysts.

Multiband Design As one of the most significant factors for determining the photocatalytic OWS efficiency, the light-response range of photocatalysts directly influences the number of available photogenerated electrons and holes. Especially, the photoexcitation efficiency in the visible-light region is used as a critical indicator for evaluating the performance of photocatalyst materials. In general, doping is a common strategy to achieve the visible-light-response photocatalyst. For example, the low-concentration Mg doping could extend the light-response range of GaN to 450 nm by introducing the acceptor states of Mg above the VB and the electronic states of N vacancy below the CB (Kibria, Chowdhury, Zhao, Trudeau, et al., 2015). As a result, the absorbed photon transformation efficiency of GaN NWs was increased to 43% at 375–450 nm, which demonstrated a feasible approach to extend the light response range of III-nitrides. However, the low density of interstitial states from dopants in the bandgap of photocatalyst can only provide a limited number of photogenerated electrons and holes from visible-light excitation. Compared to the low-concentration doping, alloying is a more efficient strategy to obtain the interstitial state-free narrow bandgap III-nitrides. The bandgap of InGaN alloy is determined by the ratio of Ga and In, which can be precisely tuned to have the bandgap in the range 3.4–0.65 eV. The visible-light-responsive InGaN photocatalysts thus prepared are widely used for photocatalytic OWS (Kibria, Chowdhury, Zhao, AlOtaibi, et al., 2015). For example, $In_{0.2}Ga_{0.8}N$ NW photocatalyst was reported to have an absorption edge of 475 nm. The apparent quantum efficiency in the visible range (400–475 nm) was ~12.3% (Kibria, Chowdhury, Zhao, AlOtaibi, et al., 2015).

Though a narrow bandgap can enable a photocatalyst to absorb more photons, the excessive narrowing would unavoidably lead to the reduction in the redox ability of the photocatalyst. In the photoexcitation process of a photocatalyst, only some short-wavelength photons with energy larger than the bandgap of the semiconductor can be absorbed. The transiently photogenerated electrons and holes would be located above the CB edge and below the VB edge, respectively. However, the photogenerated electrons and holes would rapidly relax to the conduction and VB edges, respectively. Hence, the redox ability of photocatalyst is directly determined by the potentials of its valence and CB edges. The decrease in the bandgap of a photocatalyst would unavoidably make the VB-edge potential more negative or the CB-edge potential more positive, which would undesirably weaken the oxidizing ability of photogenerated holes in the VB or the reducing ability of photogenerated electrons in the CB, respectively.

One of the most efficient strategies for maximizing both photo-redox ability and light-harvesting range is the use of multiband structures. With a well-designed material architecture, the high and low bandgap III-nitrides are used to absorb the short and long-wavelength photons, respectively. This structure can effectively absorb photons over a broad spectrum and maximize redox ability of photogenerated electrons and holes from available photons with different wavelengths. For example, the triple-band InGaN/GaN NW heterostructures were designed to possess three different bandgaps (3.40, 2.96, and 2.22 eV) (Kibria et al., 2013). When the light beam vertically irradiates the top layer of InGaN/GaN NWs, the GaN in the top layer is first to get photoexcited by photons with energy larger than 3.40 eV. Then the photons with energy less than 3.40 eV would reach the InGaN layer with a bandgap of 2.96 eV. Among those photons, the ones with an energy larger than 2.96 eV would be absorbed by the InGaN layer to produce electrons and holes. Further, the photons with energy larger than 2.22 eV reaching the InGaN dot layer would be absorbed. In the whole process, the energy loss in the photon-to-charge conversion process is effectively decreased compared to a single-bandgap photocatalyst system. The quadruple-band InGaN NWs consisting of $In_{0.35}Ga_{0.65}N$ (2.10 eV), $In_{0.27}Ga_{0.73}N$ (2.40 eV), $In_{0.20}Ga_{0.80}N$ (2.60 eV), and GaN (3.40 eV) segments were also successfully developed (Figure 11.7) (Wang, Wu, et al., 2019). Besides the quadruple-band-enhanced photon-to-charge conversion, a gradient Mg-doping along the lateral dimension of the quadruple-band NW was also designed to produce an internal electric field, which could promote an efficient charge carrier separation. Those features in the quadruple-band InGaN NWs contributed to the demonstration of a world-record STH (5.2%) in unassisted OWS by photocatalysis (Wang, Wu, et al., 2019).

Cocatalyst In general, the pristine photocatalyst often has poor surface catalytic sites for oxidizing water into oxygen or reducing water into hydrogen. Thus, some cocatalysts are introduced onto photocatalyst surface to act as water-redox sites. The water redox reaction includes both hydrogen evolution reaction (HER) and oxygen evolution reaction (OER). The cocatalyst needs to satisfy two conditions: low overpotential for water redox and strong charge-capturing ability. Hence, the cocatalyst not only catalyzes the water redox processes directly, but also promotes the charge carrier separation/transfer. Common HER cocatalysts include noble metals, such as rhodium and platinum. However, the noble metal cocatalysts easily cause the hydrogen–oxygen recombination, which is the main back reaction in the photocatalytic OWS. Thus, some strategies are developed to inhibit the hydrogen–oxygen recombination reaction. For example, a Rh/Cr_2O_3 core/shell structure was applied on InGaN NWs, which contributed to an effective OWS with stoichiometric hydrogen and oxygen production (Kibria, Chowdhury, Zhao, AlOtaibi, et al., 2015). It was considered that the Cr_2O_3 shell could allow protons to be transferred from water to the Rh surface and further reduced into hydrogen but inhibit the transfer of the oxygen produced on other regions onto Rh core, which could effectively avoid the undesirable hydrogen–oxygen

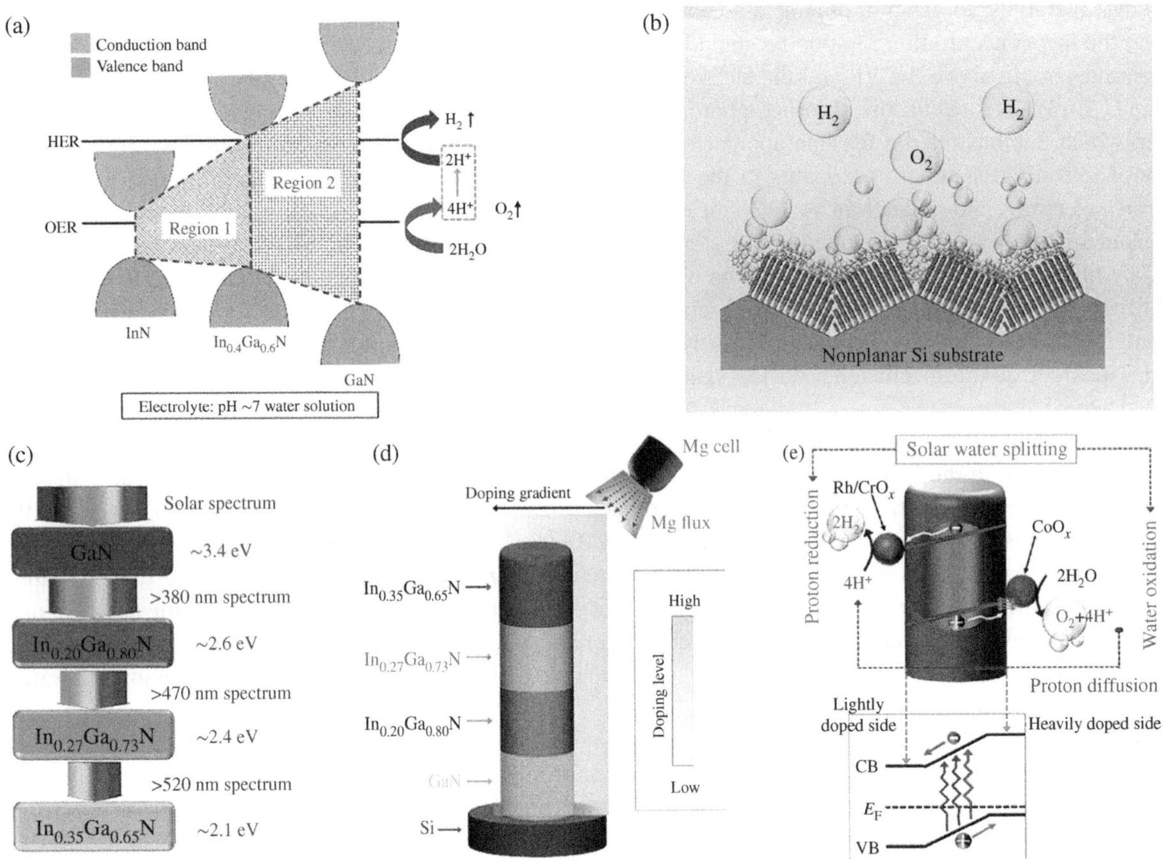

Figure 11.7 Schematic design and structural properties of monolithic multiband metal nitride nanowire arrays for photocatalytic overall water splitting in nearly pH neutral electrolyte solution. (a) Band edge positions of GaN, InN, and their InGaN compounds. In liquid solution at pH ~ 7, InGaN alloys with indium incorporation up to 40% can straddle the water redox reaction, noted as Region 2. Higher indium incorporation (Region 1) makes InGaN compounds unsuitable for overall solar water splitting. (b) Schematic illustration of photocatalytic water splitting on InGaN nanowire arrays monolithically integrated on a nonplanar Si wafer. (c) Schematic illustration of the ideal configuration for light absorption on the multiband InGaN stacks with varied indium compositions. (d) Schematic of the quadruple-band InGaN nanowire. The p-type dopant originating from the tilted Mg effusion cell (relative to the nanowire orientation) leads to the Mg doping gradient profile in the lateral direction of the nanowire, which can introduce a built-in electric field, illustrated in (e), for efficient charge carrier separation and extraction for water redox reactions. Reproduced from Wang, Wu, et al. (2019)/with permission of The Royal Society of Chemistry.

recombination reaction on Rh core. A STH conversion efficiency of 1.8% was demonstrated under concentrated sunlight illumination.

Compared to HER, OER is the rate-determining step in photocatalytic OWS, which often limits the improvement in the activity of photocatalyst materials. Though the pristine surface of III-nitride nanostructures possesses the OER sites, photogenerated holes with strong oxidizing ability can often oxidize the surface of III-nitride directly during the complex OER process. This is considered as the main reason for the photo corrosion of III-nitride nanomaterials. Though the N-terminated surface along with [001] direction can act as the protection/passivation layer for stabilizing the III-nitride surface, the most active (110) and (100) surfaces in III-nitride are co-terminated with Ga/In and N atoms (Kibria & Mi, 2016). Hence, III-nitride NWs are unavoidably oxidized in the long-term photocatalytic OWS. Some stable OER cocatalysts, such as CoO_x and NiO_x, are being developed to replace the surface sites of III-nitride as the catalytic OER sites. Loading of OER cocatalyst with effective photogenerated hole-capturing ability facilitates the separation of photogenerated holes from III-nitride photocatalyst. This can weaken or even avoid the direct oxidation of III-nitride surface. Moreover, those cocatalysts with relatively low OER overpotential commonly own more active OER sites than the surface sites of III-nitride, which can contribute to a higher STH. For example, the co-deposition of the water oxidation cocatalyst (Co_3O_4) and the water reduction cocatalyst (Rh/Cr_2O_3 core/shell) enabled InGaN NWs to achieve a higher STH of 2.7% compared to Rh/Cr_2O_3-loaded InGaN NWs (Figure 11.8) (Guan, Chowdhury, Wang, et al., 2018). More importantly, the stability of Co_3O_4

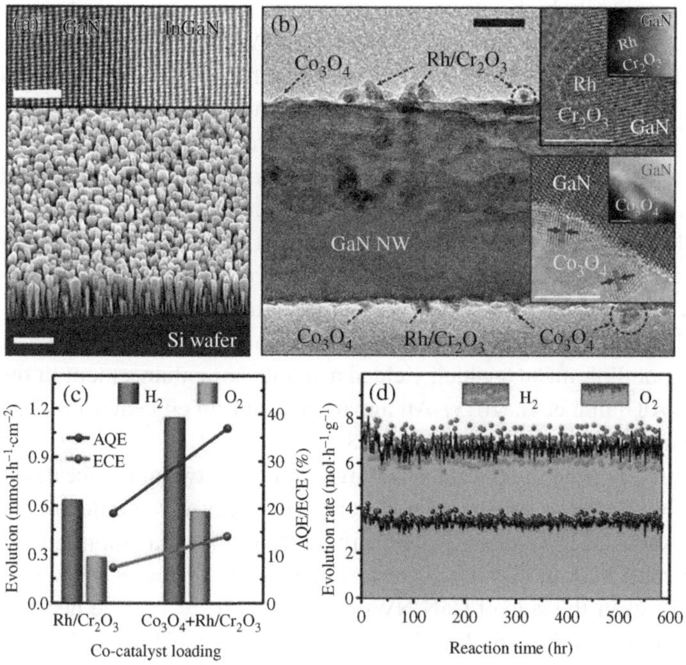

Figure 11.8 (a) SEM image of GaN/InGaN nanowire arrays on a Si wafer. Scale bar, 500 nm. HRSTEM-ZC image from a InGaN/GaN interface (scale bar 5 nm). (b) TEM image of a cocatalyst-decorated GaN nanowire with loadings of Rh/Cr_2O_3 and Co_3O_4. Scale bar, 20 nm. The inset shows STEM-BF and STEM-HAADF images of Rh/Cr_2O_3 (scale bars, 5 nm) and Co_3O_4 nanostructures (scale bar, 5 and 2 nm). (c) Rate of H_2 and O_2 evolution in pure water splitting on Ga(In)N NWs for single- and dual-cocatalyst loading. (d) Average rate of H_2 and O_2 from each cycle during a long-term repeated course of water splitting under concentrated sunlight. Reproduced from Guan, Chowdhury, Wang, et al. (2018)/with permission of American Chemical Society.

and Rh/Cr_2O_3-loaded InGaN NWs reached the world-record 580 hr. Furthermore, the co-loading of Co_3O_4 and Rh/Cr_2O_3 on InGaN NWs was also demonstrated to contribute to a stable photocatalytic OWS with a STH of 1.9% in some harsh aqueous environments like seawater (Guan, Chowdhury, Pant, et al., 2018).

Besides direct HER/OER, the cocatalyst with electron/hole-capturing abilities can also influence the photogenerated charge separation/transfer in III-nitride photocatalyst. Theoretically, the spatial separation of photogenerated electrons in HER cocatalyst and holes in OER cocatalyst can significantly inhibit the photogenerated electron–hole recombination in the photocatalyst, which is considered as the one of major challenges in the photocatalytic field. With the assistance of femtosecond transient absorption (TA) technology, ultrafast exciton and charge–carrier dynamics of InGaN/GaN NWs were studied to reveal the photogenerated charge separation/transfer. With the loading of Rh/Cr_2O_3 cocatalyst, InGaN/GaN NWs showed an additional decay peak at ~50 ps compared to the pristine InGaN/GaN NWs, which is credited to the interfacial charge transfer from InGaN/GaN NWs to Rh/Cr_2O_3 cocatalyst (Pu et al., 2015). The effective charge transfer between InGaN/GaN and Rh/Cr_2O_3 decreases the photogenerated electron–hole recombination, thus further leading to the reduced PL signal in Rh/Cr_2O_3-loaded InGaN/GaN.

Surface Tuning Though the loading of cocatalyst can significantly promote the photogenerated electron–hole separation in III-nitride photocatalyst via the interfacial charge transfer, the charge transfer from the bulk to the surface of photocatalyst is mainly influenced by the electronic properties of photocatalyst bulk and surface. The rapid charge transfer from photocatalyst to cocatalyst can effectively inhibit the undesirable bulk recombination of photogenerated electrons and holes. Especially, the band edge potentials in the bulk and surface of photocatalyst directly determine the direction and barrier of charge transfer. In general, the high downward band-bending in p-type GaN or upward band-bending in n-type GaN would inhibit the hole or electron transfer from photocatalyst to cocatalyst, respectively, which decreases the efficiency of overall photocatalytic reaction. Besides, in real reaction environment, the surface of III-nitride is covered by water, which directly influences the band edge structure of III-nitride surface. Some investigations on n-type and p-type GaN surfaces have revealed that the upward and downward band bending was enhanced on water-adsorbed n-type and p-type GaN surfaces, respectively, which suppressed the water reduction and oxidation, respectively. However, p-doped GaN NWs often contain weakly n-type or even intrinsic surfaces with the slightly higher Fermi level than the redox potential of water. Thus, the downward band bending can be effectively reduced when the surface of p-type GaN NWs is in equilibrium with water. This gives an alternative approach to balancing the charge transfer and redox reaction through tuning the Fermi level on III-nitride NW surfaces. Especially, the oxidation of water with high overpotential on most reported OER cocatalysts is commonly considered as the rate-limiting step in the photocatalytic OWS. Benefiting from the unique feature of downward band bending between p-type photocatalyst surface and water layer, the p-type III-nitride can improve the STH efficiency by producing more holes for water oxidation. Thus, the p-type Mg-dopant incorporation was proposed to optimize the surface band bending of III-nitride (Kibria et al., 2014). It was found that the Mg-dopant incorporation effectively decreased the downward band bending, which led to a nearly flat band on the photocatalyst surface. The flat band structure of Mg-doped GaN effectively balances the energy barrier of hole and electron from photocatalyst to cocatalyst, which desirably contributed to an improvement in STH efficiency by more than two orders of magnitude on p-type GaN. Moreover, this strategy was also extended to p-type visible-light-responding InGaN, which yielded a significant enhancement in the STH efficiency (Figure 11.9) (Kibria, Chowdhury, Zhao, AlOtaibi, et al., 2015). An apparent quantum efficiency of 12.3% was achieved in the visible light region (400–475 nm), which contributed to a STH of 1.8%.

Direct control of the surface atom configuration was used to tune the interfacial charge transfer between photocatalyst surface and water layer. In general, the wurtzite GaN photocatalysts grown by metal-organic chemical vapor deposition (MOCVD) tend to have a Ga-termination configuration on their top *c*-plane, which suffers from photooxidation to form Ga_2O_3 species, further inhibiting the hole transfer from GaN bulk to its surface. However, with the advanced MBE technology, the N-termination can be dominant at the top or even the side of GaN NWs (Kibria & Mi, 2016). The negative N-termination can promise a stable hole injection into the surface from bulk while maintaining the structural stability of GaN. Moreover, compared to the conventional Ga-top polarity, the reverse crystal polarization induced by the top N-polarity

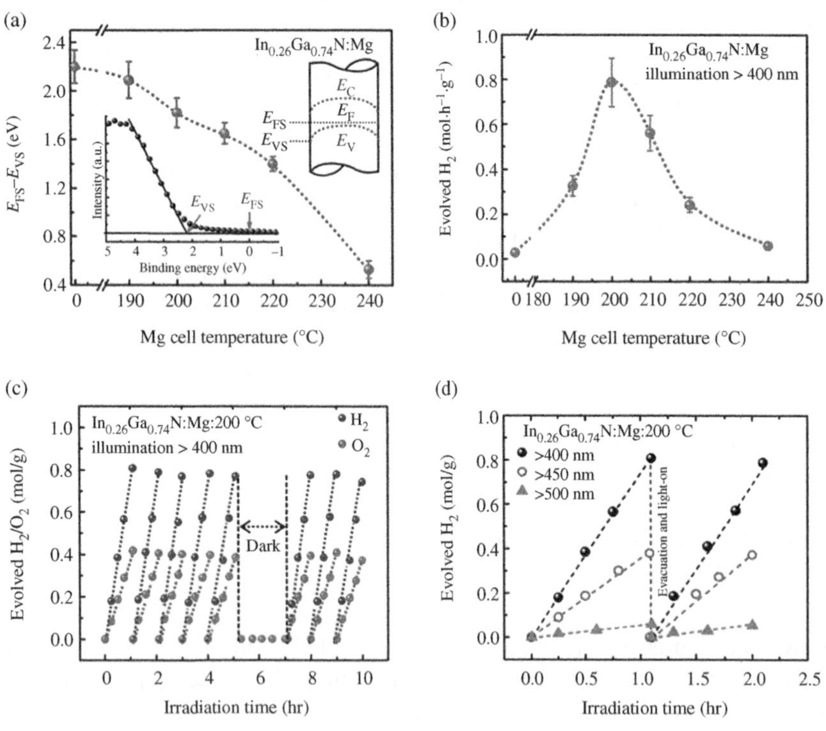

Figure 11.9 (a) $E_{FS} - E_{VS}$ for different Mg-doped $In_{0.26}Ga_{0.74}N$ nanowire samples derived from angle resolved X-ray photoemission spectroscopy (ARXPS) valence spectrum as shown in the insets. The error bars represent the accuracy of the linear extrapolation method. (b) H_2 evolution rate in overall neutral (pH ~ 7.0) water splitting for different Mg-doped $In_{0.26}Ga_{0.74}N$ nanowire arrays under visible light (>400 nm). (c) The evolution of H_2 and O_2 with irradiation time from neutral (pH ~ 7.0) water for Mg:200 °C (Mg source temperature) doped $In_{0.26}Ga_{0.74}N$ nanowire arrays under dark and visible light (>400 nm) irradiation. (d) H_2 evolution in overall water splitting as a function of irradiation time with different long-pass filters. The experiments were performed on a ~3 cm^2 wafer sample (total nanowire catalyst weight ~0.32 mg) and the estimated total incident illumination intensity on the sample is 418 mW/cm^2 (between 400 and 515 nm). The dashed lines are guides to the eye. Reproduced from Kibria, Chowdhury, Zhao, AlOtaibi, et al. (2015)/with permission of Springer Nature.

in the wurtzite GaN structure can greatly promote the spatial charge separation due to the polarity-induced internal electric field, which can also improve the photocatalytic efficiency. Besides this, the spatial charge separation could be achieved between the polar and nonpolar surfaces of GaN, which could lead to the spatial distribution of photogenerated holes and electrons on two different surfaces. The photoexcited charge carrier separation efficiency of GaN was demonstrated to have an improvement of more than 80%. This was attributed to the polarity-induced surface band binding (Li et al., 2019). Consequently, the electrons- and holes-accumulated surfaces act as the reducing and oxidizing regions on photocatalysts, respectively. Beneficial from this mechanism, the spatial assembly of HER and OER cocatalysts onto the nonpolar and polar surfaces can be exploited to further accelerate the charge transfer from bulk to surface region of the GaN photocatalyst.

Though the polarity-controlled spatial charge separation can effectively inhibit the undesirable photogenerated electron–hole recombination, a large charge-transfer distance still increases the probability of this process. Decreasing the distance (which is typically greater than 1,000 nm) between the reducing and oxidizing surfaces is an alternative strategy to reach this target. A PC diode artificial photosynthesis system based on the two-dimensional InGaN nanosheets was developed to precisely control the charge carrier flow at nanoscale, which greatly reduces the charge transfer distance (Figure 11.10) (Chowdhury et al., 2018). In the well-designed InGaN nanosheets, the gradient of Mg-dopant concentrations along the lateral dimension of nanosheets gives rise to a distinct work function difference (>300 meV) between the two parallel surfaces, thus leading to the different transfer directions for photogenerated electrons and holes. As a result, the photogenerated charge transfer distance is smaller than the thickness (~50 nm) of InGaN nanosheets. Thus, the formation of two closely adjacent anode and cathode surfaces in InGaN nanosheets greatly promotes the transportation and separation of photogenerated charge carriers. Similarly, the HER and OER cocatalysts are also deposited on the cathode and anode surfaces of InGaN nanosheets, respectively, which further accelerates the aggregation of electrons and holes on the corresponding surface. As a result, those features of InGaN diode nanosheets contribute to a STH of 3.3%.

The discussion above was focused on the advantages of nanostructured III-nitrides for solar water splitting, especially when they are grown by a state-of-the-art technology such as MBE. Herein, it is worth further noting that long-term stability is also at the core of practical applications of III-nitrides-based APIDs for hydrogen fuel production from water splitting. The recent studies about long-term stability will be discussed next.

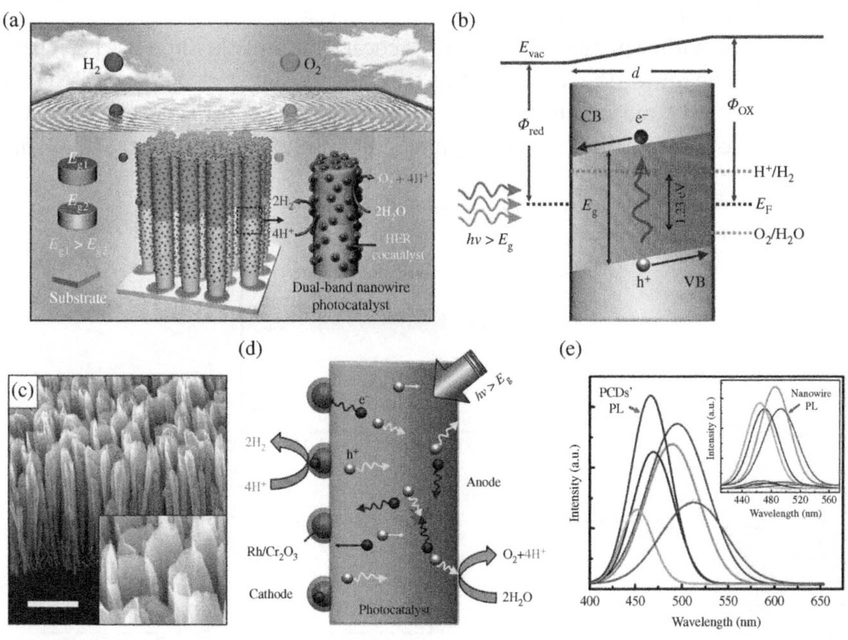

Figure 11.10 Structural and optical properties of InGaN photochemical diode (PCD). (a) Schematic illustration of wafer-level unassisted photocatalytic overall water splitting on double-band nanowire arrays. (b) Energy-band representation of the proposed photochemical diode (PCD) with radial thickness "*d*" showing the built-in electric field (band-bending) that separates the charge carriers (electrons and holes) and drives toward the opposite cathode and anode surfaces. (c) A 45° tilted SEM image of InGaN:Mg PCD nanostructures, vertically aligned on Si substrate. Scale bar, 1 μm. (d) Schematic (real space) depiction of the dynamic behaviors of charge carriers in a single-photon PCD upon photoexcitation. (e) Room temperature photoluminescence (PL) spectrum from as-grown p-InGaN PCDs for different indium incorporations (correspond to different bandgaps, which are calculated by the formula: bandgap = 1,240/wavelength). Reproduced from Chowdhury et al. (2018)/with permission of Springer Nature.

11.3.2 Long-Term Stability Studies

From the standpoint of practical applications, a superior stability of APIDs is essential for solar hydrogen production from water splitting, largely depending on photocorrosion. The corrosion of semiconductors in aqueous electrolyte is determined by many factors, such as electrolyte composition, wavelength of light, electrical bias, interfacial interactions between semiconductor and electrolyte, and presence of cocatalyst or passivation layer. To avoid cathodic or anodic corrosion during water splitting reaction, the photocathode should have a cathodic corrosion energy level larger than $0\,V_{RHE}$ to facilitate the photogenerated electrons to participate in HER rather than the cathodic corrosion and the photoanode should have an anodic corrosion energy level smaller than $1.23\,V_{RHE}$ to promote the photogenerated holes to participate in water oxidation level rather than anodic corrosion. To date, however, most semiconductor materials do not satisfy the thermodynamic requirements. For instance, the widely used Si photoelectrode can be easily oxidized during the water oxidation, and some studies reveal that Si could also get oxidized under cathodic conditions (Bae et al., 2017). To overcome these problems, various protection schemes have been demonstrated on the photoelectrodes to improve the stability. The first method is the coupling of highly active cocatalysts on the photoelectrodes to enhance the reaction kinetics of water splitting. The photogenerated charge carriers migrate toward the cocatalysts for the hydrogen evolution or OERs rather than moving to the surface of semiconductor for the corrosion reaction (Dias et al., 2016; Kuang et al., 2016). Stability of photoelectrodes improves with the presence of cocatalysts. However, the long-term stable operation still cannot be achieved due to the inevitable exposure of semiconductor surface to the electrolyte. To further improve the stability, relatively thick and uniform passivation layers such as TiO_2 (Hu et al., 2014) and Al_2O_3 (Fan et al., 2015) are deposited on the photoelectrode. Although these coatings improve the stability of these photoelectrodes significantly, one major problem is the large charge transfer resistance originating from the poor electrical conductivity of oxides and unsuitable band alignment between semiconductor and metal oxides. In this context, to achieve both high efficiency and long-term durability, a multifunctional surface protection was demonstrated by integrating *N-terminated GaN NWs* on an n^+-p Si wafer (Figure 11.11) (Vanka et al., 2019). The N-terminated surfaces of GaN behave as an effective protection layer for the underlying Si photoelectrode against photocorrosion. As a result, the Pt-decorated GaN/Si photocathodes demonstrate a superior stability of 3,000 hr with a relatively stable photocurrent density of 37–38 mA/cm^2 in 0.5 M H_2SO_4 under AM 1.5G 1-sun illumination, which is the longest reported stability for photoelectrode materials for HER.

Although GaN has been used as protection scheme with Si photoelectrode and Pt cocatalyst, less attention has been paid to fundamental understanding of the GaN composition during the water splitting reaction. Recent work demonstrated that GaN/Si photocathodes could produce hydrogen with a faradaic efficiency of ~100% without a cocatalyst (Figure 11.12) (Zeng et al., 2021). During the chronoamperometric (CA) test in 0.5 M H_2SO_4 under 1-sun illumination at $-0.6\,V_{RHE}$, the

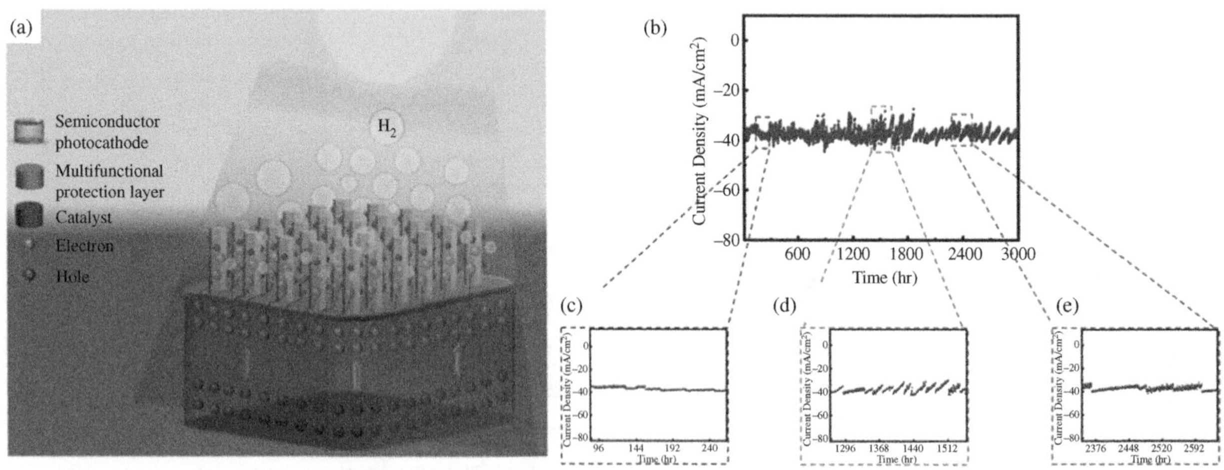

Figure 11.11 (a) Illustration of the third generation of the protection scheme, which consists of a semiconductor photoelectrode (n^+-p Si wafer), a multifunctional protective layer (n-GaN NWs), and a cocatalyst (Pt nanoparticles). The protective layer is inherently stable in the electrolyte, which not only provides long-term stability but also enhances significantly the light absorption, charge carrier separation, and extraction and reduces surface recombination further, thereby enhancing the efficiency. (b) Chronoamperometry long-term stability measurements for the platinized n-GaN NWs/n^+–p Si photocathode at $0\,V_{RHE}$ in 0.5 M H_2SO_4 under AM 1.5G 1-sun illumination. (c) Stability results of the highlighted dashed box for (c) 80–264 hr, (d) 1,270–1,539 hr, and (e) 2,350–2,640 hr of runs. Reproduced from Vanka et al. (2019)/with permission of The Royal Society of Chemistry.

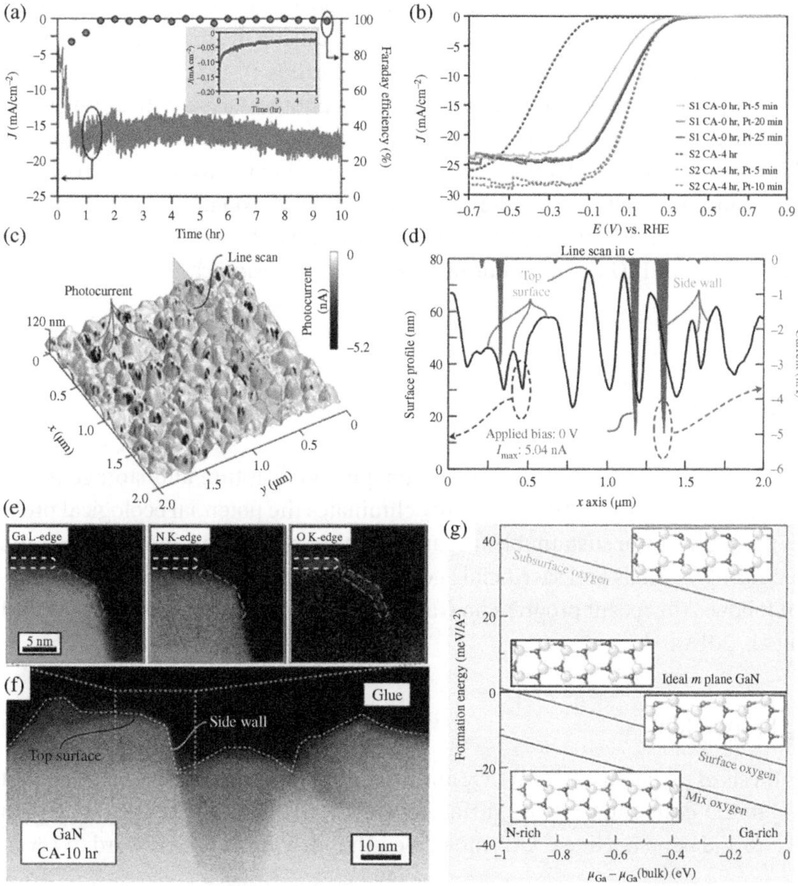

Figure 11.12 (a) CA testing of Si/GaN photocathodes for 10 hr under 1-sun illumination and constant bias at $-0.6\ V_{RHE}$, the corresponding Faradaic efficiency reveals the self-improving nature of GaN. Inset: CA testing on bare Si for 5 hr under 1-sun illumination and $-0.6\ V_{RHE}$. (b) Accelerated CA testing on Si/GaN photocathode at constant bias of $-0.6\ V_{RHE}$ under 3.5-sun illumination for 150 hr in 0.5 M H_2SO_4. (c) The topography of CA-10 sample and (d) corresponding surface profile and photocurrent extracted from the line scans. (e) STEM image of CA-10 hr surface with insets showing the EELS mappings of Ga L-edge, N K-edge, and O K-edge. (g) Calculated formation energy of *m*-plane GaN surface with different oxygen configurations within the anion- and cation-rich limits. Reproduced from Zeng et al. (2021)/with permission of Springer Nature.

measured photocurrent increased before it reached a steady state. Linear sweep voltammetry (LSV) measurements were done every hour to monitor the variations in PEC performance. The LSV curves showed a gradual anodic shift, with a gradual onset potential shift from $-0.46\ V_{RHE}$ (CA-0 hr) to $-0.08\ V_{RHE}$ (CA-10 hr). Additionally, the photocurrent density gradually increased and saturated at 25 mA/cm^2 after 3 hr. The observed self-improving photocurrent and anodic shift of the onset potential indicated an enhanced catalytic activity. Owing to the self-improving characteristics of GaN, the loading amount of noble metal catalyst (Pt) could be reduced. The CA-4 hr sample required only ~1/4 loading amount of Pt catalyst compared to as-prepared GaN to shift the turn-on voltage to $0.34\ V_{RHE}$. To trace the origin of the self-improvement in performance, atomic force microscopy measurements were performed. The surface morphology was found to be similar for CA-0 hr and CA-10 hr surfaces, which ruled out the effect of surface roughness. However, the photocurrent of CA-0 hr and CA-10 hr samples showed a substantial difference. The line profile of the photocurrent indicated that the photocurrent in CA-10 hr sample would mainly flow through the side walls of the GaN grains, whereas the top surface would barely contribute to the photocurrent. To investigate the changes after the water splitting reaction, STEM–electron energy loss spectroscopy (EELS) analysis was carried out. After 10 hr of operation, 1 nm-thin layer mixed with Ga, N, and O would format the side wall, while Ga and O covering the top surface. Theoretical calculations also suggest that the outermost layers of the nonpolar side wall favor a mixture of all three species (Ga, O, and N) whereas Ga and O are most likely stable on the polar top surfaces. Free energy of adsorption of hydrogen is smaller on GaON surface compared to GaN nonpolar side wall, meaning that HER is thermodynamically favorable on GaON surfaces. This result shows that the spontaneous partial nitrogen substitution by oxygen enhances the catalytic activity and stability of nonpolar GaN facets during the PEC water splitting. The spontaneous change in the surface chemical composition of III-nitride during the reaction will provide a new insights for designing the efficient, stable, and low-cost photoelectrode for water splitting.

Different from the PEC water splitting, the overall photocatalytic water splitting requires photocatalysts with strong resistance abilities for both photoreduction and photooxidation. Under the harsh reaction environments, such as sea water, the photooxidation-induced corrosion of III-nitrides is easy to be accelerated. In general, improving the transfer efficiency of photogenerated holes from III-nitrides to water can effectively reduce the hole-induced photooxidation. It was demonstrated that introducing some cocatalysts (such as CoO_x) onto the surface of III-nitrides would improve not only the oxidation of sea water on InGaN, but also the stability of InGaN by inhibiting the direct corrosion caused by sea water (Guan, Chowdhury, Pant, et al., 2018).

11.4 GaN-Based APID for CO_2 Reduction

Artificial photosynthesis has significant advantages compared with the conventional carbon capture and storage (CCS) approaches (Aghaie et al., 2018; Tapia et al., 2018; Yan & Zhang, 2019). It not only eliminates the potential ecological problems associated with stored CO_2 but also serves as a pathway for sustainable fuel and chemical production. Nevertheless, the grand challenge is to drive CO_2RR without electrical potentials and sacrificial agents, partly due to the stable linear configuration of CO_2 and strong C=O bonds. In what follows, the recent progress on GaN-catalyzed CO_2RR toward carbon-based fuels is described (Chu et al., 2016; Chu, Ou, et al., 2018).

11.4.1 Photochemical CO_2RR Toward CH_4 Production

III-nitrides hold great promise for bias-free synthesis of carbon fuels from CO_2 and H_2O by photocatalysis. It was found that even without cocatalysts, the reduction of CO_2 to CO dominated on the surface of GaN NWs (Figure 11.13a,b) (AlOtaibi et al., 2015). The epitaxial wurtzite GaN can adsorb and bend the linear CO_2 molecules, thus driving CO_2RR toward CO even

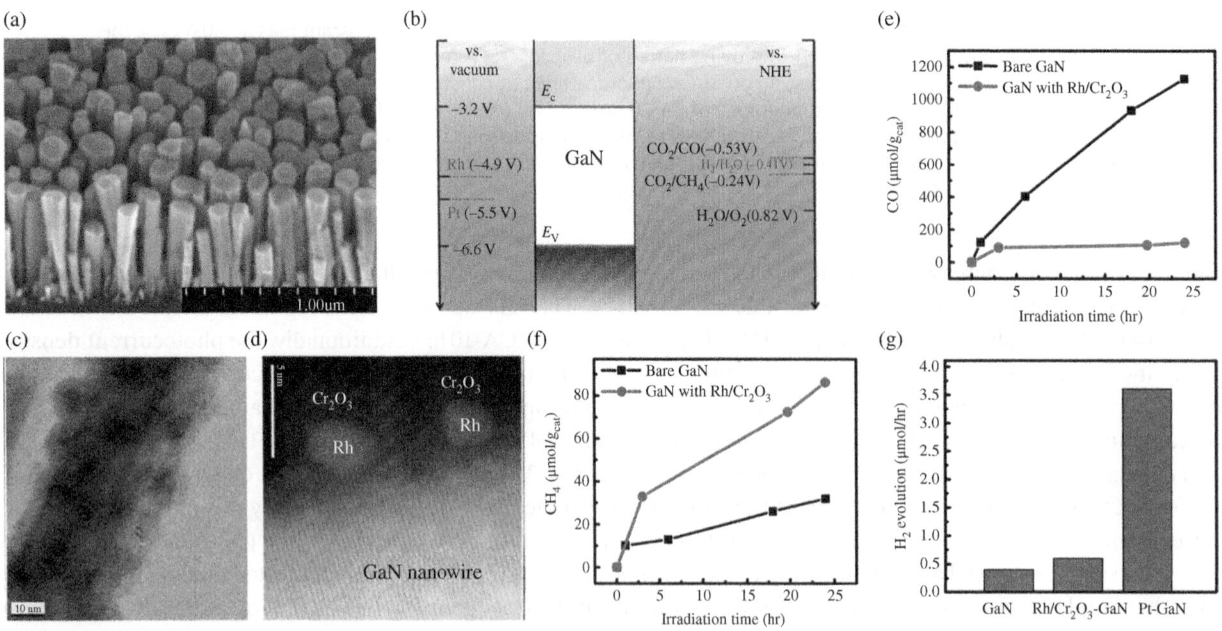

Figure 11.13 (a) A 45° tilted SEM image of GaN NWs grown on Si(111) substrate. (b) Band edge positions of GaN in comparison with the work functions of Rh and Pt with respect to the vacuum energy level. Also shown are the CO_2/CO and CO_2/CH_4 reduction potentials vs. NHE for pH = 7. (c) transmission electron microscope (TEM) image and (d) high-resolution TEM image of Rh/Cr_2O_3 core/shell decorated GaN nanowire. (e) CO and (f) CH_4 evolution on GaN NWs alone and GaN NWs decorated with core/shell Rh/Cr_2O_3. (g) H_2 evolution rate on GaN NWs and GaN NWs decorated with cocatalysts of Rh/Cr_2O_3 and Pt. Reproduced from AlOtaibi et al. (2015)/with permission of American Chemical Society.

without cocatalysts, which was also observed over GaN NWs/n$^+$-p Si during PEC process (Wang et al., 2016). The underlying mechanism requires further investigation. Compared to the dominant 2-electron product of CO, CH_4 is energy-dense ($\Delta H_c = 891$ kJ/mol) and highly compatible with the existing storage, transportation, and combustion infrastructures, thus being a more attractive product from CO_2RR (Bongartz et al., 2018; Citak et al., 2020; Wu et al., 2021; Zhang et al., 2016). However, the synthesis of methane is kinetically and thermodynamically unfavorable because of the high reaction barrier and requirement of $8H^+/8e^-$, particularly in practical APIDs without applied bias or sacrificial agents (Dimitrijevic et al., 2011; Long et al., 2017). With the help of cocatalysts, the epitaxial GaN NWs were demonstrated to facilitate unassisted conversion of CO_2 and H_2O toward CH_4 under concentrated light illumination. The formation rate of CH_4 showed a significant improvement by depositing core/shell Rh/Cr_2O_3 cocatalysts on the lateral *m*-plane surface of GaN with effective inhibition of CO evolution (Figure 11.13c–f). Pt showed a further promotion for CH_4 evolution rate, which is 4.2 times higher than that of Rh/Cr_2O_3. Herein, compared to the conventional powder systems (photocatalysts powder suspended in the electrolyte), methane was directly produced from CO_2 aqueous solution over nanostructured GaN vertically aligned on wafer-scale silicon substrate. The CH_4 evolution rate did not show observable degradation over 24 hr operation. Nevertheless, the evolution of CO was dominant over CH_4, validating the critical challenge of producing high-value chemicals during CO_2RR. Both Pt and Rh/Cr_2O_3 are well-known efficient HER catalysts, and therefore, HER is dominant over CO_2RR (Figure 11.13g), leading to very low selectivity for CH_4. To improve the overall performance of CH_4, one of the promising approaches is to develop rational cocatalysts from the scale of atoms to clusters to reduce the activation energy of CO_2RR to CH_4 by suppressing the severe HER and CO evolution reactions. Such cocatalysts can be grown using various techniques, including state-of-the-art atomic layer deposition, electrodeposition, and/or magnetron sputtering.

11.4.2 Photochemical CO_2RR Reduction Toward CH_3OH Production

Methanol is not only a liquid fuel but also a key feedstock for 30% of known chemicals such as olefins (Liu et al., 2018; Olsbye et al., 2012), biodiesel (Mohadesi et al., 2020), and methylamines (Corbin et al., 1997) in chemical industry. It is mainly produced from syngas via the well-known Fischer–Tropsch process developed for fossil fuels. As shown in Figure 11.14a–c, with the use of p-InGaN/GaN NWs decorated with Pt NPs vertically aligned on silicon, CO_2 can be converted into CH_3OH

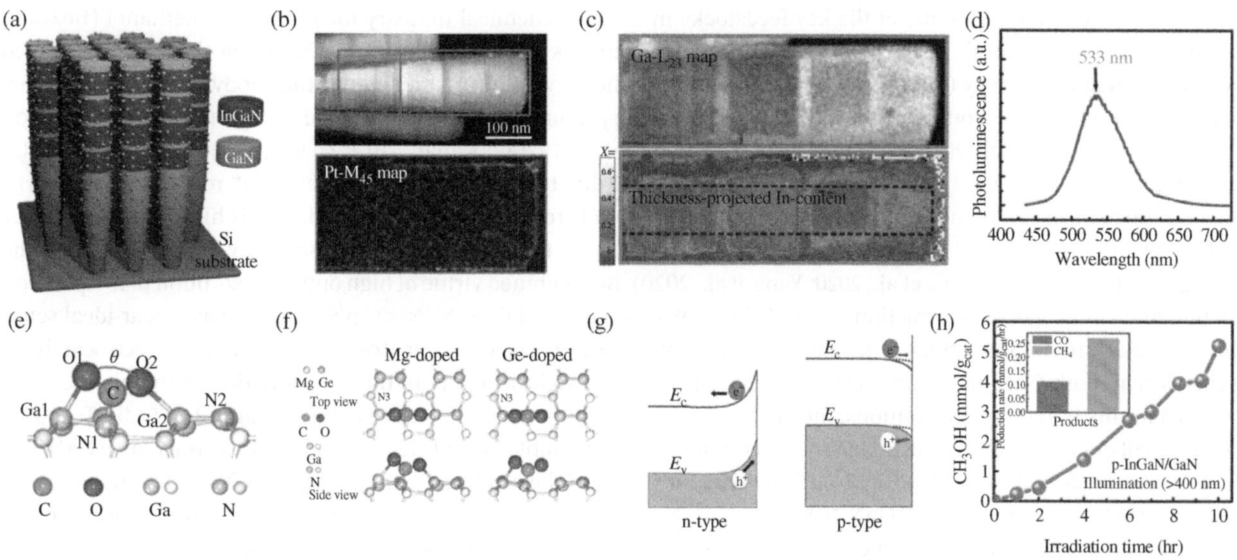

Figure 11.14 (a) Schematic illustration of p-InGaN/GaN NWs vertically aligned on silicon substrate decorated with Pt NPs. (b) The STEM-HAADF image (top) and Pt elemental mapping (bottom) extracted from the selected region (large rectangle from the middle to the right) marked in (b, top). (c) Elemental Ga map (top) from the selected region (small rectangle from the middle to the left) marked in (b, top) and thickness-projected In content map obtained from the same region (b, top). The *x* in In_xGa_{1-x} is presented in the color scale down on the left. (d) PL spectra of a p-InGaN/GaN nanowire. (e) The most stable configuration of CO_2 adsorbed on the *m*-plane surface of GaN. (f) The most relaxed and stable configurations of CO_2 molecules adsorbed on Mg- and Ge-doped GaN *m*-plane surfaces. (g) Schematic of charge carrier transfer described on semiconductors with different surface band bending. (h) Time-course CH_3OH evolution over p-InGaN/GaN NWs decorated with Pt NPs. Reproduced from AlOtaibi et al. (2016)/with permission of American Chemical Society.

using H_2 (molar ratio $CO_2/H_2 = 1:4$) as the reductant under concentrated illumination (AlOtaibi et al., 2016). In this study, distilled water was employed as the electrolyte; and atmospheric CO_2 was directly incorporated into the electrolyte by purging. Based on the PL spectroscopy measurement (Figure 11.14d), the epitaxial III-nitrides are capable of harvesting visible light (<533 nm) to drive CO_2 reduction without applied bias. Herein, the nonpolar *m*-plane surfaces of InGaN/GaN were discovered to be highly reactive with CO_2 molecules by using ab initio calculations. Particularly, upon adsorption on the surface of GaN, the inert CO_2 molecule can be heavily deformed from its linear configuration with a bent O–C–O angle of 128.27° (Figure 11.14e). The dopants were also discovered to play a critical role in affecting the behavior of CO_2RR. Mg-doped InGaN NWs showed a nearly 50-fold increase in CH_3OH evolution rate compared to the undoped ones, which could be associated with the reduction in the surface potential barriers, as well as the enhancement in the adsorption and deformation of CO_2 caused by this p-type dopant (Figure 11.14f,g). As a result, Stable and highly selective CH_3OH was produced under 10 hr of irradiation, with only 2% and 5% CO and CH_4, respectively (Figure 11.14h). In sharp contrast, the Ge-doped GaN did not show any electronic and catalytic advantages compared to the undoped samples, thus showing no enhancement in the reaction. These studies provided a fundamentalunderstanding of the causes of significantly enhanced CO_2RR activities on Mg-doped InGaN/GaN NWs and indicated the critical function of the dopants in III-nitrides. It is noted that in this study, external H_2 was essential for CO_2 reduction toward CH_3OH despite the unique capacity of III-nitrides for OWS (Kibria et al., 2013; Kibria, Chowdhury, Zhao, Trudeau, et al., 2015; Wang et al., 2011). It is thus regarded as CO_2 hydrogenation rather than a real artificial photosynthesis process. From the viewpoint of large-scale applications, it is necessary to develop catalysts for highly efficient water oxidation as well as CO_2RR simultaneously.

11.4.3 Photoelectrochemical CO_2 Reduction

In recent years, several notable achievements have been made on PEC CO_2RR over monolithically integrated GaN NWs/Si photocathodes, which are presented in the following section. Compared to photocatalysis, photoelectrocatalysis provides more options for tailoring CO_2RR toward various desired products because of the combined merits of silicon and III-nitrides and the input of applied bias, in spite of a relatively complex configuration.

11.4.3.1 Photoelectrochemical CO_2RR Toward CO/H_2 Production

The mixture of CO and H_2 is one of the key feedstocks in modern chemical industry for producing methanol (Bozzano & Manenti, 2016; Luk et al., 2017) and other important chemicals and fuels (Liu et al., 2013) via the well-established Fischer–Tropsch technology (Luk et al., 2017). In fact, from the viewpoint of kinetics and thermodynamics, CO, the key component of syngas, is favorably produced compared to other products because of the facile two-proton-coupled electron transfer process (Back et al., 2015; Fujinuma et al., 2020; Gu et al., 2019; Li & Kanan, 2012; Zhao et al., 2015; Zhu et al., 2013, 2014). Meanwhile, HER easily occurs in contrast with CO_2RR due to the low energy barrier of H^+ reduction toward H_2. Due to these features, the production of syngas (CO/H_2) from CO_2 reduction with H_2O oxidation is highly facile and may render a viable route for replacing the conventional routes built on fossil fuel reforming, especially if solar is used as the driving force (Lee et al., 2017; Xu et al., 2020; Yang et al., 2020). By combined virtue of high optical absorption of n^+–p silicon junction and strong electron extraction effect of GaN NWs, the material GaN NWs/n^+–p Si was used as a near-ideal semiconductor platform for integration with a rationally designed Cu/ZnO cocatalysts to drive CO_2RR by photoelectrocatalysis. In this design, both GaN and Si are heavily n-type doped. Under light irradiation, the accumulation of the photoexcited electrons results in flat-band conditions, thus reducing the upward bending of surface energy (Figure 11.15a,b). Together, despite small energy barrier of the n-GaN/n-Si heterointerface, the photoexcited electrons are easily extracted by GaN NW and further transferred to the deposited Cu/ZnO sites for CO_2RR. The electrons extracted by GaN NWs thus migrate easily to Cu–ZnO. Compared to bare GaN NWs/n^+–p Si, the one decorated with Cu and/or ZnO cocatalysts could provide active sites for facilitating surface chemical reactions, leading to greatly enhanced current density (Figure 11.15c). It is further observed that the synergy between Cu and ZnO was critical for activating the inert CO_2 molecules, resulting in an enhancement of CO formation (Figure 11.15d). It is particularly worth noting that the CO/H_2 ratio can be tailored in a board range by adjusting the applied bias, which benefits from the configuration of photoelectrocatalysis and is important to meet the practical requirements for various downstream products (Figure 11.15e). The reaction mechanism study showed that CO_2 was adsorbed and deformed on ZnO, and then catalyzed by Cu to CO through a series of elementary steps via the key intermediate of carboxyl (*COOH) involving multiple H^+/e^- transfers, which is not feasible with Cu or ZnO alone (Figure 11.15f). This finding validated the fact that the synergy between Cu and ZnO could play a critical role in CO_2RR. Based on this discovery, further theoretical and experimental investigations revealed the general contribution of

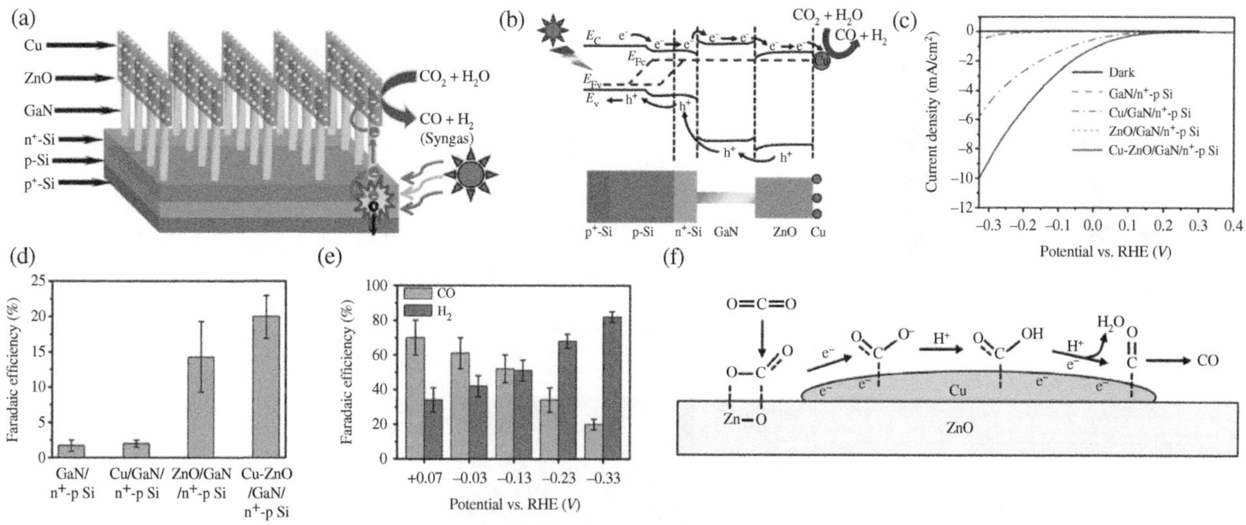

Figure 11.15 Photoelectrochemical production of syngas from CO_2 and H_2O over GaN NWs/n$^+$−p Si modified with Cu/ZnO. (a) Schematic illustration and (b) the bandgap alignment diagram of GaN NWs/n$^+$−p Si modified with Cu/ZnO. *J*−*V* (c) and Faradaic efficiencies (d) of PEC CO_2RR toward CH_4 over various photocathodes. Potential-dependent FE of CH_4 over Cu−ZnO/GaN/n$^+$−p Si (e). The synergy between Cu and ZnO for catalyzing CO_2RR toward CH_4 (f). (a, c−f) Reproduced from Chu et al. (2016)/with permission of John Wiley & Sons. (b) Adapted from Chu et al. (2016).

metal/oxide interface in offering multiple catalytic sites with complementary functions. As illustrated in Figures 11.16a,b, the metal/oxide interface showed a remarkable enhancement in CO_2 activation as compared to the metals alone. Among the tested metal/oxide interfaces, Pt–TiO_2 demonstrated an optimum performance because the Pt–TiO_2 interface allowed concurrent activation of CO_2 molecules and stabilization of the key intermediates. It is noted that the electronic modification of Pt arising from the remarkable interaction between metal and oxide is critical for selective conversion of CO_2 toward CO over Pt–TiO_2/GaN/n$^+$–p Si, which was validated using the electron localization function (ELF) for $Ti_3O_6H_6$/Pt(111) and X-ray photoelectron spectrum measurement (Figure 11.16c,d). Especially, as suggested by the ELF map, the electron redistribution between Pt and $TiO_3O_6H_6$ indicated the change in electronic properties of Pt and thereby facilitated CO_2RR.

In these two designs above, 1-D GaN NWs mainly serve as powerful electron extractors from photoexcited n$^+$–p silicon junction, which is the major absorber for a large fraction of solar photons with wavelengths up to 1,100 nm. The light-harvesting capacity of the silicon substrate is further improved by 1-D GaN due to its reduced light reflection (Kibria, Chowdhury, Zhao, AlOtaibi, et al., 2015). The exposed density of catalytic sites is also enhanced due to the unique nanostructure of 1-D NWs and their high surface area, further facilitating the CO_2RR. Most importantly, the synergy between metal sites and metal oxide supports serves as the main contributor to the improved syngas formation rate and selectivity from CO_2 and H_2O.

11.4.3.2 Photoelectrochemical CO_2RR Toward HCOOH Production

The support materials are important for CO_2RR as they influence the steric and electronic properties of the catalytic centers via metal-support interaction (Li et al., 2021; Wang et al., 2020; Wang, Kattel, et al., 2019). In our recent study, we took into consideration the surface properties of epitaxial GaN NWs for activating inert CO_2 molecules by introducing specific metallic sites (Zhou et al., 2019). The tin-based catalysts are well recognized to be highly active and selective for catalyzing CO_2RR toward HCOOH (Bejtka et al., 2019; Li et al., 2020). However, the overall performance of Sn-decorated silicon was found to be very limited, which was mainly associated with the lack of a rational catalytic architecture. In this context, by coupling MBE with electrodeposition, a rational Sn:GaN nanoarchitecture was explored. As characterized by high-resolution TEM, the edge of the GaN NW is atomically sharp and flat, thus providing an ideal supporting surface for constructing an efficient catalytic architecture at nanoscale and even atomic scale (Figure 11.17a). Sn NPs were randomly dispersed on the surface of GaN NWs. The as-designed configuration is beneficial for photon absorption, charge carrier separation, and active site exposure with high atom utilization efficiency (Figure 11.17b).

The prepared GaN:Sn nanoarchitecture also exhibited a unique CO_2 activation mechanism, which was substantially distinct from the conventional activation mechanism over bare Sn-based catalysts (An et al., 2021; Chen, Fan, et al., 2020;

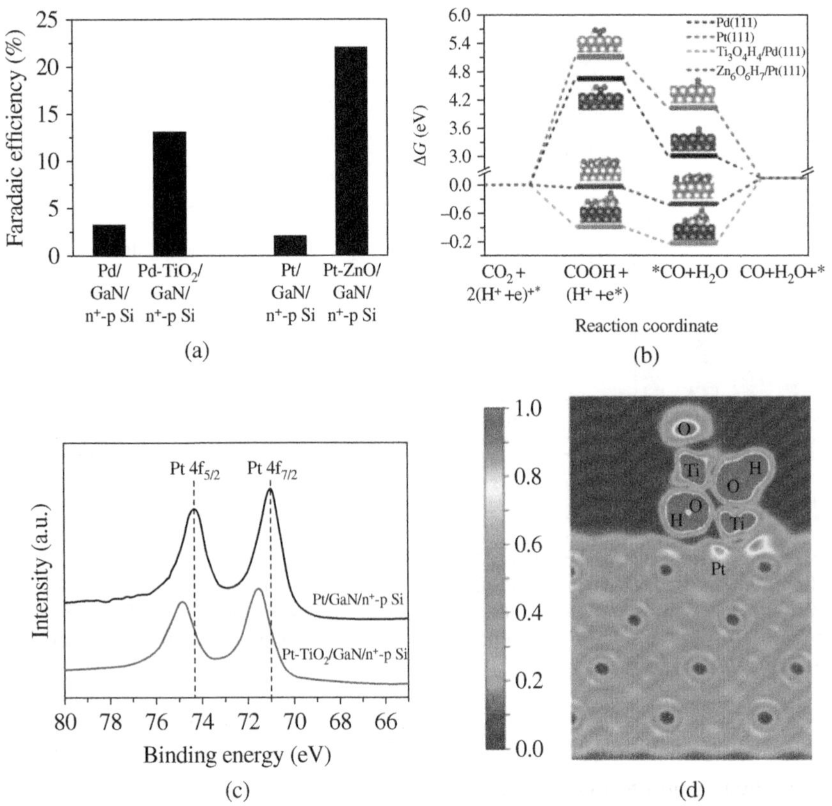

Figure 11.16 (a) Faradaic efficiencies of CO from CO_2RR over Pd/GaN/n^+−p Si, Pd−TiO_2/GaN/n^+−p Si, Pt/GaN/n^+−p Si, and Pt−ZnO/GaN/n^+−p Si (Chu, Ou, et al., 2018). (b) Calculated Gibbs energy diagrams for CO_2 reduction to CO on Pd(111), Pt(111), $Ti_3O_6H_6$/Pd(111), and $Zn_6O_7H_7$/Pt(111) surfaces at 0 V vs. RHE. (c) X-ray photoelectron spectra of Pt 4f of Pt/GaN/n^+−p Si and Pt/TiO_2/GaN/n^+−p Si. (d) Electron localization function of $Ti_3O_6H_6$/Pt(111). The probability of finding electron pairs varies from 0 (dark gray) to greater than 0 (mild shades) and to 1 (medium gray regions inside white loops). Reproduced from Chu, Ou, et al. (2018)/with permission of American Chemical Society.

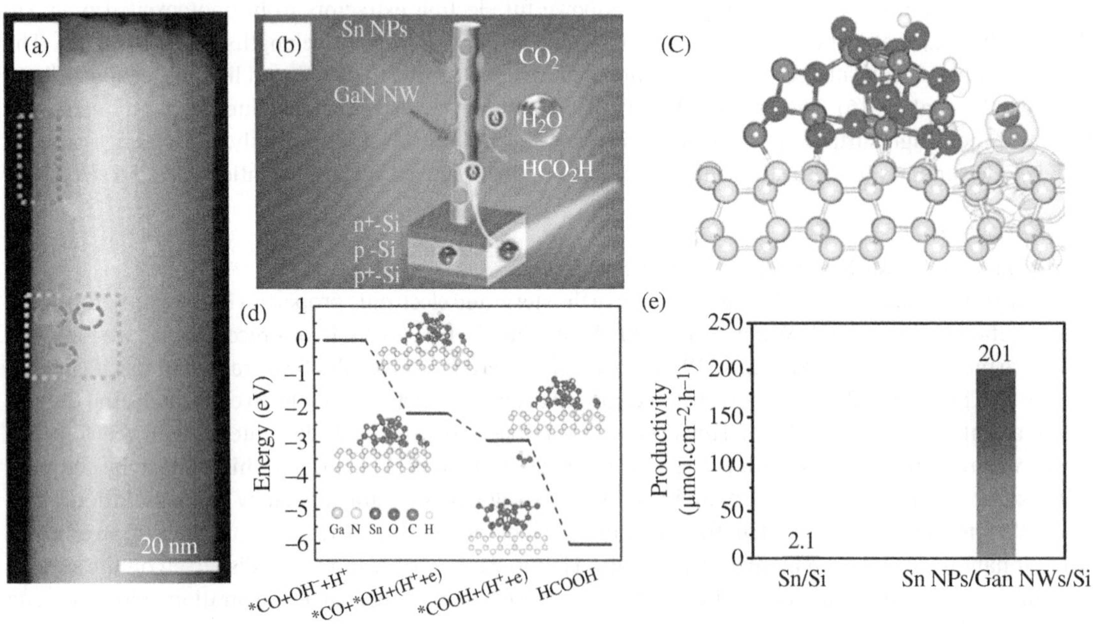

Figure 11.17 (a) A STEM-HAADF image of GaN decorated with Sn nanoparticles. (b) Schematic illustration of Sn NPs/GaN NWs/n^+−p Si for PEC CO_2RR toward HCOOH. The distinct CO_2 activation (c) and energy-favored conversion (d) toward HCOOH over the Sn:GaN interface. (e) The performance comparison between Sn/Si and Sn NPs/GaN NWs/n^+−p Si. (a–d) Reproduced from Zhou et al. (2019)/with permission of The Royal Society of Chemistry. (e) Adapted from Zhou et al. (2019).

Wu, Wu, et al., 2021). In the coordinated GaN:Sn catalytic interface, the epitaxial GaN is directly involved in CO_2 activation. By density functional theory calculations, CO_2 was found to be readily split into the key intermediate *CO via covalent-like Ga—C bond and ionic-like Sn—O bond over the Sn:GaN interface (Figure 11.17c). Subsequently, the intermediate, *CO, reacted with H_2O to produce HCOOH in an energy-favored pathway (Figure 11.17d). In this study, 1-D GaN NWs not only behaved as the bare support of catalytic centers and electron-migration channels but also played a significant role in CO_2 activation by coordinating with Sn-based catalysts. Because of this unique activation mechanism as well as the superior optoelectronic properties, the integrated device of Sn NPs/GaN NWs/Si exhibited superior performance. The productivity of HCOOH of Sn NPs/GaN NWs/Si was much higher than that of the bare Sn catalyst deposited on Si substrate, and it reached 201 $\mu mol \cdot cm^{-2} \cdot hr^{-1}$ from PEC CO_2RR in water (Figure 11.17e). Based on the discussions above, it is of scientific and practical interest to find that, apart from being light scattering alleviators and electron extractors, III-nitrides can directly influence CO_2 activation, and facilitate CO_2RR further by coupling with metallic centers. It, therefore, presents a new strategy for CO_2RR by providing more promising options for developing multifunctional catalytic nanoarchitectures by integrating them with appropriate metallic centers.

11.4.3.3 Photoelectrochemical CO_2RR Toward CH_4 Production

Due to the thermodynamic stability and sluggish kinetics of CO_2, it is extremely challenging to produce high-order chemicals like CH_4 from CO_2RR. Among the reported catalysts, copper and its derivatives are known to be the state-of-the-art catalysts for facilitating CO_2RR to produce value-added products (AlOtaibi et al., 2015, 2016; She et al., 2019; Wang et al., 2011). Hence, a monolithically integrated device of Cu/GaN/n^+–p Si was developed by photo-depositing Cu NPs onto the top of GaN NWs for PEC CO_2RR toward methane. As expected, without copper, the current density of GaN/n^+–p Si was much lower than that of Cu/GaN/n^+–p Si due to the lack of active sites for chemical reactions (Figure 11.18a). However, it was noted that in the absence of Cu catalytic centers, trace amount of CO was still evolved from GaN/n^+–p Si without formation of CH_4. It validated the potential of GaN for CO_2 activation (Figure 11.18b). An improvement in CH_4 Faradaic efficiency was achieved as Cu catalyst was incorporated. It was also noticed that the monofunctional sites and large size of the Cu-based catalyst were not efficient enough for mediating CO_2RR toward high-order chemicals. Therefore, the Faradaic efficiency of CH_4 was still as low as ~18%. The development of multifunctional catalysts by assembling copper with secondary components may favor CO_2RR by simultaneously activating CO_2 and stabilizing the reaction intermediates (Behrens et al., 2012; Yang et al., 2012).

Inspired by this thought, for the first time, a binary Cu–Fe catalyst was explored with the purpose of promoting PEC CO_2-to-CH_4 conversion. By density functional theory calculations, it was discovered that Cu and Fe in the binary catalyst were cooperative to deform the linear inert CO_2 molecule with a bent O–C–O angle of 126.05° and elongated C—O bonds. This was in stark contrast with the weak interaction of CO_2 with Cu(111) (Figure 11.19a,b). Although the potential-determining steps (i.e., *CO → *CHO) for methane synthesis remained the same, a great difference was that an appreciably reduced activation energy of 0.51 eV was achieved by copper–iron cocatalyst compared to that of 0.85 eV over pristine copper (Figure 11.19c). Moreover, the interfacial sites of $Fe_3O_6H_6$/Cu(111) were found to be directly

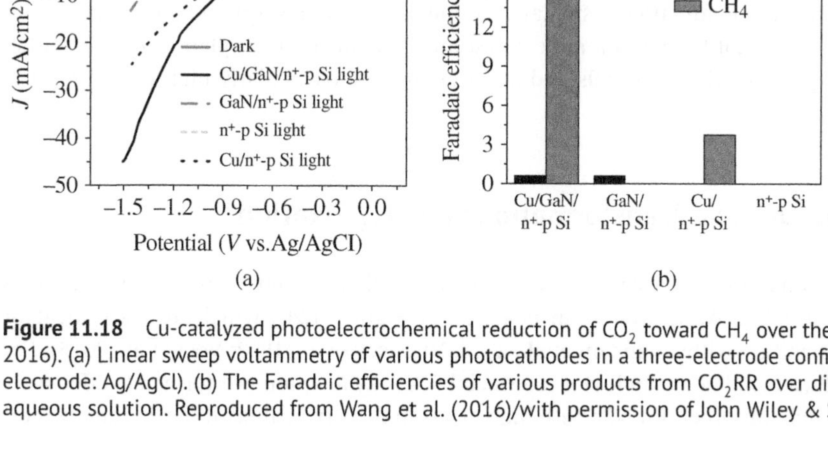

Figure 11.18 Cu-catalyzed photoelectrochemical reduction of CO_2 toward CH_4 over the GaN NWs/n^+–p Si platform (Wang et al., 2016). (a) Linear sweep voltammetry of various photocathodes in a three-electrode configuration (counter electrode: Pt, reference electrode: Ag/AgCl). (b) The Faradaic efficiencies of various products from CO_2RR over different photocathodes in 0.5 mol/L $KHCO_3$ aqueous solution. Reproduced from Wang et al. (2016)/with permission of John Wiley & Sons.

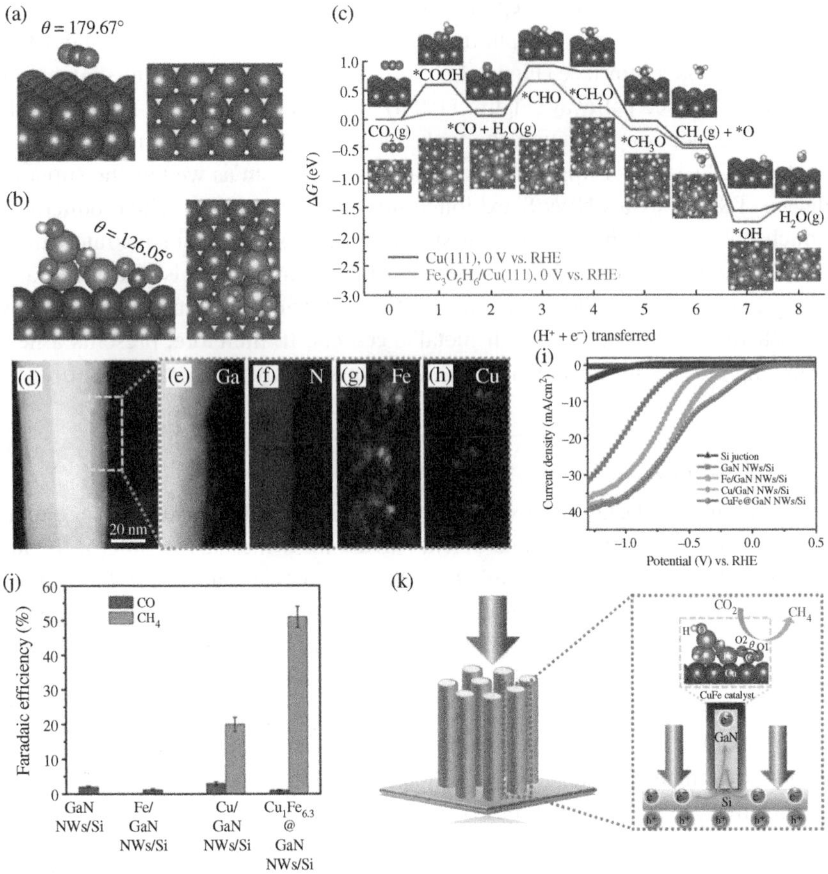

Figure 11.19 The critical role of GaN NWs in photoelectrochemical carbon dioxide reduction toward CH_4 over a binary copper–iron catalyst (Zhou et al., 2020). The optimized configurations of CO_2 upon absorption on Cu(111) (a) and $Fe_3O_6H_6$/Cu(111). (b) Large black, large gray, medium gray, medium dark, and small white spheres represent Cu, Fe, O, C, and H atoms, respectively. (c) Calculated Gibbs energy diagram for CO_2RR on Cu(111) and $Fe_3O_6H_6$/Cu(111). (e–h) The STEM-HAADF image and corresponding elemental mappings of GaN NWs decorated with CuFe. (i) *J–V* curves and Faradaic efficiencies (j) of various photoelectrodes for CO_2RR. (k) Schematics of spatial decoupling of CO_2RR from optical absorption and charge carrier separation over CuFe@GaN NWs/n^+–p Si. Reproduced from Wang et al. (2016)/with permission of National Academy of Science.

involved in binding and stabilizing all of the related reaction intermediates. As such, when electrochemically deposited onto the platform of GaN NWs/n^+–p Si (Figure 11.19d–h), the binary copper–iron exhibited an optimum current density, which primarily originated from the high-efficiency chemical reactions of CuFe@GaN NWs/n^+–p Si (Figure 11.19i). Consistent with the theoretical calculations, the Faradaic efficiency of CH_4 increased up to 51%, which is superior to that of monofunctional Cu catalysts (~19%) under the same experimental conditions (Figure 11.19j). The enhancement in CH_4 synthesis by GaN NWs was credited to their unique function in spatial decoupling of light absorption and charge carrier separation (Figure 11.19k). This study revealed that both a rational catalyst and a semiconductor platform are essential for highly efficient and selective artificial photosynthesis of fuels and other chemicals, particularly high-value products like CH_4.

11.5 Gallium Nitride-Catalyzed Organic Transformations and N_2 Fixation

In addition to artificial photosynthesis, GaN has enabled a series of achievements in light-driven organic transformations (Liu et al., 2021) and N_2 fixation (Li et al., 2017; Liu, Wang, Kong, Tan, et al., 2019) (Figure 11.20), which are also critical for carbon neutrality and environment sustainability (Nithyanandam et al., 2018; Vitousek et al., 1997; Vogt et al., 2015). For example, by coordinating GaN with Ru NPs, a metal/semiconductor interfacial Schottky junction was formed to effectively extract photoinduced electrons from GaN light absorber. Ru NPs with high electron density facilitated the extremely inert

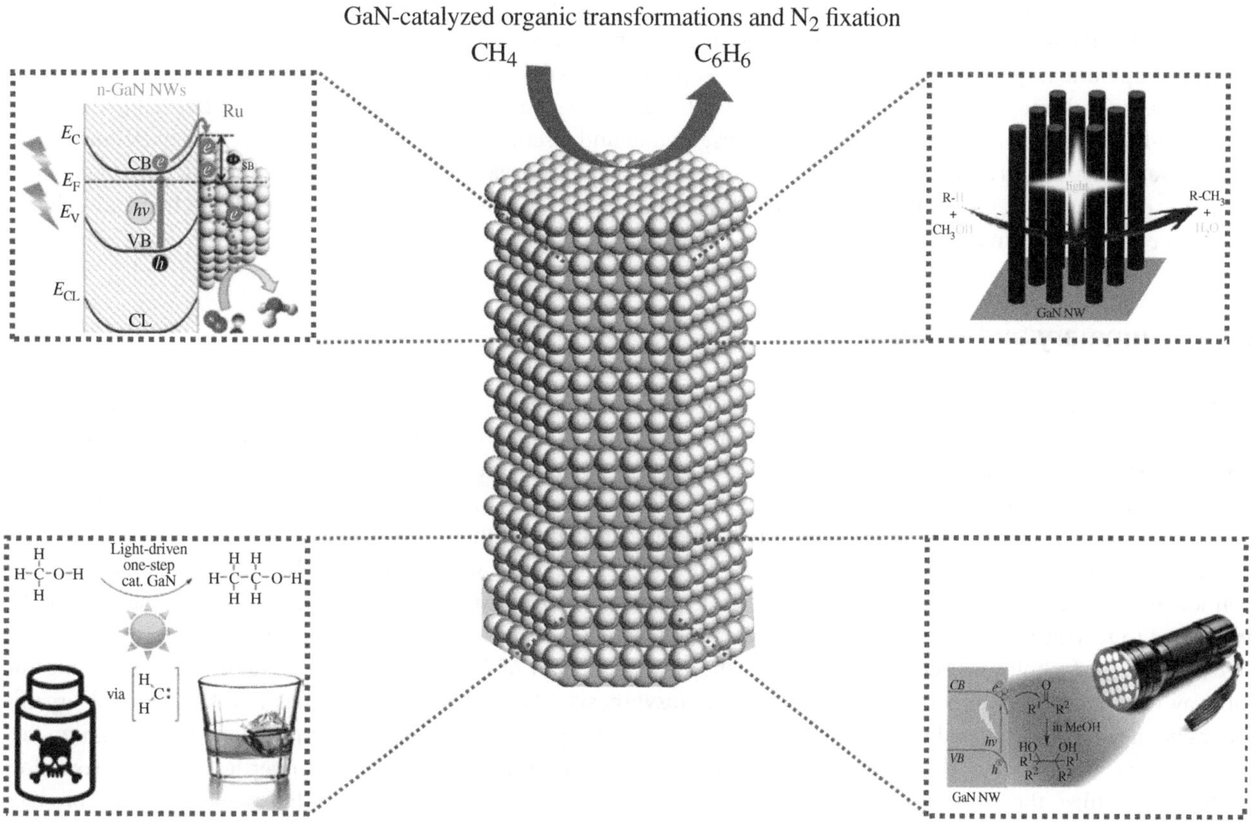

Figure 11.20 Schematic illustration of organic transformations and N_2 fixation under illumination. Adapted from Li, Fan, et al. (2014), Li, Mu, et al. (2014), Liu, Qiu, et al. (2020), Liu, Tan, et al. (2020), and Liu, Wang, Kong, Rashid, et al. (2019).

N≡N bond cleavage with the use of H_2, resulting in mild ammonia synthesis powered by sunlight (Figure 11.20, top, left). This is in stark contrast with the harsh Haber–Bosch process. Further computational and experimental investigations showed that the surface of GaN($10\bar{1}0$) wurtzite with N vacancy exhibited strong chemical adsorption of N_2 and enabled facile electron donation from the surface vacancy to N, resulting in ammonia synthesis with the use of water as the hydrogen source (Liu, Wang, Kong, Tan, et al., 2019).

In our recent studies, it has been discovered that CH_4 could be selectively converted to benzene over a photoexcited GaN surface, revealing a photon-assisted, low-temperature, and environment-friendly aromatization pathway (Li, Fan, et al., 2014; Li, Mu, et al., 2014). Furthermore, light alkanes could be thermo-catalytically converted into corresponding aromatics over commercial non-doped GaN powder with a wurtzite structure under elevated temperatures with superior reactivity, high selectivity, and stability (Li, Mu, et al., 2014). During the process of CH_4-to-C_6H_6 conversion, the C—H bond activation was discovered to be very sensitive to the surface properties of GaN, largely depending on the exposed m-plane with strong local electrostatic polarization.

For the first time, photo-driven methanol-to-ethanol transformation via methyl carbene was reported by employment of gallium nitride semiconductors (Figure 11.20, bottom, left). Under illumination, the highly negative-charged N- at the C-plane surface of GaN attacks the C—H bond of methanol, followed by the formation of methyl carbene with the elimination of a water molecule. The formed methyl carbene was a critical specie for building ethanol via C–C coupling. The carbon chain could be further elongated by tailoring the reaction temperature (Liu, Wang, Kong, Rashid, et al., 2019). Inspired by this discovery, the strategy of producing methyl carbene from CH_3OH over GaN was employed for photocatalytic methylation of simple alkanes and arenes. This is a green approach for the direct functionalization of sp^3 and sp^2 C—H bonds (Figure 11.20, top, right), which is challenging for conventional strategies. Density function theory (DFT) calculations suggested that the generated carbene species were thermodynamically favorable to be adsorbed on the nitrogen atom of the nitrogen-terminated c-plane of GaN, which is critical for the C—C bond formation (Liu, Qiu, et al., 2020).

Furthermore, the as-designed GaN with nonpolar surface was validated to be a suitable photocatalyst for light-driven C–C formation. For instance, pinacol coupling reaction has been successfully conducted by GaN-based photocatalysts under ambient temperatures, promising excellent yields and superior recyclability (Figure 11.20, bottom, right). This work shows a green and economic means for C–C coupling compared to the homogeneous photo-redox catalysis using rare and expensive Ir- and Ru-based catalysts (Liu, Tan, et al., 2020). These studies and discoveries showed the great potential of III-nitrides for solar-powered N≡N bond cleavage, C—H bond activation, C–C coupling reaction, and aromatization, which have remained extremely challenging in conventional thermocatalysis. This presents a novel strategy to help modern chemical industry and synthetic ammonia industry.

11.6 Summary and Prospects

In summary, several notable breakthroughs in artificial photosynthesis, including solar water splitting and CO_2RR, have been enabled by GaN-based APID over the recent years. These progresses have demonstrated the great potential/advantages of GaN-based nanostructures for addressing some of the key challenges of artificial photosynthesis, including (a) low optical absorption and/or insufficient redox potentials, (b) severe electron–hole recombination, (c) ineffective surface reactions, and (d) poor stability. It is especially noted that the GaN-based APIDs are mainly derived from the two common semiconductors of silicon and III-nitrides, which are widely available in industry. Thus, it can be technologically and economically viable for large-scale production of sustainable fuels and chemicals by use of sunlight, water, and CO_2. However, the research on GaN-catalyzed artificial photosynthesis is still in a relatively early stage. The precise control of photons/electrons behavior by the rational design of GaN-based APID is at the core of paving the pathway for achieving high-performance artificial photosynthesis. State-of-the-art techniques like MBE, metal–organic vapor chemical deposition, and magnetron sputtering offer viable approaches for highly controlled growth of high-quality III-nitrides at atomic scale, thus promising superior structural, optical, and electronic properties. Nevertheless, how to efficiently utilize the low-energy photons with long wavelengths, for example, infrared light, which cannot produce energetic charge carriers, remains a question in the area of solar CO_2 to fuel conversion. The combined utilization of the energetic charge carriers with the photoinduced thermal energy under concentrated illumination to drive chemical reactions may address this critical issue. In this context, tremendous efforts are needed for rationally constructing charge-thermal multifunctional III-nitride-based APID, because of the significance of the science involved and practical applications.

In addition to the semiconductor light absorber, the activity and selectivity of the GaN-based APID largely depend on cocatalysts, which affect the reaction barriers and pathways fundamentally. For solar water splitting, it is highly desirable to explore earth-abundant, efficient, and robust catalysts to replace the expensive Pt-, and/or Rh-derived metals, which also need to be coupled with III-nitrides via technically viable processes. In general, it is very difficult to produce desired products with high activity and selectivity from CO_2RR because of the inert nature of CO_2 molecules and the complex reaction networks compared to HER. To date, a wide variety of C1 products, for example, CO, H_2, HCOOH, and CH_3OH, have been reported to be produced as the dominant products with relatively low activity. The controlled synthesis of multi-carbon (C_{2+}) products from CO_2 and H_2O has remained a grand challenge because of the kinetically and thermodynamically unfavored C–C coupling process, especially in the absence of any electrical bias or any sacrificial agents. As such, the solar-to-fuel efficiencies and the activity and selectivity of desired compounds need to be significantly improved for commercial viability. The rational design of effective catalysts, coupled with well-controlled III-nitride semiconductor platform, may offer a viable path. It is particularly noted that compared to the conventional semiconductors, due to their unique surface properties, III-nitrides are able in activating the inert reactants like CO_2 with the aid of rational catalysts and/or by altering the types and concentrations of the dopants.

A deeper understanding of the mechanism at molecular level will provide fundamental guidance on the rational design of the device. With the assistance of advanced theoretical approaches, for example, molecular dynamics, first-principal calculations, as well as machine learning, in combination with state-of-the-art operando approaches, a fundamental understanding of the mechanism of artificial photosynthesis will be further developed to guide the optimization of GaN-based APID design. It is envisioned that next-generation APIDs built on III-nitrides provide a highly promising solution for producing sustainable fuels and chemicals without extra carbon emissions by using solar, carbon dioxide, water, and/or atmospheric nitrogen, thereby offering a potential path to switch the world from the fossil fuels age (FFA) to solar fuels age (SFA).

Acknowledgment

This book chapter consists of work that had been supported by the Emissions Reduction of Alberta (ERA), Department of Energy, and Natural Sciences and Engineering Research Council of Canada.

Conflict of Interest

Some IP related to this work has been licensed to NS Nanotech, Inc. and NX Fuels, Inc., which were co-founded by Z. Mi. The University of Michigan and Mi have a financial interest in these companies.

Additional Note

B. Zhou is currently a professor in the School of Mechanical Engineering at Shanghai Jiao Tong University in China.

References

Abidov, A., Allabergenov, B., Tursunkulov, O., He, L., Park, B.-H., Kim, H.-J., & Kim, S. (2014). Methanol artificial photosynthesis using iron doped TiO_2. *Journal of Biobased Materials and Bioenergy*, *8*(2), 165–169.

Aghaie, M., Rezaei, N., & Zendehboudi, S. (2018). A systematic review on CO_2 capture with ionic liquids: Current status and future prospects. *Renewable and Sustainable Energy Reviews*, *96*, 502–525.

AlOtaibi, B., Fan, S. Z., Wang, D. F., Ye, J. H., & Mi, Z. T. (2015). Wafer-level artificial photosynthesis for CO_2 reduction into CH_4 and CO using GaN nanowires. *ACS Catalysis*, *5*(9), 5342–5348.

AlOtaibi, B., Kong, X., Vanka, S., Woo, S. Y., Pofelski, A., Oudjedi, F., Fan, S., Kibria, M. G., Botton, G. A., Ji, W., Guo, H., & Mi, Z. (2016). Photochemical carbon dioxide reduction on Mg-doped Ga(In)N nanowire arrays under visible light irradiation. *ACS Energy Letters*, *1*(1), 246–252.

An, X., Li, S., Hao, X., Xie, Z., Du, X., Wang, Z., Hao, X., Abudula, A., & Guan, G. (2021). Common strategies for improving the performances of tin and bismuth-based catalysts in the electrocatalytic reduction of CO_2 to formic acid/formate. *Renewable and Sustainable Energy Reviews*, *143*, 110952.

Back, S., Yeom, M. S., & Jung, Y. (2015). Active sites of Au and Ag nanoparticle catalysts for CO_2 electroreduction to CO. *ACS Catalysis*, *5*(9), 5089–5096.

Bae, D., Seger, B., Vesborg, P. C., Hansen, O., & Chorkendorff, I. (2017). Strategies for stable water splitting via protected photoelectrodes. *Chemical Society Reviews*, *46*(7), 1933–1954.

Behrens, M., Studt, F., Kasatkin, I., Kuehl, S., Haevecker, M., Abild-Pedersen, F., Zander, S., Girgsdies, F., Kurr, P., Kniep, B.-L., Tovar, M., Fischer, R. W., Norskov, J. K., & Schloegl, R. (2012). The active site of methanol synthesis over $Cu/ZnO/Al_2O_3$ industrial catalysts. *Science*, *336*(6083), 893–897.

Bejtka, K., Zeng, J., Sacco, A., Castellino, M., Hernandez, S., Farkhondehfal, M. A., Savino, U., Ansaloni, S., Pirri, C. F., & Chiodoni, A. (2019). Chainlike mesoporous SnO_2 as a well-performing catalyst for electrochemical CO_2 reduction. *ACS Applied Energy Materials*, *2*(5), 3081–3091.

Berardi, S., Drouet, S., Francas, L., Gimbert-Surinach, C., Guttentag, M., Richmond, C., Stoll, T., & Llobet, A. (2014). Molecular artificial photosynthesis. *Chemical Society Reviews*, *43*(22), 7501–7519.

Bie, C., Zhu, B., Xu, F., Zhang, L., & Yu, J. (2019). In situ grown monolayer N-doped graphene on CdS hollow spheres with seamless contact for photocatalytic CO_2 reduction. *Advanced Materials*, *31*(42), 1902868.

Bongartz, D., Dore, L., Eichler, K., Grube, T., Heuser, B., Hombach, L. E., Robinius, M., Pischinger, S., Stolten, D., Walther, G., & Mitsos, A. (2018). Comparison of light-duty transportation fuels produced from renewable hydrogen and green carbon dioxide. *Applied Energy*, *231*, 757–767.

Bozzano, G., & Manenti, F. (2016). Efficient methanol synthesis: Perspectives, technologies and optimization strategies. *Progress in Energy and Combustion Science*, *56*, 71–105.

César, M., Ke, Y., Ji, W., Guo, H., & Mi, Z. (2011). Bandgap of $In_xGa_{1-x}N$: A first principles analysis. *Applied Physics Letters*, *98*(20), 202107.

Chen, Z., Fan, T., Zhang, Y. Q., Xiao, J., Gao, M., Duan, N., Zhang, J., Li, J., Liu, Q., Yi, X., & Luo, J. L. (2020). Wavy SnO_2 catalyzed simultaneous reinforcement of carbon dioxide adsorption and activation towards electrochemical conversion of CO_2 to HCOOH. *Applied Catalysis B: Environmental*, *261*, 118243.

Chen, F., Ji, X., & Lau, S. P. (2020). Recent progress in group III-nitride nanostructures: From materials to applications. *Materials Science & Engineering R: Reports*, *142*, 100578.

Cheng, W. H., Richter, M. H., May, M. M., Ohlmann, J., Lackner, D., Dimroth, F., Hannappel, T., Atwater, H. A., & Lewerenz, H. J. (2018). Monolithic photoelectrochemical device for direct water splitting with 19% efficiency. *ACS Energy Letters*, *3*(8), 1795–1800.

Chen, Z., Jaramillo, T. F., Deutsch, T. G., Kleiman-Schwarsctein, A., Forman, A. J., Gaillard, N., Garland, R., Takanabe, K., Heske, C., Sunkara, M., McFarland, E. W., Domen, K., Miller, E. L., Turner, J. A., & Dinh, H. N. (2010). Accelerating materials development for photoelectrochemical hydrogen production: standards for methods, definitions, and reporting protocols. *Journal of Materials Research*, *25*(1), 3–16.

Chowdhury, F. A., Trudeau, M. L., Guo, H., & Mi, Z. (2018). A photochemical diode artificial photosynthesis system for unassisted high efficiency overall pure water splitting. *Nature Communications*, *9*(1), 1707.

Chu, S., Fan, S. Z., Wang, Y. J., Rossouw, D., Wang, Y. C., Botton, G. A., & Mi, Z. T. (2016). Tunable syngas production from CO_2 and H_2O in an aqueous photoelectrochemical cell. *Angewandte Chemie International Edition*, *55*(46), 14260–14264.

Chu, S., Ou, P. F., Ghamari, P., Vanka, S., Zhou, B. W., Shih, I., Song, J., & Mi, Z. T. (2018). Photoelectrochemical CO_2 reduction into syngas with the metal/oxide interface. *Journal of the American Chemical Society*, *140*(25), 7869–7877.

Chu, S., Vanka, S., Wang, Y. C., Gim, J., Wang, Y. J., Ra, Y. H., Hovden, R., Guo, H., Shih, I., & Mi, Z. T. (2018). Solar water oxidation by an InGaN nanowire photoanode with a bandgap of 1.7 eV. *ACS Energy Letters*, *3*(2), 307–314.

Citak, F., Uslu, H., Batmaz, O., & Hos, S. (2020). Do renewable energy and natural gas consumption mitigate CO_2 emissions in the USA? New insights from NARDL approach. *Environmental Science and Pollution Research*, 63739–63750.

Corbin, D. R., Schwarz, S., & Sonnichsen, G. C. (1997). Methylamines synthesis: A review. *Catalysis Today*, *37*(2), 71–102.

Dias, P., Vilanova, A., Lopes, T., Andrade, L., & Mendes, A. (2016). Extremely stable bare hematite photoanode for solar water splitting. *Nano Energy*, *23*, 70–79.

Dimitrijevic, N. M., Vijayan, B. K., Poluektov, O. G., Rajh, T., Gray, K. A., He, H., & Zapol, P. (2011). Role of water and carbonates in photocatalytic transformation of CO_2 to CH_4 on titania. *Journal of the American Chemical Society*, *133*(11), 3964–3971.

Dong, Y., Li, K., Luo, W., Zhu, C., Guan, H., Wang, H., Wang, L., Deng, K., Zhou, H., Xie, H., Bai, Y., Li, Y., & Chen, Q. (2020). The role of surface termination in halide perovskites for efficient photocatalytic synthesis. *Angewandte Chemie International Edition*, *59*(31), 12931–12937.

Ertem, M. Z., Kharche, N., Batista, V. S., Hybertsen, M. S., Tully, J. C., & Muckerman, J. T. (2015). Photoinduced water oxidation at the aqueous GaN $(10\bar{1}0)$ interface: Deprotonation kinetics of the first proton-coupled electron-transfer step. *ACS Catalysis*, *5*(4), 2317–2323.

Fan, R., Dong, W., Fang, L., Zheng, F., Su, X., Zou, S., Huang, J., Wang, X., & Shen, M. (2015). Stable and efficient multi-crystalline n^+p silicon photocathode for H_2 production with pyramid-like surface nanostructure and thin Al_2O_3 protective layer. *Applied Physics Letters*, *106*(1), 013902.

Frischmann, P. D., Mahata, K., & Wuerthner, F. (2013). Powering the future of molecular artificial photosynthesis with light-harvesting metallosupramolecular dye assemblies. *Chemical Society Reviews*, *42*(4), 1847–1870.

Fujinuma, N., Ikoma, A., & Lofland, S. E. (2020). Highly efficient electrochemical CO_2 reduction reaction to CO with one-pot synthesized co-pyridine-derived catalyst incorporated in a nafion-based membrane electrode assembly. *Advanced Energy Materials*, *10*(39), 01645.

Fujishima, A., & Honda, K. (1972). Electrochemical photolysis of water at a semiconductor electrode. *Nature*, *238*(5358), 37–38.

Gao, X., Li, X., Tung, C., & Wu, L. (2021). Recent advances in quantum dots based artificial photosynthesis. *Chemistry*, *84*(1), 10–15.

Georgakilas, A., Dimakis, E., Tsagaraki, K., & Androulidaki, M. (2004). Promising results of plasma assisted MBE for optoelectronic applications. In M. S. Shur & A. Zukauskas (Eds.), *UV solid-state light emitters and detectors* (Vol. *144*, pp. 179–188). Springer.

Graf, M., Vonbun-Feldbauer, G. B., & Koper, M. T. M. (2021). Direct and broadband plasmonic charge transfer to enhance water oxidation on a gold electrode. *ACS Nano*, *15*(2), 3188–3200.

Gu, J., Hsu, C. S., Bai, L., Chen, H. M., & Hu, X. (2019). Atomically dispersed Fe^{3+} sites catalyze efficient CO_2 electroreduction to CO. *Science*, *364*(6445), 1091–1094.

Guan, X., Chowdhury, F. A., Pant, N., Guo, L., Vayssieres, L., & Mi, Z. (2018). Efficient unassisted overall photocatalytic seawater splitting on GaN-based nanowire arrays. *Journal of Physical Chemistry C*, *122*(25), 13797–13802.

Guan, X. J., Chowdhury, F. A., Wang, Y. J., Pant, N., Vanka, S., Trudeau, M. L., Guo, L. J., Vayssieres, L., & Mi, Z. T. (2018). Making of an industry-friendly artificial photosynthesis device. *ACS Energy Letters*, *3*(9), 2230–2231.

Habisreutinger, S. N., Schmidt-Mende, L., & Stolarczyk, J. K. (2013). Photocatalytic reduction of CO_2 on TiO_2 and other semiconductors. *Angewandte Chemie International Edition*, *52*(29), 7372–7408.

He, F., Meng, A., Cheng, B., Ho, W., & Yu, J. (2020). Enhanced photocatalytic H_2 production activity of WO_3/TiO_2 step-scheme heterojunction by graphene modification. *Chinese Journal of Catalysis*, *41*(1), 9–20.

Hu, S., Shaner, M. R., Beardslee, J. A., Lichterman, M., Brunschwig, B. S., & Lewis, N. S. (2014). Amorphous TiO_2 coatings stabilize Si, GaAs, and GaP photoanodes for efficient water oxidation. *Science*, *344*(6187), 1005–1009.

Izyumskaya, N., Avrutin, V., Ding, K., Ozgur, U., Morkoc, H., & Fujioka, H. (2019). Emergence of high quality sputtered III-nitride semiconductors and devices. *Semiconductor Science and Technology*, *34*(9), 093003.

Jain, S. C., Willander, M., Narayan, J., & Van Overstraeten, R. (2000). III-Nitrides: Growth, characterization, and properties. *Journal of Applied Physics*, *87*(3), 965–1006.

Jin, J., Yu, J., Guo, D., Cui, C., & Ho, W. (2015). A hierarchical Z-scheme CdS-WO_3 photocatalyst with enhanced CO_2 reduction activity. *Small*, *11*(39), 5262–5271.

Khaselev, O., & Turner, J. A. (1998). A monolithic photovoltaic-photoelectrochemical device for hydrogen production via water splitting. *Science*, *280*(5362), 425–427.

Kibria, M. G., Chowdhury, F. A., Zhao, S., AlOtaibi, B., Trudeau, M. L., Guo, H., & Mi, Z. (2015). Visible light-driven efficient overall water splitting using *p*-type metal-nitride nanowire arrays. *Nature Communications*, *6*, 6797.

Kibria, M. G., Chowdhury, F. A., Zhao, S., Trudeau, M. L., Guo, H., & Mi, Z. (2015). Defect-engineered GaN:Mg nanowire arrays for overall water splitting under violet light. *Applied Physics Letters*, *106*(11), 113105.

Kibria, M. G., & Mi, Z. (2016). Artificial photosynthesis using metal/nonmetal-nitride semiconductors: Current status, prospects, and challenges. *Journal of Materials Chemistry A*, *4*(8), 2801–2820.

Kibria, M. G., Nguyen, H. P. T., Cui, K., Zhao, S., Liu, D., Guo, H., Trudeau, M. L., Paradis, S., Hakima, A.-R., & Mi, Z. (2013). One-step overall water splitting under visible light using multiband InGaN/GaN nanowire heterostructures. *ACS Nano*, *7*(9), 7886–7893.

Kibria, M. G., Qiao, R. M., Yang, W. L., Boukahil, I., Kong, X. H., Chowdhury, F. A., Trudeau, M. L., Ji, W., Guo, H., Himpsel, F. J., Vayssieres, L., & Mi, Z. T. (2016). Atomic-scale origin of long-term stability and high performance of p-GaN nanowire arrays for photocatalytic overall pure water splitting. *Advanced Materials*, *28*(38), 8388–8397.

Kibria, M. G., Zhao, S., Chowdhury, F. A., Wang, Q., Nguyen, H. P. T., Trudeau, M. L., Guo, H., & Mi, Z. (2014). Tuning the surface Fermi level on p-type gallium nitride nanowires for efficient overall water splitting. *Nature Communications*, *5*(1), 3825.

Kim, W., Edri, E., & Frei, H. (2016). Hierarchical inorganic assemblies for artificial photosynthesis. *Accounts of Chemical Research*, *49*(9), 1634–1645.

Kim, D., Sakimoto, K. K., Hong, D. C., & Yang, P. D. (2015). Artificial photosynthesis for sustainable fuel and chemical production. *Angewandte Chemie International Edition*, *54*(11), 3259–3266.

Kuang, Y., Jia, Q., Ma, G., Hisatomi, T., Minegishi, T., Nishiyama, H., Nakabayashi, M., Shibata, N., Yamada, T., & Kudo, A. (2016). Ultrastable low-bias water splitting photoanodes via photocorrosion inhibition and in situ catalyst regeneration. *Nature Energy*, *2*(1), 1–9.

Kusior, A., Synowiec, M., Zakrzewska, K., & Radecka, M. (2019). Surface-controlled photocatalysis and chemical sensing of TiO_2, alpha-Fe_2O_3, and Cu_2O nanocrystals. *Crystals*, *9*(3), 163.

Lee, J. S., Won, D. I., Jung, W. J., Son, H. J., Pac, C., & Kang, S. O. (2017). Widely controllable syngas production by a dye-sensitized TiO_2 hybrid system with Re-I and Co-III catalysts under visible-light irradiation. *Angewandte Chemie International Edition*, *56*(4), 976–980.

Lei, Q., Yang, S., Ding, D., Tan, J., Liu, J., & Chen, R. (2021). Local-interaction-field-coupled semiconductor photocatalysis: Recent progress and future challenges. *Journal of Materials Chemistry A*, *9*(5), 2491–2525.

Li, L., Fan, S., Mu, X., Mi, Z., & Li, C.-J. (2014). Photoinduced conversion of methane into benzene over GaN nanowires. *Journal of the American Chemical Society*, *136*(22), 7793–7796.

Li, C. W., & Kanan, M. W. (2012). CO_2 reduction at low overpotential on Cu electrodes resulting from the reduction of thick Cu_2O films. *Journal of the American Chemical Society*, *134*(17), 7231–7234.

Li, L., Mu, X., Liu, W., Kong, X., Fan, S., Mi, Z., & Li, C. J. (2014). Thermal non-oxidative aromatization of light alkanes catalyzed by gallium nitride. *Angewandte Chemie International Edition, 53*(51), 14106–14109.

Li, L., Mu, X. Y., Liu, W. B., Mi, Z. T., & Li, C. J. (2015). Simple and efficient system for combined solar energy harvesting and reversible hydrogen storage. *Journal of the American Chemical Society, 137*(24), 7576–7579.

Li, L., Wang, Y., Vanka, S., Mu, X., Mi, Z., & Li, C.-J. (2017). Nitrogen photofixation over III-nitride nanowires assisted by ruthenium clusters of low atomicity. *Angewandte Chemie International Edition, 56*(30), 8701–8705.

Li, Z., Wu, R., Zhao, L., Li, P., Wei, X., Wang, J., Chen, J. S., & Zhang, T. (2021). Metal-support interactions in designing noble metal-based catalysts for electrochemical CO_2 reduction: Recent advances and future perspectives. *Nano Research*. https://doi.org/10.1007/s12274-021-3363-6

Li, M., Zhang, L., Fan, X., Zhou, Y., Wu, M., & Shi, J. (2015). Highly selective CO_2 photoreduction to CO over g-C_3N_4/Bi_2WO_6 composites under visible light. *Journal of Materials Chemistry A, 3*(9), 5189–5196.

Li, Z., Zhang, L., Liu, Y., Shao, C., Gao, Y., Fan, F., Wang, J., Li, J., Yan, J., Li, R., & Li, C. (2019). Surface polarity-induced spatial charge separation boosts photocatalytic overall water splitting on GaN nanorod arrays. *Angewandte Chemie International Edition, 59*(2), 935–942.

Li, L., Zhao, Z. J., Hu, C., Yang, P., Yuan, X., Wang, Y., Zhang, L., Moskaleva, L., & Gong, J. (2020). Tuning oxygen vacancies of oxides to promote electrocatalytic reduction of carbon dioxide. *ACS Energy Letters, 5*(2), 552–558.

Liu, C., Colon, B. C., Ziesack, M., Silver, P. A., & Nocera, D. G. (2016). Water splitting-biosynthetic system with CO_2 reduction efficiencies exceeding photosynthesis. *Science, 352*(6290), 1210–1213.

Liu, X., Inagaki, S., & Gong, J. (2016). Heterogeneous molecular systems for photocatalytic CO_2 reduction with water oxidation. *Angewandte Chemie International Edition, 55*(48), 14924–14950.

Liu, Y., Murata, K., Inaba, M., Takahara, I., & Okabe, K. (2013). Mixed alcohols synthesis from syngas over Cs- and Ni-modified Cu/CeO_2 catalysts. *Fuel, 104*, 62–69.

Liu, M., Qiu, Z., Tan, L., Rashid, R. T., Chu, S., Cen, Y., Luo, Z., Khaliullin, R. Z., Mi, Z., & Li, C. J. (2020). Photocatalytic methylation of non-activated sp^3 and sp^2 CH bonds using methanol on GaN. *ACS Catalysis, 10*(11), 6248–6253.

Liu, M., Tan, L., Rashid, R. T., Cen, Y., Cheng, S., Botton, G., Mi, Z., & Li, C. J. (2020). GaN nanowires as a reusable photoredox catalyst for radical coupling of carbonyl under blacklight irradiation. *Chemical Science, 11*(30), 7864–7870.

Liu, M., Tan, L., Zhou, B., Li, L., Mi, Z., & Li, C.-J. (2021). Group-III nitrides catalyzed transformations of organic molecules. *Chem, 7*(1), 64–92.

Liu, M., Wang, Y., Kong, X., Rashid, R. T., Chu, S., Li, C. C., Hearne, Z., Guo, H., Mi, Z., & Li, C. J. (2019). Direct catalytic methanol-to-ethanol photo-conversion via methyl carbene. *Chem, 5*(4), 858–867.

Liu, M., Wang, Y., Kong, X., Tan, L., Li, L., Cheng, S., Botton, G., Guo, H., Mi, Z., & Li, C.-J. (2019). Efficient nitrogen fixation catalyzed by gallium nitride nanowire using nitrogen and water. *iScience, 17*, 208–216.

Liu, X., Wang, M., Zhou, C., Zhou, W., Cheng, K., Kang, J., Zhang, Q., Deng, W., & Wang, Y. (2018). Selective transformation of carbon dioxide into lower olefins with a bifunctional catalyst composed of $ZnGa_2O_4$ and SAPO-34. *Chemical Communications, 54*(2), 140–143.

Long, R., Li, Y., Liu, Y., Chen, S., Zheng, X., Gao, C., He, C., Chen, N., Qi, Z., Song, L., Jiang, J., Zhu, J., & Xiong, Y. (2017). Isolation of Cu atoms in Pd lattice: Forming highly selective sites for photocatalytic conversion of CO_2 to CH_4. *Journal of the American Chemical Society, 139*(12), 4486–4492.

Low, J., Dai, B., Tong, T., Jiang, C., & Yu, J. (2019). In situ irradiated X-ray photoelectron spectroscopy investigation on a direct Z-scheme TiO_2/CdS composite film photocatalyst. *Advanced Materials, 31*(6), 1802981.

Luk, H. T., Mondelli, C., Ferre, D. C., Stewart, J. A., & Perez-Ramirez, J. (2017). Status and prospects in higher alcohols synthesis from syngas. *Chemical Society Reviews, 46*(5), 1358–1426.

Luo, Z., Zhou, M., & Wang, X. (2018). Cobalt-based cubane molecular co-catalysts for photocatalytic water oxidation by polymeric carbon nitrides. *Applied Catalysis B: Environmental, 238*, 664–671.

Maeda, K. (2013). Z-scheme water splitting using two different semiconductor photocatalysts. *ACS Catalysis, 3*(7), 1486–1503.

Malathi, A., Madhavan, J., Ashokkumar, M., & Arunachalam, P. (2018). A review on $BiVO_4$ photocatalyst: Activity enhancement methods for solar photocatalytic applications. *Applied Catalysis A: General, 555*, 47–74.

May, M. M., Lewerenz, H.-J., Lackner, D., Dimroth, F., & Hannappel, T. (2015). Efficient direct solar-to-hydrogen conversion by in situ interface transformation of a tandem structure. *Nature Communications, 6*(1), 1–7.

Mohadesi, M., Aghel, B., Maleki, M., & Ansari, A. (2020). The use of KOH/clinoptilolite catalyst in pilot of microreactor for biodiesel production from waste cooking oil. *Fuel, 263*, 16659.

Morkoc, H. (2001). III-Nitride semiconductor growth by MBE: Recent issues. *Journal of Materials Science: Materials in Electronics*, *12*(12), 677–695.

Navid, I. A., Vanka, S., Awni, R. A., Xiao, Y., Song, Z., Yan, Y., & Mi, Z. (2021). On the design and performance of InGaN/Si double-junction photocathodes. *Applied Physics Letters*, *118*(24), 243906.

Neatu, S., Antonio Macia-Agullo, J., Concepcion, P., & Garcia, H. (2014). Gold–copper nanoalloys supported on TiO_2 as photocatalysts for CO_2 reduction by water. *Journal of the American Chemical Society*, *136*(45), 15969–15976.

Nithyanandam, R., Mun, Y. K., Fong, T. S., Siew, T. C., Yee, O. S., & Ismail, N. (2018). Review on production of benzene from petroleum associated gas by dehydroaromatization, partial oxidation of methane and methanol-to-aromatics processes. *Journal of Engineering Science and Technology*, *13*(12), 4290–4309.

Nocera, D. G. (2012). The artificial leaf. *Accounts of Chemical Research*, *45*(5), 767–776.

Okumura, H., Malinverni, M., Martin, D., & Grandjean, N. (2016). High p-type GaN for advanced optoelectronic devices. In *2016 IEEE photonics conference* (pp. 661–662). IEEE.

Olsbye, U., Svelle, S., Bjorgen, M., Beato, P., Janssens, T. V. W., Joensen, F., Bordiga, S., & Lillerud, K. P. (2012). Conversion of methanol to hydrocarbons: How zeolite cavity and pore size controls product selectivity. *Angewandte Chemie International Edition*, *51*(24), 5810–5831.

Ong, W.-J., Tan, L.-L., Ng, Y. H., Yong, S.-T., & Chai, S.-P. (2016). Graphitic carbon nitride (g-C_3N_4)-based photocatalysts for artificial photosynthesis and environmental remediation: Are we a step closer to achieving sustainability? *Chemical Reviews*, *116*(12), 7159–7329.

Pu, Y. C., Kibria, M. G., Mi, Z. T., & Zhang, J. Z. (2015). Ultrafast exciton dynamics in InGaN/GaN and Rh/Cr_2O_3 nanoparticle-decorated InGaN/GaN nanowires. *Journal of Physical Chemistry Letters*, *6*(13), 2649–2656.

Reece, S. Y., Hamel, J. A., Sung, K., Jarvi, T. D., Esswein, A. J., Pijpers, J. J., & Nocera, D. G. (2011). Wireless solar water splitting using silicon-based semiconductors and earth-abundant catalysts. *Science*, *334*(6056), 645–648.

Roul, B., Kumar, M., Rajpalke, M. K., Bhat, T. N., & Krupanidhi, S. B. (2015). Binary group III-nitride based heterostructures: Band offsets and transport properties. *Journal of Physics D: Applied Physics*, *48*(42), 423001.

She, H., Zhou, H., Li, L., Zhao, Z., Jiang, M., Huang, J., Wang, L., & Wang, Q. (2019). Construction of a two-dimensional composite derived from TiO_2 and SnS_2 for enhanced photocatalytic reduction of CO_2 into CH_4. *ACS Sustainable Chemistry & Engineering*, *7*(1), 650–659.

Smith, P. T., Nichols, E. M., Cao, Z., & Chang, C. J. (2020). Hybrid catalysts for artificial photosynthesis: Merging approaches from molecular, materials, and biological catalysis. *Accounts of Chemical Research*, *53*(3), 575–587.

Sun, S., Watanabe, M., Wu, J., An, Q., & Ishihara, T. (2018). Ultrathin $WO_3 \cdot 0.33H_2O$ nanotubes for CO_2 photoreduction to acetate with high selectivity. *Journal of the American Chemical Society*, *140*(20), 6474–6482.

Tapia, J. F. D., Lee, J.-Y., Ooi, R. E. H., Foo, D. C. Y., & Tan, R. R. (2018). A review of optimization and decision-making models for the planning of CO_2 capture, utilization and storage (CCUS) systems. *Sustainable Production and Consumption*, *13*, 1–15.

Tian, L., Xin, Q., Zhao, C., Xie, G., Akram, M. Z., Wang, W., Ma, R., Jia, X., Guo, B., & Gong, J. R. (2021). Nanoarray structures for artificial photosynthesis. *Small*. https://doi.org/10.1002/smll.202006530

Tran, P. D., Wong, L. H., Barber, J., & Loo, J. S. C. (2012). Recent advances in hybrid photocatalysts for solar fuel production. *Energy & Environmental Science*, *5*(3), 5902–5918.

Tu, W., Zhou, Y., Liu, Q., Yan, S., Bao, S., Wang, X., Xiao, M., & Zou, Z. (2013). An in situ simultaneous reduction-hydrolysis technique for fabrication of TiO_2-graphene 2D sandwich-like hybrid nanosheets: Graphene-promoted selectivity of photocatalytic-driven hydrogenation and coupling of CO_2 into methane and ethane. *Advanced Functional Materials*, *23*(14), 1743–1749.

Tu, W., Zhou, Y., & Zou, Z. (2014). Photocatalytic conversion of CO_2 into renewable hydrocarbon fuels: State-of-the-art accomplishment, challenges, and prospects. *Advanced Materials*, *26*(27), 4607–4626.

Vanka, S., Arca, E., Cheng, S. B., Sun, K., Botton, G. A., Teeter, G., & Mi, Z. T. (2018). High efficiency Si photocathode protected by multifunctional GaN nanostructures. *Nano Letters*, *18*(10), 6530–6537.

Vanka, S., Sun, K., Zeng, G., Pham, T. A., Toma, F. M., Ogitsu, T., & Mi, Z. (2019). Long-term stability studies of a semiconductor photoelectrode in three-electrode configuration. *Journal of Materials Chemistry A*, *7*(48), 27612–27619.

Vanka, S., Zhou, B., Awni, R. A., Song, Z., Chowdhury, F. A., Liu, X., Hajibabaei, H., Shi, W., Xiao, Y., & Navid, I. A. (2020). InGaN/Si double-junction photocathode for unassisted solar water splitting. *ACS Energy Letters*, *5*(12), 3741–3751.

Verlage, E., Hu, S., Liu, R., Jones, R. J., Sun, K., Xiang, C., Lewis, N. S., & Atwater, H. A. (2015). A monolithically integrated, intrinsically safe, 10% efficient, solar-driven water-splitting system based on active, stable earth-abundant electrocatalysts in conjunction with tandem III–V light absorbers protected by amorphous TiO_2 films. *Energy & Environmental Science*, *8*(11), 3166–3172.

Vitousek, P. M., Aber, J. D., Howarth, R. W., Likens, G. E., Matson, P. A., Schindler, D. W., Schlesinger, W. H., & Tilman, D. (1997). Human alteration of the global nitrogen cycle: Sources and consequences. *Ecological Applications*, *7*(3), 737–750.

Vogt, E. T. C., Whiting, G. T., Chowdhury, A. D., & Weckhuysen, B. M. (2015). Zeolites and zeotypes for oil and gas conversion. In F. C. Jentoft (Ed.), *Advances in catalysis* (Vol. *58*, pp. 143–314). Elsevier.

Walter, M. G., Warren, E. L., McKone, J. R., Boettcher, S. W., Mi, Q., Santori, E. A., & Lewis, N. S. (2010). Solar water splitting cells. *Chemical Reviews*, *110*(11), 6446–6473.

Wang, Y., Fan, S., AlOtaibi, B., Wang, Y., Li, L., & Mi, Z. (2016). A monolithically integrated gallium nitride nanowire/silicon solar cell photocathode for selective carbon dioxide reduction to methane. *Chemistry – A European Journal*, *22*(26), 8809–8813.

Wang, J., Kattel, S., Hawxhurst, C. J., Lee, J. H., Tackett, B. M., Chang, K., Rui, N., Liu, C. J., & Chen, J. G. (2019). Enhancing activity and reducing cost for electrochemical reduction of CO_2 by supporting palladium on metal carbides. *Angewandte Chemie International Edition*, *58*(19), 6271–6275.

Wang, X., Li, K., He, J., Yang, J., Dong, F., Mai, W., & Zhu, M. (2020). Defect in reduced graphene oxide tailored selectivity of photocatalytic CO_2 reduction on Cs_4PbBr_6 pervoskite hole-in-microdisk structure. *Nano Energy*, *78*, 105388.

Wang, J., Pedroza, L. S., Poissier, A., & Fernandez-Serra, M. V. (2012). Water dissociation at the GaN($10\bar{1}0$) surface: Structure, dynamics and surface acidity. *Journal of Physical Chemistry C*, *116*(27), 14382–14389.

Wang, D. F., Pierre, A., Kibria, M. G., Cui, K., Han, X. G., Bevan, K. H., Guo, H., Paradis, S., Hakima, A. R., & Mi, Z. T. (2011). Wafer-level photocatalytic water splitting on GaN nanowire arrays grown by molecular beam epitaxy. *Nano Letters*, *11*(6), 2353–2357.

Wang, Y. J., Schwartz, J., Gim, J., Hovden, R., & Mi, Z. T. (2019). Stable unassisted solar water splitting on semiconductor photocathodes protected by multifunctional GaN nanostructures. *ACS Energy Letters*, *4*(7), 1541–1548.

Wang, Y., Vanka, S., Gim, J., Wu, Y., Fan, R., Zhang, Y., Shi, J., Shen, M., Hovden, R., & Mi, Z. (2019). An $In_{0.42}Ga_{0.58}N$ tunnel junction nanowire photocathode monolithically integrated on a nonplanar Si wafer. *Nano Energy*, *57*, 405–413.

Wang, Y., Wu, Y., Sun, K., & Mi, Z. (2019). A quadruple-band metal–nitride nanowire artificial photosynthesis system for high efficiency photocatalytic overall solar water splitting. *Materials Horizons*, *6*(7), 1454–1462.

Wen, F., & Li, C. (2013). Hybrid artificial photosynthetic systems comprising semiconductors as light harvesters and biomimetic complexes as molecular cocatalysts. *Accounts of Chemical Research*, *46*(11), 2355–2364.

Wu, Z., Wu, H., Cai, W., Wen, Z., Jia, B., Wang, L., Jin, W., & Ma, T. (2021). Engineering bismuth–tin interface in bimetallic aerogel with a 3D porous structure for highly selective electrocatalytic CO_2 reduction to HCOOH. *Angewandte Chemie International Edition*, *60*(22), 12554–12559.

Wu, L., Zhang, K., & Zhao, T. (2021). Forecasting natural gas production and consumption using grey model with latent information function: The cases of China and USA. *Scientia Iranica*, *28*(1), 386–394.

Xu, M., Li, D., Sun, K., Jiao, L., Xie, C., Ding, C., & Jiang, H.-L. (2021). Interfacial microenvironment modulation boosting electron transfer between metal nanoparticles and MOFs for enhanced photocatalysis. *Angewandte Chemie International Edition*, *60*(30), 16372–16376.

Xu, Y., Li, A., Yao, T., Ma, C., Zhang, X., Shah, J. H., & Han, H. (2017). Strategies for efficient charge separation and transfer in artificial photosynthesis of solar fuels. *ChemSusChem*, *10*(22), 4277–4305.

Xu, Y., Wang, S., Yang, J., Han, B., Nie, R., Wang, J., Wang, J., & Jing, H. (2018). In-situ grown nanocrystal TiO_2 on 2D Ti_3C_2 nanosheets for artificial photosynthesis of chemical fuels. *Nano Energy*, *51*, 442–450.

Xu, H., You, S., Lang, Z., Sun, Y., Sun, C., Zhou, J., Wang, X., Kang, Z., & Su, Z. (2020). Highly efficient photoreduction of low-concentration CO_2 to syngas by using a polyoxometalates/Ru-II composite. *Chemistry – A European Journal*, *26*(12), 2735–2740.

Yan, J., & Zhang, Z. (2019). Carbon capture, utilization and storage (CCUS). *Applied Energy*, *235*, 1289–1299.

Yang, Y., White, M. G., & Liu, P. (2012). Theoretical study of methanol synthesis from CO_2 hydrogenation on metal-doped Cu(111) surfaces. *Journal of Physical Chemistry C*, *116*(1), 248–256.

Yang, H., Yang, D., & Wang, X. (2020). POM-incorporated CoO nanowires for enhanced photocatalytic syngas production from CO_2. *Angewandte Chemie International Edition*, *59*(36), 15527–15531.

Yang, M.-Q., Zhang, N., Pagliaro, M., & Xu, Y.-J. (2014). Artificial photosynthesis over graphene-semiconductor composites. Are we getting better? *Chemical Society Reviews*, *43*(24), 8240–8254.

Yoshikawa, A., Che, S., Ishitani, Y., & Wang, X. (2009). Advances in InN epitaxy and its material control by MBE towards novel InN-based QWs. *Journal of Crystal Growth*, *311*(7), 2073–2079.

Young, J. L., Steiner, M. A., Döscher, H., France, R. M., Turner, J. A., & Deutsch, T. G. (2017). Direct solar-to-hydrogen conversion via inverted metamorphic multi-junction semiconductor architectures. *Nature Energy*, *2*(4), 1–8.

Zahran, Z. N., Tsubonouchi, Y., Mohamed, E. A., & Yagi, M. (2019). Recent advances in the development of molecular catalyst-based anodes for water oxidation toward artificial photosynthesis. *ChemSusChem*, *12*(9), 1775–1793.

Zeng, G., Pham, T. A., Vanka, S., Liu, G., Song, C., Cooper, J. K., Mi, Z., Ogitsu, T., & Toma, F. M. (2021). Development of a photoelectrochemically self-improving Si/GaN photocathode for efficient and durable H_2 production. *Nature Materials*, *20*(8), 1130–1135.

Zhang, C., Jun, K.-W., Gao, R., Kwak, G., & Kang, S. C. (2016). Efficient utilization of associated natural gas in a modular gas-to-liquids process: Technical and economic analysis. *Fuel*, *176*, 32–39.

Zhang, T., & Lin, W. B. (2014). Metal-organic frameworks for artificial photosynthesis and photocatalysis. *Chemical Society Reviews*, *43*(16), 5982–5993.

Zhang, G., Liu, B., Wang, T., & Gong, J. (2021). Arificial leaves for solar fuels. *Chinese Journal of Chemistry*, *39*(6), 1450–1458.

Zhang, B., & Sun, L. (2019). Artificial photosynthesis: Opportunities and challenges of molecular catalysts. *Chemical Society Reviews*, *48*(7), 2216–2264.

Zhao, Y., Chen, G., Bian, T., Zhou, C., Waterhouse, G. I. N., Wu, L. Z., Tung, C. H., Smith, L. J., O'Hare, D., & Zhang, T. (2015). Defect-rich ultrathin ZnAl-layered double hydroxide nanosheets for efficient photoreduction of CO_2 to CO with water. *Advanced Materials*, *27*(47), 7824–7831.

Zhao, X., Guan, J., Li, J., Li, X., Wang, H., Huo, P., & Yan, Y. (2021). CeO_2/3D g-C_3N_4 heterojunction deposited with Pt cocatalyst for enhanced photocatalytic CO_2 reduction. *Applied Surface Science*, *537*, 147891.

Zhao, S. R., & Mi, Z. T. (2017). Recent advances on *p*-type III-nitride nanowires by molecular beam epitaxy. *Crystals*, *7*(9), 17.

Zheng, J., & Shu, D. (2020). Regulation of surface properties of photocatalysis material TiO_2 by strain engineering. *Journal of Semiconductors*, *41*(9), 091703.

Zhou, B., Kong, X., Vanka, S., Cheng, S., Pant, N., Chu, S., Ghamari, P., Wang, Y., Botton, G., Cuo, H., & Mi, Z. (2019). A GaN:Sn nanoarchitecture integrated on a silicon platform for converting CO_2 to HCOOH by photoelectrocatalysis. *Energy & Environmental Science*, *12*(9), 2842–2848.

Zhou, B. W., Kong, X. H., Vanka, S., Chu, S., Ghamari, P., Wang, Y. C., Pant, N., Shih, I. S., Guo, H., & Mi, Z. T. (2018). Gallium nitride nanowire as a linker of molybdenum sulfides and silicon for photoelectrocatalytic water splitting. *Nature Communications*, *9*, 3856.

Zhou, B., Ou, P., Pant, N., Cheng, S., Vanka, S., Chu, S., Rashid, R. T., Botton, G., Song, J., & Mi, Z. (2020). Highly efficient binary copper–iron catalyst for photoelectrochemical carbon dioxide reduction toward methane. *Proceedings of the National Academy of Sciences of the United States of America*, *117*(3), 1330–1338.

Zhu, W., Michalsky, R., Metin, O., Lv, H., Guo, S., Wright, C. J., Sun, X., Peterson, A. A., & Sun, S. (2013). Monodisperse Au nanoparticles for selective electrocatalytic reduction of CO_2 to CO. *Journal of the American Chemical Society*, *135*(45), 16833–16836.

Zhu, W., Zhang, Y. J., Zhang, H., Lv, H., Li, Q., Michalsky, R., Peterson, A. A., & Sun, S. (2014). Active and selective conversion of CO_2 to CO on ultrathin Au nanowires. *Journal of the American Chemical Society*, *136*(46), 16132–16135.

Zong, X., Han, J., Ma, G., Yan, H., Wu, G., & Li, C. (2011). Photocatalytic H_2 evolution on CdS loaded with WS_2 as cocatalyst under visible light irradiation. *Journal of Physical Chemistry C*, *115*(24), 12202–12208.

Zong, X., Yan, H., Wu, G., Ma, G., Wen, F., Wang, L., & Li, C. (2008). Enhancement of photocatalytic H_2 evolution on CdS by loading MOS_2 as cocatalyst under visible light irradiation. *Journal of the American Chemical Society*, *130*(23), 7176–7177.

12

Low-Dimensional Materials for Direct Fuel Generation Assisted by Sunlight

Muhammad Shuaib Khan and Shaohua Shen

International Research Center for Renewable Energy (IRCRE), State Key Laboratory of Multiphase Flow in Power Engineering (MPFE), School of Energy and Power Engineering, Xi'an Jiaotong University, Xi'an, P. R. China

12.1 Introduction

The concomitant rise in population and life standards has established a gradual increase in energy consumption and environmental problems. More than half of world's population is urban, which is responsible for the emission of 75% of all carbon dioxide from energy usage (Low et al., 2017). Based on a careful analysis, it was projected that the global energy demand would become at least double by 2050 (Gong et al., 2019). Fossil fuels that cover up to 90% of global energy demand are mainly responsible for the emission of greenhouse gases, including carbon dioxide. Their consumption causes detrimental effects such as global warming, market uncertainty, and foreign oil dependency (Li et al., 2018; Zhang & Sun, 2019). Therefore, there is an urgent need to replace the current fossil fuels-based energy system by a system with energy derived from sustainable energy sources. The solar radiation reaches the Earth's surface at a rate of 1.3×10^5 TW, which is significantly high compared to the current power consumption of 19 TW (Armaroli &Balzani, 2007; Hisatomi et al., 2014). Due to the abundance of sunlight, the use of solar energy to produce hydrogen and useful hydrocarbons via water splitting and CO_2 reduction, respectively, is considered a feasible strategy due to its potential benefits to energy systems and environmental sustainability (Takata & Domen, 2019).

The pioneering work reported by Honda and Fujishima utilized titanium dioxide (TiO_2, a semiconductor) anode coupled with Pt cathode for photoelectrochemical water splitting, which led to the development of many photocatalytic/photoelectrochemical water splitting and CO_2 reduction systems (Fujishima & Honda, 1972). Semiconductors are characterized by an energy gap between valence and conduction bands (VB and CB, respectively) with an electrical conductivity lying between that of a conductor and an insulator. Semiconductors are used in a wide range of applications, from electronic devices to photocatalysis. When a semiconductor photocatalyst is irradiated with light, the photons with energy equal to or higher than the bandgap of semiconductor excite electrons from VB to CB leaving behind holes in VB, as shown in Figure 12.1a. The resultant photogenerated charge carriers migrate to the photocatalyst surface to take part in the redox reactions. For successful redox reactions to produce fuels, the bandgap of the semiconductor photocatalyst must straddle the redox potential of water splitting/CO_2 reduction. The redox reaction is initiated if photoexcited charge carrier injection into reactants is thermodynamically favorable. Figure 12.1b express the n-type semiconductor-based photoanode acting as a working electrode irradiated by a light source to produce electron–hole pairs. In this case, the electrons travel to the external circuit via the back contact at the anode and then to the counter electrode, usually made of a noble metal or a carbon material, and reduce water/CO_2 to produce H_2/hydrocarbon fuels. The photogenerated holes at the VB of the photocatalyst carry out the oxidation reactions (Hisatomi et al., 2014).

TiO_2 has been extensively investigated for photocatalytic water splitting; however, it possesses a wide bandgap of 3.2 eV, and only photons of ultraviolet (UV) portion of light can produce charge carriers active for redox reactions. The UV irradiations only account for 4% of the solar spectrum. To perform photocatalytic water splitting, the bottom of the CB must be located at a more negative potential than the reduction potential of water (H^+/H_2 couple, 0 V vs. NHE at pH = 0), while the top of VB must be below the oxidation potential of H_2O (O_2/H_2O couple, 1.23 V vs. NHE at pH = 0). Therefore, the minimum bandgap energy of the photocatalyst should be 1.23 eV to perform water splitting. However, the energy losses

Conversion of Water and CO_2 to Fuels using Solar Energy: Science, Technology and Materials, First Edition.
Edited by Oomman K. Varghese and Flavio L. Souza.

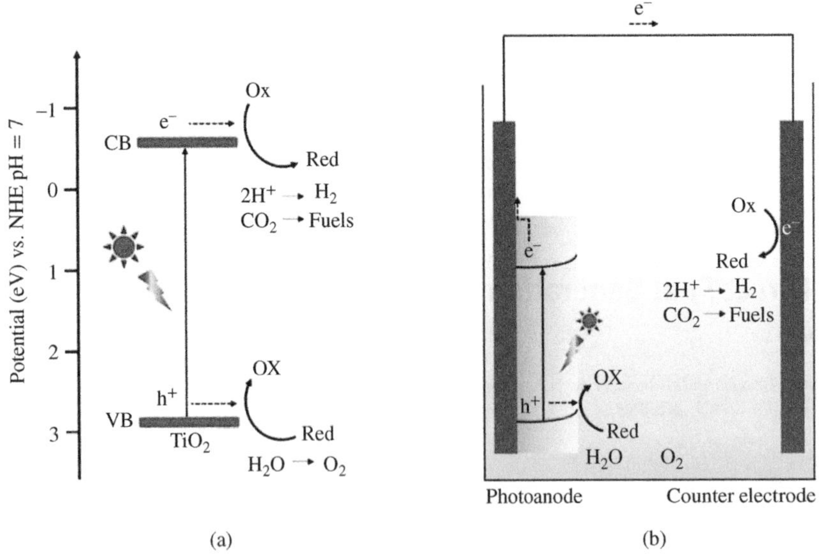

Figure 12.1 Schematic diagram of (a) Photocatalytic water splitting and CO_2 reduction based on one-step excitation; (b) Photoelectrochemical water splitting and CO_2 conversion using photoanode.

and kinetic overpotentials demand a bandgap in the range of 1.5–2.5 eV, and such photocatalysts can harvest visible light photons, which account for 44% of the solar spectrum. In addition to the basic criteria of appropriate bandgap and band edge positions for water splitting and CO_2 reduction, an efficient photocatalyst should demonstrate features including (a) high charge carrier mobility; (b) minimum charge carrier recombination rate; and (c) abundance of active sites enabling quick charge carrier exchange (Li, Yu, et al., 2015; Lin et al., 2020).

The photocatalytic and photoelectrochemical water splitting systems, due to their simple equipment design and low cost, are considered as potential scalable strategies for solar hydrogen production. Therefore, significant research efforts have been documented in the last few decades for the accelerated development of potential photocatalysts for the production of renewable solar fuels, a promising strategy that can overcome the environmental issues related to the use of fossil fuels. The low-dimensional photocatalysts can be engineered effectively to exhibit the properties required for efficient solar energy conversion.

Low-dimensional materials have at least one of the dimensions restricted to nanoscale (less than 100 nm). These materials can be categorized into zero-dimensional (0D), one-dimensional (1D), and two-dimensional (2D) materials based on their sizes in each dimension (x, y, z) measured in nanoscale range (Bukowski & Simmons, 2002; Yoffe, 1993). In 0D nanomaterials, all the three dimensions (x, y, z) are in the range of nanoscale. These nanomaterials are in the form of a dot such as nanospheres, quantum dots (QDs), and nanoparticles. In 1D nanomaterials, the growth occurs in one dimension and their lateral dimensions are in nanoscale. The examples include nanorods, nanotubes, and nanowires. In 2D nanomaterials, the growth direction is along both x, and y directions, and the third dimension is in nanoscale. The 2D nanomaterials include thin films and nanosheets. The nanoscale dimensions can modulate their properties compared to their respective bulk counterparts (Deng et al., 2016). These materials can exhibit quantum confinement effects. At nanoscale, the charge carrier pathway is significantly shortened, which can reduce the electron–hole recombination rates. Moreover, the high length-to-diameter ratio known as aspect ratio of 1D nanomaterials facilitates increased light scattering and absorption. The 2D materials possess a large surface-to-volume ratio compared to bulk counterparts, and therefore, they carry enhanced number of surface-active sites. Furthermore, majority of 2D materials, but not all, and graphene in particular, have excellent conductivity and superior electron mobility due to the presence of 2D-conjugated structure. Consequently, they can achieve superior charge carrier separation and transfer properties. Additionally, 1D and 2D materials can serve as excellent catalyst support materials (Su et al., 2018; Tan et al., 2017). The 2D materials allow the tuning of their bandgaps via controlling the number of layers, which is a useful attribute for photocatalytic applications (Khan, Zhang, et al., 2020). Figure 12.2 expresses the distinct advantages of low-dimensional materials (Low et al., 2014).

Though many low-dimensional semiconductors are capable of performing photocatalytic CO_2 reduction or water splitting, the overall energy conversion efficiency is still low, far from that required for practical applications. This is

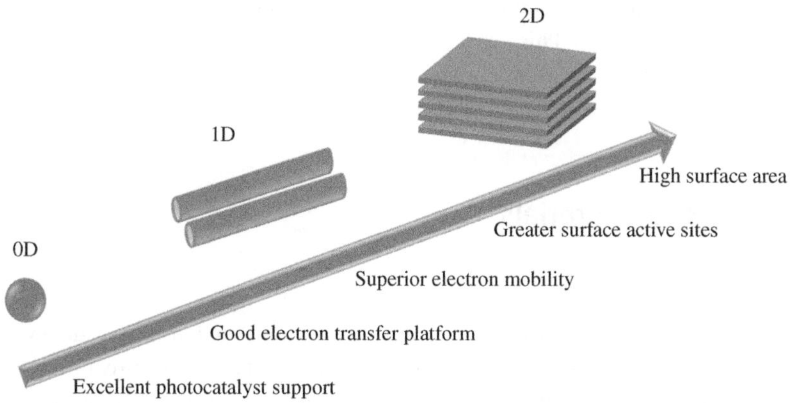

Figure 12.2 Schematic diagram describes the advantages of low-dimensional materials. Reproduced with permission (Low et al., 2014). Copyright 2015, Royal Society of Chemistry.

mainly due to several obstacles such as an increased exciton binding energy due to smaller electron screening as compared to bulk materials, flocculation of several low-dimensional materials in aqueous solution or oxidation by photogenerated holes, the presence of charge carrier recombination on surface, and insufficient redox potential of several low-dimensional materials for overall water splitting (OWS) or CO_2 reduction. A number of strategies have been developed to enhance the photocatalytic efficiencies of low-dimensional photocatalysts, and these include doping with metal or nonmetal elements, inducing defects, and coupling with metals or semiconductors. The photocatalytic properties of a material depend not only on its crystal structure, band structure, and electron affinity but also on the nature of its interface with another semiconductor or cocatalyst. Different combinations of low-dimensional materials can lead to various contact designs, such as point contact, line contact, and face contact (Figure 12.3). The face contact (2D–2D) possesses special advantages in photocatalysis as compared to point contact (0D–0D, 0D–1D, 0D–2D) and line contact (1D–1D, 1D–2D). First, it leads to an intimate interface between two semiconductors that can modulate the exciton diffusion length and energy and improve the photocatalytic quantum efficiency. Second, it is facile and efficient to form the intimate interface even when 2D semiconductors have a high lattice mismatch. Third, the large lateral size and high surface area of each 2D material contribute toward the large contact area in the face contact geometry, promoting charge separation and transfer. Fourth, face contact is highly beneficial for catalyst stability due to the alleviation of photocorrosion and agglomeration. A large number of low-dimensional materials have been combined via point contact, line contact, and face contact to build heterostructure photocatalysts with superior photocatalytic performance for water splitting and CO_2 reduction than their bulk counterparts in recent years (Tan et al., 2017).

Although the significance of dimensionality has been broadly examined in electronic and optical applications (Khan, Kim, et al., 2020; Osada & Sasaki, 2012), its utilization for catalysis has begun only recently. The significant developments in the design of low-dimensional materials and their composites have opened-up new directions for their applications, especially in photoelectrochemical processes. The expedited research on the use of low-dimensional materials in

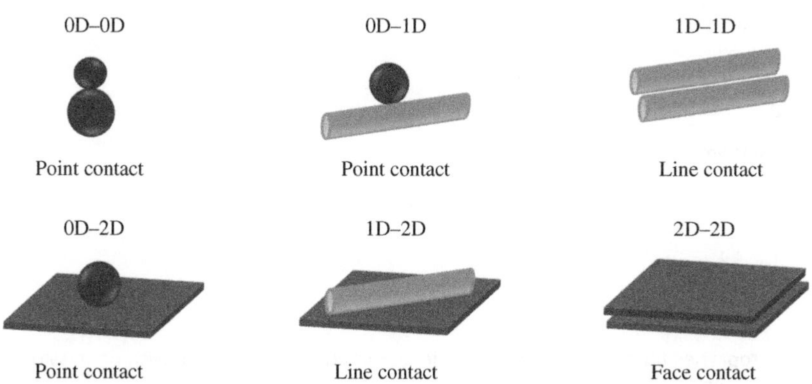

Figure 12.3 Schematic diagram of various nanoarchitectures (0D, 1D, and 2D) combined to build different types of heterostructure designs. Reproduced with permission (Low et al., 2014). Copyright 2015, Royal Society of Chemistry.

photoelectrochemical processes has not only helped the materials avoid several limitations observed in bulk materials but also revealed new challenges associated with their utilization. This chapter provides a comprehensive review of the properties and performances of low-dimensional materials used for photoelectrochemical fuel generation. At the end, we present our perspectives on the topic.

12.2 Unique Properties of Low-Dimensional Materials

12.2.1 Electronic Properties

When the bulk materials undergo dimensional reduction, the electronic density of states is modulated at nanoscale level leading to a change in electronic structure in a way that properties of low-dimensional materials are significantly different from their bulk parent materials (Figure 12.4). The performance of a catalyst is largely influenced by its electronic structure because it follows the Sabatier principle, which states that the interaction between reactants/intermediates and catalyst surface must possess moderate adsorption energy (Nilsson et al., 2005). The correlation between electronic structure and catalytic performance was observed for transition metals, where the adsorption energy is associated with the density of *D*-states at the Fermi level. To achieve a strong binding, the center of the *d*-band should be higher in relation to the energy of Fermi level. Therefore, an engineered band structure of a given catalyst enables a modulated activity (Conway & Bockris, 1957). The density functional theory (DFT) calculations allow expedited screening of many prospective materials with optimized composition, structure, and morphology to attain the preferable band structure. Simultaneously, experimental investigations focus on modulating the band structure via controlled design of low-dimensional catalysts, which can offer significantly improved activity (Voiry et al., 2018).

12.2.2 Surface Plasmon Resonance

Metal nanostructures exhibiting plasmon resonance effect have gathered significant attention in the field of photocatalytic chemistry for organic materials breakdown (Awazu et al., 2008), CO oxidation (Hung et al., 2010), water splitting, and CO_2 reduction to name a few (DuChene et al., 2018; Thimsen et al., 2011). Plasmons are combined oscillations of free charges present in a conducting material that are confined to the surfaces. When metal nanoparticles are irradiated at their plasmon frequency, resonance in absorption occurs as real part of dielectric function approaches zero and strong electric fields are produced on the surface of nanoparticles. The frequency of resultant resonance depends upon nanoparticles size, shape,

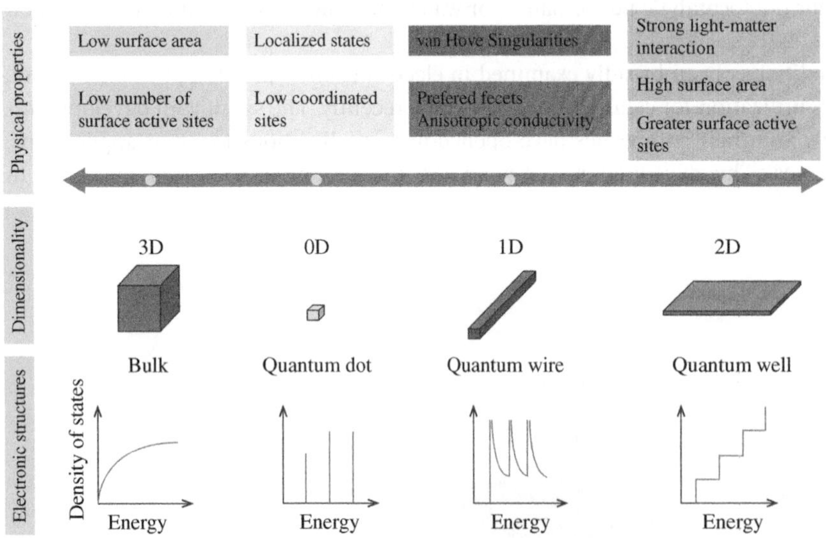

Figure 12.4 Comparison of physical and electronic properties for bulk and low-dimensional materials. Reproduced with permission (Voiry et al., 2018). Copyright 2018 Nature Publishing Group.

composition, and vicinity to other nanoparticles (Zhang et al., 2018). For instance, the plasmon resonance of gold (Au) exists in the visible spectrum, and it can be tuned to the infrared wavelength region by decreasing the nanoparticles size. Likewise, Ag shows plasmon resonance in UV, which can be modulated to visible range via synthesis of extremely small-size nanoparticles. The surface plasmon resonance (SPR) can influence photocatalytic properties via two different pathways: plasmonic charge transfer and local electric field enhancement (Linic et al., 2011).

12.2.2.1 Charge Transfer Mechanism

Tatsuma and coworker documented the mechanism of plasmon-induced charge separation in Au nanoparticles and nanoporous TiO_2 composites for experimentally observed excellent incident-photon-to-current conversion efficiency (IPCE) of 26% under visible light irradiation (Tian & Tatsuma, 2005). According to their proposed mechanism, Au nanoparticles became photoexcited owing to plasmon resonance, and excited electrons are transferred from Au nanoparticles to the conduction band of TiO_2. Simultaneously, the electron compensation in Au nanoparticles was achieved from a donor present in the solution (Figure 12.5a). Although this mechanism was reported in subsequent studies, it is worth noting that the surface plasmons consist of charge density waves on metal surfaces and have no VB and CB energy regions associated with them. The plasmonic charges exist on the Fermi level of the metal and cannot contribute to redox reaction, as explained above. In contrast, the photoresponse of few-atom metals such as Ag_8 and Au_{25} clusters expresses evident absorption and emission characteristics. Due to molecule-like features, the metal clusters are photochemically reactive and work as electron donors. Moreover, the localized surface plasmon response (LSPR) highly depends on the appropriate size and shape of metal nanoparticles. As the particle size becomes smaller than 10 nm in diameter, i.e., in the quantum size effect regime, the existing knowledge of their properties becomes unclear and at ~2 nm the plasmonic effect remains negligible. However, size-dependent LSPR in nanomaterials with quantum effect is complicated to analyze due to the presence of inconsistencies in the literature. For example, by employing femtosecond transient absorption spectroscopy with an IR probe, Furube et al. validated that the plasmon-influenced electron transfer would occur from 10-nm Au nanodots to TiO_2 nanocrystals (NCs) (Furube et al., 2007). In another study, Mubeen et al. (2011) reported a charge transfer mechanism for the same system. It was confirmed that the charge carriers were produced due to the plasmon decay process and hot electrons traveled to TiO_2 via quantum tunneling.

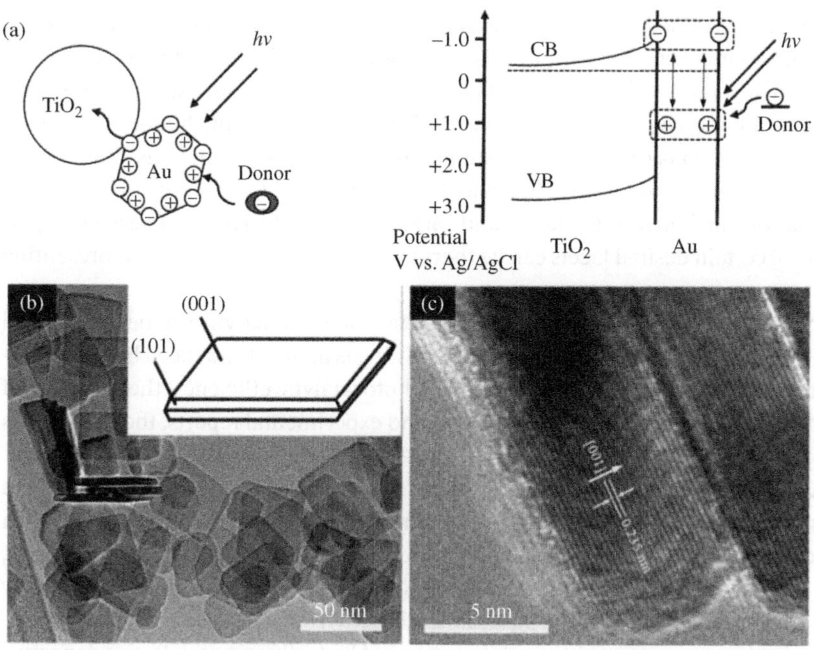

Figure 12.5 (a) Proposed mechanism for charge separation under visible light irradiation in Au NPs-TiO_2 system. Reproduced with permission (Tian & Tatsuma, 2005). Copyright 2005, American Chemical Society. (b) TEM image of TiO_2 nanosheets with well-defined rectangular outline; (c) HRTEM image represents the lattice spacing, ~0.235 nm, parallel to top and bottom facets, which corresponds to (001) planes of anatase TiO_2. Reproduced with permission (Yu et al., 2010). Copyright 2010, American Chemical Society.

12.2.2.2 Local Electric Field

As plasmon-based nanostructures perform as antennae to absorb large proportions of visible light, the resulting localized electromagnetic field can significantly improve charge carrier separation. When SPR-exhibiting nanoparticles are irradiated by light, a powerful local electric field is produced near the surface of the nanoparticles. The intensity of these plasmonic "hot spots" can approach up to 1000 times the incident electromagnetic field evaluated via electromagnetic simulations using finite-difference time-domain (FDTD) (Mizeikis et al., 2011). The resulting hot spots tend to produce 1000 times more electron–hole pairs than the incident electromagnetic field. Therefore, a significant amount of photo-triggered charge is created locally in the semiconductor due to local field enhancement of plasmonic nanoparticles. The local electric field mechanism is highly dependent on the defect states in the semiconductor material that are responsible for light absorption below its bandgap. In 2011, Mizeikis et al. reported simulation results utilizing the FDTD method for a system of spherical and hemispherical noble metal nanoparticles on a smooth TiO_2 surface to demonstrate the optical field enhancement. Their findings revealed that for hemisphere nanoparticles, the significant increase in the localized electric field was up to 10^4 at the metal/TiO_2 interface. Cronin's group reported the photocatalytic conversion of CO_2 into hydrocarbon fuels via Au nanoparticles/TiO_2 (Hou et al., 2011). When the photon energy equivalent to plasmon resonance (532 nm) was used for irradiation, a 24-fold increase in photocatalytic activity was observed due to the generation of intense local electromagnetic fields by Au nanoparticles. The strong electromagnetic fields were also responsible for sub-bandgap absorption in TiO_2, hence increasing the photocatalytic activity in the visible spectral region. The FDTD simulations helped to perform a strong evaluation of electric fields and charge at Au/TiO_2 interface.

12.2.3 Crystal Facets, Kinks, and Edges

There are various factors that control photoactivity. Among these, surface structure/properties (Linsebigler et al., 1995), reactant adsorption, and electron transfer at the interface between a catalyst and reactant present in the medium contribute significantly (Fox & Dulay, 1993). Specifically, the arrangement of atoms on the surface of a catalyst has a decisive role in manipulating photoactivity and stability. Generally, the catalyst's performance can be boosted by creating NCs with specific crystallographic facets. NC facets, represented by Miller indices {*hkl*}, can be categorized as low-index facets or high-index facets. NCs with low-index facets include those with Miller indices of 0 and 1, for instance {111} and {100} facets, whereas NCs with large-index facets include those with at least one Miller indices larger than 1, such as {210} and {310}. Metals and their respective oxide nanomaterials exposing specific facets have captured significant attention. The exposed facets regulate the morphology of NCs. For instance, a cube is enclosed by six {100} facets, while an octahedron exposes eight {111} facets. The presence of many low and high-coordination atoms at atomic steps, edges, and kinks in the high-index faceted nanocrystals often serves as highly active sites for various catalytic applications. In contrast, recent findings showed that the low-index faceted NCs could provide better stability and improved activity for various catalytic reactions. Therefore, it is important to understand the significant relationship between the facets and catalytic activity to achieve an excellent performance of NCs. Nonetheless, the formation of certain desired facets can be thermodynamically unfavorable, representing an obstacle in the nanomaterials design.

The application of this approach in nanomaterials design and its impact on the photocatalytic activity can be exemplified by titanium dioxide (TiO_2). It is well known that the TiO_2 performance for water splitting is mainly influenced by the recombination of photoexcited charge carriers. Among various methods to enhance its photocatalytic efficiency, the synthesis of TiO_2 NCs with abundant (001) facets is highly desirable. According to theoretical and experimental reports, the (001) facets are more effective for dissociative adsorption of water in contrast to the thermodynamically stable (101) facets, which favor physisorption of water molecules. Yu et al. (2010) reported the synthesis of TiO_2 nanosheets via hydrothermal method in the presence of hydrofluoric acid. These TiO_2 nanosheets consisted of a well-defined sheet-like (Figure 12.5b) structure with an average length of ~50–80 nm and thickness of ~6 nm. The lattice spacing of ~0.235 nm, parallel to top and bottom facets, corresponds to (001) planes of anatase TiO_2 (Figure 12.5c). Almost 75% of TiO_2 nanosheets were observed to be enclosed by (001) facets. Furthermore, fluorination was performed to improve the photocatalytic activity of TiO_2 nanosheets. For comparison, TiO_2 nanoparticles were also synthesized in pure water without the presence of hydrofluoric acid. Pt was deposited as cocatalyst on all the samples. Fluorinated TiO_2(001) faceted nanosheets exhibited a higher photocatalytic H_2 evolution rate compared to fluorine-free TiO_2(001) faceted nanosheets, Degussa P25, and pure TiO_2 nanoparticles (synthesized in the absence of hydrofluoric acid).

Zhou and coworkers studied the effect of Pt nanoparticles (~4–6 nm) with various morphologies (cubic, spherical, and tetrahedral) enclosed by various facets supported on carbon nanotubes for catalytic reaction. The ratio of (100)/(111) facets for cubic, spherical, and tetrahedral Pt nanoparticles decreased by 3.45 to 2.81 and 0.57. The supported Pt nanoparticles with a higher volume of (100) facets demonstrated higher catalytic activity and stability than Pt nanoparticles with (111) facets due to better adsorption of reactant molecules and prevention of deactivation induced by product adsorption.

The catalytic activity can also be improved by developing kinks and edges. In this perspective, exfoliated transition metal dichalcogenides (TMDs) are important nanomaterials due to their ability to form a variety of edges and phases that can be adopted for a specific catalytic reaction (Jaramillo et al., 2007; Tsai et al., 2015). According to DFT calculations, the electronic structures of TMDs nanosheets are highly dependent on the type of edges, phase, and composition of TMDs (Chan et al., 2014). Scanning tunneling microscopy studies demonstrated that sulfur-rich materials would selectively adsorb reactant molecules at the brim and corner of the flakes.

12.2.4 Large Surface Area and Abundant Surface-Active Sites

In addition to their novel electronic, plasmonic, distinct facets, kinks, and edges-dependent properties, low-dimensional materials also exhibit many other important features for photocatalysis, such as large surface area to volume ratio and abundant surface-active sites. This indicates that most of the atoms are present on the catalyst surface and are responsible for catalytic reactions. For instance, 1D metallic nanowires and 2D metallic nanosheets demonstrate significantly higher electron mobility, density of surface unsaturated atoms as active sites, and specific surface area than their bulk counterparts. For example, 1D nanomaterials of Si, TiO_2, and GaN are important photocatalysts for CO_2 reduction and water splitting (Wang, Pierre, Kibria, et al., 2011; Wang, Wang, Ling, et al., 2011). Due to the high surface-to-volume ratio of these nanomaterials and their composites, the distance that the charge carriers must travel to reach the surface is significantly decreased. It also decreases the charge recombination probability. As Si and TiO_2 possess a short hole diffusion length (~10–100 nm), reducing the distance from the grain interior to the electrolyte–semiconductor interface is mandatory. Oh et al. compared the performance of 1D Si nanowires and 3D planar Si as photocathodes and showed that Si nanowires could provide a larger onset potential (+0.41 V) than planar Si (+0.33 V) (Oh et al., 2012).

In low-dimensional nanomaterials, an important strategy to increase the catalytic activity is to maximize the number of catalytic active sites via surface engineering. This can be accomplished by modulating the composition, phase, defects, doping, strain engineering, and facets without introducing additional components to optimize active surface sites for photocatalytic reactions. An unquestionable advantage of low-dimensional materials is that all these modifications can be done closer to the surface and therefore, they can contribute effectively to improved catalytic response. The graphitic carbon nitride (g-C_3N_4) doped with oxygen, phosphorus, boron, and sulfur has achieved excellent activity for HER and OER. Our group utilized the synergy of boron dopants and nitrogen defects to effectively modulate the electronic band structure of g-C_3N_4 for enhancing the optical absorption in the visible region and obtaining an outstanding oxygen evolution activity (561.2 $\mu mol \cdot h^{-1} \cdot g^{-1}$) (Zhao et al., 2019). Fang and coworkers reported the plasmonic hot electron doping in MoS_2, which induced 2H to 1T MoS_2 phase transition and improved the electrical conductivity. The catalyst was constructed via deposition of Ag@Au nanorattles on MoS_2 surface for improved H_2 evolution (Kang et al., 2015).

12.2.5 Heterostructure Construction

Low-dimensional nanomaterials can assemble via van der Waals (vdW) interactions to form heterostructures or pure nanomaterials. These heterostructures can be fabricated through various techniques, such as chemical vapor deposition (CVD) and solution-phase routes. Among these two methods, CVD provides excellent control of heterostructure formation, resulting in neat interfaces between the components of heterostructure and providing exquisite control for efficient charge transfer and improved reaction kinetics. Nevertheless, mass production of heterostructure systems via CVD is prone to be expensive from a practical perspective. In contrast, solution-based processes such as hydrothermal reactions and/or electrostatic assembly can produce large amounts of heterostructure nanomaterials with sharp interfaces and high catalytic performance. The semiconductor heterostructure constructed as a combination of various low-dimensional nanomaterials can be classified into four important groups: 0D–1D, 0D–2D, 1D–2D, and 2D–2D (Figure 12.6a). These low-dimensional nanomaterial-based heterostructures have realized significant attention owing to their possible advantages of high-quality interfaces between components, extended light absorption, and perfectly aligned band offsets. The objective of low-dimensional heterostructures is to utilize most of the solar spectrum in a wide energy range, possess the ability to efficiently dissociate excitons, facilitate fast transport of charge carriers to active catalyst sites (Figure 12.6b). Precious metals (Pt, Pd, Rh, Ir) and plasmonic materials (Au, Ag, Cu) in the form of nanoparticles/QDs

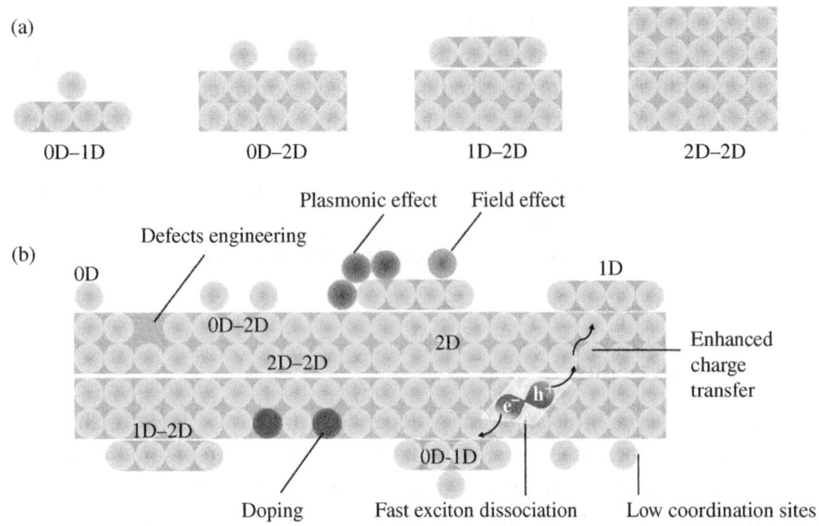

Figure 12.6 (a) Different types of low-dimensional materials-based heterostructures: 0D–1D, 0D–2D, 1D–2D, and 2D–2D; (b) Modulating the photocatalytic properties of low-dimensional materials via defect density, plasmonic effect, low coordination sites, fast exciton dissociation, improved charge transfer, and field effect for effective heterostructure construction. Reproduced with permission (Khan, Zhang, et al., 2020). Copyright 2019, Wiley-VCH.

effectively deposited on 1D and 2D semiconductors such as graphene, reduced graphene oxide (GO), transition metal oxides/nitrides, layer double hydroxide, layer perovskites, and g-C_3N_4 serve as powerful examples of 0D–1D and 0D–2D heterostructures for photocatalytic applications. In short, the fascinating low-dimensional nanomaterials are potential photocatalysts due to their unique characteristics, such as tunable bandgap, extended light absorption, large specific surface area, reduced recombination of charge carriers, and abundant active sites available on the surface.

12.3 Applications of Low-Dimensional Materials

The low dimensionality is the distinctive feature of the nanomaterials, which makes them potential materials for various applications. A commonly recognized definition of low-dimensional material is that their size should be at nanoscale (<100 nm), at least in one dimension. Researchers have developed various routes to synthesize low-dimensional materials, which are vital for modern science and technology. There exist enormous opportunities, which can be made practical by developing new types of nanostructures or miniaturizing existing microstructures into 1–100 nm systems. The importance of synthesizing smaller nanostructures can be well understood from the objectives of microelectronic and information technology where "miniaturization" means higher performance. For instance, the integrated circuits formed with multiple components per chip possess faster operations, lower costs, and lower power consumption than discrete bulk devices. It is a proven fact that a wealth of peculiar properties, including size-dependent excitation or emission, quantized conductance, Coulomb blockade, and metal–insulator transition, can be observed at nanoscale dimensions. The quantum confinement in low-dimensional materials contributes to modulation of their optical, electrical, thermoelectric, and magnetic properties. However, this chapter will focus on the emerging applications of low-dimensional materials for photocatalytic water splitting and CO_2 reduction.

12.3.1 Water Splitting

As previously mentioned, when a photoactive semiconductor capable of performing water splitting reaction is irradiated with sunlight, it undergoes three basic processes. First, it absorbs photons possessing energy equal to or higher than its bandgap to excite electrons from VB to CB generating holes in the VB (Tian et al., 2019). Second, the charge carriers diffuse to the semiconductor surface. During this process, the electrons and holes can either reach the surface successfully (Figure 12.7a, pathways 1 and 2) or undergo recombination [Figure 12.7a, pathway 3 (surface recombination) and pathway 4 (bulk recombination)]. The bulk recombination is significant in bulk materials, but it is significantly reduced in

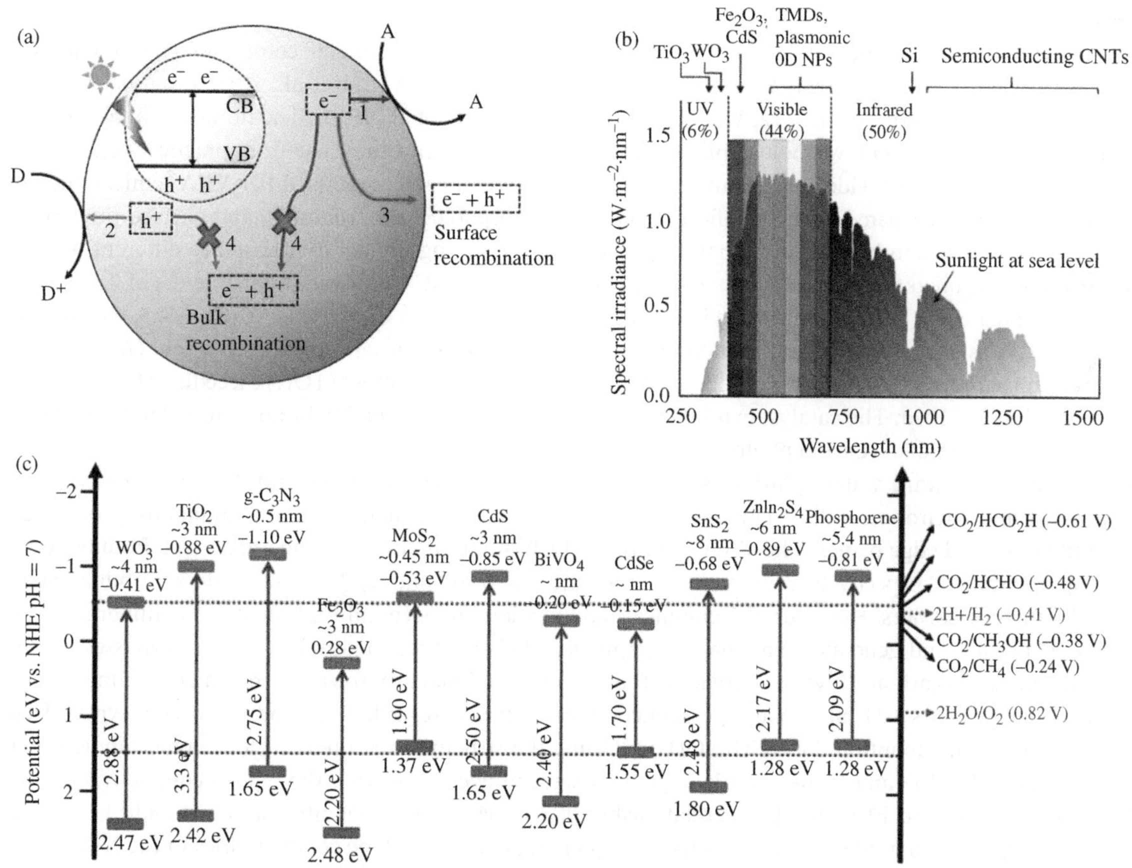

Figure 12.7 (a) Schematic illustration representing the basic steps involved in a photocatalytic reaction along with detrimental recombination pathways. A and D represent electron acceptor and electron donor, respectively; (b) Absorption region for various low-dimensional materials with respect to solar spectrum. Reproduced with permission (Voiry et al., 2018). Copyright 2018 Nature Publishing Group. (c) The respective valence band maximum and conduction band minimum for various low-dimensional materials are given. The redox potentials for water splitting and important CO_2 reduction reactions are also shown. Reproduced with permission (Khan, Zhang, et al., 2020). Copyright 2019, Wiley-VCH.

low-dimensional materials. Third, after arriving at the surface, the charge carriers are transferred to the electrolyte solution to execute the redox reactions with the adsorbed molecules (Ong et al., 2016). For an intrinsic semiconductor, its bandgap energy defines the amount of light that can be absorbed from solar spectrum. Figure 12.7b shows the solar irradiance spectrum and the absorption region demonstrated by various semiconductors (Greeley et al., 2006). The typical bandgaps and band edge positions of several important low-dimensional materials are described in Figure 12.7c for photocatalytic water splitting and CO_2 reduction.

For splitting a water molecule, two-half reactions, the hydrogen evolution reaction (HER) and oxygen evolution reaction (OER), must take place simultaneously, as given in Eqs. (12.1) and (12.2), respectively. OWS is a multielectron process that involves a standard free energy $\Delta G°$ (corresponding standard redox potential $\Delta E°$), as shown in reaction (12.3). The large energy barrier in OWS, which presents a great challenge to practical photocatalytic systems, arises from the thermodynamic requirement of positive Gibbs energy ($\Delta G° = +237.2$ kJ/mol). For photocatalytic water splitting, the VB and CB of a photocatalyst must match the redox potential of HER and OER. The CB minimum must be sufficiently negative that photogenerated electron can reduce H^+ to H_2 (−0.41 V vs. NHE at pH = 7), whereas VB maximum must be positive enough that oxidation potential of photoexcited hole can oxidize H_2O to O_2 (+0.82 V vs. NHE at pH = 7).

$$2H^+ + 2e^- \rightarrow H_2(\Delta E° = -0.41 \text{ V vs. NHE at pH 7}) \quad (12.1)$$

$$2H_2O \rightarrow O_2 + 4H^+ + 4e^-(\Delta E° = +0.82 \text{ V vs. NHE at pH 7}) \quad (12.2)$$

$$H_2O \rightarrow H_2 + 1/2O_2(\Delta G° = 237.2 \text{ kJ/mol}, \Delta E° = 1.23 \text{ V}) \quad (12.3)$$

12.3.1.1 0D Materials

Artificial photocatalytic systems usually involve bulk semiconductors (e.g., TiO_2), metal complexes (e.g., complexes of Ru, Ir, or Rh), or organic dyes (e.g., eosin Y, or fluorescein) as light absorbers (Cai et al., 2014; Cline et al., 2008; Lakshminarasimhan et al., 2012; Min & Lu, 2012; Wang et al., 2013; Willkomm et al., 2015). The limited light absorption by wide bandgap semiconductors and low photostability of molecular species limit their use from practical perspective. In order to capture solar photons in a wide energy range, inorganic QDs specifically comprised IIB–VIIA semiconductors are promising materials when prepared with specific compositions, morphologies, structures, and surface properties. An exquisite control of QD size can help tune the bandgap, therefore, providing the ability to capture different regions of the solar spectrum. In the breakthrough work reported by Han et al., CdSe NCs with varying diameters of 2.5–5.5 nm covered with dihydrolipoic acid (DHLA) were tested as a light absorber (Han et al., 2012). At pH = 4.5, the system comprised of Ni^{2+}-DHLA as catalyst for proton reduction to generate H_2 in the presence of ascorbic acid (AA) as hole scavenger. Under optimized conditions, the resultant catalyst provided a turnover number (TON) exceeding 600,000 moles of H_2 per mole of catalyst in 110 hr. The catalyst exhibited excellent H_2 evolution over 360 hr and this extended stability was associated with the role of CdSe NCs as a photosensitizer.

A cost-effective approach toward water splitting is to avoid the use of precious metals and develop photocatalysts comprised of low-cost and environmentally friendly carbon materials. GO is a material that contains oxygen-related functionalities at the basal and edge planes. The VB maximum and CB minimum of graphene are derived from bonding π and anti-bonding π^* orbitals, respectively. A monolayer graphene has zero bandgap because π and π^* orbitals touch each other at Brillouin zone corners. However, the chemical modification and size-induced quantum confinement can separate the π and π^* orbitals and generate a bandgap in graphene. GO is a p-type material as oxygen possesses higher electronegativity in contrast to carbon. However, replacing the oxygen functional group with nitrogen-containing groups on GO nanosheets edge converts GO into an n-type semiconductor. In this regard, Teng and coworkers synthesized nitrogen-doped graphene oxide-quantum dots (NGO-QDs) fulfilling the structural requirement for photocatalytic water splitting (Yeh et al., 2014). The EIS and Mott–Schottky equations were utilized to identify the type of conductivity of the NGO-QDs. As presented in Figure 12.8a, the linear relationship of $1/C^2$ vs. applied potential shows straight lines with positive and negative slopes in different potential regions corresponding to p- and n-type conductivities in the NGO-QDs. The successive introduction of nitrogen and oxygen produced n- and p-type domains in the NGO-QDs. The Fermi levels were determined from the intercepts of the extrapolated straight lines on the abscissa. The significant photoluminescence (PL) emission confirmed the development of photochemical p–n diodes within NGO-QDs where carbon sp^2 clusters worked as an interfacial junction (Figure 12.8b). The diode configuration led to an internal Z-scheme charge migration for simultaneous H_2 and O_2 production from pure water under visible light irradiation at a molar ratio of 2:1.

When the particle size of a semiconductor becomes smaller than its Bohr exciton diameter (a_B), it exhibits a quantum confinement effect. According to Marcus theory, the thermodynamic energy of confined electron–hole pairs increases, resulting in fast interfacial charge transfer (Hutchison et al., 2005). Several chalcogenides, such as ZnS, ZnSe, ZnTe, CdS, CdSe, and CdTe, demonstrate an increase in bandgap as the particle size is reduced in the quantum regime. For example, in bulk, CdSe has a moderate bandgap of ~1.74 eV and a Bohr exciton radius of 4.8 nm. At 2.0 nm, CdSe has relatively large bandgap of 2.88 eV (Zhao et al., 2013). The redox potential of both electrons and holes is increased simultaneously. However, the effective mass of a hole is relatively large compared to an electron, which makes E_{CB} more responsive toward confinement than E_{VB} (Norris & Bawendi, 1996; Sapra & Sarma, 2004). Therefore, the interfacial charge transfer in smaller QDs is more facile than in larger QDs. The simple example of CdSe QDs with diameters ~1.8–6.0 nm can best explain the quantum confinement effect that influenced H_2 evolution response. The bulk and 6.0 nm size QDs could not produce H_2 effectively, compared to smaller QDs (Figure 12.8c). When CdSe QDs passivated with 2-percaptoethanol in the presence of Na_2SO_4 as reductant are irradiated with light, electrons are excited in QDs and transferred to Cd^{2+} ions present at the surface to perform the reduction reaction for H_2 production. The photoexcited holes travel to the surface of QDs and are consumed by SO_3^{2-} (Figure 12.8d).

In addition to the QD's size, surface design with ligands such as organic molecules, metal-free inorganic ions, polymers, or molecular metal chalcogenides can significantly affect the synthesis, properties, and applications (Boles et al., 2016; He et al., 2016; Kilina et al., 2009). The presence of surface ligands can contribute to QDs solubility, lower the aggregation, reduce the surface trap sites, and modulate the exciton diffusion length and energy (Hines & Kamat, 2014; Peterson et al., 2014; Weiss, 2017). For example, semiconductor QDs employed as light absorbers were capped with ligands such as [FeFe]-hydrogenase mimics and Ni complexes that absorbed light and transferred photoelectrons to

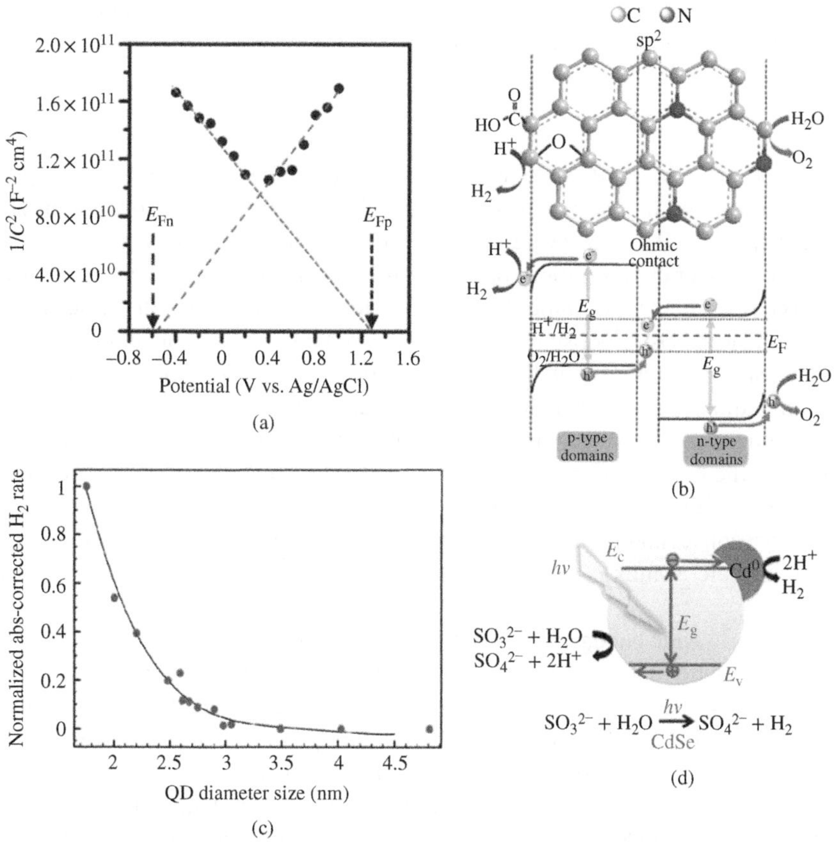

Figure 12.8 (a) The change in capacitance (C) as a function of applied potential as described in the Mott–Schottky relationship for NGO-QD electrode. The capacitance was measured by EIS. The symbols E_{Fp} and E_{Fn} demonstrate Fermi levels of the p- and n-type conductivity domains; (b) The configuration and energy band diagram of NGO-QD photochemical diode. The diode is comprised of p- and n-type domains, which are connected through sp^2 clusters to construct a Z-scheme system via an ohmic contact. Reprinted with permission (Yeh et al., 2014). Copyright 2014, Wiley-VCH. (c) The normalized photocatalytic H_2 rate relative to photocatalyst size; (d) Photocatalytic H_2 generation over CdSe QDs decorated with Cd atoms as active sites for H^+ reduction and SO_3^{2-} ions present in the suspension are oxidized by holes on the QDs surface. Reprinted with permission (Zhao et al., 2013). Copyright 2013, American Chemical Society.

HER cocatalysts (Wang, Wang, Wang, et al., 2011; Wu, Xia, et al., 2015). These ligands provide stability to colloidal QDs against agglomeration, which can influence the H_2 evolution activity. For instance, multidentate ligands such as trithiol 2,2,2-tris(mercaptomethyl) acetic acid can strongly bind to CdSe QDs, which act as stable light absorbers. The coupling of the abovementioned system and $[Co(1,2\text{-benzenedithiolato})_2]^-$ to form a catalyst allows an efficient H_2 generation in the presence of H_2Asc (Das et al., 2013). Compared to monothiol or dithiol, trithiol strongly binds to the surface of CdSe QDs and provides good dispersity and stability in H_2O, along with excellent photocatalytic performance.

Further advances in surface modification confirmed that the organic ligands affect the surface trap states, the charge separation, and control the charge carrier dynamics in semiconductor QDs by serving as an insulating layer (Wang et al., 2015). In the case of monodispersed CdSe/CdS QDs, the ligands with more thiol groups decrease the PL intensity and increase H_2 evolution response. However, the organic ligand layers possess a shielding effect that can be reduced by removing several layers or replacing them with inorganic ligands (Kovalenko et al., 2009; Nag et al., 2011). For instance, stabilized CdSe QDs after ligand removal presented better photocatalytic conversion of thiols into their respective sulfides and H_2 (Li et al., 2014). Furthermore, it was reported that bare CdS QDs provide 175 times higher H_2 production than 3-mercaptopropionic acid (MPA)-decorated CdS QDs in the company of $CoCl_2$ and Na_2SO_3 as cocatalysts and sacrificial electron donors, respectively (Chang, Orchard, et al., 2016). MPA causes a damaging effect by creating a physical barrier between CdS QDs and the cocatalyst that inhibits the mass and charge diffusion. It can be concluded that photocatalytic activity of QDs depends on different factors, such as their size, structure, and surface properties. Additionally, the photocatalytic efficiency is also governed by the nature of the cocatalyst, its ratio with respect to the QDs, and its interaction with the light absorber. Therefore, for the development of a highly efficient photocatalytic system, multiple factors such as QD's size, surface design, and surface modifications should be optimized.

12.3.1.2 1D Materials

The pioneering work of Iijima in 1991 demonstrating the synthesis of carbon nanotubes (a few micrometers in length and several nanometers in diameter) laid the foundation in materials science and technology for the development of 1D nanostructures of a large number of materials (Iijima, 1991). Over the past three decades, 1D nanostructures have received great attention, as evident from the fact that the number of publications on "1D" rose to over 200,000 (per Web of Science database) by May, 2021. The 1D nanomaterials possess high aspect ratio (length/diameter) in at least one of the dimensions (~1–100 nm). The high aspect ratio of 1D nanomaterials is responsible for their increased light absorption and scattering characteristics. In 1D nanomaterials, the dimension along the radial direction is comparable to relevant physical parameters, including wavelength of light, exciton Bohr radius, phonon mean free paths, and exciton diffusion length. The confined dimension of 1D nanomaterials generates quantized particles, such as electrons, photons, and phonons (Hu et al., 1999; Tian et al., 2014). The sunlight is absorbed across the longer dimension in 1D nanomaterials, while the charge carrier separation occurs along the short radial dimension (Lee et al., 2014). These features enable 1D nanomaterials as a potential candidate for applications related to solar energy conversion. The 1D nanomaterials of various materials, TiO_2, α-Fe_2O_3, Si, and GaN to name a few, have been studied for water splitting (Mor et al., 2015; Wang, Pierre, Kibria, et al., 2011; Wang, Wang, Ling, et al., 2011).

The transition metal oxide nanotubes, particularly wide bandgap TiO_2 nanotubes (TNTs), also gained prompt interest because of their promising properties as photocatalytic material and electron transport medium in dye-sensitized solar cells. Grimes and coworkers reported the pioneering work on the synthesis of TNTs via anodic oxidation of pure titanium sheet in aqueous solution of hydrofluoric acid (0.5–3.5 wt%) and their water splitting performance. TNTs diameter ranged from 25 to 65 nm was dependent on the anodizing voltage (Gong et al., 2001). In another study, nitrogen-doped TNT arrays were synthesized for efficient solar conversion of CO_2 and water vapor to produce methane and other hydrocarbons. The choice of cocatalyst including Pt and/or Cu nanoparticles had significant effects on the production rate of different hydrocarbons (Varghese et al., 2009). Schmuki and coworkers reported the cocatalyst-free H_2 generation using "black titania" nanotubes synthesized under high-pressure H_2 environment (Liu et al., 2014). In contrast, TiO_2 tubes converted into black titania through conventional reduction methods, such as atmospheric pressure H_2/Ar annealing, demonstrated negligible H_2 evolution performance. The high H_2 pressure annealing resulted in isolated Ti^{3+} defect structures in the anatase nanotubes, which were absent in the conventional reduction. The anodically synthesized TNTs require thermal annealing at a temperature between 400 and 600 °C in air to be crystallized in anatase phase. Furthermore, the same group applied the "water annealing" strategy on TNTs followed by hydrogenation, called as "grey" modification, to improve the H_2 evolution response of TNTs (Fu et al., 2021). The combination of water annealing and thermal annealing followed by hydrogenation resulted in a five times increase in H_2 production compared to the thermally annealed anodic TNTs (Figure 12.9a). Ampelli et al. reported an anodic synthesis of TNT arrays with a diameter of ~70 nm and wall thickness of ~8–10 nm (Ampelli et al., 2017). The Au nanoparticles were deposited via three different methods, wet impregnation, photoreduction, and electrodeposition, to increase the visible light response and plasmonic effect. The maximum H_2 evolution rate, 190 μmol in 5 hr of light irradiation, was observed for the Au-deposited TNT photocatalyst developed via electrodeposition route (Figure 12.9b). The possible reason for the high photocatalytic response was ascribed to the presence of small Au nanoparticles, which demonstrated interference fringes in the absorption spectrum from 600 to 1200 nm. It is more likely that small-sized Au nanoparticles (3–5 nm) induced a localized surface resonance and localized electromagnetic field generation response, contributing to higher activity. The photocatalytic response of Au-deposited sample prepared via wet impregnation method was lower than that of the non-modified TNT arrays due to large Au aggregates on the TNT arrays blocking the photoactive sites on the surface. Moreover, Au-TNT arrays synthesized by electrodeposition had 1.6 times higher photocurrent response than the non-modified TNT arrays. The gas-phase experiments of ethanol photo-reforming also recorded higher H_2 evolution for Au-modified TNT arrays prepared using electrodeposition.

Among transition metal oxides, hematite (α-Fe_2O_3) is a potential candidate material for solar water oxidation due to its ideal band edge position and moderate bandgap (~1.9–2.3 eV) for visible light absorption (Lin et al., 2011). Theoretically, hematite can produce a photocurrent with a density of ~12.6 mA/cm^2 and theoretical STH conversion efficiency of ~15% (Murphy et al., 2006). Additionally, α-Fe_2O_3 is a low-cost, earth-abundant, non-toxic material that possesses robust physicochemical stability. However, α-Fe_2O_3 suffers serious electron–hole recombination due to its poor intrinsic conductivity and short hole-diffusion length when employed as photoanode (Junior et al., 2021; Tilley et al., 2010). To overcome these limitations of α-Fe_2O_3, Fu et al. (2016) synthesized Nb-doped α-Fe_2O_3 nanorod photoanodes under a facile solution-based process. The optimized Nb-doped α-Fe_2O_3 nanorods exhibited four times higher photocurrent density under solar light at 1.0 V vs. Ag/AgCl and 2.5-folds higher incident photon-to-current conversion efficiency (13.7%) at 350 nm and 1.23 V vs. RHE compared to undoped α-Fe_2O_3 nanorods. The improved PEC performance was attributed to fast charge transfer ability and small diameter of Nb-doped α-Fe_2O_3 nanorods that shortened the charge transfer length to enhance the PEC activity. Wang et al. reported molecular beam epitaxy-assisted synthesis of GaN nanowires decorated with Rh/Cr_2O_3 as cocatalyst to achieve photocatalytic OWS (Figure 12.9c,d) (Wang, Pierre, Kibria, et al., 2011). The presence of Rh facilitated H_2 generation and Cr_2O_3 shell over Rh worked as a diffusion barrier for O_2 to avoid the back reaction. A highly stable process for pure water splitting into H_2 and O_2 was observed on GaN nanowires. Oh et al. demonstrated the synthesis of 1D Si nanowires via metal-catalyzed electroless etching method and compared the activity of Si nanowires and 3D planar Si as photocathodes (Oh et al., 2012). The PEC H_2 evolution response for 1D Si nanowires was superior to planar Si. The high surface area of Si nanowires maximized the interaction with reactants and achieved a high onset potential of +0.42 V compared to 0.33 V for planar Si. The lower kinetic overpotential due to large surface area, the suitable shift in flat band potential, and light trapping effect for Si nanowires were the governing factors of better PEC performance. Li and coworkers reported the controlled N doping (4%) in ZnO nanowires synthesized via hydrothermal method to implement as photoanode in PEC water splitting

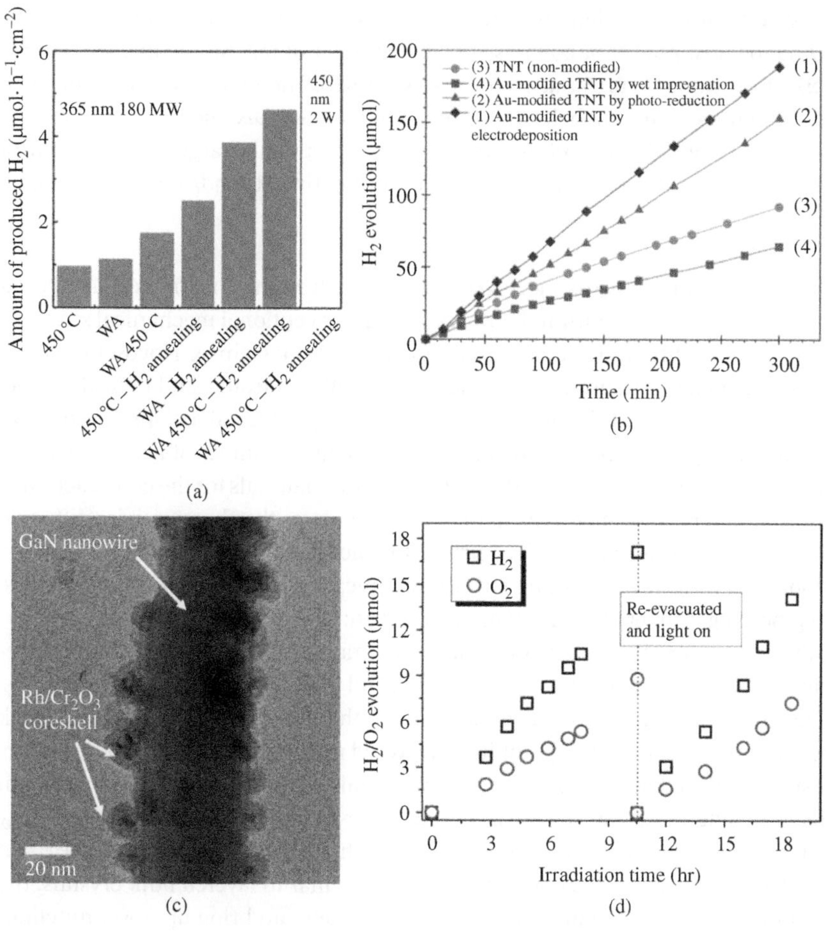

Figure 12.9 (a) Photocatalytic H_2 evolution from TiO_2 nanotubes synthesized at different conditions under 365 nm LED (180 mW/cm^2) or 450 nm (2 W) illumination in 50 vol% methanol-water electrolyte. Reprinted with permission (Fu et al., 2021). Copyright 2021. Wiley-CVH GmbH. (b) Hydrogen evolution as a function of time in gas-phase photocatalytic experiments of ethanol photo-dehydrogenation over Au-modified TNT arrays. Reprinted with permission (Ampelli et al., 2017). Copyright 2017. Elsevier. (c) Low magnification TEM image shows uniform distribution of the Rh/Cr_2O_3 nanoparticles on GaN nanowire surfaces; (d) Experimental OWS on Rh/Cr_2O_3 core-shell nanostructure deposited GaN nanowires, under a 300 W full arc xenon lamp irradiation for a duration of 18 hr. No degradation of the photocatalytic activity was observed during two cycles of reaction. Reprinted with permission (Wang, Pierre, Kibria, et al., 2011). Copyright 2011. American Chemical Society.

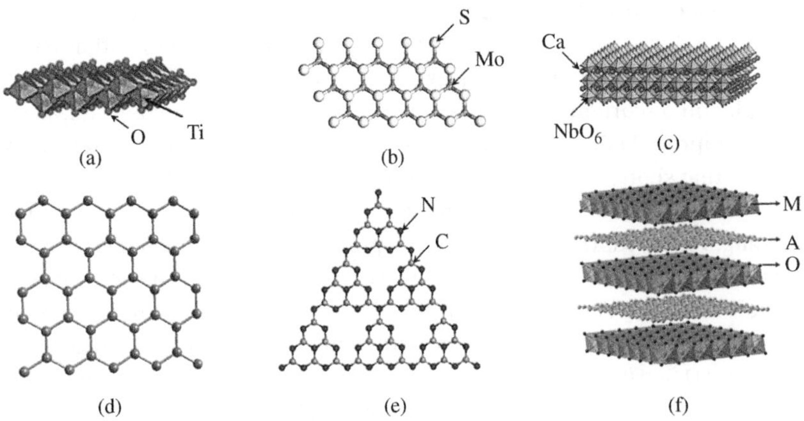

Figure 12.10 Crystal structure of (a) TiO_2. Reproduced with permission (Osada & Sasaki, 2012). Copyright 2012. Wiley-VCH. (b) MoS_2; (c) $Ca_2Nb_3O_{10}^-$. Reproduced with permission (Osada & Sasaki, 2012). Copyright 2012. Wiley-VCH. (d) Graphene; (e) tri-s-triazine structure models of g-C_3N_4; (f) LDHs. Reproduced with permission (Tan et al., 2017). Copyright 2017. American Chemical Society.

(Yang et al., 2009). The ZnO:N nanowire arrays as photoanodes demonstrated a significant increase in photoresponse in the visible region compared to non-doped ZnO nanowires. ZnO:N nanowires showed an order of magnitude increase in current density and photon-to-hydrogen conversion efficiency of 0.15% at +0.5 V (vs. Ag/AgCl). Therefore, it can be concluded that 1D nanostructures such as nanowires, nanotubes, and nanobelts are desirable nanomaterials due to their extremely large surface-to-volume ratio and significantly superior light absorption. Importantly, directional charge transport and/or precise control over crystal planes in 1D nanostructures are appealing attributes for promoting H_2 and/or O_2 evolution.

12.3.1.3 2D Materials

The 2D materials are an interesting class of materials, which possess sheet-like structure with atomic thickness and large lateral size. The presence of strong in-plane covalent bonds and ultrathin thickness provides exceptional mechanical strength, flexibility, and optical transparency for 2D materials (Chhowalla et al., 2016; Fiori et al., 2014; Geim & Novoselov, 2007). The ultrathin profile of 2D materials demonstrates many extraordinary physical, chemical, electronic, and optical characteristics that are inaccessible in their bulk counterparts. The utilization of 2D materials for photocatalytic applications has increased due to their peculiar advantages such as high surface area to volume ratio, a large number of active sites, and better electron mobility. Moreover, the 2D materials can be integrated with catalyst support materials for the construction of multicomponent photocatalysts. The ultrathin geometry of 2D materials significantly decreases the charge carrier diffusion length and improves the charge separation efficiency. Moreover, the high surface-to-volume ratio in 2D materials bestows facile modulation of properties and functionalities in terms of making surface modifications, adding functional groups, element doping, and defect/phase/strain engineering (Sun et al., 2014; Voiry et al., 2016; Zhang & Xie, 2013).

There exists an extensive library of ultrathin 2D materials with a wide composition range and crystal structure synthesized through various methods. 2D materials can be classified into two types: layered and non-layered. In layered materials, the in-plane atoms in each layer are connected through strong chemical bonds whereas these layers stack together via weak vdW forces to build bulk crystals (Nicolosi et al., 2013). For instance, graphite is a layered material in which graphene layers are weakly stacked (Freise, 1962). In contrast, non-layered materials are formed by chemical bonding in three-dimensional (3D) directions, which leads to unsaturated dangling bonds on the surface and causes a high-activity and high-energy surface. In addition to graphite, other significant 2D materials include TMDs, g-C_3N_4, h-BN, black phosphorus (BP), and layered double hydroxides (LDHs) (Osada & Sasaki, 2012; Sasaki & Watanabe, 1997). Similar to layered bulk crystals, the anticipation for the synthesis of ultrathin 2D materials from the non-layered bulk materials can bring up novel functionalities and properties, which cannot be realized from their bulk counterparts (Ambrosi et al., 2015; Chang, Hai, et al., 2016; Jin et al., 2018; Voiry et al., 2015; Zhou et al., 2009). The crystal structures of various 2D nanomaterials are shown in Figure 12.10a–f.

As previously mentioned, TiO_2 has been extensively studied as a photocatalyst due to its excellent chemical stability, non-toxicity, and low cost. In the past decades, TiO_2 nanosheets exfoliation has also gained significant consideration. For instance, TiO_2 nanosheets with higher concentrations of highly reactive [001] facets were reported to be appealing for photocatalytic processes. Han et al. synthesized TiO_2 nanosheets via hydrothermal method in the presence of hydrofluoric acid and obtained 89% of (001) facets as top and bottom surfaces (Han et al., 2009). These materials showed superior photocatalytic activity compared to Degussa P25. Among the layered perovskites, $Ca_2Nb_3O_{10}^-$ nanosheets exfoliated from bulk layered $KCa_2Nb_3O_{10}$, which belongs to Dion-Jacobson phase, were studied for various applications toward microelectronics and photocatalysis because of their homogenous structure and morphology (Haq et al., 2022; Khan, Kim, et al., 2017; Khan, Osada, et al., 2017; Wang et al., 2014). Ebina et al. (2002) utilized aggregates of restacked $HCa_2Nb_3O_{10}$ nanosheets for H_2 evolution under UV light irradiation from an aqueous solution. Compared to parent layered material, the restacked nanosheets demonstrated 10 times larger surface area and superior activity for photocatalytic H_2 generation. After RuO_x deposition during the exfoliation-restacking process, the tetrabutyl ammonium (TBA) supported $HCa_2Nb_3O_{10}$ nanosheets evolved 16- and 8-folds higher H_2 and O_2 than non-exfoliated bulk $KCa_2Nb_3O_{10}$.

The presence of cocatalysts contributes to the enhanced density of active sites and charge separation. The photocatalyst performance can also be increased by doping in order to modulate the concentration of electrons and/or expand the visible light absorption. Generally, in bulk materials, the effect of doping is not significant without a cocatalyst since a small fraction of dopants are present on the surface while the rest resides within the catalyst. In contrast, most of the dopants employed in 2D materials are distributed on the surface or close to the catalyst surface due to their ultrathin profile (~ few nm), and therefore, the dopants could enhance their activity as a cocatalyst. Ida and coworkers reported the synthesis and photocatalytic activity of Rh-doped calcium niobate (Figure 12.11a) nanosheets by exfoliating bulk layered counterpart to achieve excellent H_2 production from water/methanol system without the presence of any cocatalyst (Okamoto et al., 2011). Photogenerated holes oxidized the methanol into formaldehyde (1610 μmol), carbon dioxide (9 μmol), and formic acid (1 μmol), under 6 hr of illumination while the electrons reduced water molecules to produce H_2. The H_2 production rate from Rh(0.03)-doped $Ca_2Nb_3O_{10}$ nanosheets was 165 times higher compared to parent Rh(0.03)-doped bulk layered compound (Figure 12.11b) with a quantum efficiency of 65% at 300 nm. The H_2 production rate with undoped $Ca_2Nb_3O_{10}$ nanosheets and parent layered material were 21.0 and 1.6 μmol/hr, respectively. The H_2 produced from Rh(0.03)-doped $Ca_2Nb_3O_{10}$ nanosheets was ~1.54 times higher compared to Rh(1 wt%)-loaded $Ca_2Nb_3O_{10}$ nanosheets prepared by a conventional photo-deposition method. Furthermore, our group reported a concomitant strategy involving soft-chemical exfoliation and elemental doping on layered $KCa_2NaNb_4O_{13}$ to synthesize nitrogen (N^{3-}) and niobium (Nb^{4+}) co-doped $Ca_2NaNb_4O_{13}^-$ nanosheets (Khan, Diao, et al., 2020). The Dion-Jacobson phase, $KCa_2NaNb_4O_{13}$ ($n = 4$), is an important homologous member of layered $KCa_2Nb_3O_{10}$ ($n = 3$) perovskite, which has been utilized for various applications after exfoliation into ultrathin 2D nanosheets (Khan et al., 2021; Li et al., 2017). The optimized N^{3-}/Nb^{4+} co-doped $Ca_2NaNb_4O_{13}^-$ nanosheets due to improved light absorption after dopants exhibited 2.4 times higher photocatalytic O_2 evolution than undoped nanosheets without using a cocatalyst (Figure 12.11c).

MoS_2 is an important layered TMD for visible light photocatalysis. Bulk MoS_2 possesses an indirect bandgap of 1.2 eV, which is insufficient for performing the redox reactions for water splitting. At monolayer level, the 2D MoS_2 nanosheets possess a direct bandgap of ~1.96 eV due to quantum confinement effect. The thickness-dependent MoS_2 bandgap allows its usage as a promising photocatalyst and/or cocatalyst material for H_2 evolution (Singh et al., 2012). The role of MoS_2 as a cocatalyst is discussed via examples in the following section. Bulk g-C_3N_4 sheets possess a unique layered structure similar to graphite. However, the material has a low specific surface area <10 m^2/g. On the other hand, 2D g-C_3N_4 nanosheets offer unique physicochemical and electronic features such, as large surface-to-volume ratio, better electron mobility, strong/favorable band potentials, and a greater number of surface-active sites, making them effective for water splitting. Our group reported a facile strategy consisting of breaking and repolymerizing heptazine units in the bulk g-C_3N_4 for the synthesis of ultrathin (thickness ~ 1 nm) g-C_3N_4 nanosheets (Figure 12.11d). The experimental findings revealed that the disordered layer structure of ultrathin g-C_3N_4 resulted in strong interlayer C–N interaction and tunneling electron transfer between C and N layers (Zhao, Chen, et al., 2017). Additionally, the broken in-plane C—N bonds create more unsaturated N sites in 1 nm ultrathin g-C_3N_4 nanosheets. The subsequent g-C_3N_4 nanosheets exhibited excellent photocatalytic H_2 evolution response than their bulk counterpart (Figure 12.11e). The above studies clearly confirmed the peculiar advantages of various 2D materials compared to their bulk counterparts for photocatalytic water splitting.

12.3.1.4 Low-Dimensional Heterostructures

The quality of the interface regions between the layers in a heterostructure is a determining factor for the interfacial charge transfer efficiency. It is reported that the development of solid–solid interfaces containing a large number of defects can be controlled easily to generate quasi-continuous energy levels similar to the conductors. As a result, solid–solid interfaces exhibit excellent electrical conductivity. Among various defects, oxygen vacancies are commonly observed defects in

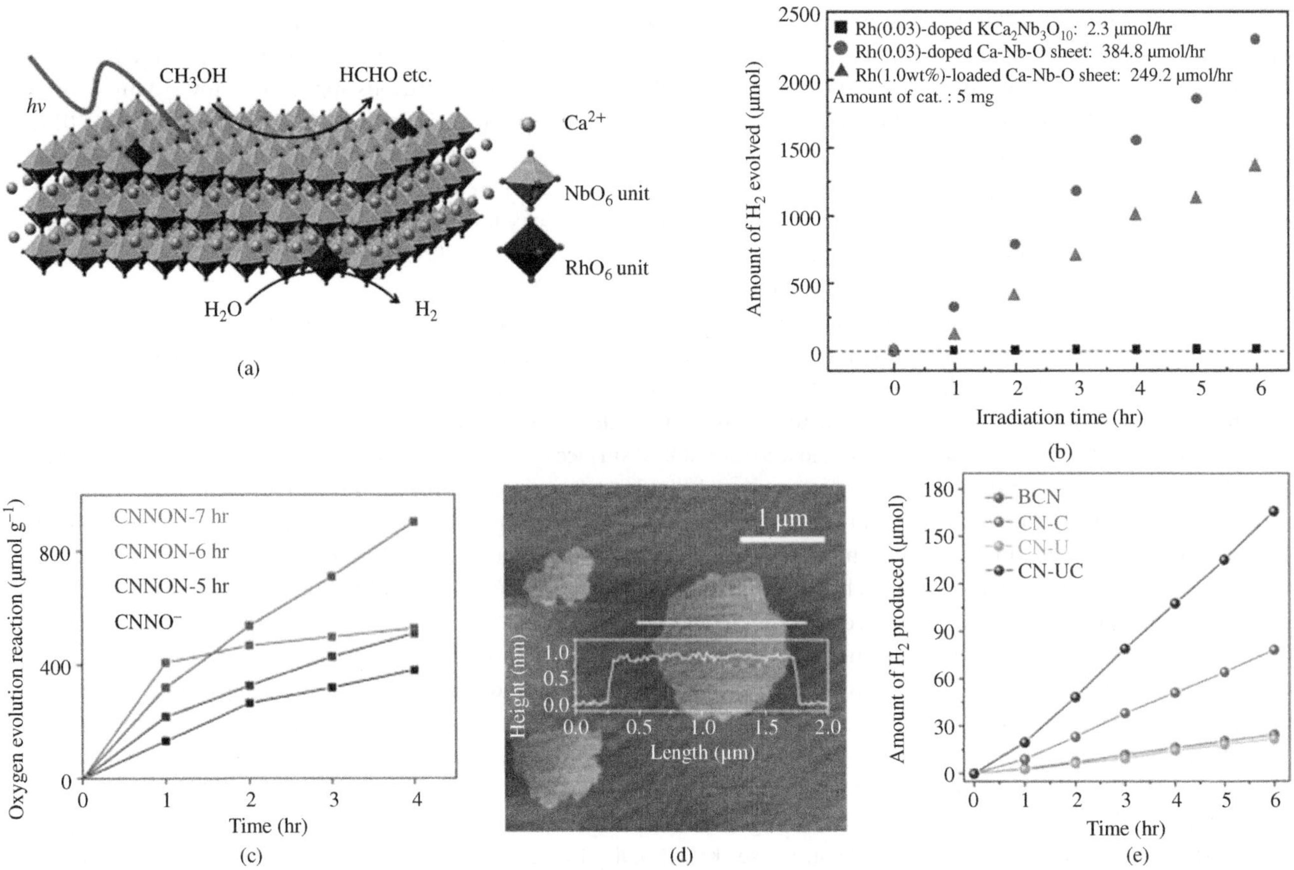

Figure 12.11 (a) Crystal structure of an Rh-doped $Ca_2Nb_3O_{10}^-$ nanosheet and its photocatalytic reaction model in water/methanol system; (b) Comparison of H_2 evolution from (●x025CF;) Rh(0.03)-doped Ca–Nb–O sheets, (■x025A0;) Rh(0.03)-doped $KCa_2Nb_3O_{10}$, and (▲x025B2;) Rh(1.0 wt%)-loaded Ca–Nb–O sheets. Reproduced with permission (Okamoto et al., 2011). Copyright 2011. American Chemical Society; (c) OER under full spectrum for undoped $Ca_2NaNb_4O_{13}^-$ ($CNNO^-$) nanosheets, N^{3-}/Nb^{4+} co-doped $Ca_2NaNb_4O_{13}^-$ ($CNNON^-$) calcined under NH_3 environment at different times (5, 6, 7 hr) without cocatalyst loading. Reproduced with permission (Khan, Diao, et al., 2020). Copyright 2019. Elsevier. (d) AFM image of ultrathin g-C_3N_4 (CN-UC); (e) Amount of H_2 gas generated by Bulk g-C_3N_4 (BCN), g-C_3N_4-calcined (CN-C), g-C_3N_4-ultrasonication (CN-U) and g-C_3N_4-ultrasonication + calcined (CN-UC) from TEOA aqueous solutions under visible light irradiation ($\lambda > 420$ nm). Reproduced with permission (Zhao, Chen, et al., 2017). Copyright 2017. Elsevier.

oxides. Xiao et al. constructed 0D/1D heterostructures comprised of g-C_3N_4 nanoparticles/TNTs with an interfacial oxygen vacancy layer (Figure 12.12a). The resultant 0D/1D heterostructure demonstrated improved visible light absorption and optimal photocurrent density of 0.72 mA/cm^2 at 1.23 V vs. reversible hydrogen electrode (RHE) under visible light (Xiao et al., 2019). The presence of an oxygen vacancy layer in 0D/1D heterostructure yielded eight-times higher photocurrent density than 0D/1D heterostructure without such a layer. As a result of oxygen vacancy layers, the Z-scheme heterostructure was obtained, which was validated through experimental and theoretical work using hydroxyl fluorescence capture technique and DFT calculations, respectively. In another work, Radhakrishnan et al. (2017) reported the use of fluorinated boron nitirde (FBN) QDs, which demonstrated broad-spectrum light absorption and high fluorescence quantum yield. The heterostructure of FBN QDs with wide bandgap TNTs arrays proved to demonstrate excellent photoelectrochemical H_2 evolution activity. The resultant OD/1D heterostructure showed high external quantum efficiency of 24% at 400 nm, which extended from ultraviolet to green region of solar spectrum. She et al. (2017) reported the photocatalytic water splitting using a 2D hybrid system comprising of α-Fe_2O_3 and g-C_3N_4 nanosheets to observe the HER half reaction and OWS. The close contact between the 2D nanostructured hybrids helped the material yield a H_2 production rate ~3×10^4 μmol·g^{-1}·h^{-1} and an external quantum efficiency of 44.3% at $\lambda = 420$ nm. The Z-scheme of this hybrid system effectively hindered charge carrier recombination and produced ~38.9 and 19.1 μmol·g^{-1}·h^{-1} of H_2 and O_2, respectively. Recently, our group reported the photocatalytic OWS via construction of 2D/2D heterostructure under Z-scheme charge transfer mechanism based on boron-doped and nitrogen deficient g-C_3N_4 nanosheets (BDCNN) as well as g-C_3N_4

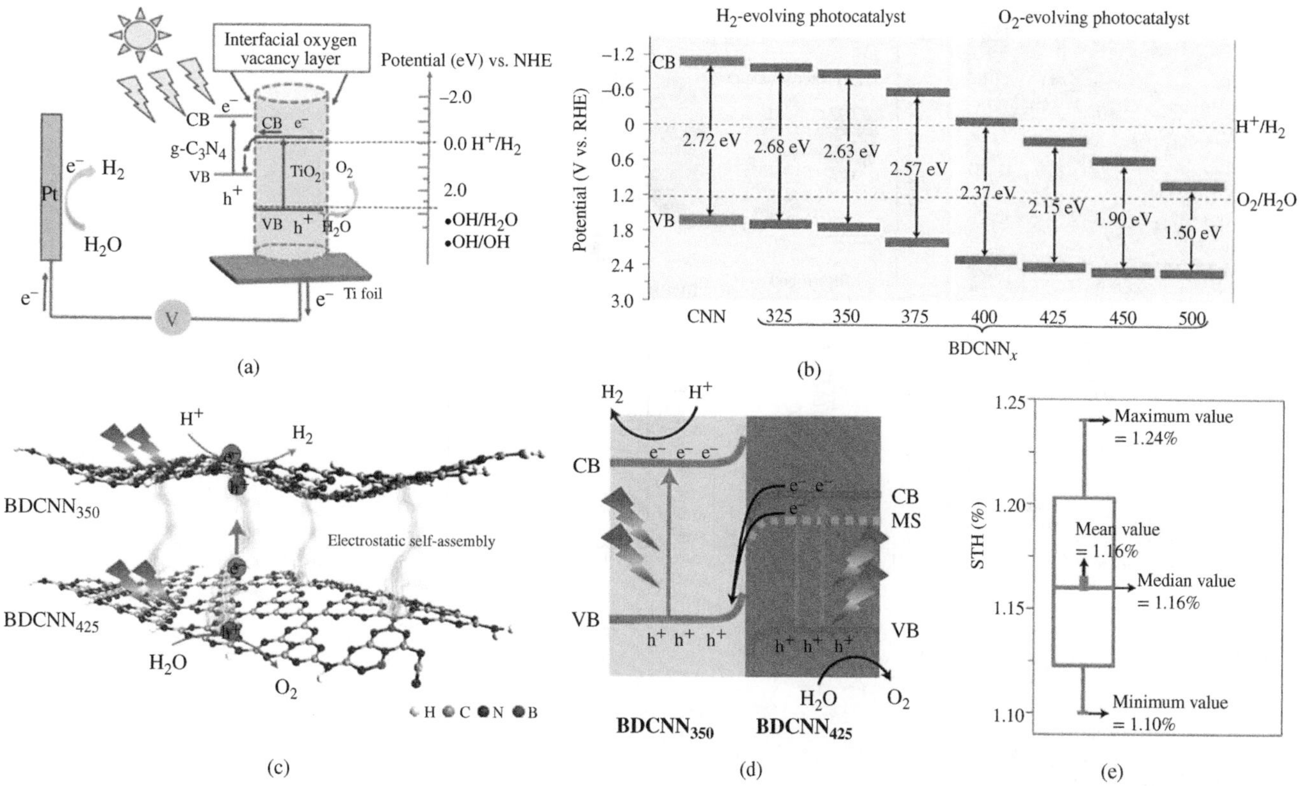

Figure 12.12 (a) Schematic diagram for charge transfer and separation for 0D/1D g-C_3N_4/TiO_2. Reproduced with permission (Xiao et al., 2019). Copyright 2019. American Chemical Society. (b) Band position diagram for g-C_3N_4 and boron-doped nitrogen deficient g-C_3N_4 nanosheets (x = 325, 350, 375, 400, 425, 450, 500 °C); (c) Graphical presentation of electrostatic self-assembly for charge transfer; (d) Perfectly aligned band structure for Z-scheme design; (e) Solar-to-hydrogen efficiency of $BDCNN_{350}$/$BDCNN_{425}$ for photocatalytic OWS. Reprinted with permission (Zhao et al., 2021). Copyright 2021. Springer Nature.

nanosheets (CNN). Figure 12.12b shows the band edge positions for CNN and various BDCNN to construct Z-scheme heterostructures (Zhao et al., 2021). An electrostatic self-assembly process resulted in a 2D/2D nanostructure with strong interfacial interactions and perfectly modulated band structure that presented excellent charge separation and transfer ability yielding stoichiometric H_2 and O_2 evolution and a solar-to-hydrogen efficiency of 1.16% under one-sun illumination (Figure 12.12c–e).

Layer-dependent bandgap provides an excellent pathway to modulate the electronic properties of various materials. In this perspective, BP is an interesting material, which possesses a layer-dependent direct bandgap in the range 0.3–2.0 eV (Liang et al., 2014; Qiao et al., 2014). In bulk form, black and red phosphorous can make a type I heterostructure with redox potentials too weak to carry out photocatalytic water splitting without a sacrificial agent. However, with quantum confinement effects, the energy band structure of black phosphorous QDs demonstrated a significant shift. Such black/red phosphorous QD heterostructures showed a staggered band alignment similar to a Z-scheme pattern (Figure 12.13a,b) for excellent charge transfer (Shi et al., 2019). The QDs-based heterostructure offered substantial light absorption and strong redox potentials for water reduction. The short lifetime of photogenerated charge carriers was reported to be a major drawback. Metal-semiconductor and p–n junction heterostructures were developed to overcome this drawback and achieve a better charge separation efficiency. The heterointerface between the components is extremely important to improve performance. Particularly, in the case of nanoparticles employed as cocatalysts decorated via conventional impregnation methods over semiconductors, the heterointerfaces usually suffer from limited interfacial contact at the molecular level, which reflects in the limited charge separation and photocatalytic activities (Bai et al., 2016; Ran et al., 2014). Sasaki's group reported an effective two-step flocculation strategy for controlled attachment of monolayers of metallic 1T-phase MoS_2 on restacked $Ti_{0.87}O_2^{0.52-}$ nanosheets via optimizing the concentration of H^+ (Nurdiwijayanto et al., 2018). The porous flaky aggregates comprising of restacked $Ti_{0.87}O_2^{0.52-}$ nanosheets decorated with monolayer MoS_2 nanosheets (Figure 12.13c) showed remarkable photocatalytic H_2 production rate (1.2 mmol·h^{-1}·g^{-1}). The exceptional

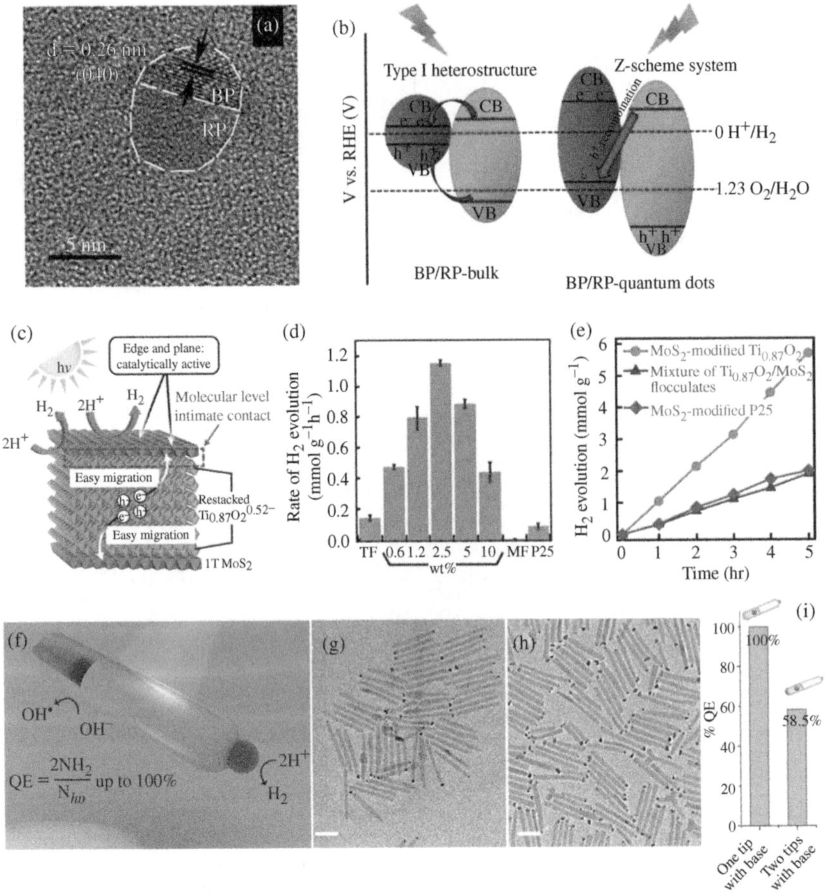

Figure 12.13 (a) HRTEM image of BP/RP-QD; (b) Schematic illustration for comparison of energy band structure of BP/RP bulk and BP/RP QDs. Reproduced with permission (Shi et al., 2019). Copyright 2019. Royal Society of Chemistry. (c) Graphical illustration represents the effective charge transfer across heterointerface and MoS_2 provides catalytic active sites for H_2 generation; (d) Comparison of photocatalytic H_2 production over TF (restacked $Ti_{0.87}O_2^{0.52-}$ nanosheets), different loading concentration of MoS_2 (0.6–10 wt%) on restacked $Ti_{0.87}O_2^{0.52-}$ nanosheets, MF (restacked MoS_2) and P25; (e) Comparison of photocatalytic activities with physically mixed restacked $Ti_{0.87}O_2^{0.52-}$ nanosheets/MoS_2 (triangle points line) and P25-TiO_2 nanoparticles modified with MoS_2 (squared points line). Reproduced with permission (Nurdiwijayanto et al., 2018). Copyright 2018. American Chemical Society. (f) Schematic for multicomponent photocatalyst describing the photocatalytic process. TEM images of CdSe@CdS nanorods with; (g) Single; (h) Double Pt tips as cocatalysts; (i) Photocatalytic quantum efficiencies for HER on single and double Pt tipped samples. Reproduced with permission (Kalisman et al., 2016). Copyright 2016. American Chemical Society.

molecular-level coupling between two dissimilar nanosheets facilitated fast electron transfer, which enabled an efficient charge separation and high catalytic activity compared to restacked $Ti_{0.87}O_2^{0.52-}$ nanosheets and P25-TiO_2 nanoparticles modified with MoS_2 (Figure 12.13d,e). Recently, Amirav and coworkers achieved a perfect photon-to-hydrogen efficiency of ~100% under visible light irradiation through Pt-tipped CdS nanorods embedded with CdSe seeds (Kalisman et al., 2016). The multicomponent photocatalyst facilitated the effective separation of charge carriers with hole accumulation on CdSe and electron transferred to Pt tips (Figure 12.13f). Different nanostructures of CdS/CdSe loaded with Pt on one side (one tip) and both sides (two tips) were also prepared (Figure 12.13g,h). The QE was reduced to 58.5% when two-tip Pt-loaded CdS/CdSe nanostructures were employed (Figure 12.13i). Moreover, the overall efficiency was strictly dependent on the extreme pH conditions, the presence of isopropanol alcohol (IPA), and hydroxyl anion-radical redox couple operating as a shuttle to relay the holes. The optimized one-tip Pt loaded CdS/CdSe nanostructured photocatalyst exhibited nearly 100% QE for H_2 evolution. These findings demonstrate that low-dimensional nanomaterials-based heterostructure photocatalysts possess significant advantages in terms of superior interfacial properties, charge separation, light energy absorption range, and control over appropriate band alignment for excellent water splitting compared to their bulk materials.

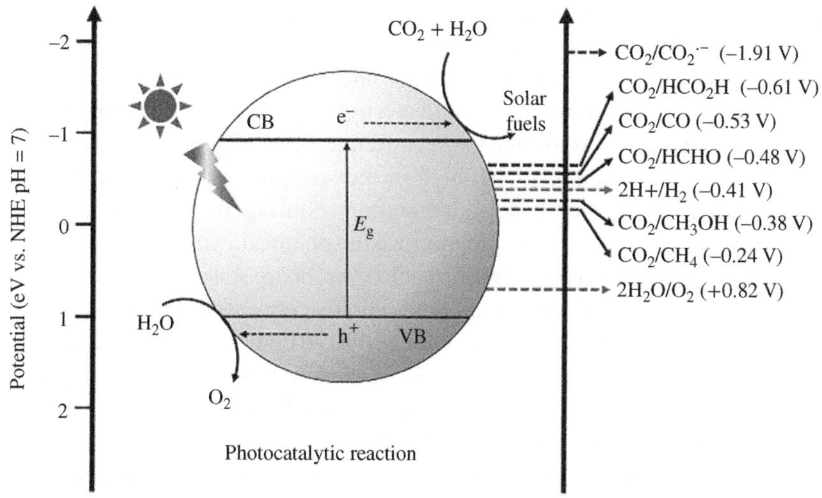

Figure 12.14 Schematic diagram for photocatalytic conversion of CO_2 into useful products and the reduction potentials for related products are also provided.

12.3.2 CO_2 Reduction

Similar to photocatalytic water splitting, the CO_2 conversion process on a semiconductor starts with the light irradiation and photon absorption with energy equal to or higher than the semiconductor bandgap to produce electron–hole pairs, which migrate to the surface for evolving the redox reaction (Figure 12.14). During the charge carrier migration, they can recombine in bulk or on the surface without being transferred to the electrolyte solution, a detrimental step in photocatalysis. The photocatalytic reduction of CO_2 is a more complex process than the water reduction process. In contrast to water reduction process involving two electrons, the conversion of CO_2 to organic products is a multielectron process and different pathways could lead to the synthesis of different products. The potential ($E°$ V vs. NHE at pH = 7) requirement of various products such as methanol (CH_3OH), formaldehyde (HCOH), carbon monoxide (CO), formic acid (HCO_2H), methane (CH_4), H_2 and O_2 production is described in Figure 12.14 and Eqs. (12.4)–(12.11). During the photocatalytic reaction, water reduction reaction competes with CO_2 reduction process, hence making CO_2 conversion more challenging. Therefore, an essential strategy is to tailor the semiconductor surface to preferentially reduce CO_2 rather than water. Furthermore, single-electron transfer to CO_2 molecule is a rate-limiting step as it requires the highest reduction potential of −1.9 V vs. NHE at pH = 7 (Rawool et al., 2021).

$$CO_2 + e^- \rightarrow CO_2^{\cdot-}, E° = -1.90\,V \tag{12.4}$$

$$CO_2 + 2H^+ + 2e^- \rightarrow HCOOH, E° = -0.61\,V \tag{12.5}$$

$$CO_2 + 2H^+ + 2e^- \rightarrow CO + H_2O, E° = -0.53\,V \tag{12.6}$$

$$CO_2 + 4H^+ + 4e^- \rightarrow HCHO + H_2O, E° = -0.48\,V \tag{12.7}$$

$$CO_2 + 6H^+ + 6e^- \rightarrow CH_3OH + H_2O, E° = -0.38\,V \tag{12.8}$$

$$CO_2 + 8H^+ + 8e^- \rightarrow CH_4 + 2H_2O, E° = -0.24\,V \tag{12.9}$$

$$2H^+ + 2e^- \rightarrow H_2, E° = -0.41\,V \tag{12.10}$$

$$2H_2O \rightarrow O_2 + 4H^+ + 4e^-, E° = +0.81\,V \tag{12.11}$$

12.3.2.1 0D Materials

The performance of photocatalytic CO_2 reduction is governed by various steps such as amount of light absorption, transport of charge carriers onto the photocatalyst surface, recombination rate of photoexcited electrons and holes, interaction of electrons and holes on photocatalyst surface with reactants/intermediates, and mass transfer of reactants to the catalyst surface. The particle size and surface properties of a photocatalyst have significant effect on modulating the aforementioned steps for improved photocatalytic response. For nanomaterials, particle size is an essential factor in specific surface area and photon conversion efficiency. Kočí et al. (2009) studied the effect of TiO_2 particle size on the photocatalytic CO_2 reduction process. They examined TiO_2 nanoparticles with diameters ranging from 4.5 to 29 nm. When the particle size was decreased to 14 nm, the yield of CO_2 photoreduction into methanol and methane became maximum. A further reduction in particle size below 14 nm resulted in lower yield due to changes in optical and electronic properties of nanomaterials. The maximum performance exhibited by 14 nm nanoparticles was ascribed to a collective effect of specific surface area, charge carrier dynamics, and light absorption efficiency. A plausible explanation for a reduced photocatalytic response for nanoparticles below 14 nm was associated with an increased possibility of agglomeration resulting in a decrease in active sites and increased bandgap due to the quantization effect. In another study, to improve the photocatalytic response of TiO_2 nanoparticles, Kwak et al. (2015) investigated the effect of different concentrations of Ni ions incorporated in TiO_2 nanoparticles. The optimized Ni-TiO_2 (0.1 mol%) presented the highest CH_4 yield than other doping concentrations of Ni ions. Ni ions incorporation decreased the PL response. The lowest PL intensity was found for Ni-TiO_2 (0.1 mol%) system. Therefore, it was concluded that addition of Ni ions decreased the photoexcited charge carrier recombination by trapping electrons from CB of TiO_2 nanoparticles. However, the higher Ni concentration (0.5 and 1.0 mol%) would not reduce electron trapping, as evident from the fact that the PL intensity increased significantly at higher Ni concentrations. The highest yield of CH_4 generated by Ni-TiO_2 (0.1 mol%) photocatalyst (14 μmol/g) was associated with large surface area and decreased electron–hole pair recombination process.

Li et al. reported the synthesis of inexpensive photocatalysts, cubic silicon carbide (3C-SiC) nanoparticles, through a facile mechanical ball milling method, which reduced CO_2 into methane with a selectivity of over 90% (Li & Sun, 2021). The large-scale production of 3C-SiC nanoparticles coated with amorphous silicon oxide (SiO_x) shell acquired a large number of surface states suitable for photoexcited charge separation. The presence of SiO_x shell provided stability against corrosion under visible light irradiation. The unique electronic structure of 3C-SiC nanoparticles allows the simultaneous water oxidation and CO_2 reduction into CH_4. Oxygen vacancies without the introduction of external impurities are considered self-doping and therefore maintain the intrinsic crystal structure. Hezam et al. (2020) reported a concentrated sunlight-assisted combustion process of CeO_2 nanoparticles to achieve better photocatalytic performance for CO_2 reduction into methanol (0.702 μmol· h^{-1}·g^{-1}) compared to CeO_2 nanoparticles synthesized via conventional solution combustion (CSC) process (0.397 μmol·h^{-1}·g^{-1}) as shown in Figure 12.15a,b. The novel sunlight-assisted method was performed under a semi-closed two-neck flask system. The evolved toxic gases (Co_x and NO_x) were bubbled in NH_3 solution for their removal to make the method green and sustainable for nanoparticle synthesis. The as-synthesized CeO_2 nanoparticles have small particle size of 12–18 nm with the presence of oxygen vacancies, large surface area, extended visible light absorption, better charge carrier separation, and superior photocatalytic activity.

Organic–inorganic hybrid perovskites are interesting materials when their size is reduced to form NCs due to increased photoluminescence quantum yield (PLQY) with concomitant advantages of large surface area and improved stability. For example, lead-based NCs have been studied for photocatalytic applications (Xu et al., 2017). Hou et al. reported the solution phase synthesis of colloidal $CsPbBr_3$ QDs (3–12 nm) for photocatalytic CO_2 reduction (Figure 12.15c,d). The bandgap energies and PL profiles depended on the quantum size effect (Hou et al., 2017). The $CsPbBr_3$ QDs achieved high selectivity of >99% for CO_2 reduction into solar fuels with additional advantages of prolonged carrier lifetime due to superior electron–hole separation and transfer. However, the lead (Pb) toxicity is an obstacle to their widespread applications. Sn-based perovskite NCs are suitable alternatives for Pb-based; however, these perovskites have disadvantages such as low stability and small PLQY (Jellicoe et al., 2016). Bismuth-based perovskite NCs are another important class of materials with their nontoxic profile and chemical stability for photocatalytic applications (Djatoubai et al., 2022a, 2022b; Zhou et al., 2018). A study conducted by Bhosale et al evaluated the photocatalytic reduction of CO_2 at the gas–solid interface for Bi-based perovskite NCs with chemical formula $A_3Bi_2I_9$ ($A^+ = Rb^+$, or Cs^+ or $CH_3NH_3^+$). The trend for CO production followed $Cs_3Bi_2I_9 \gg Rb_3Bi_2I_9 \gg MA_3Bi_2I_9 \gg TiO_2$ and that for methane followed $Rb_3Bi_2I_9 \gg Cs_3Bi_2I_9 \gg MA_3Bi_2I_9 \gg TiO_2$ (Bhosale et al., 2019). Due to superior charge transfer in $Cs_3Bi_2I_9$ NCs, they exhibited excellent photocatalytic performance with a yield of 14.9 and 77.6 μmol·h^{-1}·g^{-1} for methane and CO, respectively. The Bi-based perovskite NCs have mid-range bandgap and are active in the visible region of sunlight with promising binding energies, demonstrating their potentials for photocatalytic applications (Djatoubai et al., 2023). Wu and coworkers utilized CdSe/CdS QDs for CO_2 conversion into CO gas along with organic transformation into value-added pinacol chemical in liquid phase by simultaneous full consumption of photoexcited electrons and holes (Figure 12.15e). Under visible light irradiation, CdSe/CdS QDs generated 412.8 mmol·g^{-1}·h^{-1} of CO gas via CO_2 reduction with high selectivity of >96% in the presence of triethylamine (Guo et al., 2019). Under monochromatic LED light irradiation ($\lambda = 450$ nm, 60 mW/cm^2, and 1.0 cm^2), apparent quantum yield (AQY) of 32.7% for CO evolution was observed. Additionally, CO_2 photoreduction into CO proceeded steadily even in the presence

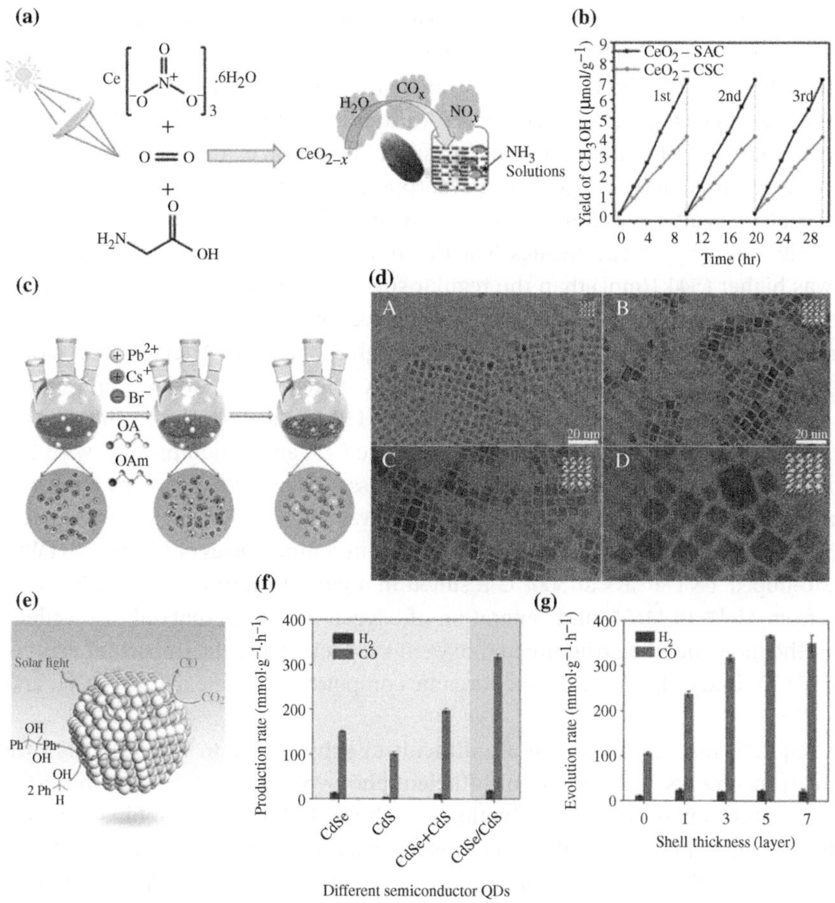

Figure 12.15 (a) Illustration of a sunlight-driven combustion (SAC) process for the development of CeO_2 nanostructures with excess oxygen vacancies; (b) Comparison of methanol evolution rate as a function of irradiation time for three cycles for SAC and conventional solution combustion (CSC) assisted synthesis of CeO_2 nanostructures. Reproduced with permission (Hezam et al., 2020). Copyright 2020. American Chemical Society. (c) Schematic diagram of solution phase synthesis of colloidal $CsPbBr_3$ QDs; (d) TEM images of $CsPbBr_3$ QDs with different particle sizes (a–d represents 3.8, 6.1, 8.5, and 11.6 nm, respectively). Reproduced with permission (Hou et al., 2017). Copyright 2017. Wiley-VCH. (e) Schematic figure of photocatalytic CO_2 reduction; (f) Comparison of CO_2 photoreduction performance of different semiconductor QDs; (g) Comparison of effect of CdS shell thickness on photocatalytic CO_2 reduction of CdSe/CdS QDs. Reproduced with permission (Guo et al., 2019). Copyright 2019. Elsevier.

of oxidative coupling of 1-phenylethanol to generate pinacol. The production yield for pinacol was up to 98%. The protons and electrons generated as a consequence of the oxidative half reaction are consumed for CO_2 reduction. The photocatalytic conversion efficiency of CdSe/CdS QDs for CO_2 reduction into CO ($318.9 \pm 5.8\ \text{mmol}\cdot\text{g}^{-1}\cdot\text{h}^{-1}$) was significantly higher than CdSe, CdS, and physical mixture of both QDs under similar reaction conditions (Figure 12.15f). Furthermore, the thickness effect of CdS shell greatly influenced the light harvesting and charge separation in heterostructure QDs as given in Figure 12.15g. In summary, the nanoparticles size, the doping concentration, surface area, and the number of active sites are imperative factors for enhancing the photocatalytic performance of 0D nanomaterials of various nanomaterials.

12.3.2.2 1D Materials

1D nanostructures of materials such as TiO_2, Si, and GaN have also been employed for CO_2 reduction due to their advantages such as high surface-to-volume ratio, which leads to short distance for charge carriers to travel to reach the surface of the photocatalyst resulting in a significant reduction in the charge carrier recombination rate. The primary step in CO_2 reduction is the adsorption of CO_2 molecules on catalyst surface. The production of defects in the photocatalyst leads to the generation of active sites that facilitate CO_2 adsorption and activation. Crystal defects can be classified as point defects, line defects, planar defects, and bulk defects. Point defects are vacancies or elemental impurities. Line defects include extended dislocations and planar defects are grain/twin boundaries and stacking faults (Adepalli et al., 2013; Naldoni et al., 2019; Thompson & Yates, 2005). Bulk defects consist of cracks, pores, and voids. Point defects in TiO_2 can be created via Ti/oxygen vacancies, Ti interstitial, and substitutional or interstitial impurity. However, oxygen vacancy generation is easier than the formation of interstitial vacancies (Zhao, Pan, and Li, 2017). The removal of one oxygen leaves an excess of two valence electrons at the vacant oxygen site. Then, the adjacent Ti^{4+} is reduced to Ti^{3+} state (Xiong et al., 2012). For adsorption and activation of CO_2 molecules, the presence of oxygen defects generates active sites. Various theoretical studies confirmed that the oxygen vacancies contribute to a stronger binding of CO_2 molecules. For instance, Thompson et al. (2003) reported the CO_2 adsorption and desorption characteristics via temperature-programmed desorption on the fully oxidized and defective TiO_2(110) surfaces. For fully oxidized surface, CO_2 is attached to five-fold coordinated Ti^{4+} atoms, whereas for defective TiO_2 surface, CO_2 is bonded to both regular sites and oxygen-vacant sites. The theoretical results confirmed that the binding energy of CO_2 to oxygen-vacant sites was higher (54 kJ/mol) than the regular sites (48 kJ/mol). Furthermore, Huygh et al. reported the finding of DFT calculations with long-range dispersion energy corrections on adsorption, dissociation, and diffusion of CO_2 on the oxidized and reduced (001) surfaces of anatase TiO_2 (Huygh et al., 2016). After the interaction of CO_2 with the oxidized surface, it takes the monodentate carbonate configuration. A small energy barrier is present for converting physiosorbed CO_2 in contrast to chemisorbed. The large energy barrier of 113.6 kcal/mol inhibits the dissociation of CO_2 on the oxidized surface. In the case of oxygen vacancy sites, highly stable CO_2 adsorption prevails with the elongation of C—O bond. The activation of C—O bonds leads to the exothermic CO_2 dissociation with an energy barrier of 22.2 kcal/mol. The final product is stabilized by taking one oxygen from CO_2 to fill the vacant site and form CO molecules. Wu and coworkers produced Mo-doped TNT arrays and calcined them in air at different temperatures (120–700 °C) (Wu, Nguyen, et al., 2015). The calcination of Mo-doped TNT arrays at 500 °C resulted in a partial rupture of nanotubes into anatase particles, reduction of Mo species from Mo^{6+} to Mo^{5+}, and generation of oxygen vacancies that led to excellent CO_2 reduction. The authors concluded that the molybdenum structure and oxygen vacancies were the important aspects for improving the photocatalytic reduction of CO_2 into CH_4, CO, and other organic compounds at a rate of 0.52, 10.41, and 13.53 μmol/g, respectively.

The reduction and oxidation potentials required for CO_2 reduction are difficult to achieve due to the necessity of a large overpotential. In Z-scheme or tandem arrangements, the presence of sufficient photovoltage can overcome this large potential difference needed for water oxidation at the anode and CO_2 reduction at cathode. PEC cells in combination with photovoltaic schemes can enhance the light absorption and photovoltage. For instance, a three-layered ZnO@ZnTe@CdTe core-shell 1D nanorod array photocathode furnished with Au nanoparticles was combined in a tandem configuration with a $CH_3NH_3PbI_3$ perovskite solar cell (Jang et al., 2016). The subsequent system achieved consumption of photons with energies higher than 1.5 eV due to concomitant usage of hybrid organic/inorganic perovskites and Au-ZnO@ZnTe@CdTe 0D–1D heterostructures. Without any external bias, CO_2 was reduced into CO with conversion efficiency of 0.35% and solar-to-fuel cumulative conversion efficiency of 0.43% when the yield of H_2, the side product, was also included in the calculation.

When 2D materials derived from their layered counterparts are rolled to form a tubular morphology, surface structure distortions are created, which generate oxygen defects due to the escape of surface oxygen atoms. The atoms in the proximity of oxygen vacancies are unsaturated in terms of coordination with a greater number of dangling bonds that can act as active sites (Sun et al., 2015). Moreover, oxygen defects can also serve as charge carrier separation centers and increase the efficiency of carrier utilization. To validate this, Di et al. (2018) used layered $Bi_{12}O_{17}Cl_2$ to synthesize oxygen-defects-incorporated $Bi_{12}O_{17}Cl_2$ superfine nanotubes, which were constructed out of $Bi_{12}O_{17}Cl_2$ bilayers to form the ultrathin tube walls. The ultrathin tube walls ensured fast bulk charge diffusion, the tubular morphology accelerated the surface charge separation, and the oxygen defects efficiently activated the CO_2 molecules adsorbed on the active sites to selectively produce CO at a rate of 48.6 $\mu mol \cdot g^{-1} \cdot h^{-1}$. This rate was 16.8 times higher than that obtained using bulk $Bi_{12}O_{17}Cl_2$ while maintaining good stability beyond 12 hr. Pan et al. documented the synthesis of In_2O_3 nanobelts coated with a carbon layer (5 nm) as presented in Figure 12.16a for direct photocatalytic reduction of CO_2 into CO and CH_4 gases at an evolution rate of 126.6 and 27.9 $\mu mol \cdot g^{-1} \cdot h^{-1}$, respectively (Pan et al., 2017). However, in the absence of carbon layer, the cocatalyst, pure In_2O_3 nanobelts, evolved H_2 gas as a major product (Figure 12.16b). The presence of carbon coating not only improved visible light absorption and separation of photoexcited charge carriers but also enhanced the chemisorption of CO_2 molecules and induced greater production of protons from water splitting to participate in CO_2 reduction.

1D Cu_2S nanomaterials decorated with cocatalysts have the ability to form a Schottky junction at the interface and facilitate the separation of photoexcited electron–hole pairs to improve CO_2 reduction. Manzi et el. (2015) reported a novel strategy for the synthesis of Pt-tipped Cu_2S nanorods via facile cation exchange method (Figure 12.16c). The Pt tips deposited over Cu_2S nanorods were efficiently utilized for the reduction of CO_2 to CO and methane at a production rate of 3.02 and 0.13 $\mu mol \cdot g^{-1} \cdot h^{-1}$, respectively. It is worth mentioning that no H_2 was quantified, suggesting that this

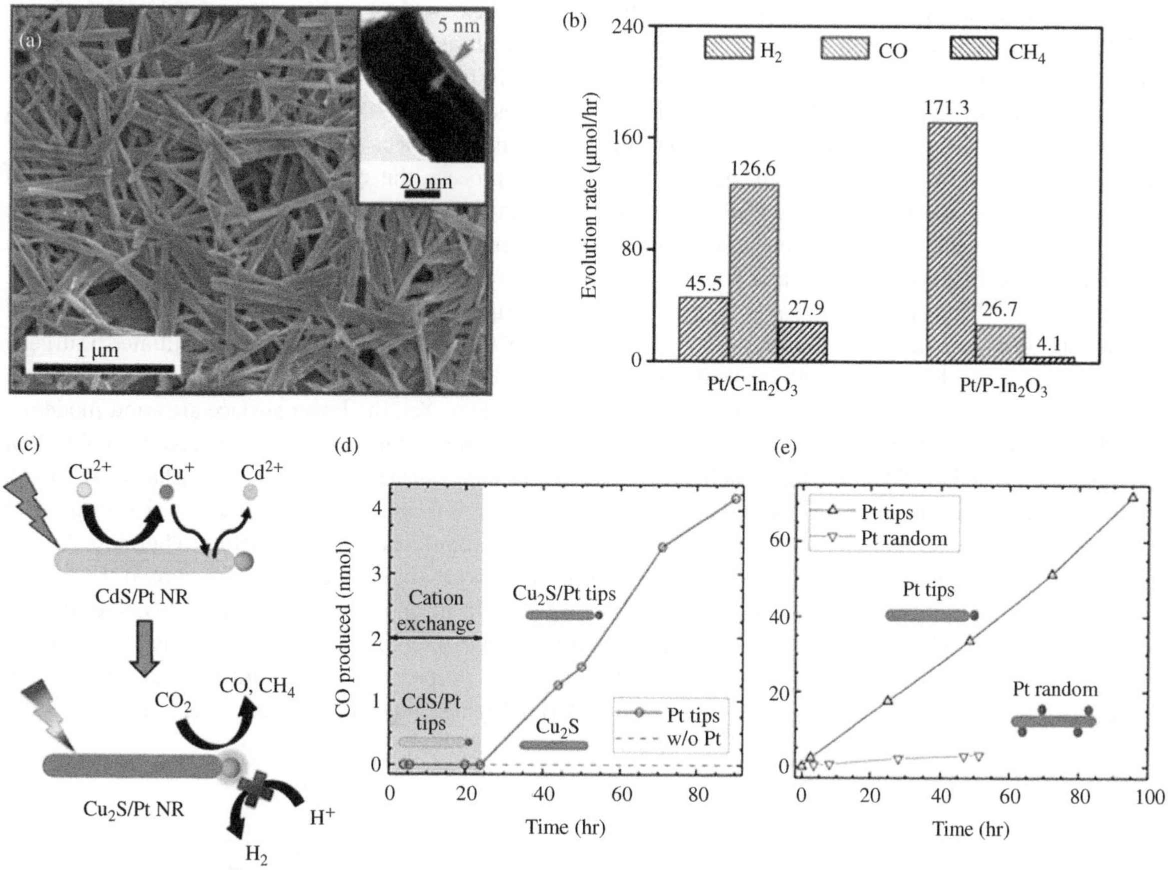

Figure 12.16 (a) SEM image of C-In_2O_3 with inset showing the TEM image; (b) Comparison of photocatalytic response for evolution of H_2, CO, and CH_4 gases over Pt/C-In_2O_3 and Pt/P-In_2O_3. Reproduced with permission (Pan et al., 2017). Copyright 2017. American Chemical Society. (c) Schematic for light-induced cation exchange process, followed by photocatalytic CO_2 reduction into CO and CH_4; (d) Pt tips/CdS and Pt tips/Cu_2S nanorods after illumination with a 447 nm laser; (e) Pt tips/Cu_2S nanorods and Pt random/Cu_2S nanorods under broad-spectrum Xe lamp. Reproduced with permission (Manzi et al., 2015). Copyright 2015. American Chemical Society.

heterojunction was selective for CO_2 reduction. Compared to Pt tipped/Cu_2S, bare Cu_2S nanorods could not present any activity due to the fast recombination of charge carriers. Additionally, the preferential Pt deposition on Cu_2S nanorod tips helped effective charge carrier separation via directional transport of electrons. Therefore, the Pt-tipped Cu_2S nanorods achieved higher activity for photoreduction of CO_2 compared to Cu_2S nanorods with randomly deposited Pt nanoparticles (Figure 12.16d,e).

12.3.2.3 2D Materials

By definition, 2D materials are crystals with thickness of one or a few atomic layers and lateral dimensions larger than 100 nm. 2D materials possess large surface area to volume ratio, which provides abundant surface-active sites. Due to ultra-thin thickness, photocatalysts with short charge carrier diffusion lengths can also perform efficiently. As dimensions have significant effects on bandgap and band edge position, these physical quantities can be controlled by tuning the number of layers. In addition, various terminating groups and dangling bonds are formed during the synthesis of 2D materials, which can be beneficial for the CO_2 adsorption that facilitate improved chemical yield. Therefore, 2D materials are an important class of nanomaterials for CO_2 photoreduction. Important 2D materials include hematene, ilmenene, MnTe, layered perovskites, and MXenes to name a few (Balan, Radhakrishnan, and Kumar, 2018; Balan, Radhakrishnan, and Peupane, 2018; Balan, Radhakrishnan, Woellner, et al., 2018; Ghimire & Klinke, 2021).

Metal oxides are important photocatalysts with strong anti-oxidant properties. The commonly used metal oxides include TiO_2, ZnO, Co_3O_4, and WO_3. The surface structure of a photocatalyst is of significant importance as it governs many physical and chemical processes occurring on the surface (Bi et al., 2011; Jiang et al., 2012). Therefore, the photocatalytic activity of semiconducting nanosheets is also greatly influenced by their exposed crystal planes. For anatase TiO_2, various facets such as {101}, {001}, and {100} have been studied extensively for photocatalysis because of their superior surface atomic structure and electronic structure (Pan et al., 2011). The {100} facets have 100% 5-coordinated Ti atoms (Ti_{5c}) that function as active sites for photocatalytic reactions. The superior photocatalytic response of TiO_2 nanostructures with {100} facets was associated with unique electronic band structure where the photogenerated electrons have sufficiently strong reductive potential to reduce H^+ into H_2. However, during the crystal growth process, the synthesis of TiO_2 nanostructures is dominated by {101} facets (~94%) and {001} facets, which are thermodynamically favorable per the minimization of surface energy (Liu et al., 2011). The pioneering theoretical work reported by Barnard et al. evaluated the effect of various adsorption configurations on the equilibrium shape and phase transition of TiO_2 nanoparticles in water. The presence of water affected the morphology of TiO_2 NCs. As a result, the size of {001} and {001⁻} truncated facets was modified in anatase and a decrease in the aspect ratio was observed in rutile NCs (Barnard et al., 2005). These theoretical findings initiated a surge of experiments on the synthesis of TiO_2 NCs with controlled morphology and surface chemistry. The synthesis of TiO_2 single crystals with {100} facets was accomplished with rods and cuboid morphology. Yet, the lower surface area and moderate percentage of {100} facets limited the practical application of TiO_2 crystals. For instance, the highest percentage for {100} facet (85.7%) was documented for TiO_2 cuboids, which have a relatively smaller surface area of 5.8 m^2/g (Zhao et al., 2011). For photocatalytic applications, the large surface area plays an important role as it can contribute to increased adsorption of reactant molecules and ensure the presence of a greater number of surface-active sites for photoreaction. Therefore, Ye and coworkers reported the synthesis of TiO_2 single crystals with an ultrathin thickness of ~2 nm and 95% of exposed {100} facets (Xu et al., 2013). The presence of a large surface area and high ratio of {100} facets in ultrathin TiO_2 nanosheets resulted in five times higher activity for both H_2 evolution and CO_2 reduction compared to reference TiO_2 cuboids with 53% of exposed {100} facets (Figure 12.17a). Large surface area provided a greater number of surface-active sites and a higher ratio of {100} facets bringing superior electronic band structure, consequently offering higher photocatalytic activity (Figure 12.17b–d).

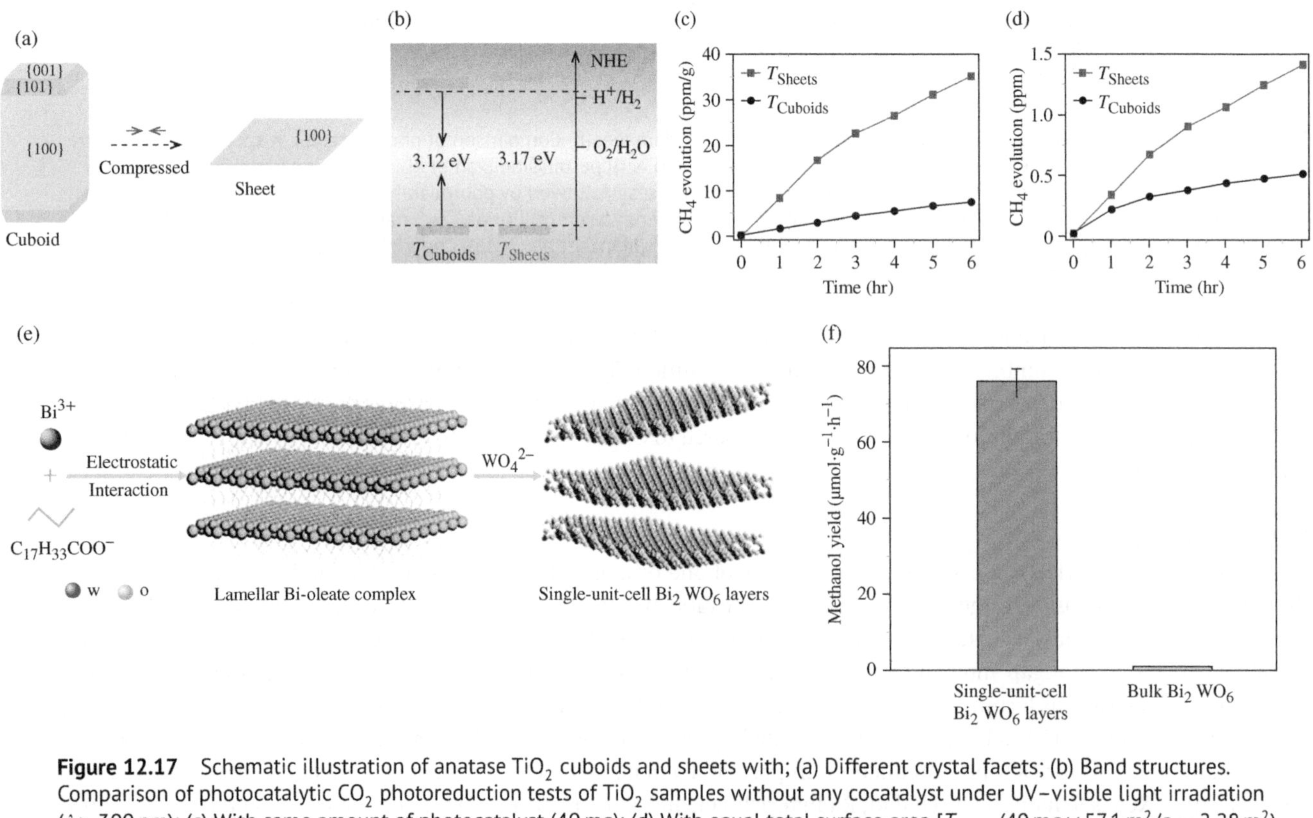

Figure 12.17 Schematic illustration of anatase TiO_2 cuboids and sheets with; (a) Different crystal facets; (b) Band structures. Comparison of photocatalytic CO_2 photoreduction tests of TiO_2 samples without any cocatalyst under UV–visible light irradiation ($\lambda > 300$ nm); (c) With same amount of photocatalyst (40 mg); (d) With equal total surface area [T_{sheets}(40 mg × 57.1 m^2/g = 2.28 m^2) and $T_{Cuboids}$(585 mg × 3.9 m^2/g = 2.28 m^2)]. Reproduced with permission (Xu et al., 2013). Copyright 2013. American Chemical Society. (e) Illustration of the formation of single-unit-cell Bi_2WO_6 layers; (f) Photocatalytic methanol evolution under 300 W Xe lamp irradiation for single-unit-cell Bi_2WO_6 layers and bulk Bi_2WO_6. Reproduced with permission (Liang et al., 2015). Copyright 2015. Wiley-VCH.

Among 2D inorganic materials, LDHs and perovskite oxides are important classes of layered materials for photocatalytic applications including CO_2 reduction. Ultrathin LDHs and perovskite oxides at atomic lengths with abundant metal sites unsaturated in coordination are promising due to a large number of surface active sites for catalytic reactions. Zhao et al. reported the bottom-up formation of ultrathin ZnAl-LDH nanosheets via inverse microemulsion or controlled hydrolysis route (Zhao et al., 2015). The oxygen vacancy-rich and Zn-coordinated unsaturated ZnAl-LDH nanosheets with three different thicknesses (~2.7, 4.08, and 210 nm) were synthesized. The experimental finding confirmed the thickness-dependent CO_2 reduction into CO. The ZnAl-LDHs nanosheets with thickness of 2.7 nm yielded maximum CO at a rate of 7.6 $\mu mol \cdot g^{-1} \cdot h^{-1}$, whereas, 4.08 nm thin nanosheets evolved 6.5 $\mu mol \cdot g^{-1} \cdot h^{-1}$ of CO, and no gas was detected with 210 nm thick platelets. Bai et al. studied the effect of different numbers of d electrons in e_g orbitals in four different MAl-LDHs (M = Mg^{2+}, Co^{2+}, Ni^{2+}, and Zn^{2+}) ultrathin nanosheets (Bai et al., 2020). In the presence of $[Ru(bpy)_3]Cl_2 \cdot 6H_2O$ as a photosensitizer, CoAl-LDH nanosheets showed excellent activity for CO_2 reduction into CO at a rate of 43.73 $mmol \cdot g^{-1} \cdot h^{-1}$ under irradiation of 600 nm monochromatic light. The suitable band edge position of CoAl-LDH nanosheets and HOMO and LUMO states of Ru complex photosensitizer offers superior CO_2 reduction even at 600 nm. Additionally, the outstanding activity of CoAl-LDH can be associated with multiple factors including a narrow bandgap of 2.45 eV, which gave good visible light absorption, significant CO_2 and H_2O adsorption due to $t_{2g}{}^6e_g{}^1$ configuration of Co^{2+} and the presence of abundant active sites owing to a large number of oxygen/metal vacancies in LDH nanosheets.

The low CO_2 photoconversion efficiency can be caused by rapid recombination of photogenerated carriers and availability of a limited number of surface-active sites. However, atomically thin 2D materials can eliminate these shortcomings. As a prototype, Liang et al. reported single unit-cell layers of Bi_2WO_6 derived from lamellar Bismuth-oleate intermediates (Figure 12.17e). The ultrathin Bi_2WO_6 layers ensured three-fold higher CO_2 adsorption ability and photo-absorption than bulk Bi_2WO_6 (Liang et al., 2015). The experimental and DFT calculations confirmed the increase in conductivity and temperature-dependent resistivities, which favor fast carrier transport. The electron–hole separation efficiency was improved as confirmed by the lifetime of charge carriers from 14.7 to 83.2 ns observed from time-resolved fluorescence spectroscopy. Hence, single unit-cell Bi_2WO_6 layers achieved methanol production at an excellent rate of 75 $\mu mol \cdot g^{-1} \cdot h^{-1}$, which was 125 times higher than bulk Bi_2WO_6 (Figure 12.17f). Surface defects demonstrate significant effect to tune the semiconductor photocatalytic properties. As in 2D nanomaterials, the surface-to-volume ratio is high, therefore, most defects are present on the surface and exhibit significant effects on photocatalysis compared to their bulk counterparts. Among various types of defects, oxygen defects are commonly studied. However, Di et al. (2019) visualized superior photocatalytic activity for CO_2 reduction in a bismuth vacancy (V_{Bi})-engineered BiOBr (V_{Bi}-BiOBr) photocatalyst. The ultrathin nanosheets of V_{Bi}-BiOBr delivered selective CO production at a rate of 20.1 $\mu mol \cdot g^{-1} \cdot h^{-1}$ in pure water in the absence of a cocatalyst, photosensitizer, or sacrificial agent. The Bi vacancies tuned the band structure, promoted charge separation, and increased CO_2 adsorption, activation, and CO desorption to obtain 3.8-fold higher CO evolution relative to BiOBr nanosheets.

12.3.2.4 Low-Dimensional Heterostructures

Copper-based semiconductors are interesting materials due to their high selectivity toward CO_2 reduction. For instance, copper oxide (CuO), and cuprous oxide (Cu_2O) possess suitable CB positions and small bandgaps, ~1.7, and ~2.2 eV, respectively, for visible light-assisted CO_2 reduction. However, the photocatalytic activity of pure copper-based nanomaterials is limited due to the high photogenerated carrier recombination rate associated with their small bandgaps (An et al., 2014; Yin et al., 2015). Furthermore, the poor photostability associated with cuprous nanomaterials as a result of photocorrosion caused by photoexcited holes is another significant factor for a poor photocatalytic response. Therefore, copper-based nanomaterials are combined with suitable cocatalysts or semiconductors in the form of heterostructures to improve electron migration for better charge carrier separation, reduce photocorrosion by efficient utilization of photoexcited holes, and decrease the activation energy barrier by modulating the reaction pathway for CO_2 photoreduction. For instance, Khatri and coworkers utilized high conductivity and a large surface area of rGO nanosheets to construct heterostructure with CuO nanorods via chemical interaction with 3-aminopropyltrimethoxysilane (APTMS) as bi-functional linker. They achieved excellent stability and photocatalytic performance for CO_2 reduction into methanol under visible light irradiation (Gusain et al., 2016). The rGO–CuO heterostructure exhibited seven times higher performance compared to pure CuO nanorods. The strong rGO–CuO interaction facilitated efficient capture and transfer of photogenerated electrons from CuO to rGO surface for CO_2 reduction and provided superior stability, up to six cycles, with negligible decrease in photocatalytic activity, which ruled out the surface carbon contamination. Similarly, when Cu_2O nanoparticles were deposited on TNTs arrays, the subsequent heterostructure yielded excellent photocatalytic CO_2 reduction into methane under visible and simulated sunlight irradiation (Li, Zhang, et al., 2015). The TNT arrays offered a large number of active sites for reactant adsorption and improved charge separation via facile transport of photogenerated charge carriers. The presence of Cu_2O nanoparticles enhanced light absorption and prevented the oxidation of hydrocarbon products. These advantages helped Cu_2O-modified TNTs arrays to achieve high efficiency for CO_2 reduction without the need for cocatalysts and sacrificial agents.

The TiO_2 NCs exposed with different facets possess distinct photocatalytic properties. Jaroniec and coworkers developed a new "surface heterojunction" concept in which the ratio of {001} and {101} facets was controlled to achieve the photoexcited charge separation on different facets in TiO_2 NCs (Figure 12.18a) and demonstrated excellent photocatalytic CO_2 reduction performance (Yu et al., 2014). Based on DFT calculations, the difference in photocatalytic activity for anatase TiO_2 was dependent on the ratio of co-exposed {001} and {101} facets due to different band structures and band edge positions of {001} and {101} facets. This heterostructure facilitated effective charge separation (Figure 12.18b). The optimal ratio of 45:55 for exposed {101} and {001} facets demonstrated superior photocatalytic reduction of CO_2 into methane. TMDs are an important class of 2D materials that could replace noble metal cocatalysts due to low cost and high stability (Li et al., 2019).

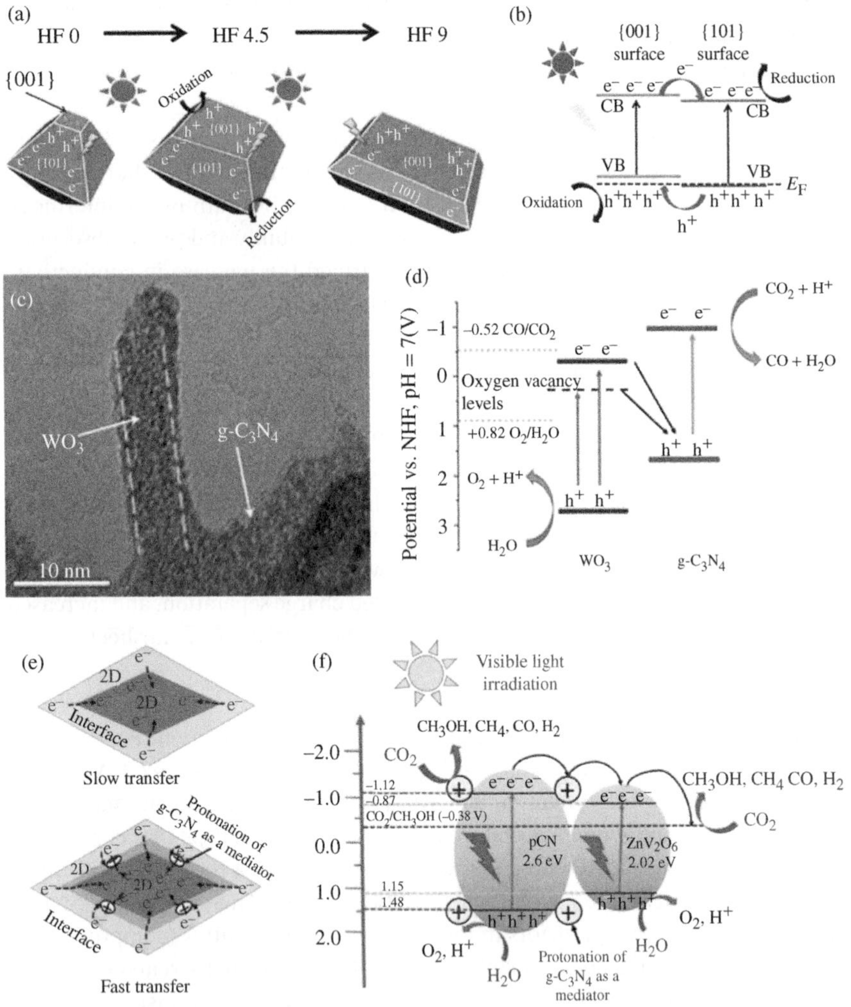

Figure 12.18 (a) Schematic illustration of the spatial separation of redox sites on the HF0, HF4.5, and HF9 samples; (b) {001} and {101} surface heterojunction. Reproduced with permission (Yu et al., 2014). Copyright 2014. American Chemical Society. (c) HRTEM image of WO_{3-x}/g-C_3N_4 composite; (d) Schematic diagram of CO_2 photoreduction into CO for WO_{3-x}/g-C_3N_4. Reproduced with permission (Huang et al., 2019). Copyright 2019. American Chemical Society. (e) Schematic illustration of contact interfaces for 2D/2D heterojunction (slow transfer) and 2D/2D heterojunction with protonation (HNO_3) as a mediator (fast transfer); (f) Schematic figure of the separation and transfer of photogenerated charges in ZnV_2O_6/pCN composite under visible light irradiation. Reproduced with permission (Bafaqeer et al., 2019). Copyright 2018. Elsevier.

Furthermore, under quantum confinement effect, the CB band position can be modulated based on number of layers. Hence, TMDs are an excellent choice as a cocatalyst. MoS_2 is the most important member of TMDs. MoS_2 has the tendency to show higher current density and lower overpotential compared to noble metals, therefore, possesses the ability to receive photoexcited electrons more effectively in composite materials. Yu and coworkers reported the 1D/2D TiO_2/MoS_2 nanostructured hybrids in which TiO_2 nanofibers were covered with MoS_2 nanosheets (Xu et al., 2018). The optimized heterostructure consisted of 10% MoS_2 loaded on TiO_2 nanofibers to achieve maximum optical absorption and increased CO_2 adsorption to realize superior photocatalytic performance for CO_2 reduction into methane and methanol at a rate of 2.86 and 2.55 $mmol{\cdot}h^{-1}{\cdot}g^{-1}$, respectively. Furthermore, MoS_2 as a cocatalyst can initiate the selective reduction of CO_2 into methanol as Mo cations at the edge sites of MoS_2 are helpful in stabilizing CH_xO_y intermediate via electrostatic attraction (Tu et al., 2017).

Pristine MXenes are metallic in nature and exhibit no semiconducting nature. However, after surface termination, MXene acquires semiconducting characteristics. MXenes are employed as a cocatalyst in combination with other semiconductors due to their lower Fermi level and metallic properties. The features such as large surface area and excellent electrical conductivity of MXenes provide them the ability to serve as a support for many photocatalysts (Wang, Wang, Liu, et al., 2020). For instance, Cao et al. constructed a heterostructure based on 2D/2D $Ti_3C_2/BiWO_6$ nanosheets. The 2D/2D combination offered short charge transport length and large interface contact area and ensured superior interfacial charge transfer abilities (Cao et al., 2018). The large surface area and porous structure assisted Ti_3C_2/Bi_2WO_6 hybrids for improved CO_2 adsorption. Under simulated sunlight irradiation, the total yield of methane and methanol was ~4.6 times higher in contrast to pristine Bi_2WO_6 nanosheets. Among all-inorganic perovskite NCs, pristine $CsPbBr_3$ has a strong CO_2 reduction potential. However, the photocatalytic activity is limited by severe electron–hole recombination. To improve the charge separation, a functional nanocomposite comprised of in-situ grown $CsPbBr_3$ perovskite NCs on 2D MXene nanosheets was reported (Pan et al., 2019). The $CsPbBr_3$/MXene nanocomposite represented substantial PL quenching and considerably shorter PL decay lifetimes than pure $CsPbBr_3$ NCs. As prepared $CsPbBr_3$/MXene nanocomposite, due to efficient charge transfer, exhibited selective reduction of CO_2 into CO and CH_4 without H_2 and other subproducts. Similarly, Wang et al. constructed Z-scheme heterostructure comprised of 0D $CsPbBr_3$ and 2D Bi_2WO_6 nanosheets (Wang, Wang, Li, et al., 2020). The 0D/2D heterostructure with strong redox potential achieved excellent photocatalytic CO_2 reduction. The total yield of CH_4/CO was 503 μmol/g, about 9.5-folds higher than pristine $CsPbBr_3$.

The g-C_3N_4, a nonmetallic semiconductor, absorbs light in the visible region of solar spectrum and is easy to synthesize from inexpensive raw materials. It shows excellent stability and nontoxicity for photocatalytic reactions. Although g-C_3N_4 is a potential photocatalyst, pristine g-C_3N_4 possesses certain drawbacks, including poor adsorption capacity for CO_2 molecules, limited visible light absorption, and easy recombination of charge carriers. In the last decade, many reports on low-dimensional g-C_3N_4 combined with various semiconductors to develop efficient heterostructure photocatalysts for CO_2 reduction were published. In a study, Zheng et al. (2019) developed a novel hybrid photocatalyst via coordinating 0D g-C_3N_4 QDs (g-CNQDs) with 2D ultrathin porphyrin MOF (PMOF). The g-CNQDs/PMOF hybrids were coordinated with Co active sites present in PMOF. As a result, the migration length for photoexcited charge carriers and gaseous substrates from g-CNQDs to Co active centers was reduced significantly. The 0D/2D hybrid photocatalyst exhibited 2.34 times superior CO production (16.1 $\mu m{\cdot}g^{-1}{\cdot}h^{-1}$) and six-folds higher CH_4 evolution (6.86 $\mu m{\cdot}g^{-1}{\cdot}h^{-1}$) compared to bare PMOF. Furthermore, 2D g-C_3N_4 nanosheets were composited with oxygen vacancy-derived WO_{3-x} nanorods (Figure 12.18c), which confirmed the separation of charge carriers via Z-scheme as analyzed from generated hydroxyl radicals (Huang et al., 2019). The presence of oxygen vacancies in WO_{3-x} nanorods generated new energy levels in bandgap and contributed to extended visible light absorption, better charge carrier separation in Z-scheme, and prolonged lifetime of electrons and holes (Figure 12.18d). Therefore, the material showed improved photocatalytic CO_2 to CO reduction performance. Furthermore, Bafqeer et al. employed protonation of g-C_3N_4 as excellent strategy for facile construction of composite structures with ZnV_2O_6 nanosheets. This system exhibited excellent electron transfer across the heterostructure interface (Figure 12.18e,f). The resultant 2D/2D heterostructure contained many high-speed charge transfer channels to transfer electrons from CB of g-C_3N_4 to the CB of ZnV_2O_6 through g-C_3N_4 (Bafaqeer et al., 2019). The protonated g-C_3N_4-based 2D/2D heterostructure showed excellent methanol yield (3742 μmol/g). Based on the abovementioned studies, it can be concluded that low-dimensional materials and their composites offer a large number of advantages compared to their bulk counterparts for facile photocatalytic/PEC reduction of CO_2 to useful solar fuels. These materials have the potential for large-scale production and utilization.

12.4 Summary and Future Perspective

In the last few decades, significant efforts have been devoted to the solar water splitting and CO_2 reduction in terms of development of low-dimensional materials. This chapter briefly describes the major advantages of low-dimensional materials related to solar conversion of water and CO_2 into solar fuels. We acknowledge that there are many significant reports and photocatalytic materials, but only a few are included as examples in this chapter. Considering the advantages of low-dimensional materials, it is most likely that these materials will be adopted for developing efficient photocatalytic systems.

The significant advances in the technologies for the synthesis of new low-dimensional materials are promising for their bright future. The ultimate goal is to achieve photocatalytic systems comprised of low-dimensional materials that can produce solar fuels at low cost and on large scales. To date, the efficiency, durability, and cost of the state-of-the-art systems are far from the requirements of a practical system. In terms of practical application for water splitting, the targeted system should be able to achieve AQY of 30% at 600 nm with an STH conversion efficiency of ~ 5% at a hydrogen cost of USD 2–3/kg. Therefore, future research needs should focus on increasing the efficiency of low-dimensional photocatalytic materials and decreasing the overall cost to generate solar fuels. For photocatalytic/PEC water splitting and CO_2 reduction, important strategies should include the design of photocatalyst that can absorb large proportion of sunlight ($E_g \sim 2$ eV) to produce photoexcited charge carriers, demonstrate fast transport of charge carriers across the interface between photocatalyst and cocatalyst/solvent, and possess enough surface-active sites for efficient redox reactions to take place.

Regarding cost, the focus of research should be on noble metal-free photocatalysts that should be abundant, stable under severe acidic or basic conditions, and non-toxic. The in-depth studies should focus on the correlation between structure, activity, and stability of the photocatalyst. Moreover, the design of multicomponent photocatalyst is needed for efficient water splitting and CO_2 reduction. A substantially better understanding of structure-activity correlation is likely to yield a better understanding of the photocatalyst design and to improve the efficiency of solar fuel generation. This will help achieve important aspects such as improved reproducibility and the capacity to modify or tailor photocatalysts. Moreover, the laboratory-based photocatalysts should be tested on outdoor solar-reactor to realize their potential in terms of practical application. In short, significant progress is yet to be achieved in the design of commercial-scale systems.

Acknowledgments

This work has been supported by National Natural Science Foundation of China (No. 5201101370), and the Fundamental Research Funds for the Central Universities.

References

Adepalli, K. K., Kelsch, M., Merkle, R., & Maier, J. (2013). Influence of line defects on the electrical properties of single crystal TiO_2. *Advanced Functional Materials, 23*(14), 1798–1806.

Ambrosi, A., Sofer, Z., & Pumera, M. (2015). 2H → 1T phase transition and hydrogen evolution activity of MoS_2, $MoSe_2$, WS_2 and WSe_2 strongly depends on the MX_2 composition. *Chemical Communications, 51*(40), 8450–8453.

Ampelli, C., Tavella, F., Genovese, C., Perathoner, S., Favaro, M., & Centi, G. (2017). Analysis of the factors controlling performances of Au-modified TiO_2 nanotube array based photoanode in photo-electrocatalytic (PECa) cells. *Journal of Energy Chemistry, 26*(2), 284–294.

An, X., Li, K., & Tang, J. (2014). Cu_2O/reduced graphene oxide composites for the photocatalytic conversion of CO_2. *ChemSusChem, 7*(4), 1086–1093.

Armaroli, N., & Balzani, V. (2007). The future of energy supply: Challenges and opportunities. *Angewandte Chemie International Edition, 46*(1–2), 52–66.

Awazu, K., Fujimaki, M., Rockstuhl, C., Tominaga, J., Murakami, H., Ohki, Y., Yoshida, N., & Watanabe, T. (2008). A plasmonic photocatalyst consisting of silver nanoparticles embedded in titanium dioxide. *Journal of the American Chemical Society, 130*(5), 1676–1680.

Bafaqeer, A., Tahir, M., & Amin, N. A. S. (2019). Well-designed ZnV_2O_6/g-C_3N_4 2D/2D nanosheets heterojunction with faster charges separation via pCN as mediator towards enhanced photocatalytic reduction of CO_2 to fuels. *Applied Catalysis B: Environmental, 242*, 312–326.

Bai, S., Wang, Z., Tan, L., Waterhouse, G. I., Zhao, Y., & Song, Y. F. (2020). 600 nm irradiation-induced efficient photocatalytic CO_2 reduction by ultrathin layered double hydroxide nanosheets. *Industrial and Engineering Chemistry Research*, *59*(13), 5848–5857.

Bai, S., Yin, W., Wang, L., Li, Z., & Xiong, Y. (2016). Surface and interface design in cocatalysts for photocatalytic water splitting and CO_2 reduction. *RSC Advances*, *6*(62), 57446–57463.

Balan, A. P., Radhakrishnan, S., & Kumar, R. (2018). A non-van der Waals two-dimensional material from natural titanium mineral ore ilmenite. *Chemistry of Materials*, *30*(17), 5923–5931.

Balan, A. P., Radhakrishnan, S., & Peupane, R. (2018). Magnetic properties and photocatalytic applications of 2D sheets of nonlayered manganese telluride by liquid exfoliation. *ACS Applied Nano Materials*, *1*(11), 6427–6434.

Balan, A. P., Radhakrishnan, S., Woellner, C. F., Sinha, S. K., Deng, L., Reyes, C. D. L., Rao, B. M., Paulose, M., Neupane, R., Apte, A., Kochat, V., Vajtai, R., Harutyunyan, A. R., Chu, C.-W., Costin, G., Galvao, D. S., Martí, A. A., van Aken, P. A., Varghese, O. K., … Ajayan, P. M. (2018). Exfoliation of a non-van der Waals material from iron ore hematite. *Nature Nanotechnology*, *13*, 602–609.

Barnard, A. S., Zapol, P., & Curtiss, L. A. (2005). Modeling the morphology and phase stability of TiO_2 nanocrystals in water. *Journal of Chemical Theory and Computation*, *1*(1), 107–116.

Bhosale, S. S., Kharade, A. K., Jokar, E., Fathi, A., Chang, S. M., & Diau, E. W. G. (2019). Mechanism of photocatalytic CO_2 reduction by bismuth-based perovskite nanocrystals at the gas–solid interface. *Journal of the American Chemical Society*, *141*(51), 20434–20442.

Bi, Y., Ouyang, S., Umezawa, N., Cao, J., & Ye, J. (2011). Facet effect of single-crystalline Ag_3PO_4 sub-microcrystals on photocatalytic properties. *Journal of the American Chemical Society*, *133*(17), 6490–6492.

Boles, M. A., Ling, D., Hyeon, T., & Talapin, D. V. (2016). The surface science of nanocrystals. *Nature Materials*, *15*(2), 141–153.

Bukowski, T. J., & Simmons, J. H. (2002). Quantum dot research: Current state and future prospects. *Critical Reviews in Solid State and Materials Sciences*, *27*(3–4), 119–142.

Cai, J.-G., Yu, Z.-T., Yuan, Y.-J., Li, F., & Zou, Z. G. (2014). Dinuclear iridium(III) complexes containing bibenzimidazole and their application to water photoreduction. *ACS Catalysis*, *4*(6), 1953–1963.

Cao, S., Shen, B., Tong, T., Fu, J., & Yu, J. (2018). 2D/2D heterojunction of ultrathin MXene/Bi_2WO_6 nanosheets for improved photocatalytic CO_2 reduction. *Advanced Functional Materials*, *28*(21), 1800136.

Chan, K., Tsai, C., Hansen, H. A., & Nørskov, J. K. (2014). Molybdenum sulfides and selenides as possible electrocatalysts for CO_2 reduction. *ChemCatChem*, *6*(7), 1899–1905.

Chang, C. M., Orchard, K. L., Martindale, B. C. M., & Reisner, E. (2016). Ligand removal from CdS quantum dots for enhanced photocatalytic H_2 generation in pH neutral water. *Journal of Materials Chemistry A*, *4*(8), 2856–2862.

Chang, K., Hai, X., Pang, H., Zhang, H., Shi, L., Liu, G., Liu, H., Zhao, G., Li, M., & Ye, J. (2016). Targeted synthesis of 2H- and 1T-phase MoS_2 monolayers for catalytic hydrogen evolution. *Advanced Materials*, *28*(45), 10033–10041.

Chhowalla, M., Jena, D., & Zhang, H. (2016). Two-dimensional semiconductors for transistors. *Nature Reviews Materials*, *1*(11), 16052.

Cline, E. D., Adamson, S. E., & Bernhard, S. (2008). Homogeneous catalytic system for photoinduced hydrogen production utilizing iridium and rhodium complexes. *Inorganic Chemistry*, *47*(22), 10378–10388.

Conway, B. E., & Bockris, J. O. M. (1957). Electrolytic hydrogen evolution kinetics and its relation to the electronic and adsorptive properties of the metal. *The Journal of Chemical Physics*, *26*(3), 532–541.

Das, A., Han, Z., Haghighi, M. G., & Eisenberg, R. (2013). Photogeneration of hydrogen from water using CdSe nanocrystals demonstrating the importance of surface exchange. *Proceedings of the National Academy of Sciences of the United States of America*, *110*(42), 16716.

Deng, D., Novoselov, K. S., Fu, Q., Zheng, N., Tian, Z., & Bao, X. (2016). Catalysis with two-dimensional materials and their heterostructures. *Nature Nanotechnology*, *11*(3), 218–230.

Di, J., Chen, C., Zhu, C., Song, P., Xiong, J., Ji, M., Zhou, J., Fu, Q., Xu, M., Hao, W., Xia, J., Li, S., Li, H., & Liu, Z. (2019). Bismuth vacancy-tuned bismuth oxybromide ultrathin nanosheets toward photocatalytic CO_2 reduction. *ACS Applied Materials & Interfaces*, *11*(34), 30786–30792.

Di, J., Zhu, C., Ji, M., Duan, M., Long, R., Yan, C., Gu, K., Xiong, J., She, Y., Xia, J., Li, H., & Liu, Z. (2018). Defect-rich $Bi_{12}O_{17}Cl_2$ nanotubes self-accelerating charge separation for boosting photocatalytic CO_2 reduction. *Angewandte Chemie International Edition*, *57*(45), 14847–14851.

Djatoubai, E., Khan, M. S., ul Haq, S., Guo, P., & Shen, S. (2022a). $BiFeO_3$ bandgap engineering by dopants and defects control for efficient photocatalytic water oxidation. *Applied Catalysis A: General*, *643*, 118737.

Djatoubai, E., Khan, M. S., ul Haq, S. U., Guo, P., & Shen, S. (2022b). Rational design of $BiFeO_3$ nanostructures for efficient charge carrier transfer and consumption for photocatalytic water oxidation. *Journal of Alloys and Compounds*, *911*, 164920.

Djatoubai, E., Khan, M. S., & ul Haq, S. (2023). Engineered cobalt single-atoms@$BiFeO_3$ heteronanostructures for highly efficient solar water oxidation. *Small*, 2206293. https://doi.org/10.1002/smll.202206293

DuChene, J. S., Tagliabue, G., Welch, A. J., Cheng, W. H., & Atwater, H. A. (2018). Hot hole collection and photoelectrochemical CO_2 reduction with plasmonic Au/p-GaN photocathodes. *Nano Letters*, *18*(4), 2545–2550.

Ebina, Y., Sasaki, T., Harada, M., & Watanabe, M. (2002). Restacked perovskite nanosheets and their Pt-loaded materials as photocatalysts. *Chemistry of Materials*, *14*(10), 4390–4395.

Fiori, G., Bonaccorso, F., Iannaccone, G., Palacios, T., Neumaier, D., Seabaugh, A., Banerjee, S. K., & Colombo, L. (2014). Electronics based on two-dimensional materials. *Nature Nanotechnology*, *9*(10), 768–779.

Fox, M. A., & Dulay, M. T. (1993). Heterogeneous photocatalysis. *Chemical Reviews*, *93*(1), 341–357.

Freise, E. J. (1962). Structure of graphite. *Nature*, *193*(4816), 671–672.

Fu, F., Cha, G., Wu, Z., Qin, S., Zhang, Y., Chen, Y., & Schmuki, P. (2021). Photocatalytic hydrogen generation from water-annealed TiO_2 nanotubes with white and grey modification. *ChemElectroChem*, *8*(1), 240–245.

Fu, Y., Dong, C.-L., Lee, W.-Y., Chen, J., Guo, P., Zhao, L., & Shen, S. (2016). Nb-doped hematite nanorods for efficient solar water splitting: Electronic structure evolution versus morphology alteration. *ChemNanoMat*, *2*(7), 704–711.

Fujishima, A., & Honda, K. (1972). Electrochemical photolysis of water at a semiconductor electrode. *Nature*, *238*(5358), 37–48.

Furube, A., Du, L., Hara, K., Katoh, R., & Tachiya, M. (2007). Ultrafast plasmon-induced electron transfer from gold nanodots into TiO_2 nanoparticles. *Journal of the American Chemical Society*, *129*(48), 14852–14853.

Geim, A. K., & Novoselov, K. S. (2007). The rise of graphene. *Nature Materials*, *6*(3), 183–191.

Ghimire, S., & Klinke, C. (2021). Two-dimensional halide perovskites: Synthesis, optoelectronic properties, stability, and applications. *Nanoscale*, *13*, 12394–12422.

Gong, D., Grimes, C. A., Varghese, O. K., Hu, W., Singh, R. S., Chen, Z., & Dickey, E. C. (2001). Titanium oxide nanotube arrays prepared by anodic oxidation. *Journal of Materials Research*, *16*, 3331–3334.

Gong, J., Li, C., & Wasielewski, M. R. (2019). Advances in solar energy conversion. *Chemical Society Reviews*, *48*(7), 1862–1904.

Greeley, J., Jaramillo, T. F., Bonde, J., Chorkendorff, I. B., & Nørskov, J. K. (2006). Computational high-throughput screening of electrocatalytic materials for hydrogen evolution. *Nature Materials*, *5*(11), 909–913.

Guo, Q., Liang, F., Li, X.-B., Gao, Y. J., Huang, M. Y., Wang, Y., Xia, S. G., Gao, X. Y., Gan, Q. C., Lin, Z. S., Tung, C. H., & Wu, L.-Z. (2019). Efficient and selective CO_2 reduction integrated with organic synthesis by solar energy. *Chem*, *5*(10), 2605–2616.

Gusain, R., Kumar, P., Sharma, O. P., Jain, S. L., & Khatri, O. P. (2016). Reduced graphene oxide–CuO nanocomposites for photocatalytic conversion of CO_2 into methanol under visible light irradiation. *Applied Catalysis B: Environmental*, *181*, 352–362.

Han, X., Kuang, Q., Jin, M., Xie, Z., & Zheng, L. (2009). Synthesis of titania nanosheets with a high percentage of exposed (001) facets and related photocatalytic properties. *Journal of the American Chemical Society*, *131*(9), 3152–3153.

Han, Z., Qiu, F., Eisenberg, R., Holland, P. L., & Krauss, T. D. (2012). Robust photogeneration of H_2 in water using semiconductor nanocrystals and a nickel catalyst. *Science*, *338*(6112), 1321–1324.

Haq, S. U., Khan, M. S., Djatoubai, E., Dong, C. L., Guo, P., Huang, Y. C., & Shen, S. (2022). A facile approach for Pt single atoms deposition on two-dimensional calcium niobate nanosheets for photocatalytic hydrogen evolution. *ACS Sustainable Chemistry & Engineering*, *10*(28), 9096–9104.

He, C., Weinberg, D. J., Nepomnyashchii, A. B., Lian, S., & Weiss, E. A. (2016). Control of the redox activity of PbS quantum dots by tuning electrostatic interactions at the quantum dot/solvent interface. *Journal of the American Chemical Society*, *138*(28), 8847–8854.

Hezam, A., Namratha, K., Drmosh, Q. A., Ponnamma, D., Wang, J., Prasad, S., Ahamed, M., Cheng, C., & Byrappa, K. (2020). CeO_2 nanostructures enriched with oxygen vacancies for photocatalytic CO_2 reduction. *ACS Applied Nano Materials*, *3*(1), 138–148.

Hines, D. A., & Kamat, P. V. (2014). Recent advances in quantum dot surface chemistry. *ACS Applied Materials & Interfaces*, *6*(5), 3041–3057.

Hisatomi, T., Kubota, J., & Domen, K. (2014). Recent advances in semiconductors for photocatalytic and photoelectrochemical water splitting. *Chemical Society Reviews*, *43*(22), 7520–7535.

Hou, J., Cao, S., Wu, Y., Gao, Z., Liang, F., Sun, Y., Lin, Z., & Sun, L. (2017). Inorganic colloidal perovskite quantum dots for robust solar CO_2 reduction. *Chemistry – A European Journal*, *23*(40), 9481–9485.

Hou, W., Hung, W. H., Pavaskar, P., Goeppert, A., Aykol, M., & Cronin, S. B. (2011). Photocatalytic conversion of CO_2 to hydrocarbon fuels via plasmon-enhanced absorption and metallic interband transitions. *ACS Catalysis*, *1*(8), 929–936.

Hu, J., Odom, T. W., & Lieber, C. M. (1999). Chemistry and physics in one dimension: Synthesis and properties of nanowires and nanotubes. *Accounts of Chemical Research*, *32*(5), 435–445.

Huang, S., Long, Y., Ruan, S., & Zeng, Y. J. (2019). Enhanced photocatalytic CO_2 reduction in defect-engineered Z-scheme WO_{3-x}/g-C_3N_4 heterostructures. *ACS Omega*, *4*(13), 15593–15599.

Hung, W. H., Aykol, M., Valley, D., Hou, W., & Cronin, S. B. (2010). Plasmon resonant enhancement of carbon monoxide catalysis. *Nano Letters*, *10*(4), 1314–1318.

Hutchison, G. R., Ratner, M. A., & Marks, T. J. (2005). Intermolecular charge transfer between heterocyclic oligomers: Effects of heteroatom and molecular packing on hopping transport in organic semiconductors. *Journal of the American Chemical Society*, *127*(48), 16866–16881.

Huygh, S., Bogaerts, A., & Neyts, E. C. (2016). How oxygen vacancies activate CO_2 dissociation on TiO_2 anatase (001). *Journal of Physical Chemistry C*, *120*(38), 21659–21669.

Iijima, S. (1991). Helical microtubules of graphitic carbon. *Nature*, *354*(6348), 56–58.

Jang, Y. J., Jeong, I., Lee, J., Lee, J., Ko, M. J., & Lee, J. S. (2016). Unbiased sunlight-driven artificial photosynthesis of carbon monoxide from CO_2 using a ZnTe-based photocathode and a perovskite solar cell in tandem. *ACS Nano*, *10*(7), 6980–6987.

Jaramillo, T. F., Jørgensen, K. P., Bonde, J., Nielsen, J. H., Horch, S., & Chorkendorff, I. (2007). Identification of active edge sites for electrochemical H_2 evolution from MoS_2 nanocatalysts. *Science*, *317*(5834), 100–102.

Jellicoe, T. C., Richter, J. M., Glass, H. F. J., Tabachnyk, M., Brady, R., Dutton, S. E., Rao, A., Friend, R. H., Credgington, D., Greenham, N. C., & Böhm, M. L. (2016). Synthesis and optical properties of lead-free cesium tin halide perovskite nanocrystals. *Journal of the American Chemical Society*, *138*(9), 2941–2944.

Jiang, J., Zhao, K., Xiao, X., & Zhang, L. (2012). Synthesis and facet-dependent photoreactivity of BiOCl single-crystalline nanosheets. *Journal of the American Chemical Society*, *134*(10), 4473–4476.

Jin, B., Huang, P., Zhang, Q., Zhou, X., Zhang, X., Li, L., Su, J., Li, H., & Zhai, T. (2018). Self-limited epitaxial growth of ultrathin nonlayered CdS flakes for high-performance photodetectors. *Advanced Functional Materials*, *28*, 1800181.

Junior, J. B. S., Souza, F. L., Vayssieres, L., & Varghese, O. K. (2021). On the relevance of understanding and controlling the locations of dopants in hematite photoanodes for low-cost water splitting. *Applied Physics Letters*, *119*, 200501.

Kalisman, P., Nakibli, Y., & Amirav, L. (2016). Perfect photon-to-hydrogen conversion efficiency. *Nano Letters*, *16*(3), 1776–1781.

Kang, Y., Gong, Y., Hu, Z., Li, Z., Qiu, Z., Zhu, X., Ajayan, P. M., & Fang, Z. (2015). Plasmonic hot electron enhanced MoS_2 photocatalysis in hydrogen evolution. *Nanoscale*, *7*(10), 4482–4488.

Khan, M. S., Diao, Z., Osada, M., & Shen, S. (2020). Nitrogen doped ultrathin calcium/sodium niobate perovskite nanosheets for photocatalytic water oxidation. *Solar Energy Materials & Solar Cells*, *205*, 110283.

Khan, M. S., Kim, H.-J., Kim, Y.-H., Ebina, Y., Sugimoto, W., Sasaki, T., & Osada, M. (2020). Scalable design of two-dimensional oxide nanosheets for construction of ultrathin multilayer nanocapacitor. *Small*, *16*, 2003485.

Khan, M. S., Kim, H.-J., Taniguchi, T., Ebina, Y., Sasaki, T., & Osada, M. (2017). Layer-by-layer engineering of two-dimensional perovskite nanosheets for tailored microwave dielectrics. *Applied Physics Express*, *10*(9), 091501.

Khan, M. S., Osada, M., Dong, L., Kim, Y. H., Ebina, Y., & Sasaki, T. (2021). Rational assembly of two-dimensional perovskite nanosheets as building blocks for new ferroelectrics. *ACS Applied Materials & Interfaces*, *13*(1), 1783–1790.

Khan, M. S., Osada, M., Kim, H.-J., Ebina, Y., Sugimoto, W., & Sasaki, T. (2017). High-temperature dielectric responses in all-nanosheet capacitors. *Japanese Journal of Applied Physics*, *56*(6S1), 06GH9.

Khan, M. S., Zhang, F., Osada, M., Mao, S. S., & Shen, S. (2020). Graphitic carbon nitride-based low-dimensional heterostructures for photocatalytic applications. *Solar RRL*, *4*(8), 1900435.

Kilina, S., Ivanov, S., & Tretiak, S. (2009). Effect of surface ligands on optical and electronic spectra of semiconductor nanoclusters. *Journal of the American Chemical Society*, *131*(22), 7717–7726.

Kočí, K., Obalová, L., Matějová, L., Plachá, D., Lacný, Z., Jirkovský, J., & Šolcová, O. (2009). Effect of TiO_2 particle size on the photocatalytic reduction of CO_2. *Applied Catalysis B: Environmental*, *89*(3), 494–502.

Kovalenko, M. V., Scheele, M., & Talapin, D. V. (2009). Colloidal nanocrystals with molecular metal chalcogenide surface ligands. *Science*, *324*(5933), 1417–1420.

Kwak, B. S., Vignesh, K., Park, N.-K., Ryu, H. J., Baek, J. I., & Kang, M. (2015). Methane formation from photoreduction of CO_2 with water using TiO_2 including Ni ingredient. *Fuel*, *143*, 570–576.

Lakshminarasimhan, N., Bokare, A. D., & Choi, W. (2012). Effect of agglomerated state in mesoporous TiO_2 on the morphology of photodeposited Pt and photocatalytic activity. *Journal of Physical Chemistry C*, *116*(33), 17531–17539.

Lee, K., Mazare, A., & Schmuki, P. (2014). One-dimensional titanium dioxide nanomaterials: Nanotubes. *Chemical Reviews, 114*(19), 9385–9454.

Li, B.-W., Osada, M., Kim, Y.-H., Ebina, Y., Akatsuka, K., & Sasaki, T. (2017). Atomic layer engineering of high-κ ferroelectricity in 2D perovskites. *Journal of the American Chemical Society, 139*(31), 10868–10874.

Li, F., Yang, H., Li, W., & Sun, L. (2018). Device fabrication for water oxidation, hydrogen generation, and CO_2 reduction via molecular engineering. *Joule, 2*(1), 36–60.

Li, H., & Sun, J. (2021). Highly selective photocatalytic CO_2 reduction to CH_4 by ball-milled cubic silicon carbide nanoparticles under visible-light irradiation. *ACS Applied Materials & Interfaces, 13*(4), 5073–5078.

Li, X., Yu, J., Low, J., Fang, Y., Xiao, J., & Chen, X. (2015). Engineering heterogeneous semiconductors for solar water splitting. *Journal of Materials Chemistry A, 3*(6), 2485–2534.

Li, X.-B., Li, Z.-J., Gao, Y.-J., Meng, Q. Y., Yu, S., Weiss, R. G., Tung, C. H., & Wu, L. Z. (2014). Mechanistic insights into the interface-directed transformation of thiols into disulfides and molecular hydrogen by visible-light irradiation of quantum dots. *Angewandte Chemie International Edition, 53*(8), 2085–2089.

Li, Y., Shi, J., Mi, Y., Sui, X., Xu, H., & Liu, X. (2019). Ultrafast carrier dynamics in two-dimensional transition metal dichalcogenides. *Journal of Materials Chemistry C, 7*(15), 4304–4319.

Li, Y., Zhang, W., Shen, X., Peng, P., Xiong, L., & Yu, Y. (2015). Octahedral Cu_2O-modified TiO_2 nanotube arrays for efficient photocatalytic reduction of CO_2. *Chinese Journal of Catalysis, 36*(12), 2229–2236.

Liang, L., Lei, F., Gao, S., Sun, Y., Jiao, X., Wu, J., Qamar, S., & Xie, Y. (2015). Single unit cell bismuth tungstate layers realizing robust solar CO_2 reduction to methanol. *Angewandte Chemie International Edition, 54*(47), 13971–13974.

Liang, L., Wang, J., Lin, W., Sumpter, B. G., Meunier, V., & Pan, M. (2014). Electronic bandgap and edge reconstruction in phosphorene materials. *Nano Letters, 14*(11), 6400–6406.

Lin, L., Hisatomi, T., Chen, S., Takata, T., & Domen, K. (2020). Visible-light-driven photocatalytic water splitting: Recent progress and challenges. *Trends in Chemistry, 2*(9), 813–824.

Lin, Y., Yuan, G., Sheehan, S., Zhou, S., & Wang, D. (2011). Hematite-based solar water splitting: Challenges and opportunities. *Energy & Environmental Science, 4*(12), 4862–4869.

Linic, S., Christopher, P., & Ingram, D. B. (2011). Plasmonic-metal nanostructures for efficient conversion of solar to chemical energy. *Nature Materials, 10*(12), 911–921.

Linsebigler, A. L., Lu, G., & Yates, J. T. (1995). Photocatalysis on TiO_2 surfaces: Principles, mechanisms, and selected results. *Chemical Reviews, 95*(3), 735–758.

Liu, G., Yu, J. C., Lu, G. Q., & Cheng, H. M. (2011). Crystal facet engineering of semiconductor photocatalysts: Motivations, advances and unique properties. *Chemical Communications, 47*(24), 6763–6783.

Liu, N., Schneider, C., Freitag, D., Hartmann, M., Venkatesan, U., Müller, J., Spiecker, E., & Schmuki, P. (2014). Black TiO_2 nanotubes: Cocatalyst-free open-circuit hydrogen generation. *Nano Letters, 14*(6), 3309–3313.

Low, J., Cao, S., Yu, J., & Wageh, S. (2014). Two-dimensional layered composite photocatalysts. *Chemical Communications, 50*(74), 10768–10777.

Low, J., Yu, J., Jaroniec, M., Wageh, S., & Al-Ghamdi, A. A. (2017). Heterojunction photocatalysts. *Advanced Materials, 29*(20), 1601694.

Manzi, A., Simon, T., Sonnleitner, C., Döblinger, M., Wyrwich, R., Stern, O., Stolarczyk, J. K., & Feldmann, J. (2015). Light-induced cation exchange for copper sulfide based CO_2 reduction. *Journal of the American Chemical Society, 137*(44), 14007–14010.

Min, S., & Lu, G. (2012). Enhanced electron transfer from the excited eosin Y to mpg-C_3N_4 for highly efficient hydrogen evolution under 550 nm irradiation. *Journal of Physical Chemistry C, 116*(37), 19644–19652.

Mizeikis, V., Kowalska, E., & Juodkazis, S. (2011). Resonant localization, enhancement, and polarization of optical fields in nano-scale interface regions for photocatalytic applications. *Journal of Nanoscience and Nanotechnology, 11*(4), 2814–2822.

Mor, G. K., Shankar, K., Paulose, M., Varghese, O. K., & Grimes, C. A. (2015). Enhanced photocleavage of water using titania nanotube arrays. *Nano Letters, 5*(1), 191–195.

Mubeen, S., Hernandez-Sosa, G., Moses, D., Lee, J., & Moskovits, M. (2011). Plasmonic photosensitization of a wide bandgap semiconductor: Converting plasmons to charge carriers. *Nano Letters, 11*(12), 5548–5552.

Murphy, A. B., Barnes, P. R. F., Randeniya, L. K., Plumb, I. C., Grey, I. E., Horne, M. D., & Glasscock, J. A. (2006). Efficiency of solar water splitting using semiconductor electrodes. *International Journal of Hydrogen Energy, 31*, 1999–2017.

Nag, A., Kovalenko, M. V., Lee, J.-S., Liu, W., Spokoyny, B., & Talapin, D. V. (2011). Metal-free inorganic ligands for colloidal nanocrystals: S^{2-}, HS^-, Se^{2-}, HSe^-, Te^{2-}, HTe^-, $TeS_3{}^{2-}$, OH^-, and $NH_2{}^-$ as surface ligands. *Journal of the American Chemical Society, 133*(27), 10612–10620.

Naldoni, A., Altomare, M., Zoppellaro, G., Liu, N., Kment, S., Zboril, R., & Schmuki, P. (2019). Photocatalysis with reduced TiO_2: From black TiO_2 to cocatalyst-free hydrogen production. *ACS Catalysis, 9*(1), 345–364.

Nicolosi, V., Chhowalla, M., Kanatzidis, M. G., Strano, M. S., & Coleman, J. N. (2013). Liquid exfoliation of layered materials. *Science, 340*(6139), 1226419.

Nilsson, A., Pettersson, L. G. M., Hammer, B., Bligaard, T., Christensen, C. H., & Nørskov, J. K. (2005). The electronic structure effect in heterogeneous catalysis. *Catalysis Letters, 100*(3), 111–114.

Norris, D. J., & Bawendi, M. G. (1996). Measurement and assignment of the size-dependent optical spectrum in CdSe quantum dots. *Physical Review B, 53*(24), 16338–16346.

Nurdiwijayanto, L., Wu, J., Sakai, N., Ma, R., Ebina, Y., & Sasaki, T. (2018). Monolayer attachment of metallic MoS_2 on restacked titania nanosheets for efficient photocatalytic hydrogen generation. *ACS Applied Energy Materials, 1*(12), 6912–6918.

Oh, I., Kye, J., & Hwang, S. (2012). Enhanced photoelectrochemical hydrogen production from silicon nanowire array photocathode. *Nano Letters, 12*(1), 298–302.

Okamoto, Y., Ida, S., Hyodo, J., Hagiwara, H., & Ishihara, T. (2011). Synthesis and photocatalytic activity of rhodium-doped calcium niobate nanosheets for hydrogen production from a water/methanol system without cocatalyst loading. *Journal of the American Chemical Society, 133*(45), 18034–18037.

Ong, W.-J., Tan, L.-L., Ng, Y. H., Yong, S. T., & Chai, S. P. (2016). Graphitic carbon nitride (g-C_3N_4)-based photocatalysts for artificial photosynthesis and environmental remediation: Are we a step closer to achieving sustainability? *Chemical Reviews, 116*(12), 7159–7329.

Osada, M., & Sasaki, T. (2012). Two-dimensional dielectric nanosheets: Novel nanoelectronics from nanocrystal building blocks. *Advanced Materials, 24*(2), 210–228.

Pan, A., Ma, X., Huang, S., Wu, Y., Jia, M., Shi, Y., Liu, Y., Wangyang, P., He, L., & Liu, Y. (2019). $CsPbBr_3$ perovskite nanocrystal grown on MXene nanosheets for enhanced photoelectric detection and photocatalytic CO_2 reduction. *Journal of Physical Chemistry Letters, 10*(21), 6590–6597.

Pan, J., Liu, G., Lu, G. Q., & Cheng, H. M. (2011). On the true photoreactivity order of {001}, {010}, and {101} facets of anatase TiO_2 crystals. *Angewandte Chemie International Edition, 50*(9), 2133–2137.

Pan, Y.-X., You, Y., Xin, S., Li, Y., Fu, G., Cui, Z., Men, Y. L., Cao, F. F., Yu, S. H., & Goodenough, J. B. (2017). Photocatalytic CO_2 reduction by carbon-coated indium-oxide nanobelts. *Journal of the American Chemical Society, 139*(11), 4123–4129.

Peterson, M. D., Cass, L. C., Harris, R. D., Edme, K., Sung, K., & Weiss, E. A. (2014). The role of ligands in determining the exciton relaxation dynamics in semiconductor quantum dots. *Annual Review of Physical Chemistry, 65*(1), 317–339.

Qiao, J., Kong, X., Hu, Z.-X., Yang, F., & Ji, W. (2014). High-mobility transport anisotropy and linear dichroism in few-layer black phosphorus. *Nature Communications, 5*(1), 4475.

Radhakrishnan, S., Park, J. H., Neupane, R., de los Reyes, C. A., Sudeep, P. M., Paulose, M., Martí, A. A., Tiwary, C. S., Khabashesku, V. N., Varghese, O. K., Kaipparettu, B. A., & Ajayan, P. M. (2017). Fluorinated boron nitride quantum dots: A new 0D material for energy conversion and detection of cellular metabolism. *Particle and Particle Systems Characterization, 36*, 1800346.

Ran, J., Zhang, J., Yu, J., Jaroniec, M., & Qiao, S. Z. (2014). Earth-abundant cocatalysts for semiconductor-based photocatalytic water splitting. *Chemical Society Reviews, 43*(22), 7787–7812.

Rawool, S. A., Yadav, K. K., & Polshettiwar, V. (2021). Defective TiO_2 for photocatalytic CO_2 conversion to fuels and chemicals. *Chemical Science, 12*(12), 4267–4299.

Sapra, S., & Sarma, D. D. (2004). Evolution of the electronic structure with size in II–VI semiconductor nanocrystals. *Physical Review B, 69*(12), 125304.

Sasaki, T., & Watanabe, M. (1997). Semiconductor nanosheet crystallites of quasi-TiO_2 and their optical properties. *The Journal of Physical Chemistry B, 101*(49), 10159–10161.

She, X., Wu, J., Xu, H., Zhong, J., Wang, Y., Song, Y., Nie, K., Liu, Y., Yang, Y., Rodrigues, M. T. F., Vajtai, R., Lou, J., Li, H., & Ajayan, P. M. (2017). High efficiency photocatalytic water splitting using 2D α-Fe_2O_3/g-C_3N_4 Z-scheme catalysts. *Advanced Energy Materials, 7*(17), 1700025.

Shi, R., Liu, F., Wang, Z., Weng, Y., & Chen, Y. (2019). Black/red phosphorus quantum dots for photocatalytic water splitting: From a type I heterostructure to a Z-scheme system. *Chemical Communications, 55*(83), 12531–12534.

Singh, N., Jabbour, G., & Schwingenschlögl, U. (2012). Optical and photocatalytic properties of two-dimensional MoS_2. *European Physical Journal B*, *85*(11), 392.

Su, T., Shao, Q., Qin, Z., Guo, Z., & Wu, Z. (2018). Role of interfaces in two-dimensional photocatalyst for water splitting. *ACS Catalysis*, *8*(3), 2253–2276.

Sun, Y., Gao, S., Lei, F., & Xie, Y. (2015). Atomically-thin two-dimensional sheets for understanding active sites in catalysis. *Chemical Society Reviews*, *44*(3), 623–636.

Sun, Y., Gao, S., & Xie, Y. (2014). Atomically-thick two-dimensional crystals: Electronic structure regulation and energy device construction. *Chemical Society Reviews*, *43*(2), 530–546.

Takata, T., & Domen, K. (2019). Particulate photocatalysts for water splitting: Recent advances and future prospects. *ACS Energy Letters*, *4*(2), 542–549.

Tan, C., Cao, X., Wu, X.-J., He, Q., Yang, J., Zhang, X., Chen, J., Zhao, W., Han, S., Nam, G. H., Sindoro, M., & Zhang, H. (2017). Recent advances in ultrathin two-dimensional nanomaterials. *Chemical Reviews*, *117*(9), 6225–6331.

Thimsen, E., Le Formal, F., Grätzel, M., & Warren, S. C. (2011). Influence of plasmonic Au nanoparticles on the photoactivity of Fe_2O_3 electrodes for water splitting. *Nano Letters*, *11*(1), 35–43.

Thompson, T. L., Diwald, O., & Yates, J. T. (2003). CO_2 as a probe for monitoring the surface defects on TiO_2 (110) temperature-programmed desorption. *The Journal of Physical Chemistry B*, *107*(42), 11700–11704.

Thompson, T. L., & Yates, J. T. (2005). TiO_2-based photocatalysis: Surface defects, oxygen and charge transfer. *Topics in Catalysis*, *35*(3), 197–210.

Tian, J., Zhao, Z., Kumar, A., Boughton, R. I., & Liu, H. (2014). Recent progress in design, synthesis, and applications of one-dimensional TiO_2 nanostructured surface heterostructures: A review. *Chemical Society Reviews*, *43*(20), 6920–6937.

Tian, N., Huang, H., Du, X., Dong, F., & Zhang, Y. (2019). Rational nanostructure design of graphitic carbon nitride for photocatalytic applications. *Journal of Materials Chemistry A*, *7*(19), 11584–11612.

Tian, Y., & Tatsuma, T. (2005). Mechanisms and applications of plasmon-induced charge separation at TiO_2 films loaded with gold nanoparticles. *Journal of the American Chemical Society*, *127*(20), 7632–7637.

Tilley, S. D., Cornuz, M., Sivula, K., & Grätzel, M. (2010). Light-induced water splitting with hematite: Improved nanostructure and iridium oxide catalysis. *Angewandte Chemie International Edition*, *49*(36), 6405–6408.

Tsai, C., Chan, K., Nørskov, J. K., & Abild-Pedersen, F. (2015). Theoretical insights into the hydrogen evolution activity of layered transition metal dichalcogenides. *Surface Science*, *640*, 133–140.

Tu, W., Li, Y., Kuai, L., Zhou, Y., Xu, Q., Li, H., Wang, X., Xiao, M., & Zou, Z. (2017). Construction of unique two-dimensional MoS_2–TiO_2 hybrid nanojunctions: MoS_2 as a promising cost-effective cocatalyst toward improved photocatalytic reduction of CO_2 to methanol. *Nanoscale*, *9*(26), 9065–9070.

Varghese, O. K., Paulose, M., LaTempa, T. J., & Grimes, C. A. (2009). High-rate solar photocatalytic conversion of CO_2 and water vapor to hydrocarbon fuels. *Nano Letters*, *9*(2), 731–737.

Voiry, D., Mohite, A., & Chhowalla, M. (2015). Phase engineering of transition metal dichalcogenides. *Chemical Society Reviews*, *44*(9), 2702–2712.

Voiry, D., Shin, H. S., Loh, K. P., & Chhowalla, M. (2018). Low-dimensional catalysts for hydrogen evolution and CO_2 reduction. *Nature Reviews Chemistry*, *2*(1), 0105.

Voiry, D., Yang, J., & Chhowalla, M. (2016). Recent strategies for improving the catalytic activity of 2D TMD nanosheets toward the hydrogen evolution reaction. *Advanced Materials*, *28*(29), 6197–6206.

Wang, B., Wang, M., Liu, F., Zhang, Q., Yao, S., Liu, X., & Huang, F. (2020). Ti_3C_2: An ideal co-catalyst? *Angewandte Chemie International Edition*, *59*(5), 1914–1918.

Wang, C., Cao, S., & Fu, W.-F. (2013). A stable dual-functional system of visible-light-driven Ni(II) reduction to a nickel nanoparticle catalyst and robust in situ hydrogen production. *Chemical Communications*, *49*(96), 11251–11253.

Wang, C., Osada, M., Ebina, Y., Li, B. W., Akatsuka, K., Fukuda, K., Sugimoto, W., Ma, R., & Sasaki, T. (2014). All-nanosheet ultrathin capacitors assembled layer-by-layer via solution-based processes. *ACS Nano*, *8*(3), 2658–2666.

Wang, D., Pierre, A., Kibria, M. G., Cui, K., Han, X., Bevan, K. H., Guo, H., Paradis, S., Hakima, A. R., & Mi, Z. (2011). Wafer-level photocatalytic water splitting on GaN nanowire arrays grown by molecular beam epitaxy. *Nano Letters*, *11*(6), 2353–2357.

Wang, F., Wang, W.-G., Wang, X.-J., Wang, H. Y., Tung, C. H., & Wu, L. Z. (2011). A highly efficient photocatalytic system for hydrogen production by a robust hydrogenase mimic in an aqueous solution. *Angewandte Chemie International Edition*, *50*(14), 3193–3197.

Wang, G., Wang, H., Ling, Y., Tang, Y., Yang, X., Fitzmorris, R. C., Wang, C., Zhang, J. Z., & Li, Y. (2011). Hydrogen-treated TiO_2 nanowire arrays for photoelectrochemical water splitting. *Nano Letters*, *11*(7), 3026–3033.

Wang, J., Wang, J., Li, N., Du, X., Ma, J., He, C., & Li, Z. (2020). Direct Z-scheme 0D/2D heterojunction of $CsPbBr_3$ quantum dots/Bi_2WO_6 nanosheets for efficient photocatalytic CO_2 reduction. *ACS Applied Materials & Interfaces*, *12*(28), 31477–31485.

Wang, P., Zhang, J., He, H., Xu, X., & Jin, Y. (2015). The important role of surface ligand on CdSe/CdS core/shell nanocrystals in affecting the efficiency of H_2 photogeneration from water. *Nanoscale*, *7*(13), 5767–5775.

Weiss, E. A. (2017). Designing the surfaces of semiconductor quantum dots for colloidal photocatalysis. *ACS Energy Letters*, *2*(5), 1005–1013.

Willkomm, J., Muresan, N. M., & Reisner, E. (2015). Enhancing H_2 evolution performance of an immobilised cobalt catalyst by rational ligand design. *Chemical Science*, *6*(5), 2727–2736.

Wu, H. B., Xia, B. Y., Yu, L., Yu, X. Y., & Lou, X. W. (2015). Porous molybdenum carbide nano-octahedrons synthesized via confined carburization in metal-organic frameworks for efficient hydrogen production. *Nature Communications*, *6*(1), 6512.

Wu, H.-Y., Nguyen, N. H., Bai, H., Chang, S. M., & Wu, J. C. (2015). Photocatalytic reduction of CO_2 using molybdenum-doped titanate nanotubes in a MEA solution. *RSC Advances*, *5*(78), 63142–63151.

Xiao, L., Liu, T., Zhang, M., Li, Q., & Yang, J. (2019). Interfacial construction of zero-dimensional/one-dimensional g-C_3N_4 nanoparticles/TiO_2 nanotube arrays with Z-scheme heterostructure for improved photoelectrochemical water splitting. *ACS Sustainable Chemistry & Engineering*, *7*(2), 2483–2491.

Xiong, L.-B., Li, J.-L., Yang, B., & Yu, Y. (2012). Ti^{3+} in the surface of titanium dioxide: Generation, properties and photocatalytic application. *Journal of Nanomaterials*, *9*, 831524.

Xu, F., Zhu, B., Cheng, B., Yu, J., & Xu, J. (2018). 1D/2D TiO_2/MoS_2 hybrid nanostructures for enhanced photocatalytic CO_2 reduction. *Advanced Optical Materials*, *6*(23), 1800911.

Xu, H., Ouyang, S., Li, P., Kako, T., & Ye, J. (2013). High-active anatase TiO_2 nanosheets exposed with 95% {100} facets toward efficient H_2 evolution and CO_2 photoreduction. *ACS Applied Materials & Interfaces*, *5*(4), 1348–1354.

Xu, Y.-F., Yang, M.-Z., Chen, B.-X., Wang, X. D., Chen, H. Y., Kuang, D. B., & Su, C. (2017). A $CsPbBr_3$ perovskite quantum dot/graphene oxide composite for photocatalytic CO_2 reduction. *Journal of the American Chemical Society*, *139*(16), 5660–5663.

Yang, X., Wolcott, A., Wang, G., Sobo, A., Fitzmorris, R. C., Qian, F., Zhang, J. Z., & Li, Y. (2009). Nitrogen-doped ZnO nanowire arrays for photoelectrochemical water splitting. *Nano Letters*, *9*(6), 2331–2336.

Yeh, T.-F., Teng, C.-Y., Chen, S.-J., & Teng, H. (2014). Nitrogen-doped graphene oxide quantum dots as photocatalysts for overall water-splitting under visible light illumination. *Advanced Materials*, *26*(20), 3297–3303.

Yin, G., Nishikawa, M., Nosaka, Y., Srinivasan, N., Atarashi, D., Sakai, E., & Miyauchi, M. (2015). Photocatalytic carbon dioxide reduction by copper oxide nanocluster-grafted niobate nanosheets. *ACS Nano*, *9*(2), 2111–2119.

Yoffe, A. D. (1993). Low-dimensional systems: Quantum size effects and electronic properties of semiconductor microcrystallites (zero-dimensional systems) and some quasi-two-dimensional systems. *Advances in Physics*, *42*(2), 173–262.

Yu, J., Low, J., Xiao, W., Zhou, P., & Jaroniec, M. (2014). Enhanced photocatalytic CO_2-reduction activity of anatase TiO_2 by coexposed {001} and {101} facets. *Journal of the American Chemical Society*, *136*(25), 8839–8842.

Yu, J., Qi, L., & Jaroniec, M. (2010). Hydrogen production by photocatalytic water splitting over Pt/TiO_2 nanosheets with exposed (001) facets. *Journal of Physical Chemistry C*, *114*(30), 13118–13125.

Zhang, B., & Sun, L. (2019). Artificial photosynthesis: Opportunities and challenges of molecular catalysts. *Chemical Society Reviews*, *48*(7), 2216–2264.

Zhang, X., & Xie, Y. (2013). Recent advances in free-standing two-dimensional crystals with atomic thickness: Design, assembly and transfer strategies. *Chemical Society Reviews*, *42*(21), 8187–8199.

Zhang, Y., He, S., Guo, W., Hu, Y., Huang, J., Mulcahy, J. R., & Wei, W. D. (2018). Surface-plasmon-driven hot electron photochemistry. *Chemical Reviews*, *118*(6), 2927–2954.

Zhao, D., Chen, J., Dong, C.-L., Zhou, W., Huang, Y. C., Mao, S. S., Guo, L., & Shen, S. (2017). Interlayer interaction in ultrathin nanosheets of graphitic carbon nitride for efficient photocatalytic hydrogen evolution. *Journal of Catalysis*, *352*, 491–497.

Zhao, D., Dong, C.-L., Wang, B., Chen, C., Huang, Y. C., Diao, Z., Li, S., Guo, L., & Shen, S. (2019). Synergy of dopants and defects in graphitic carbon nitride with exceptionally modulated band structures for efficient photocatalytic oxygen evolution. *Advanced Materials*, *31*(43), 1903545.

Zhao, D., Wang, Y., Dong, C.-L., Huang, Y. C., Chen, J., Xue, F., Shen, S., & Guo, L. (2021). Boron-doped nitrogen-deficient carbon nitride-based Z-scheme heterostructures for photocatalytic overall water splitting. *Nature Energy*, *6*, 388–397.

Zhao, H., Pan, F., & Li, Y. (2017). A review on the effects of TiO_2 surface point defects on CO_2 photoreduction with H_2O. *Journal of Materiomics*, *3*(1), 17–32.

Zhao, J., Holmes, M. A., & Osterloh, F. E. (2013). Quantum confinement controls photocatalysis: A free energy analysis for photocatalytic proton reduction at CdSe nanocrystals. *ACS Nano*, *7*(5), 4316–4325.

Zhao, X., Jin, W., Cai, J., Ye, J., Li, Z., Ma, Y., Xie, J., & Qi, L. (2011). Shape and size-controlled synthesis of uniform anatase TiO_2 nanocuboids enclosed by active {100} and {001} facets. *Advanced Functional Materials, 21*(18), 3554–3563.

Zhao, Y., Chen, G., Bian, T., Zhou, C., Waterhouse, G. I., Wu, L. Z., Tung, C. H., Smith, L. J., O'Hare, D., & Zhang, T. (2015). Defect-rich ultrathin ZnAl-layered double hydroxide nanosheets for efficient photoreduction of CO_2 to CO with water. *Advanced Materials, 27*(47), 7824–7831.

Zheng, C., Qiu, X., Han, J., Wu, Y., & Liu, S. (2019). Zero-dimensional-g-CNQD-coordinated two-dimensional porphyrin MOF hybrids for boosting photocatalytic CO_2 reduction. *ACS Applied Materials & Interfaces, 11*(45), 42243–42249.

Zhou, L., Xu, Y.-F., Chen, B.-X., Kuang, D. B., & Su, C. Y. (2018). Synthesis and photocatalytic application of stable lead-free $Cs_2AgBiBr_6$ perovskite nanocrystals. *Small, 14*(11), 1703762.

Zhou, S. S., Wang, R. Y., Han, J. B., Li, H., Gan, L., & Zhai, T. (2009). Ultrathin non-van der Waals magnetic rhombohedral Cr_2S_3: Space-confined chemical vapor deposition synthesis and Raman scattering investigation. *Advanced Functional Materials, 29*, 1805880.

Index

Note: Page numbers in *italics* refer to figures; page numbers in **bold** refer to tables.

Conversion of Water and CO_2 to Fuels using Solar Energy: Science, Technology and Materials, First Edition.
Edited by Oomman K. Varghese and Flavio L. Souza.

d

e

t